AF530832

CHEMICAL AND PROCESS EQUIPMENT DESIGN

Vessel Design and Selection

CHEMICAL AND PROCESS EQUIPMENT DESIGN

Vessel Design and Selection

By

David S. Azbel
Nicholas P. Cheremisinoff

Second Printing, 1983

230 Collingwood, P.O. Box 1425, Ann Arbor, Michigan 48106

Library of Congress Catalog Card Number 81-70863
ISBN 0-250-40478-8

Manufactured in the United States of America

Butterworths, Ltd., Borough Green, Sevenoaks
Kent TN15 8PH, England

PREFACE

The chemical and allied industries employ a multitude of unit operations in product manufacturing. Both chemical and physical mechanisms are employed in these operations, ranging from simple bulk handling and preparation of chemical feedstocks to complex chemical reactions in the presence of heat and/or mass transfer. These operations require application of scientific and engineering principles to ensure efficient, safe and economical process operations. To meet these objectives process equipment must perform its intended functions under actual operating conditions and do so in a continuous and reliable manner. Equipment must have the characteristics of mechanical reliability, which include strength, rigidity, durability and tightness. In addition, it must be designed at an optimized ratio of capital investment to service life.

Because of the diversity of products and process operations, equipment design is often unique, or case specific. For this reason, many university engineering curricula often neglect or provide only superficial treatment of design practices for equipment. In fact, most undergraduate curricula stress conceptual design fundamentals of unit operations and leave the detail design principles to job experience and training. This deficiency not only exists in the academic sphere, but in the industrial sector as well, since few textbooks provide detailed design practices and guidelines. It is the intent of the authors to correct this deficiency, at least in part, by organizing standard industrial design practices for equipment employed throughout the chemical process industry. This work will appear as a series of textbooks for the practicing engineer, providing detailed design and calculation procedures for sizing and selecting equipment.

This first work covers fundamental design principles general to all equipment. Design principles and practices for vessels are presented, realizing that vessel design is not only essential from the standpoint of process system peripherals for storage and mixing, but it is applicable to a variety of reactor designs such as tray columns, packed towers, kettles and heat exchangers. Concepts of unit operations are not discussed. Instead, emphasis is placed on computational methods, working formulas and design practices specific to shell design. Example problems and detailed solutions are included in most chapters.

The book is divided into 14 chapters. The first three chapters are general, discussing overall project planning schemes and layout, corrosion fundamentals, and material properties, respectively.

Chapter 4 discusses methods of fabrication along with their influence on design specifications. In almost the same context, Chapter 5 discusses how material properties influence design.

Chapter 6 provides detailed design methods for shells. The four major shell classifications are discussed: cylindrical, box, spherical and conical.

Chapter 7 gives design formulas and examples for shell bottoms and heads under several different operating conditions.

Chapter 8 outlines design fundamentals for reinforcing various types of openings in shells.

In Chapter 9 criteria for selecting, installing and maintaining piping, pipe connections and valves are given.

Chapter 10 gives guidelines for ensuring equipment tightness. Along similar lines, Chapter 11 provides design formulas and detailed examples for flange specification.

Chapter 12 briefly describes various types of threaded connections for shells.

Chapter 13 gives design procedures for the three major classifications of shell covers: flanged, bow and bayonet covers.

Finally, Chapter 14 gives design procedures and example problems for specifying methods of supporting process equipment.

On the personal level, we wish to thank our many friends and colleagues in industry, too numerous to name here, who contributed of their time and materials to this volume. Special thanks go to Robert L. Davis and James W. Johnson of the University of Missouri-Rolla for their suggestions and to Paul N. Cheremisinoff for his assistance in bringing this book into its final form.

David S. Azbel
Nicholas P. Cheremisinoff

David S. Azbel is Professor of Chemical Engineering at the University of Missouri-Rolla. He was previously Research Professor at the State University of New York at Stony Brook and Visiting Professor, University of Minnesota. From 1966 to 1972 he was Professor of Chemical Engineering at the All-Union Polytechnical Institute (Moscow, USSR); from 1961 to 1965, Chief of the Technology Department of the Organic and Synthetic Products Research Institute (Moscow, USSR); previously, he was project chief at the Chemical Engineering Research Institute and Petrochemical Research Institute (Irkutsk, Siberia, USSR). Dr. Azbel graduated from Moscow Institute of Chemical Engineering and received his Doctor of Science degree from the Mendeleev Institute of Chemical Technology, Moscow, USSR. Dr. Azbel's engineering, design, research and teaching span the broadest range of chemical, petroleum and nuclear experience.

Nicholas P. Cheremisinoff heads the Reactor and Fluid Dynamics Modelling Group in the chemical engineering technology division of Exxon Research and Engineering Co., Florham Park, New Jersey. He was previously a Research Scientist for a paper company. He is involved in research and engineering on atomization, fluid bed reactor design, shale oil development, coal gasification, instrumentation, and equipment design and development. Dr. Cheremisinoff received his BS, MS and PhD in chemical engineering from Clarkson College of Technology. He is the author of several Ann Arbor Science books, including *Cooling Towers: Selection, Design and Practice*; *Fluid Flow: Pumps, Pipes and Channels*; and *Gasohol for Energy Production*. In addition he is the author of many papers and is a member of several professional and honor societies including AIChE, Tau Beta Pi and Sigma Xi.

CONTENTS

LIST OF FIGURES

Chapter 4

Chapter 5

Chapter 6

Chapter 7

Chapter 8

Chapter 9

Chapter 10

Chapter 11

Chapter 12

Chapter 13

Chapter 14

LIST OF TABLES

Chapter 4

Chapter 5

Chapter 6

Chapter 7

Chapter 9

Chapter 10

Chapter 11

Chapter 12

Chapter 13

Chapter 14

Appendix

CHAPTER 1

OVERALL PROCESS SYSTEM DESIGN

1.1 INTRODUCTION

The chemical process industries (CPI), petroleum and allied industries apply physical as well as chemical methods to the conversion of raw feedstock materials into salable products. Because of the diversity of products, process conditions and requirements, equipment design is often unique, or case specific. The prime requirements of any piece of equipment are that it performs the function for which it was designed under the intended process operating conditions, and does so in a continuous and reliable manner. Equipment must have mechanical reliability, which is characterized by strength, rigidness, steadiness, durability and tightness. Any one or combination of these characteristics may be needed for a particular piece of equipment.

The cost of equipment determines the capital investment for a process operation; however, there is no direct relationship to profits. That is, more expensive equipment may mean better quality, more durability and, hence, longer service and maintenance factors. These characteristics can produce higher operating efficiencies, fewer consumption coefficients and operational expenses and, thus, fewer net production costs. The net cost of production characterizes the perfection rate of the total technological process and reflects the influences of design indices. Therefore, it is possible to compare different pieces of equipment when they are used in the manufacture of these same products.

The desirable operating characteristics of equipment include simplicity, convenience and low cost of maintenance; simplicity, convenience and low cost of assembly and disassembly; convenience in replacing worn or damaged components; ability to control during operation and test before permanent installation; continuous operation and steady-state processing of materials without excessive noise, vibration or upset conditions; a minimum of personnel for its operation; and, finally, safe operation. Low maintenance often is associated with more complex designs as well as cost. Automation of production is the most complete solution to problems associated with maintaining steady operation, easy maintenance and a minimum of operating personnel. The addition of control devices must be considered as part of the overall design and a factor that adds to the capital investment of the project. Increased automation through the use of controls increases the degree of sophistication in equipment design but lowers operational expenditures while increasing production quality. The use of automatic devices influences the form and dimensional proportions of the equipment as well as imposing additional constraints on the design. It is justified by increased production efficiencies and added security during normal and emergency operations.

Design practices often are neglected in engineering curricula. In fact, most textbooks stress conceptual design fundamentals and leave the detailed design principles to job experience and training. Consequently, equipment design is often treated as an art rather than as an exact science that applies rigorous engineering principles. This deficiency exists not only in many engineering undergraduate curricula, but also in the industrial published sector in that few texts present detailed design practices and guidelines. It is the intent of the authors to fill this void, at least in part, by organizing standard industrial design practices for equipment used throughout the CPI and other major industries. This work will take the form of a series of textbooks that provide detailed design and calculation procedures for sizing and selecting equipment. We shall depart from the standard unit operations textbooks, of which there are several classical works, by not stressing theory. Rather, we will concentrate on specific design practices, computational methods and working formulas. Hence, we hope the reader of principal interest will be the practicing engineer.

This first volume presents fundamental design principles that may be applied to all equipment. Emphasis is placed on process system peripherals, particularly vessels and their associated components. Design principles for all types of vessels, and selection, sizing and design criteria for piping system components are presented. Because practices rather than theory are stressed, only the final working formulas are presented; further, since we intend this to be a designer's guide, numerous example problems are included throughout the book.

The first chapter provides an overview of process design strategies. Fundamental definitions and a brief review of preparing process flow plans are included.

1.2. PLANNING PROJECTS AND EQUIPMENT DESIGN

There are numerous stages of activities that must be conducted before an actual process, plant or even small-scale pilot system reaches its operational stage. Figure 1.1 is a simplified flow diagram illustrating some of the major activities and their normal sequence. From the initial idea the engineer is directed to prepare a preliminary design basis. This includes a rough flow plan, a review of the potential hazards of the process and an assimilation of all available technical, economic and socioeconomic information and data. At this stage of a project often the engineer or engineers are not the final equipment designers, but merely play the devil's advocate, by establishing the equipment requirements. Dialog established between the conceptual design engineer and the process designer results in an initial process flow plan. From the flow plan, a preliminary cost estimate is prepared, many times by a different engineer whose expertise is cost estimating. Once management approval is received, the design engineer's work begins. In the initial stages the design engineer will help prepare a preliminary engineering flow plan, select the site and establish safety requirements.

This initial project stage is often considered a "predesign" period, which constitutes the basis of the conceptual design. Usually a collection of individuals are involved in discussions and planning. The cast of characters includes the project engineer, who oversees the entire project, the design engineer (with whom we are most concerned), safety engineer, environmental engineer and, perhaps, a representative from management and additional support personnel.

Once the overall process has been designed conceptually, a more detailed engineering flow plan is prepared. This flow plan serves two purposes; (1) to document the logic behind the process operation, and (2) to identify in detail major process equipment, including all control devices. A complete flow plan will also identify potential hazards and their consequences, in addition to how they are to be handled. After the environmental and safety engineers have reviewed all potential hazards related to handling toxic materials, noise, radiation, etc., recommendations are outlined for safe and standard handling and disposal practices. These recommendations often affect the overall system design, resulting in revised plans.

The next stage is the actual construction of the unit according to the revised plans. By now, the design engineer is totally involved and has selected, sized and designed most of the equipment and process piping, based (hopefully) on the standard practices outlined in this book. During the actual construction phase, the design engineer will list and review the plans with the project engineer.

At the completion of the unit or system construction, a prestartup review is conducted by the designer and his support personnel. This should include a review of all operating, as well as emergency and shutdown, procedures. The prestartup review normally involves the following personnel in addition to the designer: project engineer, trained operating personnel, operations foreman, the company environmental engineer, the division and company safety engineers and representatives from management. At this point, any

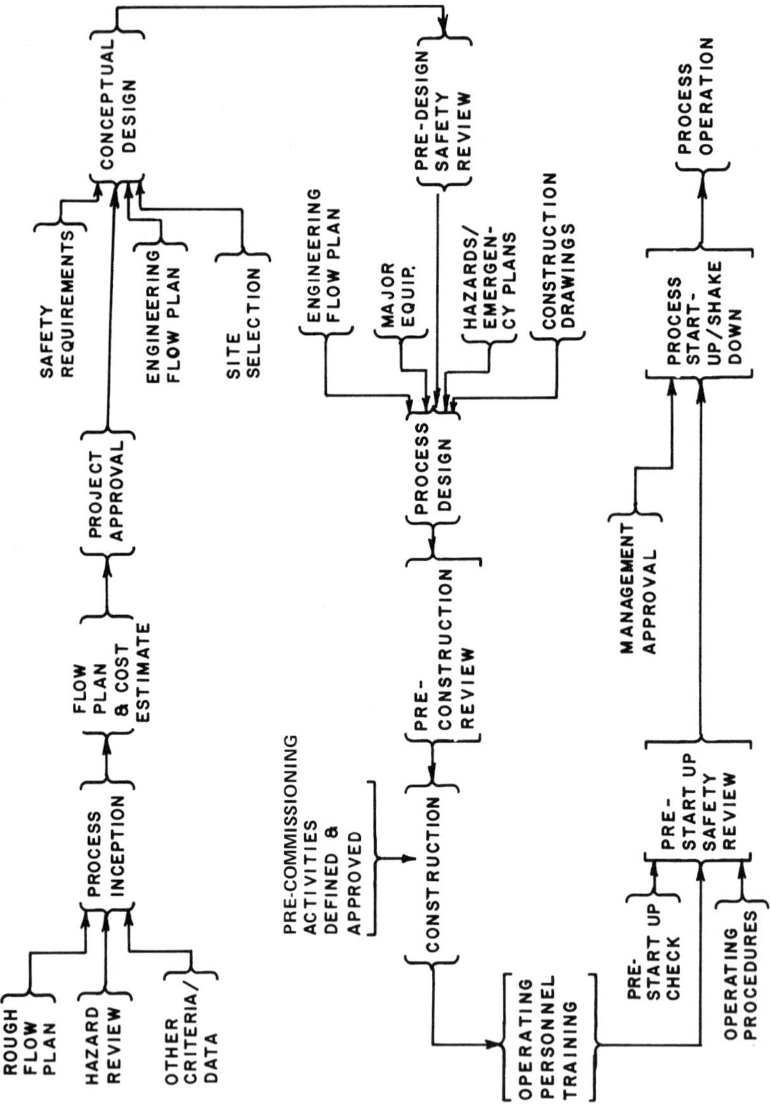

Figure 1.1. Simplified flow diagram of activities in planning and implementing process and plant design projects.

additional changes or recommendations to the process design are made. Major process revisions may be requested by the operations foreman, project engineers, design engineer, safety and environmental coordinators and/or plant operating personnel. Table 1.1 summarizes major items that are considered in the operating procedures planning.

The project planning activities may be much more complex than illustrated

Table 1.1. Major Items in Operating Guidelines Planning

PURPOSE OF PROCESS OR OPERATION

- General Discussion of Process
  - What will be done (brief summary)
  - Chemistry involved
  - Major unit operations
- Personnel Protection
  - Hazards involved–severity
  - Protective equipment–what, where, when
  - Area restrictions–what, where, when
  - Ventilation
- Startup
  - Preparation and handling
    - Feedstocks
    - Catalysts
    - Equipment
- Step-by-Step Description
  - Flow plans
  - Sketches
  - Labelled parts of units
  - Position of valves, control settings, etc.
- Sampling and Final Product Form
  - Description of equipment
  - Actions required
- Shutdown Procedure
  - Step-by-step description
    - Flow plans
    - Sketches
    - Labelled parts of unit
    - Position of valves, control settings, etc.
- Emergency Shutdown Procedure
  - Action required
  - Followup required
  - Emergency personnel/outside organizations
- Unit Cleaning Procedures
  - Description
  - Hazards or precautions
- Product or Waste Disposal

by the simple flow diagram of Figure 1.1. This depends, of course, on the magnitude of the project. Often, large complex system planning has numerous checkpoints at various stages where a continuous review of technical and revised economic forecasts is performed. Also not shown in this flow diagram is the legal framework for obtaining construction and operating permits as well as preparing the environmental impact statement and meeting local, state and federal regulations.

1.3. EQUIPMENT AND INSTRUMENTATION CODES

Process and instrumentation flow diagrams (P & I diagrams) essentially define the control and operating logic behind a process as well as provide a visual record to management and potential users. In addition, P & I diagrams are useful at various stages of a project's development by providing:

- the opportunity for safety analysis before construction begins;
- a tabulation of equipment and instrumentation for cost estimating purposes;
- guidelines for mechanics and construction personnel during the plant assembly stage;
- guidance in analyzing startup problems;
- assistance in training operating personnel; and
- assistance in solving daily operating and sometimes emergency problems.

P & I diagrams contain four important pieces of information, namely, all vessels, valves and piping, along with a brief description and identifying specifications of each; all sensors, instruments and control devices, along with a brief description of each; the control logic used in the process; and, finally, additional references where more detailed information can be obtained.

Information normally excluded from P & I diagrams includes electrical wiring (normally separate electrical diagrams must be consulted), nonprocess equipment (e.g., hoist, support structures, foundations, etc.) and scale drawings of individual components.

There are basically two parts to the diagram: the first provides a schematic of equipment and the second details the instrumentation and control devices. The P & I diagram provides a clear picture of what each piece of equipment is, including identifying specifications, the size of various equipment, materials of construction, pressure vessel numbers and ratings, and drawing numbers. Equipment and instrumentation are defined in terms of a code consisting of symbols, letters and a numbering system. That is, each piece of equipment is assigned its own symbol; a letter is used to identify each type of equipment and to assist in clarifying symbols, and numbers are used to identify individually each piece of equipment within a given equipment type. Table 1.2 illustrates common equipment symbols and corresponding letter codes.

Table 1.2. Common Equipment Symbols and Letter Codes

Equipment	Symbol	Code	Information Needs
Control valve		CV	Size, maximum flowrate, pressure drop
Piping			Material, size, wall thickness
Valves			Type: ball (B), globe (G), needle (N), etc.
Centrifugal Pump		P	Inlet/outlet pressure, flowrate
Rotameter		R	Tube, float, body, maximum flowrate
Reactor		R	Pressure vessel no., drawing no., size
Filter		FIL	Pore size
Back Pressure Regulator	KPA LOADING GAS		Range of gauge and loading source
Electric Heater			Shown on vessel with power pack and control signal
Tracing			Type: steam (S)/ electric (E)
Spring-Loaded Relief Valve		PR	Relief pressure, orifice size

When denoting instrumentation it is important that definitions be understood clearly. Terms for instruments and controls most often included on P & I diagrams are given below:

Instrument Loop–A combination of one or more interconnected instruments arranged to measure or control a process variable.

Final Control Element–A device that directly changes the value of the variable used to control a process condition.

Transducer (Converter)–A device that receives a signal from one power source and outputs a proportional signal in another power system. A transducer can act as a primary element, transmitter or other device.

Fail Closed (usually normally closed)–An instrument that will go to the closed position on loss of power (pneumatic, electric, etc.).

Fail Open (usually normally open)–An instrument that will go to the open position on loss of power (pneumatic, electric, etc.).

Fail Safe–An instrument that on loss of power (pneumatic, electric, etc.) will go to a position that cannot create a safety hazard.

Process Variable–A physical property or condition in a fluid or system.

Instrument–A device that measures or controls a variable.

Local–An instrument located on the equipment.

Remote–An instrument located away from the equipment (normally a control cabinet).

Primary Element–A device that measures a process variable.

Indicator–A device that measures a process variable and displays that variable at the point of measurement.

Transmitter–A device that senses a process variable through a primary element and puts out a signal proportional to that variable to a remotely located instrument.

Controller–A device that varies its output automatically in response to changes in a measured process variable to maintain that variable at a desired value (setpoint).

Instrumentation normally is denoted by a circle in which the variable being measured or controlled is denoted by an appropriate letter symbol inside the circle. When the control device is to be located remotely, the circle is divided in half with a horizontal line. Table 1.3 gives various instrumentation symbols and corresponding letter codes. The specific operating details and selection criteria for various process instrumentation are not discussed in this book. The reader is referred to earlier works by Cheremisinoff [1,2] for discussions on essential control and measurement instrumentation.

Piping normally is denoted by solid lines. Piping lines on the P & I diagram should be accompanied by the following identifying information:

1. line number,
2. nominal pipe size and wall thickness (or schedule number),

Table 1.3. Typical Instrument Codes and Examples

Symbol	Description	Symbol	Description
General Symbols			
	Instrument process piping		Electrically operated valve (solenoid or motor)
	Instrument air lines		
	Electrical leads		Piston-operated valve (hydraulic or pneumatic)
	Capillary tubing		
	Locally mounted instrument (single service)		3-way body for any valve
	Locally mounted transmitter		Safety (relief) valve
	Board-mounted transmitter		
	Diaphragm motor valve		Manually operated control valve
Temperature Symbols			
TW	Temperature well	TRC	Temperature recording controller
TI	Temperature indicator		
Pressure Symbols			
PI	Pressure indicator (locally mounted)	PA	Pressure alarm
PR	Pressure recorder (board mounted)	PC	Pressure controller (blind type)
Flow Symbols			
FI	Displacement-type flowmeter	FR	Flow recorder
FI	Flow indicator, differential type		

3. origin and termination,
4. design temperature and pressure,
5. specified corrosion allowance,
6. winterizing or process protection requirements (i.e., heat tracing via steam or electric),
7. insulation type and thickness,
8. test pressure (indicate hydrostatic or pneumatic), and
9. piping flexibility range (e.g., the maximum or minimum operating temperature).

1.4. VESSEL CODES AND FLANGE RATINGS

In this first volume we shall direct much of our attention to vessel design. In the United States, the primary standard for pressure vessel design is that of the American Society of Mechanical Engineers (ASME). (In subsequent chapters information on European codes for vessels shall be reviewed.) The ASME code is essentially a legal requirement. It provides the minimum construction requirements for the design, fabrication, inspection and certification of pressure vessels. The ASME code does not cover: (1) vessels subject to federal control; (2) certain water and hot water tanks, (3) vessels with an internal operating pressure not exceeding 15 psig with no limitation on size; and (4) vessels having an inside diameter not exceeding 6 inches with no limitation on pressure.

Flange ratings are also specified by the ASME. Table 1.4 gives the various flange ratings in terms of the strength of materials, as based on ASME standards. Table 1.5 gives data on flange pressure-temperature ratings. Finally, Figure 1.2 gives data on allowable stress at different temperatures for carbon steel pipe and 304 stainless steel plate.

All pressure vessels must pass appropriate hydrostatic testing before approval for service. For safety reasons, hydrostatic pressure testing is almost always recommended over a pneumatic test. The recommended

Table 1.4. Flange Ratings for Different Materials

Strength of Materials

	Carbon Steel	Stainless Steel
150 lb	@ 500°F	@ 500°F
300 lb 600 lb 900 lb 1500 lb 2500 lb	@ 850°F	@ 1000°F

Table 1.5. Typical Flange Pressure-Temperature Data

	Carbon Steel		304 SS	
°F	150 psia	300 psia	150 psia	300 psia
100	275	720	275	615
200	240	700	240	550
300	210	680	210	495
400	180	665	180	450
500	150	625	150	410
600	130	555	130	380
700	110	470	110	355
800	92	365	92	330
900	70	225	70	310
1000	40	85	40	300
1100	–	–	–	255
1200	–	–	–	155

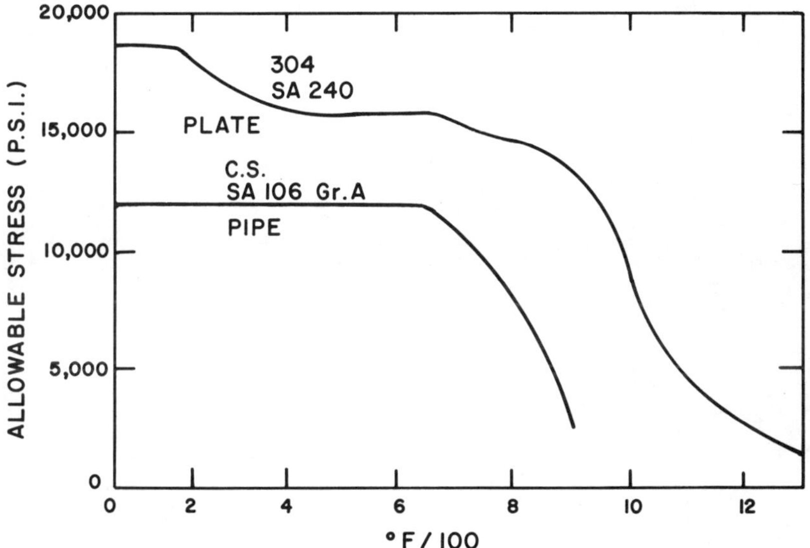

Figure 1.2. Allowable stress for different materials.

hydrostatic test is typically 150% of the temperature corrected design. The pneumatic test is typically 125% of design, as recommended by ASME. A "proof-test" is used when calculations are not possible. This requires at least twice the maximum allowable pressure and employs a brittle coat on the vessel to indicate overstress.

REFERENCES

1. Cheremisinoff, N. P. *Applied Fluid Flow Measurement* (New York: Marcel Dekker, Inc., 1979).
2. Cheremisinoff, N. P. *Process Level Instrumentation and Control* (New York: Marcel Dekker, Inc., 1981).

CHAPTER 2

DESIGN AND CORROSION

2.1 INTRODUCTION

Corrosion occurs in various forms and is promoted by a variety of causes, all related to process operating conditions. It is a continuous problem that can lead to contaminated process streams which leads to poor product quality and unscheduled equipment shutdowns, which leads to reduced production, and high maintenance and equipment replacement costs. Minimizing corrosion is a key consideration for the designer and can be accomplished in two ways: (1) proper material selection for apparatus, and (2) preventive maintenance practices. Both these approaches must be examined by the designer. This chapter reviews principles of corrosion causes and control. It is important to recognize conditions that promote rapid material degradation to compensate for corrosion in designing.

2.2 TYPES OF CORROSION

Corrosion is characterized by the controlling chemi-physical reaction that promotes each type. Each of the major types is described below.

Uniform corrosion is the deterioration of a metal surface that occurs uniformly across the material. It occurs primarily when the surface is in contact with an aqueous environment, which results in a chemical reaction between the metal and the service environment. Since this form of corrosion results in a relatively uniform degradation of apparatus material, it can be accounted for most readily at the time the equipment is designed, either by proper material selection, special coatings or linings, or increased wall thicknesses.

Galvanic corrosion results when two dissimilar metals are in contact, thus forming a path for the transfer of electrons. The contact may be in the form of a direct connection (e.g., a steel union joining two lengths of copper piping), or the dissimilar metals may be immersed in an electrically conducting medium (e.g., an electrolytic solution). One metal acts as an anode and, consequently, suffers more corrosion than the other metal, which acts as the cathode. The driving force for this type of corrosion is the electrochemical potential existing between two metals. This potential difference represents an approximate indication of the rate at which corrosion will take place. That is, corrosion rates will be faster in service environments where electrochemical potential differences between dissimilar metals are high.

Thermogalvanic corrosion is promoted by an electrical potential caused by temperature gradients and can occur on the same material. The region of the metal higher in temperature acts as an anode and thus undergoes a high rate of corrosion. The cooler region of the metal serves as the cathode. Hence, large temperature gradients on process equipment surfaces exposed to service environments will undergo rapid deterioration.

Erosion corrosion occurs in an environment where there is flow of the corrosive medium over the apparatus surface. This type of corrosion is greatly accelerated when the flowing medium contains solid particles. The corrosion rate increases with velocity. Erosion corrosion generally manifests as a localized problem due to maldistributions of flow in the apparatus. Corroded regions are often clean, due to the abrasive action of moving particulates, and occur in patterns or waves in the direction of flow.

Concentration cell corrosion occurs in an environment in which an electrochemical cell is affected by a difference in concentrations in the aqueous medium. The most common form is crevice corrosion. If an oxygen concentration gradient exists (usually at gaskets and lap joints), crevice corrosion often occurs. Larger concentration gradients cause increased corrosion (due to the larger electrical potentials present).

Cavitation corrosion occurs when a surface is exposed to pressure changes and high-velocity flows. Under pressure conditions, bubbles form on the surface. Implosion of the bubbles causes local pressure changes sufficiently large to flake off microscopic portions of metal from the surface. The resulting surface roughness acts to promote further bubble formation, thus increasing the rate of corrosion.

Fretting corrosion occurs where there is friction between two metal surfaces. Generally, this friction is caused by vibrations. The debris formed by fretting corrosion accelerates the damage. Initial damage is done by contact welding. Vibrations cause contact welds to break, with subsequent surface deterioration. Debris formed acts to accelerate this form of corrosion by serving as an abrasive. Fretting corrosion is especially prevalent in areas where motion between surfaces is not foreseen. If allowances for vibration are not made during design, fretting corrosion may be a strong candidate.

Pitting corrosion is a form of localized corrosion in which large pits are formed in the surface of a metal usually in contact with an aqueous solution.

The pits can penetrate the metal completely. The overall appearance of the surface involved does not change considerably; hence, the actual damage is not readily apparent. Once a pit forms, it acts as a local anode. Conditions such as debris and concentration gradients in the pit further accelerate degradation. There are several possible mechanisms for the onset of pitting corrosion. Slight damage or imperfections in the metal surface, such as a scratch or local molecular dislocation, may provide the environment necessary for the beginning of a pit.

Exfoliation corrosion is especially prevalent in aluminum alloys. The grain structure of the metal determines whether exfoliation corrosion will occur. In this form of corrosion, degradation propagates below the surface of the metal. Corrosion products in layers below the metal surface cause flaking of the metal.

Selective leaching occurs when a particular constituent of an alloy is removed. Selective leaching occurs in aqueous environments, particularly acidic solutions. Graphitization and dezincification are two common forms of selective leaching. Dezincification is the selective removal of zinc from alloys containing zinc, particularly brass. The mechanism of dezincification of brass involves dissolving the brass with subsequent plating back of copper while zinc remains in solution. Graphitization is the selective leaching of iron or steel from gray cast irons.

Intergranular corrosion occurs selectively along the grain boundaries of a metal. This is an electrochemical corrosion in which potential differences between grain boundaries and the grain become the driving force. Even with relatively pure metals of only one phase, sufficient impurities can exist along grain boundaries to allow for intergranular corrosion. Intergranular corrosion is generally not visible until the metal is in advanced stages of deterioration. These advanced stages appear as rough surfaces with loose debris (dislodged grains). Welding can cause local crystal graphic changes, which favor intergranular corrosion. It is especially prevalent near welds.

Stress corrosion cracking is an especially dangerous form of corrosion. It occurs when a metal under a constant stress (external, residual or internal) is exposed to a particular corrosive environment. The effects of a particular corrosive environment vary for different metals. For example, Inconel-600 exhibits stress corrosion cracking in high-purity water with only a few parts per million of contaminants at about 300°C. The stress necessary for this type of corrosion to occur is generally of the residual or internal type. Most external stresses are not sufficient to induce stress corrosion cracking. Extensive cold working or the presence of a rivet are common stress providers. Corrosion products also can build up to provide stress sufficient to cause stress corrosion cracking. The damage done by stress corrosion cracking is not obvious until the metal fails. This aspect of stress corrosion cracking makes it especially dangerous.

Corrosion fatigue is caused by the joint action of cyclically applied stresses and a corrosive medium (generally aqueous). Metals will fail due to cyclic application of stress (fatigue). The presence of an aqueous corrosive environment causes such failure more rapidly. The frequency of the applied stress affects the rate of degradation in corrosion fatigue. Ordinary fatigue is generally not frequency dependent. Low-frequency applied stresses cause more rapid corrosion rates. Intuitively, low frequencies cause extended contact time between cracks and the corrosive medium. Generally, the cracks formed are transgranular.

Hydrogen blistering is caused by bubbling of a metal surface due to absorbed hydrogen. Monatomic hydrogen can diffuse through metals, whereas diatomic hydrogen cannot. Ionic hydrogen generated by chemical processes (such as electrolysis or corrosion) can form monatomic hydrogen at a metal surface. This hydrogen can diffuse through the metal and combine on the far side of the metal forming diatomic hydrogen. The diffusion hydrogen also can combine in voids in the metal. Pressure within the void increases until the void actually grows (visibly apparent as a blister) and ultimately ruptures, leading to mechanical failure.

Hydrogen embrittlement is due to the reaction of diffused hydrogen with a metal. Different metals undergo specific reactions, but the result is the same. Reaction with hydrogen produces a metal that is lower in strength and more brittle.

Decarburization results from hydrogen absorption from gas streams at elevated temperatures. In addition to hydrogen blistering, hydrogen can remove carbon from alloys. The particular mechanism depends to a large extent on the properties of other gases present. Removal of carbon causes the metal to lose strength and fail.

Grooving is a type of corrosion particular to environmental conditions where metals are exposed to acid-condensed phases. For example, high concentrations of carbonates in the feed to a boiler can produce steam in the condenser to form acidic condensates. This type of corrosion manifests as grooves along the surface following the general flow of the condensate.

Biological corrosion involves all corrosion mechanisms in which some living organism is involved. Any organism, from bacteria and fungi to mussels, which can attach themselves to a metal surface, can cause corrosion. Biological processes may cause corrosion by producing corrosive agents, such as acids. Concentration gradients also can be caused by localized colonies of organisms. Some organisms remove protective films from metals, either directly or indirectly, leaving the actual metal surface vulnerable to corrosion. By selective removal of products of corrosion, biological organisms also can cause accelerated corrosion reactions. There are also some bacteria that directly digest certain metals (e.g., iron, copper or aluminum). Microorganisms also may promote galvanic corrosion by removing hydrogen from the surface of a metal causing a potential difference to be created between different parts of the metal.

Stray current corrosion is an electrolytic degradation of a metal caused by unintentional electrical currents. Bad grounds are the most prevalent causes. The corrosion is actually a typical electrolysis reaction.

Gaseous corrosion is a general form of corrosion whereby a metal is exposed to a gas (usually at elevated temperatures). Direct oxidation of a metal in air is the most common cause. Cast iron growth is a specific form of gaseous corrosion in which corrosion products accumulate onto the metal surface (particularly at grain boundaries) to the extent that they cause visible thickening of the metal. The entire metal thickness may succumb to this before loss of strength causes failure.

Tuberculation occurs in aqueous solutions. Mounds form over metal surfaces providing for concentration differences, favorable environments for biological growth, and an increase in acidity leading to hydrogen formation.

Deposit attack occurs when there is nonuniform deposition of a film on a metal surface. The most common form appears as unequal scale deposits in an aqueous environment. Unequal film provides for concentration cells, which degrade the metal by galvanic means.

Impingement is corrosion caused by aerated water streams constricting metal surfaces. It is similar to erosion corrosion in which air bubbles take the place of particles. The pits formed by impingement attack have a characteristic tear drop shape.

Liquid metal corrosion occurs when a metal is in contact with a liquid metal. The main type of corrosion with highly pure liquid metals is simple solution. The solubility of the solid metal in the liquid metal controls the rate of damage. If a temperature gradient exists, a much more damaging form of corrosion takes place. Metal dissolves from the higher temperature zone and crystallizes out in the colder zone. Transfer of solids to liquid metal is greatly accelerated by thermal gradients. If two dissimilar metals are in contact with the same liquid metal, the more soluble metal exhibits serious corrosion. The more soluble metal dissolves along with alloys from the less soluble metal. Metal in solutions may move by gross movement of the liquid metal or by diffusion. Depending on the system, small amounts of impurities may cause corrosive chemical reactions.

High-temperature corrosion is induced by accelerated reaction rates inherent in any temperature reaction. One phenomenon that occurs frequently in heavy oil-firing boilers is layers of different types of corrosion on one metal surface.

Causes of corrosion are the subject of extensive investigation by industry. Almost any type of corrosion can manifest itself under widely differing operating conditions. Also, different types of corrosion can occur simultaneously. It is not uncommon to see crack growth from stress corrosion to be accelerated by crevice corrosion, for example. For more detailed descriptions of the mechanisms of corrosion, the reader should consult the literature [1-10].

2.3 MATERIALS EVALUATION AND SELECTION

Materials evaluation and selection are fundamental considerations in engineering design. If done properly, and in a systematic manner, considerable time and cost can be saved in design work, and design errors can be avoided.

The design of any apparatus must be unified and result in a safe functional system. Materials used for each apparatus should form a well coordinated and integrated entity, which should not only meet the requirements of the apparatus' functional utility, but also those of safety and product purity.

Materials evaluation should be based only on actual data obtained at conditions as close as possible to intended operating environments. Prediction of a material's performance is most accurate when standard corrosion testing is done in the actual service environment. Often it is extremely difficult in laboratory testing to expose a material to all of the impurities that the apparatus actually will contact. In addition, not all operating characteristics are readily simulated in laboratory testing. Nevertheless, there are standard laboratory practices that enable engineering estimates of the corrosion resistance of materials to be evaluated.

Environmental composition is one of the most critical factors to consider. It is necessary to simulate as closely as possible all constituents of the service environment in their proper concentrations. Sufficient amounts of corrosive media, as well as contact time, must be provided for test samples to obtain information representative of material properties degradation. If an insufficient volume of corrosive media is exposed to the construction material, corrosion will subside prematurely.

The American Society for Testing Materials (ASTM) recommends 250 ml of solution for every square inch of area of test metal. Exposure time is also critical. Often it is desirable to extrapolate results from short time tests to long service periods. Typically, corrosion is more intense in its early stages (before protective coatings of corrosion products build up). Results obtained from short-term tests tend to overestimate corrosion rates which often results in an overly conservative design.

Immersion into the corrosive medium is important. Corrosion can proceed at different rates, depending on whether the metal is completely immersed in the corrosive medium, partially immersed or alternately immersed and withdrawn. Immersion should be reproduced as closely as possible since there are no general guidelines on how this affects corrosion rates.

Oxygen concentration is an especially important parameter to metals exposed to aqueous environments. Temperature, temperature gradients and concentration gradients in solutions should also be reproduced as closely as possible. Careful attention should be given to any movement of the corrosive medium. Mixing conditions should be reproduced as closely as possible.

The condition of the test metal is important. Clean metal samples with uniform finishes are preferred. The accelerating effects of surface defects lead to deceptive results in samples. The ratio of the area of a defect to the total surface area of the metal is much higher in a sample than in any metal in service. This is an indication of the inaccuracy of tests made on metals with improper finishes. The sample metal should have the same type of heat treatment as the metal to be used in service. Different heat treatments have different effects on corrosion. Heat treatment may improve or reduce the corrosion resistance of a metal in an unpredictable manner. For the purpose of selectivity, a metal stress corrosion test may be performed. General trends of the performance of a material can be obtained from such tests; however, it is difficult to reproduce the stress that actually will occur during service.

For galvanic corrosion tests it is important to maintain the same ratio of anode to cathode in the test sample as in the service environment.

Evaluation of the extent of corrosion is no trivial matter. The first step in evaluating degradation is the cleaning of the metal. Any cleaning process involves removal of some of the substrate. In cases in which corrosion products are strongly bound to the metal surface, removal causes inaccurate assessment of degradation due to surface loss from the cleaning process. Unfortunately, corrosion assessments involving weight gain measurements are of little value. It is rare for all of the corrosion products to adhere to a metal. Corrosion products that flake off cause large errors in weight gain assessment schemes.

The most common method of assessing corrosion extent involves determining the weight loss after careful cleaning. Weight loss is generally considered a linear loss by conversion. Sometimes direct measurement of the sample thickness is made. Typical destructive testing methods are used to evaluate loss of mechanical strength. Aside from inherent loss of strength due to loss of cross section, changes brought about by corrosion may cause loss of mechanical strength. Standard tests for tensile strength, fatigue and impact resistance should be run on test materials.

There are several schemes for nondestructive evaluation. Changes in electrical resistance can be used to follow corrosion. Radiographic techniques involving X-rays and gamma rays have been applied. Transmitted radiation as well as back scattered radiation have been used.

Radiation transmission methods, in which thickness is determined by (measured as) the shadow cast from a radioactive source, are limited to pieces of equipment small enough to be illuminated by small radioactive sources. There are several schemes for highlighting cracks. If the metal is appropriate, magnetic particles can be used to accentuate cracks. Magnetic particles will congregate along cracks too small to be seen normally. An alternate method involves a dye. A dye can be used that will soak into cracks preferentially.

Because of the multitude of engineering materials and the profusion of material-oriented literature it is not possible to describe specific engineering practices in detail in a single chapter. However, we can outline general criteria for parallel evaluation of various materials that can assist in proper selection. The following is a list of general guidelines that can assist in material selection:

1. Select materials based on their functional suitability to the service environment. Materials selected must be capable of maintaining their function safely and for the expected life of the equipment, and at reasonable cost.

2. When designing apparatus with several materials, consider all materials as an integrated entity. More highly resistant materials should be selected for the critical components and for cases in which relatively high fabrication costs are anticipated. Often, a compromise must be made between mechanically advantageous properties and corrosion resistance.

3. Thorough assessment of the service environment and a review of options for corrosion control must be made. In severe, humid environments it is sometimes more economical to use a relatively cheap structural material and apply additional protection, rather than use costly corrosion-resistant ones. In relatively dry environments many materials can be used without special protection, even when pollutants are present.

4. The use of fully corrosion-resistant materials is not always the best choice. One must optimize the relation between capital investment and cost of subsequent maintenance over the entire estimated life of the equipment.

5. Consideration should be given to special treatments that can improve corrosion resistance (e.g., special welding methods, blast peening, stress relieving, metallizing, sealing of welds). Also, consideration should be given to fabrication methods that minimize corrosion.

6. Alloys or tempers chosen should be free of susceptibility to corrosion and should meet strength and fabrication requirements. Often a weaker alloy must be selected than one that cannot be reliably heat treated and whose resistance to a particular corrosion is low.

7. If, after fabrication, heat treatment is not possible, materials and fabrication methods must have optimum corrosion resistance in their as-fabricated form. Materials that are susceptible to stress corrosion cracking should not be employed in environments conducive to failure. Stress relieving alone does not always provide a reliable solution.

8. Materials with short life expectancies should not be combined with those of long life in nonreparable assemblies.

9. For apparatuses for which heat transfer is important, materials prone to scaling or fouling should not be used.

10. For service environments in which erosion is anticipated, the wall thickness of the apparatus should be increased. This thickness allowance should ensure that various types of corrosion or erosion do not reduce the apparatus wall thickness below that required for mechanical stability of the operation. Where thickness allowance cannot be provided, a proportionally more resistant material should be selected.

11. Nonmetallic materials should have the following desirable characteristics: low moisture absorption, resistance to microorganisms, stability through temperature range, resistance to flame and arc, freedom from outgassing, resistance to weathering, and compatibility with other materials.

12. Fragile or brittle materials whose design does not provide any special protection should not be employed under corrosion-prone conditions.

Thorough knowledge of both engineering requirements and corrosion control technology is required in the proper design of equipment. Only after a systematic comparison of the various properties, characteristics and fabrication methods of different materials can a logical selection be made for a particular design. Tables 2.1 through 2.5 can assist in this analysis. Table 2.1 lists general physical and material characteristics, as well as characteristics of strength, that should be considered when comparing different metals and/or nonmetals for a design. Table 2.2 is a listing of fabrication parameters that should be examined in the materials comparison process. In addition to the characteristics listed in Tables 2.1 and 2.2, an examination of design limitations and economic factors must be made before optimum material selection is accomplished. Design limitations or restrictions for materials might include:

- size and thickness
- velocity
- temperature
- composition of constituents
- bimetallic attachment
- geometric form
- static and cyclic loading
- surface configuration and texture
- special protection methods and techniques
- maintainability
- compatibility with adjacent materials

Economic factors that should be examined may be divided into three categories: (1) availability, (2) cost of different forms, and (3) size limitations and tolerances [3]. More specifically, these include:

Table 2.1. Parameters to Analyze in Materials Selection

Metals	Nonmetals
General Physical Characteristics	
1) Chemical composition (%) 2) Contamination of contents by corrosion products 3) Corrosion characteristics in: Atmosphere Water Soil Chemicals Gases Molten metals 4) Creep characteristics @ temperature range 5) Crystal structure 6) Damping coefficient 7) Density (g/cm^3) 8) Effect of cold working 9) Effect of high temperature on corrosion resistance 10) Effect on strength after exposure to: Hydrogen High temperatures 11) Electrical conductivity (mho/cm) 12) Electrical resistivity (Ω/cm) 13) Fire resistance 14) Hardenability 15) Maximum temperature not affecting strength (°C) 16) Melting point (°C) 17) Corrosion factor (rapidity of corrosion) 18) Susceptibility to corrosion: General Hydrogen damage Pitting Galvanic Corrosion fatigue Fretting Stress corrosion cracking Corrosion/erosion Cavitation damage Intergranular Selective attack High temperature 19) Thermal coefficient of expansion (°C^{-1}) 20) Thermal conductivity (W/m °C) 21) Wearing quality: Inherent Via heat treatment Via plating	1) Anisotropy characteristics (main and cross-direction) 2) Area factor (in.^{2}/lb/mil) 3) Burn rate (in./min) 4) Bursting strength (Mullen points) 5) Change in linear dimensions @ 100°C for 30 min. (%) 6) Clarity 7) Color 8) Creep characteristics @ temperature range–creep apparent modulus (10° lb_f/in.^{2}). 9) Crystal structure 10) Crystalline melting point 11) Damping coefficient 12) Decay characteristics in: Atmosphere Alcohols Chemicals Gases High relative humidity Hydraulic oils Hydrocarbons Solvents Sunlight Water 13) Deflection temperature (°C) 264 (lb_f/in.^{2}) fiber stress 66 (lb_f/in.^{2}) fiber stress 14) Density (g/cm^3) 15) Dielectric constant 16) Dielectric strength: short time/step-by-step 17) Dissipation factor (1 MΩ) 18) Effect on decay from: high temperature/low temperature/exposure to heat 19) Electrical loss factor (1 MΩ) 20) Electrical resistivity arc/sec insulation (96 hρ 90% RH and 35°C)MΩ 21) Combustion properties/fire resistance 22) Flammability 23) Fillers 24) Gas permeability (cm^3/100 in.^{2}/mil thick/24 hr/atm at 25°C): CO_2, H_2, N_2, O_2 25) Heat distortion temperature at 264 lb_f/in.^{2}) (°F) 26) Thermal coefficient of expansion (in.$^{-1}$ °F)

Table 2.1, continued

Metals	Nonmetals
General Physical Characteristics	
	27) Thermal conductivity (Btu/ft^2 h °F in.$^{-1}$)
	28) Light transmission, total white (%)
	29) Maximum service temperature (°C)
	30) Melt index (dg/min.)
	31) Minimum and maximum temperatures not affecting strength (°C)
	32) Softening temperature (°C)
	33) Stiffness–Young's modulus
	34) Susceptibility to various forms of deterioration:
	General
	Cavitation/erosion
	Erosion
	Fatigue
	Fouling
	Galvanic (metal-filled plastics)
	Impingement
	Stress cracking and crazing
	35) Thermal conductivity (W/m°C)
	36) Wearing quality:
	Inherent
	Via treatment
Strength and Mechanical Characteristics	
1) Bearing ultimate (N/mm^2)	1) Abrasion resistance
2) Complete stress-strain curve for tension and compression	2) Average yield (lb$_f$/in.^{2})
3) Compression modulus of elasticity (kg/mm^2)	3) Bonding strength (lb/thickness)
4) Fatigue properties	4) Brittleness
5) Hardness (Vickers)	5) Bursting pressure (lb$_f$/in.^{2})
6) Impact properties (Charpy kg/cm^2 @ 20°C):	6) Compressive strength:
Notch sensitivity	Flatwise (lb$_f$/in.^{2})
Effect of low temperature	Axial (lb$_f$/in.^{2})
Maximum transition temperature (°C)	at 10% deflection (lb$_f$/in.^{2})
7) Poisson's ratio	7) Deformation under load
8) Response to stress-relieving methods	8) Elongation (%)
9) Shear modulus of elasticity (kg/mm^2)	9) Elongation at break (%)-75°F (24°C)
10) Shear ultimate (Pa)	10) Fatigue properties
11) Tension modulus of elasticity (Pa)	11) Flexibility and flex life
12) Tension-notch sensitivity	12) Flexural strength (N/mm^2)
13) Tension yield	13) Hardness (Rockwell)
	14) Impact strength, Izod (ft lb^{-1} in.$^{-1}$ notch)
	15) Inherent rigidity
	16) Modulus of elasticity (lb$_f$/in.^{2} or kg/mm^2)
	In compression
	In flexure
	In tension
	In shear

Table 2.1, continued

Metals	Nonmetals
Strength and Mechanical Characteristics	
	17) Resistance to fatigue 18) Safe operating temperature (°C) 19) Shear ultimate (Pa) 20) Tear strength: Propagating (g/mil) Initial (lb/in.) 21) Tensile strength ($lb_f/in.^2$ or kg/mm^2) 22) Vacuum collapse temperature

Table 2.2. Fabrication Parameters to Analyze in Materials Selection

General Subject	Parameter
Metals	
Brazing and soldering	Compatibility Corrosion effect Flux and rod
Formability at elevated and room temperature	Aging characteristics Annealing procedure Corrosion effect of forming Heat treating characteristics Quenching procedures Sensitivity to variation Tempering procedure Effect of heat on prefabrication treatment
Formability in annealed and tempered states	Apparatus stress × local stream curve Characteristics in: Bending Dimpling Drawing Joggling Shrinking Stretching Corrosion effect of forming Elongation × gauge length Standard hydropress specimen test True stress-strain curve Uniformity of characteristics
Machinability	Best cutting speed Corrosion effect of: Drilling Milling

Table 2.2, continued

General Subject	Parameter
Metals	
	Routing Sawing Shearing Turning Fire hazard Lubricant or coolant Material and shape of cutting tool Quality suitability for: Drilling Routing Milling Sawing Shearing Turning
Protective coating	Anodizing Cladding Ecology Galvanizing Hard surfacing Metallizing Need of application for: Storage Processing Service Paint adhesion and compatibility Plating Prefabrication treatment Sensitivity to contaminants Suitability Type surface preparation
Quality of finish	Appearance Cleanliness Grade Honing Polishing Surface effect
Weldability	Arc welding Atomic hydrogen welding Corrosion effect of welding Cracking tendency Prefabrication treatment effects Electric flash welding Flux Friction welding Heat zone effect Heli-arc welding Pressure welding Spot welding

Table 2.2, continued

General Subject	Parameter
Metals	
	Torch welding Welding rod
Torch cutting	Cutting speed
Nonmetals	
Molding and injection	Compression ratio Compression molding pressure ($lb_f/in.^2$) Compression molding temperature (°C) Injection molding pressure ($lb_f/in.^2$) Injection molding temperature (°C) Molding qualities Mold (linear) shrinkage (in./in.) Specific volume (lb^3)
Lamination	Lamination pressure ($lb_f/in.^2$) Lamination temperature (°C)
Formation at elevated temperatures	
Machinability	Adverse effects of: Drilling Milling Sawing Shearing Turning Best cutting speed Fire hazard Machining qualities Material and shape of cutting tool
Protective coating	Cladding Painting Plating Sensitivity to contaminants Suitability Type surface preparation
Quality of finish	Appearance Cleanliness Grade Polishing Surface and effect
Joining	Adhesive joining Bonding Cracking tendency Heat zone effect Welding

Table 2.3. General Properties of the Corrosion Resistance of Metals to Various Chemicals [11]

R = Recommended
M = Moderate Service
L = Limited Service
U = Unsatisfactory
Blank = No information

	Carbon Steel	Cast Iron and Ductile Iron	304 Stainless Steel	316 Stainless Steel	347 Stainless Steel	Nickel-Resist Iron	Carpenter 20; Durinet 20	Worthite	Duriron-Durichlor	Monel	Inconel	Hastelloy B	Hastelloy C	Hastelloy D	Chlorimet 3	Aluminum and Alloys	Copper and Cu Alloys	Brass	Lead	Nickel
Aluminum Chloride	U	U	U	U	U	L	R	U	R	U	L	R	M	R	R	U	L	L	U	L
Aluminum Hydroxide			R	R			R	R	M	M					R	R	R	R		R
Aluminum Sulfate	U		L	L	R	L	R	R	R	L	L	R	R	R	R	R	L	L	R	L
Alums, Dilute	U	U	R	L	L	L	R	R	R	R	R	R	R	R	R	R	L	L	R	R
Amines (various)	R	R	R	R	R	R	R	R	R		R	R	R	R	R	L	L	L		R
Ammonia Gas	R		R	R	R	R	R		R			R	R	R	R	U	U			R
Ammonium Carbonate	M	M	M	M	R		R		R	M	M	M	M	M	R	L	L	L		R
Ammonium Chloride	U	U	U	L	L	R	R	R	R	M	M	L	M	M	R	U	U	U	L	M
Ammonium Hydroxide	R	R	R	R	R	R	R	R	M	U	R	M	M	M	R	L	U	U	M	U
Ammonium Nitrate	U		R	R	R	L	R	R	R	R	L	U	R	U	R	U	U	U		
Ammonium Sulfate	R		L	L	M	R	R	R	R	M	R	R	R	R	R	L	L	L	L	M
Benzene	M	M	R	R	R	R	R	R	R	M	R	M	M	M	R	R	R	R	M	M
Calcium Carbonate	R		R	R		R	R	R	R	R	R	R	R	R	R	U	R	R	R	R
Calcium Chlorate			L	R	U		M	L	M	M	M		M	M	M		L	L		R
Calcium Chloride	R	R	L	L	L	L	M	R	R	M	M	M	R	M	R	L	M	M	U	R
Calcium Hydroxide	R		R	R	R	R	R	R	M	R	R	R	R	R	R	L	M	L	L	R
Calcium Hypochlorite	L		U	L	L	L	L	L	R	U	U	U	L	U	U	U	U	U	U	U
Calcium Sulfate	L		R	R			R	R	R	M		M	M	M	R	L	R	R		
Carbon Dioxide (dry)	R	R	R	R	R	R	R	R	R	R	R	R	R	R	R	R	R	R	R	R
Carbon Dioxide (wet)	L	L	R	R	R	L	R	R	R	L	R	R	R	R	R	L	L	L	L	R

Table 2.3, continued

R = Recommended
M = Moderate Service
L = Limited Service
U = Unsatisfactory
Blank = No information

	Carbon Steel	Cast Iron and Ductile Iron	304 Stainless Steel	316 Stainless Steel	347 Stainless Steel	Nickel-Resist Iron	Carpenter 20; Durinet 20	Worthite	Duriron-Durichlor	Monel	Inconel	Hastelloy B	Hastelloy C	Hastelloy D	Chlorimet 3	Aluminum and Alloys	Copper and Cu Alloys	Brass	Lead	Nickel
Chlorine (wet)	U	U	U	U	U	U	U	U	L	U	U	U	R	U	M	U	U	U	L	L
Chromic Acid Solution	L	M	U	L	U	U	M	R	R	U	L	U	M	U	M	M	U	U	M	U
Copper Chloride	U	U	U	U	U		U	R	U	U	U	R	R	R	L	U	U	U	U	U
Copper Sulfate	U	U	R	R	R	L	R	R	R	L	L	R	R	R	R	U	U	U	L	L
Fatty Acids	U	U	R	R	R	L	R	R	R	M	R	R	R	R	R	R	L	L	U	R
Ferrous Chloride	U	U	U	U	U	R	L	U	R	U	U	L	R	R	R	U	U	U	U	U
Ferrous Sulfate	U	U	M	U	R	L	R	R	R	U	U	M	M	M	R	M	L	L	L	U
Hydrochloric Acid (conc.)	U	U	U	U	U	U	U	U	M	U	U	M	L	U	M	U	U	U	U	U
Hydrochloric Acid (dilute)	U	U	U	U	U	L	U	U	R	L	L	L	L	L	R	U	U	U	L	L
Hydrogen Chloride (dry gas)	R	L	L	R	L	U	U	U	R	L	L	R	R	U	R	U	U	U	L	L
Hydrofluoric Acid	L	U	U	U	U	U	M	U	U	M	L	M	M	M	R	U	U	U	U	L
Hydrocarbons (aliphatic)	R	R	R	R	R	R	R	R	R	R	R	R	R	R	R	R	R	R	R	R
Hydrogen Peroxide (conc.)	U	U	M	U	L	U	M	L	M	L		L	R	L	R	R	U	U	L	
Hydrogen Sulfide (dry)	M		L	R	L	R			R	L	M	L	R	L	R	U	U	U	R	R
Hydrogen Sulfide (wet)	L		L	R	L	L	R	R	M	U	M	L	M	L	R	U	U	U	U	R
Nitrating Acid (>15% H_2SO_4)	R	R	L	L	L		R	R	R	U		U	R	U	R	U	U	U		U
Nitrating Acid (<15% H_2SO_4)	U		L	L	L		R	R	R	U		U	R	U	R	U	U	U		U
Nitrating Acid (<15% HNO_3)	U		L	U	L		R		R	U		U	R	U	R	U	U	U		U
Nitrating Acid (<1% acid)	U		L	R	L		R		R	U		U	R	U	R	U	U	U		U
Nitric Acid (conc.)	U	U	M	M	R		R	R	R	U	L	U	M	U	R	L	U	U	L	U

Nitric Acid (dilute)	U	U	R	R	R	U	R	R	R	U	U	U	R	U	R	U	U	U	U	U
Nitrous Acid	U		R	R	R		R	R	R	R					R		R	R		U
Phenol (conc.)	M	M	L	R	L	R	R	R	R	M	R	M	M	M	R	L	U	U	L	M
Phosphoric Acid (100%)	U	U	L	L	R		R		R	L	L	M	R	R	R	U			M	M
Phosphoric Acid (hot >45%)	U	U	U	U	R	U	R	R	R	U	U	L	L	L	R	U	U	U	M	U
Phosphoric Acid (cold >45%)	U	U	L	R	R	L	R	L	R	L	L	M	R	M	R	U	U	U	M	L
Phosphoric Acid (<45%)	U	U	L	R	R	L	R	L	R	L	L	M	R	M	R	L	U	U	M	L
Potassium Carbonate	L		R	R		L	R	R	R	R	M	R	R	R	R	U	L	L	R	R
Potassium Chlorate	R		R	R			R	L	R	M	M	U	R	U	R	M	L	L		M
Potassium Chloride	R		L	L		R	R	R	R	M		R	R	R	R	L	L	L		M
Potassium Permanganate	M	M	M	M	M		R	R	R	L	M	U	R	U	R	R	R	R	U	M
Sodium Bicarbonate	L		R	R	R	R	R	R	R	R	R	R	R	R	R	R	L	L	R	R
Sodium Bisulfate	U	U	L	R	R		R	R	R	M		R	R	R	R	U	L	L	R	
Sodium Bisulfite	U		R	R			R	R	L						R	R	L	L	R	
Sodium Carbonate	M	M	M	M	R		R	R	R	M	L	M	R	R	R	U	M	M	L	M
Sodium Chlorate	L		R			R	L	L	R		M	U	R	U	R	M	M	M		
Sodium Chloride	M	M	L	L	L	L	R	R	M	M	L	M	M	L	R	L	M	M	M	M
Sodium Hydroxide (conc.)	L	L	L	R	R	R	R	R	L	R	R	R	M	R	R	U		L	U	R
Sodium Hydroxide (dilute)	R	R	M	M	M	M	R	R	M	L	L	R		R	R	U	R	R	U	L
Sodium Hydrosulfite							R		R	U	U	L	R	L	R	R	L	L		U
Sodium Hypochlorite	U	U	L	L	L	L	L	L	R	U	U	U	L	U	R	U	U	U	U	U
Sodium Nitrate	R	M	R	R	R	L	R	R	R	M	R	U	M	U	R	R	M	M	U	M
Sodium Phosphate	R		R	R	M		R		R	M		R	R	R	R	U	L	L		M
Sodium Silicate	R		R	R	M	R	R		R	M	M	R	R	R	R	L	L	L		R
Sodium Sulfate	R	R	L	L	R	R	R	R	R	M	R	R	R	R	R	R	R	R	R	R
Sodium Sulfide	M	R	L	R	R	R	R	R	R	M	M	R	R	R	R	U	U	U	L	M
Sodium Sulfite	M	R	L	L	L	R	R	R	U	L	L	U	R	U	R	L	L	L	R	L
Stearic Acid	L	R	R	R	R	L	R	R	R	R	R	R	R	R	R	L	L	L	L	R
Sulfur Dioxide (dry)	R	R	R	R	L	R	R	R	R	U	U	U	R	U	R	R	R	R	R	U
Sulfur Dioxide (wet)	U		L	R			R		L	U	U	U	R	U	R		U	U	R	U
Sulfur Trioxide	R		R	R		R	R	R	L	U	U				R	R	L	L		U
Sulfuric Acid (fuming 98%)	L	L	U	U		L	R	R	U	U	U	R	U	R	R	L	U	U	U	U

Table 2.3, continued

R = Recommended
M = Moderate Service
L = Limited Service
U = Unsatisfactory
Blank = No information

	Carbon Steel	Cast Iron and Ductile Iron	304 Stainless Steel	316 Stainless Steel	347 Stainless Steel	Nickel-Resist Iron	Carpenter 20; Durinet 20	Worthite	Duriron-Durichlor	Monel	Inconel	Hastelloy B	Hastelloy C	Hastelloy D	Chlorimet 3	Aluminum and Alloys	Copper and Cu Alloys	Brass	Lead	Nickel
Sulfuric Acid (hot, conc.)	U		U	U			M		R	U	U	U	U	R	M	U	U	U	L	U
Sulfuric Acid (cold, conc.)	M	M	R	R	L		R	R	R	M		R	R	R	R	L	U	U	R	
Sulfuric Acid (75-95%)	M	L	U	U	L	L	R	R	R	U	L	R	R	L	R	U	U	U	R	U
Sulfuric Acid (10-75%)	U	U	U	U	L	L	R	R	R	L	L	R	R	R	R	U	U	U	R	U
Sulfuric Acid (<10%)	U	U	U	L	L	L	L	R	R	L	L	R	R	R	R	L	U	U	R	L
Sulfurous Acid	U	U	L	R	L	U	R	R	U	U	U	U	M	M	R	L	U	U	L	U
Toluene	R	R	R	R	R	R	R	R	R	R	R	R	R	R	R	R	R	R	R	R
Water (fresh)	R	R	R	R	R	R	R	R	R	R	R	R	R	R	R	R	R	R	R	R
Water (distilled)	U	U	R	R	R	U	R		R	U	R	R	R	R	R	U	R	R	U	L
Zinc Chloride	U	U	U	L	U	L	R	R	R	M		M	M	M	R	U	U	U	L	M
Zinc Sulfate	U	U	M	M	M	R	R	R	R	M	M	M	M	M	R	L	M	M	M	M

Table 2.4. General Properties of the Corrosion Resistance of Nonmetals to Various Chemicals [11]

R = Recommended
M = Moderate Service
L = Limited Service
U = Unsatisfactory
Blank = No information

	Carbon Ceramics			Rubbers			Plastics															
	Carbon and Graphite	Glass (Pyrex®)	Chemical Porcelain	Natural Rubber	Neoprene	Butadiene	Asphaltic Bitumastic	Acrylic (Lucite, Plexiglas)	Polyethylene	Polyvinylchloride	Saran	Kel-F	Teflon® a	Penton	Polystyrene (Styron)	Haveg 41	Heresite	Molded Phenol Formaldehyde (Durez)	Epoxy Resins	Nylon	Durcon 6	Woods–Maple, Oak, Pine
Aluminum Chloride	R	R	R	R	R	R	R		R	R	R	R	R	R	R	R	R	R	R	R		
Aluminum Hydroxide		R	R						R	R		R	R	R						R	R	
Aluminum Sulfate	R	R	R	R	R	R	R	R	R	R	R	R	R	R	R	R	R	R		R	R	R
Alums, Dilute	R	R	R	R	R	R	R	R	R	R	R	R	R	R	R	R	R	R	R		R	R
Amines (various)	R	R	R	L						U		R	R			R	R			R	R	
Ammonia Gas	R	R	R	L	R	R		R	R	L	U	R	R	R	U			R	R		R	
Ammonium Carbonate	R	R	R	R			R		R	R	R	R	R	R	R			R	R	R	R	
Ammonium Chloride	R	R	R	R	R	R	R	R	R	R	R	R	R	R	R	R	R	R	R	U	M	
Ammonium Hydroxide	R	L	R	L	L		L	R	R	R	U	R	R	R	R	R	R	U	R	R	R	
Ammonium Nitrate	R	R	R	L	L	R	R	R	R	R	R	R	R	R	R			R	R	R	R	
Ammonium Sulfate	R	R	R	R	R	R	R	R		R	R	R	R	R		R	R	R	R		U	
Benzene	R	R	R	U	U		U	U	L	U	L	R	R	M	U	L	R	R	R	R	R	
Calcium Carbonate	R	R	R	R	L					R	R	R	R	R						R	R	
Calcium Chlorate	R	R	R				R			R		R		R					R		R	
Calcium Chloride	R	R	R	R	R	R	R	R	R	R	R	R	R	R	R	R	R	R	R	U	R	
Calcium Hydroxide	R	R	R	R	R		R			R	R	R	R	R					R	R	R	
Calcium Hypochlorite	R	R	R	R	U	R	L		R	R	R	R	R	R	R	L	L	U	R	U	R	U

Table 2.4, continued

R = Recommended M = Moderate Service L = Limited Service U = Unsatisfactory Blank = No information	Carbon Ceramics			Rubbers			Plastics															
	Carbon and Graphite	Glass (Pyrex®)	Chemical Porcelain	Natural Rubber	Neoprene	Butadiene	Asphaltic Bitumastic	Acrylic (Lucite, Plexiglas)	Polyethylene	Polyvinylchloride	Saran	Kel-F	Teflon® [a]	Penton	Polystyrene (Styron)	Haveg 41	Heresite	Molded Phenol Formaldehyde (Durez)	Epoxy Resins	Nylon	Durcon 6	Woods–Maple, Oak, Pine
Calcium Sulfate	R	R	R	R	R		R			R	R	R	R	R					R	R	R	
Carbon Dioxide (dry)	R	R	R	R	R		R			R	R	R	R	R		R	R	R	R	R	R	
Carbon Dioxide (wet)	R	R	R	R	R	R	R	R	R	R	R	R	R	R	R	R	R	R		R	R	R
Chlorine (wet)	R	R	R	R	U	L	U		R	L	L	R	R	M	R	L	L	L	L	U	U	
Chromic Acid Solution	U	R	R	U	U	U	U	U	R	L	R	R	R	L	U	U	R	U	U	U	R	
Copper Chloride		R	R	L	R	R	R		R	R	R	R	R	R		R			R	U	R	
Copper Sulfate	R	R	R	R	R	R	R	U	R	R	R	R	R	R	R	R	R	R	R	U	R	
Fatty Acids	R	R	R	U	U	R	U		U	L	R	R	R	U	R	R	R	R	R	R	R	
Ferrous Chloride	R	R	R	R	R	R	R			R	L	R	R	R	R	R	R	R	R	U	R	
Ferrous Sulfate	R	R	R	R	R	R	R		R	R	L	R	R	R	R	R	R	R	R	U	M	
Hydrochloric Acid (conc.)	R	R	R	R	L	R	R	R	R	R	R	R	R	R	U	R	R	R	R	U	R	U
Hydrochloric Acid (dilute)	R	R	R	R	L	R	R	R	R	R	R	R	R	R	R	R	R	L	R	U	R	R
Hydrogen Chloride (dry gas)	R	R	R				U			R		R	R						R		R	
Hydrofluoric Acid	U	U	U	U	U		U	R	R	L	U	R	R	M	R	U	U	L	R		R	
Hydrocarbons (aliphatic)	R	R	R	U	R	R		R				R	R		R	R	R	L			M	
Hydrogen Peroxide (conc.)		R	R	L	U	R	L	R		R		R	R	R	U	R	R	R		U	R	
Hydrogen Sulfide (dry)		R	R	R		R	R			R		R	R		R			R	R	U	R	
Hydrogen Sulfide (wet)		R	R							R		R	R	R							R	

Nitrating Acid (>15% H_2SO_4)	U	R	R	U			U					R	R			U	R		U		R	
Nitrating Acid (<15% H_2SO_4)	U	R	R	U			U					R	R			L	R				R	
Nitrating Acid (<15% HNO_3)	U	R	R	U			U					R	R			L	R				R	
Nitrating Acid (<1% acid)	U	R	R	U			U					R	R			L	R				M	
Nitric Acid (conc.)	U	R	R	L	U	U	U		U	U	L	R	R	U	U	U	R	R	U	U	R	
Nitric Acid (dilute)	L	R	R	U	U	R	U	R	L	R	R	R	R	M	R	L	R	L	U		R	
Nitrous Acid		R	R	L	U	R	R					R	R	R							R	
Phenol (conc.)	R	R	R	L	L	R	U	R		L	L	R	R	M		L	R	L		U	R	
Phosphoric Acid (100%)	M	L	R	R	R	R	R		R	R	L	R	R		U	R	R	U		L	R	
Phosphoric Acid (hot >45%)	R	L	U	L	L		U		L	R		R	R		U	R	R	U	R		R	
Phosphoric Acid (cold >45%)	R	R	L	L	L		R		R	R	R		R	R	R	R	R		R		R	
Phosphoric Acid (<45%)	R	R	L	R	L				R	R	R		R	R		R	R			U	R	
Potassium Carbonate	R	L	R	R	R		R		R	L	R		R	R					R	R	R	
Potassium Chlorate	R	R	R							R	R		R	R						U	R	
Potassium Chloride	R	R	R	R	R		R		R	R	R		R	R		R			R	R	R	
Potassium Permanganate	R	R	R	L	U		U		R	R	L		R	R	R			R	R	U	R	
Sodium Bicarbonate	R	R	R	R	R	R	R	R		R	R		R	R	R	R		R	R	R	R	
Sodium Bisulfate	R	R	R	R	R	R				R	R		R	R	R	R	R	R	R		R	
Sodium Bisulfite	R	R	R	R	L	R	R		R	R			R	R				R	R	U	R	R
Sodium Carbonate	R	L	R	R	R	R		U	R	R	R	R	R	R	R	R	R	R	R	R	R	M
Sodium Chlorate	R	R	R	R	R	R	R	R	R	R	R		R	R	R			R		U	R	R
Sodium Chloride	R	R	R	R	R		R	R	R	R	R	R	R	R	R	R	R	R	R	R	R	R
Sodium Hydroxide (conc.)	R	L	U	R	R	R	R	R	R	R	U	R	R	R	R	U	U	U	R	R	R	U
Sodium Hydroxide (dilute)	R	L	L	R	R	R	R	R	R	R	L	R	R	R	R	U	L	U	R	R	R	U
Sodium Hydrosulfite		R	R	R		R							R	R				R			R	
Sodium Hypochlorite	U	R	R	R	U	R	L			R	R		R	R	R	U	R	L	U	U		
Sodium Nitrate	R	R	R	R	R	R	R			R	R		R	R	R	R	R	R	R	U	R	
Sodium Phosphate		L	R	R	L	R	R	R		R			R		L	L		R	R	R	R	
Sodium Silicate		L	R	R	R	R	R			R			R		R			R		R	R	
Sodium Sulfate	R	R	R	R	R		R		R	R	R		R	R		R	R		R	R	R	
Sodium Sulfite	R	L	R	R	R	R			R	R	R		R		R	R	R		R	R	M	U
Sodium Sulfide	R	R	R	R	U	R	R			R	R		R	R		R	R	L	R	R	R	

Table 2.4, continued

R = Recommended
M = Moderate Service
L = Limited Service
U = Unsatisfactory
Blank = No information

	Carbon Ceramics			Rubbers			Plastics															
	Carbon and Graphite	Glass (Pyrex®)	Chemical Porcelain	Natural Rubber	Neoprene	Butadiene	Asphaltic Bitumastic	Acrylic (Lucite, Plexiglas)	Polyethylene	Polyvinylchloride	Saran	Kel-F	Teflon®[a]	Penton	Polystyrene (Styron)	Haveg 41	Heresite	Molded Phenol Formaldehyde (Durez)	Epoxy Resins	Nylon	Durcon 6	Woods–Maple, Oak, Pine
Stearic Acid	R	R	R	L	U		U		R	R	R		R	R		R	R		R	R	R	
Sulfur Dioxide (dry)	R	R	R	L	U	L	R			R	L		R	R		R			R	L	R	
Sulfur Dioxide (wet)	R	R	R	L	U					R	L		R	R						U	R	
Sulfur Trioxide		R	R	R	U	R	U			R			R						R	R	R	
Sulfuric Acid (fuming 98%)	U	R			U		R	U		U	L		R	U	U			U	U		R	U
Sulfuric Acid (hot, conc.)		R		U	U								R	L		U			U		R	U
Sulfuric Acid (cold, conc.)	U	R	R	U	U			L		R	L		R	L		L			U	U	R	U
Sulfuric Acid (75-95%)	R	R	R		U		R		R	R	L		R	M	L	L		L	L		R	
Sulfuric Acid (10-75%)	R	R	R	R	L	R	R	R	R	R	R	R	R	R	R	R		L	R	L	R	L
Sulfuric Acid (<10%)	R	R	R	R	R	R	R	R	R	R	R	L	R	R	R	R		R		L	R	L
Sulfurous Acid	R	R	R	L	R		R		U	R	L		R	R	R	R		R	R	U	M	R
Toluene	R	R	R	U	U	U		U	L	U	L	L	R	M	U	L		L		R	R	R
Water (fresh)	R	R	R	R	R	R	R	R		R	R		R	R	R			R			R	R
Water (distilled)	M	M	R	R	R			R		R			R	R							R	
Zinc Chloride	R	R	R	R	R	R	R		R	R	R		R	R	R	R		R	R	L	R	
Zinc Sulfate	R	R	R	R	R		R		R	R	R		R	R	R	R		R	R	R	R	

[a]Registered trademark of E. I. du Pont de Nemours and Company, Inc., Wilmington, Delaware.

Table 2.5. Corrosion Rates of Steel and Zinc Panels Exposed for Two Years [11]

No.	Location	mils/yr Steel	mils/yr Zinc	Environment[a]
1.	Norman Wells, NWT, Canada	0.06	0.006	R
2.	Phoenix, AZ	0.18	0.011	R
3.	Saskatoon, Sask., Canada	0.23	0.011	R
4.	Vancouver Island, BC, Canada	0.53	0.019	RM
5.	Detroit, MI	0.57	0.053	I
6.	Fort Amidor, Panama C.Z.	0.58	0.025	M
7.	Morenci, MI	0.77	0.047	R
8.	Ottawa, Ont., Canada	0.78	0.044	U
9.	Potter County, PA	0.81	0.049	R
10.	Waterbury, CT	0.89	0.100	I
11.	State College, PA	0.90	0.045	R
12.	Montreal, Que., Canada	0.94	0.094	U
13.	Melbourne, Australia	1.03	0.030	I
14.	Halifax, NS, Canada	1.06	0.062	U
15.	Durham, NH	1.08	0.061	R
16.	Middletown, OH	1.14	0.048	SI
17.	Pittsburgh, PA	1.21	0.102	I
18.	Columbus, OH	1.30	0.085	U
19.	South Bend, PA	1.32	0.069	SR
20.	Trail, BC, Canada	1.38	0.062	I
21.	Bethlehem, PA	1.48	0.051	I
22.	Cleveland, OH	1.54	0.106	I
23.	Miraflores, Panama C.Z.	1.70	0.045	M
24.	London (Battersea), England	1.87	0.095	I
25.	Monroeville, PA	1.93	0.075	SI
26.	Newark, NJ	2.01	0.145	I
27.	Manila, Philippine Islands	2.13	0.059	U
28.	Limon Bay, Panama C.Z.	2.47	0.104	M
29.	Bayonne, NJ	3.07	0.188	I
30.	East Chicago, IN	3.34	0.071	I
31.	Cape Kennedy, FL ½ mile	3.42	0.045	M
32.	Brazos River, TX	3.67	0.072	M
33.	Pilsey Island, England	4.06	0.022	IM
34.	London (Stratford), England	4.40	0.270	I
35.	Halifax, NS, Canada	4.50	0.290	I
36.	Cape Kennedy, FL 180 ft.	5.20	0.170	M
37.	Kure Beach, NC 800 ft.	5.76	0.079	M
38.	Cape Kennedy, FL 180 ft.	6.52	0.160	M
39.	Daytona Beach, FL	11.7	0.078	M
40.	Widness, England	14.2	0.400	I
41.	Cape Kennedy, FL 180 ft.	17.5	0.160	M
42.	Dungeness, England	19.3	0.140	IM
43.	Point Reyes, CA	19.8	0.060	M
44.	Kure Beach, NC 80 ft.	21.2	0.250	M
45.	Galetea Point, Panama C.Z.	27.3	0.600	M

[a]R rural; RM rural-marine; SI semi-industrial; SR semi-rural; M marine; I industrial; IM industrial-marine; U urban

1. Availability - In required quantities (single, multiple, limited, unlimited)
   - In different forms (bar, casting such as sand, centrifugal, die, permanent mould, etc., extrusion, forging, impact extrusion, pressing, sintered, powder pressing)
   - In metallized and pretreatment forms (galvanized, plastic coated, plated, prefabrication treated)
   - In cladded forms
   - Uniformity of material
   - Freedom from defects
   - Delivery time
2. Cost in different forms - Bar, shape, plate, sheet
   - Casting (sand, centrifugal, die, permanent mould, etc.)
   - Extrusion
   - Forging
   - Impact extrusion
   - Pressing
   - Sintered
   - Powder pressing
3. Size limitations and tolerances in different forms
   - Gauge
   - Length
   - Weight
   - Width

Tables 2.3 through 2.5 give general corrosion-resistance ratings of different materials. Table 2.3 lists various metals and Table 2.4 gives ratings for various nonmetals. Table 2.5 gives typical corrosion rates of steel and zinc panels exposed to the atmosphere in various locations about the U.S. Figure 2.1 also illustrates relative corrosion rates of steel and zinc in major areas of the world.

2.4 DESIGN GUIDELINES

Often, complex apparatus and systems, process piping arrangements and even support structures utilize different metals, alloys or other materials. These are often employed in corrosive or conductive environments and, in practice, the contact of dissimilar materials cannot be avoided totally. It is important that the designer minimize the damaging effects of corrosion by optimizing the compatibility of materials either by selection or arrangement in the overall design. Compatible materials are those that will not cause an uneconomic breakdown within the system, even though they are utilized together in a particular medium in appropriate relative sizes and compositions. In addition to material influences on each other by virtue of inherent or induced differences of electric potentiality, adverse chemical reactions can occur as a result of changes in materials caused by environmental variations. All these possibilities must be examined thoroughly by the designer.

The following general considerations should be followed in designing all types of process equipment:

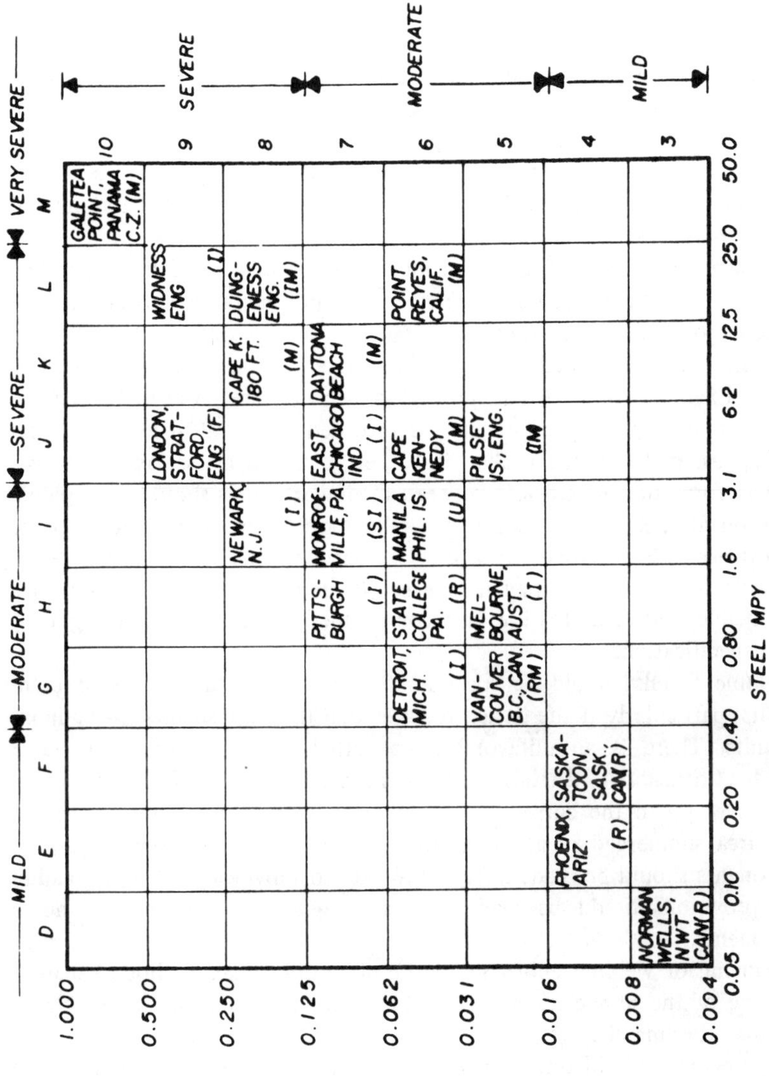

Figure 2.1 Comparison of corrosion rates of zinc and steel in various parts of the world [11].

1. Dissimilar metals should be in contact (either directly or by means of a conductive path such as water, condensation, etc.) only when the functional design so dictates.

2. Scales of Galvanic Potentials are useful indicators of galvanic corrosion; however, information is needed on the amount of current flowing between dissimilar metals.

3. To ensure compatibility, detailed engineering descriptions of all materials and their metallurgical properties are needed. General information (e.g., mild steel) does not provide sufficient data to establish compatibility in conductive or corrosive media.

4. Galvanic corrosion of dissimilar metals can be minimized by controlling humidity near such bimetallic connections. In general, continuously dry bimetallic joints do not corrode.

5. Avoid faying surfaces of dissimilar metals by separating them completely. Examples of poor and proper connections are given in Figure 2.2. Note that dielectric separation can be provided in several manners, e.g., insulating gaskets (synthetic rubber, PTFE, etc.), spreadable sealants, coatings.

6. The formation of crevices between dissimilar metals should be avoided. Corrosion at such connections is generally more severe than either galvanic or crevice corrosion alone. Also, crevices between metals and certain types of plastics or elastomers may induce accelerated rates of combined crevice and chemical attack. Testing is recommended prior to establishing final design specifications.

7. Noble metals should be specified for major structural units or components, particularly if the design requires that these are smaller than adjoining units. There is an unfavorable area effect of small anode and large cathode. Corrosion of a relatively small anodic area can be 100–1000 times more severe than the corrosion of bimetallic components, which have the same area submerged in a conductive medium. Hence, less noble (anodic) components should be made larger or thicker to allow for corrosion. In addition, provision should be made for easy replacement of the less noble components.

8. Brazing or welding alloys should be more noble (i.e., cathodic) than at least one of the joined metals. Also, these alloys should be compatible to both the other metals.

9. Fasteners made of dissimilar metal should be insulated completely from both metals of the joint (or at least the one that is least compatible with the metal of the fastener).

10. Clad metals are candidates for galvanic corrosion along exposed edges. An example is copper/aluminum clad to aluminum.

11. Proper system and sequences of welding attachment of bimetallic pads for structures and equipment should be specified to avoid distortion and input stresses.

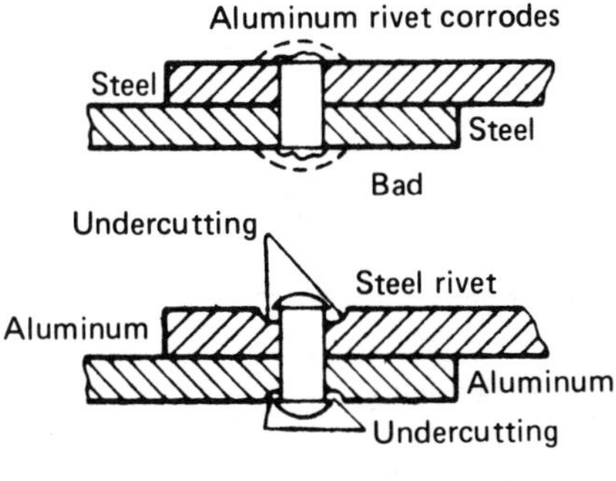

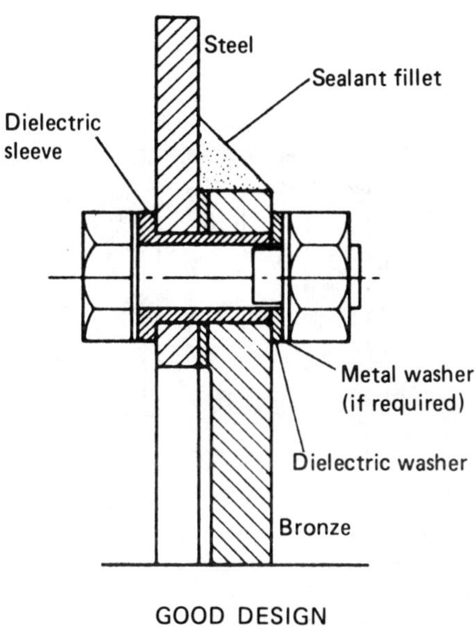

Figure 2.2. Examples of poor and proper connections of dissimilar metals.

12. In aluminum castings, integral corrosion-resistant steel inserts may be used. An example is shown in Figure 2.3.

13. Sources of mercury (e.g., mercury thermometers) should be avoided in the vicinity of aluminum and copper alloy equipment.

14. Avoid coupling carbon or graphite components with other metals in conductive environments.

15. Designs that establish large temperature gradients in equipment resulting in adverse polarization of metals should be avoided.

16. If dielectric separation of fasteners in noncompatible joints cannot be implemented readily, fasteners should be coated with a zinc chromate primer and exposed ends encapsulated. This is illustrated in Figure 2.4.

17. Use sealing (encapsulating or enveloping type with shrinkable plastic) on bimetallic joints if geometrical arrangements prohibit access to such joints for replacement.

The following general guidelines are most applicable to piping system designs.

1. Ensure effective separation between piping sections of dissimilar metals. Examples of this are illustrated in Figure 2.5. As shown, dielectric non-absorbent gaskets of adequate thickness can be inserted between dissimilar

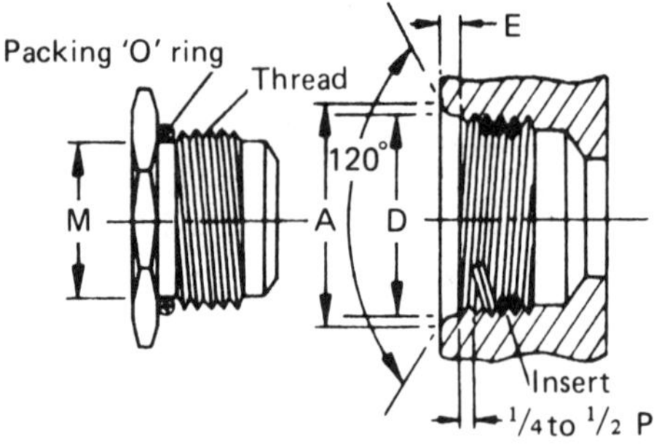

Figure 2.3. Example of a corrosion-resistant steel insert used in an aluminum casting.

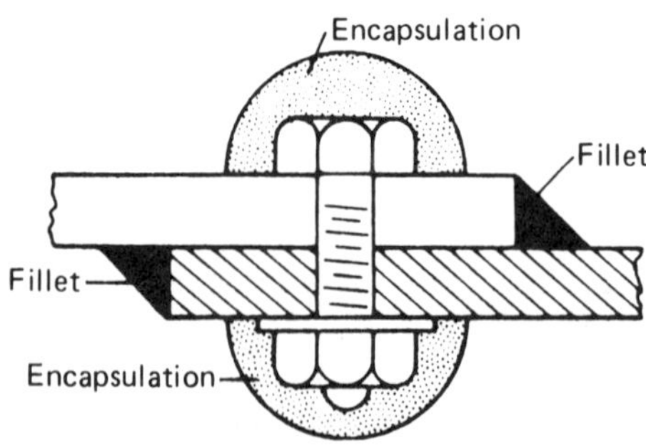

Figure 2.4. Encapsulation of exposed metal connections.

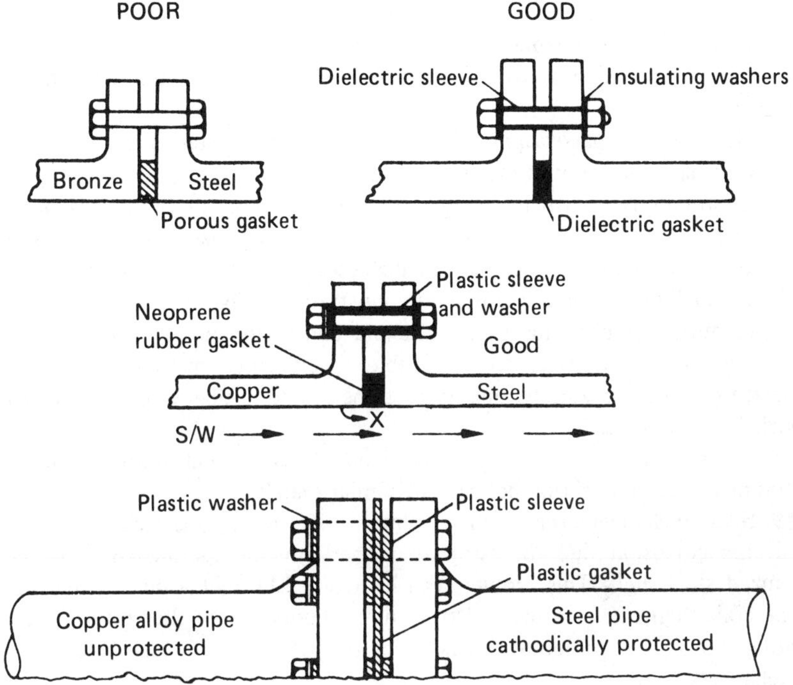

Figure 2.5. Gasket insertion between pipe flanges for sealing purposes and to minimize galvanic corrosion between dissimilar piping metals.

pipe connections. Note that graphite packings and gaskets should not be used for dielectric separation except for steam service or similar applications at elevated temperatures, as with nonconductive media.

2. Piping should not be directly attached to dissimilar metal structures via conductive materials.

3. Graphite and carbon packing should not be used in pipe systems containing conductive media upstream of heat exchangers and other critical equipment. Graphite particles can deposit onto tube bundles in heat exchangers and promote galvanic corrosion. Where possible, use insert seals and packing.

4. Avoid fitting copper alloy pipes upstream of carbon steel equipment. Copper salts from copper-base pipes can dissolve in solution and pose problems to carbon-steel components and vessels downstream, due to the formation of carbonate and bicarbonate salts. If the use of copper alloy pipes is unavoidable, sacrificial sections of mild steel pipe can

be inserted between such connections. These sacrificial sections should be easily accessible to enable replacement and thus should be provided with adequate wall thickness to meet a well-planned maintenance program frequency.

5. Pickling and passivation of Monel and stainless steel pressure vessels should be specified to prevent deep pitting.

6. In situations in which piping protrudes partitions or bulkheads of dissimilar metals, proper precautions should be taken against galvanic corrosion. Possible solutions include the use of dielectric gaskets or sleeves and the use of plastic adhesive tapes. Examples are illustrated in Figure 2.6.

7. In buried pipeline installations, avoid contact of piping with structures of dissimilar metals. Also, where possible, specify uniform quality, grade and surface conditions. Various quality sections should not be welded together in buried installations.

8. Tinning of copper piping or components is a good approach toward minimizing galvanic action between dissimilar metals.

9. Heat exchangers that utilize copper coils are potential candidates for galvanic corrosion due to dissolved copper salts interacting with the galvanized steel shell. This problem can be avoided by nickel plating the coils. The coils then can be separated from direct contact with the vessel via insulation. Also, it is preferable to conduct the water on the tube side of heat exchangers.

The above factors represent considerations that the design engineer must account for to ensure compatibility between components and equipment materials. In addition to these, there are geometric considerations that can minimize corrosion problems if accounted for in design. The following are general guidelines pertaining to geometry in a design aimed at minimizing corrosion. The overall design approach involves the selection of the optimum geometry for a piece of equipment that is less likely to undergo certain types of corrosion, either directly or indirectly. Such shapes, forms, combinations of forms and their method of attachment, along with their fabrication technique and treatment, should be specified so as not to aggravate corrosion.

1. For structures and equipment, the utility should be located where it cannot be affected by natural and climatic conditions. This includes (1) corrosive pollution that may be airborne, (2) prevalent winds, and (3) surface water currents from near or remote sources.

2. Undrainable traps that accumulate liquids and absorbent solid wastes should be avoided. Structures should be designed to be self-draining.

3. Provisions should be made for the removal of moisture or other corrosive media from critical areas.

4. Laps and crevices should be avoided if possible. If they cannot, then effective seals should be used (particularly in areas of heat transfer) between metal and a porous material or where aqueous environments contain inorganic chemicals or dissolved oxygen.

5. Laps should be faced downward on exposed surfaces.

6. Effort should be made to design shapes that will reduce the effects of high fluid velocity, turbulence and the formation of gas bubbles.

7. Asymmetrical shapes of unequal thickness should be avoided for galvanizing. Extremes in weight and cross sections of design members also should be avoided.

8. Impellers should have shapes that minimize high turbulence formation and reduce low-pressure buildup at their tips, which can lead to cavitation.

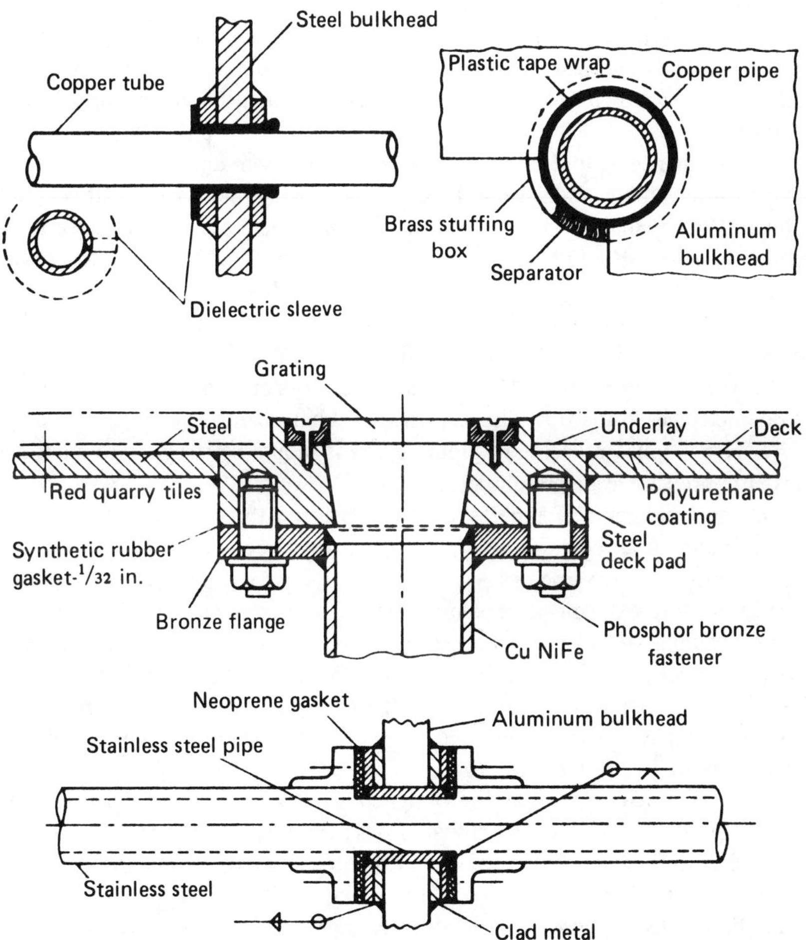

Figure 2.6. Examples of minimizing galvanic corrosion when piping penetrates partitions and bulkheads.

For piping arrangements and vessels, the following geometric considerations are recommended.

9. Piping systems should be designed for an economic flow velocity. For relatively clean fluids, a recommended velocity range where minimum corrosion can be expected is 2 to 10 fps. If piping bores exist, maximum fluid velocities may have a mean velocity of 3 fps for a 3/8-in. bore to 10 fps for an 8-in.-diameter bore. Higher flow velocities are not uncommon in situations that require uniform, constant oxygen supply to form protective films on active/passive metals.

10. Condensate filters, deaerators, traps, drains and other means should be provided for the removal of rust, debris and any other contaminants from the system that may promote corrosion.

11. The interior of piping systems should be streamlined for easy drainage. Stubs and dead ends should be avoided, and pipelines should be sloped continuously downstream to their outlets. Elbows should be sloped for drainage purposes.

12. Turbulence, rapid surging, excessive agitation and impingement of fluids onto piping walls should be avoided. Throttle valves, orifices and similar flow-regulating devices should be employed only where necessary. Control devices should be selected partly on the basis of minimum resistance to the flow. For example, a venturi tube is preferable to an orifice plate [12]. In general, one should avoid using flow-controlling devices in close proximity to bends or changes of flow direction downstream.

13. When using soft metals such as lead, copper and their alloys, avoid sudden changes in the flow direction, such as sharp bends.

14. To minimize nucleation, piping systems should be designed to maintain absolute pressure as high as possible.

15. The bend radii of pipes should be designed to be as large as possible. A minimum of three times the pipe diameter is recommended to maintain safe, economic flow velocities.

16. For heat exchangers, coolers, heaters, condensers and related equipment, welding of tubes in tube sheets is recommended over the rolling-in system. Tubes should extend beyond the tube sheets. Also, the design should be such that cooling water starvation at the periphery of the tube bundle is avoided (this is illustrated in Figure 2.7). Heat exchanger tubes should be slanted to provide proper drainage.

17. Heat exchangers should be designed to provide a realistic amount of excess auxiliary exhaust steam with reasonable velocity steam inlets and exhausts. Also, steam baffles should be slanted away from exchanger bracing and other critical areas.

18. For tanks and vessels, welded units are preferred over riveted or bolted designs. Fastener joints provide sites for crevice corrosion. Undrainable horizontal flat tops of tanks should be avoided unless proper drainage schemes are included in the design. Tank bottoms should be sloped toward

drain holes to eliminate the collection of liquids and sludge after emptying. Examples of good and poor designs for this latter case are shown in Figure 2.8.

19. Inlet pipes to vessels should be directed toward the center. Also, the inlet pipe should protrude into the tank and as close as possible to the normal

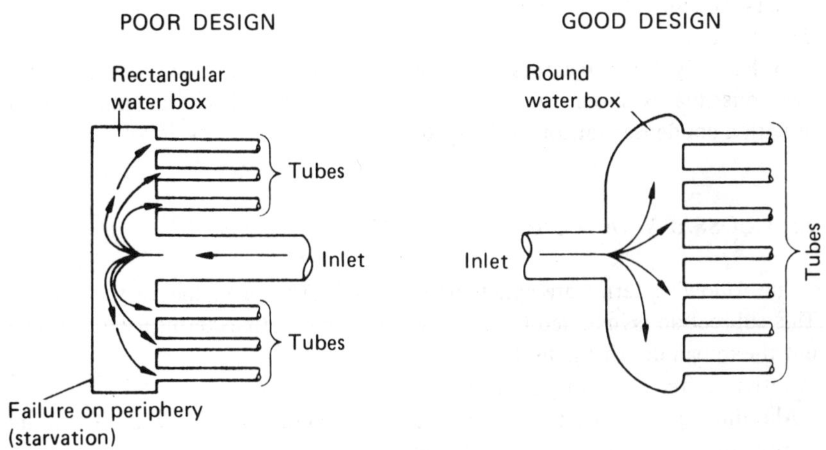

Figure 2.7. Poor and good designs for heat exchanger inlets.

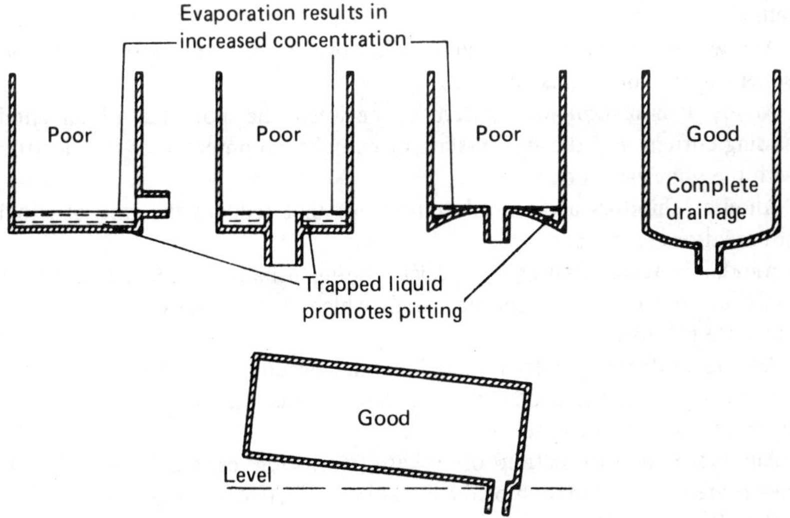

Figure 2.8. Poor and good designs for vessel drainage.

level. This minimizes splashing during filling. Splashing causes precipitation to accumulate on walls, producing fouling and potential corrosion sites.

20. With side inlets and outlets on vessels, protrusions should be avoided that can cause additional turbulence.

21. Tanks designed to hold hygroscopic corrodants should be well sealed to prevent their breathing damp air.

22. Proper vessel designs should avoid discharges from high-positioned coolers directed down a pipe. This situation reduces pressure in coolers via siphon action.

23. Partially filled reactors and storage vessels containing vapors of corrosive constituents should be vented or provided with either vacuum removal or with a condenser return to the system.

2.5 GLOSSARY OF CORROSION TERMS

The following terms are widely used in the corrosion engineer's vocabulary. This subsection is included for the newcomer and defines terms used throughout the remainder of the text.

Active—a free corroding condition.

Aluminizing—a process for impregnating a metal's surface with aluminum to provide protection against oxidation and corrosion.

Anchor Pattern (Surface Profile)—the shape and amplitude of the profile of blast-cleaned or grooved steel, which influences the bond between metallic or paint films and the substrate.

Anion—negatively charged ions that migrate toward the anode of a galvanic cell.

Anode—the electrode at which oxidation of the surface or some component of the solution is occurring.

Anode Polarization—the difference between the potential of an anode passing current and the steady-state or equilibrium potential of the electrode with the same electrode reaction.

Anodic Inhibitor—a chemical constituent that reduces the rate of anodic or oxidation reaction.

Anodic Metallic Coating—a special coating usually comprised, either entirely or in part, of an anodic metal, which is electrically positive to the substrate to which it is applied.

Anodic Protection—a technique for reducing corrosion of a metal surface via passing sufficient anodic current to it to cause its electrode potential to enter into the passive state.

Anodizing—the formation of a hard, corrosion-resistant oxide film on metals via anodic oxidation of the metal in an electrolytic solution.

Base Potential—the potential toward the negative end of a scale of electrode potentials.

Blast Peening—a treatment for relieving tensile stress via inducing beneficial compressive stress in the surface by kinetic energy of rounded abrasive particles.

Breakaway Corrosion—a sudden increase in corrosion rate, particularly under conditions of high-temperature dry oxidation.

Cathode—the electrode of an electrolytic cell where reduction takes place. During corrosion, this is the area at which metal ions do not enter the solution. During cathodic reactions, cations take up electrons and discharge them, hence reducing oxygen. That is, there is a reduction from a higher to a lower state of valency.

Cathodic Inhibitor—a chemical constituent that reduces the rate of cathodic reaction.

Cathodic Protection—a means of reducing the corrosion rate of a metal surface by passing sufficient cathodic current to it to cause its dissolution rate to become very low.

Cation—positively charged ions that migrate to the cathode in a galvanic cell.

Caustic Embrittlement—a form of stress corrosion cracking that occurs in steel exposed to alkaline solutions.

Composite Plate—an electrodeposit that consists of two or more layers of metals deposited separately.

Corrosion Fatigue Limit—the maximum stress that a metal can endure without failure. This is determined in a stated number of stress applications under defined conditions of stressing and corrosion.

Corrosion Potential—the potential of a corroding surface in an electrolyte relative to some reference electrode.

Corrosion Rate—the rate at which corrosion occurs. It is usually reported in units of inches of penetration per year (ipy), mils of penetration per year (mpy), milligrams of weight loss per square decimeter per day (mdd); microns per year (μm/yr), or millimeters per year (mmpy). Note that 1μm = 0.0395 mils.

Couple—an electrical contact made between two dissimilar metals.

Critical Humidity—the relative humidity (RH) at and above which the atmospheric corrosion rate of a metal increases significantly.

Current Density—the average current flowing in an electrolyte (common units are amperes per square foot (A/ft^2), amperes per square decimeter (A/dm^2), amperes per square centimeter (A/cm^2), or milliamperes per square centimeter (mA/cm^2) of either cathode or anode surface.

Deactivation—in corrosion control refers to the removal of a constituent of a liquid that is active in promoting corrosion.

Deposit Attack—to localized corrosion under, and resulting from, a deposit on a metal surface.

Dielectric Strength—the magnitude of electrical nonconductance of a material.

Differential Aeration—the stimulation of corrosion at a localized area by differences in oxygen concentration in the electrolytic solution that is in contact with the metal surface.

Diffusion Coating—application of a metallic coating. The chemical composition of the metal is modified by diffusing the coating into the substrate at the metal's melting temperature.

Electrogalvanizing—the process of galvanizing by electroplating.

Electrolysis—a reaction in which chemical change results in an electrolyte being produced from the passage of electric current.

Electrolyte—chemical constituent, usually a liquid, containing ions that migrate in an electric field.

Electrolytic Cleaning—method of degreasing/descaling metal surfaces via electrolysis. The metal is utilized as an electrode.

Electrophoretic Plating—the production of a layer of deposit as a result of discharge of colloidal particles in solution onto an electrode.

Electroplating—the process of electrodeposition onto a metallic substrate of a thin adherent layer of a metal or alloy having desirable chemical, physical and/or mechanical properties.

Exfoliation—also called lamination, refers to the falling away of metal in layers.

Filiform Corrosion—type of corrosion that takes place under a film in the form of randomly distributed hairlines.

Flame Plating—the deposition of a hard metal coating onto a substrate via application of molten metal at supersonic velocities.

Flash Corrosion—light surface oxidation of cleaned metals that are exposed to the environment for short times.

Fouling—deposition of scale materials on metal surfaces.

Galvanizing—the method of coating iron or steel with zinc by immersion of the metal in a bath of molten zinc.

Green Rot—a corrosion product particular to nickel alloys and greenish in color that normally results from carburization and oxidation of certain nickel alloys at temperatures around 1000°C (1832°F).

Hermetic Seal—an impervious seal made by the fusion of metals of ceramics, which prevents the passage of gas or moisture. The seal can be achieved by brazing, soldering, welding, fusing glass or ceramics.

Ion Erosion—the deterioration of materials caused by ion impact.

Iron Rot—the deterioration of wood caused by contact with iron.

Laminar Scale—rust formation in heavy layers.

Localized Attack—corrosion in which one area of the metal surface is primarily anodic and another predominantly cathodic.

Metal Cladding—the combination of two or more metal compounds bonded metallurgically face to face.

Metallic Coatings—coatings that consist fully or partially of metal applied to metals or nonmetals for the purpose of protection or to improve certain properties.

Metallizing—also called metal spraying; refers to the application of a metal coating to a surface (either metallic or nonmetallic) by means of a spray of molten particles.

Mill Scale—an oxide layer on metals produced by metal rolling, hot forming, welding or heat treatment.

Noble—positive direction of the electrode potential.

Noble Potential—a potential that is more cathodic (i.e., positive) than the standard hydrogen potential.

Oxidation—the loss of electrons by a constituent in a chemical reaction.

Parting—the selective attack of one or more constituents of a solid solution alloy.

Passivation—a reduction of the anodic reaction rate of an electrode involved in an electrochemical reaction, such as corrosion.

Passivity—a condition of a metal or alloy in which the material is normally thermodynamically unstable in a given electrolytic solution but remains visibly unchanged for a prolonged period. The electrode potential of a passive metal is always appreciably more noble than its potential in the active state.

Peen Plating—the deposition of the coating metal (in powder form) on the substrate via a tumbling action in the presence of peening shot.

Pickling—a form of chemical and electrolytic removal of mill scale and corrosion products from the surfaces of metals in an acidic solution. Electrolytic pickling may be anodic or cathodic, depending on the polarization of the metal in the solution.

Plasma Plating—deposition on critical areas of metal coatings resistant to wear and abrasion; normally this is done by means of a high velocity and high-temperature ionized inert gas jet.

Rash Rusting—also called peak spotting; refers to a local corrosion due to inadequate coating of the peaks of a rough surface.

Reduction—the reverse of oxidation; a chemical change of state in which one constituent gains electrons.

Rust—a corrosion product consisting mainly of hydrated iron oxide; the term is used to describe the corrosion products of iron and ferrous ions.

Rust Creep—also called underfilm corrosion; refers to corrosive action that results in damaged or uncoated areas and extends subsequently under the surrounding inert protective coating.

Scaling—the formation of thick corrosion products as layers on a metal surface; in piping systems it is usually the deposition of water-insoluble constituents on a metal surface.

Season Cracking—stress corrosion cracking of brass.

Sherardizing—the process of coating iron or steel with zinc by heating the product to be coated in zinc powder at a temperature below the melting point of zinc.

Stress-Accelerated Corrosion—increased corrosion rate caused by applied stresses.

Surface Preparation—the cleaning of a surface prior to treatment.

Surface Treatment—any suitable means of cleaning and treating a surface that produces a desired surface profile that has required coating characteristics.

Tuberculation—the formation of localized corrosion products scattered over the surface in the form of knob-like tiles.

Vacuum Deposition—also vapor deposition or gas plating; the deposition of metal coatings by means of precipitation (sometimes in vacuum) of metal vapor onto a treated surface. The vapor may be produced by thermal decomposition, cathode sputtering or evaporation of the molten metal in air or an inert gas.

Weld Decay—localized corrosion of weld metal.

REFERENCES

1. Butler, G., and H. C. K. Ison. *Corrosion and Its Prevention in Waters* (London: Leonard Hill Publishers, 1966).
2. Evans, U. R. *The Corrosion and Oxidation of Metals: Scientific Principles and Practical Applications* (London: Edward Arnold Publishers, Ltd., 1960).
3. Pludek, V. R. *Design and Corrosion Control* (New York: John Wiley & Sons, Inc., 1977).
4. Tomashov, N. D. *Theory of Corrosion and Protection of Metals: The Science of Corrosion* (New York: Macmillan Publishing Co., Inc., 1966).
5. Fontana, M. G., and N. D. Greene. *Corrosion Engineering* (New York: McGraw-Hill Book Co., 1978).
6. Fontana, M. G., and R. W. Staehle. *Advances in Corrosion Science and Technology,* Vol. 2 (New York: Plenum Publishing Corp., 1972).
7. Staehle, R. W. "Comments on the History of Engineering and Science of Stress Corrosion Cracking," in *Proc. Fundamental Aspects of Stress Corrosion Cracking* (Houston, TX: National Association of Corrosion Engineers, 1969).
8. LaQue, F. L., and H. R. Copson. *Corrosion Resistance of Metals and Alloys* (New York: Van Nostrand Reinhold Co., 1963).
9. Harada, Y. "High Temperature Corrosion in Heavy Oil Firing Boilers," *Proc. Fifth Int. Cong. on Metallic Corrosion* (Houston, TX: National Association of Corrosion Engineers, 1974).
10. Gilbert, P. T. *Corrosion Problems of the Petroleum Industry* (London: Society of Chemical Industry, 1960).
11. Cheremisinoff, P. N., and R. A. Young. *Pollution Engineering Practice Handbook* (Ann Arbor, MI: Ann Arbor Science Publishers, Inc., 1975).
12. Cheremisinoff, N. P. *Applied Flow Measurement: Fundamentals and Technology* (New York: Marcel Dekker, Inc., 1979).

CHAPTER 3

PROPERTIES OF MATERIALS AND THEIR SELECTION

3.1 GENERAL PROPERTIES AND SELECTION CRITERIA

Proper material selection for chemical and process equipment is one of the first important problems encountered by the designer. Among the many parameters that must be considered are structural strength specifications, heat resistance, corrosion resistance, physical properties, fabrication characteristics, composition and structure of material and cost.

The properties that materials must have for a particular application depend largely on the environment in which they are to be used in. Material selection begins from determination of equipment, operating conditions, temperature, pressure, and various components in the process.

No materials have properties that fulfill all requirements. For example, good heat conductivity is a desirable property for the fabrication of heat exchanger surfaces, but not for insulation purposes. Obviously, both positive and negative properties can coexist in a single material. A corrosion resistant material may be insufficient for heat resistance or mechanical strength. Strong materials may be too brittle, e.g., ferrosilicon. Also, materials that have good mechanical and chemical properties may be too expensive.

The initial cost of a material does not provide the entire economic picture. At first, strong materials that are expensive may be more favorable than less expensive ones. The cost of processing cheap materials is sometimes high, thus creating abnormally high fabrication costs. For example, the cost of a ton of granite is a dozen times cheaper than that of nickel chromium/molybdenum steel. However, granite absorption towers are more expensive than steel towers of the same volume because of the high costs associated with processing granite. Furthermore, granite towers are much heavier than the steel ones; therefore, they require stronger, and thus more expensive, foundations.

Because any material may be characterized by some desirable and nondesirable properties with respect to a specific application, the selection of materials is reduced to a reasonable compromise. In so doing, one strives to select materials so that properties correspond to the basic demands determined by the function and operating conditions of the equipment, tolerating some of the undesirable properties.

The basic requirement for materials intended for fabricating chemical apparatuses is mostly corrosion resistance because this determines the durability of equipment. Often, corrosion data are reported as a weight loss per unit of surface area per unit of time. It is easy to transfer from such data to the penetration rate using the following relation:

$$V = 8.76 \frac{G}{\rho} \text{ mm/yr} \tag{3.1}$$

where G = weight loss at uniform corrosion (kg/m^2 hr)
V = corrosion rate (mm/yr)
ρ = density of material (kg/dm^3)

Materials must have high chemical resistance as well as durability. For example, if the material dissolves in the product, the product quality may deteriorate, or materials may act as catalysts promoting side reactions and thus decreasing the yield of the primary product. Usually there are several materials suitable for use under the process conditions. In such cases the material is selected by additional considerations. For example, if a vessel must be equipped with a sight glass, the material for fabricating this item must be transparent and safe. In this case, plexiglas may be used if the vessel operates at low temperatures. For higher temperatures, glass is used; however, glass is brittle and very sensitive to drastic temperature changes. Therefore, the accessories must be designed so that the glass could not be broken occasionally and the poisonous or aggressive liquid allowed to escape. In this application, double glasses or valves must be provided for an emergency to shut off the accessory from the working space of the vessel. Consequently, the poor construction property of glass may cause additional complications in the design. At very high temperatures sight glasses are made from mica. For high-pressure drops they are made from rock crystal (an excellent but very expensive material). An example is shown in Figure 3.1.

Gauges used for measuring the liquid level in vessels may be of semi-transparent and even nontransparent materials. Figure 3.2 illustrates a simple level gauge on a steel vessel used for liquid ammonia storage. As shown, a narrow strip of insulation is taken away from the vessel's shell to expose the bare metal. Consequently, the heat transfer coefficient from the

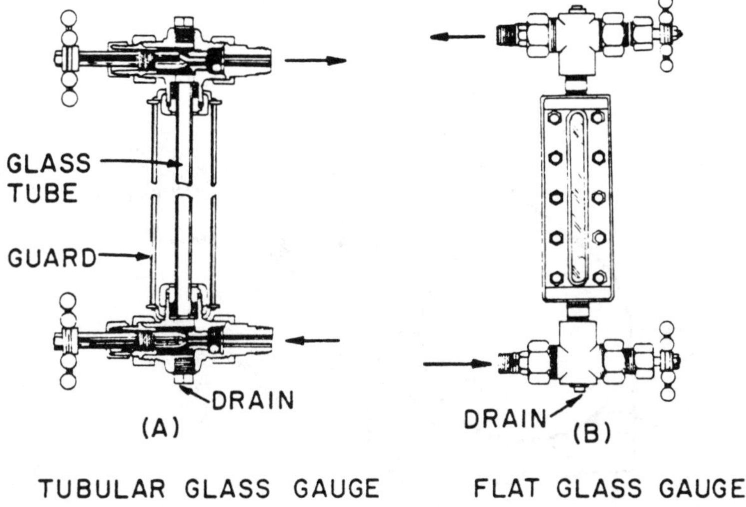

Figure 3.1. Typical glass sight gauges.

liquid ammonia is high, and the heat transfer coefficient from air to the wall is low. The wall temperature close to the liquid ammonia will be almost the same as that of ammonia, and the unprotected part of the vessel that is contacted with liquid will be covered with a layer of frost indicating the height of the liquid level.

Let us now consider the basic materials used in the fabrication of chemical equipment from the point of view of a designer. The principal construction materials for welded, forged and cast chemical vessels are: cast irons, gray cast iron, white cast iron, malleable cast irons, nodular cast iron, austenitic cast iron, high-silicon cast iron, low-carbon steels (mild steel), high-carbon steels, low-carbon/low-alloy steels, high-carbon/low-alloy steels, high-alloy steels (corrosion-resistant, heat resistant and high-temperature), nickel and nickel alloys. Each of these is described below.

3.2 PROPERTIES OF CAST IRONS

Three main factors that determine the properties of cast iron are:

1. the chemical composition of the cast iron,
2. the rate of cooling of the casting in the mold, and
3. the type of graphite formed.

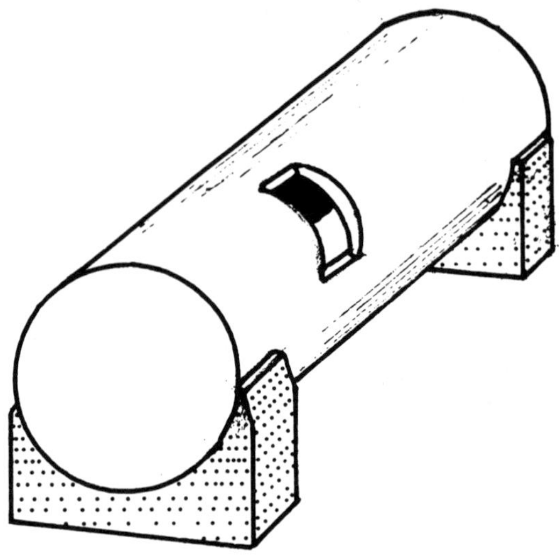

Figure 3.2. Liquid-level gauge for an ammonia tank.

Most commercial cast irons contain 2.5–4% carbon, and it is the occurrence of some of this carbon as free graphite in the matrix that is the characteristic feature of thin material. About 0.8–0.9% carbon is in a bound form as cementite (iron carbide).

The cast irons usually have a ferrite-pearlite structure, which determines its mechanical properties. The ferrite content determines the cast iron's viscosity, while the pearlite content determines its rigidity and strength.

Because cast iron has a carbon content approximately equivalent to its eutectic composition, it can be cast at lower temperatures than steel and flows more readily than steel because of its much narrower temperature solidification range. The presence of the graphite flakes in cast iron decreases its shrinkage on solidification much less than that of steel. These factors contribute to the fabrication of cast iron as sound castings in complex shapes and with accurate dimensions at low cost.

The physical properties of cast irons are characterized by the following data:

- density $\rho = 7.25$ kg/dm^3
- melting temperature $t_m = 1250 - 1280$°C
- heat capacity $C_p = 0.13$ kcal/kg°C
- heat conductivity $\lambda = 22 - 28$ kcal/m °C hr
- coefficient of linear expansion $\alpha = 11 \times 10^{-6}$

The cast irons do not possess ductility. They cannot be pressed or forged even while heated; however, their machining properties are considered good. Typical mechanical properties of various types of cast iron are given in Table 3.1.

3.2.1 Gray Cast Iron

Gray cast iron is the most commonly used cast iron and is the least expensive. It is the easiest to cast and machine. The tensile strength of gray cast iron ranges from 155 to 400 N/mm^2 (10 to 26 ton/in.^{2}). The tensile modulus ranges from 70 to 140 kN/mm^2 and the hardness from 130 to 300 DPN.

In nearly all standards for gray cast iron the grades are designated according to the tensile strength, not composition. In the British standard BS1452, for example, there are seven grades from 155 to 400 N/mm^2 (10 to 26 tonf/in.^{2}). This is the tensile strength measured on a test bar having a diameter of approximately 30 mm (1.2 in.). The actual strength of a casting will differ from that of the test bar according to the cross-sectional area (Table 3.2).

Castings are designed to be loaded in compression because the compressive strength of gray iron is about three times that of its tensile strength.

The recommended maximum design stress in tension is one-quarter the ultimate tensile strength (for cast irons a value up to 185 N/mm^2 (12 tonf/in.^{2})). The fatigue strength is one-half the tensile strength. Notched

Table 3.1. Typical Mechanical Properties of Various Types of Cast Iron [1]

Material	Specification	Tensile strength (tonf/in.^{2})	Tensile strength (N/mm^2)	Elongation (%)
Gray Cast Iron	BS1452 Grade 10	10	155	–
	14	14	215	–
	26	26	400	–
Nodular Cast Iron	BS2789 SNG 24/17	24	370	17
	32/7	32	500	7
	47/2	47	730	2
Malleable Cast Iron				
Blackheart	BS310 B290/6		290	6
	B340/12		340	12
Whiteheart	BS309 W340/3		340	3
	W410/4		410	4
Pearlite	BS3333 P440/7		440	7
	P540/5		540	5
	P690/2		690	2

Table 3.2. Typical Data Showing the Effect of Strength on Gray Iron Castings [1]

Gray iron to BS1452	Tensile strength, N/mm^2, of casting with section thickness of:				
	10 mm	20 mm	75 mm	100 mm	150 mm
Grade 20	350	280	280	230	220
14	230	200	150	140	120
12	200	170	120	110	110
10	170	140	110	90	75

specimens show the same value as unnotched specimens. For 220 N/mm^2 (14 $tonf/in.^2$) grades and above, the fatigue strength of unnotched specimens is approximately one-third the tensile strength. There is some notch sensitivity, although much less than is found in steel.

3.2.2 White Cast Iron

White cast iron is very hard (from 400 to 600 DPN) and brittle. All white cast irons are very difficult to machine and usually are finished by grinding. Table 3.3 gives properties of the four principal types of white cast irons.

3.2.3 Malleable Cast Irons

This type of cast iron is made by high-temperature heat treatment of white iron castings. The mechanical properties of malleable cast irons are given in Table 3.1; usually they are applied to the fabrication of conveyor chain links, pipe fittings and gears.

Table 3.3. Properties of White Iron [1]

	Unalloyed White Iron	Low-Alloy White Iron	Martensitic White Iron (Ni-hard)	High-Carbon, High-Chromium, White Iron
Composition (%)				
Carbon	3.5	2.6	3.0	2.8
Silicon	0.5	1.0	0.5	0.8
Nickel			3.5	–
Chromium		1.0	2.0	27
Hardness (DPN)	600	400	600	500
Tensile strength, (N/mm^2)	270	300	330	420

3.2.4 Nodular Cast Iron

Nodular cast iron (also referred to as ductile cast iron) is manufactured by inoculating the molten metal with magnesium or cesium. It is characterized by a homogeneous structure, higher than usual abrasion-resistance and strength for dynamic loads, and by easy machining. A wide variety of grades are available, with typical tensile strengths ranging from 380 to 700 N/mm^2 (25 to 40 tonf/in.^{2}), elongations from 17 to 2%, and hardness from 150 to 300 DPN (see Table 3.1). The tensile modulus is approximately 170 kN/mm^2. The design stress is half the 0.1% proof stress, and the fatigue design stress is one-third the fatigue limit. The nodular cast iron is used for many applications such as valves in pipelines for petroleum products, underground pipelines and so on.

3.2.5 Austenitic Cast Iron

Austenitic cast irons (either flake graphite irons or nodular graphite irons) are produced by mixing in nickel from 13–30%, chromium from 1–5% and copper from 0.5–7.5 (to lower nickel-containing grades to augment the corrosion resistance at lower cost).

The main advantages of austenitic cast irons are corrosion and heat resistance. For corrosion resistance, the flake and nodular are similar, but the mechanical properties of nodular cast irons are superior. Some of the commercially available austenitic cast irons are given in the Tables 3.4 and 3.5.

3.3 APPLICATION REQUIREMENTS OF CAST IRONS

3.3.1 Abrasion Resistance

The white cast irons and their low alloys have good abrasion resistance properties [2,3]. White cast irons are used for grinding balls, segments for mill liners and slurry pumps. In the ceramic industry they are used for muller tyres and augers; in the pulp and paper industry for attrition mill plates and chip feeders; and in the paint industry for balls for grinding pigments.

3.3.2 Corrosion Resistance

The corrosion resistance of unalloyed and low-alloy flake, nodular, malleable and white cast iron is comparable to mild- and low-alloy steel. However, these cast irons have a major advantage over steel; namely, greater cross section or wall thickness than steel. Consequently, they have a

Table 3.4. Properties of Spheroidal

BS3468 Designation	ASTM[a] A439 Designation	Trade Names	C (% max.)	Si (%)	Mn (%)
AUS202 Grade A	D-2	SG Ni-resist type D2	3.0	1.0 to 2.8	0.7 to 1.5
AUS202 Grade B	D-2b	SG Ni-resist type D2b	3.0	1.0 to 2.8	0.7 to 1.5
AUS203	D-2c	SG Ni-resist type D2c	3.0	1.0 to 2.8	1.8 to 2.4
–	ASTM A571		2.6	2.5	3.75 to 4.5
AUS204	–	SG Nicrosilal	3.0	4.5 to 5.5	1.0 to 1.5
AUS205	D-3	SG Ni-resist type D3	2.6	1.5 to 2.8	0.5 max.
–	D-4	SG Ni-resist type D4	2.6	5.0 to 6.0	1.0 max.
–	D-5	SG Ni-resist type D5, Minovar	2.4	1.0 to 2.8	1.0 max.

[a]There are slight differences between BSS and ASTM compositions.
[b]Properties in brackets are indicative, not mandatory.
[c]0.2% proof stress.

Table 3.5. Properties of Flake

BS3468 Designation	ASTM A436 Designation	Trade Names	C (% max.)	Si (%)	Mn (%)
AUS101 Grade 1	Type 1	Ni-resist type 1	3.0	1.0 to 2.8	1.0 to 1.15
AUS101 Grade B	Type 1b	Ni-resist type 1b	3.0	1.0 to 2.8	1.0 to 1.5
AUS102 Grade A	Type 2	Ni-resist type 2	3.0	1.0 to 2.8	1.0 to 1.5
AUS104	–	Nicrosilal	1.6 to 2.2	4.5 to 5.5	1.0 to 1.5
AUS105	Type 3	Ni-resist type 3	2.6	1.0 to 2.0	0.4 to 0.8
–	Type 4	Ni-resist type 4	2.6	5.0 to 6.0	0.4 to 0.8
–	Type 6		3.0	1.5 to 2.5	0.8 to 1.5

Graphite-Grade Cast Irons

Ni (%)	Cr (%)	UTS (N/mm^2 min.)	0.5% Proof Stress (N/mm^2 min.)	Elongation (% min.)	Hardness (HB)	Elastic Modulus (kN/mm^2)
1.80 to 22.0	1.0 to 2.5	370	230	8.0	201 max.	(115)[b]
18.0 to 22.0	2.0 to 3.5	370	230	6.0	205 max.	(151)
21.0 to 24.0	0.5 max.	370	230	20.0	170 max.	(110)
21.0 to 24.0	0.20 max.	430		30	170 max.	
18.0 to 22.0	1.0 to 2.5	370	230	10.0	230 max.	
28.0 to 32.0	2.5 to 3.5	370	230	7.0	201 max.	(110)
28.0 to 32.0	4.5 to 5.5	400			202–273	
34.0 to 36.0	0.10 max.	380	210[c]	20.0	131–185	

Graphite-Grade Cast Irons

Ni (%)	Cr (%)	Cu (%)	UTS (N/mm^2 min.)	Elongation (% min.)	Hardness (HB)	Elastic modulus (kN/mm^2)
13.5 to 17.5	1.0 to 2.5	5.5 to 7.5	140	2.0	212 max.	(90)
13.5 to 17.5	2.0 to 3.5	5.5 to 7.5	180	–	248 max.	(105)
18.0 to 22.0	1.0 to 2.5	0.5 max.	140		212 max.	(90)
18.0 to 22.0	1.8 to 4.5	0.5 max.	190	2.0	248 max.	(110)
28.0 to 32.0	2.5 to 3.5	0.5 max.	170	–	212 max.	(105)
29.0 to 32.0	4.5 to 5.5	0.5 max.	170	–	149–212	
18.0 to 22.0	1.0 to 2.0	3.5 to 5.5	170	–	124–174	

longer life, although they corrode at the same rates. For example, the corrosion rate for boiling concentrated sulfuric acid in large cast iron pots is very high, but a reasonable life is obtained by making the bottom of the pots three inches thick. Secondly, although the matrix of a cast iron pipe may rust, for example, the graphite network prevents disintegration of the pipe and permits its duty for a longer time than a steel pipe.

The austenitic cast irons are in widespread use in many industries (food, pharmaceutical, petroleum, chemical, petrochemical, pulp and paper, etc.) in mildly corrosive and erosive situations where the life of unalloyed or low-alloy cast iron or steel is short, but the high cost of stainless steel and nonferrous alloys cannot be justified.

Other austenitic cast iron applications can be found in food and dairy production, where the metallic contamination of the product must be eliminated.

3.3.3 Temperature Resistance

The persistent increase in volume of cast iron items in high-temperature situations becomes the limiting factor in the use of unalloyed cast irons, especially in flake graphite castings [4,5].

The addition in a casting of about 1% of chromium can control the growth in the temperature range from 400 to 600°C. Above 600°C, scaling due to surface oxidation becomes an undesirable phenomenon in the use of unalloyed cast irons. For achieving dimensional stability and long life, silal cast irons (containing 5.5–7% silicon) may be used for temperatures up to 800°C, in cases where it is not subjected to thermal shock. For thermal cycling and thermal shock situations for temperatures up to 950°C, the 30% nickel austenitic cast irons are preferred; and above this temperature, where there is no thermal shock, the 28% chromium cast iron is recommended.

Table 3.6 gives commonly used maximum working stresses for various grades of cast irons up to 600°C.

Gray cast irons do not have the abrupt ductile to brittle fraction transition down to −40°C as takes place in steels. Special austenitic nodular cast iron similar to the AUS 203 grade, but with a higher manganese content of about 4%, has been obtained for cryogenic purposes for temperatures down to −253°C.

3.3.4 Welding Cast Iron

Welding is sometimes used to repair broken and defective castings [6]. This process is more difficult than welding steel because the high carbon content in cast iron may lead to brittle structures on cooling, thus

Table 3.6. Maximum Working Stresses for Various Grades of Cast Iron up to 600°C

	Maximum safe working stresses (N/mm^2)				
	350°C	400°C	450°C	500°C	600°C
Gray Iron, Grade 17	45	30	15	–	–
Nodular Iron: SNG 27/12	60	40	20		
Blackheart, Malleable, B340/12	45	30	20		
Pearlitic, Malleable, P33/4	75	45	30		
Austenitic Nodular, AUS203	120	100	75	45	20

causing cracking. However, special techniques have been developed for fusion and nonfusion welding. Some of these are outlined in Table 3.7. The engineer should consider carefully whether the properties of varieties of cast irons will suit his demands before specifying more expensive materials.

3.4 PROPERTIES OF STEELS

The second group of structural materials in the iron base category is steels. They have obtained an exclusive importance because of their strength, viscosity, and their ability to withstand dynamic loads. Also,

Table 3.7. Rods and Electrodes for Fusion-Welding Cast Iron

Filler Rod or Electrode	Welding Method	Suitability	Comments
Cast Iron (e.g. BS1453)	Oxyacetylene	Gray irons but not malleable	Matching color Not machinable
Mild Steel	Metal arc	Cavity filling of thin castings	Not machinable
Nickel iron (55/45)	Metal arc	For high-strength gray and nodular irons	Reasonable machinability
Nickel Copper (70/30)	Metal arc	For all gray irons	Easy to machine
Brass (60 Cu/40 Zn)	Oxyacetylene	Gray and malleable irons	Often called 'bronze' welding
Tin Bronze (7 Sn/93 Cu)	Oxyacetylene	Gray and malleable irons	

they are beneficial for producing castings, forgings, stamping, rolling, welding, machining and heat treatment works. Steels change their properties over a wide range depending on their composition, heat treatment and machining.

Most steels have a carbon content of 0.1–1%, but in structural steels this does not exceed 0.7%. With higher carbon contents, steel increases in strength but decreases in plasticity and weldability. In the carbon steels designed for welding, the carbon content must not exceed 0.3%; in the alloy steels it must not exceed 0.2%. When the carbon content in the steels exceeds the abovementioned value, they are susceptible to air hardening. Hence, high stresses may be created and hardening fractures in welding zones may be formed. The steels with low carbon content (below 0.2%) are easily stamped and stretched, easily cemented and nitrated, but badly machined. The physical properties of low-carbon, low-alloy steels are characterized by the following data:

- density = 7.85 kg/dm^3
- melting temperature t_m = 1400–1500°C
- thermal conductivity λ = 40–50 kcal/m°C hr

3.4.1 Low Carbon Steels (Mild Steel)

Mild steel (<0.25% carbon) is the most commonly used, readily welded construction material, and has the following typical mechanical properties (Grade 43A in BS4360; weldable structural steel):

- Tensile strength, 430 N/mm^2
- Yield strength, 230 N/m^2
- Elongation, 20%
- Tensile modulus, 210 kN/mm^2
- Hardness, 130 DPN

No one steel exceeds the tensile modulus of mild steel. Therefore, in applications in which rigidity is a limiting factor for design (e.g., for storage tanks and distillation columns), high-strength steels have no advantage over mild steel. Stress concentrations in mild steel structures are relieved by plastic flow and are not as critical in other, less-ductile steels.

Low-carbon plate and sheet are made in three qualities: fully kilned with silicon and aluminum, semikilned (or balanced), and rimmed steel. Fully killed steels are used for pressure vessels. Most general-purpose structural mild steels are semikilned steels. Rimming steels have minimum amounts of deoxidation and are used mainly as thin sheet for consumer applications.

The strength of mild steel can be improved by adding small amounts (not exceeding 0.1%) of niobium, which permits the manufacture of semikilned steels with yield points up to 280 N/mm^2. By increasing the manganese content to about 1.5% the yield point can be increased up to 400 N/mm^2. This provides better retention of strength at elevated temperatures and better toughness at low temperatures.

3.4.2 Corrosion Resistance

Equipment from mild steel usually is suitable for handling organic solvents, with the exception of those that are chlorinated, cold alkaline solutions (even when concentrated), sulfuric acid at concentrations greater than 88%, and nitric acid at concentrations greater than 65% at ambient temperatures [7].

Mild steels are rapidly corroded by mineral acids even when they are very dilute (pH less than 5). However, it is often more economical to use mild steel and include a considerable corrosion allowance on the thickness of the apparatus. Mild steel is not acceptable in situations in which metallic contamination of the product is not permissible.

3.4.3 Heat Resistance

The maximum temperature at which mild steel can be used is 550°C. Above this temperature the formation of iron oxides and rapid scaling makes the use of mild steels uneconomical. For equipment subjected to high loadings at elevated temperatures, it is not economical to use carbon steel in cases above 450°C because of its poor creep strength. (Creep strength is time-dependent, with strain occurring under stress.)

3.4.4 Low Temperatures

At temperatures below 10°C the mild steels may lose ductility, causing failure by brittle fracture at points of stress concentrations (especially at welds) [8,9]. The temperatures at which the transition occurs from ductile to brittle fraction depends not only on the steel composition, but also on thickness.

Stress relieving at 600–700°C for steels decreases operation at temperatures some 20°C lower. Unfortunately, suitable furnaces generally are not available, and local stress relieving of welds, etc., is often not successful because further stresses develop on cooling.

3.4.5 High-Carbon Steels

High-carbon steels containing more than 0.3% are difficult to weld, and nearly all production of this steel is as bar and forgings for such items as shafts, bolts, etc. These items can be fabricated without welding. These steels are heat treated by quenching and tempering to obtain optimum properties up to 1000 N/mm^2 tensile strength.

3.4.6 Low-Carbon, Low-Alloy Steels

Low-carbon, low-alloy steels are in widespread use for fabrication-welded and forged pressure vessels. The carbon content of these steels is usually below 0.2%, and the alloying elements that do not exceed 12% are nickel, chromium, molybdenum, vanadium, boron and copper. The principal applications of these steels are given in Table 3.8.

3.4.7 Mechanical Properties

The maximum permissible loading of low-alloy steels according to the ASME code for pressure vessels is based on proof stress (or yield point), which is applicably superior to those of carbon steels. The cost of a pressure vessel in alloy steel may be more expensive than in carbon steel. However, consideration should be given to other cost savings resulting from thinner-walled vessels, which provide fabrication savings on weldings, stress relieving, transportation, erection and foundation. Table 3.9 compares mild- and low-alloy steels used for fabricating spherical gas storage tanks.

3.4.8 Corrosion Resistance

The corrosion resistance of low-alloy steels is not significantly better than that of mild steel for aqueous solutions of acids, salts, etc. The addition of 0.5% copper forms a rust-colored film preventing further steel deterioration; small amounts of chromium (1%) and nickel (0.5%) increase the rust

Table 3.8. Applications of Low-Carbon, Low-Alloy Steels [10]

0.5 Mo 1.25 CrMo 2.25 CrMo 6 to 12 CrMoVW	High creep strength for: 1. pressure vessels such as boilers operating at elevated temperatures; and 2. oil refinery vessels such as crackers and reformers with high hydrogen pressures.
5 to 9% Cr	For oil refinery applications involving high-sulfur process streams, e.g., pipe stills.
CuCr (Corten)	Rust-resisting steels for structural applications.
2 to 9% Ni	For cryogenic applications.

Table 3.9. Comparison of Mild and Low-Alloy Quenched and Tempered Steels [11][a]

	Low-Alloy Steel, a	Low-Alloy Steel, b	Mild Steel
Relative total weight of shell plate	1	1.35	3.2
Relative total weight of other steel required for supporting structures, etc.	1	2	20
Diameter of vessels, (m)	39	31	14
Capacity of each vessel, (m^2)	30,000	15,000	1500
Shell thickness, (mm)	35	37	35

[a]Steel a was quenched and tempered to a tensile, strength of 830 N/mm^2 and a yield strength of 670 N/mm^2, and steel b to 620 and 450 N/mm^2, respectively. It was considered that the maximum thickness of metal that could be welded onsite was 38 mm.

resistance of copper steels still further. Low alloy steels have good resistance to corrosion by crude oils containing sulfur. This is illustrated by the data in Figure 3.3.

In operations involving hydrogen at partial pressures greater than 35 kgf/cm^2 and temperatures greater than 250°C, carbon steels are decarborized and fissured internally by hydrogen [13]. Small additions of molybdenum prevent hydrogen attack at temperatures up to 350°C and pressures up to 56 kgf/cm^2. For higher temperatures and pressures chromium/molybdenum steels (2.25 Cr, 0.5 Mo) are used. Figure 3.4 shows operating limits for steels in atmospheres containing hydrogen.

3.4.9 Oxidation Resistance and Creep Strength

Chromium is the most effective alloying element for promoting resistance to oxidation. Table 3.10 gives temperatures at which steels can be used in air without excessive oxidation. In atmospheres contaminated with sulfur, lower maximum temperatures are necessary.

In fractionation columns for petroleum products, where the oxygen content is restricted, higher temperatures can be used without excessive waste of the metal.

The creep strength of steels is a factor limiting the maximum temperatures for such high-pressure equipment as shells and stirrers of high temperature reactors. Table 3.10 presents creep data for temperatures ranging from 400 to 600°C. The stress for 1% creep in 100,000 hours (which is a design criterion) is accepted to be not less than two-thirds of the creep stresses.

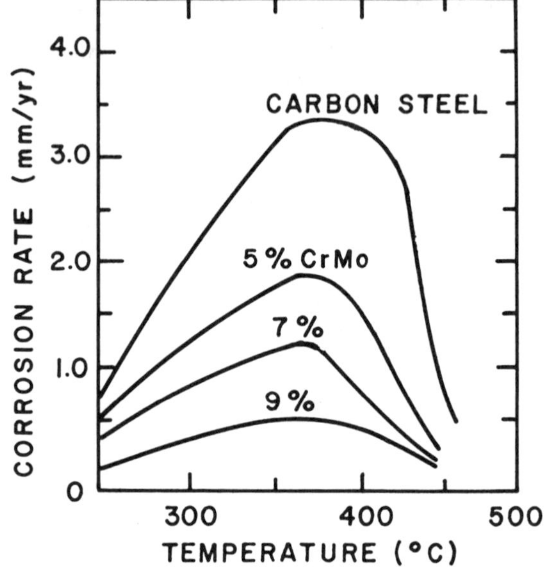

Figure 3.3. Effect of temperature on corrosion rates of steels in crude oil containing sulfur [12].

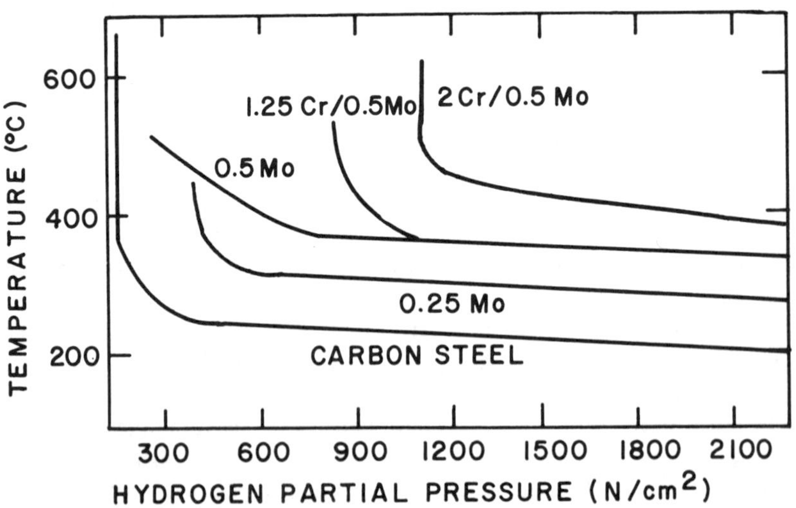

Figure 3.4. Operating limits for steels in atmospheres containing hydrogen.

Table 3.10. Alloying Effects that Improve Creep Properties [14]

Steel	Stress for rupture, N/mm², in 100,000 hr at						
	400°C	460°C	480°C	500°C	560°C	580°C	600°C
Carbon (silicon-killed)	180	75	48	27			
1% Cr Mo				145	40		
21% Cr Mo				168	70	52	
½% Cr Mo V				185	70		
12% Cr Mo V				215	87	70	55

3.4.10 Low-Temperature Ductility

Nickel is the alloying element used for improving low-temperature ductility. The addition of 1.5% nickel to 0.25% Cr/0.25% Mo steels provides satisfactory application for moderately low temperatures down to about -50°C.

Heat treatment by quenching and tempering improves the low temperature ductility of steels such as 0.5 Cr, 0.5% Mo, 1% Ni Type V. For lower-temperature application (below −196°C), up to 9% nickel is used as the sole alloying element.

3.4.11 High-Carbon, Low-Alloy Steels

High-carbon (about 0.4%), low-alloy steels that are not weldable usually are produced as bars and forging for such items as shafting, high-temperature bolts and gears, and ball bearing components. These steels can be less drastically quenched and tempered to obtain tensile strengths of at least 1500 N/mm^2, thus minimizing the danger of cracking [15].

3.5 PROPERTIES OF HIGH-ALLOY STEELS

Stainless and heat-resisting steels containing at least 18% by weight chromium and 8% nickel are in widespread use in industry. The structure of these steels is changed from magnetic body centered cubic or ferritic crystal structure to a nonmagnetic, face-centered cubic or austenitic crystal structure.

3.5.1 Chromium Steels (400 Series), Low-Carbon Ferritic (Type 405): 12–13% Chromium

The main use of this type steel is for situations in which the process material may not be corrosive to mild steel, yet contamination due to rusting is not tolerable and temperatures or conditions are unsuitable for aluminum. However, prolonged use of these steels in the temperature range of 450 to 550°C causes low-temperature embrittlement of most ferritic steels with more than 12% chromium [16].

3.5.2 Medium Carbon Martensitic: 13–17% Chromium (Types 403, 410, 414, 416, 420, 431, 440)

These steels resist oxidation scaling up to 825°C but are difficult to weld and, thus, are used mainly for items that do not involve welded joints [17]. They are thermally hardened and useful for items that require cutting edges and abrasion resistance in mildly corrosive situations. However, they should not be tempered in the temperature range of 450 to 650°C. This reduces the hardness and wear resistance and also lowers the corrosion resistance because of the depletion of chromium in solution through the formation of chromium carbides.

3.5.3 Medium Carbon Ferritic: 17–30% Chromium (Types 430 and 446)

The 17% ferritic steels are easier to fabricate than the martensitic grades. They are used extensively in equipment for nitric acid production. The oxygen- and sulfur-resistant 30% chromium steel can be used at temperatures up to 1150°C but only for lightly loaded and well-supported furnace items because of its poor creep and brittlement properties when equipment is down to ambient temperatures [18].

3.5.4 Chromium/Nickel Austenitic Steels (300 Series)

The excellent corrosion resistance over a wide range of operating conditions and readily available methods of fabrication by welding and other means of shaping metals make these steels the most extensively used throughout the chemical and allied industries.

The formation of a layer of metal oxide on the surface of this steel provides better corrosion resistance in oxidizing environments than under reducing conditions. Common steels 304, 304L, 347, 316 and 316L are used for equipment exposed to aqueous solutions of acids and other low-temperature corrosive conditions. For high-temperature regimes involving

Table 3.11. AISI Classifications of Wrought Stainless and Heat-Resisting Steels (based on AISI type numbers) [19]

Type number	Chemical composition, %					
	C	Mn (max.)	Si (max.)	Cr	Ni	Trace Elements
	200 Series-Economy steels with Ni partly replaced by Mn					
201	0.15 max	5.50- 7.50	1.00	16.00-18.00	3.50- 5.50	N: 0.25 max.
202	0.15 max.	7.50-10.00	1.00	17.00-1900	4.00- 6.00	N: 0.25 max.
301	0.15 max.	2.00	1.00	16.00-18.00	6.00- 8.00	
302	0.15 max.	2.00	1.00	17.00-19.00	8.00-10.00	
302B	0.15 max.	2.00	2.00-3.00	17.00-19.00	8.00-10.00	
303	0.15 max.	2.00	1.00	17.00-19.00	8.00-10.00	Mo: 0.60 max.
303Se	0.15 max.	2.00	1.00	17.00-19.00	8.00-10.00	Se:0.15 min.
304	0.08 max.	2.00	1.00	18.00-20.00	8.00-12.00	
304L	0.03 max.	2.00	1.00	18.00-20.00	8.00-12.00	
305	0.12 max.	2.00	1.00	17.00-19.00	10.00-13.00	
308	0.08 max.	2.00	1.00	19.00-21.00	10.00-12.00	
309	0.20 max.	2.00	1.00	22.00-24.00	12.00-15.00	
309S	0.08 max.	2.00	1.00	22.00-24.00	12.00-15.00	
310	0.25 max.	2.00	1.50	24.00-26.00	19.00-22.00	
310S	0.08 max.	2.00	1.50	24.00-26.00	19.00-22.00	
314	0.25 max.	2.00	1.50-3.00	23.00-26.00	19.00-22.00	
316	0.08 max.	2.00	1.00	16.00-18.00	10.00-14.00	Mo: 2.00-3.00
316L	0.03 max.	2.00	1.00	16.00-18.00	10.00-14.00	Mo: 2.00-3.00
317	0.08 max.	2.00	1.00	18.00-20.00	11.00-15.00	Mo: 3.00-4.00
321	0.08 max.	2.00	1.00	17.00-19.00	9.00-12.00	Ti: 5 × C, min.
347	0.08 max.	2.00	1.00	17.00-19.00	9.00-13.00	Cb-Ta: 10 × C, min.
348	0.08 max	2.00	1.00	17.00-19.00	9.00-13.00	Cb-Ta: 10 × C, min. Ta: 0.10 max., Co: 0.20 max.

Table 3.11, continued

Type number	Chemical composition, %					
	C	Mn (max.)	Si (max.)	Cr	Ni	Trace Elements
			Ferritic Steels			
F405	0.08 max.	1.00	1.00	11.50-14.50		
F430F	0.12 max.	1.25	1.00	14.00-18.00		Mo: 0.60 max.
F430Se	0.12 max.	1.25	1.00	14.00-18.00		Se: 0.15 min.
F446	0.20 max.	1.50	1.00	23.00-27.00		N: 0.25 max.
			Martensitic Steels			
M403	0.15 max.	1.00	0.50	11.50-13.00		
M410	0.15 max.	1.00	1.00	11.50-13.50		Al: 0.10-0.30
M414	0.15 max.	1.00	1.00	11.50-13.50	1.25- 2.50	
M416	0.15 max.	1.25	1.00	12.00-14.00		Mo: 0.60 max.
M416Se	0.15 max.	1.25	1.00	12.00-14.00		Se: 0.15 min.
M420	Over 0.15	1.00	1.00	12.00-14.00		
M430	0.12 max.	1.00	1.00	14.00-18.00		
M431	0.20 max.	1.00	1.00	15.00-17.00	1.25- 2.50	
M440A	0.60-0.75	1.00	1.00	16.00-18.00		Mo: 0.75 max.
M440B	0.75-0.95	1.00	1.00	16.00-18.00		Mo: 0.75 max.
M440C	0.95-1.20	1.00	1.00	16.00-18.00		Mo: 0.75 max.

oxidation, carborization, etc., the 309 and 310 compositions may be recommended because of their higher chromium content and, thus, better resistance to oxidation [20].

Type 304–19/10 (chromium nickel) provides a stable austenitic structure under all conditions of fabrication. Carbon (0.08% max.) is sufficient to have reasonable corrosion resistance without subsequent corrosion resistance for welded joints. Type 304 is used for food, dairy and brewery equipment, and for chemical plants of moderate corrosive duties.

Type 304L–This is used for applications involving the welding of plates thicker than about 6.5 mm.

Type 321–This is an 18/10 steel that is stabilized with titanium to prevent weld decay or intergranular corrosion. It has similar corrosion resistance to types 304 and 304L but a slightly higher strength than 304L; also, it is more advantageous for use at elevated temperatures than 304L.

Type 347–This is an 18/11 steel that is stabilized with niobium for welding. In nitric acid it is better than Type 321; otherwise, it has similar corrosion resistance.

Type 316–This has a composition of 17/12/2.5 chromium/nickel/molybdenum. The addition of molybdenum greatly improves the resistance to reducing conditions such as dilute sulfuric acid solutions and solutions containing halides (such as brine and sea water).

Type 316L–This is the low-carbon (0.03% max.) version of type 316 that should be used where the heat input during fabrication exceeds the incubation period of the 316 (0.08% carbon) grade. For example, it is used for welding plates thicker than 1 cm.

Type 309–This is a 23/14 steel with greater oxidation resistance than 18/10 steels because of its higher chromium content.

Type 315–This has a composition that provides a similar oxidation resistance to type 309 but has less liability to embrittlement due to sigma formation if used for long periods in the range of 425 to 815°C. (Sigma phase is the hard and brittle intermetallic compound FeCr formed in chromium rich alloys when used for long periods in the temperature range of 650 to 850°.)

Alloy 20–This has a composition of 20% chromium, 25% nickel, 4% molybdenum and 2% copper. This steel is superior to type 316 for severely reducing solutions such as hot, dilute sulfuric acid.

3.5.5 Precipitation Hardening Stainless Steels

These steels do not have AISI numbers and are referred to by trade name. Examples are given in Table 3.12. They can be heat-treated to give the following mechanical properties:

- Ultimate tensile strength, 1235 N/mm^2
- 0.2% proof stress, 1080 N/mm^2
- Elongation, 10%
- Hardness, 400 DHN

Properties are higher than those of austenitic steels and they retain a general level of corrosion resistance considerably better than that of chromium martensitic steels. They can be supplied as forgings, castings, plate, bar and sheets and can be readily welded and formed before hardening. A typical application is for gears in pumps used for metering chemicals where their hardness prevents wear and galling in contact with Type 316 bodies.

3.5.6 Chromium/Nickel/Ferrite/Austenite Steels

These steels also are not yet included in the AISI system. Trade names and typical compositions are given in Table 3.13. These steels can be welded successfully because they are not predisposed to excess grain growth at elevated temperatures. However, the general level of their corrosion resistance is usually inferior to that of austenitic steels, although they have good resistance to stress corrosion cracking. For example, using austenitic steels in hot, slightly acid solutions containing chlorides causes rapid cracking in a few weeks, whereas the ferrite/austenite steels may last many years.

Table 3.12. Examples of Precipitation Hardening Stainless Steels

	Composition, %								
Name	C	Si	Mn	Cr	Ni	Mo	Cu	Al	N_2
Armco 17/7 PH	0.07	0.3	0.6	17.0	7.0	–	–	1.2	–
Armco 14/8 PH–Mo	0.03	0.3	0.6	14.5	8.5	2.5	–	1.2	–
Allegheny Ludlum AM350	0.09	0.5	1.0	17.5	4.2	2.75	–	–	0.1
Allegheny Ludlum AM355	0.13	0.25	1.0	15.5	4.2	2.75	–	–	0.1
Firth Vickers 520(S)	0.05	0.3	1.0	16.0	5.5	1.75	2.0	–	–

Table 3.13. Compositions of Ferrite/Austenite Stainless Steels

	Composition (% wt)							
Name	C	Si	Mn	Cr	Ni	Mo	Nb	Cu
Firth Vickers 702	0.02	0.5	0.6	15.7	2.5	1.0	0.5	
Sandvik 3RE60	0.02	1.6		18.5	4.7	2.7		
Langley Ferralium				25	5	1.5		2

3.5.7 Maraging Steels

For corrosion resistance, these steels (18% nickel, 9% cobalt, 3% molybdenum, 0.2% titanium and 0.02% carbon) are similar to the 13% chromium steels and, therefore, are suitable for mildly corrosive situations. Because of their very high strength after heat treatment (yield strength–1390 N/mm^2, elongation–15%, impact strength) maraging steels find some use in very high-pressure equipment.

3.6 APPLICATIONS OF HIGH-ALLOY STEELS

With austenitic stainless steels a high carbon content may cause the formation of chromium carbides at grain boundaries, consequently producing intergranular corrosion. This is most likely to occur during welding (called "weld decay"). This phenomenon may be avoided by using either a low-carbon steel (grade L) (i.e., less than 0.03% carbon), or a steel containing titanium or niobium, such as Types 321 and 347.

Intergranular corrosion depends on the length of time the steel is exposed to the sensitizing temperature (500–750°C), even if made from low-carbon or titanium- or niobium-stabilized steel.

Equipment fabricated from such a steel may undergo corrosion by condensation of even mild corrosives unless it is possible to keep it above the dew-point or to neutralize acidic condensates. This kind of corrosion can be prevented by a preliminary heat-treating at temperatures of 815–915°C. The niobium-stabilized steels respond best to this treatment.

Stress corrosion cracking, usually occurring at temperatures above 80°C, takes place in equipment made from austenitic stainless steel but does not affect ferritic steels in this way. Stress cracking most often occurs in solutions of chlorides. Concentrations of a few parts per million can cause severe cracking, even in a medium that would not be considered corrosive, for example in water main lines. Stress corrosion cracking can be caused by some thermal insulating materials, but can be prevented by cladding the insulation with aluminum. This eliminates rain from washing chlorides into contact with the steel.

Residual stresses occur from welding and other fabrication techniques even at very low stress values. Unfortunately, stress relief of equipment is not usually a reliable or practical solution. Careful design of equipment can eliminate crevices or splash zones in which chlorides can concentrate. The use of high-nickel stainless steel alloy 825 (40% nickel, 21% chromium, 3% molybdenum and 2% copper) or the ferritic/austenitic steels would solve this problem.

3.6.1 Oxidation Resistance

The ferritic chromium steels (chromium is the principal alloying element) are the most economical for very lightly loaded high-temperature situations. However, they are inadequate when creep must be accounted for. Austenitic steels are often recommended for such conditions. The 17% chromium alloys (Type 430) resist scaling up to 800°C; and 25% alloy (Type 446) up to 1100°C [21].

3.6.2 Mechanical Properties at Elevated Temperatures

The austenitic steels containing nickel are used for load-bearing applications, pressure vessels operating above 550°C, as well as for light-load cyclic operation because they have a more adherent scale than chromium steels and generally do not become brittle under high-temperature service. The 18/10 alloys are suitable for use up to 800°C in air; the 25/10 Type 310 alloys are suitable for use up to 1100°C. When using Type 316 alloy at high temperatures, care should be taken that the atmosphere is not stagnant as catastrophic oxidation of molybdenum may occur.

For high-pressure, high-temperature situations where steels are required with certified creep strength properties, the AISI austenitic steels are given the suffix H (e.g., 347H, 316H etc.).

Below creep range temperatures, costs can be reduced by using nitrogen-containing (for example BS1501 Part 6, Grades 304S65 and 316S66) or worm-worked grades, as these have higher proof strength than ordinary grades [22].

3.6.3 Mechanical Properties at Low Temperatures

The austenitic steels can be used at very low temperatures (low-alloy ferritic steels containing 9% nickel down to −196°C) without the risk of brittle fracture [23].

3.7 CORROSION-RESISTANT NICKEL AND NICKEL ALLOYS

Nickel alloys have two main properties: good resistance to corrosion and high-temperature strength. There are alloys for medium- and low-temperature applications and for high-temperature conditions in which creep resistance is of main importance [24].

The standard quality of commercially pure nickel (nickel 99% minimum, carbon 0.15% maximum; nickel 200/201) can be readily welded and fabricated in all wrought forms and as castings. Because of high costs, it is restricted to special applications for which nickel alloys are not adequate (for example, for equipment used in the production of caustic soda where it is not subject to stress corrosion cracking in hot caustic soda solutions) [25,26].

Unalloyed nickel is used where it is necessary to eliminate iron and copper contamination (nickel 200 up to 300°C and nickel 201 above 300°C).

3.7.1 Nickel/Copper (Alloy 400)

Alloy 400 has good mechanical properties and is easy to fabricate in all wrought forms and castings. K-500 is a modified version of this alloy and can be thermally treated and is suitable for items requiring strength, as well as corrosion resistance. Alloy 400 has immunity to stress corrosion cracking and pitting in chlorides and caustic alkali solutions.

Alloy 400 is also adequate for equipment processing of dry halogen gases and chlorinated hydrocarbons and can be used in reducing environments.

3.7.2 Nickel/Molybdenum

This alloy has a nominal composition of 65% nickel, 28% molybdenum and 6% iron. It is generally used in reducing conditions. It is intended to work in very severely corrosive situations after post-weld heat treatment to prevent intergranular corrosion. These alloys have outstanding resistance to all concentrations of hydrochloric acid up to boiling-point temperatures and in boiling sulfuric acid solutions up to 60% concentration.

3.7.3 Nickel/Molybdenum/Chromium

The composition of this alloy (54% nickel, 15% molybdenum, 15% chromium, 5% tungsten and 5% iron) is less susceptible to intergranular corrosion at welds. The presence of chromium in this alloy gives it better resistance to oxidizing conditions than the nickel/molybdenum alloy, particularly for durability in wet chlorine and concentrated hypochlorite solutions, and has many applications in chlorination processes. In cases in which hydrochloric and sulfuric acid solutions contain oxidizing agents such as ferric and cupric ions, it is better to use the nickel/molybdenum/chromium alloy than the nickel/molybdenum alloy.

3.7.4 Nickel/Chromium/Molybdenum/Iron

Because the composition of this alloy (47% nickel, 22% chromium, 7% molybdenum and 17% iron) has a higher iron content it cannot withstand such aggressive corrosion conditions as nickel/molybdenum and nickel/

molybdenum/chromium alloys. It is, however, less expensive. The nickel makes these alloys immune to stress corrosion cracking and also superior to stainless steels with respect to pitting in chloride solutions. Because of these properties, their greater cost over stainless steel is justified.

3.7.5 Nickel/Chromium/Molybdenum/Copper

These alloys (50/60% nickel, 20/30% chromium, 5/8% molybdenum, and 5/7% copper) have very good resistance to hot sulfuric acid solutions and similar environments. They are only available as castings but can be hardened by heat treatment. The castings are suitable for parts requiring cutting edges and good wear resistance under corrosion conditions, but should not be used in contact with halogens, halogen acids, and halogen salt solutions.

3.7.6 Nickel/Silicon

Nickel/silicon alloy (10% silicon, 3% copper, and 87% nickel) is fabricated only as castings and is rather brittle, although it is superior to the iron/silicon alloy with respect to strength and resistance to thermal and mechanical shock. It is comparable to the iron/silicon alloy in corrosion resistance to boiling sulfuric acid solutions at concentrations above 60%. Therefore, it is chosen for this and other arduous duties where its resistance to thermal shock justifies its much higher price compared with iron/silicon alloys.

3.8 HEAT-RESISTANT NICKEL ALLOYS

3.8.1 Nickel/Chromium

The high-chromium casting alloys (50% nickel, 50% chromium and 40% nickel, 60% chromium) are designated for use at temperatures up to 900°C in furnaces and boilers fired by fuels containing vanadium, sulfur and sodium compounds (e.g., residual petroleum products). Alloys with lower chromium contents cannot be used with residual fuel oils at temperature above 650°C because the nickel reacts with the vanadium, sulfur and sodium impurities to form compounds that are molten above 650°C [27].

3.8.2 Nickel/Chromium/Iron

Alloy 800 (32% nickel, 20% chromium and 46% iron) is used for furnace equipment such as muffles, trays and radiant tubes and in oil and petrochemical plants as furnace coils for the reforming and pyrolysis of hydrocarbons. Higher-strength versions of alloy 800 were developed to meet this situation (802 has a higher carbon content; alloy 807 has a higher hot strength by adding cobalt and tungsten). For 807, the stress to produce rupture in 100,000 hr at 900°C is 13.8 N/mm^2 compared with 8.5 N/mm^2 for alloy 800.

3.9 COPPER AND COPPER ALLOYS

The outstanding properties of copper-base materials are high electrical and thermal conductivity, good durability in mildly corrosive chemical environments and excellent ductility for forming complex shapes. As a relatively weak material, copper is often alloyed with zinc (brasses), tin (bronzes), aluminum and nickel to improve its mechanical properties and corrosion resistance.

The classification system used in the U.S. for copper and copper alloys is given in Table 3.14.

Some different grades of copper are described in Table 3.15.

The specific gravity of the soft pure metal is 8.94. Additional properties are:

- heat capacity C_p = 0.093 kcal/kg°C
- melting temperature t_m = 1083°C
- thermal conductivity λ = 334 kcal/m °C hr
- linear expansion coefficient $\alpha = 1.65 \times 10^{-5}$
- Young's modulus E = 1,080,000 kg/cm^2
- molding temperature 1150°C

Table 3.14. Classification Used for Copper Alloys in the United States [28]

Series	Constituents
100	Not less than 99.4% copper
200	50–99% copper plus zinc and minor elements
300	Zinc and lead alloys
400	Zinc and tin alloys
500	Tin and phosphorus or phosphorus and zinc alloys
600	Aluminum, aluminum and zinc, or zinc and manganese alloys
700	Nickel, nickel and zinc, or zinc and lead

Table 3.15. Various Grades of Copper [28]

Alloy	Alloying Elements (% by wt)	Thermal Conductivity (Btu/ft^2·hr °F/ft)	Ultimate Tensile Strength (ton/in.^{2})	
			(annealed)	(hard worked)
Tough-Pitch High-Conductivity (B.S. 1036)	0.04 oxygen	229	14	20
Tough-Pitch (B.S. 1038, 9)	0.10 oxygen	228	14	20
Tough-Pitch Arsenical (B.S. 1173)	0.10 oxygen 0.3-0.5 As	102	14	21-23
Deoxidized (B.S. 1172)	0.02-0.08 P	126-175	14	20
Deoxidized Arsenical (B.S. 1171)	0.02-0.08 P 0.3-0.5 As	102	14	21-23
Oxygen-Free High-Conductivity	Trace	229	14	20
Silver-bearing	0.05 Ag 0.04 oxygen	228	14	22
Chromium Copper	0.5 Cr	100	15	29
Tellurium Copper	0.75 Te	208	15	
Beryllium Copper	About 2% Be with other elements (Co, Ni, Cr)	50	33	75

The approximate tensile strength is 14 ton/in.^{2} at ordinary temperatures, and its strength decreases with increasing temperature. Typical mechanical properties of copper as a function of temperature are given in Table 3.16.

Copper retains high impact strength and increases its tensile strength under low temperatures, including cryogenic applications. Typical data are given in Table 3.17.

Along with high mechanical properties, copper improves its conductivity in the range of lower temperatures (at −160°C ~400, −190°C ~450, and −252°C ~1600 kcal/m °C hr). It softens in the temperature range of 200 to 220°C as shown in Figure 3.5. The casting properties of copper are rather fair, but copper can be readily stretched, flattened, rolled, welded and brazed. For chemical plant work, welded or brazed joints have become almost universal.

Copper does not form protective oxide films. Therefore, its corrosion resistance is poor against most acids and salts. Many gases—halogens, sulfurous anhydride, sulfur vapors, hydrogen sulfide, carbon dioxide, ammonia—destroy copper. However, copper is highly corrosion resistant to alkali solutions.

3.9.1 Brasses

These are alloys containing more than 50% copper used to overcome the softness, low tensile strength and high casting temperature of the pure

Table 3.16. Mechanical Properties vs Temperature for Copper

	Temperature (°C)				
	20	100	200	300	500
Tensile Strength (kg/cm^2)	2300	2200	1800	1500	840
Elongation (%)	49	48	46	32	18
Brinell Hardness (HB)	42	41	39	37	35

Table 3.17. Mechanical Properties vs Low Temperature for Copper

	Temperature (°C)					
	+20	−10	−40	−80	−120	−180
Tensile Strength (kg/cm^2)	2200	2250	2370	2730	2900	4100
Yield Strength (kg/cm^2)	600	620	650	700	750	800
Elongation (%)	48	40	47	47	45	38
Necking (%)	76	78	77	74	70	77

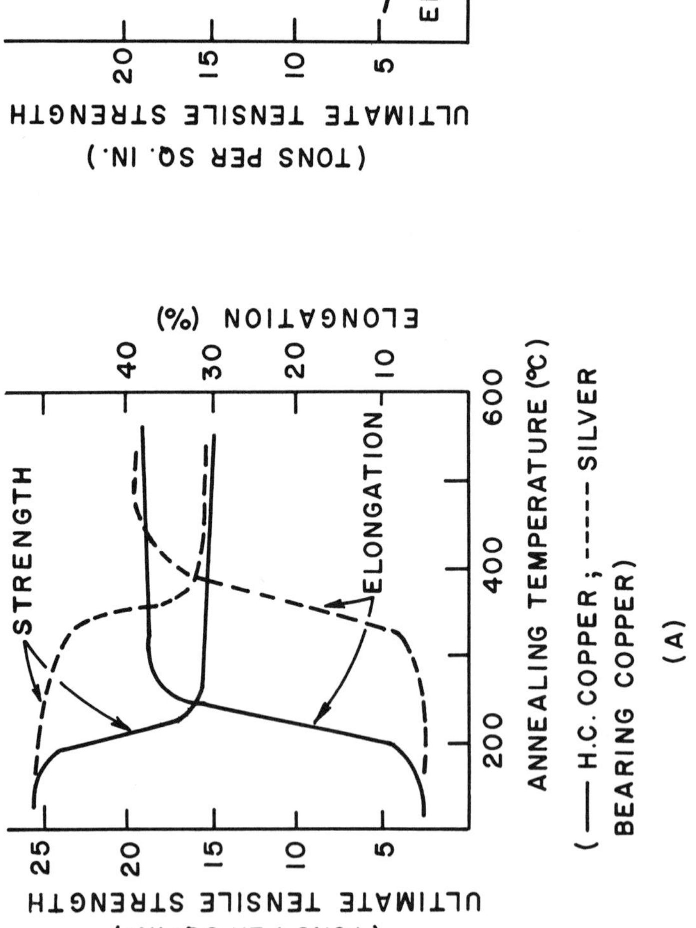

Figure 3.5. Effect of temperature on the tensile strength of copper: (A) effect of annealing on strength and ductility; (B) hardened high-conductivity copper [29].

metal. The compositions and properties of commonly used brasses are presented in Table 3.18.

These annealed brasses are used for fabrication of pressure vessels. They are characterized by the following physical properties:

- density = 8.5 kg/dm^3
- melting temperature t_m = 940°C
- heat capacity C_p = 0.092 kcal/kg°C
- heat conductivity λ = 90–100 kcal/m°C
- temperature elongation $\alpha = 2 \times 10^{-5}$

The strength and ductility of brasses are well maintained over a range of 300° to −180°C, and castings are easy to make as well as to machine. Brass behaves similarly to copper in chemical plant environments, with somewhat greater rates of attack.

3.9.2 Tin Bronzes

This is the name given to copper-tin alloys containing additional alloying elements (Table 3.19). Small amounts of phosphorus are added to deoxidize the metal and in residual amounts to harden the finished alloy. Mixtures treated in this way are referred to as phosphor-bronzes. These have the best corrosion resistance of the alloys listed in Table 3.19 and are used in applications involving contact with dilute acid solutions where bronzes containing zinc (as an alternative to phosphorus, i.e., the gunmetals) would not be sufficiently durable. The phosphor-bronzes have a low coefficient of friction and good resistance to wear. They are most often used for gears and bearings. Lead-bearing alloys corrode more rapidly than those containing only tin and copper; however, apart from this, all bronze alloys can be used with confidence wherever copper can resist corrosion.

3.9.3 Aluminum and Manganese Bronzes

The aluminum-bearing (5–12% of aluminum) alloys retain high strength, good corrosion resistance and good oxidation resistance at temperatures up to 400°C. The aluminum manganese bronzes are noted for high strength and good corrosion resistance at temperatures on the order of 400°C. These bronzes are available only as castings. They have good machining qualities combined with easy welding. With regard to corrosion resistance they appear to behave at least as well as the true bronzes.

3.9.4 Silicon Bronzes

Containing up to 3% silicon, silicon bronzes are characterized by high mechanical and antifriction properties. They are made in all wrought forms,

Table 3.18. Properties of Common Brasses [28]

Material	Nominal Composition, %					Mechanical Properties		
	Cu	Zn	Sn	Al	Others	Hardness (DPN)	0.1% Proof Stress (N/mm^2)	UTS (N/mm^2)
70/30 Brass	70	30			Arsenic 0.02/0.06	78	80	325
Aluminum Brass	76	22		2	Arsenic 0.02/0.06	75	90	355
60/40 Brass	60	40				90	115	360
Naval Brass	62	37	1			95	135	390
High-Tensile Brass (manganese bronze)	59	38			Iron 1 Manganese 0.5 Aluminum 0.2	125	145	470

Table 3.19. Properties of Tin Bronzes and Gunmetals [30]

Materials	Nominal Composition, %					Mechanical Properties		
	Cu	Sn	Zn	Pb	Others	Hardness (DPN)	0.1% Proof Stress (N/mm^2)	UTS (N/mm^2)
5% Phosphor Bronze, Wrought	95	5			Phosphorous	75	125	340
Admiralty Gunmetal, Castings	88	10	2			85	125	300
Leaded Gunmetal 85/5/5/5, Castings	85	5	5	5		65	95	220
Nickel Gunmetal	86	7	2		Nickel 5	80	155	300
Phosphor-Bronze, General-Purpose Castings	90	9.5			Phosphorus 0.1/0.4	85	125	270
Bronze for Bearings	89	10			Phosphorus 0.5			

such as plates, sheets and castings. The silicon bronzes are well molded, cold- and hot-pressure shaped (rollings, forging, stamping, etc.) and welded.

These alloys have corrosion resistance similar to that of copper, with mechanical properties equivalent to mild steel. Because silicon bronzes do not generate sparks under shocks, they can be used in the fabrication of explosion-proof equipment. Compared to tin bronzes, the tinless bronzes have a higher shrinkage (1.7–2.5% against 1.3–1.5% of tin bronzes) and less fluid-flow, which is an important consideration in designing.

3.9.5 Cupro-nickels

The cupro-nickel alloys (5–30% of nickel) are perhaps the best of all for strength and resistance to corrosion. Table 3.20 gives typical properties.

3.9.6 Corrosion Resistance

Copper-base alloys perform best under reducing conditions and in the absence of aeration. Copper and its alloys are resistant to dilute solutions of several mineral acids such as sulfuric and hydrochloric, and to a wide range of organic acids such as acetic and formic. Aluminum bronze is suitable in slightly oxidizing situations. Copper-base alloys are resistant to most alkaline solutions, but never should be exposed to strong oxidizing acids such as nitric and chromic, as well as aqueous ammonia. Copper-base alloys are also resistant to most neutral salts, except to those forming soluble complexes [31].

3.10 MECHANICAL PROPERTIES OF LEAD AND LEAD ALLOYS

Lead is the softest and most easily worked metal used in plant construction. The main difficulty in design is that the metal has a very low creep stress,

Table 3.20. Mechanical Properties of Annealed Cupro-Nickel Alloys [30]

Materials	Nominal Composition, %					Mechanical Properties		
	Cu	Ni	Fe	Mn	Others	Hardness (DPN)	0.1% Proof Stress (N/mm^2)	UTS (N/mm^2)
90/10 Copper-Nickel-Iron	88	10	2	1		70	110	310
80/20 Copper-Nickel-Manganese	80	20		0.5/0.50		75	110	340

even at ordinary temperatures, with or without work-hardening effects. In the form used for chemical plants, the purity of the metal is almost complete; small amounts of alloying additions in lead are intended to improve its mechanical properties without any significant decrease in corrosion resistance.

There are three standard leads available in the U.S. for process plant construction. These are described in Table 3.21.

Lead has the following physical properties:

- density $\rho = 11.35$ kg/dm^3
- melting point $t_m = 327°C$
- heat capacity $C_p = 0.031$ kcal/kg °C

Table 3.21. Standard U.S. Leads [32]

Analyses (%)	Chemical Lead	Acid Lead	Copper Lead
Silver: max.	0.020	0.002	0.020
min.	0.002		
Copper: max.	0.080	0.080	0.080
min.	0.040	0.040	0.040
Arsenic, Antimony and Tin, max.	0.002	0.002	0.015
Zinc, max.	0.001	0.001	0.002
Bismith, max.	0.005	0.025	0.10
Lead, min.	99.90	99.90	99.85

Table 3.22. Mechanical Properties of Sheet Lead [32,33]

Ultimate Tensile Strength (kg/cm^2)	130–180
Elongation (%)	40–50
Necking (%)	100
Brinell Hardness (HB)	4–4.6

Table 3.23. Mechanical Properties of Annealed Lead vs Temperature [32,33]

	Temperature (°C)				
	20	80	150	200	265
Ultimate Tensile Strength (kg/cm^2)	135	80	50	40	20
Elongation (%)	31	24	23	20	18
Necking (%)	100	100	100	100	100

- thermal conductivity λ = 30 kcal/m °C hr
- temperature elongation = 3.9 × 10^{-7}

The mechanical properties of lead are given in Tables 3.22 and 3.23.

Lead alloys have higher strength and lower melting points than pure lead and, therefore, have a lower service temperature (less than 100°C).

Dispersion-strengthened lead (DSL), obtained by a uniform dispersion of lead oxide through the lead particle matrix, has the traditional corrosion resistance of lead but much greater stiffness. DSL is fabricated as pipe and other extruded items, but has a limited application for process plant construction because the welding technique does not provide adequate strengths in joints.

The recommended maximum design stresses for a life of 5 to 10 years based on long-time creep tests are given in Table 3.24.

Another important factor in the selection of a lead alloy is fatigue strength, which may arise from high-frequency vibration from pumps and stirrers or from differential expansion from heat and cooling cycles. The marked increase of fatigue strength obtained by alloying with copper, silver and tellurium can be seen from Table 3.25.

Table 3.24. Maximum Stresses in Pipe Wall of Lead Alloys [33]

	Maximum Stress, S, N/mm^2, in Pipe Wall			
Temperature (°C)	**99.99% Lead**	**Copper, Tellurium and Silver Leads**	**8% Antimonial Lead**	**DSL**
20	2.21	2.42	3.50	10.34
60	1.24	1.38	1.24	10.34
100	0.66	0.86	0.76	9.62
150		0.52		3.50

Table 3.25. Fatigue-Strength Data of Lead Alloys [33]

	Endurance Limit, +N/mm^2, for 20 × 10^6 Cycles	
Lead	**20°C**	**80°C**
99.99	3.17	2.10
99.99% +0.06 copper	4.06	3.00
99.99% +0.005% silver +0.005% copper	4.17	3.05
99.99% +0.06% copper +0.04 tellurium	7.70	5.10
DSL	13.8	12.50

3.10.1 Corrosion Resistance

The corrosion resistance of lead is due to the formation of a thin surface film of an insoluble lead salt that protects the metal from sulfuric acid and related compounds of any strength at ordinary temperatures. Even when the temperature increases to nearly 100°C the rates of corrosion are still low. However, strong, hot sulfuric acid attacks lead rapidly, especially if the acid is flowing.

Nitric acid in any concentration attacks lead steadily, but mixtures of nitric and sulfuric–nitration acids–are not as active and can be handled in lead.

Phosphoric acid made by the "wet process," in which phosphate rock is treated with sulfuric acid, is highly inert toward lead in any concentration for temperatures up to 150°C. However, in the "dry process," where hydrogen phosphate (H_3PO_4) is made directly from phosphorus or phosphorus pentoxide (P_2O_5), a chemical reaction with lead occurs.

Lead chloride is freely soluble in hot aqueous solutions, but lead fluoride is almost insoluble in dilute HF solutions. When the HF concentration reaches about 40%, steel is preferred.

Organic chlorinations are handled in lead where the presence of iron might produce catalyst substitution in an undesirable position. Hence, lead is the material most frequently specified for chlorinators.

Chromic acid and its salts normally are prepared in lead. Lead is especially suitable for organic oxidations because its inertness avoids any interference from reactions.

Neutral or weak acid-salt solutions usually can be handled in lead plants, with the exception of those few heavy metals that may form lead alloys by substitution. The alums and sulfates generally have little action.

3.11 ALUMINUM AND ALUMINUM ALLOYS

The main criteria in the selection of aluminum and its alloys for chemical plants are corrosion resistance, ease of fabrication and price. High-quality aluminum grades are used for chemical and process plant applications.

Physical properties of aluminum are characterized by the following data:

- density $\rho = 2.7$ kg/dm^3
- melting point $t_m = 657$°C
- heat capacity $C_p = 0.218$ kcal/kg °C
- thermal conductivity $\lambda = 188$ kcal/m °C hr
- thermal elongation coefficient $\alpha = 2.4 \times 10^{-5}$

The positive properties of aluminum are its high heat conductivity (4.5 times higher than that of steel), low specific gravity, high ductility providing good rolling, and cold and hot stamping. The negative properties are its poor castability, poor cutting and low strength. The most important specifications of aluminum as a structural material are given in Tables 3.26–3.28.

Table 3.26. Mechanical Properties of Aluminum

	Mild, Annealed Aluminum	Hardened Aluminum
Ultimate Tensile Strength (kg/cm^2)	700–1000	1500–2000
Yield Strength (kg/cm^2)	300–400	1400–1800
Elongation (%)	30–40	4–8
Necking (%)	70–90	50–60
Brinell Hardness (HB)	15–25	40–55

Table 3.27. Mechanical Properties of Aluminum Annealed at 370°C

	Temperature, °C						
	20	75	135	310	400	510	600
Ultimate Tensile Strength (kg/cm^2)	1160	1000	765	260	125	55	35
Elongation (%)	19	24	32	39	42	45	48
Necking (%)	79	83	88	97	99	99	100

Table 3.28. Allowable Tensile and Compression Stresses for Mild Aluminum (annealed) vs Metal Operating Temperature

Aluminum Temperature (°C)	Allowable Tensile Stress (kg/cm^2)	Allowable Bending Stress (kg/cm^2)
30	150	250
31-60	140	225
61-80	130	200
81-100	120	175
101-120	105	150
121-140	90	125
141-160	75	100
161-180	60	75
181-200	45	50

The low strength of aluminum can be considerably improved by alloying with magnesium, silicon, manganese, copper, etc. However, the alloys have substantially the same modulus of elasticity (70 kN/mm^2).

3.11.1 Aluminum Alloy Compositions

Aluminum alloys can be divided into three classes: aluminum of various degrees of purity; wrought alloys; and casting alloys.

3.11.2 Aluminum of Commercial 99% Minimum Aluminum Purity

This is used widely for chemical and process plant applications where its resistance to corrosion and its high thermal conductivity are desirable characteristics. The strength of aluminum can be increased by cold working, as shown in Table 3.29.

The addition of alloying elements to the commercially pure metal results in an increase in its strength and usually has some favorable effects on other characteristics.

3.11.3 Nonheat-Treatable Magnesium and Manganese Alloys

These represent the best compromise between corrosion resistance and strength. It is the most useful class of alloys for chemical and process plant construction. Examples are given in Table 3.30.

Table 3.29. Effect of Purity on the Properties of Aluminum

			Typical Mechanical Properties	
BSI Designation	Purity (% min.)	Degree of Cold-Work	Tensile Strength (N/mm^2)	Hardness (DPN)
1	99.99	0[a]	54	16
		H4[b]	85	25
		H8[c]	108	30
1A	99.8	0	69	18
		H4	100	31
		H8	130	40
1B	99.5	0	85	21
		H4	108	31
		H8	140	39

[a]0 = fully softened.
[b]H4 = partially cold-worked.
[c]H8 = fully cold-worked.

Table 3.30. Typical Properties of Fully Annealed Nonheat-Treatable Aluminum Alloys

BS Designation	Main Alloying Elements		Typical Mechanical Properties	
	Magnesium	Manganese	Tensile Strength (N/mm^2)	Hardness (DPN)
N3[a]		1/1.5	108	29
N4	1.7/2.8		185	45
N6	4.5/5.5		280	65
N8	4.0/4.9	0.5/1.0	295	70

[a]N signifies a nonheat-treatable alloy.

3.11.4 Heat-Treatable Alloys

These are produced by adding small amounts of copper, magnesium and/or silicon, which can increase their strength much more by heat-treatment than by cold-working (Table. 3.31).

Differences in composition require different temperatures for the high-temperature solution treatment, as well as variations both in time and temperature of aging. Heat-treatable plate and sheet alloys are not widely used for process plant construction because heat treatment has to be applied after welding to restore the mechanical properties.

Table 3.31. Effect of Heat Treatment on Heat-Treatable Aluminum Alloys

BSI Designation	Main Alloy Elements				Heat Treatment Condition	Typical Mechanical Properties	
	Copper	Magnesium	Silicon	Manganese		Tensile Strength (N/mm^2)	Hardness (DPN)
H30[a]	–	0.4/3.4	0.6/1.3	0.4/1.0	M	215	70
					TF[b]	310	100
H15	3.8/4.8	0.2/0.8	0.5/0.90	0.3/1.2	M[c]	400	125
					TF	430	160
H9	–	0.4/0.9	0.3/0.7		M	155	50
					TF	200	80

[a]H denotes a heat-treatable alloy.
[b]TF is solution heat treatment followed by precipitation hardening.
[c]M as cast.

3.11.5 Casting Alloys

These are used as corrosion-resistant materials. Examples are given in Table 3.32. Some can be strengthened by heat treatment. The alloys containing substantial amounts of silicon have the best foundry characteristics and a high resistance to corrosion, but are not readily machined.

3.11.6 Temperature Effects

Tensile strength diminishes rapidly with increasing temperature above 200°C. The high-magnesium alloys N5, N6 and N8 should not be used above 65°C because higher temperatures make them susceptible to stress corrosion cracking.

Aluminum and its alloys are excellent for low temperatures as well as for cryogenic applications because their tensile strength and ductility are increased at low temperatures.

They are widely employed in the manufacture, storage and distribution of liquified gases, particularly on sea and road tankers. The most popular alloy for cryogenic applications is 4.5% magnesium alloy (N8). Table 3.33 gives the boiling points of the most common cryogenic liquids and the minimum temperatures at which various materials can be used.

3.11.7 Corrosion Resistance

Clean metallic aluminum is extremely reactive. Even exposure to air at ordinary temperatures is sufficient to promote immediate oxidation. This reactivity is self-inhibiting, however, which determines the general corrosion behavior of aluminum and its alloys due to the formation of a thin, inert, adherent oxide film. In view of the great importance of the surface film, it can be thickened by anodizing in a bath of 15% sulfuric acid (H_2SO_4) solution or by cladding with a thin layer of an aluminum alloy containing 1% zinc.

Aluminum is suitable for contact with cold dilute sulfuric acid solutions, especially if an anodizing treatment has been applied. As the H_2SO_4 concentration increases, rates of attack also increase. This is illustrated in Figure 3.6. Aluminum may be used for handling oleum; however, the passivity of the metal with sulfuric acid is soon destroyed as the temperature rises, and the boiling solutions of an acid of any concentration attacks rapidly.

Aluminum and stainless steel are used almost interchangeably for any strength of *nitric acid*. Figure 3.7 compares the rate of attack of cold nitric acid on stainless steel and aluminum. Figure 3.7 shows that higher rates of aluminum corrosion occur up to about 80% nitric acid (HNO_3), but aluminum is still to be preferred over stainless steel for any concentration above 80%.

Table 3.32. Various Aluminum Casting Alloys

BSI Designation	Alloy Elements: Magnesium	Silicon	Manganese	Heat Treatment Condition	Mechanical Properties[a]: Tensile Strength (N/mm^2)	Hardness (DPN)
LM5	3.0/6.0	–	0.3/0.7	M[b]	125	60
LM6	–	10.0/13.0	–	M	170	55
LM9	0.2/0.6	10.0/13.0	0.3/0.7	TE[c]	185	70
				TF[d]	245	100
LM18	–	4.5/6.0	–	M	280	40
LM25	0.2/0.45	6.5/7.5	–	M	130	60
				TE	155	70
				TF	245	100

[a]As sand castings. The new heat treatment designations and their former equivalents:
[b]M = as cast.
[c]TE = precipitation treated (P).
[d]TF = solution and precipitation treated (WP).

Table 3.33. Aluminum Alloys Recommended for Cryogenic Applications

Cyrogenic Liquid	Boiling Point (°C)	Material	Minimum Temperature of Use (°C)
Propane	–42	Carbon steels	–50
Carbon Dioxide	–78	2.25% nickel steel	–65
Acetylene	–84	3.5% nickel	–100
Ethylene	–104	Aluminum/magnesium alloys	–270
Methane	–161	Austenitic stainless steel	–270
Oxygen	–182	Nickel alloys	–270
Argon	–186	Copper alloys	–270
Nitrogen	–196		
Hydrogen	–253		

Aluminum does not have the mechanical reliability of stainless steel, especially at higher temperatures. However, it is not nearly as susceptible to "weed decay" or localized attack around the welds.

3.11.8 Organic Acids

These are the second great field of application for aluminum alloys, with the exception of aluminum magnesium alloys. One restriction always applies

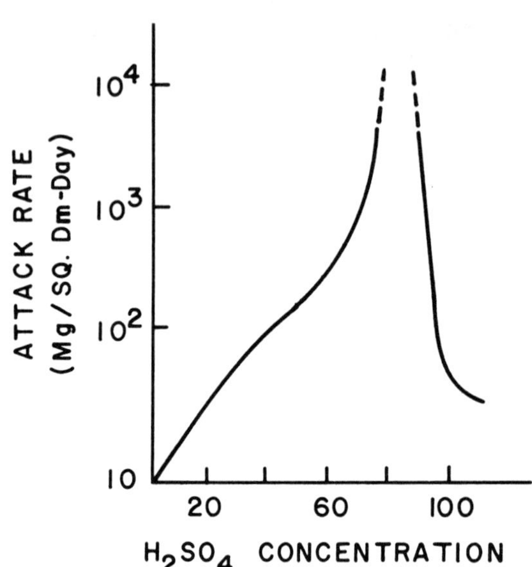

Figure 3.6. Effect of sulfuric acid on aluminum.

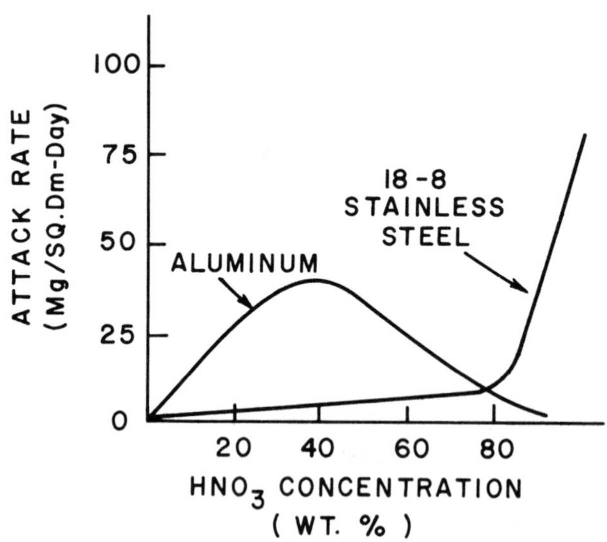

Figure 3.7. Effect of nitric acid on stainless steel and aluminum.

in their use—the rapid increase in corrosion rate when perfectly anhydrous acids are being handled at high temperature. The resistance to most acid-reacting organic compounds increases with the acid concentration.

It is possible to carry out such oxidation processes as the conversion of acetaldehyde to acetic acid, or methyl alcohol to formaldehyde in aluminum plants, thus avoiding boiling anhydrous acids. The metal is especially valuable for handling delicate chemicals, which must not acquire metallic taste or color. For these reasons, aluminum has found extensive use in the food, dairy, brewing and fishing industries.

Neutral salts and aqueous solutions of various acids generally follow the acid action. Aluminum has no apparent action on microbiological processes (i.e., the production of antibiotics by deep-vessel fermentation). Fermentation tanks, as well as various absorbing and extracting units, can be made from aluminum.

Since aluminum is not attacked by hydrogen sulfide (H_2S) solutions, it is used widely as a material in refineries for the handling of hydrocarbons made from "sour" crudes. In the strongly oxidizing conditions of manufacturing hydrogen peroxide, aluminum is one of the few materials that does not undergo decomposition.

Steam-heated aluminum castings are used for the melt spinning of nylon and polyester fibers and have been used for storage of raw materials during manufacturing, as well as for storage of acetic acid in cellulose acetate plants.

3.12 MISCELLANEOUS PRECIOUS METALS

Titanium, tantalum and zirconium are used for construction in process plants. The principal physical and mechanical properties of these three metals are given in the Table 3.34.

Table 3.34. Properties of Titanium, Tantalum and Zirconium [34]

	Density (g/cm^2)	Melting Point (°C)	Coefficient of Expansion $\times 10^{-6}$ (°C)	Thermal Conductivity (W/m °C)	Yield Strength (N/mm^2)	Tensile Modulus (N/mm^2)	Hardness (DPN)
Titanium	4.5	1668	9.0	15	345	103,000	150
Tantalum	16.6	2996	6.5	55	240	185,000	170
Zirconium	6.5	1852	7.2	17	290	80,000	180

3.12.1 Titanium

Titanium is a white metal and, when cold, is brittle and may be powdered. At a red heat it may be forged and drawn. The tensile strength of titanium is almost the same as that of steel, while its specific gravity (4.5) is almost two times less than that of steel. Hence, its specific strength (tensile strength/specific gravity) is 1000, which is considerably higher than that of 18/8 steel, which has a value of 700. Titanium is now available as plate, sheet and tube, and its use in chemical plant construction is considered common.

The mechanical properties of titanium are greatly affected by small amounts of oxygen and nitrogen. The properties of the commercially pure grade metal and its alloys are given in the Table 3.35.

The alloys with aluminum, vanadium and tin have considerably greater strength but lower corrosion resistance. These alloys are used as rotating components in centrifuges, where the strength to weight ratio is important. About 0.2% palladium alloy gives better corrosion resistance than the first four grades in Table 3.35.

According to the ASME code (Section VIII, Div. 1) titanium may be used up to 300°C. The fatigue strength of the metal is about half the tensile strength. Typical values for the tensile strength of titanium and its alloys at temperatures up to 500°C are given in Table 3.36. The corrosion resistance of all grades of commercial pure titanium is similar. This protection relies on a surface film of metallic oxide. Therefore, titanium is most useful in oxidizing environments.

Table 3.35. Mechanical Properties of Titanium and Alloys (ASTM B265/337/338) [35]

Grade	Alloying Element (% max.)	UTS (N/mm^2 min.)	Elongation (%)	Hardness (DPN)
1	Oxygen 0.18	240	24	150
2	0.25	345	20	180
4	0.40	550	15	260
5	Aluminum 6 Vanadium 4	900	10	
6	Aluminum 5 Tin 2.5	820	10	
7	Palladium 0.15/.25 Oxygen 0.25	345	20	180
8	Palladium 0.15/0.25 Oxygen 0.35	450	18	210

Table 3.36. Effect of Elevated Temperatures on Strength of Titanium and Alloys [35]

Material ASTM Grade	Tensile Strength, N/mm^2, at Test Temperature of					
	Room Temperature	100°C	200°C	300°C	400°C	500°C
1	310	295	220	170	130	
2	480	395	295	205	185	
3	545	460	325	250	200	
4	760	585	425	310	275	
5	1030	890	830	760	690	655
6	960	820	690	620	535	460
7	480	395	295	205	185	
8	545	460	325	250	200	

Titanium is generally suitable for use in boiling nitric acid, aqua regia, nitrites, nitrates, chlorides, sulfides, phosphoric acid, chromic acid and organic acids. The main advantage of titanium over stainless steel is that it is not affected by pitting or stress corrosion cracking in solutions containing chloride ions and has better resistance to erosion. It can be more easily protected anodically (less than 50 W for a surface equal to 100 m^2).

The corrosion resistance of unalloyed titanium in hydrochloric or sulfuric acids can be increased significantly by anodic protection, which maintains the oxide film so that the corrosion will be negligible even in severely reducing conditions.

If the metal is exposed to highly oxidizing conditions in the complete absence of water, a violent reaction may occur (for example, in completely dry chlorine). In this case, 0.015% water is added as the minimum for passivation of titanium.

3.12.2 Tantalum

Tantalum is a light bluish metal: ductile, malleable and, when polished, resembles platinum. The metal is characterized by high strength and infusibility. Its melting point is 3000°C. The metal has high ductility, good forging, flattening and stamping. Excellent welds can be made by the TIG process; however, as tantalum reacts with oxygen and nitrogen at temperatures above 300°C, careful shielding with argon of all areas likely to exceed this temperature is vital for success [36,37].

Tantalum has a degree of corrosion resistance similar to that of glass; therefore, it can be used in environments for which glass is required but without the risk of fracture and for purposes of heat transfer. The thermal conductivity of the metal is similar to that of nickel and nickel alloys.

In tantalum equipment very high flowrates can be admitted before erosion and cavitation occur, and a much higher thermal flux can be achieved. Therefore, the higher cost of tantalum sometimes can be justified.

The same volume of metal tantalum is 30 times more expensive than titanium, but it has the range of corrosion resistance more comparable with the precious, rather than the base, metals. It is only 3% of the cost of platinum and 8% of the cost of gold.

In many applications tantalum can be substituted for platinum and gold, and there are some environments in which tantalum is more corrosion resistant than platinum. Table 3.37 lists the main chemicals for which tantalum is not a suitable substitute for platinum and, conversely, those for which tantalum is better than platinum. Tantalum is rapidly embrittled by nascent hydrogen even at room temperature. Therefore, it is very important to avoid the formation of galvanic couples between tantalum and other metals.

3.12.3 Zirconium

Of high purity, zirconium is a white, soft, ductile and malleable metal. At 99% purity, when obtained at high temperatures it is hard and brittle. The rapid development of production techniques of zirconium has resulted because of its suitability for nuclear engineering equipment.

Table 3.37. Comparative Corrosion Resistance of Tantalum and Platinum

Chemical	Tantalum	Platinum
Acetylene	G[a]	NR
Alkalis	NR[b]	G
Bromine (wet or dry)	G	NR
Bromic Acid	G	NR
Cyanides	G	NR
Fluorine	NR	G
Compounds Containing Fluorine	NR	G
Ethylene	G	NR
Lead Salts	G	NR
Lead Oxide	NR	G
Metals (molten)	G	NR
Mercury	G	NR
Mercury Compounds	G	NR
Oleum	NR	G
Phosphoric Acid	NR	G
Sulfur Trioxide	NR	G

[a]G = good.
[b]NR= not recommended.

Zirconium has outstanding resistance to hydrochloric acid and is a cheaper alternative to titanium for this duty. It is superior to titanium in resistance to sulfuric acid. Zirconium has excellent resistance to caustic alkalies in all concentrations and is superior to both titanium and tantalum in this respect.

3.12.4 Precious Metals

The precious metals are many times the cost of the base metals and, therefore, are limited to specialized applications or to those in which process conditions are highly demanding (e.g., where conditions are too corrosive for base metals and temperatures too high for plastics; where base metal contamination must be avoided, as in the food and pharmaceutical industries; or where plastics cannot be used because of heat transfer requirements; and for special applications such as bursting discs in pressure vessels). The physical and mechanical properties of precious metals and their alloys used in process plants are given in Table 3.38.

3.12.5 Silver

Silver is a white metal; it is softer than copper and harder than gold. One use of the pure metal (about 99.99%) is as a liner bonded to stronger or cheaper metals. The metallic bond is usually of high thermal conductivity.

Both steel and copper vessels may be lined with thin silver sheets in the same way as for homogeneous lead lining. As silver is extremely resistant to most organic acids at all concentrations and temperatures, it is used widely for handling foodstuffs and pharmaceutical products where nontoxicity and

Table 3.38. Properties of Precious Metals [38]

	Platinum	Gold	Silver	10% Rh/Pt	20% Rh/Pt	10% Ir/Pt	20% Ir/Pt	70% Au 30% Pt 1% Rh
Density (g/cm^3)	21.45	19.3	10.5	20.0	18.8	21.6	21.7	20
Melting point (°C)	1769	1063	961	1850	1900	1800	1815	1250
Thermal Conductivity (W/m °C)	70	290	418					
Young's Modulus (kN/mm^2)	170	70	70	195	215			
Tensile Strength Annealed (N/mm^2)	140	110	140	325	415	370	695	925
Hardness (DPN) Annealed	40	20	26	75	90	120	200	250

discoloration are essential. Silver is inert to hot alkaline solutions and very resistant to fused alkalis in the absence of oxidizing agents and to all neutral salt solutions.

3.12.6 Gold

Gold can be used only in very small portions or very thin coatings because of its cost. Most of the applications for which it was used in the past have now been accomplished with tantalum at a much lower cost. A gold/platinum/rhodium alloy is used in the manufacture of rayon-spinning jets in the production of rayon fibers. This alloy presents the combination of strength, corrosion resistance and abrasion resistance necessary to prevent changes in hole dimensions.

3.12.7 Platinum

Platinum, plus other platinum group metals (Pt, Pd, Ir, Os, Rh, Ru), within the range of 99.8–99.99% Pt content are almost completely inert to chemical reagents under oxidizing conditions over a wide range of temperatures. At high temperatures under reducing conditions, however, it is attacked by all base metals, by molten silver and gold, and elemental silicon, boron, arsenic, phosphorus, bismuth and sulfur. One particular case is the handling of molten glass in the manufacture of glass wool or glass threads for weaving. Again, platinum electrodes are used for electrolytic production of highly oxidizing materials, such as ammonium persulfate and chlorine from brine solutions.

3.13 METALLIC COATINGS

Metallic coating involves the deposition of metals and alloys onto other metals ranging in thickness from a few microns to several millimeters. This method allows for the possibility to obtain the properties of the coating at low cost, compared with making the items entirely from the coating composition. Coating permits the use of metals or alloys that are too brittle or too weak in the solid form (for example, chromium and zinc). The main reasons for applying a coating are prevention of corrosion, oxidation and abrasion. Coatings are produced by four main methods: electrodeposition, spraying, dipping and diffusion.

3.13.1 Electrodeposition

Nickel, chromium and zinc are commonly used as electrodeposits. Chromium, the hardest of these coatings, is applied for abrasion resistance and low coefficient of friction. Nickel and zinc electrodeposits are used for resistance to corrosion, the latter for mildly corrosive conditions [39,40].

3.13.2 Dip Coating

Dip coating involves immersion of steel or copper in a bath of molten coating metal (zinc, tin and/or aluminum). Hot dip-galvanized (zinc-coated) steel should not be used in circuits containing copper equipment. This can result in galvanic corrosion at the copper/galvanized junctions, as well as cause overall galvanic corrosion of the zinc by copper redepositing from the water or process stream. Galvanized equipment is not recommended for use with liquors above 60°C. Above this temperature there is a reversal of the polarity of the zinc/steel couple, and the coating ceases to be protective where flaws appear in the coating. Impervious coatings of tin for mild corrosive conditions can be formed on steel and copper by dipping in a molten bath of tin.

Aluminum is the highest melting point metal (660°C) applied by hot dipping. Aluminized steel can be used at temperatures up to 550°C without appreciable oxidation. This steel has very good resistance to gases and vapors containing small quantities of sulfur dioxide and hydrogen sulfide [41,42,43].

3.13.3 Sprayed Coatings

Zinc, aluminum, nickel alloys, cobalt alloys and tungsten carbide are applied for sprayed coatings, which are slightly porous. Flame-sprayed zinc coatings are used for corrosion protection of steel and provide similar properties for galvanized coatings.

Sprayed aluminum coatings used on steel for protection against atmospheric corrosion are preferred over zinc for use in areas with considerable contamination of the atmosphere by sulfur oxides [44]. Sprayed aluminum also is used for the protection of steel at elevated temperatures up to 550°C. For temperatures of 550–900°C, aluminum is converted to a high-melting point aluminum/iron compound by heating the coated equipment to 800/900°C and maintaining it at that temperature for 15 minutes. For protection up to 1000°C, a sprayed coating of nickel chromium and nickel and cobalt alloys is applied. Nickel or cobalt alloys containing small amounts of boron or silicon can be deposited with very simple equipment, requiring very little heating of the base metal.

3.13.4 Diffusion Coatings

The purpose of diffusion coatings is not to produce a coating of another metal on the substrate, but to change the composition of the surface layers of the substrate by alloying with the diffusing metal chosen (zinc, aluminum, chromium and silicon). The surface properties after such treatment depend not only on the metal diffused, but also on the composition of the substrate. The diffusion coating causes very little change in the dimensions of the piece being treated, which is important for items machined to fine limits, such as nuts and bolts [45,46].

Zinc diffusion is used for protection against atmospheric corrosion. Aluminum diffusion is used to improve the oxidation resistance of low-carbon steels.

Chromium diffusion applied to a low-carbon steel produces a surface that has the characteristics of ferritic stainless steel, such as AISI446 to a depth about 0.1 mm. When diffusion is applied to a high-carbon steel, a surface rich in chromium carbides is formed. This has a hardness greater than 1000 VHN, which provides good resistance to abrasion.

Nickel alloys and stainless steels such as AISI310 (25Cr/20Ni) diffusion treated with chromium enhance resistance to sulfur gases at high temperatures. The chromium-rich surface prevents the formation of nickel sulfide.

The use of equipment close to the temperature at which the material was diffusion treated will result in continuing diffusion of chromium, aluminum etc., into the substrate, thus depleting chromium with consequent loss in oxidation and corrosion resistance. For aluminum, this effect is noticeable above 700°C in steels, and above 900°C in nickel alloys. For chromium, the effect is pronounced above 850°C for steels and above 950°C for nickel alloys.

Silicon used for diffusion treatment of carbon steels enhances corrosion resistance to sulfuric acid. Such a treatment has the surface durability of iron/silicon alloys without their marked brittleness.

3.14 CARBON, GRAPHITE AND GLASS

3.14.1 Carbon and Graphite

Structural carbon shapes fabricated by heating coke with a mixture of tar and pitches are porous and are made impermeable by impregnation with a resin (usually a phenolic resin). Cashew nut shell liquid resin is used when resistance to alkalis and acids is required.

Graphite is used widely in process plants for its high thermal conductivity (about six times that of stainless steel). Typical properties of impregnated carbon and graphite are given in Table 3.39.

Impregnated carbon and graphite can be used up to 180°C, and porous graphite can be used up to 400°C in oxidizing environments and 3000°C in a reducing atmosphere. Carbon and graphite bricks and tiles are used for

Table 3.39. Properties of Carbon and Graphite [47]

	Carbon	Graphite
Density (g/cm^3)	1.8	1.8
Tensile Strength (N/mm)	28	10
Compressive Strength (N/mm^2)	135	70
Tensile Modulus (kN/mm^2)	10	3
Thermal Conductivity (W/m °C)	4	70
Linear Coefficient of Expansion $(°C^{-1})$	3×10^6	4×10^6

lining process vessels and are particularly suitable for applications involving severe thermal shock.

Tube and shell heat exchangers, small distillation columns, reactors, valves, pumps and other items are available in impregnated graphite. Graphite can be joined only by cementing, which embrittles on aging. It is prone to mechanical damage, particularly when subjected to tensile stresses.

3.14.2 Glass

By virtue of its chemical and thermal resistances, borosilicate glass has superior resistance to thermal stresses and shocks, and is used in the manufacture of a variety of items for process plants. Examples are pipe up to 60 cm in diameter and 300 cm long with wall thicknesses of 2–10 mm, pipe fittings, valves, distillation column sections, spherical and cylindrical vessels up 400-liter capacity, centrifugal pumps with capacities up to 20,000 liters/hr, tubular heat exchangers with heat transfer areas up to 8 m^2, maximum working pressure up to 275 kN/m^2, and heat transfer coefficients of 270 $kcal/hz/m^{2}°C$ [48,49].

Borosilicate glass has the following properties:

- density ρ = 2.7–31 kg/dm^3
- heat capacity C_p = 0.1–0.3 kcal/kg°C
- melting point t_m = 1000–1200°C
- thermal conductivity λ = 0.4–1.0 kcal/m°C hr
- linear coefficient of expansion $\alpha = 5 \times 10^{-6}$
- tensile strength 500–900 kg/cm^2
- compression strength 6,000–13,000 kg/cm^2
- modulus of elongation E = 620,000 kg/cm^2
- Poisson's ratio $\tilde{\mu}$ = 0.27–0.29.

Because borosilicate is a brittle material, its design stress is restricted to less than 7 N/mm^2. Borosilicate glass is attacked by hydrofluoric acid even when a solution contains only a few parts per million of fluoride ions, and at elevated temperatures by strong solutions of phosphoric acid (85% acid at 100°C is the approximate limit). It is attacked also by strong bases such as sodium hydroxide and potassium hydroxide solutions, in which the effect is linear with time. As for all other materials, borosilicate represents almost the final resort in corrosion resistance. It has a very high degree of passivity to sensitive chemicals.

3.15 CEMENTS, BRICKS AND TILES

3.15.1 Cements

Cements are used mainly for jointing brickworks, drainage gullies, pipes and storage tanks. Portland cement has very poor resistance to acids but good resistance to alkaline liquors.

High-alumina cement is very quick setting, but its acid resistance is only slightly better than Portland cement, and it is rapidly attacked by alkalis. Super-sulfated cement is used for liquors high in sulfates. It is resistant to acidic conditions down to a pH of 3.5 and has alkali resistance similar to Portland cement [50].

3.15.2 Bricks and Tiles

Lining vessels and equipment using this type of construction consists of a membrane, acid-resistant bricks or tiles, and chemical-resistant mortar for joining the bricks (tiles) together.

Membranes are applied directly to the metal or concrete surface to protect from corrosion by any liquor that penetrates the brick lining through pores and cracks. Membranes consist of sheet material bonded to the metal or concrete, e.g., flexible PVC sheet, or it may be formed in situ (e.g., polyester resin reinforced with glass fiber, or synthetic rubber sheet, lead, polyisobutylene, polyethylene and asphalt).

Brick lining (40–65 mm thick) is used for reducing the temperature at the membrane. This protects the membrane from deteriorating with free access of the process liquors and prevents its erosion and other mechanical damage [51,52].

Red and blue acid-resistant bricks are resistant to all inorganic and organic chemicals, except for hydrofluoric acid and hot concentrated caustic alkalis. Acid-resistant fireclay bricks are used for conditions involving elevated temperatures and corrosive condensates. Highly vitrified materials such as chemical stoneware, porcelain and basalts are used for extremely severe duties or where contamination of the process liquors is undesirable.

Chemical-resistant cements are used for all acids up to temperatures of 170°C, but they are attacked rapidly by even dilute alkaline solutions.

Sulfur mortars (mixture of sulfur and inert fillers with small amounts of organic plasticizers) are used for shrinkage mitigation and for eliminating thermal shocks for temperatures up to 80°C. These mortars have poor resistance to alkalis and nonpolar organic solvents.

Phenolic mortars have excellent resistance to acids, particularly for dilute nitric acid (up to 50%) and sulfuric acid (60–90%) but can only tolerate very dilute alkaline solutions at low temperatures.

Furnace mortars are used over a very wide range of conditions. They are resistant to nonoxidizing acids, alkalis, and solvents up to 190°C. Carbon fillers should be used for conditions involving strong alkalis and compounds containing fluorine.

Maximum chemical resistance of polyester mortars is obtained from cements made from isophthalic or bisphenol resins. The bisphenolic resins are preferred for alkaline and hypochlorite solutions.

Epoxy resin cements are specifically intended for resistance to caustic alkalis and organic solvents, but they also have fair acid resistance. They have excellent bond strength to other materials including ceramic and concrete. The corrosion resistances of the cements described above are given in Table 3.40.

Bricks or tiles that line steel vessels have the tendency to crack when the vessel is heated because of the differences in the coefficients of thermal expansion between ceramic and steel.

Table 3.40. Chemical Resistance of Bedding and Jointing Cements

	Dilute Acids	Concentrated Acids	Caustic Alkalis	Mineral Oils	Animal and Vegetable Oils	Oxidizing Acids
Rubber-Latex	G[a]	P[b]	G	F[c]	P	P
Silicate	VG[d]	VG	P	G	G	VG
Sulfur	G	G	P	G	P	G
Phenolic	VG	VG	P	G	G	F
Furane	VG	VG	P	G	G	P
Polyester	VG	G	VG	VG	VG	VG
Epoxy	VG	G	VG	VG	VG	F

[a]G = good.
[b]P = poor.
[c]F = fair.
[d]VG = very good.

3.16 PLASTIC AND THERMOPLASTIC MATERIALS

Plastics are highly resistant to a variety of chemicals. They have a high strength per unit weight of material; therefore, they are of prime importance to the designer of chemical process equipment. Their versatility in properties has provided new and innovative designs of equipment. They are excellent substitutes for expensive nonferrous metals.

Plastics are high-molecular-weight organic compounds of natural or mostly artificial origin. In fabrication, plastics are added with fillers, plasticizers, dyestuffs and other additives, which are necessary to lower the price of the material, and give it the desired properties of strength, elasticity, color, point of softening, thermal conductivity, etc.

Plastics are subdivided into two types: thermoplastic and thermosetting. The thermoplastics can be softened by heat and hardened again by subsequent cooling. This process is reversible and can be repeated many times. By contrast, the thermosetting resins are first softened and melted and, at subsequent heating to a definite temperature, they are irreversibly hardened, becoming insoluble in most organic solvents [53].

Plastics are particularly resistant to inorganic chemicals but are often inferior to metals in resistance to organic chemicals. Table 3.41 gives general resistance properties and typical uses of thermoplastics. The strength characteristics obtained from short-time tests are not suitable for design purposes because all plastics gradually elongate or creep when subjected to sustained loads for long periods, even at ambient temperatures.

The physical and mechanical properties of the principal thermoplastics of interest for process plant applications are listed in Table 3.42. Table 3.43 gives typical hydrostatic design stresses for different types of thermoplastic pipe. Plastics widely employed in piping systems are described briefly below.

3.16.1 Polyolefins

Polyethylene and polypropylene are semitransparent plastics made by polymerization. They are produced from ethylene and propylene in a variety of grades. Their mechanical properties are determined mainly by density (degree of crystallinity) and molecular weight, characterized by the Melt Index (MI).

The effect of these two parameters on mechanical and physical properties of polyethylene and polypropylene are shown in Tables 3.44 and 3.45. The copolymer grade is usually propylene with a little ethylene (5%), which considerably improves the impact strength while causing only a slight loss in stiffness.

Table 3.41. General Properties and Uses of Thermoplastic Materials [54]

Plastic Material	First Introduced	Strength	Electrical Properties	Acids	Bases	Oxidizing Agents	Common Solvents	Product Manufacturing Methods	Common Applications
Acetal Resins	1960	H[a]		P[b]	P	P	G[c]	Injection, blow or extrusion molded	Plumbing, appliance, automotive industries
Acrylic Plastics	1931		G	P	P	F-P	F-P	Injection, compression, extrusion or blow molded	Lenses, aircraft and building glazing, lighting fixtures, coatings, textile fibers
Arc Extinguishing Plastics	1964		E[d]					Injection or compression molded and extruded	Fuse tubing, lightning arrestors, circuit breakers, panel boards
Cellulose Plastics Cellulose Acetate	1912	M[e]					P	All conventional processes	Excellent vacuum-forming material for blister packages, etc.
Cellulose Acetate Butyrate		H				F	F	Molded with plasticizers	Excellent moisture resistance-metallized sheets and film, automobile industry
Cellulose Nitrate	1889	M						Cannot be molded	Little use today because of fire hazard
Cellulose Propionate		H						All conventional processes	Toys, pens, automotive parts, radio cases, toothbrushes, handles

Table 3.41, continued

Plastic Material	First Introduced	Strength	Electrical Properties	Acids	Bases	Oxidizing Agents	Common Solvents	Product Manufacturing Methods	Common Applications
Ethyl Cellulose		H+						All conventional processes	Military applications, referigerator components, tool handles
Chlorinated Polyether	1959	M+	A+[g]	VG	VG	VG	VG	Injection, compression, transfer or extrusion molding	Bearing retainers, tanks, tank linings, pipe, valves, process equipment
Fluorocarbon (TFE)	1930	M	A	VG	VG	VG	VG	Molding by a sintering process following preforming	High-temperature wire and cable insolution, motor-lead insulation; chemical process equipment
Fluorinated Ethylene Propylene (FEP)		M+	A	G	G	G	G	Injection, blow molding, and extrusion and other conventional methods	Autoclavable laboratory ware and bottles
Glass-Bonded Mica	1919	M	G	VG	G	G	G	Moldable with inserts like the organic plastics	Arc chutes, radiation generation equipment, vacuum tube components, thermocouples
Hydrocarbon Resins	1960	M	A	A	A	A	A	Molding with transfer and compression process, casting	Used as lamination resins for various industrial laminates

Methylpentene Polymers (TPX)	1965	M+	E	F	F	F	F	Most conventional processes	Used for electrical and mechanical applications
Parylene (polyparaxylene)	1960							A monomer of the organic compound is vaporized and condensed on a surface to polymerize	Coating material for sensing probes
Phenoxy Plastics	1962	H-M	F	F	F	F	F	Injection, blow and extrusion molding, coatings and adhesives	Adhesives for pipe-bonding compounds, bottles
Polyamide Plastics Nylon	1938	H-M	A	P	P	P	VG	Injection, blow and extrusion molding	Mechanical components (gears, cams, bearings), wire insulation pipe fittings
Polycarbonate Plastics	1959	H-M	VG	G-F	G-F	F	F	All molding methods, thermoforming, fluidized bed coating	Street light globes, centrifuge bottles, high-temperature lenses, hot dish handles
Polychlorotrifluro-ethylene (CTFE)	1938	H	E	VG	VG	VG	VG	Molded by all conventional techniques	Wire insulation, chemical ware, pipe lining, pipe, process equipment lining
Polyester-Reinforced Urethane	1937	H		G	G	G	G	Compression molded over a wide temperature range	For heavy-duty leather applications–industrial applications
Polyimides	1964	H-M	E	G-VG	P	P	G-VG	Molded in a nitrogen atmosphere	Bearings, compressors, valves, piston rings

Table 3.41, continued

Plastic Material	First Introduced	Strength	Electrical Properties	Acids	Bases	Oxidizing Agents	Common Solvents	Product Manufacturing Methods	Common Applications
Polyolefin Plastics Ethlene Vinyl Acetate (EVA)	1940	H		G	G	G-F	G-F	Most conventional processes	Molded appliance and automotive parts, garden hose, vending machine tubing
Polyallomers	1962	H	G-VG	F	F	F	F	Molding processes, all thermoplastic processes	Chemical apparatus, typewriter cases, bags, luggage shells, auto trim
Polyethylene	1939	H	VG	G-VG	G-VG	P	P	Injection, blow, extrusion and rotational	Pipe, pipe fittings, surgical implants, coatings, wire and cable insulation
Polypropylene Plastics	1954	H-M	VG	VG	VG	F	F	Same as PVC	Housewares, appliance parts, auto ducts and trim, pipe, rope, nets
Polyphenylene Oxide	1964	M	F-G	E	E	VG	VG	Extruded, injection molded, thermoformed and machined	Autoclavable surgical tools, coil forms, pump housings, valves, pipe
Polysulfone	1965	M	VG	VG	VG	F	P-F	Extrusion and injection molded	Hot water pipes, lenses, iron handles, switches, circuit breakers

Polyvinylidene Fluoride (VF_2)	1961	H	VG+	G-VG	G	G-VG	G-VG	Molded by all processes, fluidized bed coatings	High-temperature valve seats, chemical resistant pipe, coated vessels, insulation
Styrene Plastics ABS Plastics	1933	M-H	VG+	G-VG	G-VG	F-G	F	Thermoforming, injection, blow, rotational and extrusion molds	Business machine and camera housings, blowers, bearings, gears, pump impellers
Polystyrene	1933	M-H	VG+	G	G	F	P-F	Most molding processes	Jewelry, light fixtures, toys, radio cabinets, housewares, lenses, insulators
Styrene Acrylonitrile (SAN)		H	VG	VG	G	G	G	Most molding processes	Lenses, dishes, food packages, some chemical apparatus, batteries, film
Urethane	1955	M-H+		G-VG	G	G	F-G	Extruded and molded	Foams for cushions, toys, gears, bushings, pulleys, shock mounts
Vinyl Plastics,	1835								
Compolymers of Vinyl Acetate and Vinyl Chloride,	1912	M+		G	G-F	G	G	All molding processes	Floor products, noise insulators
Polyvinyl Acetate	1928	M		P-G	G	G	P	Coatings and adhesives	Adhesives, insulators, paints, sealer for cinder blocks

Table 3.41, continued

Plastic Material	First Introduced	Strength	Electrical Properties	Acids	Bases	Oxidizing Agents	Common Solvents	Product Manufacturing Methods	Common Applications
Polyvinyl Aldehyde	1940	H		VG	VG	VG	VG	Most molding processes	Used for coatings and magnet wire insulation, interlayer of safety glasses
Polyvinyl Chloride (PVC)	1940	M-H		VG	G	G	G	Extrusion, injection, rotational, slush, transfer, compression, blow mold	Pipe conduit and fittings, cable insulation, downspouts, bottles, film
PVC Plastisols	1940	M-H		VG	G	G	G	Slush and rotationally molded, foamed, extruded	Used in coating machines to cover paper, cloth and metal
Polyvinylidene Chloride	1940	H		VG	VG	VG	VG	Same as PVC	Auto seatcovers, film, bristles, pipe and pipe linings, paperboard coatings

[a] H = high.
[b] P = poor.
[c] G = good.
[d] E = excellent.
[e] M = moderate.
[f] A = average.

Table 3.42. Mechanical Properties of Thermoplastics [55]

	Polyethylene	Polypropylene	PVC[a]	ABS[a]	PTFE[a]	Acrylics	Nylon 66	Acetal
Specific Gravity	0.93	0.90	1.4	1.05	2.20	1.2	1.15	1.4
Tensile Strength (N/mm^2)	10	34	55	34	21	70	83	70
Elongation at Break (%)	500	300	15	50	300	5	100	70
Tensile Modulus (N/mm^2)	170	1360	3100	2100	415	3450	2760	2900
Thermal Expansion ($10^{-3}/°C$)	15	11	7	10	10	7	10	10

[a]PVC = polyvinychloride; ABS = acrylonitrile butadiene styrene; PTFE = polytetrafluorenethylene.

Table 3.43. Hydrostatic Design Pressures for Thermoplastic Pipe for Temperatures up to 130°C [55]

	Maximum Hydrostatic Design Pressures, N/mm^2, for life of 10 years						Minimum Temperature of Usage[a]
	20°C	40°C	60°C	80°C	100°C	120°C	°C
Low-Density Polyethylene	2.7	1.7					−40
High-Density Polyethylene	4.8	3.9					−20
Polypropylene							
Homopolymer	6.9	4.1	2.2	1.2	1.0		−1
Copolymer	5.9	3.2	1.7	0.7			−5
CPVC[b]	13.8		6.9	3.5	1.1		−1
High-Impact PVC[b]	6.2	3.1					−20
ABS[b]	11.0		5.5	1.6			−30
PVF[b]	9.0	7.7	5.9	4.8	3.4	2.4	

[a]Temperature below which the pipe is very brittle.
[b]PVC = polyvinylchloride; ABS = acrylonitrile butadiene styrene; CPVC = chlorinated PVC; PVF = polyvinylfluoride.

Table 3.44. Effect of Density on Poleythylene Polymers [55]

	Polyethylene Density, g/cm^3					Polypropylene (0.90)	
	0.92	0.93	0.94	0.95	0.96	Copolymer	Homopolymer
Crystallinity (%)	55	62	70	77	85		
Tensile Modulus, (N/mm^2)	170	275	450	625	850	1300	1500
ASTM Type[a]	I		II		III		
Melting Point (°C)	110				135	165	175
Heat Distortion Temperature (°C)	40	50	60	70	85	100	105

[a]D1248: Polyethylene plastics molding and extrusion materials.

Low-density polyethylene, a tough, relatively flexible material, is used at temperatures up to 50°C and does not become brittle until the temperature falls to −40°C.

This material is resistant to most hot, highly oxidizing inorganic acids,

Table 3.45. Effects of Degree of Crystallinity and Molecular Weight [55]

Increase in Density (crystallinity)	Decrease in Melt Index	Basic Properties Unaffected by Density and MI
Increases	Increases	
Elastic modulus Stress at yield Surface hardness Softening and melting temperature Impermeability to gases	Resistance to environmental stress cracking Tensile strength Elongation at break	Corrosion resistance, thermal properties, electrical properties, refractive index, decomposition temperature
Decreases	Decreases	
Toughness Liquid absorption	Temperature at which brittleness occurs	

alkalis and aqueous solutions of inorganic salts. Even though polyethylene is not actually dissolved by any organic substances at temperatures below 50°C, it is not a practical material for use in contact with many solvents (among them are chlorinated solvents and aromatics) because of its marked swelling. Although polar liquids such as alcohols, esters, amines and phenols are not solvents suitable for use with polyethylene and do not cause swelling, they may cause environmental stress cracking. This occurs when polyethylene is stressed by either external loads or internal molding strains in contact with these solvents.

High-density polyethylene ($\rho \cong 0.94$–0.96 g/cm^3) has up to five times the stiffness of low-density polyethylene at ambient temperatures and can be used at much higher temperatures. Its chemical resistance is similar to that of the low-density grades, but the resistance to swelling by solvents is higher.

High-density polyethylene is very susceptible to environmental stress corrosion cracking, especially if used at the high end of the temperature range.

The chemical resistance of polypropylene is similar to that of polyethylene. Its standard grade is suitable for use at temperatures up to 90°C. All commercial grades of polypropylene contain an antioxidant, whose concentration in the polymer determines the maximum working temperature of equipment made from its high grades. The stiffness of polypropylene can be improved by incorporation of asbestos and glass fibers. Compared with polyethylenes, polypropylene has the great advantage of not being subject to environmental stress cracking.

The copolymer is safer to use in most process plant equipment where creep strength is low. However, for long life under pressure, particularly at the higher end of the temperature range, the homopolymer is preferred.

3.16.2 Polyvinyl Chloride (PVC)

PVC, the polymerization product of chlorine-substituted ethylene derivatives, is probably the most widely used plastic for process plant construction. It is available in four different types: rigid, high impact, high temperature and plasticized.

3.16.3 Rigid PVC (UPVC)

This is used most often in process plants. It is a tough, low-cost material with probably the widest range of chemical resistance of any of the low-cost plastics. On a volume basis, PVC is more favorable than polypropylene because the modulus of PVC is considerably higher than that of polypropylene, so it will form more rigid structures when used at the same thickness. On a weight basis it is not as favorable as PVC because it has a specific gravity of 1.4 compared with 0.92 for polypropylene.

The maximum temperature at which PVC can be used unsupported is 60°C. However, it can be used up to 100°C by external reinforcement with glass-reinforced polyester or epoxy resin. At low temperatures, rigid PVC becomes brittle and, in this case, a high-impact PVC should be used. Rigid PVC not suffering from environmental stress cracking is very resistant to most nonhighly oxidizing inorganic acids and alkalis. It is attacked by halogen gases, particularly when they are damp. PVC is much better than polyolefins for resistance to mineral oils, petrol and paraffin, but absorbs and is swollen by aromatic hydrocarbons, chlorinated solvents, esters and ketones.

This plastic is used widely in extruded forms such as sheet, pipe and rod, and as injection moldings such as pipe fittings [56-58].

3.16.4 High-Impact PVC

These compounds are particularly advantageous for equipment exposed to low temperatures. This plastic containing chlorinated ethylene does not differ significantly in chemical resistance from rigid PVC, but does have a slightly lower creep strength.

3.16.5 Chlorinated PVC (CPVC)

This is used at temperatures up to 100°C and has similar chemical resistance impact. Presumably it will supersede normal PVC because of its superior tolerance to higher temperatures.

3.16.6 Plastic PVC

Produced from PVC compounds containing plasticizers such as dioctyl phthalate, this is a flexible sheeting suitable for lining tanks made from steel and concrete. The maximum temperature at which this lining can be used is 60°C.

Some solvents, including aromatic and chlorinated hydrocarbons, ketones and ethers, will soften the sheet by acting as additional plasticizers.

3.16.7 Acrylonitrile-Butadiene-Styrene (ABS)

ABS, a tough rigid plastic, is widely used as sheet, pipe and pipe fittings. ABS's chemical resistance is slightly inferior to rigid PVC. It has good impact and low-temperature properties similar to those of high-impact PVC, but has better resistance than rigid PVC to high temperatures (up to 80°C).

3.16.8 Fluorinated Plastics

Fluorinated polymers stand out sharply against other construction materials for their excellent corrosion resistance and high-temperature stability. In this respect they are not only superior to other plastics but also to platinum, gold, glass, enamel and special alloys. The fluorinated plastics used in process plants are polytetrafluorethylene (PTFE), fluorinated ethylene/propylene (FEP), polytrifluoromonochlorethylene (PTFCE) and polyvinyl fluoride (PVF). They are much more expensive than other polymers and so are only economical in special situations [59].

In addition to excellent corrosion resistance at temperatures up to 250°C, PTFE has a very low coefficient of friction. Therefore, it is widely used for metal lining, self-lubricating bearings, seals, etc. PTFE has a very low mechanical strength and so should be loaded only lightly. Its creep strength can be increased by using fillers (glass and carbon fibers). PTFE can be used in contact with all chemicals except molten alkali metals, fluorine and chlorine trifluoride. Unlike PTFE, fluorinated ethylene propylene (FEP) can be extruded and injection molded. Therefore, FEP is available in a greater variety of forms than PTFE and can be used up to 200°C. The PTFE and FEP coatings do not protect metal from corrosion as they are porous.

PTFCE (or CTFC) can be applied as a porous-free coating where the highest corrosion resistance is required.

3.16.9 Polyvinyl Fluoride (PVF)

PVF is a rigid plastic with corrosion resistance over a much wider range of conditions (it can be used with most chlorinated solvents up to 100°C) and temperature (−60° to 150°C).

3.16.10 Acrylics

Acrylics are used in applications in which transparency is necessary and breakage is likely to occur. It is tougher than glass and can be used in the temperature range from −40°C to +85°C.

Acrylics are chemically resistant at room temperature to dilute acids, except hydrofluoric and hydrocyanic, all alkalis and mineral oils. They are attacked by chlorinated solvents, aromatic hydrocarbons, ketones, alcohols, ethers and esters [60].

3.16.11 Chlorinated Polyether

Chlorinated polyether used in temperatures up to 120°C in forms of coatings, extruded pipe and sheet linings is an intermediate between the low-cost commodity polymers such as PVC and polyethylene, and fluorinated polymers [61].

3.16.12 Nylon (Polyamide)

Nylons, long-chain synthetic polymeric amides, are fabricated in many different types. The most commonly used forms are described in Table 3.46.

High stiffness, fair resistance to corrosion and good frictional properties sometimes make nylons more attractive than metals. The stiffness of the polymer may be increased by glass filling. Nylons containing molybdenum

Table 3.46. Properties of Different Nylons [62]

	Type Nylon				
	66	Glass Filled 66	8	610	11
Density (g/cm^3)	1.14	1.4	1.13	1.09	1.04
Melting point (°C)	260		220	220	190
Moisture Absorption (%) Equilibrium at 100% RH	8	5.5	11	3.5	1.5
Elastic Modulus (N/mm^2)					
Dry	3000	7500	2750	1750	1250
Wet	1200	2500	700	1100	900
Tensile Strength (N/mm^2)	85	170	80	60	60
Elongation at Break (%)	60	3	100	100	100
Heat Distortion Temperature (°C) (204 psi ASTM D648)	75	245	70	55	50

disulfide or graphite can be used over the temperature range of −50° to +75°C. Nylons are resistant at room temperature to very dilute acids, alkalis up to 20% concentration, esters, alkyl halides, mercaptans, glycols, aldehydes (except formaldehyde), mineral oils and petroleum fuels. They are attacked by halogens and halogenated compounds, as well as phenols and cresols.

3.16.13 Miscellaneous Engineering Plastics

To overcome the disadvantages of nylon as an engineering material—high water absorption and poor creep strength at elevated temperatures—many newer polymers were developed. Table 3.47 lists polymers that are among the most commercially important: acetal, polycarbonate, polyphenylene oxide and polysulfone.

3.16.14 Acetal Resin

This has the same stiffness as dry nylon but absorbs less than one-fifth the amount of water and retains its strength and shape in humid conditions at considerably higher temperatures. Acetal has poor resistance to aqueous solutions of strong acids and should not be used at a pH below 3. The acetal's copolymers are satisfactory in most alkaline conditions. All grades have very good resistance to mineral oils and petrols and to most organic solvents including chlorinated compounds.

Table 3.47. Properties of Different Engineering Plastics

	Acetal	Polycarbonate	Polyphenylene Oxide	Polysulfone	Nylon 66
Tensile yield stress (N/mm^2)	70	60	60	70	70
Elongation at fracture (%)	60	80	25	75	60
Elastic Modulus (dry) (N/mm^2)	2800	2500	2500	2500	3000
Heat Deflection Temperature (°C) at 1.8 N/mm^2 (264 psi)	110	135	130	175	75
Water Absorption, 24 hr at 100% RH	0.22	0.15	0.07	0.22	1.5
Equilibrium Immersion	0.8	0.35	0.15		8
UL Temperature Index (°C)	80	115	105	140	65

3.16.15 Polycarbonate

This has a very high resistance to impact damage, even at subzero temperatures. It has good creep strength in dry conditions up to 115°C but degrades by continuous exposures to water hotter than 65°C. It is resistant to aqueous solutions of acids, aliphatic hydrocarbons, paraffins, alcohols (except methanol), animal and vegetable fats and oils, but is attacked by alkalis, ammonia, aromatic and chlorinated hydrocarbons.

3.16.16 Polyphenylene Oxide

A stable material in humid conditions at temperatures up to 105°C, polyphenylene oxide is resistant to most aqueous solutions of acids and alkalis but is attacked by many organic solvents, particularly by aromatics and chlorinated aliphatics.

3.16.17 Polysulfone

Used at temperatures up to 140°C, polysulfone has good resistance to aqueous solutions of acids and alkalis; it is satisfactory with aliphatic solvents but is swollen by aromatics and stress cracked by several organic solvents, including acetone, ethyl acetate, trichlorethylene and carbon tetrachloride.

3.17 THERMOSETTING PLASTICS

The thermosetting plastics–phenolics, polyesters and epoxies–are used at higher temperatures (about 150°C) and pressures than thermoplastics. They are finding ever increasing applications for process plant equipment.

Thermosetting resins are available as powders and liquids. The powders can be molded in a similar manner to thermoplastics and are used in process plants where large numbers of small items are required (for example, packing rings).

Liquid resins are usually reinforced with fibers (glass, asbestos), because of their brittleness. They are almost always used for process plant construction. As liquid resins they can be catalyzed to cure at room temperature and low pressures. Relatively cheap wooden molds are required to build quite large items such as tanks and ducting on a one-off basis. The principal materials in this group of plastics are described below.

3.17.1 Phenolic Resins

Phenoplasts manufactured on the basis of thermoreactive phenolformaldehyde resin, harden on heating up to 120-170°C and then become insoluble. They are usually reinforced with asbestos fiber.

The phenolic/asbestos laminates (used up to 200°C) have excellent resistance to most mineral and organic acids but are attacked by strong oxidizing agents such as nitric and concentrated sulfuric acids and strong alkalis such as sodium and potassium hydroxide. Tanks, scrubbers, columns, pumps, pipes, etc., are fabricated from phenolic/asbestos laminates.

3.17.2 Polyester Resins

Polyester resins, reinforced with glass fibers, are used widely in the construction of process equipment. Some physical and mechanical properties are presented in Table 3.48. Table 3.49 lists various materials used as filler and the properties they impart to different plastics.

There are four types of polyester resins available: orthophthalic, isophthalic, bisphenol and chlorinated. The general purpose orthophthalic resins do not have adequate resistance to corrosion for most process plant requirements. Isophthalic resins can be used up to 60°C. They are corrosion resistant to most organic and mineral (up to 25%) acids, salts that are not strongly alkaline, aliphatic solvents and petroleum products, but not to aromatics, strongly oxidizing acids, hot distilled water and strong alkalis [63,64].

The bisphenol polyesters, being considerably more expensive than isophthalics, have better chemical resistance to strong alkaline solutions and oxidizing solutions. For about the same price as the bisphenols, the

Table 3.48. Various Properties of Fiberglass Resins

Property	Polyester/ Glass Mat	Polyester/ Woven Glass Cloth	Epoxy/ Woven Glass Cloth	Epoxy/ Filament Wound Glass Rovings
Glass (%w/w)	25	60	60	70
Specific Gravity	1.5	1.7	1.7	1.8
Thermal Expansion, $\times$ 10^{-6} °C	25	12	12	11
Modulus of Elasticity (kN/mm^2)	7	15	18	20
Flexural Strength (N/mm^2)	125	345	415	70

Table 3.49. Various Filler Materials and

Filler Material	Chemical Resistance	Heat Resistance	Electrical Insulation	Impact Strength	Tensile Strength
Alumina Powder					
Alumina Trihydrate			X		
Asbestos	X	X	X	X	
Bronze					
Calcium Carbonate		X			
Calcium Silicate		X			
Carbon Black		X			
Carbon Fiber					
Cellulose				X	X
Alpha Cellulose			X		X
Coal (powdered)	X				
Cotton (chopped fibers)			X	X	X
Fibrous Glass	X	X	X	X	X
Graphite	X				X
Jute				X	
Kaolin	X	X			
Mica	X	X	X		
Molybdenum Disulfide					
Nylon (chopped fibers)	X	X	X	X	X
Orlon®	X	X	X	X	X
Rayon			X	X	X
Silca, Amorphous			X		
TFE-Fluorocarbon					
Talc	X	X	X		
Wood Flour			X		X

chlorinated resins are nonflammable. The addition of fire-retardant compounds–chlorine or bromine–can reduce the flammability of other polyester resins.

3.17.3 Epoxy Resins

The fully glass-reinforced epoxy laminates have much higher strength and corrosion resistance to solvents and alkalis than do polyester laminates, but are more expensive. They are used in the process industries for manufacturing piping to be used for very high working pressures by filament winding techniques with glass rovings.

3.17.4 Furane Resins

Furane resins are superior to polyesters and epoxies for resistance to ketones, chlorinated solvents and carbon disulfide. However, as they are

Their Property Contributions to Plastics [54]

Dimensional Stability	Stiffness	Hardness	Electrical Conductivity	Thermal Conductivity	Moisture Resistance	Handleability
			X	X		
	X				X	X
X	X	X				
	X	X	X	X		
X	X	X				X
X	X	X				
X	X		X	X		X
			X	X		
X	X	X				
X						
					X	
X	X	X				
X	X	X			X	
X	X	X	X	X		
	X					
X	X	X			X	X
X	X	X			X	
	X	X			X	X
X	X	X				X
X	X	X		X	X	
X	X	X				
					X	X
X	X	X				
X	X	X			X	X
X						

very brittle resins, they are difficult to make into self-supporting laminates, so often are used as an inner layer to glass-reinforced polyester laminates [65].

3.17.5 Rubber Linings

Rubber, natural and synthetic, has been used extensively for many years in chemical process plants. Rubber is a product obtained by thermal processing (vulcanization) of a mixture of raw natural synthetic caoutchouc with sulfur.

Soft rubber is obtained by adding 2–4% sulfur; by adding extra sulfur (25–40%), the rubber can be made into ebonite, which is a hard, brittle material, having a wider range of chemical resistance than soft rubber. Soft ordinary rubber is chemical and erosion resistant, but its thermal resistance is not high (about 80°C).

Neoprene, which is basically polychloroprene, has better resistance to heat than does natural rubber (up to 105°C), and has better resistance to mineral and vegetable oils and fatty acids. It is attacked by aromatic and halogenated hydrocarbons. Butyl rubber, basically polyisobutylene, is used as a heat-resistant, impermeable material. It is better than natural rubber when in contact with oxidizing acids, such as dilute solutions of nitric acid. It is attacked by free halogens, chlorinated and aromatic hydrocarbons and petroleum oils.

Nitrile rubbers, copolymers of butadiene and acrylonitrile, are used for resistance to swelling by mineral oils and fuels enhanced by formulations with a high acrylonitrile/butadiene ratio. They have poor resilience and low-temperature properties, and should not be used with ketones, phenols or aromatic hydrocarbons [66-69].

Hypalon, chlorosulfated polyethylene, is particularly noted for its resistance to strong oxidizing materials such as sodium hypochlorite, chromic and nitric acids. It has good resistance to mineral and vegetable oils but is not recommended for use with aromatic and chlorinated hydrocarbons.

Fluorinated rubbers, copolymers of hexafluoropropylene and vinylidene-fluorides, have excellent resistance to oils, fuels and lubricants at temperatures up to 200°C. They have better resistance to aliphatic, aromatic and chlorinated hydrocarbons and most mineral acids than other rubbers, but their high cost restricts their engineering applications. Cheremisinoff et al. [54] provide extensive physical and mechanical properties data on engineering plastics. A glossary of terms concerned with fabrication and properties of plastics is given in the last section of this chapter.

3.18 ORGANIC COATINGS AND PAINTS

Organic coatings are applied mainly to mild steel structures and equipment. They are also used on aluminum, zinc-sprayed and galvanized steel, but to a lesser extent. The applications for organic coatings can be divided into three areas: corrosion by atmospheric pollution, protection from splash by process liquors, and linings for immersion in process liquors [70-74].

Application of protective paints consists of surface preparation of steel, priming coat and finishing coats. Wherever possible, steel should be blast-cleaned before painting. Primers thoroughly wet the metal to promote adhesion of finishing paints and carry inhibitive pigments. For example, red lead oxide will minimize the spread of rust on metal surfaces. The total thickness of finishing coats must be at least 0.125 mm for adequate protection and life. Four coats of paint usually are necessary to achieve this.

Paints based on phenolic resins are oil modified to permit drying at ambient temperatures. They are very suitable for most industrial atmospheres. Paints with a higher standard of chemical resistance are required for equipment that is splashed by corrosive process liquors.

Chlorinated rubber paints and vinyl paints have excellent resistance to high concentrations of acids and alkalis at temperatures up to 80°C. High-build chlorinated rubber paints, which give a thickness of 0.12 mm per coat are commonly used for process plant equipment.

Epoxy resin paints, inferior to chlorinated rubber for resistance to strong acids, are excellent for dilute acids and strong alkalis. They produce a harder, more abrasion-resistant coating than does chlorinated rubber and are much better for resistance to fats, oils and many organic solvents. Table 3.50 gives data on the chemical resistance of epoxy resin coatings to different materials.

Chemical-resistant finishing paints are frequently used under immersion conditions; however, they are not used in cases in which pinholes in the coating might lead to catastrophic corrosion of the underlying metal. Coatings for immersion require special equipment for application (ovens for high-temperature curing, tanks for dipping or special guns for spraying), but they provide the solution of complete elimination of porosity.

Stoved phenolics have outstanding acid resistance (up to 200°C in dry conditions and up to 100°C in wet conditions), except to strong oxidizing acids. They are unsuitable for use with alkaline solutions above pH 10, wet chlorine or hypochlorite solutions. Phenolics/silicon formulations can be used for steam up to 180°C without a significant effect on heat transfer rates.

Polyester/glass-flake linings can be applied onsite because they cure at ambient temperature. Their corrosion resistance depends on the type of polyester resin used.

PVC plastisol coatings are tough, with an abrasion resistance similar to rubber. They are resistant to acids and alkalis but usually are not suitable for solvents because of extraction of the plasticizers in the coating.

Most plastics are now available as powders and can be applied as coatings by fluidized bed or spraying techniques. Nylon 11 and polyethylene have proved most useful for chemical plant applications.

3.19 GLOSSARY OF FABRICATION AND PLASTICS TERMS

A-stage — Initial or early stage in the reaction of some thermosetting resins in which the material is still soluble in certain liquids and fusible; referred to as resol.

Acid-acceptor — Chemical that acts as a stabilizer by chemically combining with an acid that may be present initially in trace quantities in a plastic; also may be formed via decomposition of the resin.

Table 3.50. Chemical Resistance of Epoxy Resin Coatings [54]

Material	Concentration (%)	Temperature (°C)							
		5	15	27	38	49	60	71	82
Acetic Acid	1-5	G[a]	G	F[b]	F	F	F	F	P[c]
	5-10	F	F	P	P	P	NR[d]	NR	NR
	10-50	NR	NR	NR	NR	NR	NR	NR	NR
Acetone	1-5	G	G	G	F	F	F	P	P
	10-20	F	F	F	P	P	NR	NR	NR
Alcohols (ethyl)	–[e]	X[f]	X	G	G	F	F	P	P
Alum Sulfate	–	X	X	X	X	X	X	X	X
Ammonium Chloride	–	X	X	X	X	X	X	X	X
Ammonium Fluoride	–	X	X	X	X	X	X	X	X
Aromatic Solvents	–	X	X	X	X	G	G	G	F
Beer	–	X	X	X	X	X	X	X	X
Black Liquor	–	X	X	X	X	X	X	X	X
Boric Acid	1-5	X	X	X	G	G	G	F	F
Calcium Chloride	1-50	X	X	X	X	X	X	G	G
Carbon Tetrachloride	–	X	X	G	G	G	G	F	F
Chromic Acid	1-5	F	F	NR	NR	NR	NR	NR	NR
Citric Acid	1-5	F	F	F	G	G	G	G	F
Cooking Oils	–	X	X	X	X	X	X	X	G
Copper Salts	–	X	X	X	X	X	X	X	X
Esters	–	X	X	X	X	X	X	X	X
Ethers (ethyl ether)	–	X	X	X	X	X	G	G	G
Formaldehyde	1-35	X	X	X	X	X	X	X	X
Ferric Chloride	–	X	X	X	X	G	G	G	F
Ferrous Salts	–	X	X	X	X	X	X	X	X
Gasoline	–	X	X	X	X	X	X	X	G
Glycerin	–	X	X	X	X	X	X	X	X
Hydrochloric Acid	1-5	X	X	X	G	G	F	F	F
Hydrofluoric Acid	1-5	G	G	G	P	P	NR	NR	NR
Kerosene	–	X	X	X	X	X	X	X	G
Lactic Acid	1-10	X	X	X	G	G	G	F	F
Lead Acetate	–	X	X	X	X	X	X	X	G
Manganese Salt	–	X	X	X	X	X	X	X	G
Methyl Ethyl Ketone	1-5	G	G	G	F	F	NR	NR	NR
Mineral Spirits	–	X	X	X	X	G	G	G	F
Naptha	–	X	X	X	X	X	X	X	X
Nitric Acid	1-5	F	F	P	P	P	P	NR	NR
	10-20	P	P	P	NR	NR	NR	NR	NR
Oxalic Acid	Saturated	G	G	G	F	F	NR	NR	NR
Phosphoric Acid	20	F	F	P	P	NR	NR	NR	NR
Potassium Hydroxide	–	X	X	X	X	X	X	X	X
Salt Brine	–	X	X	X	X	X	X	X	G
Soaps	–	X	X	X	X	X	X	X	G
Detergents	–	X	X	X	X	X	X	X	G
Sodium Chromate	–	X	X	X	X	X	X	X	G

Table 3.50, continued

Material	Concentration (%)	Temperature (°C)							
		5	15	27	38	49	60	71	82
Sodium Dichromate	–	G	G	G	G	G	G	F	F
Sodium Fluoride	–	X	X	X	X	X	X	X	X
Sodium Hydroxide	1-10	X	X	X	X	X	X	X	G
	50	X	X	X	G	G	G	G	F
Sodium Hypochlorite	3	G	G	F	P	P	NR	NR	NR
Sodium Phosphate	–	X	X	X	X	X	X	X	X
Sodium Sulfate	–	X	X	X	X	X	X	X	X
Sodium Sulfite	–	X	X	X	X	X	X	X	X
Sodium Thiosulfate	–	X	X	X	X	X	X	X	X
Sulfite Liquor	–	X	X	X	X	X	X	X	X
Sulfuric Acid	1-5	X	X	X	G	G	F	F	P
	10-20	X	X	P	P	NR	NR	NR	NR
Vegetable Oils	–	X	X	X	X	X	X	X	X
Water (fresh)	–	X	X	X	X	X	X	X	X
Water (distilled)	–	X	X	X	X	X	G	G	G
White Liquor	–	X	X	X	X	X	X	X	G

[a]G = good.
[b]F = fair.
[c]P = poor.
[d]NR = not recommended.
[e]– = all conditions.
[f]X = excellent.

Acrylic plastics — Group of plastics based on resins generated from the polymerization of acrylic monomers (e.g., ethyl acrylate and methyl methacrylate).

Adherend — A component or body held to another body by an adhesive.

Adhesion — Condition in which two surfaces are bonded together by interfacial forces caused by valence forces or interlocking forces or both (see mechanical adhesion and specific adhesion).

Adhesion, mechanical — Bonding between two surfaces caused by interlocking action of molecules.

Adhesion, specific — Adhesion between surfaces whereby valence forces predominate that are similar to those promoting cohesion.

Adhesive	Material that holds parts together by surface attachment. Examples include glue, mucilage, paste and cement. Various forms of adhesives include liquid or tape adhesives (physical type) and silicate or resin adhesives (chemical type).
Adhesive, assembly	Adhesive for bonding materials together, e.g., boat, airplane, furniture, etc.; term commonly used in wood chemistry to distinguish between 'joint glues' and veneer glues. Term applied to adhesives employed in fabricating finished goods, differs from adhesives used in fabricating sheet materials such as laminates or plywood.
Aging	The effect of exposure of plastics to the environment for a length of time. The specific effect and degree depend on the moisture in, and temperature and composition of, the environment, in addition to the length of exposure.
Alkyd plastics	Group of plastics composed of resins based on saturated polymeric esters whereby the recurring ester groups are an integral part of the primary polymer chain and the ester groups exist in cross-links that are present between chains.
Allyl plastics	Group of plastics composed of resins formulated by addition polymerization of monomers containing allyl groups (e.g., diallyl phthalate).
Amino plastics	Group of plastics generated by the condensation of amines (e.g., urea and melamine with aldehydes).
Anneal	As applied to molded plastics, the process of heating material to a specified temperature and slowly cooling it to relieve stresses.
Assembly	The positioning or placing together in proper order layers of veneer or other materials, with adhesives, for purposes of pressing and bonding into a single sheet or unit.
Assembly time	Refers to the elapsed time after an adhesive is applied until pressure effects curing.
B-stage	Intermediate-stage reaction step for various thermosetting resins. During this stage the material swells when in contact with certain liquids and

becomes soft when heat is applied. The material may not dissolve or fuse entirely. Resin in this stage is referred to as resitol.

Back-pressure-relief port	Opening from an extrusion die used for excess material to overflow.
Binder	Part of adhesive composition responsible for adhesive forces.
Blanket	Veneers laid up on a flat table. Complete assembly is positioned in a mold at one time; used primarily on curved surfaces to be molded by the flexible bag process.
Blister	Elevation of the surface of a plastic caused by trapped air, moisture, solvent; can be caused by insufficient adhesive, inadequate curing time, excess temperature or pressure.
Blocking	Adhesion between layers of plastic sheets in contact; condition arises during storage or use when components are under pressure.
Bloom	Visible exudation or efflorescence on the surface of a plastic; caused by plasticizer, lubricant, etc.
Bolster	Spacer or filler material in a mold.
Bond	The attachment at the interface or exposed surfaces between an adhesive and an adherend; to attach materials together with adhesives.
Bulk density	Density of a molding material in loose form, such as granular, nodular, etc., with units in g/cm^3 or lb/ft^3.
Bulk factor	Ratio of the volume of loose molding compound to the volume of the same amount in molded solid form; ratio of density of solid plastic component to apparent density of loose molding compound.
Case harden	Process of hardening the surface of a piece of steel to a relatively shallow depth.
Cast film	Film generated by depositing a layer of liquid plastic onto a surface and stabilizing by evaporating the solvent, by fusing after deposition or by cooling. Cast films are generated from solutions or dispersions.

Catalyst	Material used to activate resins to promote hardening. For polyesters, organic peroxides are used primarily. For epoxies, amines and anhydrides are used.
Cavity	Portion of a mold that forms the outer surface of the molded product.
Cell	Single cavity caused by gaseous displacement in a plastic.
Cellular plastic	A plastic that suffers a density decrease by the presence of numerous cells dispersed throughout the material.
Cellular striation	Layering of cells within a cellular plastic.
Cellulosic plastics	Group of plastics composed of cellulose compounds, for example esters (e.g., cellulose acetate) and ethers (e.g., ethyl cellulose).
Centrifugal casting	Process in which tubular products are fabricated through the application of resin and glass strand reinforcement to the inside of a mold that is rotated and heated. The process polymerizes the resin system.
Chalking	Dry, chalk-like deposit on the surface of a plastic.
Chase	Main portion of the mold containing the molding cavity, mold pins, guide pins, etc.
Chemically formed plastic	Cellular plastic whereby the material's structure is formed by gases generated from the chemical reaction between its constituents.
Clamping plate	Mold plate that matches the mold and is used to fasten the mold to the machine.
Closed-cell foam	Cellular plastic composed predominantly of non-interconnecting cells.
Cohesion	Forces binding or holding a single material together.
Cold flow	Creep: the dimensional change of a plastic under load with time followed by the instantaneous elastic or rapid deformation at room temperature; permanent deformation caused by prolonged application of stress below the elastic limit.

Cold molding	The fashioning of an unheated mixture in a mold under pressure. The article is then heated to effect curing.
Cold pressing	Bonding process whereby an assembly is subjected to pressure without applying heat.
Cold slug	Material to first enter an injection mold.
Cold-slug well	Section provides opposite sprue opening of the injection mold, used for trapping cold slug.
Condensation	Chemical reaction whereby two or more molecules combine and separate out water or other substance. When polymers are formed, it is referred to as polycondensation.
Consistency	Resistance of a material to flow or to undergoing permanent deformation under applications of shearing stresses.
Contact molding	Process whereby layers of resin-impregnated fabrics are built up one layer at a time onto the mold surface, forming the product. Little or no pressure is required for laminate curing.
Copolymer	Formed from two or more monomers (see polymer).
Core	Portion of the mold that forms the inner surfaces of the molded product.
Core and separator	Center section of an extrusion die.
Core pin	Pin for molding a hole.
Core-pin plate	Plate that holds core pins.
Crazing	Tiny cracks that develop on a laminate's surface. Caused by mechanical or thermal stresses.
Creep	See cold flow.
Cross linking	Generation of chemical linkages between long-chain molecules; can be compared to two straight chains joined together by links. The rigidity of the material increases with the number of links. The function of a monomer is to provide these links.
C-stage	Final reaction stage of various thermosetting resins. In this stage material is insoluble and infusible. Resin in fully cured thermosetting

	molding is in this stage and is referred to as resite.
Cull	Remaining material in the transfer vessel after the mold has been filled.
Cure	Process in which the addition of heat, catalyst or both, with or without pressure, causes the physical properties of the plastic to change through a chemical reaction. Reaction may be condensation, polymerization or addition reactions.
C-veil	Thin, nonwoven fabric composed of randomly oriented and adhered glass fibers of a chemically resistant glass mixture.
Degradation	Deleterious change in a plastic's chemical structure.
Delamination	Separation of a laminate's layers.
Deterioration	Permanent adverse change in the physical properties of a plastic.
Diaphragm gate	Gate employed in molding tubular or annular products.
Die-adaptor	Piece of an extrusion die that serves to hold die block.
Die block	Part of extrusion die that holds the core and forming bushing.
Die body	Part of an extrusion die used to separate and form material.
Dilatant	Property of a fluid whose apparent viscosity increases with shear rate.
Dished	Displays a symmetrical distortion of a flat or curved section; as viewed, it appears concave.
Dispersant	In an organosol, the liquid constituent that displays solvating or peptizing action on the resin; subsequent action aids in dispersing and suspending resin.
Dispersion	Heterogeneous mixture in which finely divided material is distributed throughout the matrix of another material. Distribution of finely divided solids in a liquid or a solid (e.g., pigments, fillers).
Doping	Coating a mandrel or mold with a material that prevents the finished product from sticking to it.

Dowel	Pin that maintains alignment between the various sections of a mold.
Draft	Angle of clearance between the molded article and mold, allowing removal from the mold.
Dry spot	Incompleted area on laminated plastics; the region in which the interlayer and glass are not bonded.
Durometer hardness	A material's hardness as measured by the Shore Durometer.
Ejector pin	Pin or dowel used to eject molded articles from a mold.
Ejector-pin-retainer plate	Receptacle into which ejector pins are assembled.
Elasticity	Property of materials whereby they tend to retain or recover original shape and size after undergoing deformation.
Elastomer	A material under ambient conditions which can be stretched and, on release or with applied stress, returns with force to its approximate original size and shape.
Epoxy plastics	Group of plastics composed of resins produced by reactions of epoxides or oxiranes with compounds such as amines, phenols, alcohols, carboxylic acids, acid anhydrides and unsaturated compounds.
Ethylene plastics	Group of plastics formed by polymerization of ethylene or by the copolymerization of ethylene with various unsaturated compounds.
Evenomation	Softening, discoloration, mottling, crazing, etc. Process of deterioration of a plastic's surface.
Exotherm	Indicates that heat is given from a reaction between a catalyst and a resin.
Expandable plastics	Plastics that can be transformed to cellular structures by chemical, thermal or mechanical means.
Extender	A material which, when added to an adhesive, reduces the amount of primary binder necessary.
Extraction	Transfer of materials from plastics to liquids with which they are in contact.

Extrusion	Process in which heated or unheated plastic compound is forced through an orifice, forming a continuous article.
Filament winding	Process in which continuous strands of roving or roving tape are wound, at a specified pitch and tension, onto the outside surface of a mandrel. Roving is saturated with liquid resin or is preimpregnated with partially cured resin. Application of heat may be required to promote polymerization.
Filler	Inert material that is added to a plastic to modify the finished product's strength, permanence and various other properties; an extender.
Fin	Portion of the 'flash' that adheres to the molded article.
Finishing	Removal of any defects from the surfaces of plastic products.
Fisheye	A clump or globular mass that does not blend completely into the surrounding plastic.
Flash	Excess material that builds up around the edges of a plastic article; usually trimmed off.
Foamed plastic	Cellular structured plastic.
Force plate	A plate used for holding plugs in place in compression molding.
Furane plastics	Group of plastics composed of resins in which the furane ring is an integral portion of the polymer chain; made from polymerization or polycondensation of furfural, furfural alcohol and other compounds containing furane rings; also formed by reaction of furane compounds with an equal weight or less of other compounds.
Fusion	As applied to vinyl dispersions, the heating of a dispersion, forming a homogeneous mixture.
Fusion temperature	Fluxing temperature; temperature at which fusion occurs in vinyl dispersions.
Gel	State at which resin exists before becoming a hard solid. Resin material has the consistency of a gelatin in this state; initial jelly-like solid phase that develops during the formation of a resin from a liquid.

Gel coat	Specially formulated polyester resin that is pigmented and contains fillers. Provides a smooth, pore-free surface for the plastic article.
Gel point	Stage at which liquid begins to show pseudo-elastic properties.
Gelation	Formation of a gel.
Glass	Inorganic product of a fusion reaction. Material forms on cooling to a rigid state without undergoing crystallization. Glass is typically hard and brittle and will fracture conchoidally.
Glass transition	Transition region or state in which an amorphous polymer changed from (or to) a viscous or rubbery condition to (or from) a hard and relatively brittle one. Transition occurs over a narrow temperature region; similar to solidification of a glassy state. This transformation causes hardness, brittleness, thermal expansibility, specific heat and other properties to change dramatically.
Gum	Class of colloidal substances prepared from plants. Composed of complex carbohydrates and organic acids that swell in water. Also, a number of natural resins are gums.
Halocarbon plastics	Group of plastics composed of resins generated from the polymerization of monomers consisting of a carbon and a halogen or halogens.
Hardener	Compound or mixture that, when added to an adhesive, promotes curing.
Heat treat	Refers to annealing, hardening, tempering of metals.
Hot soils	Soils having a resistivity of less than 1,000 ohm-cm; generally very corrosive to base steel.
Hydrocarbon plastics	Plastics composed of resins consisting of carbon and hydrogen only.
Inhibitor	Material that retards chemical reaction or curing.
Isocyanate plastics	Group of plastics produced by the condensation of organic isocyanates with other plastics. Examples are the urethane plastics.
Isotactic	Type of polymeric molecular structure that contains sequences of regularly spaced asymetrical

atoms that are arranged in similar configuration in the primary polymer chain. Materials having isotactic molecules are generally in a highly crystalline form.

Isotropic — Refers to materials whose properties are the same in all directions. Examples are metals and glass mats.

Laminate — Article fabricated by bonding together several layers of material or materials.

Laminated, cross — Laminate in which some of the layers of materials are oriented at right angles to the remaining layers. Orientation may be based on grain or strength direction considerations.

Laminated, parallel — Laminate in which all layers of materials are oriented in parallel with respect to grain or strongest direction in tension.

Lignin plastics — Group of plastics composed of resins formulated from the treatment of lignin with heat or by reaction with chemicals.

Line pipe — Pipeline used for transportation of gas, oil or water; utility distribution pipeline system ranging in sizes ⅛-42 in. o.d. inclusive. Fabricated to American Petroleum Institute (API) and American Water Works Association (AWWA) specifications.

Lyophilic — Referring to vinyl dispersions, having affinity for the dispersing medium.

Lyophobic — Referring to vinyl dispersions, no affinity or attraction for dispersing medium.

Mechanical tubing — Welded or seamless tubing manufactured in a large range of sizes of varied chemical compositions (sizes range $\frac{3}{16}$-10 ¾ in. o.d. inclusive for carbon and alloy material); usually not fabricated to meet any specification other than application requirements; fabricated to meet exact outside diameter and decimal wall thickness.

Mechanically foamed plastic — Cellular plastic whose structure is fabricated by physically incorporated gases.

Melamine plastics — Group of plastics whose resins are formed by the condensation of melamine and aldehydes.

Metastable	Unstable state of plastic as evidenced by changes in physical properties not caused by the surroundings. Example is the temporary flexible condition some plastics display after molding.
Mold base	Assembly of all parts making up an injection mold, excluding cavity, cores and pins.
Molding, bag	Process of molding or laminating in which fluid pressure is applied, usually by means of water, steam, air or vacuum, to a flexible film or bag that transmits the pressure to the material being molded.
Molding, blow	Method of forming plastic articles by inflating masses of plastic material with compressed gas.
Molding, compression	Process of shaping plastic articles by placing material in a confining mold cavity and applying pressure and usually heat.
Molding, contact pressure	Method of molding or laminating whereby pressure used is slightly greater than is necessary to bind materials together during molding stage (pressures generally less than 10 psi).
Molding, high pressure	Molding or laminating with pressures in excess of 200 psi.
Molding, injection	Process of making plastic articles from powdered or granular plastics by fusing the material in a chamber under pressure with heat and forcing part of the mass into a cooler cavity where it solidifies; used primarily on thermoplastics.
Molding, low pressure	Molding or laminating at pressures below 200 psi.
Molding, transfer	Process of molding plastic articles from powdered, granular or preformed plastics by fusing the material in a chamber with heat and forcing the mass into a hot chamber for solidification. Used primarily on thermosetting plastics.
Monomer	Reactive material that is compatible with the basic resin. Tends to lower the viscosity of the resin.
Nonrigid plastic	Plastic whose apparent modulus of elasticity is not greater than 10,000 psi at room temperature in accordance with the Standard Method of Test for Stiffness in Flexure of Plastics (ASTM Designation: D747).

Novolak	Phenolic-aldehydic resin that remains permanently thermoplastic unless methylene groups are added.
Nylon plastics	Group of plastics comprised of resins that are primarily long-chain synthetic polymeric amides. These have recurring amide groups as an integral part of the principal polymer chain.
Organosol	Suspension of finely divided resin in a volatile organic slurry.
Phenolic plastics	Group of plastics whose resins are derived from the condensation of phenols (e.g., phenol and cresol, with aldehydes).
Piling pipe	Round-welded or seamless pipe for use as foundation piles where pipe cylinder acts as a permanent load-carrying member; usually filled with concrete. Used below the ground in foundation work in the construction industry for piers, docks, highways, bridges and all types of buildings. Fabricated to ASTM piling specifications (ASTM A-252).
Plastic	According to ASTM, a material containing an organic substance of large molecular weight that is sold in its finished state and, at some stage, is manufactured into finished goods and can be shaped to flow.
Plastic, semirigid	Plastic having apparent modulus of elasticity in the range of 10,000-100,000 psi at 23°C, as determined by the Standard Method of Test for Stiffness in Flexure Plastics (ASTM Designation: D747).
Plastic welding	Joining of finished plastic components by fusing materials either with or without the addition of plastic from another source.
Plasticate	Softening by heating or kneading.
Plasticity	Property of plastics that permits the material to undergo deformation permanently and continuously without rupture from a force that exceeds the yield value of the material.
Plasticize	Softening by adding a plasticizer.

Plasticizer	Material added to a plastic to increase its workability and flexibility. Plasticizers tend to lower the melt viscosity, the glass transition temperature and/or the elastic modulus.
Plastisol	Suspension of finely divided resin in a plasticizer.
Polyamide plastics	See nylon plastics.
Polyester plastics	Group of plastics composed of resins derived principally from polymeric esters that have recurring polyester groups in the main polymer chain. These polyester groups are crosslinked by carbon/carbon bonds.
Polyethylene	Plastic or resin made by the polymerization of ethylene as the sole monomer.
Polymer	Material produced by the reaction of relatively simple molecules with functional groups that allow their combination to proceed to high-molecular weights under suitable conditions; formed by polymerization or polycondensation.
Polymerization	Chemical reaction that takes place when a resin is activated.
Polypropylene	Plastic or resin derived from the polymerization of propylene as the principal monomer.
Polystyrene	Plastic derived from a resin produced by the polymerization of styrene.
Polyvinyl acetate	Resin derived from the polymerization of vinyl acetate.
Polyvinyl alcohol	Polymer derived from the hydrolysis of polyvinyl esters.
Polyvinyl chloride	Resin derived from the polymerization of vinyl chloride.
Polyvinyl chloride-acetate	Copolymer of vinyl chloride and vinyl acetate.
Pot life	Time period beginning once the resin is catalyzed and terminating when material is no longer workable; working life.
Preform	Coherent block of granular plastic molding compound or of fibrous mixture with or without resin. Prepared by sufficiently compressing material, forming a block that can be handled readily.

Prepolymer	An intermediate chemical structure between that of a monomer and the final resin.
Pressure tubing	Tubing used to convey fluids at elevated temperatures and/or pressures. Suitable for head applications, it is fabricated to exact o.d. and decimal wall thickness in sizes ranging from ½-6 in. o.d. inclusive and to ASTM specifications.
Primer	Coating that is applied to a surface before application of an adhesive, enamel, etc. The purpose is to improve bonding.
Promoted resin	Resin with an accelerator added but not catalyst.
Reinforced plastic	According to ASTM, those plastics having superior properties over those consisting of the base resin, due to the presence of high-strength fillers embedding in the composition. Reinforcing fillers are fibers, fabrics or mats made of fibers.
Resin	Highly reactive material which, in its initial stages, has fluid-like flow properties. When activation is initiated, material transforms into a solid state.
Roller	A serrated piece of aluminum used to work a plastic laminate. Purpose of device is to compact a laminate and to break up large air pockets to permit release of entrapped air.
Roving	Bundle of continuous, untwisted glass fibers. Glass fibers are wound onto a roll called a 'roving package.'
Saran plastics	Group of plastics whose resins are derived from the polymerization of vinylidene chloride or the copolymerization of vinylidene chloride and other unsaturated compounds.
Shelf life	Period of time over which a material will remain usable during storage under specified conditions such as temperature and humidity.
Silicone plastics	Group of plastics whose resins consist of a main polymer chain with alternating silicone and oxygen atoms and with carbon-containing side groups.
Softening range	Temperature range in which a plastic transforms from a rigid solid to a soft state.

Solvation	Process of swelling of a resin or plastic. Can be caused by interaction between a resin and a solvent or plasticizer.
Standard pipe	Pipe used for low-pressure applications such as transporting air, steam, gas, water, oil, etc. Employed in machinery, buildings, sprinkler and irrigation systems, and water wells but not in utility distribution systems; can transport fluids at elevated temperatures and pressures not subjected to external heat applications. Fabricated in standard diameters and wall thicknesses to ASTM specifications, its diameters range from 1/8 to 42 in. o.d.
Stress-crack	Internal or external defect in a plastic caused by tensile stresses below its short-time mechanical strength.
Structural pipe	Welded or seamless pipe used for structural or load-bearing applications in aboveground installations. Fabricated in nominal wall thicknesses and sizes to ASTM specifications in round, square, rectangular and other cross-sectional shapes.
Structural shapes	Rolled flanged sections, sections welded from plates and specialty sections with one or more dimensions of their cross section greater than 3 in. They include beams, channels and tees, if depth dimensions exceed 3 in.
Styrene plastics	Group of plastics whose resins are derived from the polymerization of styrene or the copolymerization of styrene with various unsaturated compounds.
Styrene-rubber plastics	Plastics that are composed of a minimum of 50% styrene plastic and the remainder rubber compounds.
Syneresis	Contraction of a gel, observed by the separation of a liquid from the gel.
Thermoelasticity	Rubber-like elasticity that a rigid plastic displays; caused by elevated temperatures.
Thermoforming	Forming or molding with heat.
Thermoplastic	Reverse of thermoset. Materials that can be reprocessed by applying heat.

Thermoset	Those plastics that harden on application of heat and cannot be reliquefied, resin state being infusible.
Thixotropy	Describes those fluids whose apparent viscosity decreases with time to an asymptotic value under conditions of constant shear rate. Thixotropic fluids undergo a decrease in apparent viscosity by applying a shearing force such as stirring. If shear is removed, the material's apparent viscosity will increase back to or near its initial value at the onset of applying shear.
Tracer yarn	Strand of glass fiber colored differently from the remainder of the roving package. It allows a means of determining whether equipment used to chop and spray glass fibers is functioning properly and provides a check on quality and thickness control.
Urea plastics	Group of plastics whose resins are derived from the condensation of urea and aldehydes.
Urethane plastics	Group of plastics composed of resins derived from the condensation of organic isocyanates with compounds containing hydroxyl groups.
Vacuum forming	Fabrication process in which plastic sheets are transformed to desired shapes by inducing flow; accomplished by reducing the air pressure on one side of the sheet.
Vinyl acetate plastics	Group of plastics composed of resins derived from the polymerization of vinyl acetate with other saturated compounds.
Vinyl alcohol plastics	Group of plastics composed of resins derived from the hydrolysis of polyvinyl esters or copolymers of vinyl esters.
Vinyl chloride plastics	Group of plastics whose resins are derived from the polymerization of vinyl chloride and other unsaturated compounds.
Vinyl plastics	Group of plastics composed of resins derived from vinyl monomers, excluding those that are covered by other classifications (i.e., acrylics and styrene plastics). Examples include PVC, polyvinyl acetate, polyvinyl butyral, and various

	copolymers of vinyl monomers with unsaturated compounds.
Vinylidene plastics	Group known as saran plastics.
Weathering	Exposure of a plastic to outdoor conditions.
Yield value	Also called yield stress; force necessary to initiate flow in a plastic.

NOMENCLATURE

C_p heat capacity (kcal/kg-°C)
E modulus of elongation (kg/cm^2)
G weight loss at uniform corrosion (kg/m^2-hr)
t_m melting temperature (°C)
V corrosion rate (mm/yr)
λ thermal conductivity (kcal/m-°C-hr)
$\tilde{\mu}$ Poisson's ratio
ρ density (kg/dm^3 or g/cm^3)

REFERENCES

1. British Cast Iron Research Association. "Engineering Data on Cast Iron," Aevechurch, Birmingham, England (1968).
2. Gilbert, G. N. J. *Foundry Trades.* 667-672 (May 19, 1966); 713-723 (May 26, 1966).
3. Richardson, R. C. D. In: *Proc. Inst. Mech. Eng.* 182:410 (1967/1968).
4. Palmer, K. B. "Iron and Steel," 44:39-46 (February 1971).
5. ASTM. "Report on Properties of Cast Iron at Elevated Temperatures," ASTM Special Report N248 (1959).
6. "Code of Practice for the Repair and Reclamation of Gray Iron Castings by Welding and Allied Methods," *Proc. Inst. Brit. Foundrymen,* 43:A114-122 (1950).
7. National Association of Corrosion Engineers, "Corrosion Data Survey," Houston, TX (1979).
8. Diamant, R. M. E. "The Prevention of Corrosion, Business Book," London (1971).
9. Boyd, G. M., *Brittle Fracture of Steel Structures* (London: Butterworths, 1970).
10. Lessels, J., and J. Sinclair. *J. Iron and Steel Inst.* 205:249-256 (1967).
11. Sutherland, J. D., and I. Scott. *Welding and Metal Fabrication.* 26-30 (January 1967).
12. Geerlings, H. G., and J. C. Jongebreuer. "Corrosion in Oil Refinery Equipment," paper presented at the 1st International Congress on Metallic Corrosion, London, 1961.

13. Nelson, G. A. *Hydrocarbon Proc.* 45(5):201-204 (1966).
14. DePaul, R. A., et al. "Welding Research Supplement," *Welding J.* 44:409s-416s (October 1970).
15. "Mechanism of Temper Embrittlement," paper presented at the ASTM Symposium, Philadelphia, PA, October, 1970.
16. Moller, G. E. *Materials Protection* 62-67 (May 1966).
17. Truman, J. E. *Corrosion Technol.* 2:243 (1955).
18. Colberk, E. W., and R. P. Garner. *J. Iron and Steel Inst.* 139 (1939).
19. Keating, F. H. *Chromium Nickel Austenitic Steels* (London: Butterworths, 1956).
20. Asbough, W. G. *Material Protection* 18-23 (May 1965).
21. Farber, M., and D. M. Ehrenburg. *J. Electrochem. Soc.* 99:427 (1952).
22. Eigelstein, H. C., and E. N. Skinner. "The Effect of Composition on the Scaling of Iron-Chromium-Nickel Alloys Subjected to Cyclic Temperature Conditions," ASTM, STP No. 165 (1954).
23. McConnell, J. H., and R. R. Brady. *Chem. Eng.* 67:125-128 (July 11, 1960).
24. Rosenberg, J. J. "Nickel and Its Alloys," U.S. National Bureau of Standards Monograph (1968).
25. Spahn, H., and H. Grafen. *Technische Mitteilungen* 61(3):140-145 (1968).
26. Corrosion Data Survey, NACE, Houston, TX (1979).
27. Greenert, J. W. *Corrosion* 18(3):95t-102t (1962).
28. "Copper and Copper Alloy Data," Conseil International pour le Developpment du Cuivre, Geneva (1979).
29. Baker, *Proc. Chem. Eng. Group* 19:41 (1937).
30. Carey, R. J. T. *Aust. Eng.* 54-69 (June 7, 1954).
31. Moore, D. C. *Metals and Materials* 251-257 (June 1970).
32. Newson, E. R. *Chem. Proc.* 33-39 (December 1970).
33. Roll, K. H. *Chem. Eng.* 60:264-280 (February 1953).
34. "British Standard Code of Practice CP3003: Lining of Vessels and Equipment for Chemical Processes, Part 9: Titanium."
35. Gleekman, L. W. "Trends in Materials Application–Non-Ferrous Metals," Chem. Eng. Casebook, 111-118 (October 12, 1970).
36. Taylor, D. F. *Ind. Eng. Chem.* 32:439 (1950).
37. Bishop, C. R., and H. Stern. *Corrosion* 379t-385t (August 1961).
38. "Bishop Standard Code of Practice CP3003: Lining of Vessels and Equipment for Chemical Process, Part 8: Precious Metals."
39. "BS4641: Electroplated Coatings of Chromium for Engineering Purposes."
40. "BS4758: Electroplated Coatings of Nickel for Engineering Purposes."
41. "BS729: Hot Dip Galvanized Coatings on Iron and Steel Articles."
42. "ASTM A123: Zinc (Hot Galvanized) Coatings on Products Fabricated from Rolled, Pressed and Forged Steel Shapes, Plates, Bars and Strips."
43. "British Standard Code of Practice CP2008: Protection of Iron and Steel Structures from Corrosion, 69."
44. "BS2569: Sprayed Metal Coatings. Part I: Protection of Iron and Steel by Aluminum and Zinc Against Atmospheric Corrosion."
45. "BS729. Part 2: Sheradised Coatings."
46. "New Chromizing and Chromium-Aluminizing Processes," *Mach. Prod. Eng.* (May 3, 1967).

47. Hilliard, A. *Brit. Chem. Eng.* 4(3):138-143 (1959).
48. "High Strength Glass Ceramics," *Eng. Mat. Design* 14 (January 1960).
49. Brain, R. C. *Chem. Proc. Eng.* 101-102 (February 1970).
50. Carnet, I. *Mat. Protection* 56-58 (March 1967).
51. "British Standard Code of Practice: 3003: Lining of Vessels and Equipment for Chemical Processes, Part 10: Brick and Tile."
52. "Acid-Proof Vessel Construction with Membrane and Brick Linings," NACE Committee Report, *Corrosion* 195t-204t (March 1957).
53. *Modern Plastics Encyclopedia* published annually (New York: McGraw-Hill Book Co.).
54. Cheremisinoff, N. P., and P. N. Cheremisinoff. *Fiberglass-Reinforced Plastics Deskbook* (Ann Arbor, MI: Ann Arbor Science Publishers, Inc., 1978).
55. Vincent, P. I. *Eng. Mat. Design* 816-821 (December 1961).
56. Mottram, S., and D. A. Lever. *The Industrial Chemist* (February, March, April, May, 1957).
57. Czehorsky, G. *Kunstoff* 60: 354-358 (June 1970).
58. "British Standard Code of Practice CP3003: Lining of Vessels and Equipment for Chemical Processes, Part 4: Plasticized PVC Sheet."
59. Allan, A. J. G., and F. M. Chapman. *Mat. Design Eng.* 106-108 (October 1958).
60. "ASTM D78-: Methacrylate Molding and Extrusion Compounds."
61. *Mat. Protection* 79-82 (August 1966).
62. Powell, P. C. *Design Eng.* 72-75 (October 1971).
63. Mallinson, J. H. *Chemical Plant Design with Reinforced Plastics* (New York: McGraw-Hill Book Co., 1969).
64. Fenner, O. H. *Mat. Protection* 23-25 (May 1968).
65. Munger, C. G. *Mat. Protection* 19-24 (January 1969).
66. Saxman, T. E. *Mat. Protection* 43-45 (October 1965).
67. Butt, L. T. *Rubber and Plastics Age* 42(5):528-530 (1961).
68. Evans, L. S. *Rubber and Plastics Age* (November 1963).
69. "British Standard Code of Practice CP3003: Lining of Vessels and Equipment for Chemical Processes, Part 1: Rubber."
70. "British Standard Code of Practice CP2008: Protection on Iron and Steel Surfaces from Corrosion."
71. *Steel Structures Painting Manual* (Pittsburgh: Steel Structures Painting Council, 1964).
72. "Curing of Interior Tank Linings," NACE Technical Committee Report, *Mat. Protection* 87-89 (January 1964).
73. "British Standard Code of Practice CP3003: Lining of Vessels and Equipment for Chemical Processes, Part 6: Phenolic Resins."

CHAPTER 4

FABRICATION METHODS AND THEIR INFLUENCE ON DESIGN

4.1 OVERVIEW OF FABRICATION AND DESIGN PRINCIPLES

The technological properties of a structural material determine the method of fabrication of pieces of equipment and their components. Cast iron apparatus is founded; steel components also may be founded but more often are fabricated from rolled steel; pipe or profile iron is made by cutting, bending and welding half-finished articles; ceramic vessels are formed and burnt. Pieces of equipment fabricated by different methods require different design criteria, even if they are to be used for the same application.

As an example, Figure 4.1 illustrates two alternative methods for shaping tray towers. One tower is made from sheet metal by welding. Its trays, furnished by a number bubble-caps, are fastened to the shell by welding, as are all the shell nozzles. The second tower is made from cast iron and consists of several flanged sections. The trays are cast separately and fastened between the flanged sections. Each tray has only one bubble-cap. All nozzles are molded together with the tower.

Design variations between the two towers are established by the different methods of welding and casting. It is impossible to cast the tower in one piece; it must be made in several sections.

A thicker shell than is needed for the application has to be fabricated because of the casting method. A decrease in the number of bubble-caps was established by a need for simplifying the shape of a casting pattern, thus lowering its price. It is difficult both to cast a thin-walled nozzle with a thick-walled section of tower and to avoid its breaking during transportation.

As another example, assemblies for heating chambers of two vaporizers (from steel and cast iron) are shown in Figure 4.2 (A) and (B). The tubular plate of the steel vaporizer is welded to its shell and the tubes rolled in the

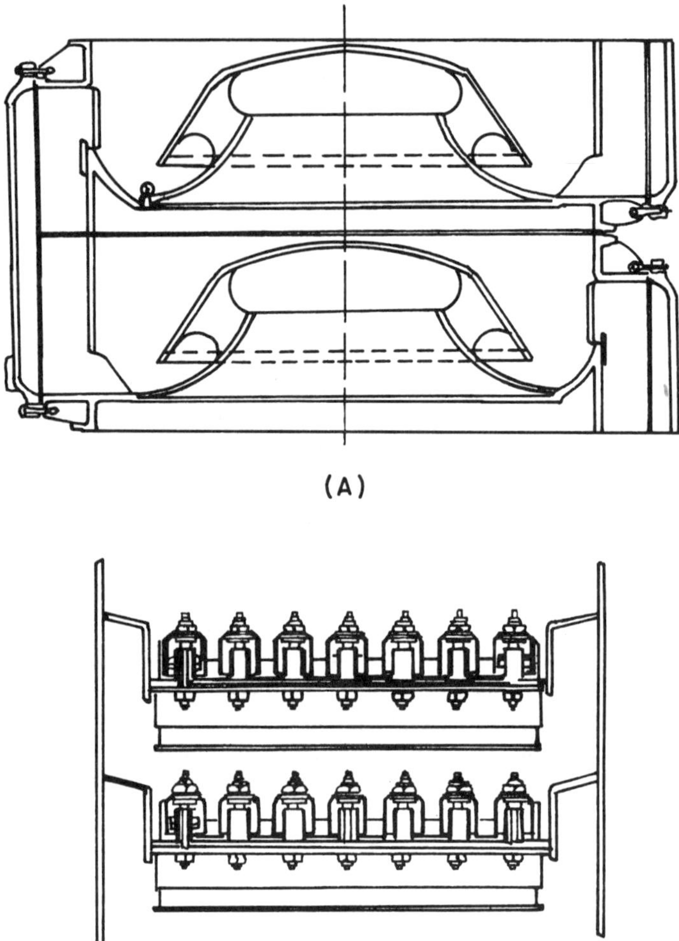

Figure 4.1. Schematic section through a bubble-cap tray tower: (A) cast iron tower; (B) steel tower.

plate. The design of a similar assembly of a cast iron vaporizer is more complicated: the plate is fastened with bolts to the shell and the tubes are tightened with special profile nuts. Such complications are inevitable because cast iron cannot be welded; also the tubes cannot be rolled out because of the metal's brittleness.

Apparatuses fabricated from the same structural material but by different

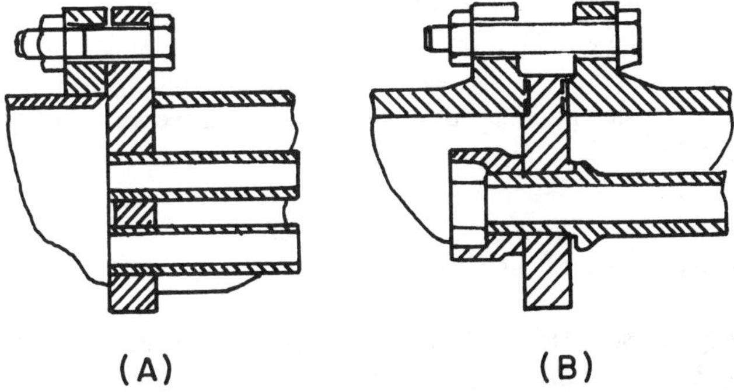

Figure 4.2. Tube joints: (A) steel heating chamber; (B) cast iron heating chamber.

techniques also will differ from each other. Figure 4.3(A) shows the old design of a vessel head riveted to the shell of a receiver. The lower head was made concave to support the rivets during clinching (riveting). For the welded version, there is no need for a concave head; both heads are fastened with outside convexity. This design permits butt-welded joints that improve and lax requirements for the final design (Figure 4.3(B)).

It follows from these examples that the design of apparatuses and their assemblies depend on the method of fabrication. Consequently, the designer

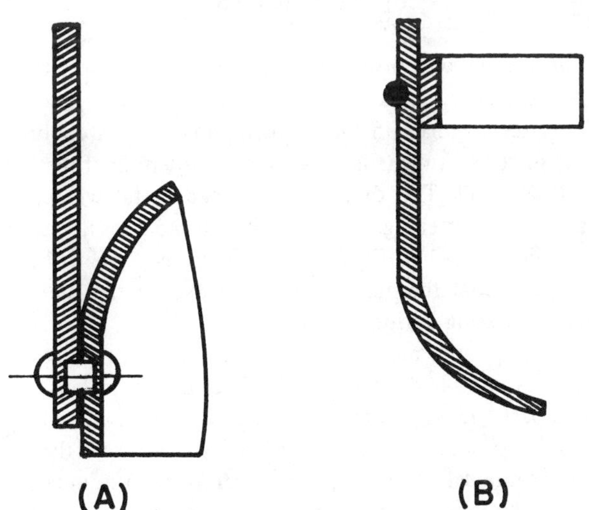

Figure 4.3. Riveted and welded head connections.

must be well versed in the principal fabrication methods and consider all modern techniques and their limitations.

Assuming our reader is familiar with machine-building technology, only data and conclusions about this subject that are necessary for a design are presented below.

4.2 DESIGN OF CAST EQUIPMENT

Casting processes permit a very wide latitude in chemical engineering because one can obtain castings of cast iron and steel that have very complicated shapes. The designer must be familiar with concepts and details before producing a successful design. There are many variables in castings: properties of the material, the simplicity of a pattern, ease of molding and subsequent machining of castings, cost considerations, etc.

Castings must be designed so that the material of the article after cooling has the desired requirements in all its parts. Therefore, one of the most important criteria in casting designs is the ability to provide a uniform cooling rate. As a first approximation, it is possible to assume that the casting cooling rate depends on the ratio of casting perimeter to cross section area. Care must be taken to ensure the same thickness of all outer walls of the casting to obtain equal cooling rates. The inside wall, wells and ribs generally are harder to cool at the same rate as outside walls. To maintain constant rates, the thickness, S_1 should be less than that of an outside wall, S':

$$S_1 = 0.7S' \text{ to } 0.8S' \tag{4.1}$$

Violating the "equal-wall-principle" can cause the formation of residual thermal stresses, cracks, pores or even blow holes.

Thermal stresses originate in the following manner. In cooling a nonequal-walled piece, such as a tube (Figure 4.4), the hardening of its thin wall occurs before the thick wall. The thin wall hardened metal will prevent the free shrinkage and size shortening of the thick wall. As a result, stresses will bend the article, as illustrated in Figure 4.4. If there is a great difference in wall thickness, the residual thermal stresses may exceed the material's ultimate strength and the casting during cooling.

Local thickening and metal accumulation are extremely objectionable. They form pores and blow holes as a result of discontinuous metal supply in the thickened part and a more rapid cooling of the thin wall [1-7].

It is desirable to specify a minimum wall thickness of the casting, which should be determined from strength considerations. Minimum wall thickness provides a good metal structure as well as economical efficiency of the casting. It is determined by metal flowability, its capability to fill up the

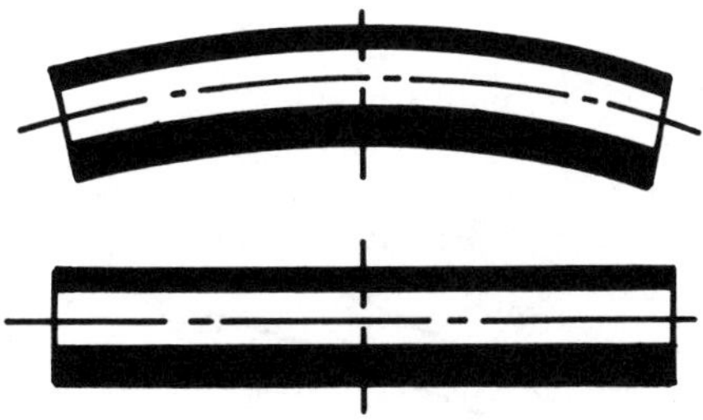

Figure 4.4. Deformation of nonequal-walled tube after cooling.

mold, features of the molding and the size of the casting. Recommended minimum wall thicknesses for sand castings are given in Table 4.1. Note, however, that sand casting does not permit as small a thickness as other casting processes (e.g., die casting, casting in shell molds, chill casting, injection molding) [6].

The wall thickness of a founding in sand castings may be determined by using the plot in Figure 4.5, where the wall thickness is given with respect to the casting overall size, N. N (in meters) is calculated from the following formula:

$$N = \frac{2L + b + h}{3} \tag{4.2}$$

where L = casting length
b = casting width
h = casting height

Increasing the wall thickness of castings more than 30–40 mm decreases their strength, porosity and graphite grain growth. Fabricating vessels from

Table 4.1. Minimum Wall Thickness in Sand Castings [6]

Material	Minimum Wall Thickness (mm)		
	Small	**Average**	**Large**
Gray Cast Iron	3-5	8-10	12-15
Malleable Cast Iron	2.5-4	6-8	
Cast Steel	6	10-12	20
Alloys Nonferrous Metals	2-3	5-7	10-20

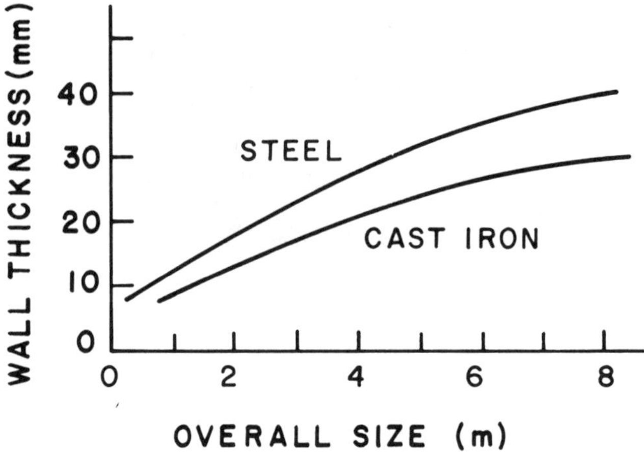

Figure 4.5. Wall thickness vs conditional overall size [7].

cast iron with walls thicker than 50–60 mm is not recommended. However, they can be made from steel, with a considerably greater wall thickness, and their mechanical properties can be considerably improved by forging in simple shapes. Ribs and hollow or box sections are used to decrease the weight of cast apparatus and to make the construction either stronger or more rigid.

The hollow or box sections should not be combined with ribbed sections since this complicates the casting unnecessarily. In selection, it is necessary to account for ribbed constructions, which are better cast. Their cooling conditions are more favorable than box sections. However, box sections are more rigid and their shapes are streamlined, smooth and provide a more agreeable shape (Figure 4.6).

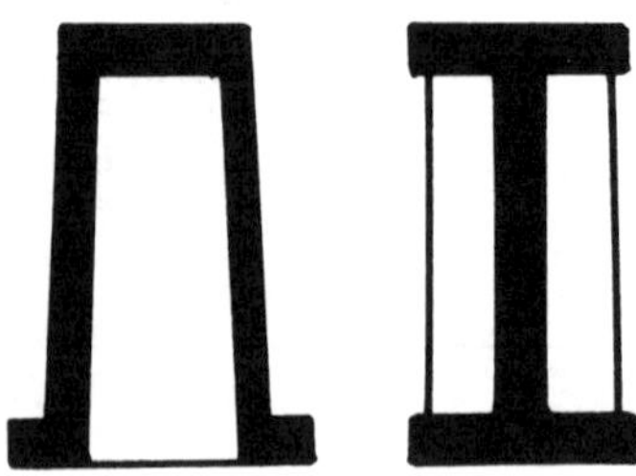

Figure 4.6. Ribbed and box sections.

In designing chemical equipment it is impossible from a practical standpoint to make all parts of a casting the same thickness. Parts of equipment are loaded differently and because of full-strength considerations must be fabricated in different wall thicknesses. In this case, castings should be designed so that the casting massiveness and its wall thickness gradually increase in the direction of improving heat evacuation.

For obtaining a sound casting without residual stresses, it is necessary to maintain strictly a smooth change of sections (Figure 4.7) and to avoid wall connections without fillets (Figure 4.8(A) and (B)).

It is not desirable to design conjugating walls with a thickness ratio greater than 2:

$$\frac{S}{S_1} \leqslant 2 \tag{4.3}$$

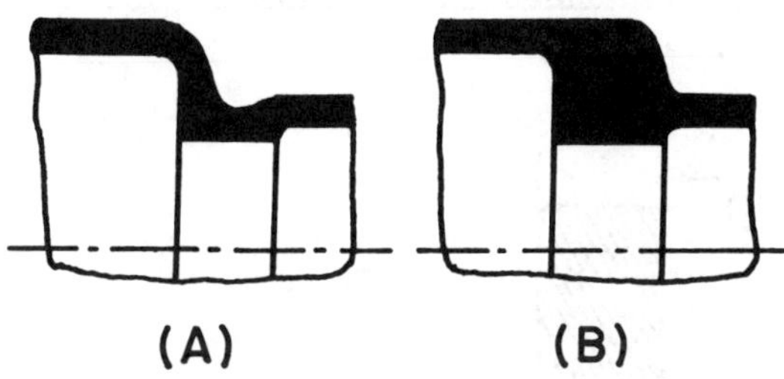

Figure 4.7. Smoothing the wall thickness for improving the casting. (A) good design; (B) poor design.

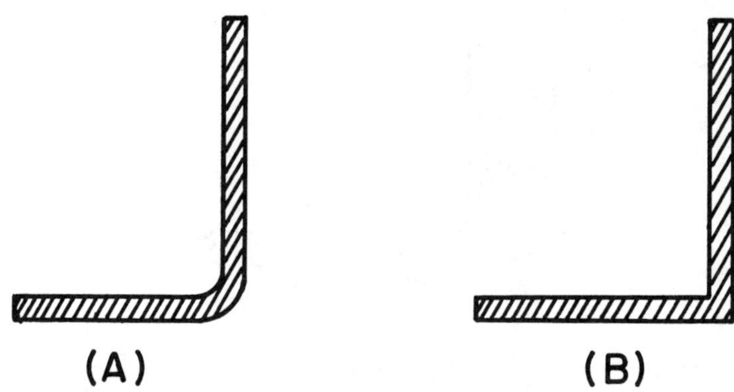

Figure 4.8. Conjugation of cast walls: (A) good design; (B) poor design.

Junctions of sections of different thicknesses are achieved using a taper (Figure 4.9(A)) with the following ratio:

$$\frac{S_1 - S}{\ell} = \frac{1}{4} \text{ to } \frac{1}{5} \tag{4.4}$$

If $S/S_1 < 2$, it is possible to have a corner junction (Figure 4.9(B)) that can be combined with a paper junction (Figure 4.9(C)).

The selection of the curvature radius for junctions of walls at a right angle

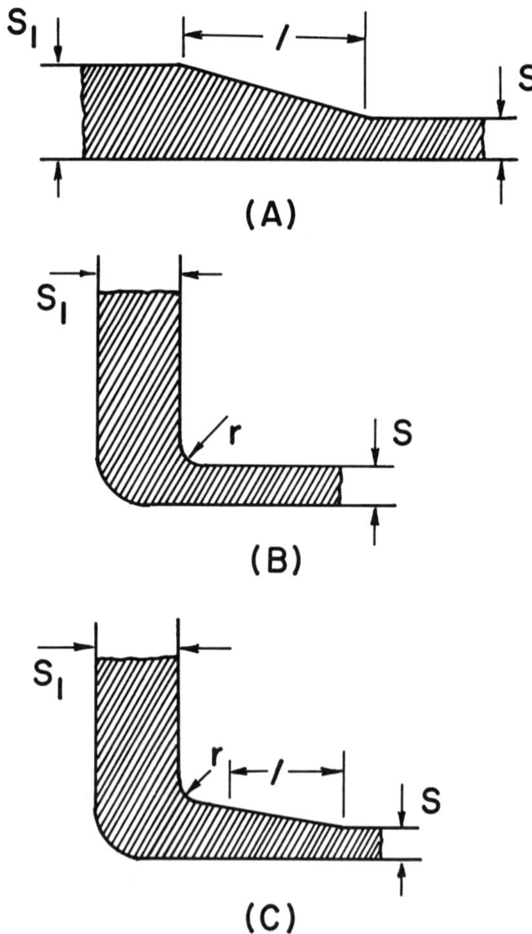

Figure 4.9. Design of junctions between walls of different thickness.

with different wall thicknesses as, for example, the junction of a pipe with a flange, is contradictory. On the one hand, increasing the radius decreases the stress concentration and reduces the probability of crack formation. A large fillet may produce a large accumulation of metal and a high probability of cavity formation.

If the junction radius ranges from ⅙ to ⅓ of the arithmetic mean of the wall thickness,

$$r = \frac{S_1 + S}{6} \text{ to } \frac{S_1 + S}{12} \tag{4.5}$$

Then both problems are minimized. Greater size radii for junctions at right angles and at sharp angles will cause unacceptable local wall thickening.

Intersections should be designed so that the concentration of metal is at a minimum. Figures 4.10(A) and (C) illustrate good designs. Figures 4.10(B) and (D) illustrate poor design.

Particular attention must be given to junctures of sections of different thicknesses. An abrupt change from a thin section to a heavy one (Figure 4.11(A)) will produce a high stress concentration at the juncture. This may result in shrinkage cracks that often occur during cooling or in cracking when a tensile load is applied. A large fillet (Figure 4.11(B)) is apt to produce large local accumulation of metal that will be still liquid when the outside fibers have solidified. This results in pore formation inside the casting, or even a blow hole.

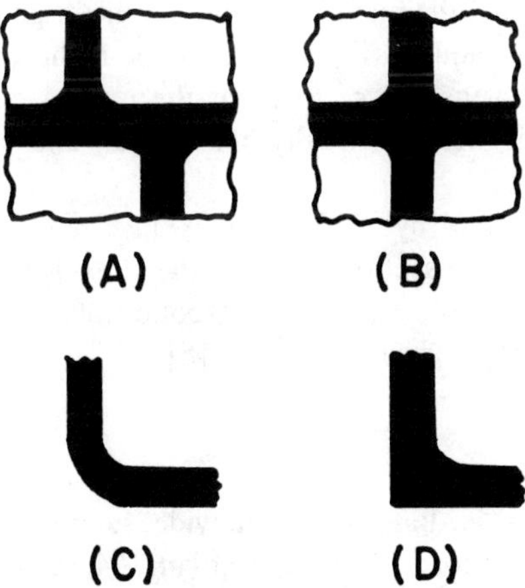

Figure 4.10. Treatment of intersections.

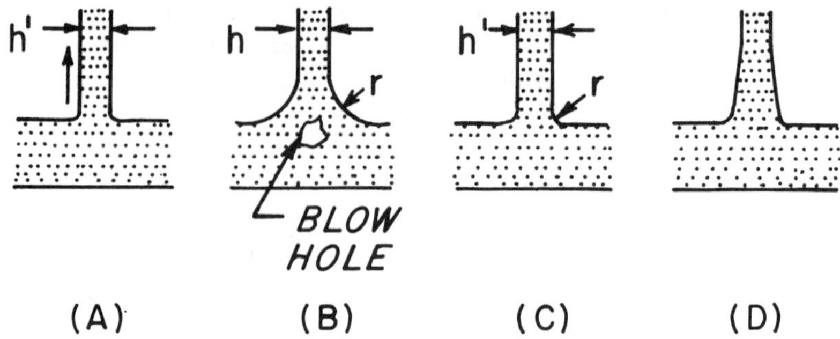

Figure 4.11. Junctures of different sections.

A moderate fillet, Figure 4.11(C) (approximately r = 0.5 h′) will eliminate excessive stress concentration and prevent the formation of a crack. A gradual taper, Figure 4.11(D), will provide even better results.

When heavy sections, or bosses, are surrounded by light sections, it is important to secure proper feeding of the boss by placing ribs of sufficient cross sections on the light parts. The main sections should be made sufficiently thick to permit the addition of a gate of a size that secures good feeding of all sections. The ribs improve evacuation of heat from a massive part of the casting and provide a more even load distribution over the entire volume of material. Ribs are added to make a construction either stronger or more rigid. While ribs always attain the latter objective, they may fail in regard to the former. The ribs strengthen the part if their addition lowers the maximum stress of it. If the maximum stress in the original design is lowered, but the stress in the added rib is equal to or higher than in the original design, the rib is either useless or weakens the part. A crack may develop in the high-stressed ribs that eventually will spread and cause the casting to fail.

The rigidity measured by deflection is inversely proportional to the moment of inertia, I, of a section, whereas the stress is inversely proportional to the section modulus, Z. The resulting conditions can be illustrated best by the numerical example given by Maleev [8]:

Example 4.1

A cast-steel plate 20 in. long and 15 in. wide, supported at the ends, carries a uniformly distributed load of 8000 lb (Figures 4.12 and 4.13(A)). We wish to determine the influence of conventional ribs on the rigidity and strength of the plate. The bending moment of the load is

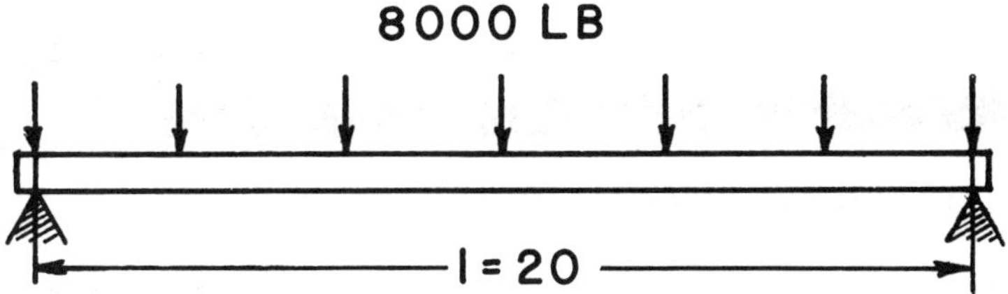

Figure 4.12. Loading of the plate.

$$M = \frac{1}{8} F\ell = \frac{1}{8} \times 8000 \times 20 = 20{,}000 \text{ lb-in.}$$

Naturally, the ribs must be in the direction of the greatest dimension, or lengthwise. Dividing the width into three strips, each 5 in. wide, and adding narrow, ¼-in. ribs to double the thickness of the plate, h = ¾ in.; the section Figure 4.13(B) is obtained.

In Table 4.2, I, Z, W and S are given of the plate *a* and of the plate *b* after the small ribs are added.

The rigidity, as shown by the increased moment of inertia, $I_2 = 0.71$ as compared with I = 0.53, is increased by about 34% (Table 4.2). However, the strength is decreased 32%, as the maximum stress is increased from t = 14,200 to $t_2 = 21{,}000$ psi.

The next type of rib, Figure 4.13(C), was obtained by changing the overall height, h_2, to obtain at least the same strength. By increasing h_2 to 3 in., the plate became 4.13 times more rigid with a slight decrease in stress, from 14,200 to 13,700 psi (Table 4.2). However, such thin and high ribs are very difficult to cast and cause additional stresses from the abrupt change of the thickness from ¾ in. to ¼ in.

Table 4.2. Influence of Ribs on Rigidity and Strength of the Plate [8]

Section of Plate	Moment of Inertia, I ($in.^4$)	Section Modulus, Z ($in.^3$)	Weight, W (lb)	Highest Stress, τ (psi)	Relative Rigidity	Relative Strength	Relative Weight
a	0.53	1.41	63.0	14,200	1	1	1
b	0.71	0.95	66.2	21,000	1.34	0.68	1.05
c	2.19	1.46	72.5	13,700	4.13	1.04	1.15
d	1.90	1.69	75.6	11,800	3.58	1.20	1.20
e	2.20	1.96	62.1	10,200	(4.15)	(1.39)	(0.99)

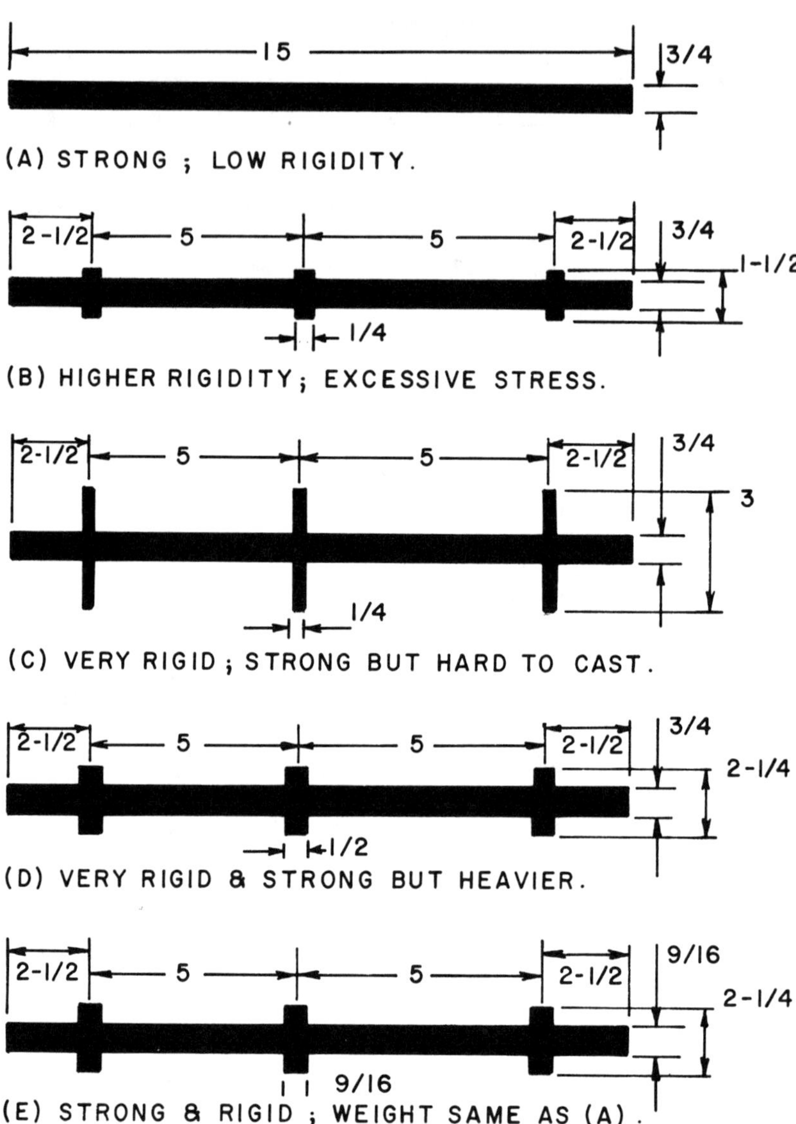

Figure 4.13. Various ribs added to a plate.

A better rib design is shown in Figure 4.13(D) and in Table 4.2. The height of each rib is equal to the thickness of the plate; the width of the ribs is twice that of the two previous attempts. Table 4.2 shows that rigidity is only slightly smaller than in Type C, but the strength is increased about

20% above that of the flat plate. The weight, W, is not appreciably affected by the ribs.

Finally, Figure 4.13(E) and Table 4.2 give the data for a plate that has even greater rigidity and greater strength and is slightly lighter than the flat plate. In this case, the body of the plate is made thinner, and the ribs are of the same thickness as the plate. This plate will have very small internal casting stresses and little stress concentration. The relative figures for this plate are enclosed in parentheses to indicate that the results depend not only on the ribs, but on a change of the plate thickness as well. The thickness of external ribs at their junction with a wall is taken equal to 0.7–0.9 minimum external wall thickness. The height of the rib is no more than 5-6 times the wall thickness.

The shape of rib has to be simple. A curved reinforced rib, Figure 4.14(A) is better than a straight one, Figure 4.14(B), as it is less rigid and causes less stress concentration. However, the maximum bending stress in rib A will be higher than in rib B. Care must be taken to put in sufficiently heavy ribs.

In the event of impact loading, the decrease of failure may be achieved by increasing the resilience of casting parts involved, which, in turn, requires greater deformations. Since ribs act in the opposite direction, it is obvious that they are to be avoided where impact loads are anticipated.

Fluctuating loads have a similar effect, although not as pronounced as shock loads; therefore, all conclusions about the latter relating to ribs apply to fluctuating loads as well. Ribs should be used chiefly for static loads. Wide and low ribs are safer than thin and high ones.

In casting design one must account for the fact that cast iron, and especially ferrosilicon and similar alloys, resist compression loads but act poorly where bending and tensile loads are applied. Gray iron resists four times greater compression loads than tensile ones. Therefore, the ribs have to be located at the side of compressed fibers, for example, in vessels under

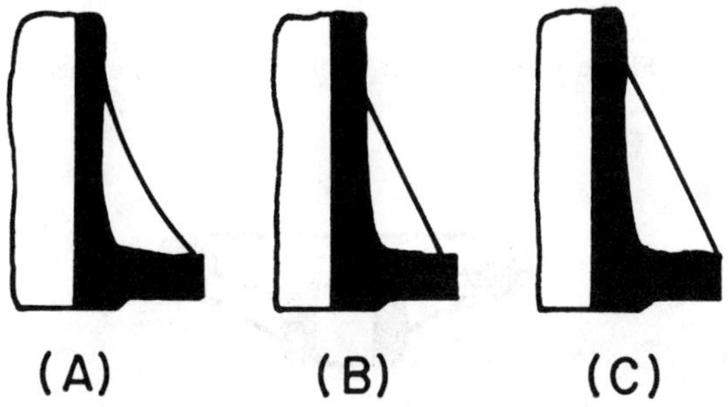

Figure 4.14. Cast ribs: (A) concave; (B) straight; (C) straight expanded to rim.

internal pressure (Figure 4.15(A)) and in vessels under external pressure (Figure 4.15(B)). Care should be taken to avoid loading the vessels with bending moments.

Ribs that are allocated incorrectly may produce defective castings. For example, placing radial ribs on heads of large vessels will increase their rigidity and prevent free shrinkage during cooling. This may cause excessive stresses and shrinkage cracks. The radial allocation of wheel spokes, pulleys and flywheels is apt to produce an unsound casting. To increase elasticity, it is desirable that all radial ribs, spokes, etc., be changed by tangent, curved or conical shapes. Figure 4.16 illustrates the pilot-wheel with conically situated spokes.

The intersection of ribs crosswise is less favorable than that of T- or V-shape. Intersecting ribs (Figure 4.17(A)) should be avoided as they are apt to produce unsound castings. Six ribs (Figure 4.17(B)) make a poor design, which can be improved materially by the addition of a ring rib (Figure 4.18), which permits the escape of metal accumulation.

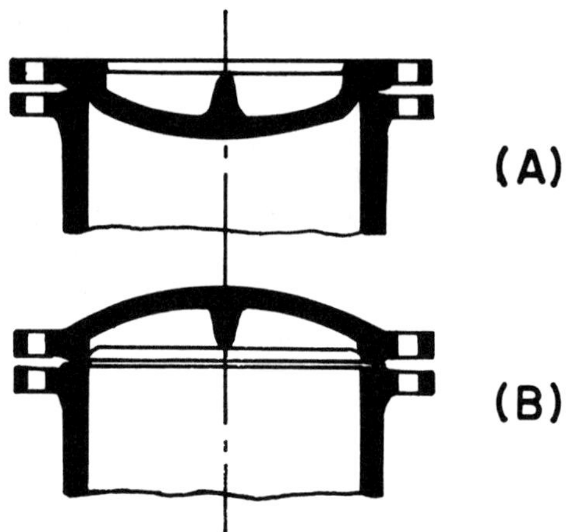

Figure 4.15. Layout of ribs on the cast iron head.

Figure 4.16. Pilot-wheel with conically situated spokes.

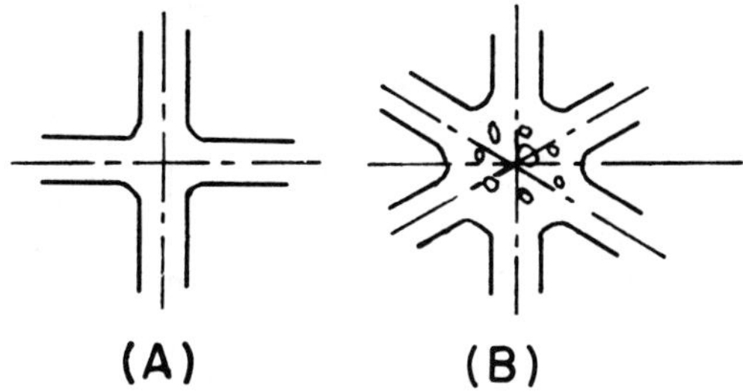

Figure 4.17. Junction of intersecting ribs.

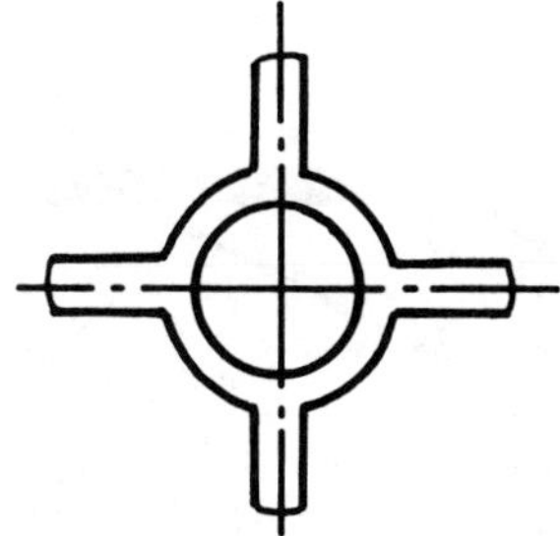

Figure 4.18. Junction of radial ribs by ring rib.

The correctness and smoothness of junctures is checked by the inserted circle method, which is explained in Figure 4.19.

The smaller the difference in the circles, the better the casting. The diameter ratio in a good design must meet the following criterion:

$$\frac{D}{d} \leqslant 1.2 \tag{4.6}$$

Figures 4.7 and 4.19 illustrate how minor changes in the casting shape can significantly decrease local accumulation of the metal.

To decrease stress and metal accumulation, it is desirable not to bring the ribs inside of an apparatus up to its angles. That is, at the point of juncture of three walls, it should leave a hole not less than 30–50 mm in diameter (Figure 4.20).

Holes in medium and large castings are provided if their diameters are more than 50 mm. For the lesser holes, it is more economical to drill them in a

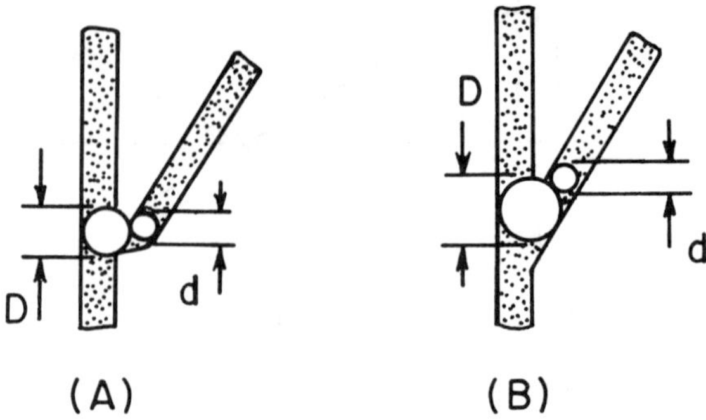

Figure 4.19. Checking the casting design by the inserted circle method: (A) good design; (B) poor design.

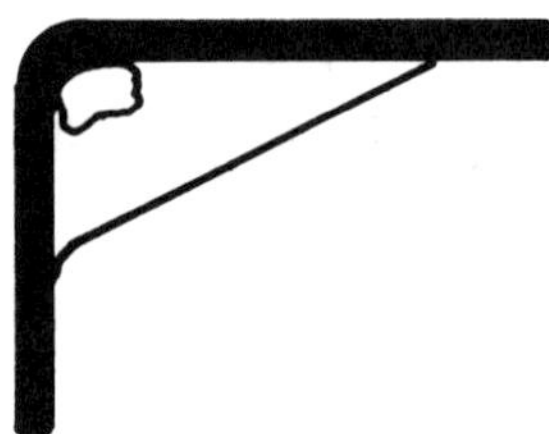

Figure 4.20. Hole in the rib.

finished casting. To lower the stress concentration around a hole, it must be reinforced by a one- or two-sided bead.

Rectangular or square openings, such as windows in compressor pedestals, must have well-rounded corners (Figure 4.21(A)) and should be reinforced with beads.

Too small a fillet will produce a stress concentration that will cause cracks at the corners under fluctuating loads (see Figure 4.21(B)). Long nozzles should not be cast together with thick-walled cylinders—they can be broken during cooling or transportation. Such nozzles must be provided with ribs or be changed by bosses. Their thickness should not exceed 50–60% of the wall thickness. Location of the parting plane must be selected in a manner such that cores are not required because they complicate and raise the price of the molding.

Height of the chamber, H (Figure 4.22) formed as a result of metal displacement by the part of the pattern fixed on its base in the lower mold box

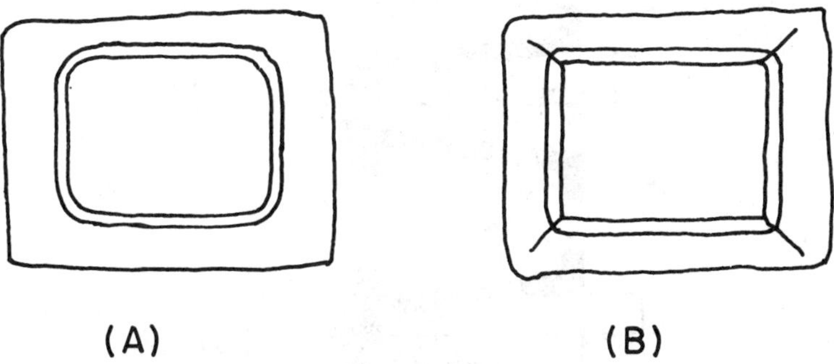

Figure 4.21. Shaping the rim of a rectangular opening.

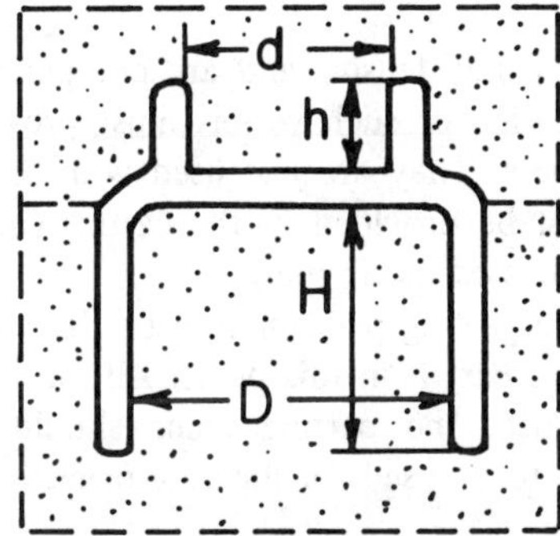

Figure 4.22. The sizes of the chamber in the mold without bosses.

may reach up to $H \leqslant D$ for hand molding and $H \approx D$ for machine molding. The height of a chamber formed by a part of the pattern that is suspended downward is $h \leqslant 0.3d$.

Standardization possibilities and machine molding of bosses in designing boss castings must be given careful consideration. Where possible, core marks must have a cross section equal to the core section. This relaxes requirements for exact and sure fixing of a core mark necessary for preventing displacement during casting and fabricating defective pieces as a result of casting unevenness. In all cases, one should avoid console cores fastening. The cores must be fixed on two sides, as shown in Figure 4.23(A).

In casting design it is necessary to avoid foundry gaggers. They are one of the basic sources of an unsound casting because of pores and liquation. The

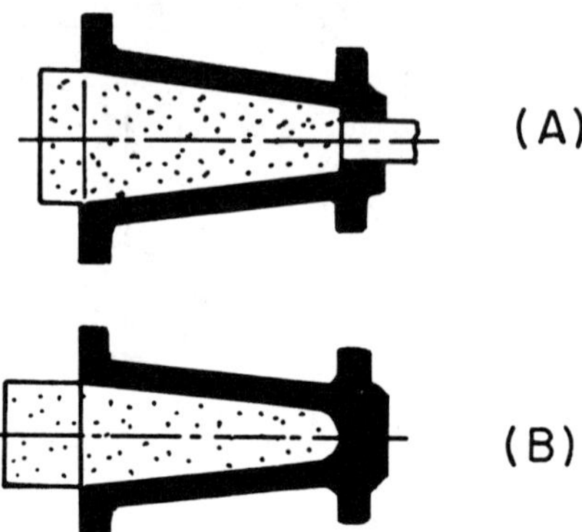

Figure 4.23. Core fixing: (A) on both sides; (B) console fixing.

gaggers decrease the casting density and are not permissible in fabricating pressure vessels. The design of such vessels must provide the boss fixing in the mold without gaggers. They are not used with ferrosilicon and similar castings. Openings must be provided in the casting (no less than 50 mm in diameter) to withdraw bosses and frames.

Wherever possible, the parting plane should be flat and located so that the largest dimension of the part is in this plane. All mold cavities should be as shallow as possible. Also, the parting plane should be kept away from surfaces that are likely to be used for locating or clamping when the part is machined.

Location of the parting plane may considerably influence the cast finish. For example, at displacement of pattern parts during molding, thick seams may result that require trimming. It is possible to avoid seam formation by the location of other parting planes (Figure 4.24).

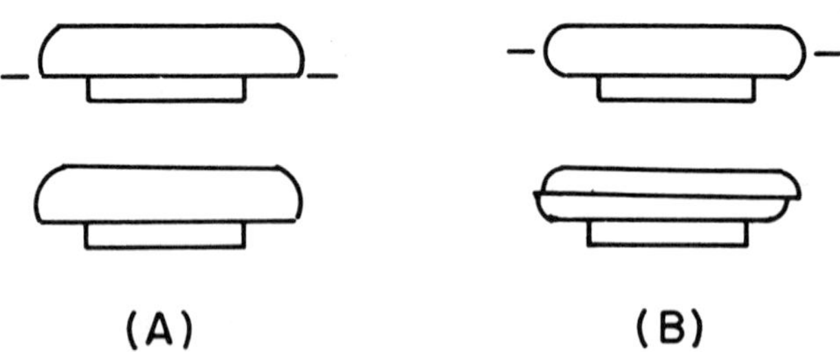

Figure 4.24. Selection of the location of a parting plane: (A) good design; (B) poor design. Lines show the parting plane.

To lower the price of castings and decrease defective products, the designer should consider the fabrication features of patterns, core boxes and molding, and the casting process itself. Although casting provides fabrication parts of complicated shapes, it is always necessary to strive for simple configurations and shapes. The pattern fabrication (Figure 4.25(A)) with only two radii is easier than that with four radii of curvature (Figure 4.25(B)).

Patterns and bosses contoured by planes and surfaces of revolution are more simple in fabrication and not expensive. Notice that large, flat surfaces are extremely unfavorable for castings. It is better to manufacture them with short ribs or pits, and round bottom as cones with a large central angle (Figure 4.26(A)), thus avoiding flat castings (Figure 4.26(B)), because blow holes may be formed in horizontal planes.

To permit easy withdrawal of a pattern, all surfaces perpendicular to the parting plane—walls, ribs, etc.—require draft (Figures 4.27 and 4.28). External draft is used at surfaces that, on cooling, will shrink away from the mold.

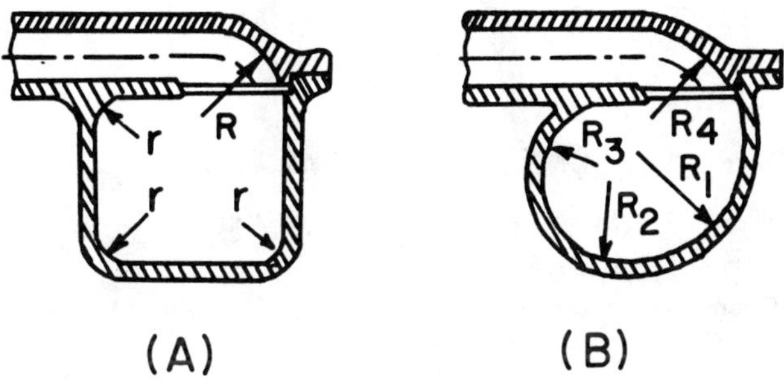

Figure 4.25. Two pattern version of a (A) simple form; (B) more complicated form.

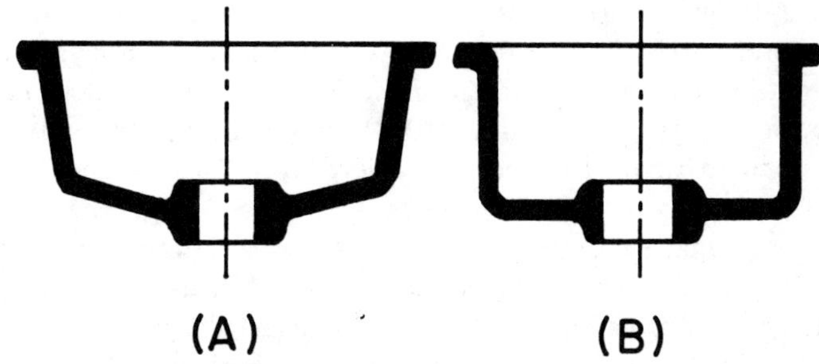

Figure 4.26. Bottoms: (A) casting with cone bottom; (B) casting with flat bottom.

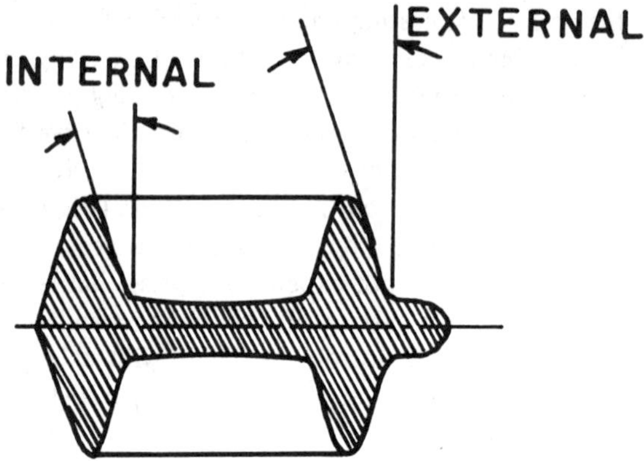

Figure 4.27. External and internal draft.

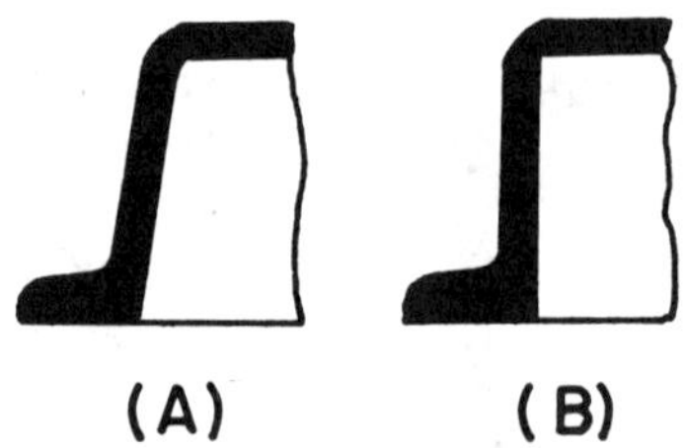

Figure 4.28. Castings with a draft: (A) good design; (B) poor design.

Internal draft is used on surfaces that, on cooling, tend to tighten the mold. The pattern must have a shape that enables easy withdrawal from the mold (Figure 4.28(A)). It is not desirable to provide for ribs, partitions, projections and notches (Figure 4.29). These make the molding process very difficult.

If ribs, projections, etc., are necessary, they should be arranged so that they do not prevent the withdrawal of the pattern. Figures 4.30(A) and (C) illustrate good designs, and Figures 4.30(B) and (D) illustrate poor designs.

Checking the pattern shape may be achieved if we imagine a beam directed parallel to the pattern motion (Figures 4.31(A) and (B)). The absence of shadows shows that the pattern shape favors the pattern withdrawal as illustrated in Figure 4.31(A); and the shaded sections (Figure 4.31(B)) show that the pattern cannot be withdrawn if there is no detachable part.

To avoid scrapping of the heavy piece through breakage of the extension, the light and protruding extensions should not be cast in one piece with

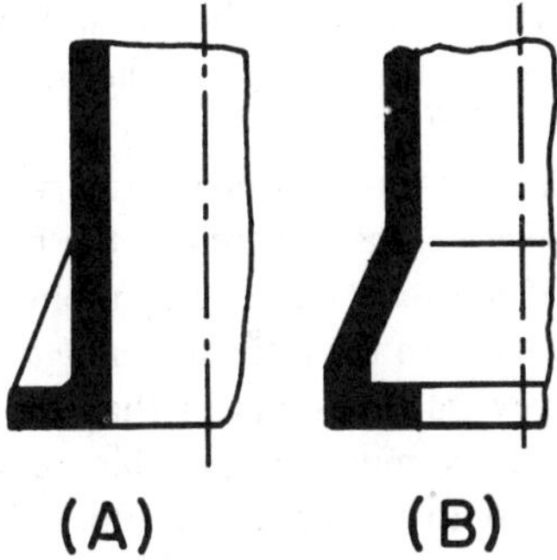

Figure 4.29. The edge of an apparatus of silicon cast iron: (A) easy withdrawal pattern. (B) difficult withdrawal pattern.

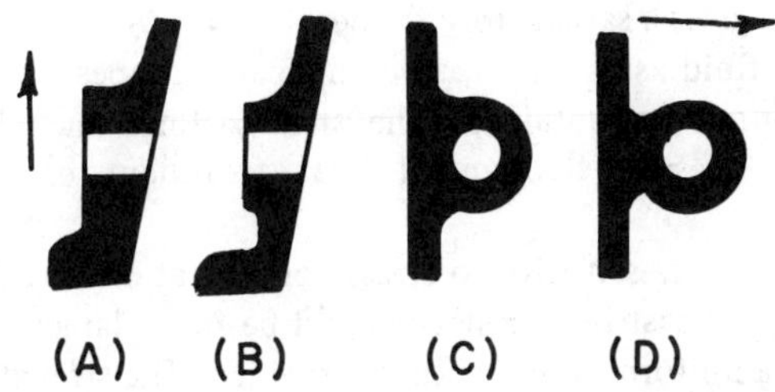

Figure 4.30. The notches do not permit easy withdrawal of the pattern: (A,C) good designs; (B,D) poor designs. The arrows show the direction of pattern withdrawal.

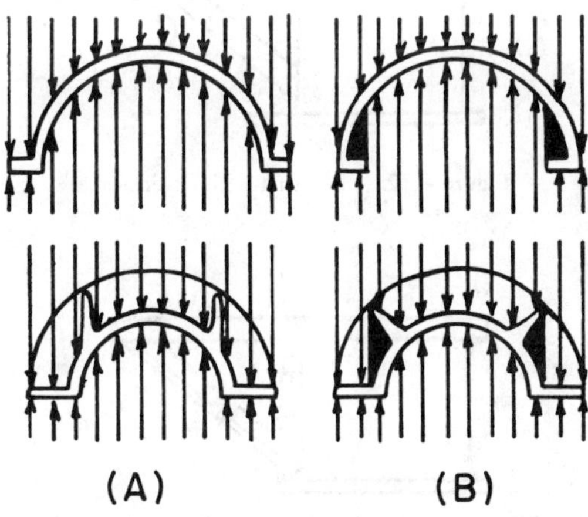

Figure 4.31. Checking the possibility of pattern withdrawal: (A) good; (B) poor.

heavy sections or parts. Figure 4.32 shows an incorrect design; that is, brace c is out of proportion with the rest of the bed. Figure 4.33 shows the correct, two-piece design.

In most cases, straight lines look better than curved lines and are cheaper to obtain in pattern. However, the parabola is the proper shape for a heavy pedestal (Figure 4.34), which resists a bending moment, because its strength is more nearly uniform. As both the ultimate strength and yield point of cast iron in compression are several times greater than those in tension, cast iron parts should be designed as much as possible to avoid their loading in tension. Sections subjected to tensile stress either should be reinforced or relieved of stress by appropriate tie rods (Figure 4.35), or clamp-shaped devices such as forged-steel bearing caps, b (Figure 4.36).

For the same reason, heavy-loaded items, for example, a punch box section frame, should be made with a thick wall (Figure 4.37). Stress concentration in section A may cause it to break, whereas section B, with a gradual change of wall thickness, gives a satisfactory frame.

Steel is not as fluid as cast iron, so complicated shapes, sharp corners and thin sections cannot be obtained. The steel castings must be annealed to relieve internal stresses. Subsequent heat treatment can increase their strength.

Light alloys are more sensitive to stress concentrations produced by fillets and notches than is cast iron. Fillets should be made larger, notches should be avoided, and a uniform wall thickness provided. The strength of aluminum

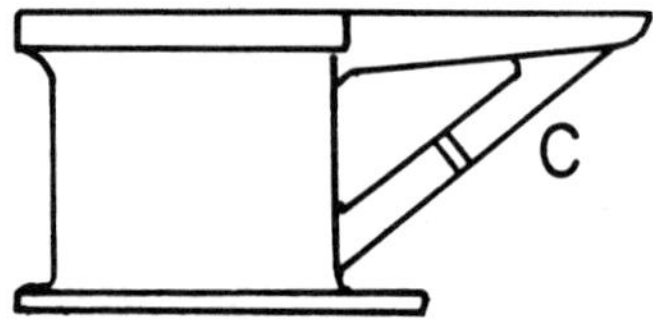

Figure 4.32. Bed cast with extension.

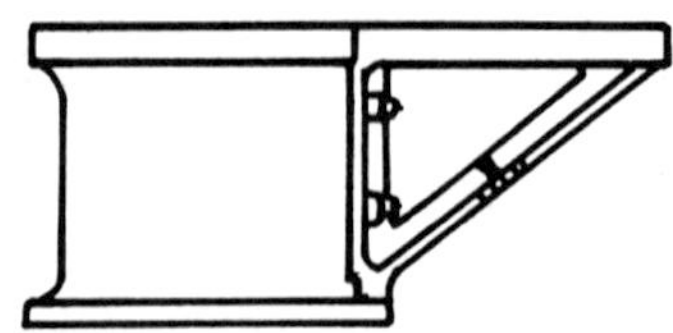

Figure 4.33. Bed with a separate extension.

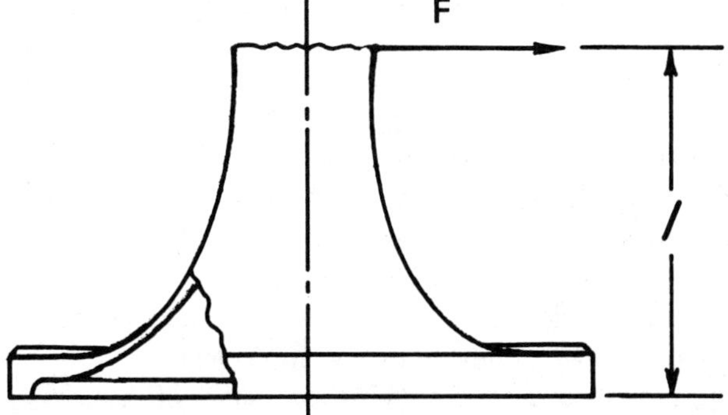

Figure 4.34. Pedestal loaded in bending.

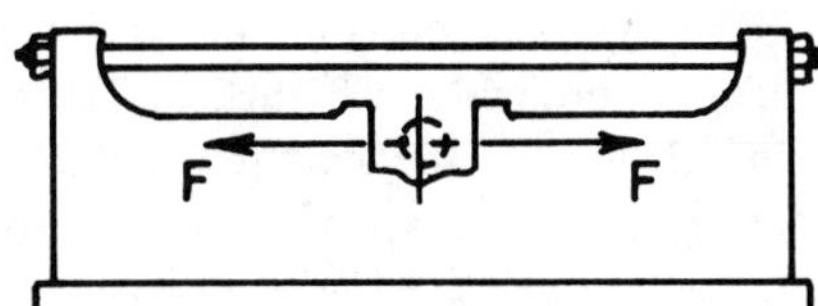

Figure 4.33. Bed plate with tie rod.

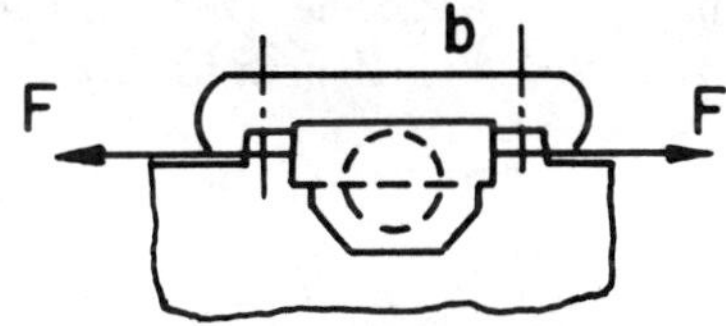

Figure 4.36. Steel bearing cap.

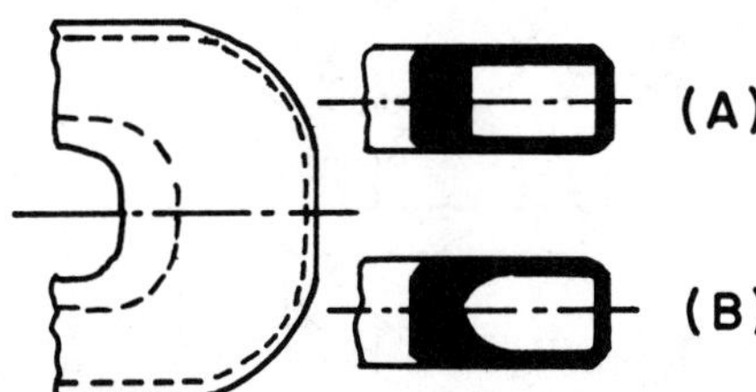

Figure 4.37. Punch-press frame.

sections in tension can be increased by cast-in steel reinforcing rings or anchors.

In designing a casting part, care should be given to any machining that must be performed. Machining allowance (the amount of excess material) is required to permit a proper machining or finishing of the surfaces of the casting. In determining the machining allowance, there must be a tolerance on both the casting and the machining dimensions. Because different amounts of shrinkage are caused by variations in pouring temperature, incomplete closing of the molds and rapping the pattern to make it easier to remove it from the mold, castings require relatively large tolerances.

Machining allowance vary with the size and shape of the casting. A cylinder 2 in. in diameter and 4 in. long, for instance, might be cast 2¼ in. in diameter and 4¼ in. long to permit ⅛ in. to be removed in machining its surfaces [9]. Should the cylinder be 12 in. long, however, a cast diameter of 2½ in. might be required because of the possibility of warpage of the casting. For the part illustrated in Figure 4.38, ¼ in. allowance on all diameters and ⅛ in. allowance on lengths probably will suffice. Therefore, the diameters will become 10¼ in. and 4¼ in., and the lengths 1⅛ in. and 3 in.

Most metals contract during cooling, so provision must be made for this change in dimension. The shrinkage allowance depends on the size of the casting and the material of which it is made, and is generally specified as a definite amount per foot of length or diameter. The usual allowance for cast iron is ⅛ in./ft; for steel, ¼ in./ft; for aluminum, 3/16 in./ft. So that the casting may have the required dimensions, the 10¼ in. diameter must be increased to approximately 10⅜ in.; the 4¼ in. diameter to 4 19/64 in.; the 3 in. length to 3 1/32 in.; and the 1⅛ in. length to 1 9/64 in.

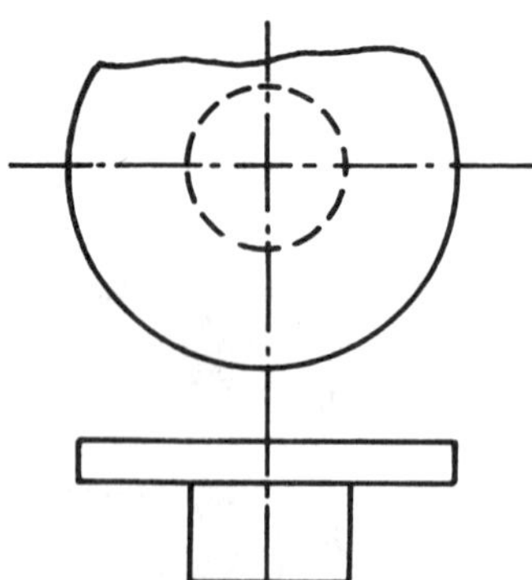

Figure 4.38. Finished or machined part for which a casting is to be made.

To remove the pattern without injury to the mold, it should be tapered in the direction in which it is drawn or removed from the mold. The necessary increase in size of the upper portion of the pattern (i.e., the draft allowance) depends on the size of the casting and the manner in which the pattern is drawn from the mold. The pattern (Figure 4.39) will be drawn in a direction parallel to the axis of the component cylinders.

Therefore, it will be necessary to change these cylinders into a frusta of cones. Draft allowance is generally ¼ in./ft of depth, so the upper diameters of the two component cones are $1/32$ in. and $1/16$ in. larger than the lower diameters. The final dimensions are given in Figure 4.39.

Somewhat controversial is the best casting to be used–either a heavy casting in one piece or its dismembering as several small ones. In most cases, it is more expeditious to make use of dismembering heavy and complicated castings in several small and simple ones and then to assemble them after casting and machining. As a result of such subdivision, the quality of castings increases and the amount of defective parts becomes lower. Each part can be cast from a more suitable material and made thin-walled, which considerably decreases the casting weight and material consumption. The cost of machining and assembly increases in this case.

The net cost of casting depends on material cost, plant equipment, casting technique, weight and complexity of casting and number of parts. Assuming the cost of a ton of casting from gray iron as a unit, it is possible to estimate relative net costs of castings from other metals.

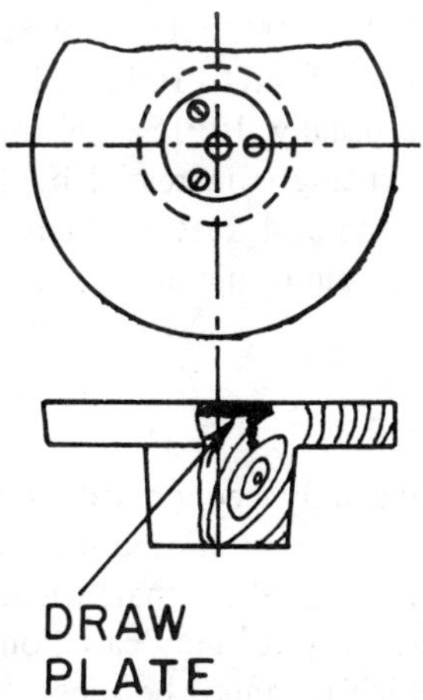

Figure 4.39. Typical pattern.

Table 4.3. Relationship Between Net Cost and Complexity of Castings

Complexity	Castings of Gray Iron	Castings of Steel	Castings of Bronze
Simple	1.0	1.0	1.0
Mean Complex	1.15-1.20	1.30-1.40	1.0
Complex	1.40-1.60	1.70-1.80	1.05
Very Complex	1.70-2.0	2.20-2.30	1.10

Castings usually are divided into four groups: simple, mean complex, complex, and very complex. Most parts of chemical apparatus are of mean and complex groups. Table 4.3 presents the net cost relationship versus casting complexity.

The higher increase in net cost of steel castings with complexity compared to that of cast iron depends on poorer cast properties of steel—its lower fluidity and worse mold filling. On the other hand, the capability of bronze to fill extremely narrow channels and molds of complicated shapes makes the difference in net cost of items of various complexity insignificant.

Heavy castings are cheaper than light ones. Assuming the cost of 1 ton of material in a 3000-kg casting as unity, we will obtain the following relationship for cost versus casting weight for mean and complex castings: 300 kg-1.00; 100-1000 kg-1.05; 10-100 kg-1.20; 1-10 kg-1.40; 0.21-1 kg-1.80.

Correspondingly, for steel castings we have: all castings heavier than 3000 kg-0.95; 1000-3000 kg-1.00; 500-1000 kg-1.08; 200-500 kg-1.20; 50–200 kg–1.35: 10-50 kg-1.65; up to 10 kg-2.15.

The cost of patterns is usually 10-15% of the casting cost. For up to 10 castings, the cost per casting is increased by 10% and by 5% for up to 25 pieces. For many parts the cost does not increase. The net cost of casting depends very much on the founding method:

- Sand mold casting–1
- Shell molding–2
- Casting by the lost-wax process–10

The last two methods are only economical in mass production and for precision casting, which eliminates aftertreatment.

The preceding data on net cost show that, wherever possible, the designer must use the cheapest material (e.g., gray cast iron). Steel castings should be used only for cases in which it cannot be substituted by cast iron. Because of pattern rapping, misalignment of bosses and various other reasons, the

Table 4.4. Allowances in Sizes of Castings from Steel and Cast Iron with Use of Wooden Patterns

Maximum Casting Size (mm)	Nominal Size, mm						
	Up to 50	50–120	120–260	260–500	500–800	800–1,250	1,250–2,000
Up to 500	1.0	1.5	2.0	2.5	–	–	–
500–1,250	1.2	1.8	2.2	3.0	4.0	5.0	–
1,250–3,150	1.5	2.0	2.5	3.5	5.0	6.0	7.0
3,150–6,300	1.8	2.2	3.0	4.0	5.5	6.5	8.0
6,300–10,000	2.0	2.5	3.5	4.5	6.0	7.5	9.0

casting will never follow the drawing exactly. Hence, machining allowances must be made to produce a successful design. Reference surfaces should be noted in drawings for checking the castings and finished products. Allowances in sizes for steel and cast iron castings used in the Soviet Union and parts of Europe are given in Table 4.4.

4.3 DESIGN OF WELDED EQUIPMENT

Process equipment is fabricated by a number of well-established methods such as casting, welding, forging, machining, brazing and soldering, and sheet-metal forming. Each method has certain advantages for particular equipment. However, welding is the most important and is widely used for fabrication of steel apparatuses. This method of construction is virtually unlimited with regard to size and is applied extensively for fabrication and erection of large-sized process equipment in the field, which is often made by the method of subassembly [10-24].

Three factors influence the economics of welding processes: (1) the size of weldment required; (2) the number of weldments required; and (3) the relative complexity of the weldment [25]. The size of the required weld has a marked effect on the process that is used.

The design may be broken down into a number of subassemblies of perhaps similar form if a large weldment is made from a thin section. An example of this kind of construction is an all-welded steel flat-bottomed cylindrical vessel. If the fabrication of the vessel is made in the field, then manual arc welding by highly skilled craftsmen is likely to be the major process. In these circumstances capital investment charges may be comparatively low, but overall production costs will be high because labor cost will be very high. If the vessel construction is streamlined by prefabrication of relatively small subassemblies in a comfortable workshop not subject to climatic conditions, and by employing manipulators and fixtures that enable less skilled operators to use rapid-production automatic and semiautomatic high speed welding processes, then the skilled men and slower manual processes need be used only for difficult parts of the final assembly. This means relatively large initial capital investment charges, but comparatively low overall labor costs; so the total production cost tends to fall with increasing number of orders. If the component is large with relatively thick sections, then subassemblies are likely to be difficult to arrange. The problem is then one of adapting the process to the fabrication techniques most suited to the circumstances. If it is not possible to install some form of automatic arc welding equipment or an electro-slag welding plant, then costly positioned manual arc welding may have to be used. The cost of the manual process is not limited to the deposition costs but may have to allow for more careful treatment and a greater risk of defective welding. Suitable large manipulating equipment is a great help for manual welding of larger components and can reduce labor costs greatly.

The greater the number of finished parts required, the more sophisticated the methods become and the lower the production cost per part. With the increasing number of parts, it is possible to devise fixtures to simplify and speed up the manual welding and then, possibly, to arrange an automatic process, employing an existing machine.

For complex shape components requiring expensive manufacturing techniques, it is worthwhile to use one of the more costly welding processes to aid in fabricating the shape in such a way that overall fabrication is simplified.

In fabrication of process equipment from steel, the following welding methods are used [26-29]:

1. Automatic and semiautomatic hidden-arc welding
2. Manual electric-arc welding
3. Electro-slag welding
4. Automatic and semiautomatic carbon dioxide–gas-shielded welding
5. Automatic, manual and machine argon-arc welding
6. Acetylene welding.

Automatic and semiautomatic hidden-arc welding is now the basic method for the fabrication of process equipment from carbon, low-alloy and high-alloy steels.

Compared to manual electric-arc welding, the automatic hidden-arc welding possesses a number of significant advantages; the major ones are:

1. the higher productivity of welding–4-5 times higher than manual electric-arc welding;
2. the higher equality of welded joints; and
3. butt-welding without beveling, which considerably simplifies preliminary operations, saves weld metal and decreases the cost of equipment.

The hidden-arc welding process involves submerging the arc beneath a blanket of granulated mineral flux, where it generates heat for the electrode and deposits weld metal under a protective layer formed from a portion of melted flux. Several inches of weld metal can be deposited in one pass, which greatly decreases the welding time involved. As the arc is covered, there is no arc flash, and smaller amounts of smoke and obnoxious fumes are produced compared to other welding processes.

Hidden-arc welding is used primarily for the fabrication of straight and circumferential butt-joints of vessel shells from sheet steel of any thickness up to 3200 mm in diameter, as well as corner welds of flanges and other connections for apparatuses of medium and large sizes, tee welds and other standard part joints for chemical equipment.

The semiautomatic hidden-arc welding of steels is similar to automatic welding and differs by a manual motion of arc along the seam. Compared to automatic welding, semiautomatic welding has the following advantages:

1. Due to the high passability and maneuverability of a semiautomatic device, it is possible to weld in locations that are inaccessible for automatic hidden-arc welding.
2. One can butt-weld thin sheets (2–3 mm) and corner welds with a 3–4 mm fillet, which is inaccessible or difficult for automatic hidden-arc welding.
3. The cost of welding is lower due to lower energy consumption (approximately 30–40% less), and there is less consumption (approximately 15–20%) of welding (filler) wire compared to automatic welding.

Semiautomatic hidden-arc welding is used for fabricating small-sized and thin-walled equipment, and for welding to apparatus flanges, nozzles, tubes and other parts where the use of automatic welding is either impossible or ineffective.

Manual electric-arc welding of steel is used widely in the fabrication of process equipment. This method is used preferentially for tube joints, accessories, internal appliances of apparatuses, supporting constructions and different kinds of steel structures completing the process equipment.

Electro-slag welding is used widely in fabricating high pressure, thick-walled apparatuses, thick and large flanges, and other items that have a large initial thickness. Compared to other welding methods, electro-slag welding has the following advantages:

1. It consumes a small amount of energy, approximately 1.5–2.0 times less than for automatic hidden-arc welding and 4 times less than manual arc-welding.
2. 150–200 mm of weld metal can be deposited in one pass with the use of one electrode; when more electrodes are used, the welding of any metal thickness is possible.
3. Welding of any metal thickness is possible without bevelling; this makes preliminary operations easier and decreases the laboriousness and cost of apparatus fabrication.
4. Better conditions for evacuation of gases and slags provides a high quality of welded joints.

The productivity and economy of electro-slag welding depends significantly on the thickness of welded parts. If their thickness is less than 50 mm, the laboriousness and cost of welded joints made by electro-slag welding is higher than with electric-arc welding; however, if the part thickness is more than 100 mm, electro-slag welding is considerably more economical.

Automatic and semiautomatic carbon dioxide-gas shielded welding is used widely for low-carbon, high- and low-alloy steels. This method provides a high quality of welded joints. Automatic, manual and machined argon-arc welding is used mostly for welding apparatuses from high-alloy steels, as well as for welding tubes to the tube plate.

Acetylene welding of steels is used in fabricating process apparatuses in limited quantities, because of its defects. The considerable heating of the main metal determined by a longer heat action results in higher deformation of parts to be welded and enables superheating and grain raise in the weld joint. Compared to other welding methods, acetylene welding is less productive and not economical. However, it may be applied for light gauges of metal (20 gauge or less) which are difficult to weld by the arc-welding process. In common with every other type of fabrication, selection of a particular welding method is a matter of balanced consideration of design requirements, metallurgy and economics.

The influence of welding on the apparatus design depends primarily on the design of a welded joint. This is determined by the strength and tightness that can be obtained by the material in and around the joint, accessibility to the welding process and accommodation to the side-welding effects.

Each of the above factors has to be considered in every application. Unfortunately, each will not have the same relative importance in every case. In one case, limited weld strength will determine location; in some cases, the process may be the determining influence; and in many cases incidental effects, such as residual stresses or deformations could be the main influence.

The designer has a free hand in selecting joint locations if full parent metal strength is assured in a weld joint, provided that the selected position is accessible and side effects can be absorbed. In this case, the selection of joint positions is determined primarily by economics.

If full parent strength is not attainable in a weld joint, the problem arises of locating welds in relatively low-stressed positions or increasing the sectional size in the weld vicinity to give more load carrying area. In locating the weld joints, it is necessary to account for rigidity; therefore, the complete weldment must be studied as a unit, not as a pin-jointed structure.

Correct joint preparation is important to the particular process to be used. Table 4.5 gives a summary of the type of process, the type of joint preparation and a note on relevant limitations [25]. We deal particularly with the problems associated with progressive fusion welds of the arc type because of their widespread importance.

Table 4.5. Various Types of Joints [25]

Welding Process	Type of Weld					Joint Preparation										Materials Limitations			Geometry Limitations				
	Progressive Fusion	Progressive Fusion of Filler	Unit Fusion	Unit Fusion of Filler	Unit Solid Phase	Square Butt	Fillet/Corner Joints	Lap Joint	Modified	Open-Square Butt	Local Spot	T-Joint	Single and Double U Butt or J Butt	Single or Double V Butt	Single V Butt	Limited Range	Wide Range	Very Wide Range	Machine Capacity	Positional Limitation	Maximum Size of Weldment Limitation	Maximum Thickness Limitation	Long Seams Desirable
Braze Welding		✓					✓							✓			✓					✓	
Brazing																							
Dip				✓				✓									✓				✓		
Furnace				✓				✓										✓			✓		
Manual		✓						✓									✓					✓	
Electric Resistance																							
Butt					✓	✓										✓			✓				
Flash butt					✓	✓										✓			✓				
Projection			✓					✓			✓					✓			✓			✓	
Roller seam	✓		✓					✓			✓					✓			✓			✓	
Spot			✓					✓			✓					✓			✓			✓	
Stud			✓									✓				✓			✓				
Electron Beam	✓					✓												✓	✓	✓			
Electro-Slag		✓								✓						✓				✓			

Table 4.5, continued

Welding Process	Type of Weld					Joint Preparation										Materials Limitations			Geometry Limitations				
	Progressive Fusion	Progressive Fusion of Filler	Unit Fusion	Unit Fusion of Filler	Unit Solid Phase	Square Butt	Fillet/Corner Joints	Lap Joint	Modified	Open-Square Butt	Local Spot	T-Joint	Single and Double U Butt or J Butt	Single or Double V Butt	Single V Butt	Limited Range	Wide Range	Very Wide Range	Machine Capacity	Positional Limitation	Maximum Size of Weldment Limitation	Maximum Thickness Limitation	Long Seams Desirable
Friction					√	√						√					√						
Metal Arc																							
Flux-coated (deep penetration)		√				√			√			√				√							
Flux-coated (normal)		√				√	√						√	√		√							
Inert gas-shielded (self-adjusting arc)		√					√		√							√							√
Spot			√					√			√					√						√	
Stud			√									√				√			√				
Submerged arc (powdered flux)		√							√							√				√			√
Oxyacetylene		√				√	√							√			√					√	
Tungsten Arc		√				√	√							√			√						

A variety of types of welded joints are used in fabricating equipment, depending on the service, the thickness of the metal, fabrication procedures and code requirements. There are five basic forms of welded joints: edge, butt, lap, corner and tee. Figure 4.40 illustrates these forms.

Figure 4.40(A) shows a plain edge joint. The dotted lines shown are referred to as leaf-and-edge joints and are obtained by bending the plates over after welding. Figures 4.40(B) and (C) show a single and a double-edged butt joints, respectively; (D) and (E) are lap joints; (F) and (G) are corner joints; (H) and (I) are tee joints.

With the exception of the joint (A), the other joints can be obtained by applying either a *butt weld*, where the weld metal is deposited between the plate edges to be joined (Figure 4.40 (B,C,G)), or *fillet weld,* where the weld metal is deposited in the corner formed by two surfaces (Figure 4.40 (D, E or H)). These welds are classified as: *reinforced* (Figure 4.40 (B,F)), *flush* (Figure 4.40 (D,E,G)) and *concave* (Figure 4.40(I)), if we consider them from the standpoint of strength. The dimension "α" (Figure 4.40) is called the *throat* and "β" the *leg* of a weld. A butt weld is considered the best welding joint. The selection of the edge form depends on the welding procedure, the plate thickness and whether both sides of the joint or only one side can be welded. Figure 4.41 illustrates various forms of edges of the plates for manual welding. These are used for square, butt-welded joints, single-welded butt joints with and without backing strips. The forms of edges of the plates used for automatic welding are shown in Figure 4.42. The gap 1 between sheets is taken as a function of sheet thickness S as shown in Table 4.6.

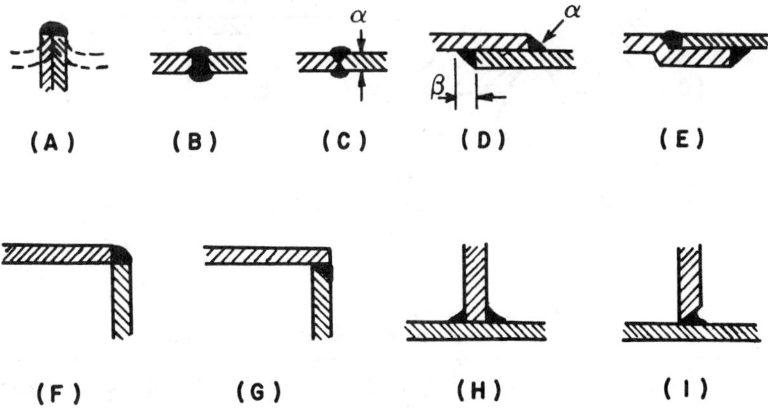

Figure 4.40. Forms of welded joints. Weld shown in black.

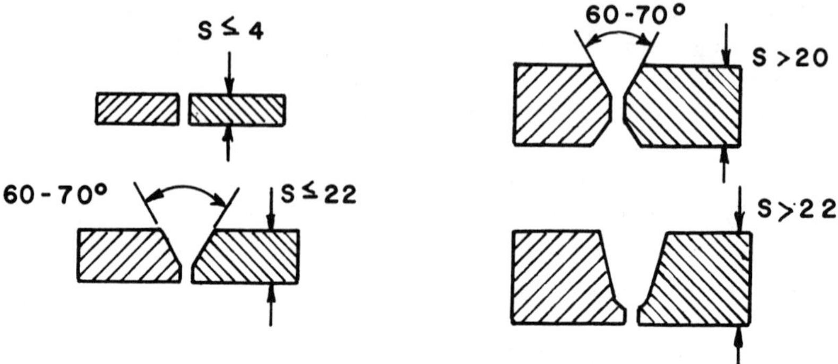

Figure 4.41. Forms of edging of plates for manual welding.

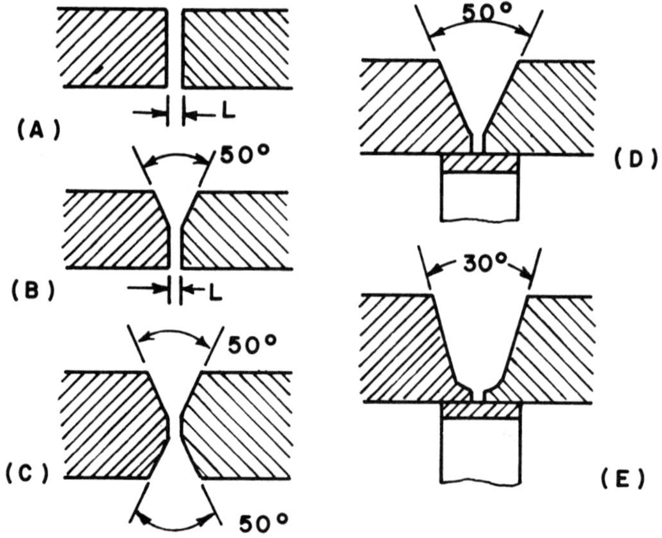

Figure 4.42. Forms of edges of plates for automatic welding.

Table 4.6. Sheet Thickness as a Function of Gap Distance

S_1, mm	Up to 8	10–12	14–18	20–30	30–40	40–50
ℓ_1, mm	0–1	1–2	2–3	3–4	4–5	7–8

For welding vessels, longitudinal joints of shell plates should not be thicker than 6-8 mm. Thicker shell plates require beveling, as shown in Figure 4.42(B). Beveling of plates is necessary if it is desirable to increase the role of the weld metal in the joint formation; for example, in welding stainless steel with carbon steel to avoid weld cracks. The double-welded butt joint (Figure 4.42(C)) is used widely in weld metals in joints, as well as for welding sheets more than 50 mm thick. In the latter, it is desirable to use electro-slag welding without beveling with a gap of 30 mm between the plates. The edges of the plates (Figures 4.42(D) and (E)) are used only for transverse joints, single-welded butt joints with backing strip and one-sided access to the joint. The edge (Figure 4.42(E)) is used for plates up to 18 mm.

The minimum size of fillet welds, K, (Figure 4.43) used for tee and lap welds, depends on the plate thickness to be welded. Sufficient penetration may be achieved by following the recommended sizes of fillets, K, given in Table 4.7.

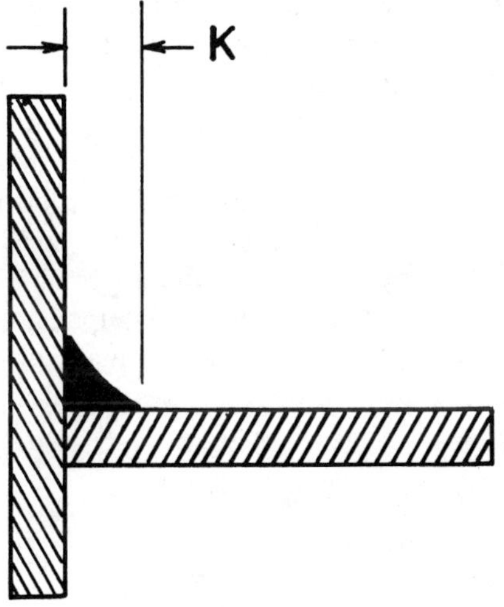

Figure 4.43. Fillet weld.

Table 4.7. Recommended Fillet Sizes

S_1, mm	Up to 4	4–8	9–15	16–25	25
K_{min}, mm	3	4	6	8	10

The maximum fillet size is as follows:

$$K \leqslant 1.2\ S_{min} \tag{4.7}$$

where S_{min} is the thickness of the thinnest plate to be welded. The length of the longitudinal fillet weld should not be less than 50K because of incomplete fusion at the beginning and end of the weld.

Bevelless automatic hidden-arc welding shells with tube plates (Figure 4.44(A)) provide better results than beveled shells (Figure 4.44(B)). There is no need to turn a groove in the tube plate and groove the shell as the plate becomes weakened and welding becomes more complicated. In addition, the cost of machining increases.

The design of a weld joint should provide a sound seam. Wherever possible avoid overhead welds. Automatic hidden-arc welding is only possible in downhand and horizontal positions. Any configuration of weld joints is possible when using manual arc-welding.

Butt-welded sheets should have a uniform thickness, wherever possible. If their thickness differs more than 20–25%, it is necessary to make the edges of the sheets the same. The slope of the thicker sheet should be 1:5 (Figure 4.45). The weld joint must provide a free shrinkage of deposited weld metal and have a gap between sheets.

In welding processes, access to the weld position with welding equipment is essential. Figure 4.46 illustrates this for manual welding. The example given does not exhaust the complexity of the problem. It depends to a certain extent on the size of the weldment. A very small one is manipulated successfully by hand or by a small fixture to bring welds into workable positions. With very large sizes, welding equipment usually has to be manipulated about the work. In this case, automatic or even semiautomatic progressive welding processes can be realized economically to long lengths of readily accessible joints.

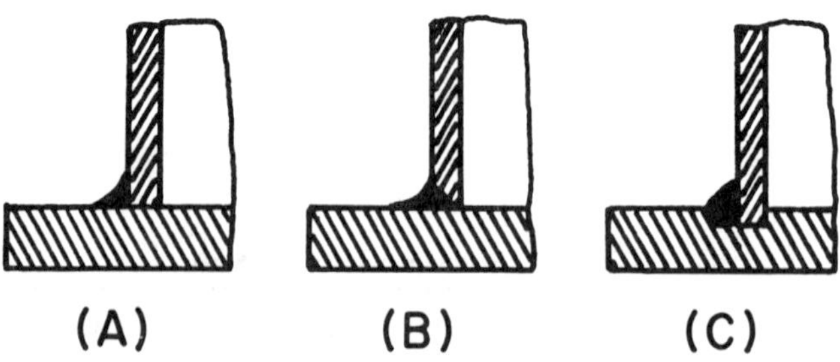

Figure 4.44. Welded joints of shell and tube plate.

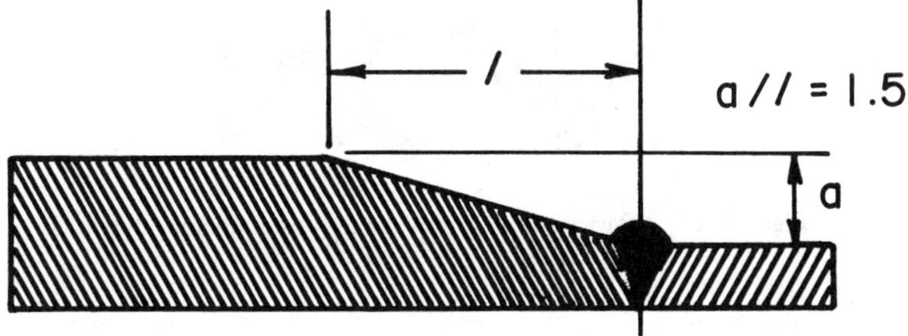

Figure 4.45. Single-welded butt joint for sheets of different thicknesses.

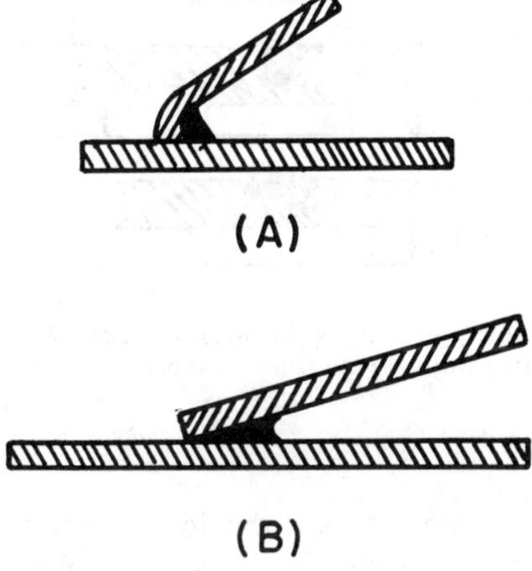

Figure 4.46. Joint accessibility for welding.

Certain welding processes require certain types of weld joints, each with their own problems of accessibility and accommodation. For example, automatic and semiautomatic hidden-arc welding processes require protection of the weld metal and flux and must be maintained in one place, as shown in Figure 4.47. The first design version in Figure 4.47 is less successful, but the second one requires a temporary welded strip to keep the flux in place. In some cases where the weld joint is undercut and welding is impossible, for example, in small-diameter tubes and vessels or transverse assembling seams, there is a need to insert a backing strip ring or provide a half-lap joint (Figure 4.48).

Figure 4.47. Devices that prevent the flux from draining down.

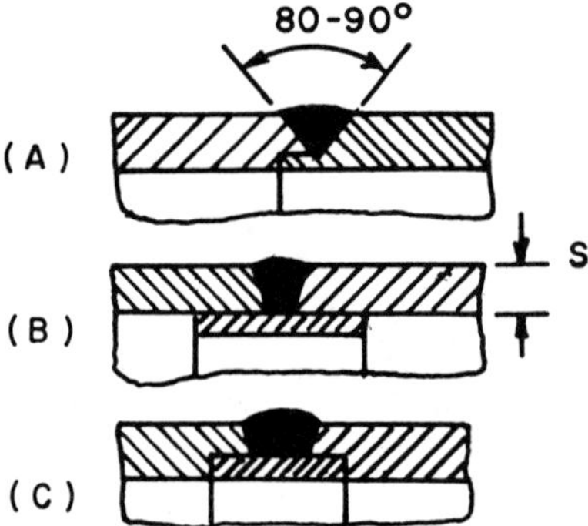

Figure 4.48. Weld joints preventing weld metal passage: (A) half-lap joint; (B) butt-joint with backing ring strip; (C) the same for tubes.

Strength factors of a welded joint should be considered in the same way as in any other fabricating systems, that is, the weakest links in the system have to be found and their effects evaluated. The mechanical properties of a continuous welded joint are not uniform. The following three zone conditions of a material must be considered:

1. unaffected parent material;
2. heat-affected zones of parent material; and
3. weld deposit or bonding zone of the weld.

Because of low welding temperatures, in some cases condition (2) will be nonexistent; however, conditions (1) and (3) always will be present.

The tensile strength of a welding joint gives a value at which complete rupture would take place under steady load conditions. It provides a safer indication in welding than some other applications. The mode of fracture of welded material in a tensile test can give a good guide to the material's probable responses to constrained loading conditions. A brittle type of test fracture is a danger sign. The mode of development of a fracture often may be correlated with weldability. A fracture showing marked lipping is a good sign. A transcrystalline cleavage fracture means high static strength but little resistance to notches, etc. A rough shearing-type fracture indicates a heterogeneous material is liable to crack during welding.

A relatively low yield stress indicates low strength but good shock resistance to welding stresses. For welding applications, it is recommended that the yield stress should be at least 10% lower than the fracture stress.

Shearing strength is one of the most important factors in all lap joints. However, in many applications the shearing strength of the weld metal is not likely to be a critical factor.

A low elastic modulus in a material can help absorb strain and shock-loading. In cases in which the elastic behavior of the weld material is similar to that of the adjacent parent material, its effect usually is not great. The effect is pronounced if the elasticities of adjacent materials are very different. It is then possible for stress concentrations to build up, which can lead to a failure under shock loading, or even steady load in some cases; or it can lead to premature fatigue failure. A high modulus of elasticity usually implies relatively high residual stress in almost any kind of fusion weld.

For withstanding the welding stresses, as well as normal service, it is desirable that every weld joint provide a reasonable degree of plasticity. Light mechanical shocks to the apparatus may cause severe shocks to the area containing weld joints. If they are not shock resistant, the joints must be placed in the middle of slender parts, rather than toward a corner near a rigid member. A material that has poor shock resistance is also sensitive to the presence of a notch because welds do not arrest a crack once one is started.

Estimation of fatigue in welded applications is very important wherever vibration and other fluctuation loads are applied. Fatigue is not an inherent weakness of weld joint metal and, therefore, is not the main source of trouble. Rather, the effect of transferring stress from one kind of metallurgical structure into another, weld defects and stress concentration due to weld contour are major problems.

In dissimilar structures existing side-by-side, stress passes from one to the other and lowers the fatigue resistance due to elastic response. Weld defects due to stress concentrations will affect fatigue strength adversely, the amount of which depends on the type and location of the defects. If a weld with reinforcement is left as welded, there must be a notch effect at the discontinuity that lowers the fatigue strength. This is dependent on the sharpness of the changes of the section and the nature of the weld metal.

A welded structure at elevated temperatures is liable to the same limitation of creep resistance as any other structure used; therefore, it is necessary to know the creep properties of the weld joint metal. The designer should account for the state of the weld with respect to structure and residual stresses. As a result of the interaction of these factors in the case of very large welds, the resistance to creep may be lowered. To increase creep resistance it is advisable to bring the weld structure into as stable a condition as possible by heat treatment before use at elevated temperatures.

Four effects are related to the application of welding in process equipment fabrication:

1. transient stresses,
2. residual stresses,
3. transient distortion, and
4. permanent distortion.

The severity of each depends on the welding method used, the design configuration of the weldment and the local circumstances.

It is often necessary to disperse weld joints. Figure 4.49 illustrates the location of longitudinal and transverse weld joints of a vessel shell. The distance between the weld joints "a" must be no less than 35 or 40 mm. Note that it is not possible to connect several parts with one weld joint (Figure 4.50).

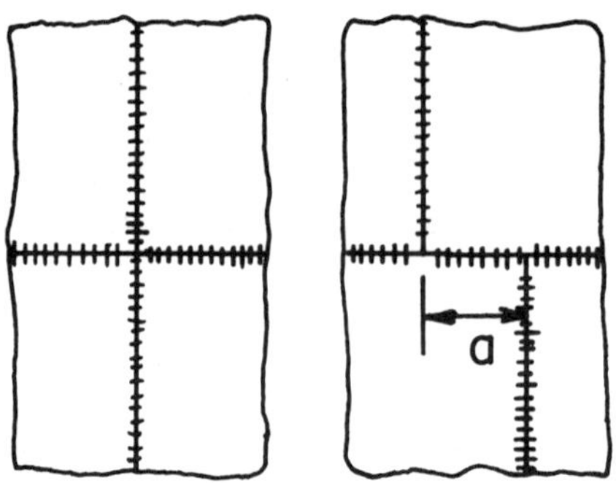

Figure 4.49. Location of weld joints.

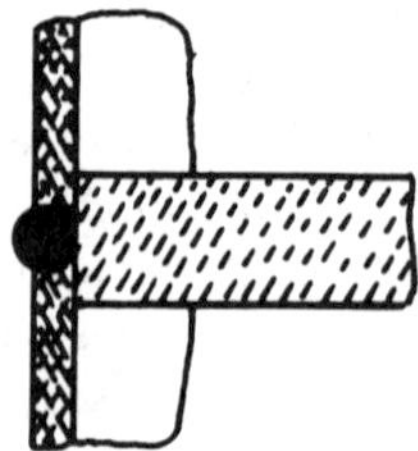

Figure 4.50. Poor design of a weld connection of three parts.

Longitudinal weld joints should not be interrupted by holes and nozzles. There should be regular inspections of weld joints in process equipment. The weld joints should not be overlapped by any kind of supports. If the pickup overlaps the transverse weld joint, it should be designed as shown in Figure 4.51.

Stress concentration can arise within a material in service:

1. if it is deliberately designed so that some parts of the structure have different specific stress intensities from others immediately adjacent to parts; or if certain parts are subjected to combined stresses imposing more strained conditions than on adjacent parts; or
2. accidentally, when it is difficult to consider the general behavior of the materials of construction, because of the effects of the geometry whose significance could not be foreseen, or as a result of damage or defect.

If different stress intensities of one type of stress exist in one member in adjacent zones, the transverse contractions or expansions relative to the stress planes will differ and each zone will impose a transverse restrain on the other, which is proportional in intensity to the difference in strain behavior.

If the zone subject to a basic principal tensile stress has further lateral tensile stress added to the principal stress, then it is most likely a source of trouble.

Residual stresses and distortion must be at a minimum. This can be achieved by good design (i.e., by decreasing the volume of weld metal, avoiding local metal accumulation and joint intersection, locating the weld joint symmetrically with respect to the axis). The influence of thermal stresses is especially unfavorable for construction formed from flat sheets, which can result in high warping. Such designs should be avoided wherever possible. A considerable decrease in stress may be achieved by incorporating

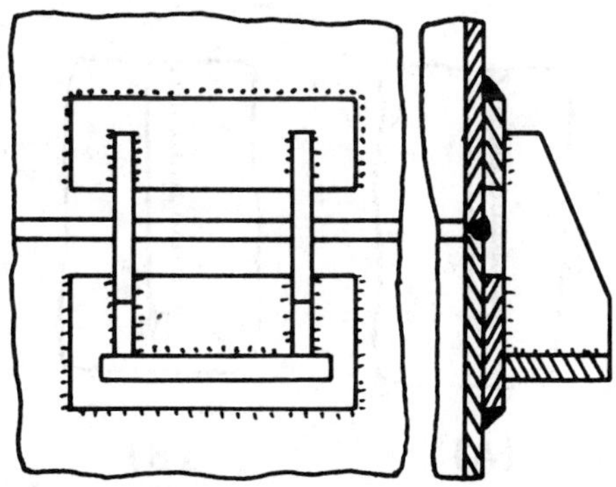

Figure 4.51. Pickup design providing inspection of a transverse butt weld.

flexible elements in welded connections, for example, a butt-flanged connection or a nozzle with a shell (Figure 4.52). There are two versions of this design: (A) flanged opening in the shell, and (B) flanged rim of the nozzle. Beading of a nozzle is more convenient than that of the shell; however, version A is preferable because the walls of a shell almost always are thicker than those of a nozzle. With a shell flanging, the rim of an opening becomes thinner, approaching the nozzle thickness, which is favorable for welding.

It is not recommended to weld nozzles, rims and other parts at the ends of an apparatus. In such cases, the stresses cannot be distributed evenly over the section of the material and can cause weld cracks and rim warping. Welded rim ends should be welded from both sides (Figure 4.53). Shrinkage during cooling can cause distortion of welded parts, but can be avoided by separating the weld joint of the weld in parts (Figure 4.54). Considerable welding strains determine the sequence of the apparatus fabrication. If, for

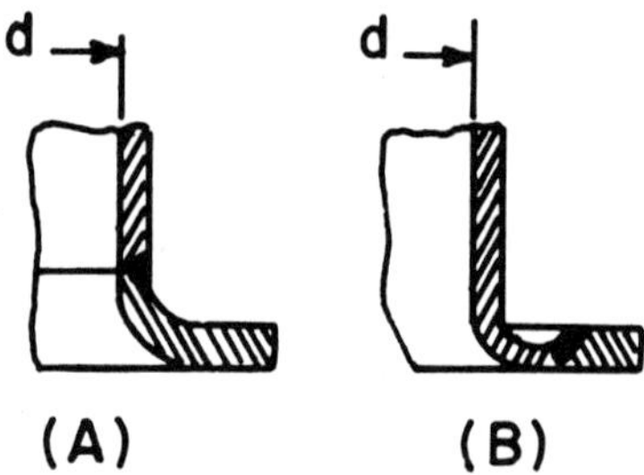

Figure 4.52. Welded joint of a flanged nozzle.

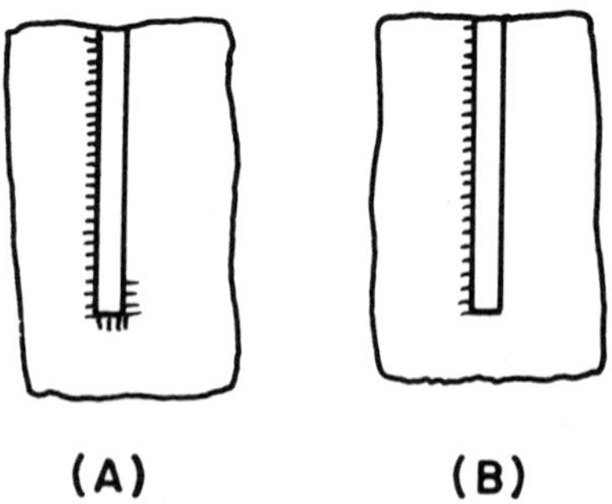

Figure 4.53. Fixing the welded ribs.

example, the connection is made by rolling out, riveting, caulking, etc., then these connections must be made after welding, or thermal strains will break the tightness and strength of such connections. Due to an inevitable warping, the machining of welded vessels, checking and flange turning must be done after welding if there is a need for finished parts. The location of a weld joint then must be separated from a finished surface as far as possible to avoid damaging during welding (Figure 4.55).

Because the weld metal is always more brittle than the base metal, the weld joint must be carried out of the bending stress zone. Construction should be avoided where such stresses exist (Figure 4.56). Any residual stress left in a weld joint may react with applied working stresses as additive stress and bring the material into its plastic flow range or to its fracture stress if it has limited plasticity. In the case of an angle residual tensile stress orientation to the working tensile stress, a restraint on the freedom to deform under the working stress will arise, causing the apparent yield stress to increase, and thus reducing the ductility of the material. For sensitive materials,

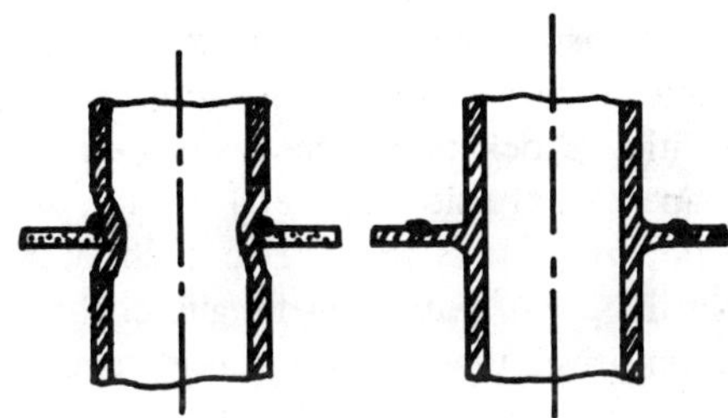

Figure 4.54. Separation of weld joint for strain decrease.

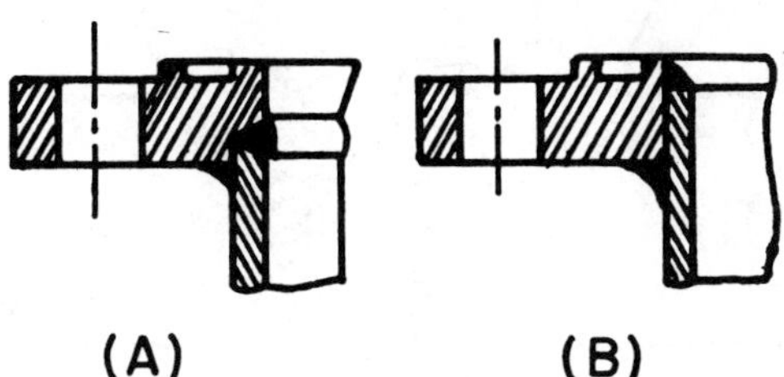

Figure 4.55. Separation of the finished surface from the weld joint: (A) good design; (B) poor design.

Figure 4.56. Poor weld joint designs where bending stresses arise.

ductility may be considerably diminished and brought to embrittlement under shock loading. It is important to account for the effects of any residual stress in the entire range of service temperatures to which materials are to be subjected. Changes in temperature can lead to changes in resistance. Some steels may become more sensitive with a fall in temperature, whereas others become more sensitive with rise in temperature. Lap joints and T-joints are vulnerable to severe stress concentrations and never should be used to withstand either shock or fatigue loading conditions. A weld that has a reinforced shape or extrusion of metal outside the uniformity of the main contour is a potential stress raiser. Hence, the weld profile should be machined smooth in the event that either fatigue or shock loading is critical. If the material is sensitive, then even a gentle reinforcement of a weld (Figure 4.57) can be dangerous to the sharpness of corners. These are likely zones for orientation of fatigue failure.

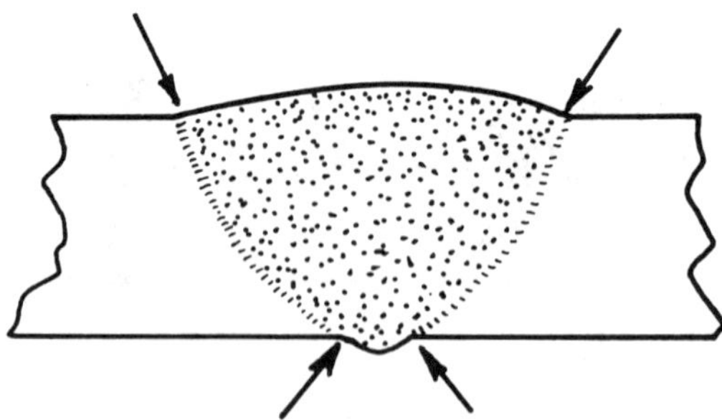

Figure 4.57. Possible stress concentration positions on a normal fusion weld profile.

There may be dangers also in the interior of a weld joint (Figure 4.58), even if full allowance is made for the external shape. These kinds of joints are sometimes used for relatively unimportant work but are generally not recommended. In the case of bad workmanship full penetration is not achieved, and there is the chance of weld discontinuity, which may bring about notch induced failure. Even if measures are taken to avoid apparent geometric discontinuities in a weld joint, stress concentrations may arise if adjacent zones of the material do not react in the same manner as to the applied stress.

From economic considerations, it is desirable to minimize the amount of welding by combining bending and stamping with weld joints (Figure 4.59); combining continuous and intermittent weld joints; and using a minimum

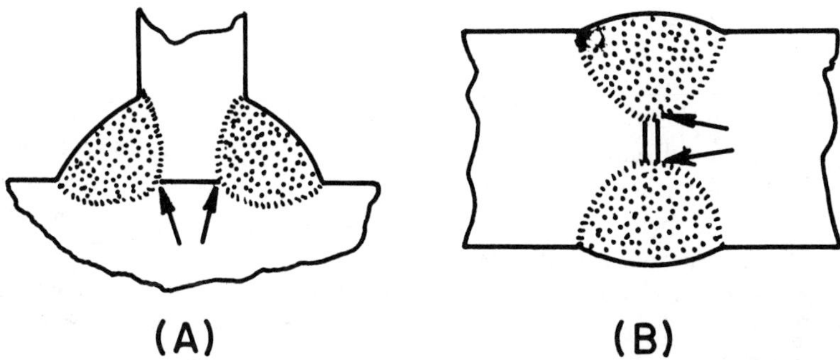

Figure 4.58. Joint conditions likely to give internal stress concentrations where indicated by arrows: (A) normal fillet-welded T-joint; (B) Double-V-joint with unblended roots.

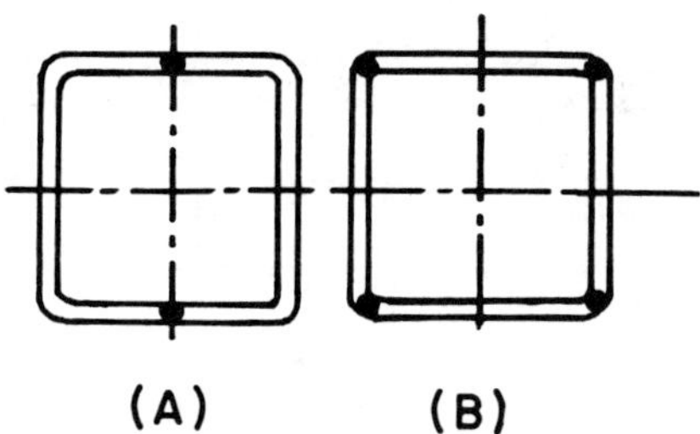

Figure 4.59. Decreasing the weld joint length by combination bending and welding: (A) good design; (B) poor design.

angle of beveling. Unfortunately, the possibilities of decreasing the metal weld volume are quite limited in fabrication since weld joints not only must be strong, but tight as well. In welding small parts, such as strips and studs for bubble caps in distillation columns, pressure contact welding is used. This will not only decrease the length of weld joints but also increase the accuracy of assembly.

The joint should be arranged to minimize any tendency to pry off the elements to be welded. A rule of thumb in designing a welded joint is to provide strength by length rather than by size, as it is necessary to use more weld metal for developing the required strength when a large short weld is applied rather than a small long weld.

One of the most important principles in the design of welded joints is to arrange the welds to be symmetrical in location, size and length. For members such as structural steel angles (Figure 4.60), the length of a weld is divided in accordance with the following equation:

$$L_1 \times a = L_2 \times b \tag{4.8}$$

where a and b are dimensions from the centroidal axis of the section. Welded joints subjected to tension are divided into three classes according to their internal stresses (Figure 4.61). In class 1, the resistance of the weld is exclusively in shear; in class 2, the resistance is in shear and tension; in class 3, the resistance is in pure tension.

The approximate relative strengths of welds in percentage of the base metal strength for various classes of joints and types of weld are presented in Table 4.8. These data should be used only for orientation in selecting the

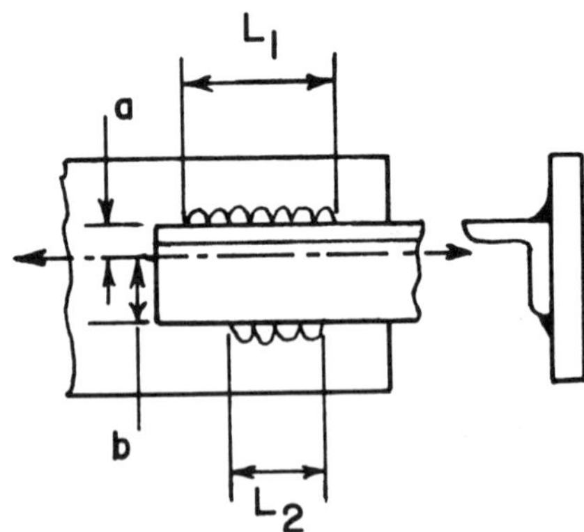

Figure 4.60. Symmetry applied to welding an angle.

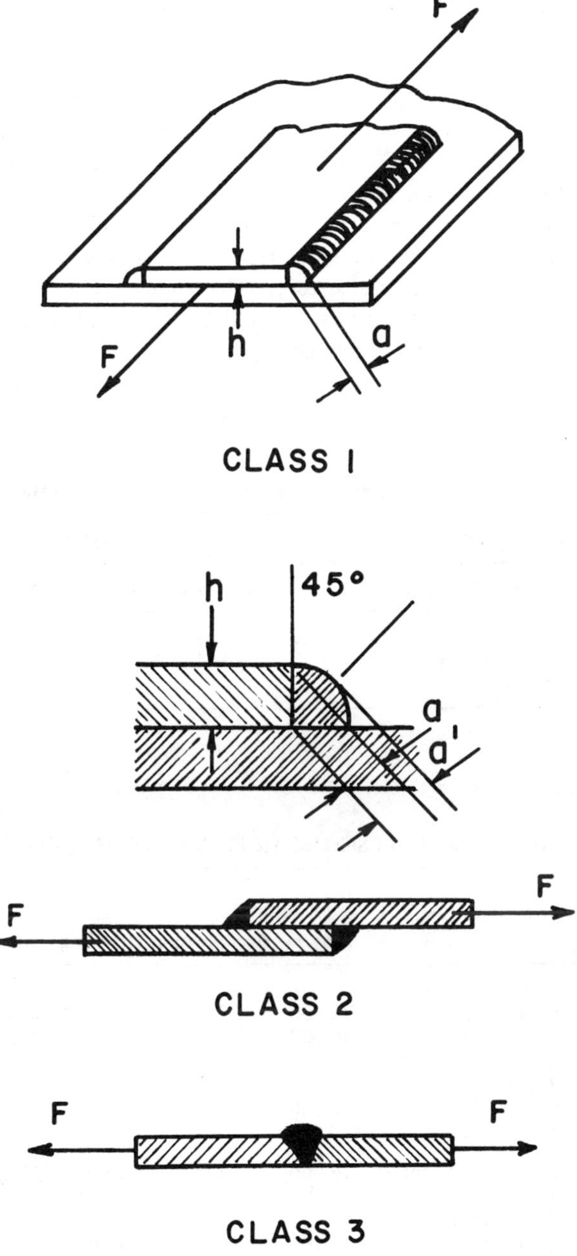

Figure 4.61. Force action in various classes of welds.

type of joint and weld [8]. For properly designed and skillfully made welds, the strength is almost equal to that of the base metal.

Table 4.9 gives the design stresses for shielded flush welds. The throat dimension a' of a reinforced fillet (Figure 4.61) should not be taken more than 1.2a, and for a concave weld $a' = 0.5a$ [30]. The effective length of each weld should be taken ⅝ in. shorter than the total length to allow for imperfections caused by starting and stopping a weld bead.

Welds should be proportioned to be stiffer than the adjacent material, and thus throw the deflection into the latter in all cases when the joints are exposed to bending or reverse stresses. The endurance limit of the weld metal in the case of repeated stresses should be taken no higher than 18,000 psi. A safety factor of 2 or 2.5 may be applied.

Table 4.8. Ratio of Static Strength of Shielded Welds to Base Metal [8]

Joint Classification		Weld Type			
Class Type	**Condition**	**Leak Proof**	**Concave**	**Flush**	**Reinforced**
1	Shear	0	0.40	0.90	1.10
2	Shear and tension	0	0.30	0.75	0.95
3	Tension	0	0.20	0.60	0.75

Table 4.9. Strength of Shielded-Arc Flush Steel Welds (psi) [8]

		Tension	Compression	Bending	Shear	Shear and Tension
Limit Stress	Elastic Limit					
	Base metal	32,000	35,000	35,000	20,000	–
	Deposited metal	40,000	44,000	44,000	24,000	–
	Endurance limit of deposited metal	22,000	–	26,000	–	–
Recommended Design Stress	Static load (safety factor 2)	16,000	18,000	18,000	11,000	11,000
	Load varies from 0 to F (safety factor 2)	14,500	16,000	16,000	10,000	10,000
	Load varies from +F to –F (safety factor 2.75)	8,000	8,000	9,000	5,000	5,000

Under static loads initial stresses are not dangerous. This is true when the yield point exceeds the maximum stressed point, causing the metal to creep and the load to redistribute more evenly over the cross section. After several local overloadings, the stress distribution in the structure approaches the distribution that would exist without initial stresses. For variable loads the initial stress actually may reduce the safety factor. In the case of impact the presence of initial stresses is extremely undesirable. The following examples described by Maleev [8] illustrate this point:

Example 4.2

We wish to determine the load that an arc-welded joint of class 1 can carry safely if it joins ⅜-in.-thick plates. Each fillet is 4½ in. long and reinforced. Figure 4.61 shows the system. The load varies from 0 to a certain maximum value.

The cross-sectional area of a flush fillet resisting shear would be $a \times (\ell - 0.625)$, and the area of a fully reinforced fillet may be taken as $1.2a \times (\ell - 0.625)$. Thus, the total cross-sectional area for both fillets is

$$A = 0.707\,h \times 1.2\,(\ell - 0.625)\,2$$
$$= 0.707 \times 0.375 \times 1.2\,(4.5 - 0.625) \times 2 = 2.46 \text{ in.}^2$$

Taking the working stresses from Table 4.2 as $S_s = 10{,}000$, the safe load is

$$F_1 = A_1 S_s = 2.46 \times 10{,}000 = 24{,}600 \text{ lb}_f$$

Example 4.3

We wish to determine the loads that welded joints of class 2 and 3 (see Figure 4.61) can carry safely, using the conditions and data from the previous example and comparing them with the strength of the strip.

In the joint of class 2, the section area in tension is equal to that in shear. Therefore, the strength in shear is the deciding factor of strength. The area of the weld is

$$A_2 = 0.375\,(4.5 - 0.625) \times 2 = 2.91 \text{ in.}^2$$

and the safe load is

$$F_2 = A_2 \times S_s = 2.91 \times 10{,}000 = 29{,}100 \text{ lb}$$

Hence, a class 2 joint is considerably stronger than a class 1 joint, although they are not quite comparable.

In the joint of class 3 the dangerous section with a reinforced joint has an area

$$A_3 = 0.375 \times 1.2 \times 3.875 = 1.74 \text{ in.}^2$$

From Table 4.2, the allowable stress in tension S = 14,500 psi:

$$F_3 = A_3 S = 1.74 \times 14{,}500 = 25{,}200 \text{ lb}_f$$

A class 1 joint is the weakest one, although it is only 2% below class 3. If the base metal is S.A.E. 1010 steel with an endurance limit of 24,000 psi, the safe load for a ⅜-in. × 4½-in. strip, using a safety factor n = 2, is

$$F = \frac{0.375 \times 4.5 \times 24{,}000}{2} = 20{,}200 \text{ lb}_f$$

Thus, all welded joints theoretically are stronger than the strip.

Example 4.4

We wish to compare the strength of the joints in the two preceding examples with class 1 and class 2 joints made with flush welds and a class 3 joint with a reinforced weld ground flush afterwards; all welds are to be made by the oxyacetylene process.

In a class 1 joint with flush weld, the dangerous cross section is

$$A_1 = 0.707 \times 0.375 \times 2 \times 4.5 = 2.38 \text{ in.}^2$$

and with a stress coefficient of 0.85 for gas welds, the safe load is

$$F_1 = A_1 S_s \times 0.85 = 2.38 \times 10{,}000 \times 0.85 = 20{,}200 \text{ lb}_f$$

For a class 2 joint, the area of dangerous cross section is

$$A_2 = 0.375 \times 4.5 \times 2 = 3.37 \text{ in.}^2$$

and the safe load for a gas weld is

$$F_2 = A_2 S_s \times 0.85 = 3.37 \times 10{,}000 \times 0.85 = 28{,}600 \text{ lb}_f$$

In a class 3 joint, the dangerous section is

$$A_3 = 0.375 \times 4.5 = 1.685 \text{ in.}^2$$

and the safe load, using a coefficient 0.85, is

$$F_3 = 1.685 \times 14{,}500 \times 0.85 = 20{,}750 \text{ lb}_f$$

Thus, with these flush welds, all joints have a strength equal to, or higher than, the strip.

In fabricating process equipment from aluminum and its alloys, the following welding methods are employed [23,24]:

1. automatic hidden-arc welding;
2. automatic argon-arc welding with a consumable electrode;
3. manual argon-arc welding with a nonconsumable (tungsten) electrode; and
4. manual electric-arc welding.

The automatic hidden-arc welding of aluminum is a highly efficient method that is used widely. This method is applied for welding without beveling, straight and circumferential weld joints of large vessels from sheet thicknesses more than 4 mm. It does not use expensive inert gas and provides good quality and high corrosion resistance of weld joints.

Argon-arc welding of aluminum and its alloys is a universal welding process giving high-quality weld joints (single- and double-welded butt joints, lap joints, corner joints and tee joints) for tubes and sheets ranging from 0.03 mm to 10 mm thick. The welding is made with preliminary metal heating to 150–200°C.

Manual electric-arc welding is of limited usefulness in fabricating process equipment because of low quality. In fabrication from copper and brass, automatic and manual electric-arc welding is used. Welding copper and brass is possible only in lower positions because of the flowability of these metals. For single-welded butt joints, good weld formation is achieved by using copper, graphite or asbestos backing strips. Copper or brass sheets more than 5 mm thick require preliminary heating up to 200–300°C, and afterwards undergo welding, forging and rapid cooling with water. The manual electric-arc welding of brass sheets more than 10 mm thick also may be accomplished with carbon electrodes and by heating up to 300 to 350°C. After welding, the component must be annealed at 300 to 310°C. For nickel sheets up to 2 mm thick, it is recommended to make a butt joint with argon-arc welding. This provides high quality of connection. For thicker sheets, automatic hidden-arc welding with ceramic flux, manual electric-arc welding and acetylene welding are employed. Automatic welding of nickel sheets is made with joint heating up to 300–350°C to decrease weld stresses and to increase the mechanical strength of the weld joint.

4.3.1 Influence of Machining on Half-Finished Parts

Half-finished parts must have a shape convenient for making alterations in machines. For castings, it is necessary to provide special ribs to permit further work in a chuck (Figure 4.62) on machines. These ribs may be removed after machining. The rough and clean bases must be selected properly for easy marking and fixing. For cast vessels it is convenient to take parting planes as a base. For steel apparatuses, one should take the face edge of shell.

The shape and size of half-finished parts should be considered so that the machined surface is minimal. The surface to be machined should be clearly separated from the surface that is nonmachined. Therefore, all surfaces to be machined must have a projection over nonmachined ones as, for example, in cast flanges (Figure 4.63), or parts machining with another surface roughness.

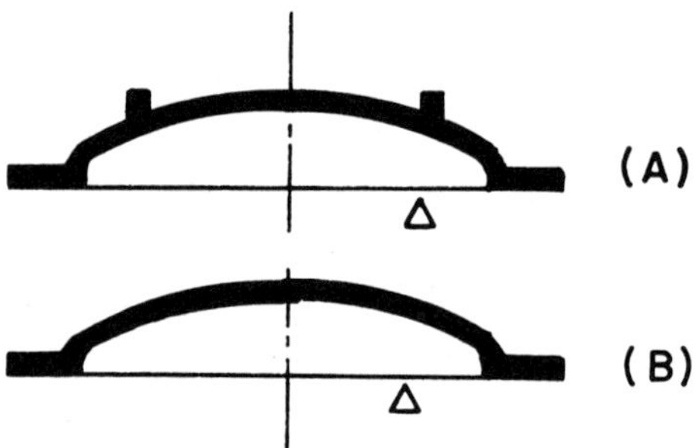

Figure 4.62. Ribs on a shell head for assistance during machining.

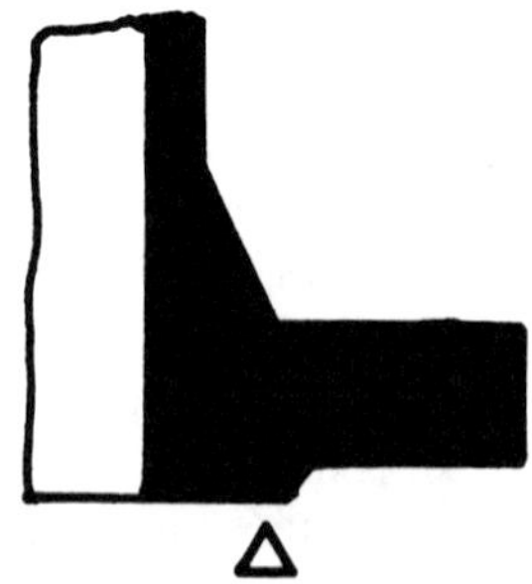

Figure 4.63. Flange projection for machining.

The thickness of similar lugs should not be small (i.e., not less than the thickness provided by the casting tolerance). Otherwise, warpage of the pattern and casting may prevent lug formation. The ribs must not reach the machining surface to prevent a shock load on the cutting tool when turning (Figure 4.16(C)).

Machining surfaces should be mutually situated either parallel or perpendicular. Only when absolutely necessary (Figure 4.64) is it permissible to make an angle arrangement different from 0, 30, 45, 60 or 90°. It is sometimes profitable to use countersinks instead of total flange machining (Figure 4.65). For machining massive parts by speed methods sufficient balance should be provided to avoid vibration.

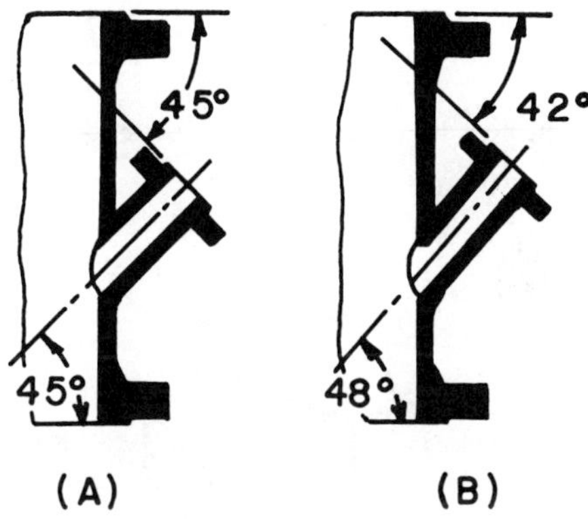

Figure 4.64. Surfaces must be situated at angles convenient for machining: (A) good design; (B) poor design.

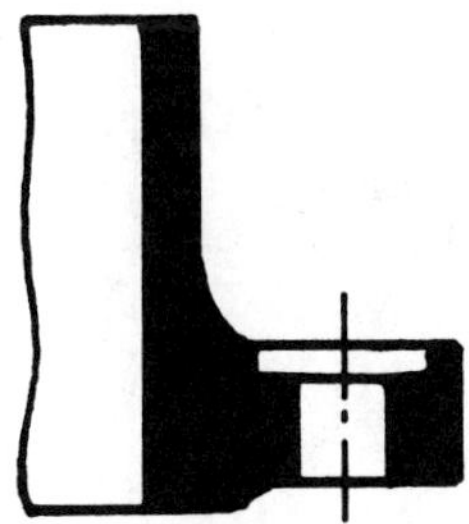

Figure 4.65. Flange machining by countersink.

Wherever possible, parts must provide for the passing-through of machining of surfaces and planes (Figure 4.66). In this relation, version A is better than B in Figure 4.66.

One also may provide a slot for passing out a cutting tool (Figure 4.67). Three versions of boring a vessel for centering a head extension are given in Figure 4.68. Versions A and B are equally correct, but version C is not recommended.

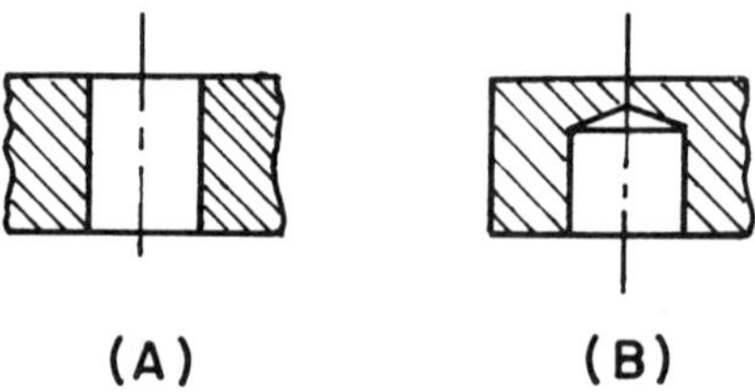

Figure 4.66. The hole drilled through (A) is better than the blind hole (B).

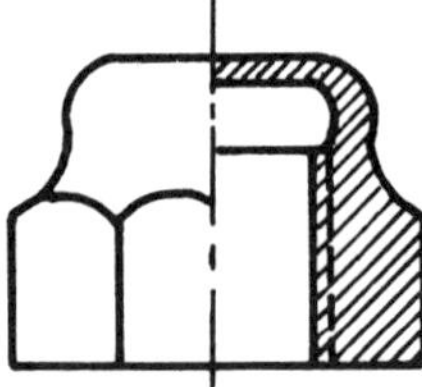

Figure 4.67. A slot for passing out the cutting tool from a nut screw cap.

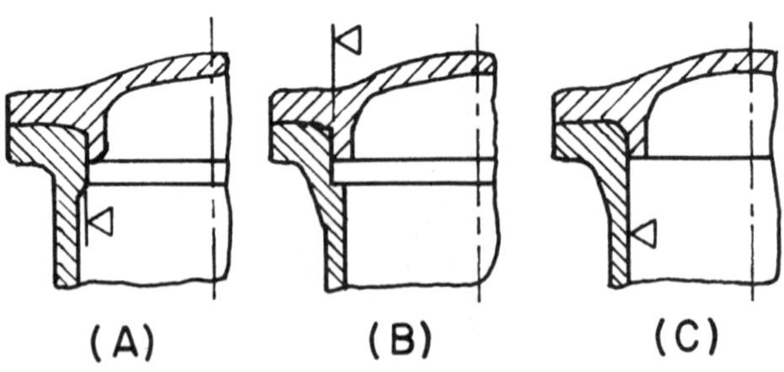

Figure 4.68. Versions of boring of a vessel for centering the head: (A,B) are good designs; (C) poor design.

As already noted, a part must be accessible for machining. For example, it is impossible to drill a hole for a screw in a pulley hub (Figure 4.69) since the pulley rim prevents this. Therefore, it is necessary to drill an angular hole, or to drill the pulley and then a hub (Figure 4.69).

The shape of a part must not make the task of machining difficult. A drill should be started wherever possible on a surface perpendicular to the axis of the hole to be drilled. The surface at the drill exit should be perpendicular to prevent the tool from breaking (Figure 4.70).

Holes should be drilled through. Where blind holes are used, sufficient thickness should be left at the bottom of the hole on the opposite side of the wall. When using ordinary drills and drilling methods, a depth equal to 6–8 times the diameter of the drill is permissible. With the use of special drills, the depth of drilling can be increased. It is recommended that half-finished parts have a shape providing minimum drilling depth (Figure 4.71).

Internal threads for steel bolts and screws subject to a tension load should provide an engagement equal to the thread diameter (1½ times the thread diameter for cast iron or brass, and 2 times the thread diameter for aluminum or zinc). In threads tapped in blind holes (Figure 4.72), the depth of the hole should be deeper than the depth of the thread by a distance of at least 5 times the pitch of the thread. A thread relief should be provided in the hole with a width that is at least 3 times the thread pitch, and a diameter that exceeds the maximum major diameter. The thread engagement for standard studs should be at least 1.25 times the diameter of the thread for steel, 1.5 times the diameter for cast iron, and as much as 2.5 times the diameter for other materials. It is not recommended to align the parts by studs because the accuracy of stud fabrication is lower than that of smooth

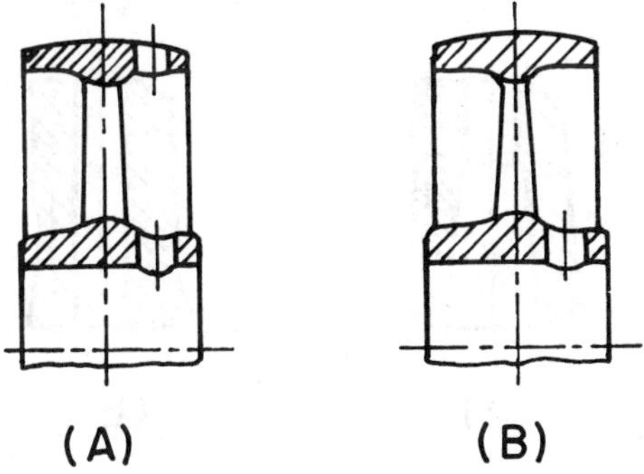

Figure 4.69. Drilling on the pulley hub: (A) good design; (B) poor design.

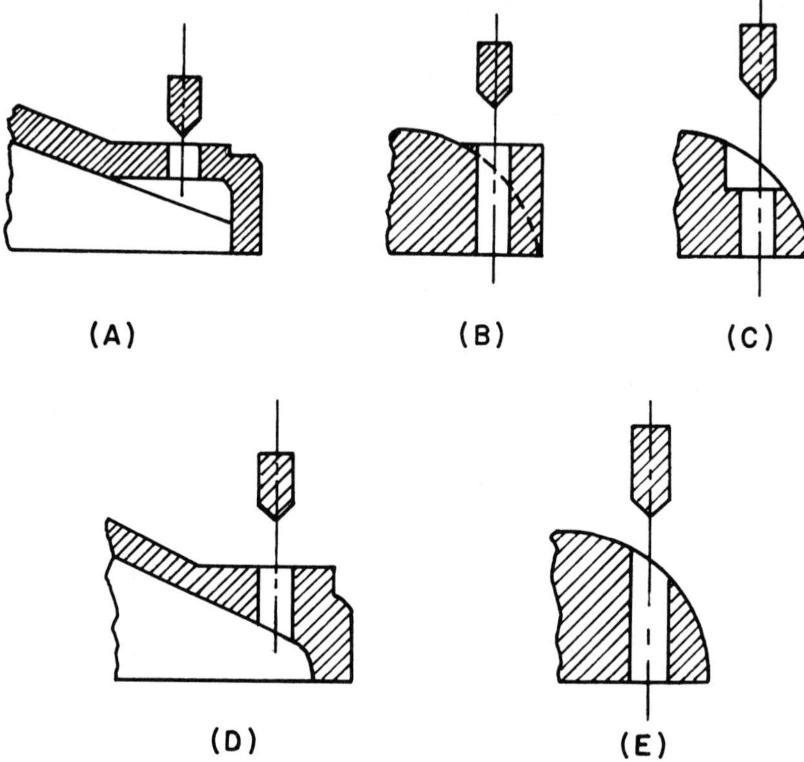

Figure 4.70. Examples of drilling schemes: (A,B,D) good designs; (C,E) poor designs.

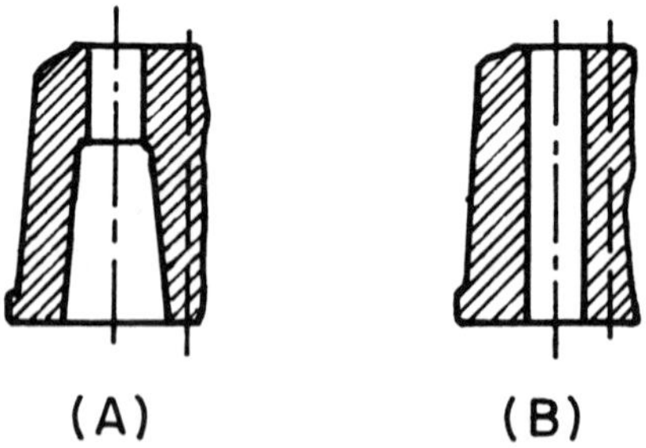

Figure 4.71. Decreasing the drilling depth: (A) good design; (B) poor design.

finished parts; thus the alignment must be done by a special cylindrical driving band (Figure 4.73).

Precise holes made with a reamer should be done all the way through and be provided with an entering chamfer. Shaped surfaces, which must be machined with a master form or pattern gauge, should be provided only in extreme need. It is better to use conventional methods of machining (e.g., milling and turning).

A part to be produced by milling must be rigid enough to withstand the cutting force without deflection. The part should be designed so that the maximum number of surfaces can be machined in one pass. Removal of the part and placing it in another position will take more time and allow opportunity for error.

All internal and fillet radii should be the same size on parts produced with an end mill. The fillet radius produced by the radius on the corner of the cutter should not be too large as it can interfere with the cutting action. Parts must be so designed that the portion produced by the bottom of the

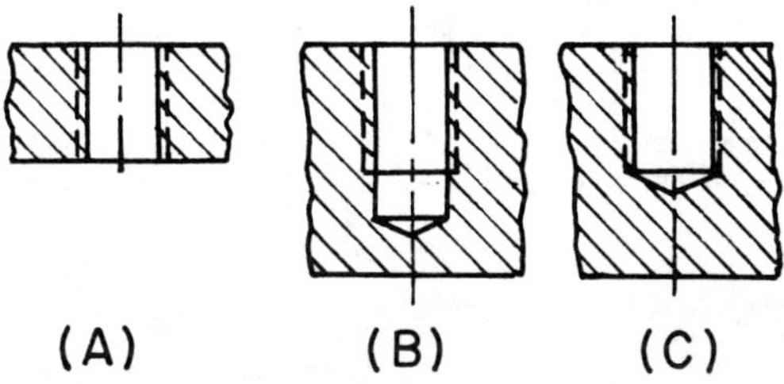

Figure 4.72. Hole tapped for a stud: (A,B) good designs; (C) poor design.

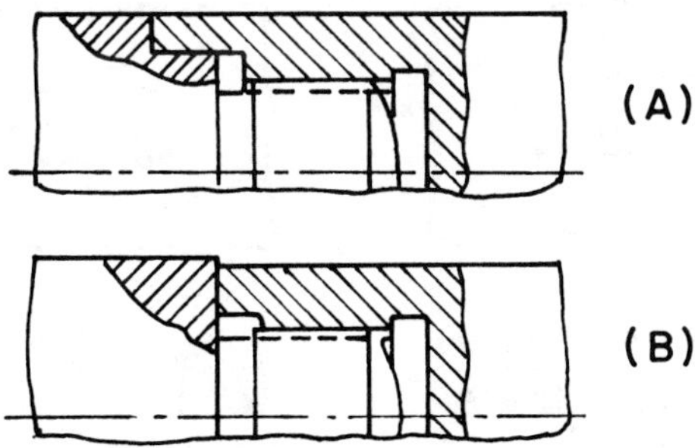

Figure 4.73. Example of threads that do not provide alignment: (A) alignment with a driving band, good design; (B) alignment by a thread, poor design.

end mill is at a 90° angle to that produced by the side, as shown in Figure 4.74 (A).

Slots should be made with a regular milling cutter (Figure 4.75 (A)) rather than with an end mill (Figure 4.75 (B)) because such slots are more accurate and cheaper.

Turning is a widespread method for parts fabrication. It is versatile with respect to both the operations that can be performed and the size of the parts that can be produced. Parts that have a length in excess of 15 times their diameter require special precautions because of the deflection caused by the force of the cutting. The concentricity of the various machined

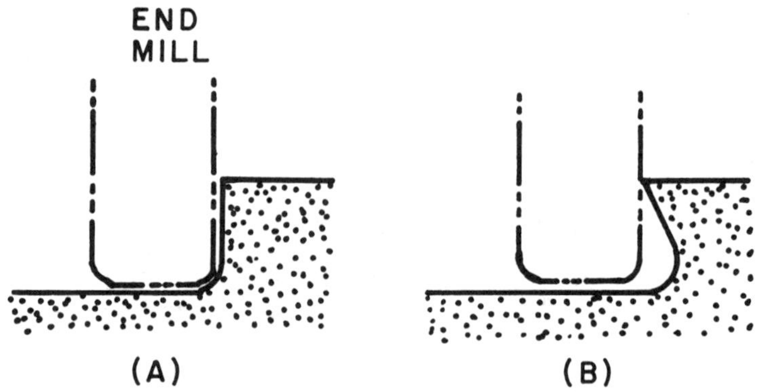

Figure 4.74. (A) Preferred configuration when using an end mill; (B) unsatisfactory configuration.

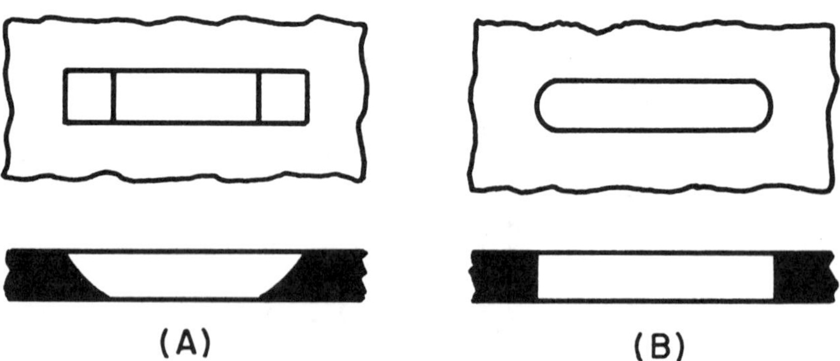

Figure 4.75. (A) Preferred slot; (B) unsatisfactory slot.

diameters can be held closer when they are all machined in a single setup. All fillet radii that can be made by the same cutting tool should have the same dimension. Corners should have a chamfer rather than a radius.

Shafts and other parts bounded by surfaces of revolution should be made smooth, without sharp changes in diameters. In the finished shafts, the adjacent diameters should not differ by more than 3–5 mm. Square ends should be avoided on shafts (preferring in this case the use of keys). If there is a need for coaxial collars of large diameters, removable thrust rings (Figure 4.76) are recommended as opposed to turning them from a thick round blank.

The grooves and slots should be designed so that they can be machined with the use of an engine or vertical lathe, as well as with a milling machine. As far as is practicable, the depth of grooves and slots should be located in the same plane. For fulfillment, for example, of the last condition, the transverse slot depth in tube plates should match the depth of the rim bore for a gasket (Figure 4.77).

In designing equipment, we are concerned not only with determining dimensions to assure proper assembly and function, but with providing for tolerances great enough to permit economical manufacture [31-34].

To lower the cost of fabrication, select maximum tolerances (the total variations that are permitted) and a minimum degree of surface finish. Wherever possible, the surface should be left unfinished. It should never be machined to present a better appearance because the increase in class of accuracy and degree of surface finish are related to the cost of equipment.

Surfaces of a part may be divided in two groups: functional and nonfunctional. The nonfunctional surfaces are those that usually do not come in

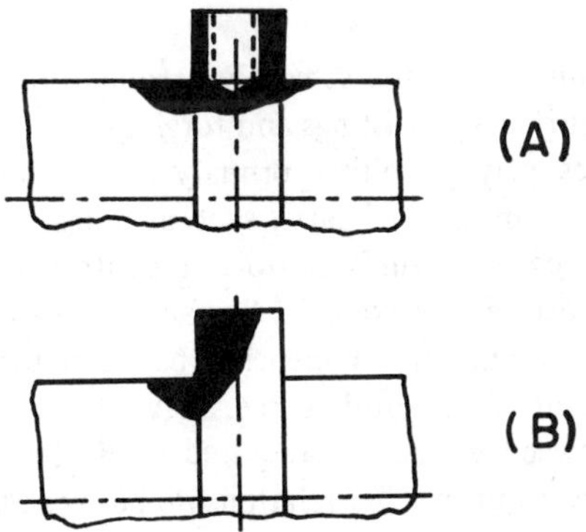

Figure 4.76. (A) Solid and (B) removable collars.

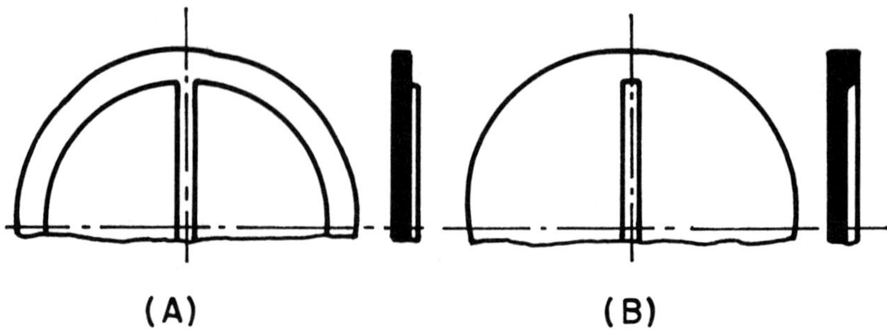

Figure 4.77. Slot boring for a gasket in a tube plate: (A) good deisgn; (B) poor design.

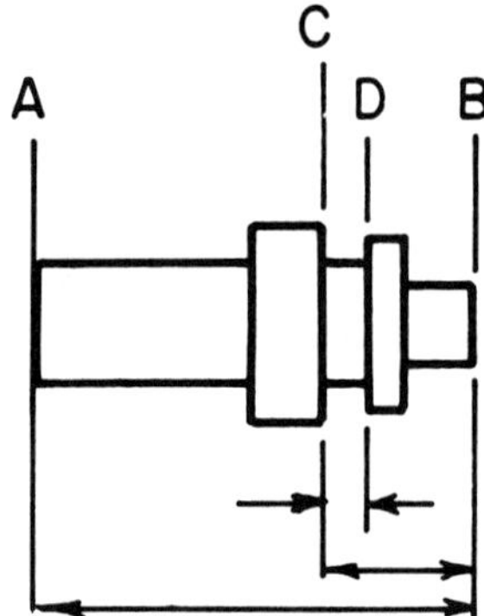

Figure 4.78. Relation of dimensions.

contact with other surfaces in the assembly or operation of the machine, for example, the rough surfaces of castings and forgings.

Functional surfaces may be either primary or secondary: the first ones control the position of motion of parts, such as holes, which have locations in two directions. Secondary surfaces do not control position as a primary surface, such as a spotface, for example. Surfaces have a direct or first-degree relationship if there is only one dimension between them; for example, a second degree, as A and C, and third degree, as A and D in Figure 4.78.

A relationship between surfaces is required when they are functional surfaces. One should try to utilize the relationship between surfaces that results from fabricating operations for primary functional surfaces that require an accurate relationship.

Whenever a close relationship must be maintained between two surfaces, they should be directly related. If they have an indirect relationship, several dimensions must be held much more accurately than would otherwise be necessary. For example, in Figure 4.79 dimension C should be used rather than A and B if the size of the slot must be strictly maintained. If the width of slot C is critical and must be accurate, dimensions A, B or D do not necessarily have to be closely held.

Parts fabrication to exact dimensions is not economically feasible; in most instances it is necessary to keep only a few dimensions strictly held. The tolerance must have an acceptable magnitude of variation from the exact size. Because tolerance has a great effect on cost, one must be certain that a large tolerance will not cause difficulty in assembly or operation.

Tolerances may be specified in several ways (Figure 4.80). If the tolerance were not necessary, it would be referred to as the design size (Figure 4.80(A)). The unilateral system (tolerance all in one direction) is illustrated in Figure 4.80(B); the bilateral system (tolerance allowed in both directions) in Figure 4.80(C); and in Figure 4.80(D) the tolerance is indicated by specifying the maximum and minimum sizes that are acceptable. The method shown in Figure 4.80(C) may be used to indicate the desired size.

The interpretation of the dimension (Figure 4.81(A)) allows the conditions in Figure 4.81(B) and (C) as variation in size, but not the lack of parallelism (Figure 4.81(B)) or the lack of flatness (Figure 4.81(C)).

Geometric tolerances applied regardless of feature size are used to avoid extremely close tolerances. In such a system a note may specify that a surface is to be "flat within 0.002 total." The interpretation is shown in Figure 4.82(A). The entire surface must lie between two parallel planes 0.002 in. apart. To avoid this condition (Figure 4.81(B)), a tolerance on parallelism (Figure 4.82(B)) would be used, the interpretation of which is

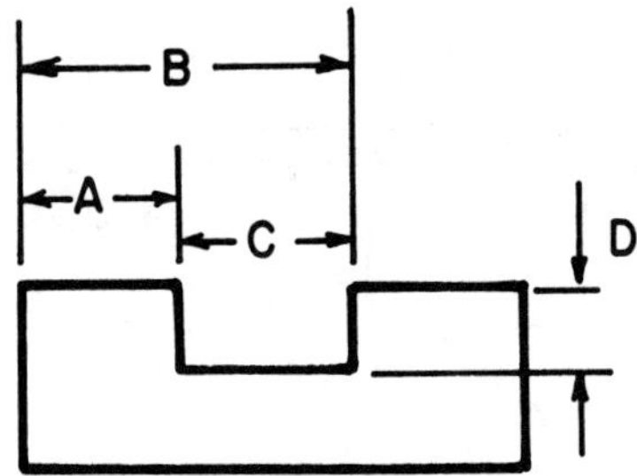

Figure 4.79. Effect of dimensioning on tolerance required.

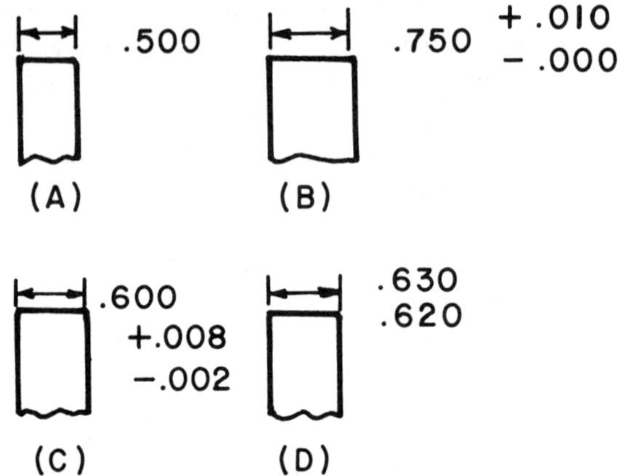

Figure 4.80. Tolerance systems.

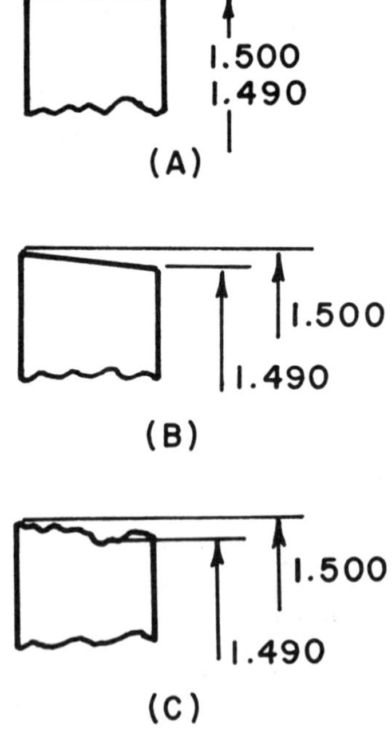

Figure 4.81. Interpretation of a dimension.

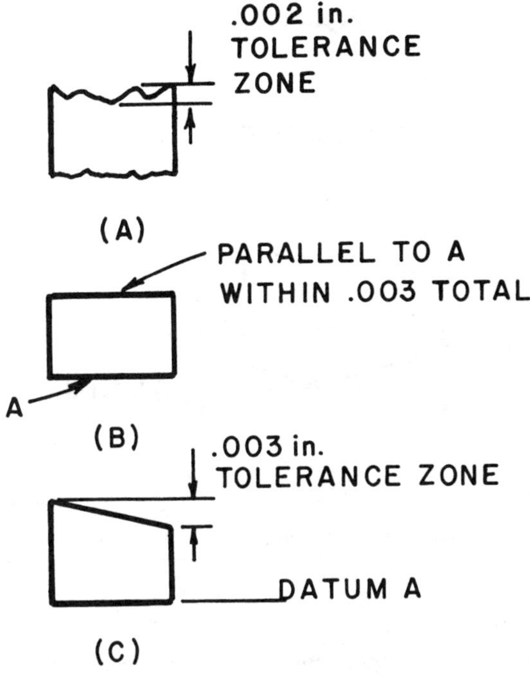

Figure 4.82. Geometric tolerance.

shown in Figure 4.81(C). The surface should be located between two parallel planes 0.003 in. apart and parallel to the datum plane, A, established by the high points of the surface, as defined in Figure 4.83. When a system of dimensions and tolerances (Figure 4.84(A)) is used for locating several holes, it is possible to obtain the situation shown in Figure 4.84(B), where two extreme (tolerance is exaggerated) positions of the upper left-hand hole are presented by the two dotted circles. Following this figure, one must make allowance for the size of the holes required, which will be rather involved. Consequently, the hole will have to be oversized by a considerable amount if reasonable tolerances are to be permitted.

The design may permit considerable tolerance in the initial location of a part attached by bolts in the holes. Once located, however, it has to be limited to only a small variation from that position. In such a situation the method shown is obviously undersizable, and it can be avoided by using true-position dimensioning when the dimensions locating the holes have no tolerance. The holes then are permitted to vary from this exact location by

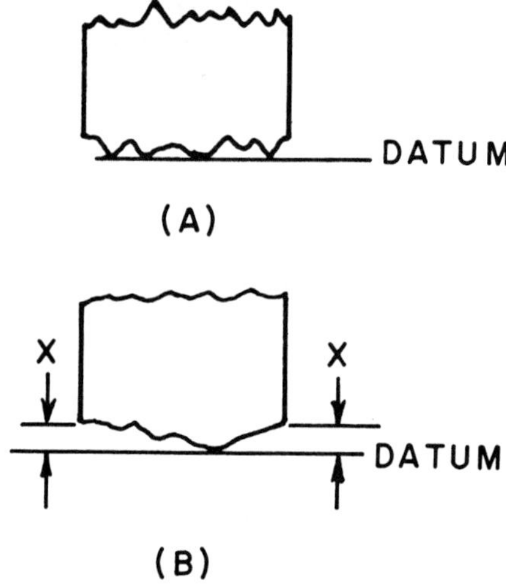

Figure 4.83. Interpretation of datum plane.

a specified amount, as shown in Figure 4.85. In this figure the shaded circles denote a tolerance zone of 0.010 diameter, within which the center of the hole has to be located and the dotted circles show the two extreme positions of the upper left-hand hole. Dealing with a true position when the holes are smallest, assembly would be possible with larger holes when the distance between them exceeds the allowed tolerance. Generally, this additional tolerance is acceptable, and the true position is considered to apply to the maximum material condition (MMC). But if this additional tolerance is not to be permitted, the notation RFS is used, which means that the true position is to be held regardless of a feature size. Locating holes in parts that must be assembled also may be done by placing a note on each set of holes stating that one must match the other. With this method one can achieve the desired result with less effort on the part of fabricating. Critical portions of the tooling of both parts can be made simultaneously by assuring proper matching without having to work to very close tolerances.

A layout should be carefully drawn to scale showing the mean dimensions and locations in the final stages of a design. The layout illustrates a clearance where required, and if the parts were made to the mean dimension the actual clearance would be close to that shown. The allowed tolerances may combine to create an interference. Tolerance also can combine to create an excessive clearance. Both these unsatisfactory conditions are referred to as tolerance stacks. Calculations should be made wherever an interference or excessive clearance appears possible to determine whether a tolerance stack exists.

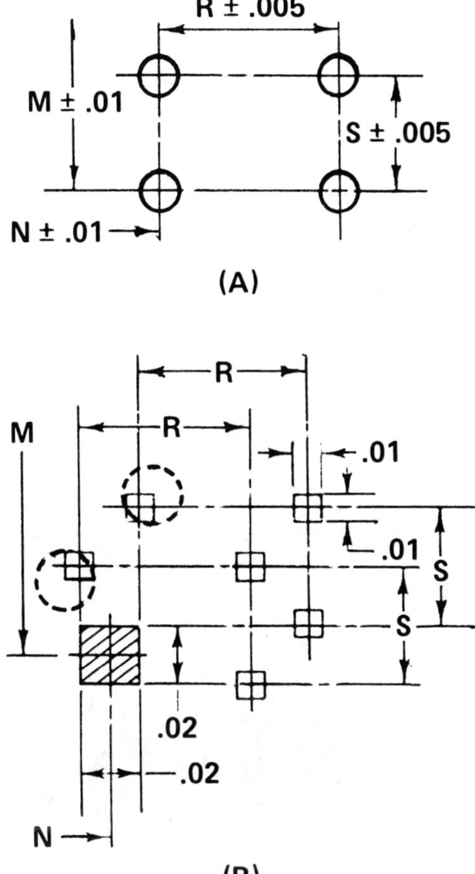

Figure 4.84. Interpretation of location dimensions.

Parr [35] gives a procedure that facilitates such calculations. It is explained below using Figure 4.86.

1. After checking the layout, it seems that there could be an interference between the gear and the loss.

2. The dimensions are changed, where necessary, to the form with a balance bilateral tolerance. The source of these dimensions are the detail drawings of the various parts.

3. The dimension desired is that between surfaces H and E. To obtain this it will be necessary to trace the combination of dimensions that determine it: H to B, B to A, A to C, C to D, D to G, G to F, and F to E.

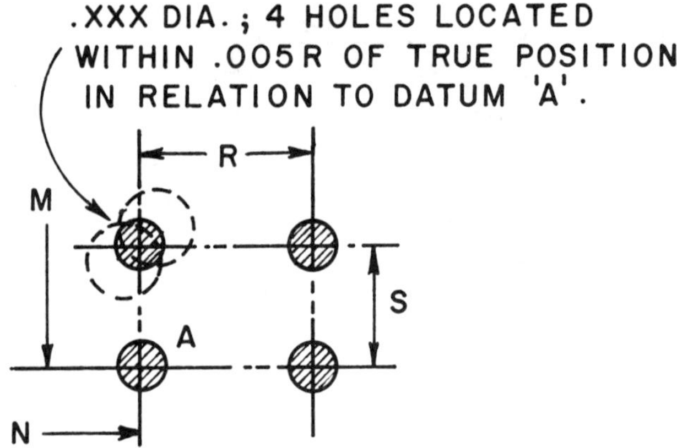

Figure 4.85. True-position dimensioning.

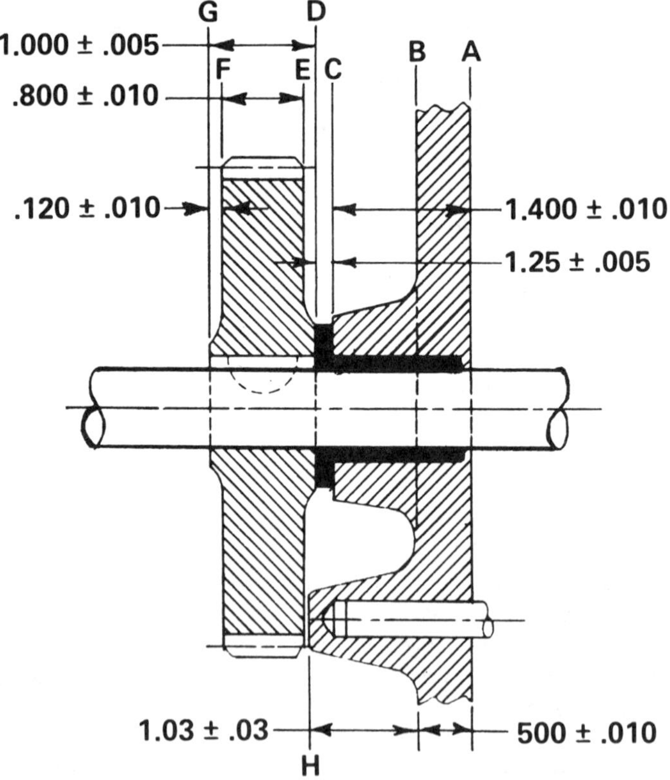

Figure 4.86. A tolerance stack situation.

4. In the above tracing, if the apparent movement is to the right the dimension is considered positive; if it is to the left, it is considered negative. Thus, H-B and B-A are positive; A-C and C-D are negative.

5. A table such as Table 4.10 is prepared. In the first column, the dimension is identified by part description or letters as used here.

6. The dimension H-B is to the right. Therefore, it is placed in the (+) column. The tolerance is placed in the tolerance column. B-A is to the right. A-C is to the left; therefore, it is placed in the (−) column, and its tolerance in the tolerance column. This continues to F-E.

7. The figures in the three columns are added.

8. The negative total is larger, indicating a clearance. The amount is equal to the difference between the negative total and positive total, or 0.075. This is the mean clearance. Figure 4.87(A) will help clarify this.

9. The total tolerance is applied to the mean clearance. The (+) tolerance will decrease the clearance (Figure 4.87(B)). And the (−) tolerance will increase the clearance (Figure 4.87(C)).

10. The maximum clearance will be 0.16 and the "minimum clearance" will be an interference of 0.005. Thus, there is a tolerance stack between the gear and the boss. This unsatisfactory condition can be corrected in several ways. The most obvious and easiest would be to reduce the height of the boss. Assume that this is not possible, that the gear must remain in the position shown, and that the tolerances are not to be reduced. There is probably

Table 4.10. Tolerance Stack

Identification	Dimension +	Dimension −	Tolerance ±
H-B	1.030		0.030
B-A	0.500		0.010
A-C		1.400	0.010
C-D		0.125	0.005
D-G		1.000	0.005
G-F	0.120		0.010
F-E	0.800		0.010
	+2.450	−2.525	±0.080

Mean Clearance	Interference	Maximum Clearance
−2.525	−0.075	−0.075
+2.450	+0.080	−0.080
−0.075	+0.005	−0.155

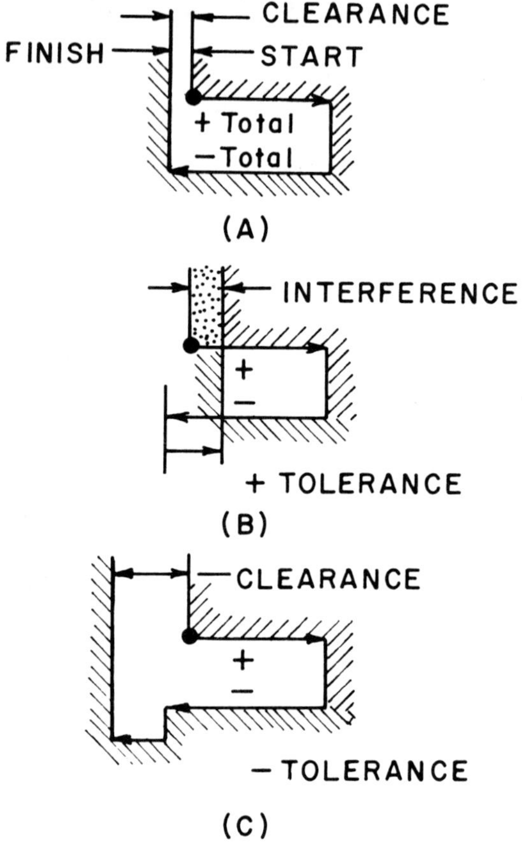

Figure 4.87. Tolerance accumulation.

no reason for the gear being dimensioned in the manner shown; that is, another method of dimensioning that would produce essentially the same part would be acceptable. The mean distance between surfaces D and E is 1.00 − (0.800 + 0.120) = 0.080. If this dimension were used rather than the 0.120 between F and G, and the tolerance were the same as for the 0.120 dimension, the calculations for possible tolerance stack would be as shown in Table 4.11. The minimum clearance is 0.010. There is no tolerance stack. This clearly shows the importance of carefully considering the manner in which a part is dimensioned.

In any piece of process equipment there are certain relations between the dimensions of the component parts that are essential if the equipment is to

Table 4.11. Tolerance Stack

Identification	Dimension +	Dimension −	Tolerance ±
H-B	1.030		0.030
B-A	0.500		0.010
A-C		1.400	0.010
C-D		0.125	0.005
D-E		1.080	0.010
	+1.530	−1.605	±0.065

Mean Clearance	Minimum Clearance	Maximum Clearance
−1.605	−0.075	−0.075
+1.530	+0.065	−0.065
−0.075	−0.010	−0.140

function properly. In some cases in the mechanical design of process equipment, the parts must have a certain amount of looseness, or play, between them when assembled. In other cases the parts must be so made that force must be used to assemble them. In a mixer, for example, the shaft should rotate freely, but without any perceptible shaking in the bushings. On the other hand, the bushings should fit tightly in the frame. The difference in size between the bore of the bushing and the diameter of the shaft is referred to as clearance; the difference between the outer diameter of the bushing and the hole in the frame is referred to as interference, since the bushing is larger than the hole. Both of these intentional differences in sizes of mating parts are termed allowances. Clearance is a positive, and interference a negative, allowance.

The American Standards Association has adopted eight classifications of fits (ASA Bulletin B 4a-1925). A class 1 loose fit has a large allowance; it provides for considerable freedom where accuracy is not essential. A class 2 free fit has a liberal positive allowance and is used as a running fit for speeds higher than 600 rpm. A class 3 fit has a medium positive allowance and is used for more accurate machine tools and automotive parts. A class 4 snag fit has a zero allowance and necessitates considerable precision; it is the closest fit that can be assembled by hand. A class 5 wringing fit has a zero to negative allowance and gives practically metal-to-metal contact. A class 6 tight fit has slight negative allowance, requires light pressure to assemble and is used for gears, pulleys and extremely long contacts. A class 7 medium force fit has negative allowance; it requires considerable pressure to assemble and the parts are considered permanently assembled. A class 8 heavy force and shrink fit has considerable negative allowance and is used for press fits in steel or for shrink fits where heavy force fits are impractical.

There are four classes of screw thread fits (ASA Bulletin B 1.1 1935). A class 1 fit is quite loose and is recommended where shake or play is not objectionable, and where rapid assembly is desired. A class 2 fit is used for a great bulk of commercial screw threaded products, and represent a high quality fit. Classes 3 and 4 represent extremely close fits, and are used only where freedom from vibration and play are absolutely essential.

The tolerance on interchangeable parts has a variation in the amount of clearance or interference. The maximum material condition takes place when both the shaft and the hole have the maximum amount of material—when the shaft is at its largest size and the hole is at its smallest size. The minimum material condition is the reverse—the shaft is at its smallest size and the hole is at its largest size.

For either a clearance fit or an interference fit, the tightest fit will be the maximum material condition, and the loosest fit will be the minimum material conditions. Thus, the difference between the two diameters at maximum material condition determines the allowance. The following example illustrates clearance setting.

Example 4.5

A shaft is to fit in a hole reamed to 1.4995/1.5005. The minimum clearance is to be 0.0005, and the tolerance on the shaft is ±0.001. We wish to determine the size of the shaft and the maximum clearance.

Solution

The minimum clearance will occur when the hole is its smallest dimension and shaft is its largest dimension. The largest shaft dimension is, therefore, the smallest hole dimension minus the minimum clearance, that is, 1.4995 − 0.0005 = 1.4990. The smallest shaft will be 0.002 less than the largest. Thus, the shaft size is 1.499/1.497. The maximum clearance will occur with the smallest shaft and the largest hole. Thus, 1.5005 − 1.4970 = 0.0035.

To reduce costs when determining the sizes involved in a fit, it is necessary to consider the use of standard sizes. For various reasons, such as providing shoulders for location, the shaft requires machining. In this case the size of the hole should be such that it can be fabricated with standard tools. The size of the shaft should be computed to obtain the desired allowance. If the shafts do not require machining for reasons other than to provide the proper fit, it is convenient to use a standard shaft and to calculate the size of the hole to provide the required allowance.

It should be remembered that when a clearance fit is used, the center of the hole and the center of the shaft will never coincide when in operation.

When using a setscrew (Figure 4.88) or a tapered key, the actual eccentricity should equal one-half the clearance between the actual parts. The center of the hub has to follow a circular path around the center of the shaft, the radius of which would equal the eccentricity. It is necessary to determine whether the eccentricity of this kind is objectionable and the amount of eccentricity that can be tolerated. After careful consideration of all factors involved, it is possible to use standard values for determining the limits of both the shaft and hole for various types of fits.

The amount of interference depends on the purpose of the fit. Small interference is used for long fits or fits involving thin sections. Large interference fits are applied for permanent assembly or where the fit has to transmit a force. When the design is such that force cannot be used, a shrink fit may be applied by heating the outer part of a hole to increase its size to the extent that the parts can be assembled. After returning to the temperature at which the interference was measured, a fit of that amount of interference will exist. A medium fit should have an average interference of about 0.0005 in. for each inch of hole diameter. For small interference fits half of this amount is used. Twice this amount is appropriate for fits that require a larger interference. In selecting a suitable fit, the tensile stress induced should be less than the yield stress in tension (depending on the nature of the stress, it need not be appreciably less). The following example illustrates determining limits.

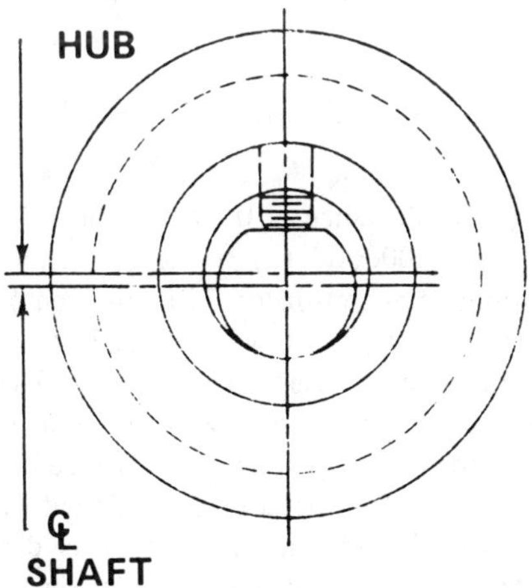

Figure 4.88. Effect of clearance on eccentricity.

Example 4.6

We wish to determine the limits for the shaft and hub, the stress in the hub, and the temperature required for a shrink fit intended for permanent assembly, where the nominal diameter of the shaft is 4 in. and both shaft and hub are steel.

Solution

For a permanent assembly, an appropriate average interference is 0.001 in./in. of diameter, or 0.004. The hole will have to be bored and the shaft ground; therefore, reasonable tolerance for each is 0.0005. Make the hole 4.0000/4.0005. The shaft should then be 4.0045/4.0040. The maximum interference is 0.0045 and the minimum is 0.0035. The average is the 0.004 required. The stress is found from the following equation:

$$\tau_t = \frac{E\delta}{D} = \frac{30{,}000{,}000 \times 0.0045}{4} = 33{,}800 \text{ psi}$$

where τ_t = tensile stress (psi)
E = modulus of elasticity (psi)
δ = interference (in.)
D = nominal diameter of hole (in.)

The maximum interference is used to give the greatest stress. Any steel will have sufficient strength, and some types of iron could be used.

The temperature is found using the following equation:

$$T = \frac{\delta}{D\alpha} = \frac{0.006}{4 \times 0.000007} = 214^\circ\text{F}$$

The 0.006 includes an allowance of 0.0015 to provide for cooling and to facilitate assembly. The 0.000007 (α) is the coefficient of expansion for steel. Assuming a room temperature of 70°F, the temperature of the part need not be greater than 284°F, which will not damage the part.

The interference fits may be used to avoid the inaccuracies caused by clearance, but they also can be the source of inaccuracies. For example, by using a single keyway, the center of the hub is displaced relative to the center of the shaft because there is less metal on the side with the keyway that resists the pressure caused by the interference. This displacement may be avoided by using two keys or by machining a flat on the side opposite the keyway equal in width to the keyway.

Pressing a thin-walled tubular part into a more substantial part reduces the inside diameter of the inner part. Such a reduction can be corrected by machining the inside diameter after assembly. When a portion of the tubular part is pressed into another part, the condition shown in Figure 4.89 results, and correction has to be made by machining the inside diameter after assembly. To avoid inaccuracies caused by interference fits, the interference should be kept to a minimum and the parts should be symmetrical in the area of the fit. Tolerance produced by different manufacturing processes for an average part are presented in Table 4.12. It should be noted that the values given in the Table 4.12 depend on a number of factors: machine condition, material being machined, operator skill and tooling. Figure 4.90 illustrates how cost is

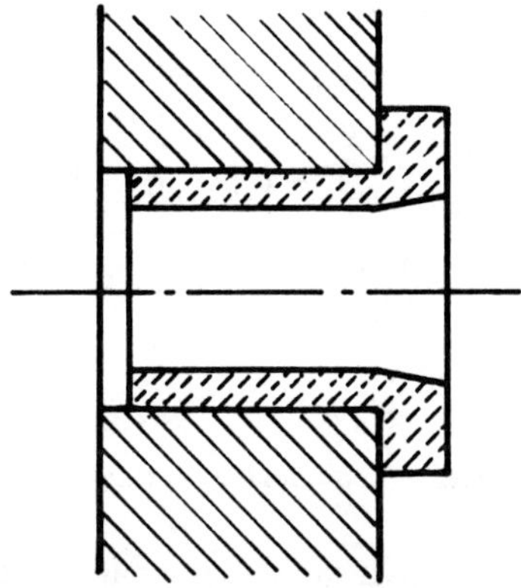

Figure 4.89. Effect of press fit on a part with a thin wall.

Table 4.12. Typical Tolerances for Different Manufacturing Processes [35]

Manufacturing Process	Tolerance
Turning	0.003
Milling	0.004
Shaping	0.005
Planing	0.005
Drilling (0.5 dia.)	0.008
Reaming (0.5 dia.)	0.0015
Boring	0.003
Grinding	0.001
Broaching	0.002
Honing	0.0005
Lapping	0.0001

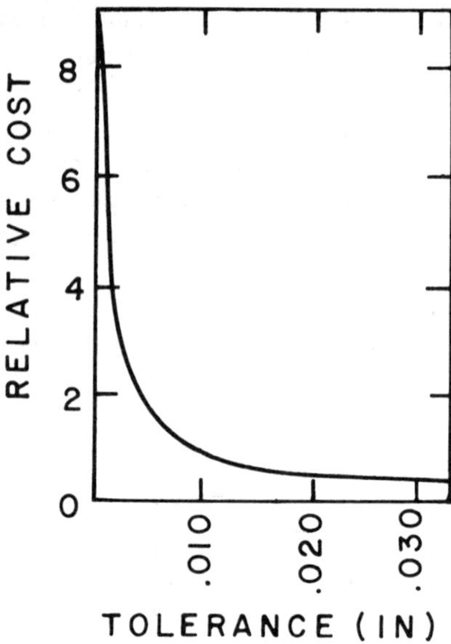

Figure 4.90. Relation between tolerance and cost [35].

affected by tolerance and gives the relative relationship between machining costs and tolerance.

The degree of roughness of part surfaces influences the function of the part regardless of whether it is lubricated. The performance of the mating parts improves as surface roughness decreases. This is especially true for fatigue strength of heat-treated parts, which is greatly increased with increased smoothness. Table 4.13 gives roughness-height values for various manufacturing processes which can be expected in production from an average machine. The minimum allowable surface roughness for a particular application cannot be determined by calculation. However, typical values for various applications may be helpful in obtaining approximations. Typical values are given in Table 4.14.

In maintaining a close tolerance, a smooth surface is produced. A smooth surface requires little more effort to hold a close tolerance also. This relationship can be seen in Figure 4.91. It would be meaningless to demand, for example, a tolerance of 0.002 in. between surfaces whose roughness height was 250 μin. because the peak-to-valley distance on one of the surfaces would be 5 times as great as the tolerance. The approximate relationship of surface roughness to cost is shown in Figure 4.92. The number of different processes involved in producing smooth surfaces, the cost of machines, the

Table 4.13. Typical Surface Roughness of Manufacturing Processes [35]

Roughness (μin.)	Manufacturing Process
	(Machined surfaces)
125	Shaping, planing, drilling
63	Milling, turning, boring
32	Broaching, reaming
16	Grinding
8	Honing
2	Superfinishing, lapping
	(Natural surfaces)
500	Sand cast
250	Forged
63	Die cast
32	Cold drawn

Table 4.14. Typical Applications of Surface Roughness [35]

Surface Roughness (μin.)	
250	Clearance surfaces. Maximum for final machining for quality machine work except large rough parts.
125	Mating surfaces where there is little or no motion between them and where bearing loads are light.
63	Press fits. Rolling surfaces such as cams with rolling followers. Gear-locating faces. Bearings where the motion is intermittent and load is moderate.
32	Gear teeth. Sliding or rotating bearings with moderate to heavy loading.
16	Precision journal bearings. Gear teeth subject to heavy loading. Friction surfaces such as clutch plates.
8	Surfaces for hydraulic dynamic seals. Crank pins and production gauges.
4 and less	Primarily precision measuring instruments.

process time, and the case in handling explain why there is greater cost associated with producing smooth surfaces.

4.3.2 Sheet Materials

Many manufacturing articles of process equipment are made from sheet metal. Such articles are usually lighter in weight and often less expensive

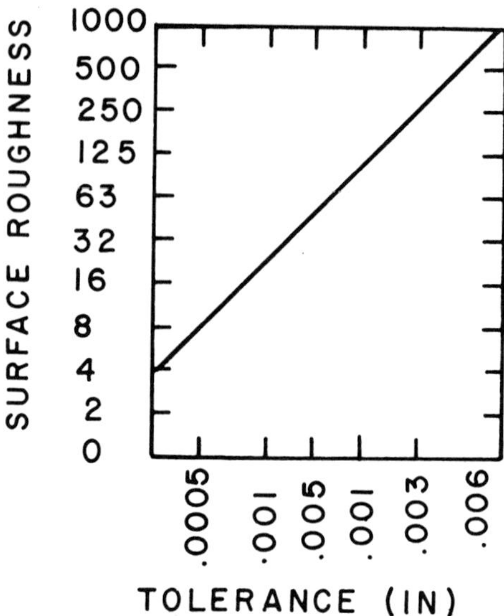

Figure 4.91. Relation between surface roughness and tolerance [35].

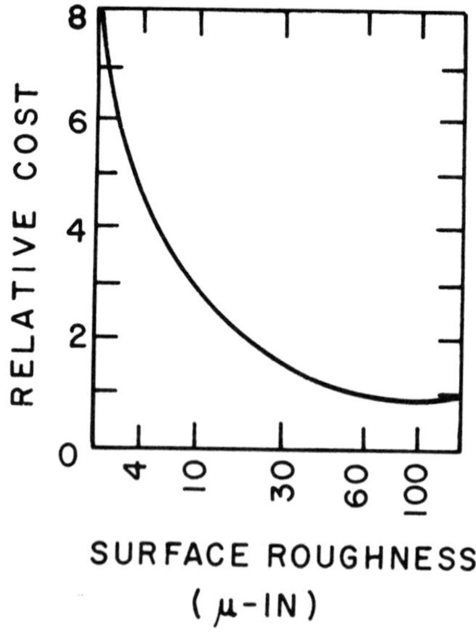

Figure 4.92. Relation between surface roughness and cost [35].

than castings or forgings. Such important parts as shells and heads for vessels, pipes, elbows, hoppers, etc., may be made from sheet iron and steel, galvanized iron, copper, aluminum or brass.

The shape to which the sheet material is cut must be simple. The parts should be cut by power squaring shears and the amount of scrap should be minimized. Inner corners must be avoided where possible as shown in Figure 4.93. A small projection is shown in Figure 4.94.

In general, it is easier to cut a rectangular sheet and weld it to two strips. This will simplify cutting and decrease the amount of scrap. It is possible to obtain parts of a more complicated shape by a cutting torch. It is easier to manufacture circular pipes than square ones.

In sheet bending, elastic and plastic deformation can arise. The minimum bend radius under cold conditions depends on the sheet thickness, S, and is determined as follows:

$$R = K'S \tag{4.10}$$

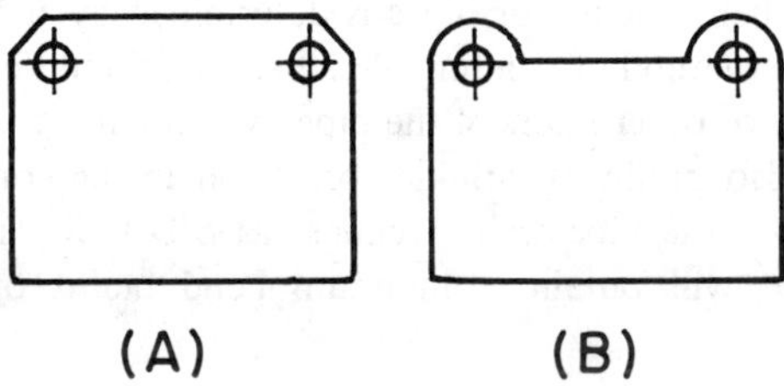

Figure 4.93. Shape of a part cut from sheet: (A) good design; (B) poor design.

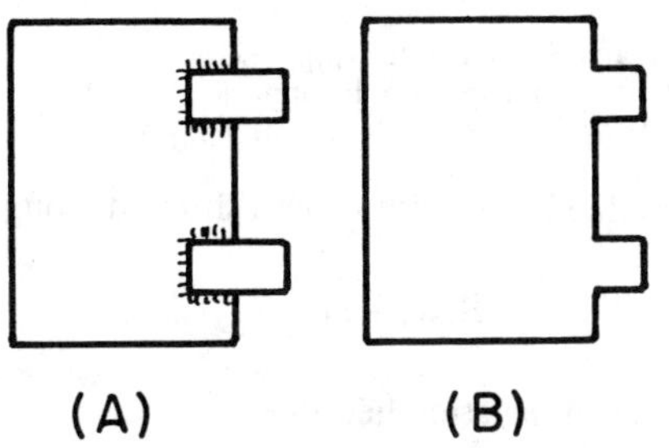

Figure 4.94. Welded strip simplifies the cut: (A) good design; (B) poor design.

Table 4.15. K′-Coefficient Values

Material	Across Fibers	Along Fibers
Steel	0.5	1.2
Brass (Hard)	0.5	1.2
Mild Brass and Aluminum	0.3	0.5
Copper	0.25	0.4

where K' is a coefficient, depending on material and direction of plate rolling. Table 4.15 shows typical values of K'.

If the direction of sheet rolling is unknown, the larger K' value should be selected. Cold bending of sheets with a small radius is accompanied by cold-hardening. It is desirable that bent sheets be heat-treated to improve their structure. To avoid cracking, bends should be made across the grain of the metal. The grain will be parallel to the long edge of the stock. The minimum bend radius in which there is no cold-harden is $R = 205$. Hence, the minimum diameter of a cold-rolled shell should be $D \geqslant 40S$.

The minimum bend radius for pipes is determined by their diameter, wall thickness and the method of bending. Pipe bending should be in the direction of the elongation of outer fibers of the pipe, which tends to decrease the wall thickness. This also produces wrinkle formation in the compressed zone of the pipe. Denoting the pipe wall thickness as $S/D = K'$ and assuming that $R_{min} = 20S$, we will obtain a minimum bend radius of a pipe (Figure 4.95 (A)):

$$R = D\,(20K' + 0.5) \tag{4.10}$$

The wall thickness, S, usually is taken in the limits from 0.04 to 0.1D. To avoid defective bends, use the following minimum bend radius for seamless pipes when fabricating with mandrel:

At $D < 51$ mm and $S > 1.2$ mm, $R_{min} = 2D$
At D from 51 to 73 mm and $S > 1.6$ mm, $R_{min} = 3D$
At D from 37 to 103 mm and $S > 2$ mm, $R_{min} = 4D$

For bending of a welded pipe with a seam situated along neutral line:

$$R_{min} = 6D \tag{4.11}$$

and for a seam situated along stretched side:

$$R_{min} = 9D \tag{4.12}$$

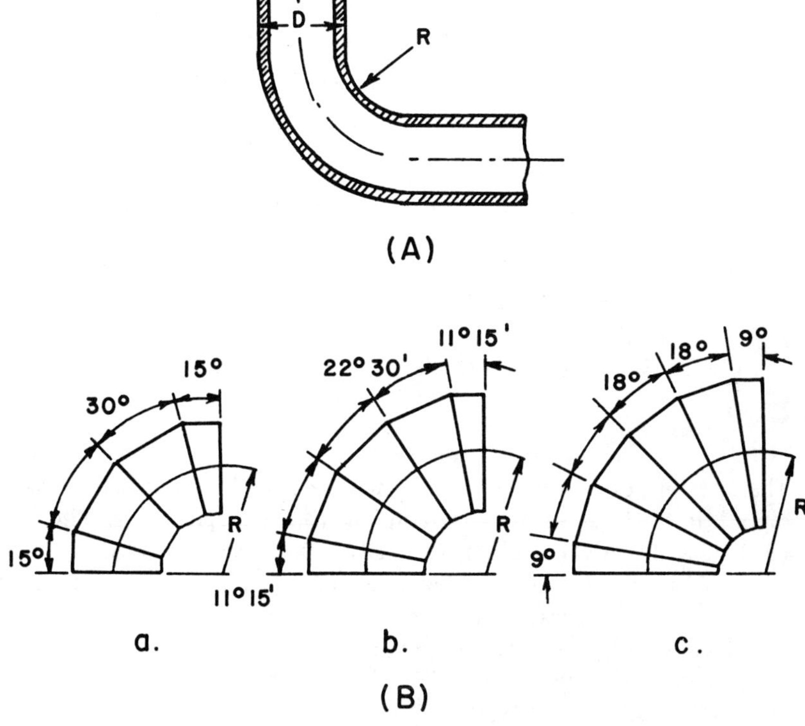

Figure 4.95. (A) Bend of a pipe; (B) normalized welded bends.

The large radius bendings cause an increase in the apparatus' external dimensions. To decrease them use special-shaped and stamped parts with bevelled edges for welding. This improves the design considerably and lowers its price.

Bending large pipes is very difficult, because the sizes of bends become excessively large. Therefore, the bends for large pipes usually are welded (Figure 4.95 (B)).

4.3.3 Assembly/Disassembly Considerations

When designing, consider the influence of assembly and disassembly for the process equipment. There are three reasons for considering assembly: reliability, the cost of initially assembling the parts, and the cost of assembly when maintenance is performed. The design should provide safeguards against incorrect assembly.

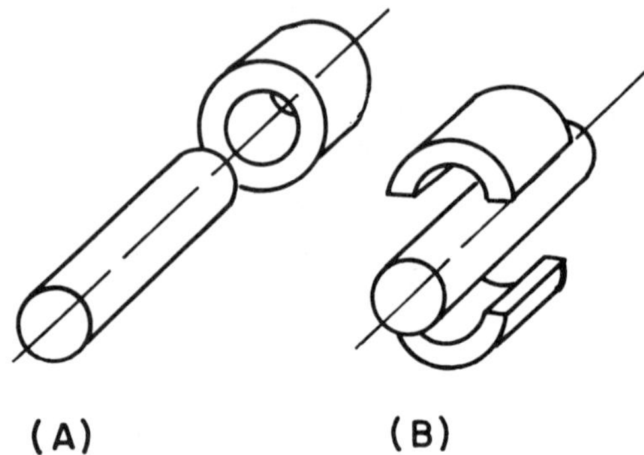

Figure 4.96. Methods of assembly: (A) axial method; (B) radial method.

The assembly of process equipment may be done by the radial and axial methods. In the axial assembly method, one part is slipped on the other and moved along the axis. For example, a ball bearing is slipped on the shaft. In the radial assembly method, an exterior embracing part is provided by a radial parting plane. Because of this there is no need to push the interior part through the exterior one. An example providing radial assembly is a split-sliding bearing. Both these methods of assemblies are shown in the Figure 4.96.

The axial method is convenient because it provides good coaxial alignment of parts and simplifies the design and fabrication. However, the assembly itself to some extent becomes more complicated. The radial assembly method is used when the interior part is very long, heavy and awkward, or when there are some parts that prevent disassembly. All parts assembled by the axial method must be slipped on the covered part from one side.

It is always desirable to design equipment so that it would be possible to preassemble and adjust subassemblies and sets (for example, a bubble-cup tray) and then couple them together. In so doing, it is important to keep the number of "type-size" parts used in the apparatus as small as possible. This is convenient both for assembling and for decreasing the amount of spare parts needed. Standard parts, such as bolts, nuts and screws should be used wherever possible because drawings, tooling and manufacturing paper work are saved, as well as assembly and maintenance operation.

It should be noted that hands and tools are required to accomplish assembly and hence sufficient access to components must be provided. The distance between adjacent nuts, or nuts and walls, for example, must be sufficient to allow the nut to be put on and to apply a wrench to it (Figure 4.97).

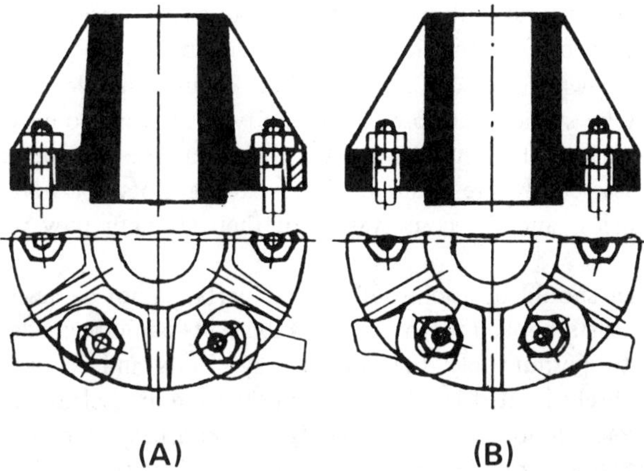

Figure 4.97. Relative position of bolts: (A) good design; (B) poor design because it does not consider the casting draft.

If bolts are used it is necessary to provide a support for a bolt head. For example, it is important to insert and support the bolt in the opening in a frame (Figure 4.98); the high washer makes it easier to reach the bolt head.

Free access to all holders and fasteners, as well as to the parts that must be aligned during operation (for example, stuffing boxes), should be provided. The closeness of adjacent parts or the configuration of apparatus parts may

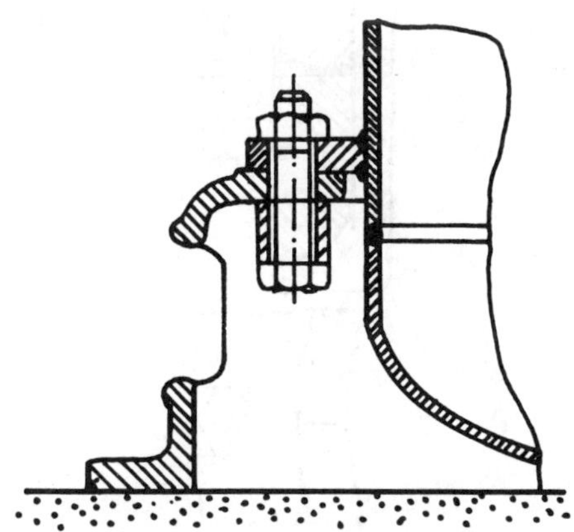

Figure 4.98. Opening in a frame for bolt inserting and support.

prevent insertion and withdrawal of bolts. In such cases, slots are provided and bolts are inserted in the slots from the side (Figure 4.99).

Bolt withdrawal from such a slot need not require complete removal of the nut. It may be sufficient to provide one to two turns and to move the bolt in a radial direction. This withdrawal often is used in cases in which rapid assembly and disassembly are desired. However, the slots for bolt withdrawal weaken flanges, and for conservation of their strength they must be reinforced.

The design should be foolproof so that parts cannot be assembled incorrectly. For parts hooked up in two ways the bolts could be made of different diameters to permit only the proper way. Bolts and screws should not require a closely limited length to provide the required clearance, especially where the end cannot be checked. Designs should be avoided where the axes of fasteners are not parallel to each other (Figure 4.100).

The design of assemblies and subassemblies must provide a simple and convenient assembly. This possibility has to be checked with great care. The drain nozzle of an apparatus (Figure 4.101) is made removable to allow for apparatus disassembly. However, the dimensions chosen do not permit the job to be done because of the bent part. To do it, one needs to make either the nozzle diameter larger, or to increase the bend radius of the pipe to withdraw it through the nozzle. The parts (such as shafts, sleeves, glands of stuffing boxes, sealing rings, etc.) that are withdrawn must be so designed to

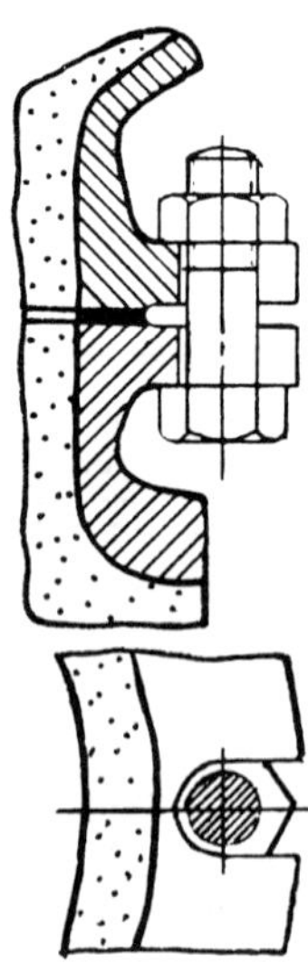

Figure 4.99. Inserted bolts.

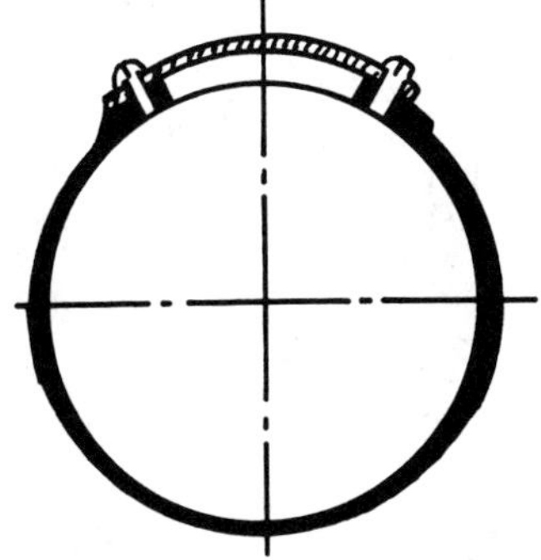

Figure 4.100. Example of a design with nonparallel axes of fasteners.

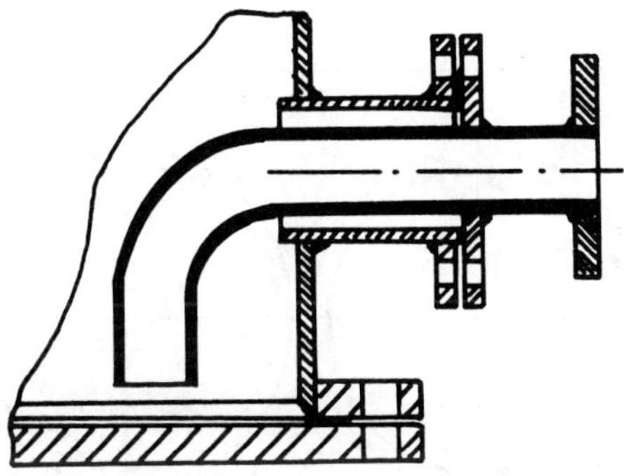

Figure 4.101. A drain nozzle.

be conveniently clamped when disassembled. Threaded holes for screws and studs should be provided that will simplify the withdrawal process. Pins must be so designed to be easily detached by a screw or be knocked out.

Proper (Figure 4.102(A)) and poor designs (Figure 4.102(B)) of stuffing boxes are shown. The gap formed by conical machining of the sleeve provides for the insertion of a screwdriver, allowing easy detachment of the gland. The proper example (Figure 4.103) is a spring catch lock of a small apparatus. The spring catch lock is made as a split snap elastic ring. It can be inserted easily into a groove, but withdrawal is very difficult. Simply bending the ring ends will provide easy withdrawal. For reasons of disassembly, it is necessary to insert a ball bearing with an interference fit. It is sufficient to use a push fit. In the latter case the ball bearing withdrawal is much easier.

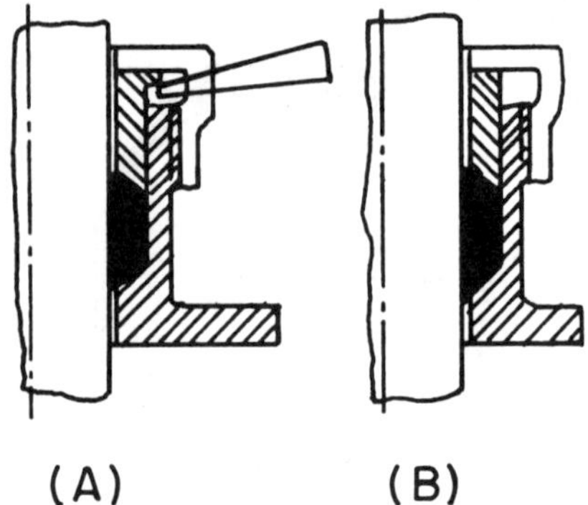

Figure 4.102. The design of small stuffing boxes: (A) good design; (B) poor design because the gland withdrawal is difficult.

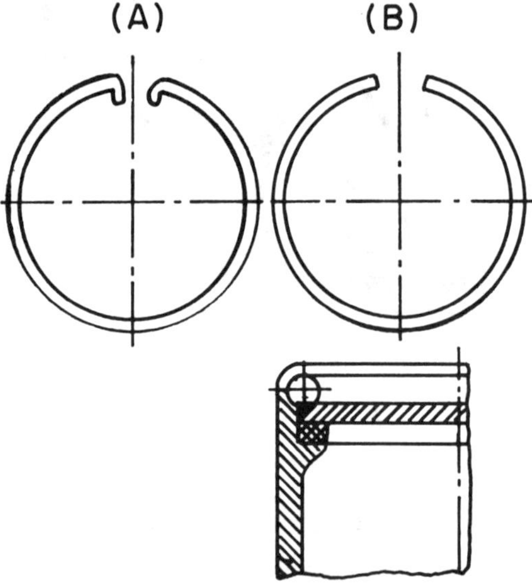

Figure 4.103. Spring catch lock for a head: (A) good design; (B) poor design.

The shape of adjacent parts must simplify the assembly: the mate part must be bevelled to simplify the assembly (Figure 4.104). The sizes and shape of chambers and fillets must be so designed that the parts cannot be pinched. For low-stressed parts, it is possible to use undercuts (Figure 4.105(D)).

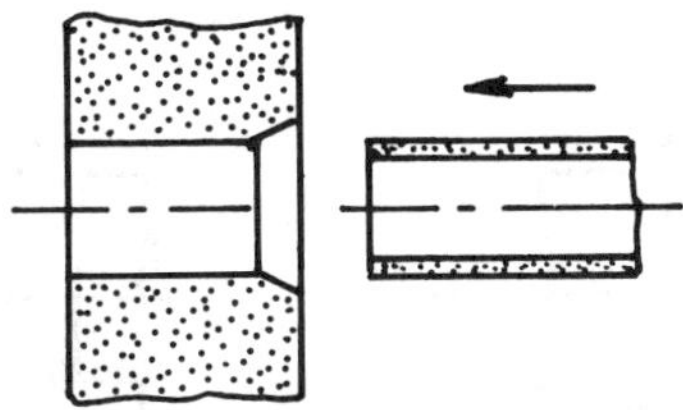

Figure 4.104. Chamfer on the tube plate to simplify the assembly.

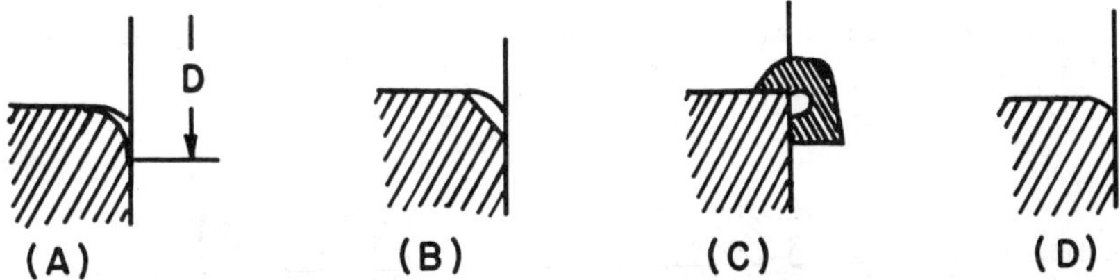

Figure 4.105. Mate parts: (A,B,C) good designs; (D) poor design.

The design must provide a good alignment when assembled. The alignment of bolted parts are made either by a centering projection or by register pins. It is important to provide parts that are mated only in one surface, that is, statically determinate. Extra mate surfaces complicate the fitting and the design. Figure 4.106(A) gives an example in which it is difficult to machine the shaft end of the collar and cone, which are supported simultaneously with the force against the boss. A proper design is shown in Figure 4.106(B).

Another example of a statically determinate design is shown in Figure 4.107(A). In the proper design the full gasket strain should be left as a gap providing the possibility of further tightening of the connection. In version (B), (Figure 4.107(B)), further gasket tightening is impossible. The statically indeterminate designs are sensitive to temperature variations.

The lengthened parts must be fixed only on one end to provide for free expansion of the other one. To simplify an axial assembly, it is desirable that the extensions on the interior part enter the exterior one, not simultaneously but stepwise. In designing, one must keep in mind that disassembling an apparatus, especially in a functioning unit, is much more difficult than assembling on the construction site, where special devices and tools are required.

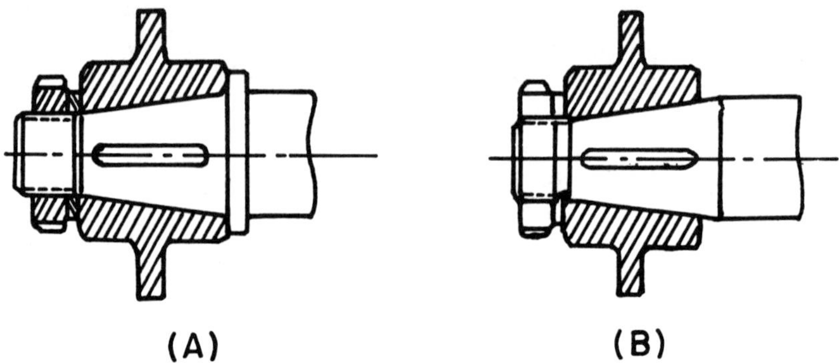

Figure 4.106. The fit of a boss on a conical shaft: (A) poor design; (B) good design.

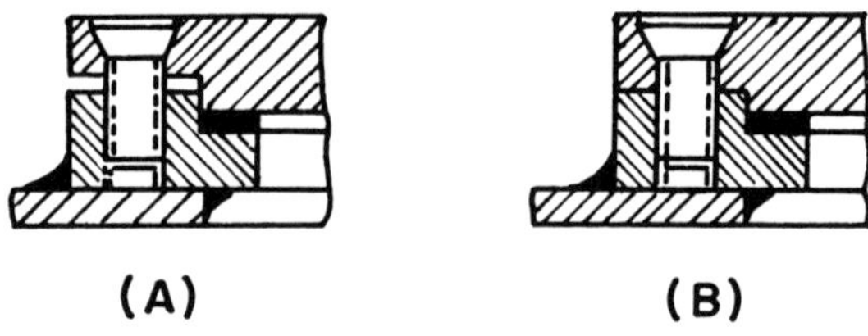

Figure 4.107. Assembly of a blank: (A) good design; (B) poor design.

NOMENCLATURE

A	cross-sectional area (in.^{2})
a, b	dimensions (see Figure 4.60) (in.)
D	diameter (in.)
d	distance or dimension (in.)
E	modulus of elasticity (psi)
F	force (lb_f)
h	height (in.)
I	moment of inertia (in.^{4})
K	maximum fillet size (in.)
K′	coefficient (see Equation 4.10)
L	length (in.)
M	bending moment (lb_f-in.)
MMC	maximum material condition

N casting size (see Equation 4.2) (m)
n safety factor
R bend radius (in.)
r junction radius (in.)
S, S′ thickness (in.)
T temperature (°F)
W weight (lb_m or kg)
Z section modulus ($in.^3$)
α coefficient of thermal expansion (in./°C)
δ interference (in.)
τ_t tensile stress (psi)

REFERENCES

1. Angus, H. T. *Cast Iron: Physical and Engineering Properties* (London: Butterworths, 1976).
2. Lux, B., et al., eds. *The Metallurgy of Cast Iron* (Switzerland: Georgi Publishers, Inc., 1977).
3. Clarke, C. D. *Molding and Casting: Its Technique and Application,* Standard Arts, Maryland (1972).
4. Foundry Service Ltd. *Foundryman's Handbook* (Elmsford, NY: Pergamon Press, Inc., 1976).
5. Syevia, J. G. *Cast Metal Technology*, A-W (Reading, MD: Addison-Wesley Pub. Co., 1972).
6. Ammen, C. W. *The Complete Handbook of Sand Casting* (Summit, PA: TAB Books, 1979).
7. Flinn, R. A. *Fundamentals of Metal Casting* (Reading, MD: Addison-Wesley Publ. Co., 1963).
8. Maleev, V. L. *Machine Design* (Scranton, PA: International Textbook Company, 1931).
9. Hesse, H. C. *Engineering Tools and Processes* (New York: Van Nostrand Reinhold Company, 1944).
10. Althouse, A., et al. *Modern Welding* (South Holland, IL: Goodheart Wilcox Co., 1980).
11. *Classification and Application of Welded Joints for Machinery and Equipment* (Miami, FL: American Welding Society, 1977).
12. *Welding Handbook* (Columbus, OH: American Welding Association, 1972).
13. AWS Committee on Machinery and Equipment. "Standard of Classification and Application of Welded Joints for Machinery and Equipment," (Miami, FL: American Welding Society, 1977).
14. Blodgett, O. W., *Design for Welding* (Miami, FL: American Welding Society, 1968).
15. Brumbaugh, J. *Welder's Guide* (Indianapolis, IN: Theodore-Audel, 1973).
16. Cary, H. B. *Modern Welding Technology* (Englewood-Cliffs, NJ: Prentice-Hall Inc., 1979).

17. Griffin, I. H., and E. M. Roben. *Basic Welding Techniques* (New York: Van Nostrand-Reinhold Co., 1977).
18. Hicks, J. G. *Welded Joint Design* (New York: Halsted Press, 1979).
19. Houldcraft, P. W. *Welding Processes* (London: Oxford University Press, 1975).
20. Jefferson, T. B., ed. *Welding Encyclopedia* (Zurich, IL: Jefferson Publishers, 1974).
21. Kennedy, G. A. *Welding Technology* (Indianapolis, IN: Bobbs-Merrill Co., 1977).
22. Masubuchi. *Analysis of Design and Fabrication of Welded Structures* (Elmsford, NY: Pergamon Press, Inc., 1980).
23. Baird, R. J. *Oxyacetylene Welding* (South Holland, IL: Goodheart-Wilcox Co., 1980).
24. Bell, J., et al. *Welding Practices* (Philadelphia: International Ideas, 1977).
25. Tweeddale, J. G. *Welding Fabrication,* Vol. 3 (New York: Elsevier-North Holland, Inc., 1969).
26. *Welding and Brazing of Carbon Steels*, Book 1, *Arc Welding*, American Society for Metals, Ohio (1976).
27. *Welding and Brazing of Carbon Steels*, Book 2, *High-Deposition Rate and Special-Application Welding*, American Society for Metals, Ohio (1976).
28. *Welding and Brazing of Carbon Steels*, Book 3, *Resistance Welding*, American Society for Metals, Ohio (1976).
29. *Welding and Brazing of Carbon Steels*, Book 4, *Gas Welding and Brazing*, American Society of Metals, Ohio (1976).
30. *Hutte*, Vol. II (New York: McGraw-Hill Book Co., 1942), p. 797.
31. Hill, R., and C. Jensen. *Modern Engineering Tolerancing* (New York: McGraw-Hill Book Co., 1976).
32. Lieblich, J. H. *Dimensions and Tolerances,* Global Eng. (1973).
33. Ligget, J. V. *Fundamentals of Position Tolerance* (New York: American Society of Mechanical Engineers, 1970).
34. Madsen, D. A. *Geometric Dimensioning and Tolerancing,* Goodheart (1977).
35. Parr, R. E. *Principles of Mechanical Design* (New York: McGraw-Hill Book Co., 1970).

CHAPTER 5

INFLUENCE OF MATERIAL PROPERTIES ON DESIGN

5.1 DESIGN FROM HIGH-ALLOY STEELS

Apparatuses of austenitic steels and alloys are widely employed in the chemical and allied industries. Excessive overheating of austenitic steels, even if they contain titanium or niobium, results in a burning out of alloy components and a reduction in chemical resistance.

Welded joints should be designed so that the parts to be connected are heated simultaneously to the melting point. This condition is achieved automatically when parts of the same thickness are butt-welded. For welding parts of different thicknesses the thicker part should be made bevelled to match the thinner part with a slope of ¼ to ⅕ [1-4].

Some design solutions will illustrate this principle. Butt-welding of a tubular shaft end, for example, is considerably better than that of a conventional method. This is illustrated in Figure 5.1.

Similarly, the welding of a blade to a hollow shaft is more favorable than to a solid one, as shown in Figure 5.2. The same considerations are valid for connecting any different-walled parts, for example, welding a shell to a tube plate, flanges to thick-walled pipes, etc. The conventional method of welding a shell to the tube and plate (Figure 5.3) used in fabricating carbon steel apparatuses cannot be recommended for austenitic steels. The plate thickness decrease is achieved by turning on two grooves (Figure 5.3(A)). The other method sometimes used (Figure 5.3(B)) is more expensive, as machining produces more waste from trimmings. For connecting a thin-walled shell to a plate, the design shown in Figure 5.4 is recommended.

A circular slot (3–4 mm wide) is turned onto the plate and then the thin rim of the plate is butt-welded to the shell. In welding parts of different thicknesses, an intermediate ring (Figure 5.5) or connection of a thin-walled shell with a massive tube plate (Figure 5.6) is employed. A deep circular slot

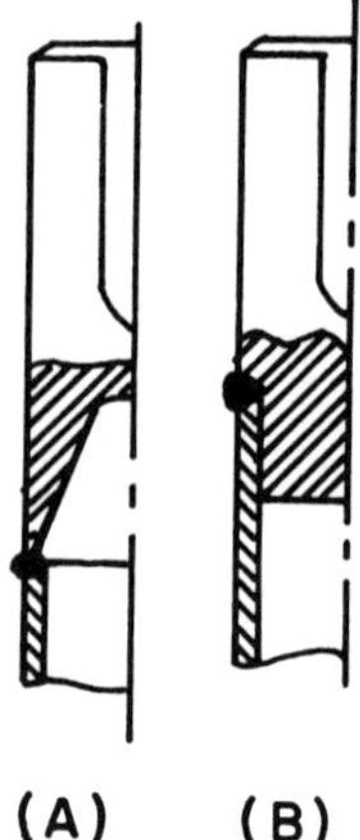

Figure 5.1. Welding of a tubular shaft end: (A) good design; (B) poor design.

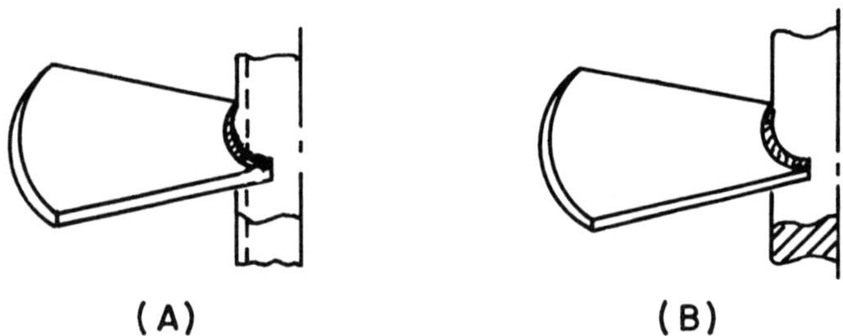

Figure 5.2. Blade welding to a shaft: (A) good design; (B) poor design.

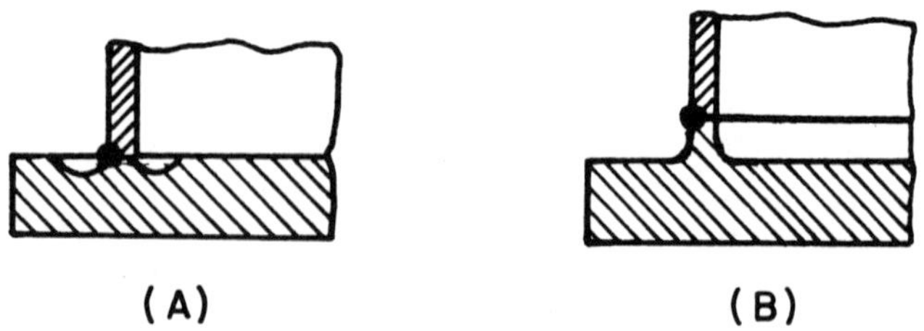

Figure 5.3. Welding a shell to a plate: (A) good design; (B) poor design.

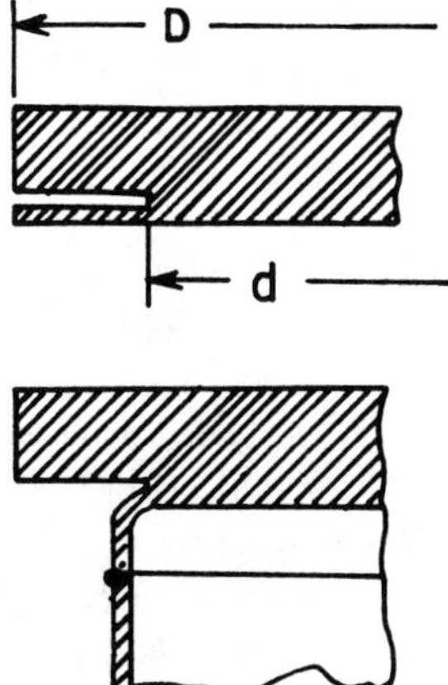

Figure 5.4. Welding a shell to a flanged plate.

Figure 5.5. Welding parts of different thicknesses with the use of an intermediate ring.

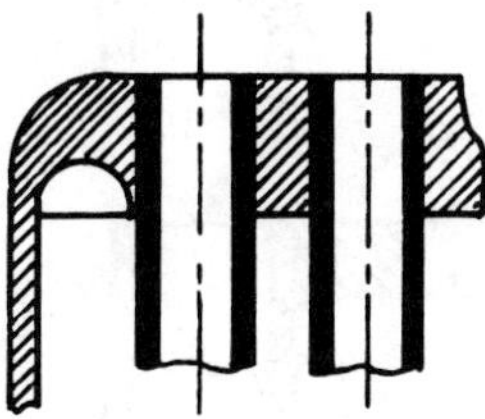

Figure 5.6. Welding a shell to a tube plate.

is turned onto the plate with the radius no less than 1.5 times the shell wall thickness. The upper end of the plate is rounded. Owing to its flexibility, such a connection decreases the boundary stresses as well.

All the thick-walled parts are connected with thin-walled ones in the same manner (Figures 5.7 and 5.8).

If it is not possible to make a butt joint, the piece should be turned on a slot as a massive part to decrease its thickness (Figure 5.9). Such a design decreases the risk of burning out the alloy components. However, border stresses may develop as a result of shrinkage after welding.

Lap joints are unsatisfactory for two reasons: First, the thickness of a lap joint is equal to the sum of the thicknesses of two sheets, which is equivalent to the metal accumulation. This is dangerous because there is a possibility of burning through a thin sheet. Second, in such joints the strip of metal

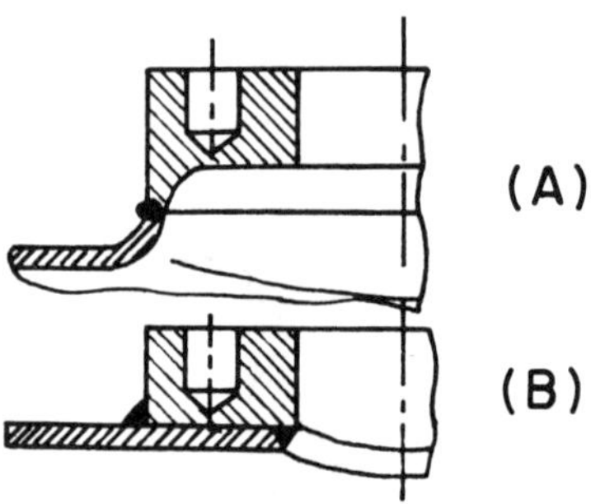

Figure 5.7. Welding of a boss to a shell: (A) good design; (B) poor design.

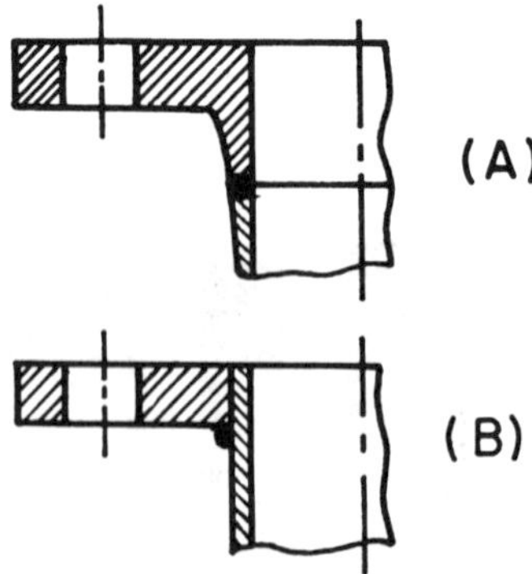

Figure 5.8. Welding a flange to a shell: (A) good design; (B) poor design.

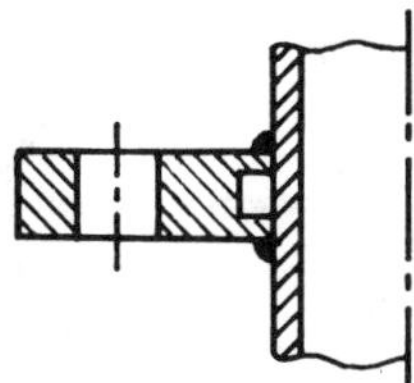

Figure 5.9. Slotting for equalizing of welded parts.

between welds would be under tension, and every tension in austenitic steels decreases their strength because of a recombination of the crystalline lattice as result of deformations. Proper and poor designs are shown in Figure 5.10.

It is important to maintain the composition of steel and, therefore, not to overheat many times in the same place. The weld joints of equipment of stainless steel should be separated from each other. In connecting shells, the distance between longitudinal seams should not be less than five times the width of a weld and not less than 50 mm; therefore, piling up welds is not permissible (Figure 5.11(B)). The version shown in Figure 5.11(A) is preferable.

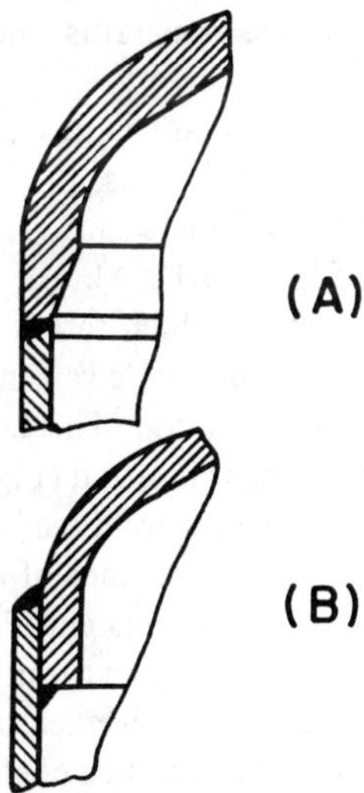

Figure 5.10. Welded assembly of head to shell: (A) good design; (B) poor design.

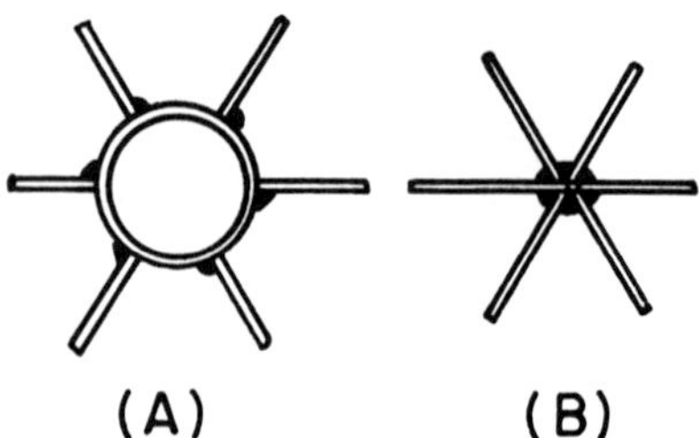

Figure 5.11. Rib welding: (A) good design; (B) poor design.

Welding of austenitic with nonaustenitic steels is difficult due to the danger of diluting the seam with nonaustenitic metal and the formation of fragile transition layers between the austenitic seam and nonaustenitic steel. The influence of the difference between these two types of steels on the quality of equipment will become clear if we compare some physical properties of carbon and stainless steel, for example, the 18-8 type.

The coefficient of thermal expansion of 18-8 steel is $\alpha = 0.0000173$, or 1.5 times larger than that of carbon steel ($\alpha = 0.0000112$), and the coefficient of thermal conductivity of chrome-nickel steel is equal to 12–16 kcal/m °C hr, which is 3-4 times less than for carbon steels ($\lambda = 40 - 50$ kcal/m °C hr). This results in lower heating along the width of the seam during welding, which, together with a larger coefficient of thermal expansion, may cause excessive stresses that break the apparatus and deteriorate the corrosion resistance of the steel.

To prevent the formation of excessive stresses, flexible elements are used that play the role of compensators. Because of their deformation, the stresses are decreased in the zones around the seams. The typical design solution to this problem is shown in Figure 5.12(A). As shown, the nozzle is butt-welded to a flanged rim of a shell, which serves as a compensator. The usual design for an apparatus of carbon steel given in Figure 5.12(B) is not valid for stainless because of crack formation. The same principle is used in connections in which a reinforced ring is inserted (Figure 5.13).

The use of flexible connections near weld seams is quite reasonable because they help to maintain the corrosion resistance of an apparatus.

Relatively low heat conductivity of austenitic steels should be considered when designing heat exchangers. If one heat transfer coefficient is small compared to the other, then the thermal resistance of a wall, S/λ, has almost no influence on the value of the heat transfer coefficient, k. For example, let $\alpha_1 = 10$ and $\alpha_2 = 10{,}000$ kcal/m^2 °C hr, and wall thickness S = 0.5 cm. For a wall of carbon steel, the heat transfer coefficient will be

Figure 5.12. Welding a nozzle to a shell: (A) good design; (B) poor design.

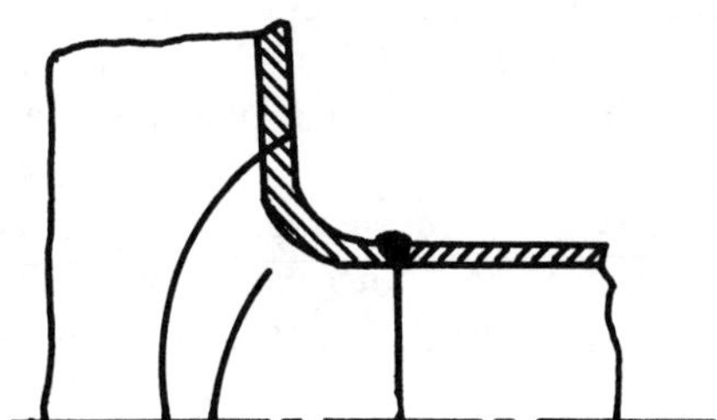

Figure 5.13. Welding in a reinforcing ring.

$$k = \frac{1}{\frac{1}{10} + \frac{1}{10{,}000} + \frac{0.005}{50}} \approx 10 \text{ kcal/m}^2 \text{ °C hr} \qquad (5.1)$$

For a wall of chrome-nickel steel, we will obtain approximately the same value:

$$k = \frac{1}{\frac{1}{10} + \frac{1}{1000} + \frac{0.005}{12}} \approx 10 \text{ kcal/m}^2 \text{ °C hr} \qquad (5.2)$$

The situation changes greatly when designing high-efficiency equipment, where change of phases takes place on both sides of the heat transfer surfaces and both heat transfer surfaces and their heat transfer coefficients, α_1 and α_2, are comparable to each other, for example, 10,000 kcal/m^2 °C hr. Then, for a wall of carbon steel, we obtain

$$k = \frac{1}{\frac{1}{10{,}000} + \frac{1}{10{,}000} + \frac{0.005}{50}} \approx 3{,}300 \text{ kcal/m}^2 \text{ °C hr}$$

and for a wall of austenitic steel,

$$k = \frac{1}{\frac{1}{10{,}000} + \frac{1}{10{,}000} + \frac{0.005}{12}} \approx 1{,}600 \text{ kcal/m}^2 \text{ °C hr}$$

Thus, for the heat transfer surface made from austenitic steel, the heat transfer coefficient decreases by about 2. Consequently, the low heat conductivity of austenitic steels may prevent the design of highly efficient heat exchanger equipment. The high cost of austenitic steels (mostly because of the high content of chrome and nickel) demands that the designer be economy conscious.

Only parts contacting aggressive substances should be fabricated from chrome-nickel austenitic steels. All other parts beyond the working space of the apparatus, for example, flanges, lugs, etc., should be manufactured from conventional steel. Flanges are preferred to be loose, using beading or bead rings (Figure 5.14).

Welding austenitic steels with carbon steels is possible because the melting points of both types are similar. However, such welding is difficult due to the danger of diluting the seam with nonaustenitic metal and the potential formation of fragile transition layers between the austenitic seam and carbon steel. To prevent dilution of the seam, electrodes and wire are used with an increased content of nickel, chromium and other alloying elements, and an attempt is made to provide minimum melting of the carbon steels. During manual arc welding of parts of heterogeneous steels with electrodes of

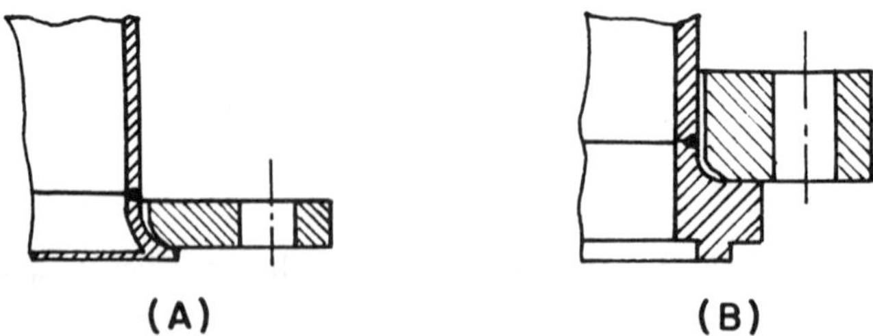

Figure 5.14. Flanges for apparatuses of austenitic steel: (A) on beading; (B) on bead ring.

ordinary nonalloyed steels, cracks generally do not appear in the metal of the seam. However, this does not eliminate the danger of local destruction of the zone near the seam.

The appearance of fragile layers between the austenitic seam and the non-austenitic steel can occur from the following conditions: as a result of diffusion of carbon in the steel into the metal of the seam, and/or insufficient mixing of the metal of the seam with the nonaustenitic steel.

Diffusion of carbon into the seam is accompanied by decarbonizing of the nonaustenitic steel in the zone near the seam, causing its weakening. However, destruction can occur not only in this zone, but also on the austenitic-carbide layer in the austenitic seam because of carbon enrichment. To prevent carbon diffusion, it is recommended to weld combined joints with electrodes using an increased nickel content.

If there is an insufficient amount of alloying components in the filler metal, breaks may develop on the fragile layer, not only during the welding of heterogeneous steels but also during seam welding of a layer of austenitic metal on the surface of carbon steels [4]. This is illustrated in Figure 5.15(A).

One way to prevent breaks is by preliminary seam welding of the intermediate layer of noncarbon soft iron (Armco iron) on the surface of steel being subjected to seam welding with austenitic metal. This is shown in Figure 5.15(B). In welding heterogeneous steels, there is always the danger of a decrease in alloying substances in the welding zone (shown in Figure 5.16 by the dotted line). This will occur only when the steel alloy's sheet thickness is sufficient that corrosion can be avoided in the place opposite the weld. The deterioration of corrosion resistance in such connections may arise as a result of residual welding stresses.

For thin austenitic metal sheets (3-5 mm thick), a reduction in corrosion resistance almost always occurs in places opposite the weld joints with parts

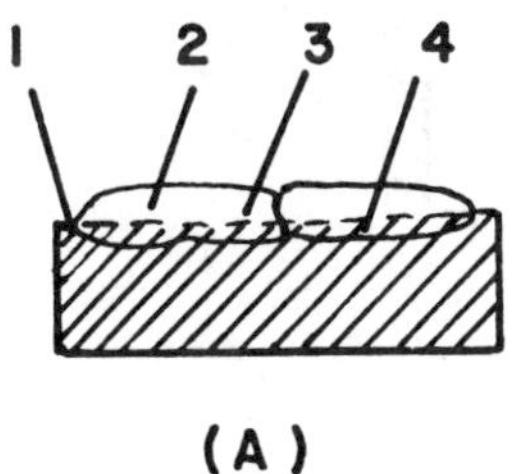

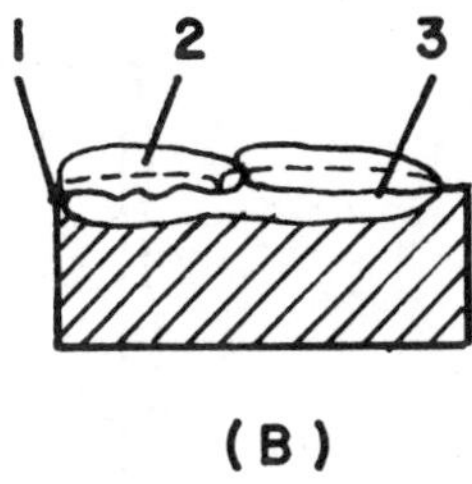

Figure 5.15. Beading of a layer of austenitic steel on the surface of carbon steel: (A) without intermediate layer (breaks along the martensite layer); (B) with an intermediate layer of noncarbon soft iron; 1-carbon steel; austenitic beading; martensite layer; 4-breaks; 5-intermediate layer (soft Armco iron).

of carbon steel. Hence, such assemblies as that shown in Figure 5.17, in which a carbon steel is welded on the outside of stainless pipe, is completely unsatisfactory because the metal of the pipe opposite the weld will corrode at a faster rate than the pipe itself, eventually breaking the connection.

To avoid deterioration of the base metal, it is sufficient that the seam between the austenitic and carbon steel be moved as far as possible from the corrosion site (generally a distance larger than the depth of diffusion). This is realized by inserting between austenitic and carbon steel parts an additional piece of austenitic steel. Thus, the seam between the carbon and austenitic steels would be located far away from the region subjected to the aggressive

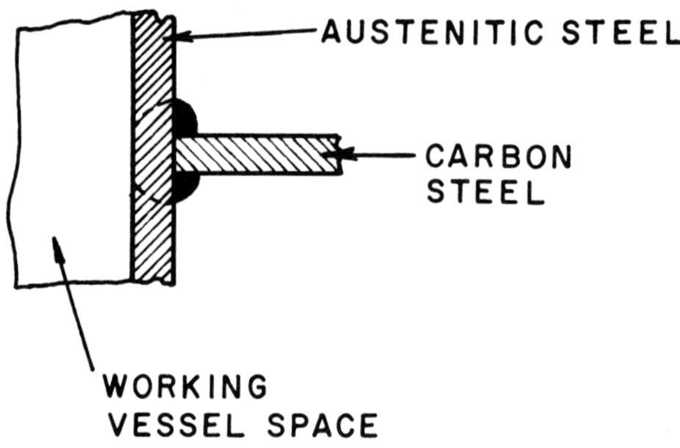

Figure 5.16. Metal zone deteriorated by diffusion of alloy elements. (Shown by dotted line.)

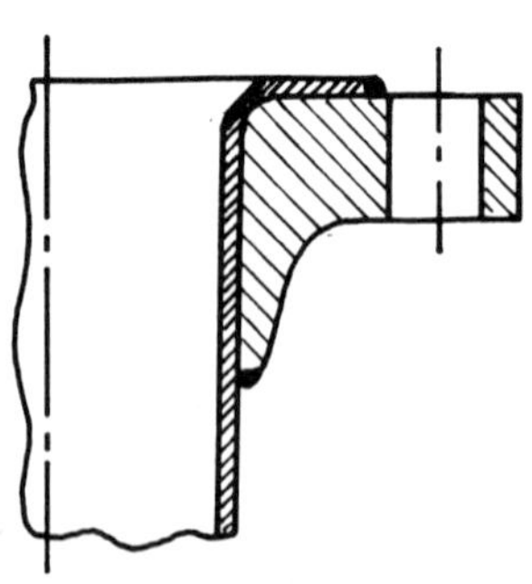

Figure 5.17. Poor flange design.

media, and the strength of the equipment would not be violated. Figure 5.18 shows some examples of such a design: welding a support shell to the column of austenitic steel (Figure 5.18(A)); welding a jacket (Figure 5.18(A)) and a shell of heat exchanger to the tube plate (Figure 5.18(C)).

The same purpose is served by a pad of austenitic steel between a vessel made from stainless steel and a lug from carbon steel. This enables the load to distribute over the larger surface of the shell, as shown in Figure 5.18(D).

After welding parts of austenitic steels, the seams should be cleaned carefully, and the apparatus itself should be polished and pickled. This increases the corrosion resistance by a factor of 3-5.

In the overwhelming majority of cases, the application of welding austenitic steels can be managed without heat treatment. However, when it is necessary to remove welding stresses to avoid corrosional cracking of weld joints or to increase durability properties of the steel and seam, heat treatment is necessary. Elimination of the inclination of austenitic seams to intercrystalline corrosion is attained by hardening the welded structures.

If it is possible to subject only the small parts to hardening without difficulties, then the entire structure does not require treatment. It has been determined that a considerable increase in corrosional stability can be attained without hardening from high temperatures (1050-1150°C). It is sufficient to heat the weld joint to a temperature of 850-900°C to hold it at this temperature for 2-3 hours and cool it in the air to give the metal even greater stability against corrosion. This is obtained as a result of hardening. Such heat treatment is called stabilizing or diffusion annealing.

Welding evokes the appearance of residual stresses, which may cause one of the most dangerous forms of destruction of welded structures of austenitic steels in certain aggressive media (corrosion under stress or corrosional cracking). To avoid this, it is necessary to remove the stress in the structure by heat treatment. Heating to a temperature of 650°C is sufficient to remove stresses in parts of ordinary carbon steel. Austenitic steels possess higher indices of yield; therefore, with heating to 650°C, internal stresses caused by welding are maintained. Furthermore, at such a temperature the formation of chromium carbides results on the boundary of grains causing the appearance of the inclination of austenitic seams (resulting in intercrystalline and knife corrosion). Therefore, to remove stresses from a part made from austenitic steels, heat treatment at higher temperatures (800-850°C) is used. Consequently, stabilizing annealing increases the stability of the basic metal and the weld seam, not only against intercrystalline corrosion, but also against corrosion under stress. Considering this, parts of particular importance are subjected to stabilizing annealing. If it is not possible to subject the entire part to annealing, one must resort to local heat treatment of weld seams. Thus, for instance, pipelines of austenitic pipes are treated this way during assembling of chemical and other process plants.

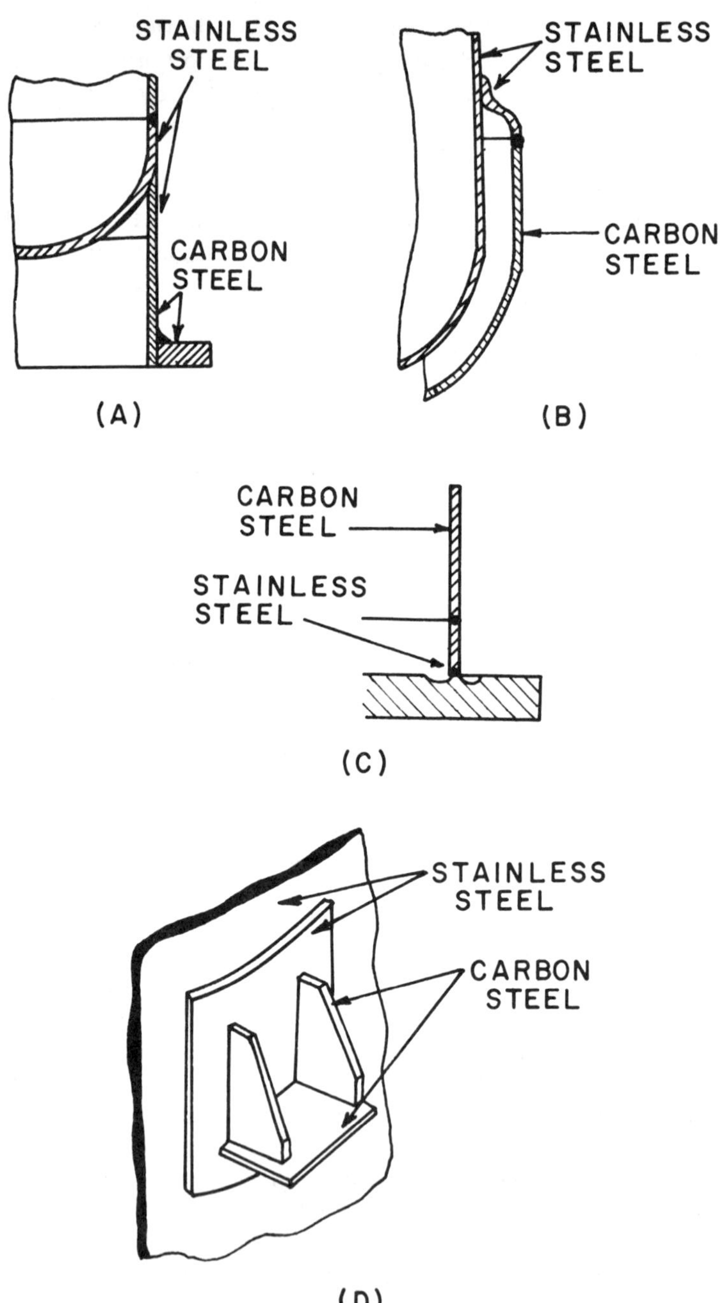

Figure 5.18. Examples of the use of intermediate parts when welding austenitic steels with carbon steel.

5.2 DESIGN OF APPARATUSES FROM COPPER

Pieces of equipment made from copper are fabricated from rolled copper sheets or tubes. Copper has excellent ductility and, hence, can be readily formed into complex shapes. Since it is relatively inexpensive, it has enjoyed wide use as half-finished parts, as well as in the form of large apparatuses and their connections. The shells of apparatuses are easily rolled, with heads stamped or beaten out. To increase the rigidity of thin-walled shells, they are bead formed by a creasing machine (Figure 5.19). Bead forming also is used for fixing parts inside the equipment.

Permanent connections of copper parts are made by soft and hard soldering, welding and riveting. Soft soldering is always made with the use of lap joints. The butt joints generally are not strong enough because soft soldering strength is low. For example, butt-soldering will be 2200/320 = 7 times weaker than the copper sheet. From considerations of full strength and taking an equal safety factor for sheet and solder, one obtains the proper amount of overlapping. It is approximately equal to 7 times the sheet thickness, b = 7S (Figure 5.20).

Soldered joints made from soft solder can be used up to +120°C. At this temperature the strength of the connection decreases considerably. Taking this into account, as well as a possible seam inequality, it is recommended that the allowable stress for soldered seams be no more than 25 kg/cm^2. The soldered joints are excellent at low temperatures down to cryogenics. Of the three examples given in Figure 5.20, example (A) is more favorable because of its symmetry and the absence of additional bending moments. However, its adjustment is difficult for large-diameter vessels. The seam (B) is used for connections with limited parting, and it assists parts alignment in assembling. The assembling force creates a bending

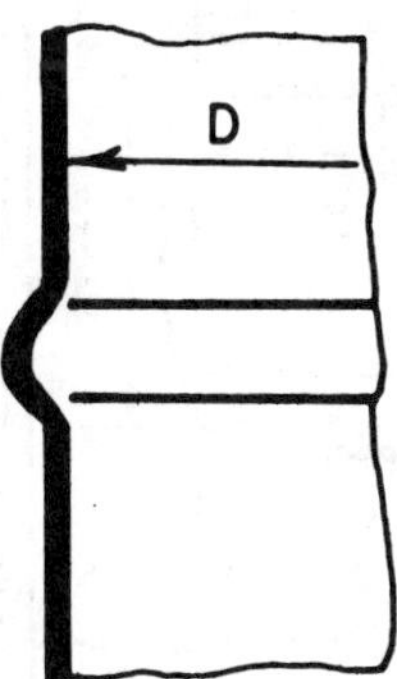

Figure 5.19. Bead forming of copper shell.

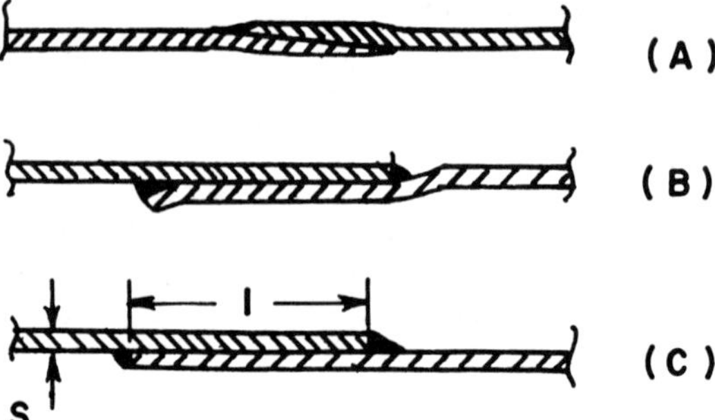

Figure 5.20. Examples of soldered joints.

moment, which gives rise to additional stresses in the seam; therefore, such connections should be avoided.

In designing any loaded soldered connection, the connection area must be determined according to the condition that the seam should withstand all acting forces. For example, in the design of the seam between a shell and a boss (Figure 5.21), the following criteria should be fulfilled.

$$\pi d \ell \sigma_{allow.} \geqslant \frac{\pi d^2}{4} P \qquad (5.3)$$

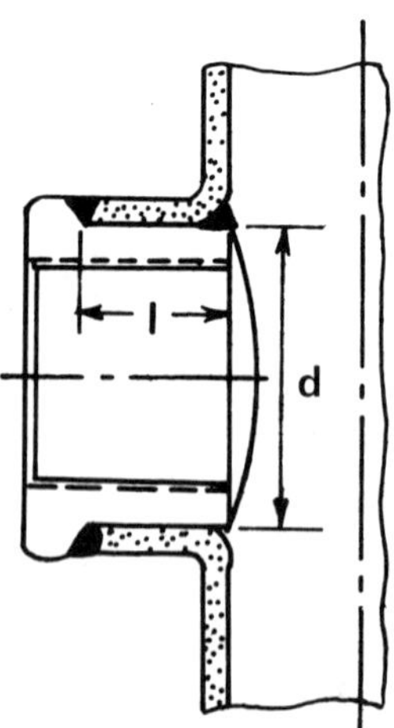

Figure 5.21. Soldering of a boss.

where $\sigma_{allow.}$ = allowable stress for the solder. The value to be determined is ℓ.

The increase of strength and seam unloading are achieved by the use of locks and folders, which play the role of rigid elements at the same time. Figure 5.22 shows three examples of folders, and Table 5.1 gives the widths of folders for different sheet thicknesses [5-7].

Soft soldering is the basic method of making tube connections in heat exchanger plates. For installations working under pressure (in cryogenics units, pressures can be up to 200 atm), tube diameters must be small (from 5×1 to 10×1.5) because of copper's low strength. It is difficult to roll out such small tubes, and generally a tight fit in tube sheets is not achieved because of the metal's softness. A recommended design for soldered connections of tubes is given in Figure 5.23.

The tube sheet shown is provided with grooves. Tin-coated tubes are drawn through the sheet so that they extend 5–10 mm over the sheet. Then the sheet with tubes is heated and poured by solder. To increase strength, solder pouring sometimes is done on both sides of the sheet. Soft soldering is used in many cases, including in the tightening of riveted connections.

The greatest advantage of soft soldering is its low melting temperature, which minimizes the formation of thermal stresses that can cause part warping. Soft soldering devices that do not permit load increases on

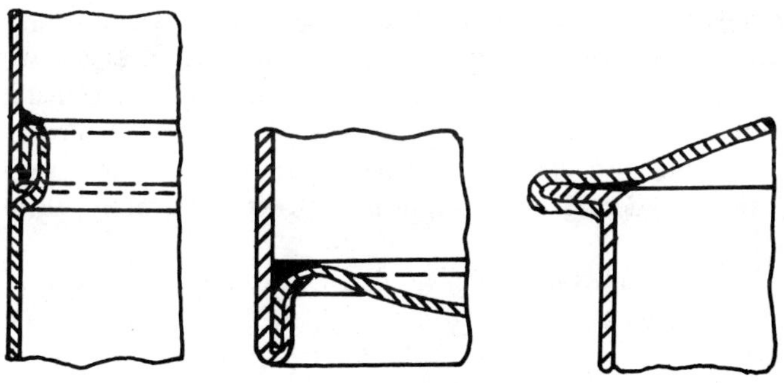

Figure 5.22. Examples of folders.

Table 5.1. Typical Sheet Thickness and Folder Widths

Sheet Thickness (mm)	0.75	1.25	1.5	1.75	2.00	2.50
Folder Width (mm)	7	13	14		16	19

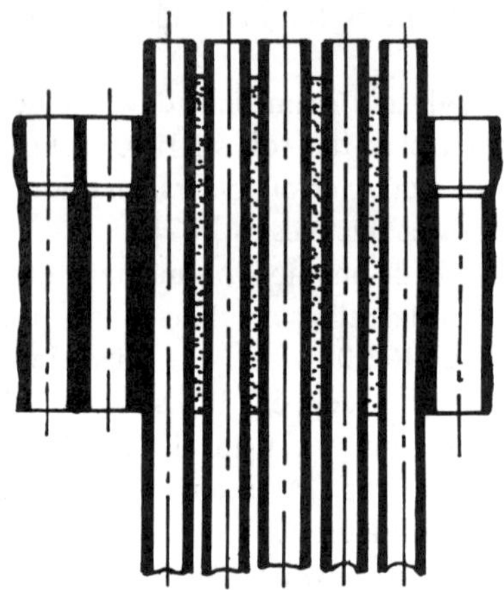

Figure 5.23. Soldering copper tubes to a tube sheet.

the seam above the design rating should be used (for example, stresses arising from uneven heating during starting or assembling of the equipment, etc.). This is helpful in the use of lens or bellows compensators.

Hard solders and brass are used for connecting parts that are heavily loaded statically or dynamically. Brass melts at 910°C whereas most solders have a melting temperature of 840°C. Silver solders containing 10–70% silver are used for vital soldering, where not only tightness and strength are important, but surface finish of seams as well, and where there is also a danger of burning through thin parts. The tensile strength for mild brass is higher than that of copper at the same temperatures. The relative elongation for all three solders is quite sufficient (20–40%); therefore, sheets can be butt-welded.

In some cases hard soldering is needed. One example is in connecting tubes for heat exchangers for oxygen units (Figure 5.24). The connecting tubes are soldered in the upper head of the condenser-vaporizers of copper oxygen columns. The hard solders are diffused into the material of connecting parts; therefore, the connections are very strong. In hard soldering, the possibility of warping connecting parts exists because of residual thermal stresses in the seam. It is important that parts that are to be connected be well heated and not be moved out of their positions. For soldering thin sheets, a connection with set teeth is used, as illustrated in Figure 5.25.

The stretched end of one sheet is cut at a length equal to 5 times the sheet thickness. The teeth are set and the other sheet also stretched, but not

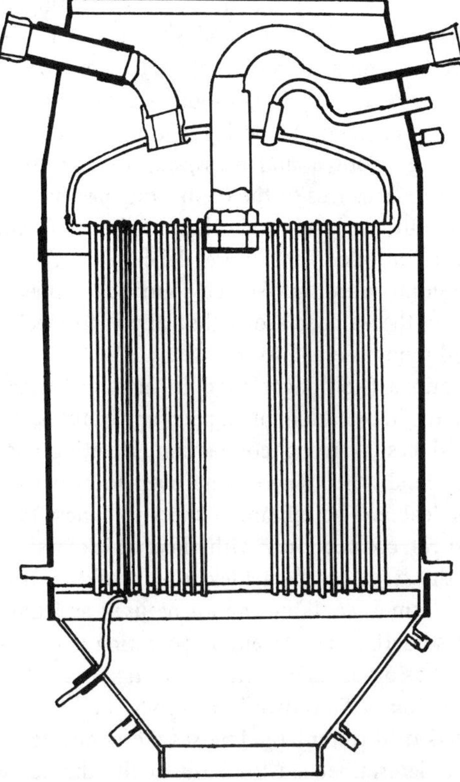

Figure 5.24. Tube soldering by silver solder.

Figure 5.25. Design of a longitudinal seam of a shell.

until it is inserted among the teeth. After hammering, the seam is hard soldered, then cleaned, rolled out and smoothed. The shells for oxygen columns are rolled once more to restore them to their correct shape.

For important seams, such as longitudinal seams on vessels, hard soldering or welding is recommended. Soldering permits parts to be connected to different metals; e.g., copper and brass, copper and steel, steel and brass, etc., provided that the surfaces of the connecting parts are moistened.

Welding successfully replaces the other methods of connecting parts of copper equipment, but demands more delicacy in fabrication because of difficulties in avoiding warping. Welded joints of copper equipment should be butt-welded for the same reasons that apply to steel. The safety factor for copper-welded joints is equal to $\psi = 0.8$.

Riveted connections (no longer used in manufacturing steel equipment) are still used in the fabrication of copper apparatuses. Riveting eliminates warping and produces a strong connection. Riveting with copper rivets is done under cold conditions. There is no need to heat rivets because copper is sufficiently malleable. The minimum sheet thickness for riveting is 3 mm. In the case of sheet connections, with sheet thicknesses at 7 mm or more, the tightness of the seam can be achieved by caulking. If the sheet thickness ranges from 3 to 7 mm, caulking cannot be used and the tightness must be achieved by soft soldering. For strength, soldering is not taken into account in the design. It is assumed that all loads are taken by the rivets. The design elements of riveted joints are shown in Figure 5.26.

If sheets or parts of different thicknesses are connected, all design sizes of a seam should be determined on the basis of the thinner sheet. Parts whose total thickness is more than four times the diameter of a rivet should not be connected because, in this case, the stem of the rivet may be curved and the quality of the connection would be spoiled. The strength factor of a row seam is $\psi = 0.6$; for a two-row seam it is $\psi = 0.7$.

The allowable shear stresses for copper rivets are given in Table 5.2. The allowable bearing stresses are equal to double the shear stress at the same temperature. The detachable connections of copper apparatuses are unique because copper is too mild and expensive to subject flanges to large loads. Therefore, flanges for these systems usually are made from regular carbon steels as lap-joint flanges fastened on a lap-joint stub or bead ring, which, in turn, are soldered to the shell with hard solder. The inner flange edge should be rounded so the beading or the bead ring will not be spoiled. To protect the flange from damage during disassembling, it should be riveted to the beading as shown in Figure 5.27(A), or provided with bead forming (Figure 5.27(B)).

For providing smooth sealing surface, countersunk riveting should be performed. If the beading thickness is not enough to countersink the head of the rivet, counterboring should be done to build it in as part of the sheet

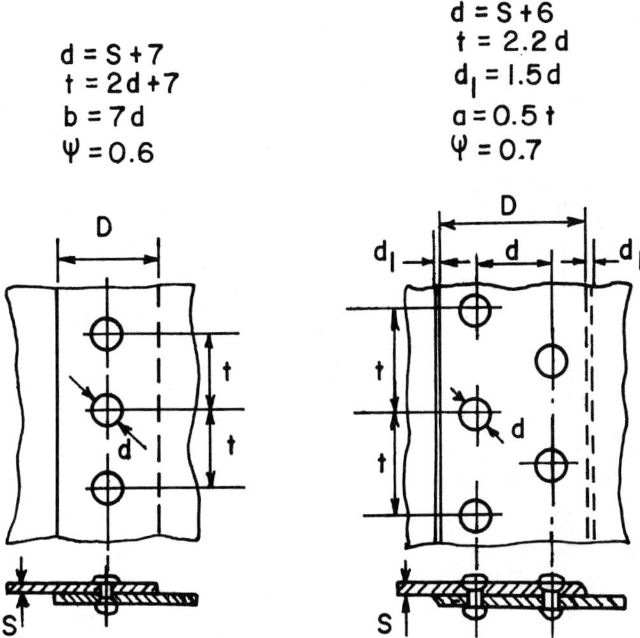

Figure 5.26. Design elements of riveted seams for copper apparatuses.

Table 5.2. Allowable Shear Stress for Copper Rivets

Wall Temperature (°C)	120	121-140	141-160	161-180	181-200	201-220	221-240	241-250
Allowable Shear Stress (kg/cm^2)	360	350	340	320	300	280	260	240

(Figure 5.28). The connection bosses of bronze or brass are bended round by copper, then soldered as shown in Figure 5.29.

The contact area of a sheet and a boss should be sufficient from the standpoint of equality of acting forces of pressure and elasticity in the solder. Nozzles are either soldered or welded. Sometimes they are connected by a socket joint. Bosses and other parts covered by copper are fabricated from copper alloys. Figure 5.30 shows an example of fastening a plate in a

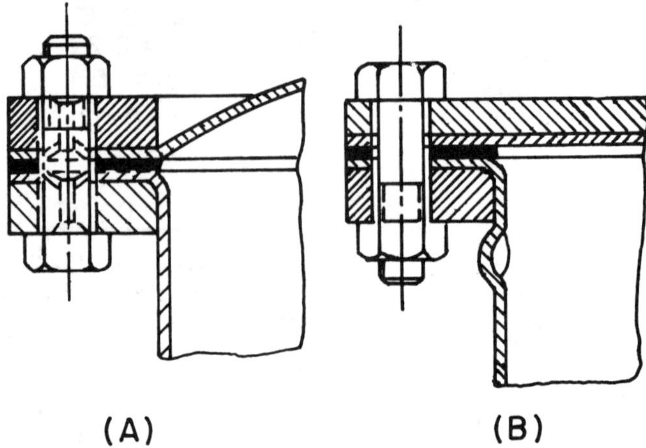

Figure 5.27. Fastening of a flange to the shell: (A) by riveting; (B) by bead forming.

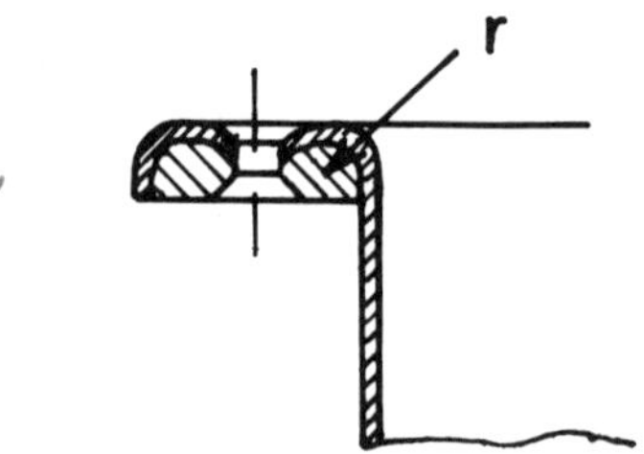

Figure 5.28. Building in the head of a rivet.

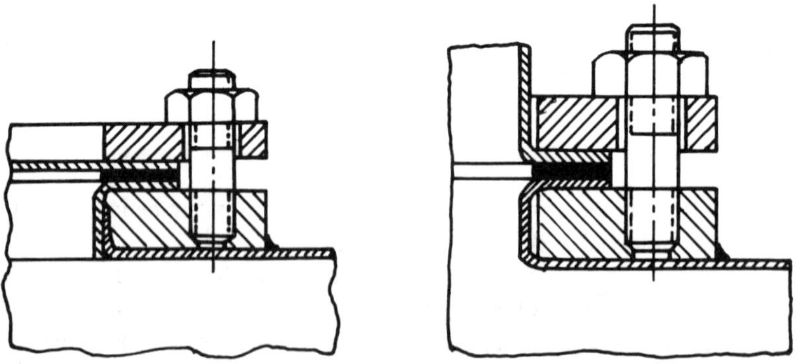

Figure 5.29. Design of a boss for a copper apparatus.

small-diameter distillation column. The shell is bead formed with the split ring inserted in the lower part of the two beads and then soldered to the shell with a soft solder. The tray is put on this ring and then inserted into the other ring above, and the entire assembly is soldered together. Bead soldering is possible only for shells up to 3.0 mm thick.

Beads formed for shells that work under pressure (e.g., shells of lower oxygen columns (P = 6 atm)) should be unloaded. This shell is inserted inside a more thick-walled shell (6–8 mm thick) and the same pressure is created in the gap between two shells. Typical examples of bubble caps are shown in Figures 5.31, 5.32 and 5.33.

Figure 5.30. Fastening a tray by the use of rings.

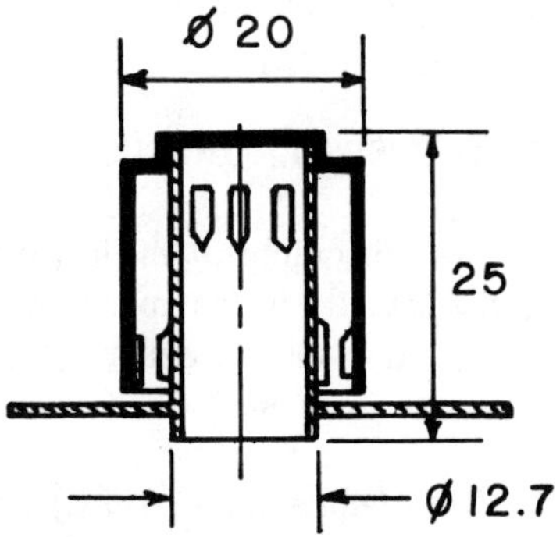

Figure 5.31. Bubble cap soldered to a tray.

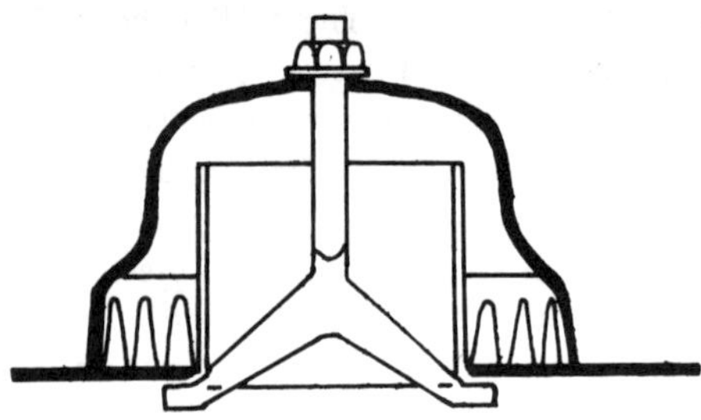

Figure 5.32. Removable bubble cap for a copper column.

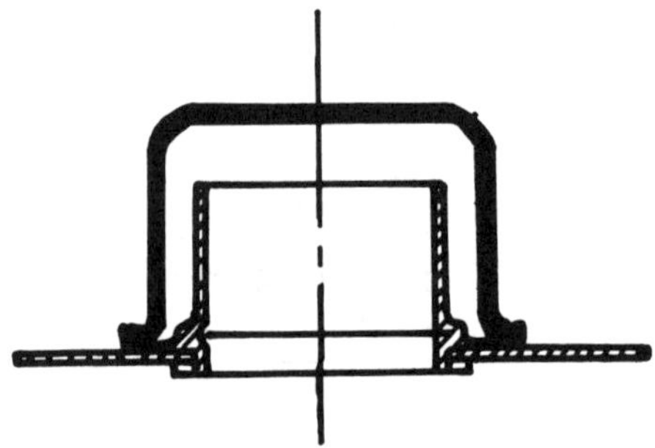

Figure 5.33. Fixing a bubble cap with a tab washer.

Figure 5.31 presents a bubble cap of a cryogenic column fixed to the tray by soldering. An example (Figure 5.32) is a bubble cap that could be disassembled by unscrewing only one nut.

5.3 DESIGN OF APPARATUSES FROM ALUMINUM [8-11]

The design characteristics of equipment made from aluminum are determined by the metal's fluidity, its instantaneous oxidizability and its low mechanical strength. Aluminum's instant oxidizability makes soldering practically impossible. Hence, the main method of aluminum parts connection is welding.

Because of its low melting temperature (657°C) and the low viscosity of the melted metal, it is important to melt the two parts to be connected simultaneously. Therefore, all connections should be butt-welded. The parts to be connected must have equal thickness at linkup, consequently, the welded joints are designed in the same manner as for vessels of austenitic steels. For example, all nozzles are welded to a shell beading. This is done not because we wish to compensate for thermal stresses, but to provide butt-welding. Flanges usually are fabricated from carbon steel. Lap-joint flanges and lap-joint stubs are used for aluminum equipment. Because aluminum is easily flanged, there is no need for beading rings. Steel parts, for example support lugs, cannot be welded or soldered to aluminum apparatus. Hence, these components also are fabricated from aluminum. However, it is preferential to fabricate them from steel and then weld them to a detachable steel ring, which is fastened with bolts tightly pressing against the vessel's shell. This is shown in Figure 5.34. Sight glasses generally are fastened as ordinary blind flanges onto the nozzle beading, as shown in Figure 5.35.

5.4 DESIGN OF CERAMIC-LINED EQUIPMENT

Enameling is a very delicate process. The quality of enameling depends on many factors, but is largely established by the properties of the base metal.

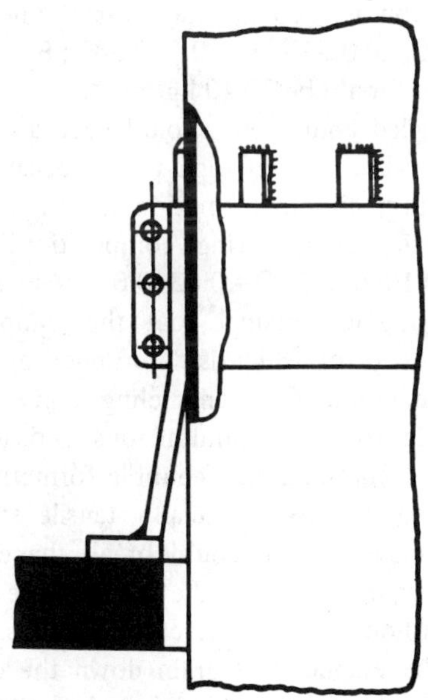

Figure 5.34. Lugs for aluminum apparatuses.

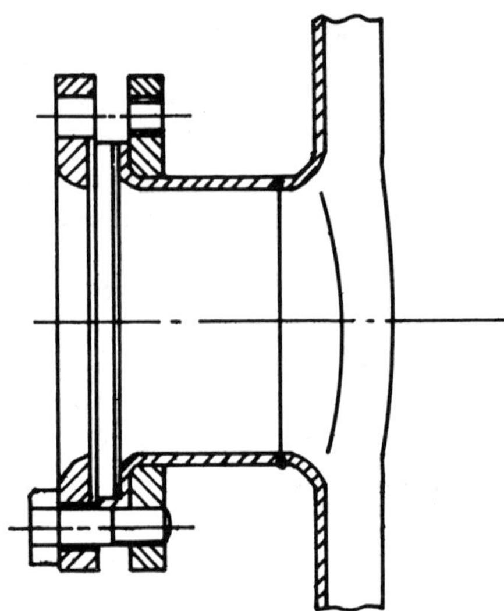

Figure 5.35. Sight glass for an aluminum apparatus.

Cast iron used for fabrication of enameled equipment must be strong and tight. It should have little inclination to grain growth, and its coefficient of thermal expansion should be close to that of the enamel. The following composition of cast iron is recommended for enameling: C = 3.5–3.5%; Si = 1.4–1.8%; Mn = 0.55–0.65%; P = 0.3–0.38%; S = 0.10–0.12%; and the cross-breaking strengths should be 37-40 kg/mm^2.

Steel used for enameled equipment should have a carbon content of no more than 0.12% and should be of high purity because the gases released from slag insertion cause bubble formation in the enamel. Steel used in enameling should have the following composition: C = 0.005–0.09%; Mn = 0.35–0.4%; P = 0.10–0.02%; S = 0.03–0.04%; Si–traces.

For obtaining a high-grade enamel coat the equipment should have a special shape, the necessity of which is determined by the following conditions: preparation of equipment for enameling; conservation of even thickness of coating during roasting; simultaneous and uniform heating and roasting of enamel; and elimination of bubble formation during roasting as well as dangerous deformations (especially tensile strain due to residual thermal and other stresses, which could break the enamel coating). For obtaining a high-grade enamel coating, the shape of the equipment should be simple and smooth without sharp rims, corners and cavities. If there is insufficient rounding, the enamel will drain down the corners or crack. All parts of equipment should be prepared carefully before enameling. The seams should be polished and the corners rounded; the greater the radius of rounding, the better the quality of the coating. Consequently, the equipment should be designed such that its internal surfaces have rounded shapes. This provides easy access for mechanical work and inspection. Local cavities are unwanted in apparatuses subjected to enameling. Some enamel accumulates after cooling, which can cause cracks due to thermal stresses.

Larger parts enclosed by flat sheets are very unfavorable. They tend to undergo warping during roasting, as well as dent formation where accumulated enamel would be broken away after cooling. Planes should be reinforced by ribs. It is better to avoid abrupt planes where possible by employing rounded surfaces. High-grade and uniform enamel roasting, as well as coating conservation during cooling, is provided by a uniform rate of heating and cooling of all the equipment parts. This condition is easily fulfilled when the walls of an apparatus have the same thickness. Therefore, in designing, care should be taken to avoid using massive parts and other large metal accumulations. Difficulties arise in the design of flanges, lugs, supports, etc. Lugs and similar parts are welded on steel vessels subjected to enameling by an intermittent weld (Figure 5.36) without strengthening plates.

Apparatuses with tubular lug-supports should be welded to short tubes (150-200 mm diameter) before enameling. The lugs are then welded to these tubes afterward. The main cause of bubble formation in enamel is gas release from seams during roasting, and especially dangerous is the root of a seam where slag is always accumulated, providing a source of air holes. In Figure 5.37 an example is shown where bubble formation develops against the root of the weld. Welds should be cut out carefully and welded onto the surface from the inside, and the seam itself should be polished. A proper design for enameling is shown in Figure 5.38. An example of a poor design is shown in Figure 5.39(B).

The bubbles were formed as a result of air penetration from the gap between the tube and flange through microcracks and pores in the seam. The change of the lower continuous seam in the intermittent seam provided a free air exit from the gap outside and eliminated spillage completely (Figure 5.39(A)).

Elimination of residual stresses in enamel vessels is achieved by the following measures: equal wall thickness, reasonable seams location, and butt-welding of nozzles to beaded shells. Nozzles up to 200 mm in diameter often are fabricated in a conical shape, as shown in Figure 5.40. The beading is important to coat the place where the enamel is connected.

Designs that could develop local edge stresses during cooling should be avoided (Figure 5.41) because they are very dangerous for brittle enamel. Also to be avoided is the use of ribs because they prevent free vessel shrinkage

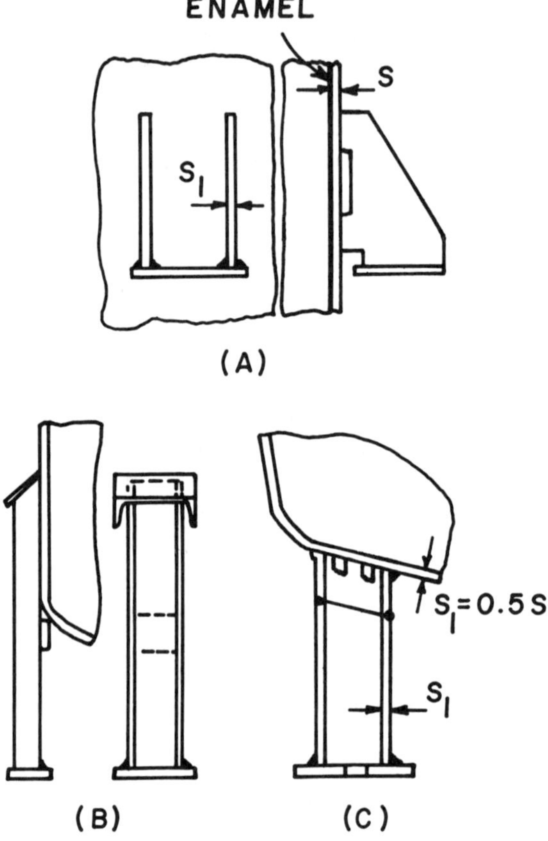

Figure 5.36. Examples of lug welding to an enameled apparatus.

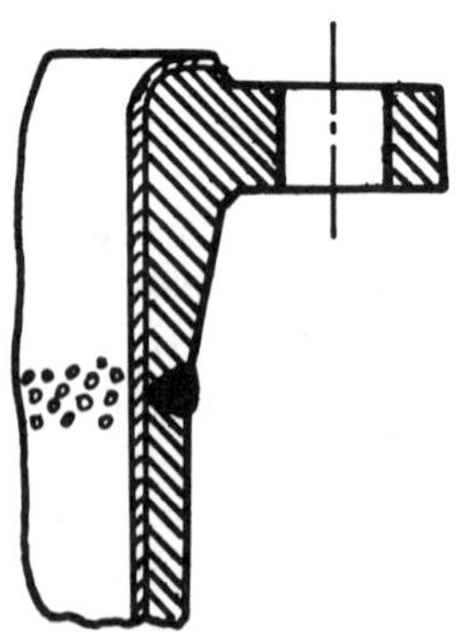

Figure 5.37. Defective enameling as a result of gas formation during roasting.

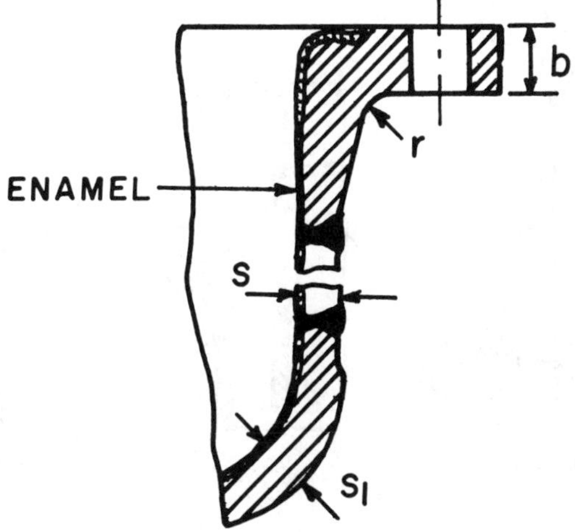

Figure 5.38. Proper design of apparatus for enameling.

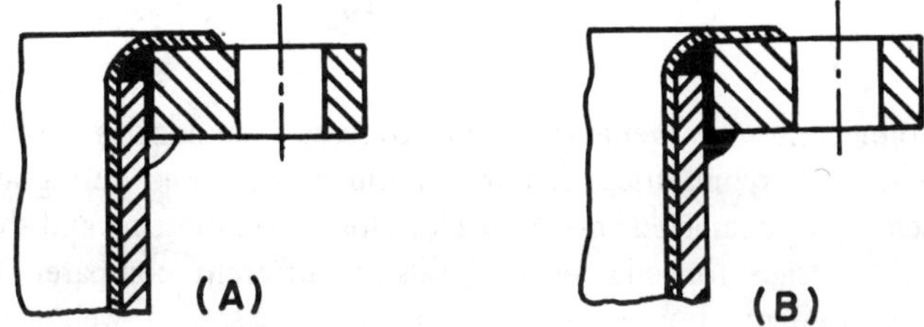

Figure 5.39. Welding of a flat flange: (A) good design; (B) poor design.

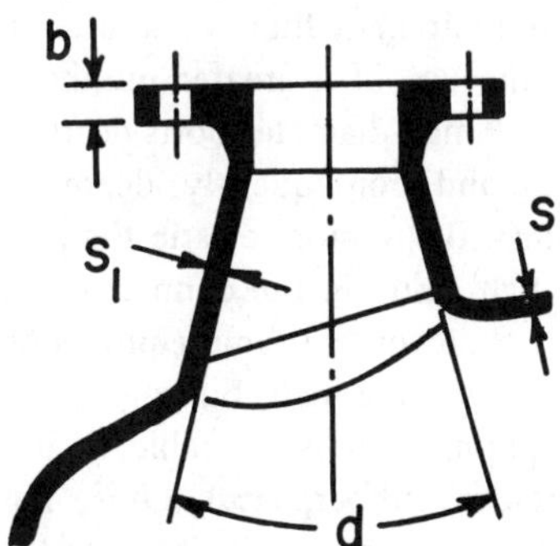

Figure 5.40. Conical nozzle of a vessel to be enameled.

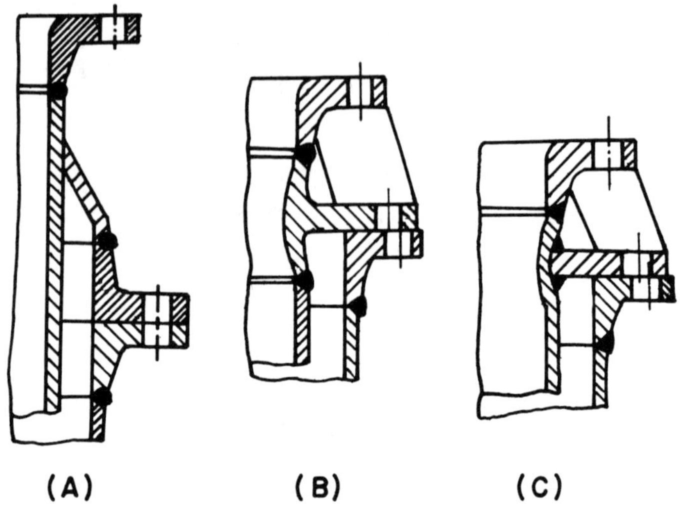

Figure 5.41. Connection of a jacket to an enameled vessel: (A) good design; (B,C) poor designs.

during cooling and may generate local stresses. The nozzles on the head should be located symmetrically. Care should be exercised during assembly and disassembly of enameled vessels to avoid local overstressing and shocks.

The flange design for enameled vessels is different compared to non-enameled equipment. For a weight decrease of cast and welded flanges, complex shapes must be used to approach equal wall thicknesses (Figure 5.42(A)).

The flanges of main parting are especially complicated in fabrication (Figure 5.42(B)); however, such a complication is justified because it reduces the amount of defective equipment that is manufactured. The other feature of such connections is the use of a greater number of thin bolts or studs (12–16 mm in diameter). Small-diameter bolts decrease the distance between the gasket and bolt circle and, consequently, decrease the bending moments applied to the head flange. Being more elastic than thick bolts or studs they provide more uniform tightening in the connection. The gasket thickness is taken as 3-4 times more than for the same equipment without enamel. This results in a decrease in the seating load, bolting force and bending moments. To tighten enameled equipment it is desirable to use soft gasket materials (polyethylene, fluoroplastics, rubber, etc.). Loose flanges are quite suitable for enameled vessels. After roasting, enameled vessels may undergo warping. The danger of warping should be considered in designing, especially with

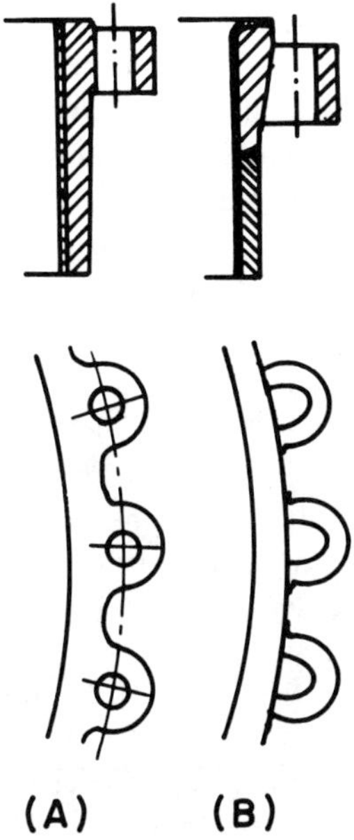

Figure 5.42. Main parting flanges for an enameled vessel: (A) cast; (B) welded.

mixers. Consequently, a sufficient gap should be provided between the impeller and the walls of the vessel, as well as the shaft and guiding sleeve. To decrease the frequency of defective parts, all small parts and assemblies, stuffing boxes, thermometer sockets, devices for sampling, etc., should be detachable.

The difficulty of enameling and the possibility of covering only streamlined and smooth surfaces considerably limits design possibilities, forcing the designer to fulfill heat transfer surfaces either as jackets or field tubes, and columns fabricated either from packed or stuffed grid plates. The high corrosion resistance of enamel vessels has justified their use in the manufacturing of a variety of high-purity products [12-14].

5.5 DESIGN OF PARTS FROM MOLDING MATERIALS

Equipment fabrication from prefabricated sheets of plastic materials generally is not difficult. For thermosetting materials (e.g., Faolite) the process is reduced to formation items from mild sheet and their hardening by heat treatment at 100-160°C.

Thermoplastics (polyvinyl chloride plastic, acrylics, etc.) are stamped from sheets heated to a softening temperature, then welded or glued. After stamping, the parts usually are machined. The fabrication of vessels from sheets is simple and does not require special limitations.

The fabrication of parts by stamping of molding material is specific and demands careful design considerations. In parts designing it is necessary to provide shapes that produce the following features:

1. high quality fabrication and a minimum of defective items;
2. minimum consumption of material;
3. ease of parts removal from lower and top stamping dies; and
4. minimum final finishing of a part.

The defective products result in incomplete material filling of stamping dies, inadequate pressing, cracking and warping (due to poor selection of part shape), and nonuniform hardening [15-17].

Smooth shape and equal wall thickness are very important to obtain high-grade parts. Therefore, the first criterion for obtaining quality parts is good design. The wall thickness should not be excessive. Excessive wall thickness results in additional consumption of material, which increases the time a part is punched in the machine. The minimum wall thickness that provides filling the mold is 2–4 mm. The minimum wall thickness for medium parts is approximately equal to 1 mm. The maximum wall thickness must not exceed 10–12 mm. With thick walls in regions of material accumulation, incomplete hardening occurs even if the part is kept in the mold for a long time. This tends to decrease the strength of a part. It is best to maintain the small wall thickness over the entire piece. If different wall thicknesses are inevitable, then the transition from one wall thickness to another should be as smooth as possible. The walls of plastic parts must be coupled smoothly.

Sharp corners on walls result in crack formation and stress concentration; therefore, they should be avoided. The following radii of curvature are recommended (Figure 5.43) in vertical projections and internal radii in horizontal projections:

$$R \geqslant 0.5 - 15 \text{ mm} \tag{5.4}$$

and for outside horizontal projections:

$$R = 5 - 10 \text{ mm} \tag{5.5}$$

For loaded parts, maximum possible radii of curvatures should be taken. To decrease bending stresses in the corners, it is permissible to make a smooth thickening by bringing it to the thickness of the wall (Figure 5.44).

Replacing massive thick-walled parts with thin-walled ones with ribs results

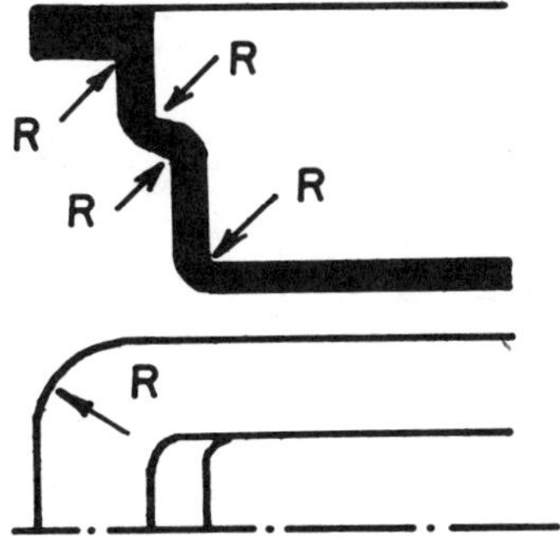

Figure 5.43. Radii of curvature for a plastic part.

Figure 5.44. Reinforcement of an angle.

in a savings of molding material. This increases quality and avoids warping of a part. An example of changing from a massive part to a thin one with ribs is given in Figure 5.45.

The rib thickness should be equal to the wall thickness and the edges of walls and the coupling walls and ribs are rounded. The ribs should be located in the direction of the top die so the part can be removed from the mold. The ribs should not prevent machining, and thus should not be brought to the machining surface (Figure 5.46(A)).

The wall draft located in the direction of pressing is used to permit easy withdrawal of a part. The draft is made no less than 1:200: the internal draft is recommended to assume more than the external one. Examples of proper and poor designs are given in Figure 5.47.

Increased rigidity and minimal warping can be achieved by changing flat surfaces into spherical or shaped surfaces (Figure 5.48). For example, parts with high walls should be fabricated with bent-in bottoms (Figure 5.47). Sometimes both ribbing and shaping are used to increase rigidity.

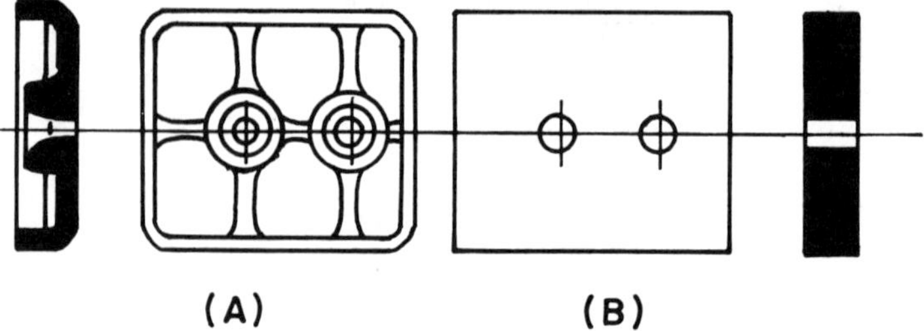

Figure 5.45. Replacement of a massive part (B) by a ribbed one (A).

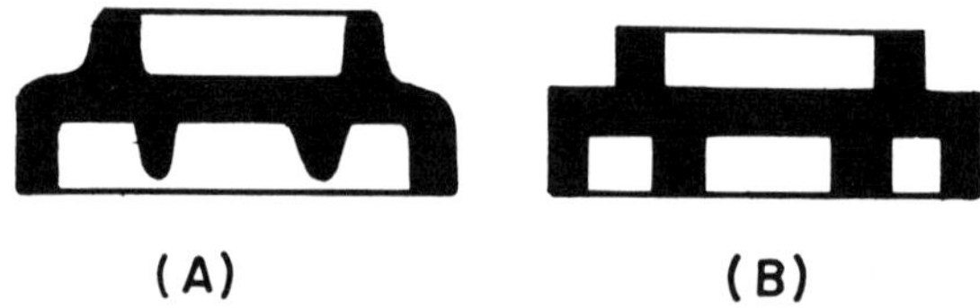

Figure 5.46. Decreasing the rib height for ease of machining: (A) good design; (B) poor design.

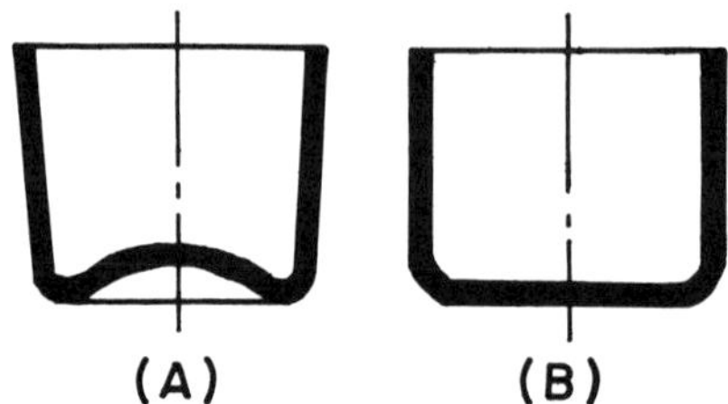

Figure 5.47. Draft made for easy part withdrawal: (A) good design; (B) poor design.

The shape of a part's edges should be considered carefully. Too thin an edge can be broken and would violate the principle of equal thicknesses of walls (Figure 5.49(C,D)).

If the part must be elastic, it is best to make it both without a bead and without edge thickening (Figure 5.49(A)), or to make a bead of minimum thickness (Figure 5.49(B)). The bead should extend around the perimeter

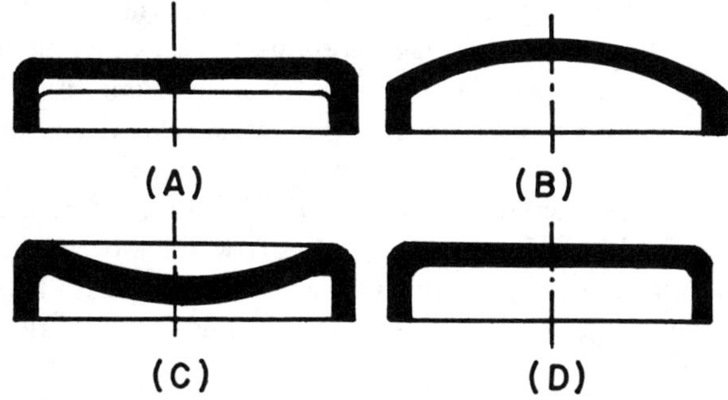

Figure 5.48. Flat surfaces such as A and D are poor designs. B and C represent acceptable designs that minimize warping.

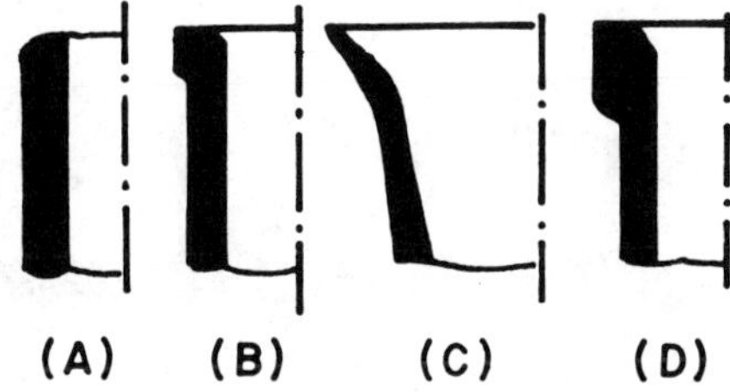

Figure 5.49. Examples of part edges: (A,B) good designs; (C,D) poor designs.

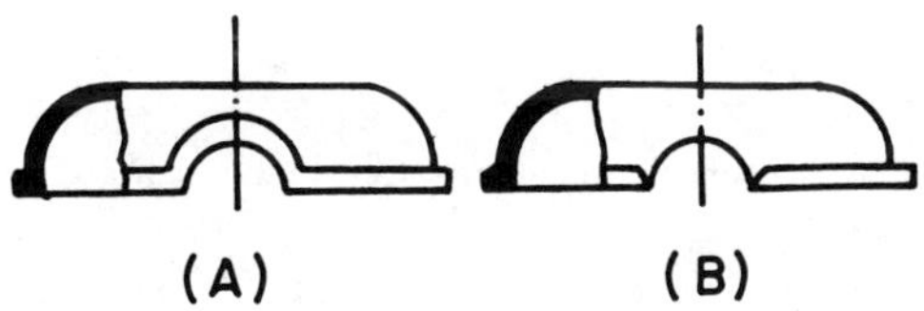

Figure 5.50. Design of an edge: (A) good design; (B) poor design.

of the part (Figure 5.50(A)). Bead interruption is not desirable because it may result in crack formation (Figure 5.50(B)).

It is not recommended that the part be supported around the entire base. Because of warping, the supported surface always becomes somewhat curved. Tightening with bolts may warp parts and induce bending stresses. Curved surfaces will hinder the support stability in this case. The supporting surface is better handled with separate lugs or projections, which should have a shape sufficient to withstand the stresses resulting from fastening. Stress concentration should be avoided, and the design should account for the maximum material volume (Figure 5.51(A,B)).

Holes for bolts and screws should be drilled with a clearance. This will provide easy fit and assembly even if some part warping occurs after stamping. Holes in parts of molding materials may be fabricated by two methods: molding a through-hole by pressing, or molding a blind hole with subsequent drilling after pressing.

Concentric vertical through-holes may be pressed up to 10 diameters long. Vertical blind holes may be pressed up to 7.5 diameters long. Vertical blind holes with eccentric hole geometries should be done up to 2 diameters long. If a larger length is required, the stud base should be thickened as shown in Figure 5.52. The studs should have a slope of 1:50.

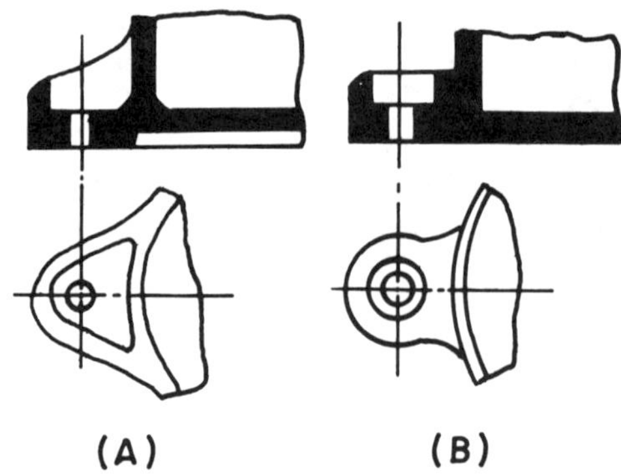

Figure 5.51. Supporting projections: (A) good design; (B) poor design.

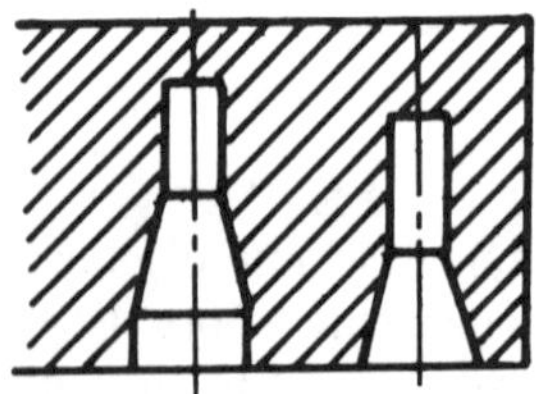

Figure 5.52. The shape of holes.

Undercutting should be avoided because it complicates the design of press molds and decreases the efficiency of pressing. For example, the head design according to Figure 5.53(A) is considerably simpler than that of Figure 5.53(B); however, they work equally well in equipment. For the same reason, holes should be avoided on the sides of the part. Such holes complicate the press mold fabrication, which, in this case, should be made detachable. Bringing the hole up to the upper edge of the part simplifies the forming considerably (Figure 5.54).

Often it is possible to avoid undercutting by fabricating a part in two sections, as shown in Figure 5.55(A,B). The designs in Figure 5.55(C,D) are not recommended. In changing the shape of the part one also can avoid undesirable undercutting. The thread in plastic parts at diameters greater than 5 mm is formed by pressing. The threaded connections may be given any profile. To avoid breaking the connection, it should be started at some distance from the edge of the part. If a metal screw must be screwed into the threaded hole it should be pressed in the plastic part in a threaded metal sleeve.

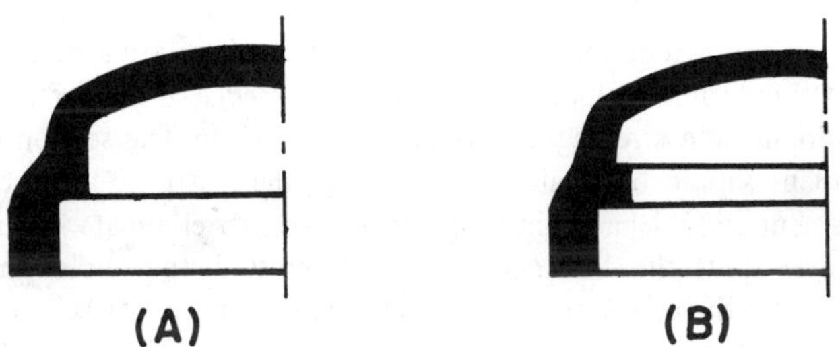

Figure 5.53. Head designs: (A) good design; (B) poor design.

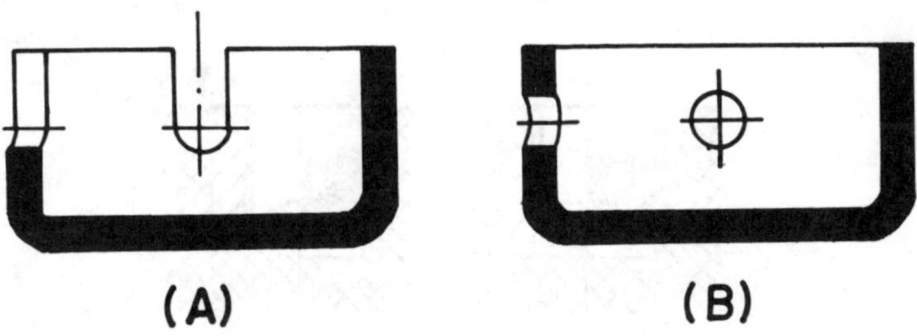

Figure 5.54. Holes in the side of a part: (A) good design; (B) poor design.

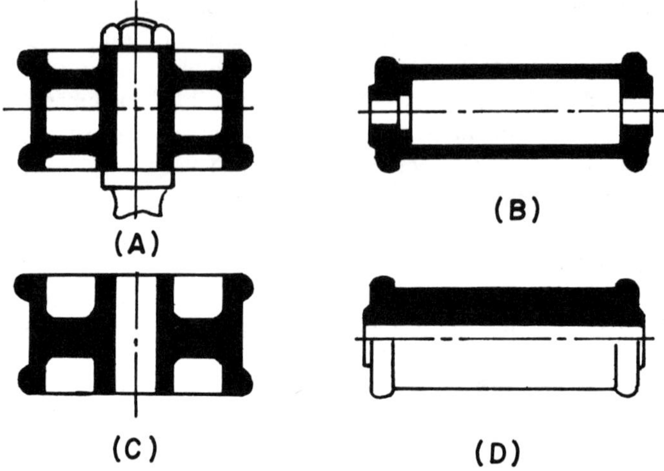

Figure 5.55. Avoiding undercutting by the use of section parts: (A,B) good designs; (C,D) poor designs.

Reinforcement is required to increase the strength of a plastic part. Reinforcement should not be located very close to the edge of the part, otherwise cracks, chipping and swelling of the plastic may occur. The section of metal reinforcement should be smaller than that of the plastic to avoid cracking. Reinforcement should have knurling, or grooves, to eliminate turning over and withdrawal. If the threaded sleeves are pressed, then holes should be blind to prevent the plastic from being penetrated (Figure 5.56).

The sizes of threaded sleeves are given in Figure 5.57. The heads of pressed studs should be larger than the stud diameter. The head should overlap the breech in the press mold to fit the stem with no less than 0.5 mm on each side; that is, $D \geqslant d + 1$ mm; otherwise the plastic will pour the thread

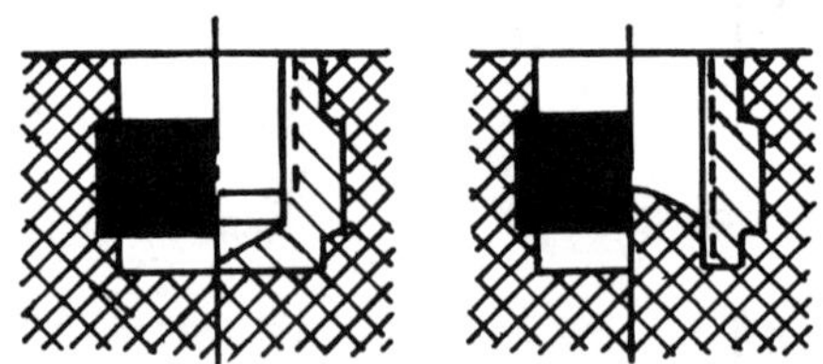

Figure 5.56. Pressing of threaded sleeves: (A) good design; (B) poor design.

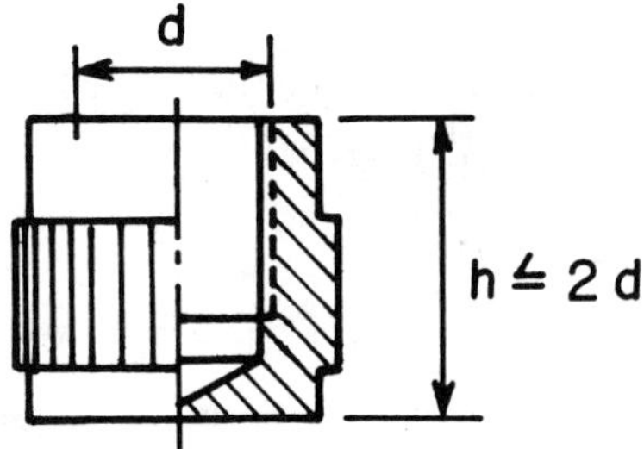

Figure 5.57. The sizes of threaded sleeves.

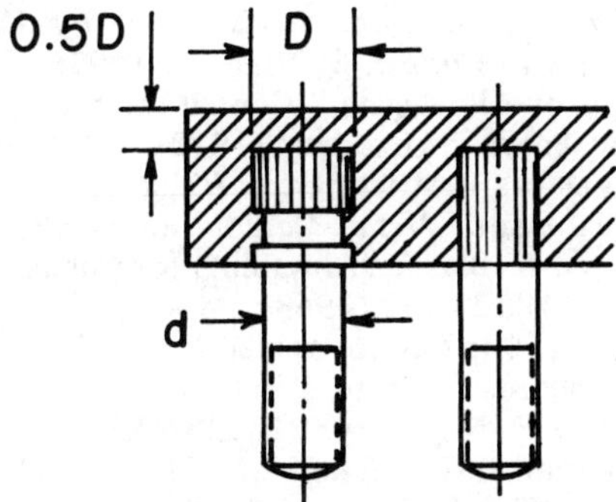

Figure 5.58. Pressing of studs.

(Figure 5.58). The length of threaded studs should not be less than 1.5 times the diameter or the stud will stand crooked. Plastic flash should be accessible for elimination by grinding or machining.

NOMENCLATURE

b	overlap length (in.)
d	diameter (in. or mm)
F	effective thread length (m)
h	height (mm)
K	overall heat transfer coefficient (kcal/m^2 C° hr)
ℓ	length (in.)
P	pressure (kg/cm^2)

S wall thickness (cm)
α coefficient of thermal expansion (in./(in. °F))
λ thermal conductivity (kcal/m °C hr)
σ_{allow} allowable stress (N/m^2)
ψ safety factor or strength factor

REFERENCES

1. Jefferson, T. B. *Welding Encyclopedia* (Zurich, IL: Jefferson Publishers, 1974).
2. Kennedy, G. A. *Welding Technology* (Indianapolis, IN: Bobbs-Merrill Publishers, 1977).
3. Hicks, J. G. *Welding Joint Design* (New York: Halstead Publishers, 1979).
4. Medovar, B. I. *Welding Austenitic Steels and Alloys,* Kiev, USSR (1964).
5. Elliot, Z. *Working with Copper,* Sterling, NY (1976).
6. "Copper and Copper Alloy Data," Conseil International pour le Development du Cuigze, Geneva, Switzerland (1970).
7. Moore, D. C. "Factors Restricting the Wider Use of Copper and Copper Alloys in Chemical Plant," *Metals Materials* (June 1970), pp. 251-257.
8. Blewett, R. V., and S. Black. "Aluminum for Chemical & Process Plant," *Brit. Welding J.* 10(4):124-131 (1963).
9. "British Standard 1500; Fusion Welded Pressure Vessels, Part 3: Aluminum," British Standards Institute, London, England.
10. Blewett, R. V., and E. N. Skerry. "The Performance of Aluminum Welds in Corrosion Environments," *Metallurgia,* 71:72-81 (1965).
11. "British Standard 2901: Parts 1 and 2," British Standards Institute, London, England.
12. Bates, F. *Enameling: Principles and Practices* (New York: TX Crowell Publishers, 1974).
13. Clarke, G., et al. *The Technique of Enameling* (New York: Van Nostrand Reinhold, 1977).
14. Herbert, M. *Metalwork and Enameling* (New York: Dover Publishing, 1971).
15. *Modern Plastic Encyclopedia* (New York: McGraw-Hill Book Co., 1981).
16. Vincent, P. I. *Thermoplastics in Design, Vol. 4, Engineering Materials and Design* (New York: McGraw-Hill Book Company, 1961), pp. 816-821.
17. Czehorsky, G. "The Use of Heat Resistant PVC in the Construction of Chemical Plant and Equipment," *Kunstoffe*, 60:354-358 (1970).

CHAPTER 6

PRINCIPLES OF SHELL DESIGN

6.1 INTRODUCTION

The chemical process industries (CPI) are concerned with the conversion of one material into another. Thus they require materials to be handled and stored in vessels of varied construction, depending on the state of the material, its physical and chemical properties and the required operations to be performed. Most process equipment simply includes vessels with various modifications that enable them to perform certain required functions. For example, an autoclave is a high-pressure vessel equipped with a mixer and heating element; a distillation or absorption column is a vessel that contains a series of trays; a heat exchanger is also a vessel that contains suitable provisions for the transfer of heat through tube walls.

In the design of a vessel as a unit, a number of criteria must be considered. Some of these criteria are: (1) the properties of the material used, (2) the induced stresses, and (3) the elastic stability. Others will be discussed later.

To simplify designs and lower the cost of fabrication, the components of the vessel are constructed in the form of simple geometric shapes–spherical, cylindrical and conical–and the shells of most vessels are a combination of them.

From the point of view of material savings and uniform distribution of stresses induced in the material of the walls from loads, a spherical shape is the most favorable. A spherical geometry provides minimum surface per unit volume, and its wall thickness is minimum for a given pressure. However, the fabrication of spherical shells is more complicated and expensive than cylindrical ones, and they are awkward for distributing internal working devices and for providing even flow of reagents.

For this reason, spherical vessels are used mostly for the storage of gases and liquids. For large-volume, low-pressure storage operations, spherical vessels are more economical; however, at higher storage pressures the volume of gas is reduced and the cylindrical type of storage vessel becomes more economical. For handling small amounts of gases, cylindrical storage vessels are more economical than spherical ones because the cost of fabrication becomes the controlling factor. Spherical vessels also are used in cases in which it is necessary to maintain the ratio of external surface to internal volume of the vessel at a minimum, for example, as storage tanks for liquid oxygen. An example is shown in Figure 6.1.

Most apparatuses are designed as cylindrical shells fabricated of rolled sheet metal or cylindrical cast shells. Apparatuses with flat walls are used less frequently, usually only in equipment working under low pressure drops. This type of apparatus is not economical because of its poor resistance to pressure and high consumption of metal per unit of useful volume. Apparatuses with flat walls also are inconvenient from the standpoint of manufacturing: they are warped in welding, and their casting is limited and involves the fabrication of defective parts. However, in many cases it is impossible to manage without flat elements as they often serve important roles in making up parts of an apparatus (such as tube sheets of heat exchangers, trays for columns, etc.). In these cases, they should be reinforced by ribs, tie rods, tubes, etc.

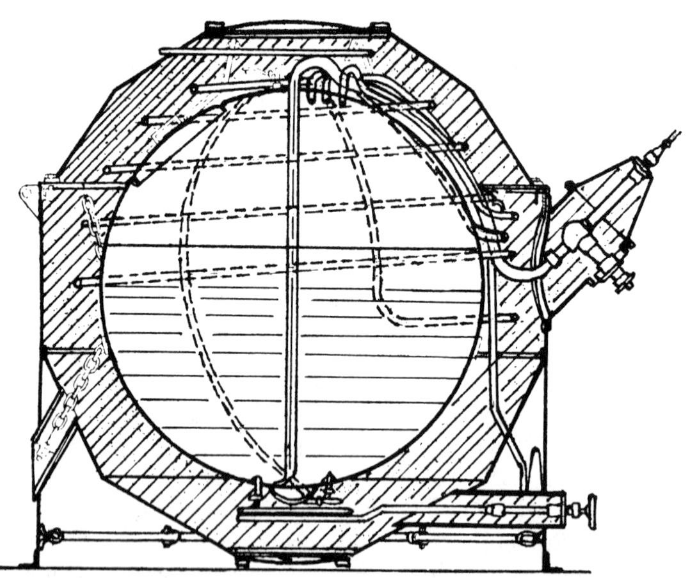

Figure 6.1. Storage tank for liquid oxygen.

The most common types of vessels are classified according to their:

1. Function: storage vessels, reactors, mixers
2. Structural material: steel, cast iron, copper, plastics, etc.
3. Method of fabrication: welded, cast, brazed, forged, etc.
4. Geometry: cylindrical, spherical, conical, combined
5. Scheme of loading: working under atmospheric internal, or external, pressures
6. Wall temperature: heated, unheated
7. Corrosion action: moderate or high corrosion effect
8. Location in space: vertical, horizontal, sloped
9. Method of assembly: detachable, nondetachable
10. Wall thickness: thin-walled ($d_e/d < 1.5$), thick-walled ($d_e/d \geqslant 1.5$)

From the point of view of structural design, the shells of vessels and apparatuses are parts of envelopes of rotation connected with plates loaded symmetrically with respect to the axis. For simplification, the stressed state of material in such construction is considered biaxial, which is permissible because of low values of radial stresses originating in thin-walled apparatuses [1].

The stressed state of shell materials in similar structures is the sum of two components: (1) the stressed state originated by uniformly distributed forces on the surface as a result of gas or liquid pressure, empty weight, etc., and (2) the stressed state originating under action of forces and moments distributed around a contour (i.e., end effect).

The first is determined either by membrane or moment theory, as may be required. The results obtained by the use of membrane theory are accurate for engineering design and widely used for structural design [2]. The moment theory usually is not applied for determination of the stressed state from forces distributed uniformly on a surface. The design equations from moment theory are complex and the design often laborious; however, the difference in results obtained by the use of these two methods is insignificant.

End forces and moments originate in sections where there is an abrupt change in load, wall thickness or in the properties of the structural material, as well as in places where fixing or applying additional connections occurs. The stressed state induced by end forces and moments are determined by the use of moment theory [3]. The stress and deformation induced by end effects have local characteristics and influence only the zones located close to the place where end forces and moments are applied. However, end stresses may reach high values, and design measures must be taken to decrease them. The end stresses are especially dangerous for apparatuses fabricated from brittle materials, such as silicon cast iron, thermosetting plastics, ceramic, etc., as well as for apparatuses operating under alternating loads. When designing the stressed state of material from distributed forces, the stressed state induced by end effects should be taken into account.

There are two fundamentally different approaches to determining structural reliability of the parts of equipment. According to the theory of elasticity, strength is determined by the ultimate stress with which the part can pass without rupture. The theory of plasticity demands that the strength should be determined by the ultimate load that the part can pass without residual deformation.

The first design method is based on the assumption that the material of the part is in an elastic state everywhere, and nowhere does the stress exceed the yield point. In the complex-stress state of the material, the combined stress is determined by the use of one of these strength theories, and then compared to the stress.

The materials of parts of process apparatuses generally are not loaded uniformly. If the structural design is based on the ultimate stress originating at the most loaded place of construction, inevitably it would cause excessive consumption of structural materials. Therefore, for designs based on the membrane theory, average stresses are assumed, paying no attention to the existence of local stresses of significant intensity. The disadvantage of such an approach to strength evaluation is that the actual safety factor in different parts of the structure remains unclarified.

In heavily loaded parts of structures fabricated from plastic material, partial transition into the elastic-plastic state occurs because total utilization of the strength of the material takes place under such loads where plastic deformations extend on all dangerous portions. This design method allows for a more objective evaluation of the maximum load the structure can withstand without destruction. However, this method is valid only for designing equipment of plastic materials, whereas the design method based on ultimate stresses may be used for equipment designs from any material.

It should be remembered that the stresses exceeding the yield point result in a material, which is cold-hardened accompanied by a decrease in corrosion resistance. These should not be used in the parts that contact corrosive substances.

There are always changes in temperature and pressure during the operation of equipment. Lowering temperature decreases the plasticity of the material and may cause cracks and failure. Likewise, an increase in pressure or temperature may become dangerous because of stress increases in the parts of equipment and decreases in material strength.

6.2 CYLINDRICAL SHELLS

Cylindrical shells are characterized by their great structural strength, simplicity of fabrication and reasonable consumption of material. They are employed widely throughout the manufacturing of process equipment.

Depending on their function, cylindrical vessels may be vertical or horizontal. Preference often is given to the vertical type, especially for thin-walled vessels operating under low internal pressure. Additional bending stresses are eliminated in the shell from gravity forces of the vessel itself, and from the fluid inside the vessel.

Cylindrical shells of elastic materials (steels, nonferrous metals, and most of their alloys, PVC, etc.) operating under internal pressure up to 10 MN/m^2 are fabricated mostly of rolled, and subsequently welded, sheets. Except for welding, the joint connections of cylindrical shells of copper or brass sheets are made by soldering with hard and soft solders. Cylindrical shells of elastic materials (mostly of steel) for high-pressure vessels (higher than 10 MN/m^2) are fabricated mostly of forged pieces, with subsequent heat treating and machining outside and inside the shell. Besides the forged shells for high-pressure vessels, compound designs are used (sandwich and wound structures, etc.).

Cylindrical shells of brittle materials (cast iron, bronzes, quartz glass, etc.) for vessels operating under low pressure (up to 0.8 MN/m^2) are molded with subsequent machining of internal surfaces, (and sometimes without machining). The cast shells usually are fabricated together with a bottom. In some cases, cast shells are fabricated of elastic materials (steels, nonferrous metals and their alloys, etc.). The relation between the height, H, of the shell and its internal diameter, d, for vertical vessels, as well as between the length, L, of the cylindrical shell and its internal diameter, d, for horizontal vessels, is determined by process requirements, which are in the limits of

$$\frac{H}{d} \leqslant 30 \text{ to } \frac{H}{d} \leqslant 10$$

The fabrication of cylindrical shells by the rolling of sheets is a widespread method for manufacturing low-pressure (up to 1.6 MN/m^2) and medium pressure (from 1.6 to 10.0 MN/m^2) vessels.

In designing welded and soldered cylindrical shells from rolled sheets, the following should be considered:

1. The total length of seams should be minimized; therefore, the sheets should be chosen of large sizes conforming to reasonable cutting.
2. The shells might be rolled either along the long or the short side of the sheet. In cutting, it is recommended that a minimum number of longitudinal seams in the shell be provided.
3. The welding of longitudinal and circumferential seams must be butt-welded.
4. The longitudinal seams of adjacent courses should be shifted from each other at a distance equal to more than three times the shell thickness, and not less than 100 mm between center lines of the seams.
5. All joints should be available for inspection and, if necessary, for repair.

6. The longitudinal seams in horizontal vessels should not be located in the lower part of the vessel if they are not attainable for inspection.
7. It is not recommended to provide holes for tubes, manholes, etc. on the seams (especially on longitudinal seams).

Shell thicknesses of vessels operating at pressures up to $P = 0.7\ kg/cm^2$ are not designed. Instead, they are taken from manufacturing considerations, and durability is determined by the corrosion resistance of the material. The minimum thickness of a shell rolled from sheet metal and welded is taken for carbon and low-alloy steels, 4 mm; for austenitic steel, 3-4 mm; for copper shells, 2.5-30 mm; for cast shells, 20-25 mm.

Before a storage tank can be designed, the optimum height to diameter ratio should be established.

The volume of sheet material required for the fabrication of a storage tank is determined by the following formula (Figure 6.2):

$$A = \pi d_e H S_1 + \frac{\pi d_e^2}{4}(S_K + S_\partial) = \pi d_e H S_1 + \frac{\pi d_e^2}{4} S_2 \tag{6.1a}$$

The tank capacity is

$$V = \frac{\pi d_e^2}{4} H \tag{6.1b}$$

The tank diameter is

$$d_e = \sqrt{\frac{dV}{\pi H}} \tag{6.2}$$

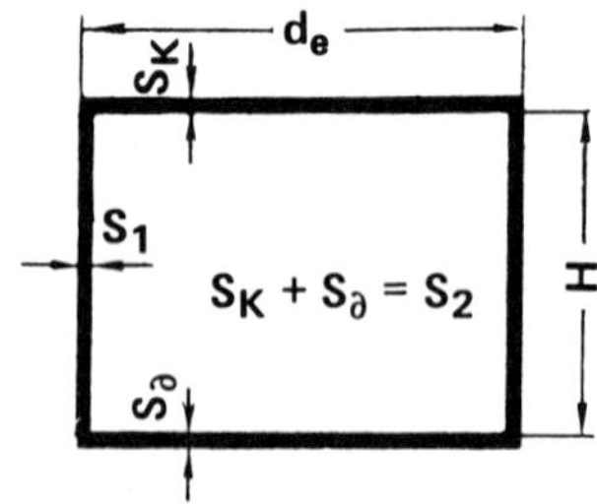

Figure 6.2. Sketch for designing a cylindrical tank.

Substituting d_e in Equation 6.1a, we obtain

$$A = 2S_1\sqrt{\pi VH} + \frac{VS_2}{H}$$

For determining the minimum volume of steel, we find the first derivative of A/H and set it equal to zero:

$$\frac{dA}{dH} = S_1\sqrt{\pi V/H} - \frac{VS_2}{H} = 0$$

$$VS_2/H = S_1\sqrt{\pi VH}$$

The optimum tank height is then

$$H_{opt} = \sqrt[3]{\frac{V}{\pi}\left(\frac{S_2}{S_1}\right)^2} \tag{6.3}$$

Substituting H_{opt} in Equation 6.3, we obtain the optimum tank diameter:

$$D_{opt} = 2\sqrt[3]{\frac{V}{\pi}\frac{S_1}{S_2}} \tag{6.4}$$

In addition to tanks of constant shell thickness, tanks with courses of different thicknesses also are used. The upper courses are fabricated of minimum sheet thickness (4 mm for steel). Middle and lower ones are made of thicker sheet thickness according to the design. The limiting tank volume beyond which the tank should be fabricated of different thickness sheets is determined as follows: the hydrostatic head of a liquid column, H, on the tank bottom is $P = H\rho$ (where ρ is liquid density). The shell thickness of a tank with external diameter, d_e, is

$$S_1 = \frac{H\rho d_e}{2\sigma_{allow}} \tag{6.5}$$

Substituting the optimum values H_{opt} and D_{opt} and solving, an equation for V is obtained:

$$V = \pi S_1^2 \sqrt{\frac{\sigma_{allow}^3}{\rho^3 S_2}} \qquad (6.6)$$

Taking shell thickness, $S_1 = 0.4$ cm, head and bottom thickness, 0.4 cm, that is $S_2 = 0.8$ cm, liquid density 1 kg/dm^3 and allowable stress in the shell, $\sigma_{allow} = 1400$ kg/cm^2, we obtain the optimum tank volume for a constant shell thickness, $V_{max} \approx 1000$ m^3.

6.2.1 Design of Steel Rolled-Up Welded Cylindrical Shells Operating Under Internal Pressure

The structural design of cylindrical shells is based on the equations of membrane theory. For the equations of this theory to be in a convenient form for engineering design, it is necessary to:

1. select the strength theory best reflecting the material behavior;
2. establish a safety factor and allowable stresses;
3. consider the weakening of construction induced by welding and other connections; and
4. consider the wall thinning down as a result of negative tolerances of a sheet and corrosion effects.

The energy theory of strength–Huber/von Mises/Hencky theory [4-6]–assumes that the critical state of a material depends on the amount of potential energy accumulating at deformation in the unity of material volume that is numerically equal to the specific work of deformation expressed by all three main stresses. According to this theory, the design stress is

$$\sigma_{design} = \sqrt{\sigma_T^2 + \sigma_v^2 + \sigma_R^2 - 2(\sigma_T\sigma_v + \sigma_T\sigma_R + \sigma_v\sigma_R)} \leqslant \sigma_{allow} \qquad (6.7)$$

where σ_T, σ_v and σ_R are the circumferential, longitudinal and radial stresses, respectively. As the dangerous state of plastic materials is approached, a definite amount of deformation potential energy, associated only with the change of the specimen, is assumed (the volume of a specimen at a given amount of deformation does not change). Hence, Poisson's coefficient $\tilde{\mu} = 0.5$ under this condition.

Consider the case of a thin-walled cylindrical shell operating under internal pressure, P (Figure 6.3). The radial stress, σ_R, is assumed to be equal to zero. The longitudinal force resulting from an internal pressure acting on a thin cylinder of thickness, t, length, ℓ, and diameter, d, is (force tending to rupture vessel longitudinally)

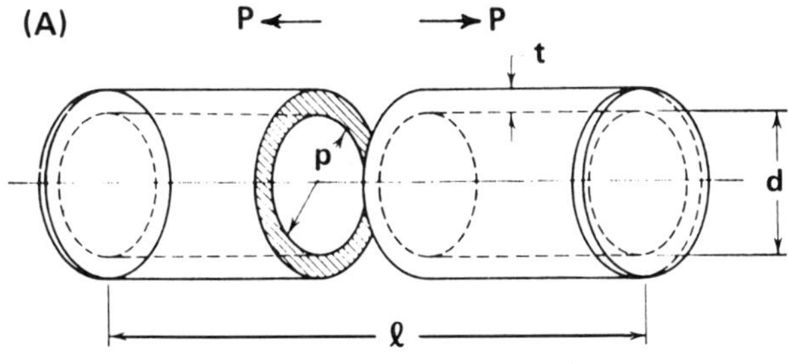

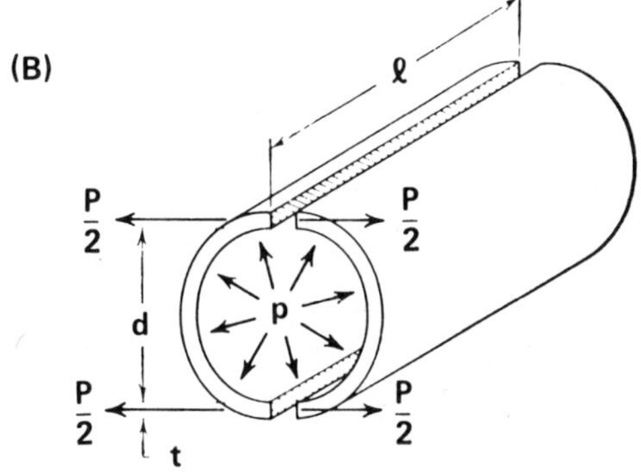

Figure 6.3. (A) Longitudinal forces acting on thin cylindrical shell under internal pressure; (B) circumferential forces acting on thin cylindrical shell under internal pressure.

$$P' = \frac{P\pi d^2}{4}$$

and the area of metal resisting longitudinal rupture is

$$a = t\pi d$$

Therefore, the induced stress is

$$\sigma = \frac{P'}{a} = \frac{P\pi d^2/4}{t\pi d} = \frac{Pd}{4t} \tag{6.8}$$

or

$$t = \frac{Pd}{4\sigma} \tag{6.8a}$$

The circumferential force tending to rupture the shell shown in Figure 6.3(B) is

$$P' = Pd\ell$$

and the area of metal resisting force is

$$a = 2t\ell$$

The stress in this case will be

$$\sigma = \frac{P'}{a} = \frac{Pd\ell}{2t\ell} = \frac{Pd}{2t} \tag{6.9}$$

or

$$t = \frac{Pd}{2\sigma} \tag{6.9a}$$

According to Equations 6.8 and 6.9, Equation 6.7 may be rewritten as follows:

$$\sigma_{design} = \sqrt{\left(\frac{PR_2}{2t}\right)^2 + \left(\frac{PR_2}{t}\right)^2 - 2 \times 0.5\left(\frac{PR_2}{2t}\right)\left(\frac{PR_2}{t}\right)} \leqslant \sigma_{allow} \tag{6.10a}$$

where

$$R_2 = \frac{d_c}{2} = \frac{d + (t + c)}{2}$$

where R_2 is the radius of the central surface. Hence,

$$\sigma_{design} \approx 0.87 \frac{PR_2}{t} \leqslant \sigma_{allow} \tag{6.10b}$$

According to the energy theory of strength, considering the combined stress of material the design stress for plastic materials is 13% less compared with the maximum value of the main stress. From Equation 6.10a we obtain

$$\sigma_{allow} = \frac{Pd_c}{\frac{2}{0.87}t} = \frac{Pd_c}{2.3\ t} \tag{6.10c}$$

and

$$t = \frac{Pd_c}{2.3\ \sigma_{allow}} \tag{6.11}$$

Introducing the joint efficiency factor, ψ, and an allowance factor, c, for corrosion, erosion and negative tolerance of the shell thickness, we obtain the following design formulas:

$$t = \frac{Pd_c}{2.3\ \sigma_{allow}\psi} + c \tag{6.11a}$$

and

$$\sigma_{allow} = \frac{Pd_c}{2.3\ \psi(t - c)} \tag{6.11b}$$

Taking into account Equations 6.10b and 6.11a, we obtain the final design formula for determination of the shell thickness:

$$t = \frac{Pd}{2.3\ \psi\sigma_{allow} - P} + c \tag{6.12}$$

This formula may be used for design where

$$\beta = \frac{d_e}{d} \leqslant 1.5 \tag{6.13}$$

Sizing the shell thickness from Equation 6.12 can be done after appropriate selection of values for σ_{allow}, ψ and C.

The most important problem in design is the proper selection of the allowable stress to provide safe operation of equipment and to eliminate excessive consumption of structural material.

The allowable stress is controlled by a number of factors, such as:

1. mechanical properties of material–its strength and ductility (or brittleness);
2. character of load and its constancy in time;
3. temperature (its influence on strength and ductility of material);
4. design method with which the load can be estimated and reliability of stresses computed from these loads; and
5. the uniformity of material, the hazard if the failure occurs and such other considerations as local stress concentration, impact shock, fatigue and corrosion.

The static nature of mechanical loads and a wide temperature range of shells are characteristic for process apparatus. This last case leads to the fact that the same material, e.g., carbon steel, might be impermissibly brittle when in cryogenic temperatures, and be too weak and prone to fatigue under high temperatures.

In shell design, the criterion for determining allowable stresses in moderate temperatures, is the tensile strength, σ_T. In this case,

$$\sigma_{allow} = \frac{\sigma_T}{\eta_T} \tag{6.14}$$

where η_T is a safety factor.

For shells that are designed on the basis of ultimate stresses, it is evident that the allowable stress, σ_{allow}, should not exceed the yield point, σ_y. Consequently, the following inequality should be fulfilled:

$$\sigma_{allow} < \sigma_y$$

Increasing temperature has different influences on various materials. Non-ferrous metals and some plastics undergo a decrease in ultimate strength and yield point, σ_y, with increasing temperature. As shown in Figure 6.4, the ultimate strength of a low-carbon steel, σ_T, first increases and then rapidly decreases with temperature rise. The yield point, σ_y, decreases continuously

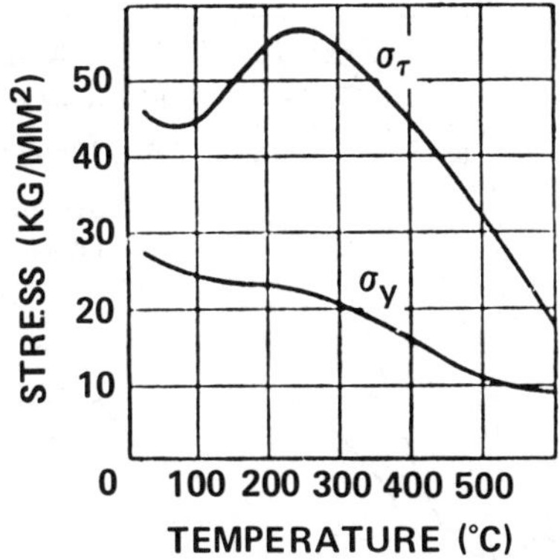

Figure 6.4. Influence of temperature on the ultimate strength, σ_T, and yield point, σ_y.

with heating. Therefore, determination of yield point with respect to ultimate strength, σ_T, may bring about an incorrect increase in the safety factor when heating low-carbon steel, although in reality the opposite occurs.

The reason for the use of the ultimate strength as a criterion for allowable stress is the lack of a plastic zone for brittle materials, such as gray cast iron. This is illustrated in Figure 6.5. This plot indicates a no yield point, and that fracture occurs at the ultimate strength. For brittle materials similar to gray cast iron, the allowable stress should be based on the ultimate strength.

Most nonferrous materials have no well-defined yield point. In this case, the value of a yield point corresponding to some specified permissible strain may be obtained. For example, in the 0.2% offset method a line is drawn parallel to the modulus line from the 0.2% elongation point, and the intercept of the stress-strain curve with this line is taken as the yield strength of the material.

In cases in which failure can be expected as a result of plastic deformation, the yield point should be used as a basis for determining the allowable operating stresses. In Figure 6.6, the points on the curve correspond as follows: A, proportionality limit; B, yield point; and C, ultimate strength. However, if the vessel is designed to meet code requirements, the procedure specified in the codes must be used [2]. The pressure-vessel codes still use a safety factor of four based on the ultimate strength for determining the allowable stresses for pressure vessels. For evaluation of safety factors the following relation is recommended:

$$\eta_y = \frac{\sigma_y}{\sigma_{allow}} \tag{6.14a}$$

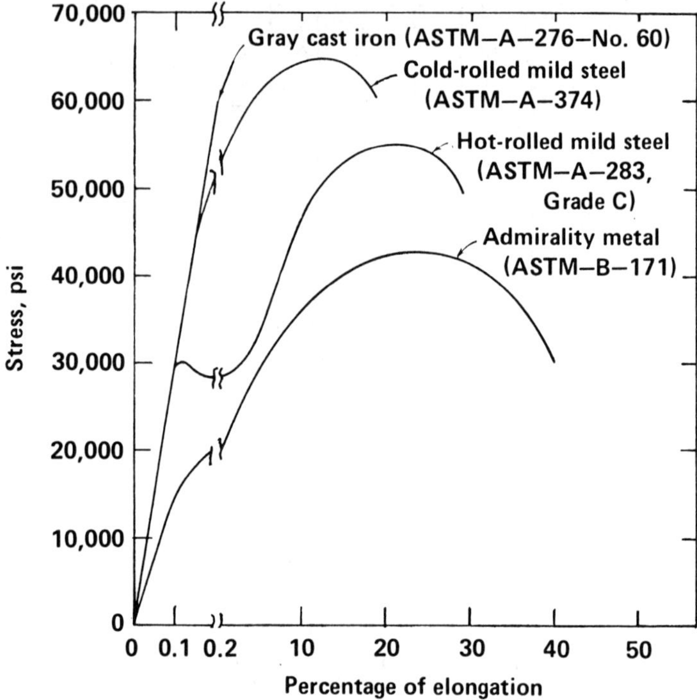

Figure 6.5. Typical stress-strain curves for various metals.

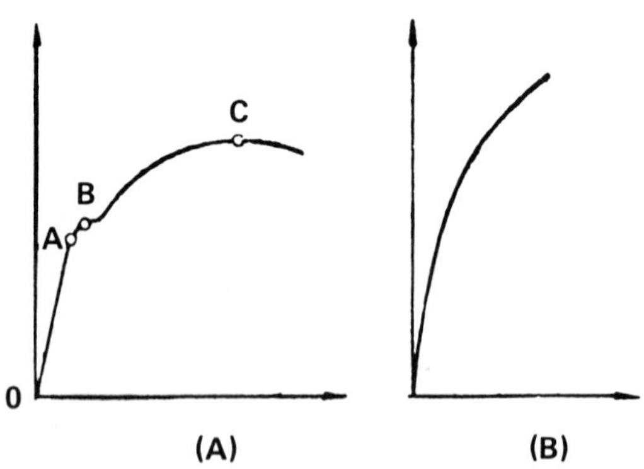

Figure 6.6. Stress-strain diagrams for (A) steel and (B) cast iron.

Since temperature increases are accompanied by an increase in the metal's ductility, the yield point, σ_y, decreases continuously. Hence, one may use this ratio for the determination of allowable stresses under higher temperatures by substituting into the equation a value for the yield point of the material for a given operating temperature, σ_y^t:

$$\sigma_{allow} = \frac{\sigma_y^t}{\eta_y} \tag{6.15}$$

To this point the assumption has been made that the strain under load does not vary with time, which essentially is true for ferrous materials under load at temperatures up to 650°F. Beyond the temperature at which the material "creeps" under load, an increase of strain occurs with time, but not exceeding stresses beyond the yield point. The higher the service temperature, the more rapid the elongation of the loaded parts. At the same time the strain does not become elastic but residual. The yield point at high temperatures cannot be used as a criterion for allowance of the stressed state because the creep of the material may cause failure as a result of deformation increases over a long time period.

Thus, the basic criterion for shells operating at moderate temperatures is the size stability of loaded elements. Hence, at a high temperature range we should not use this criterion; rather, we should strive for an increase in sizes, ensuring that creep does not exceed a permissible limit. The creep rate usually is taken to tighten the hot bolt connections, corresponding to approximately 10^{-7} mm/mm hr. The allowable residual strains because of creep are 1%, for parts of carbon steel and 1.5% for heat-resistant alloy steels.

The important feature of behavior of loaded material under creep is relaxation, that is, stress decreases with time as a result of plastic deformation. The creep rate is extremely dependent on two factors: temperature and the stress in metal. Figure 6.7 shows typical creep rate curves for a material placed under constant axial load at constant temperature [3]. Extension A of the diagram indicates elastic strain, which occurs on application of the initial load. In the time interval, B, "the first stage of creep" begins, followed by, the constant-rate period, C, the "second stage of creep," which is the region that limits the service life of equipment. The "third-stage of creep," D, is an increasing rate period that ends in fracture of the specimen if the test is continued. When the test is interrupted before fracture, the specimen undergoes an elastic contraction (extension E). The amount of permanent strain is indicated by extension, F. Typical creep curves for a high-alloy steel are presented in Figure 6.8. As shown, the change in stress greatly increases the creep rate and shortens the service life of the material. By obtaining

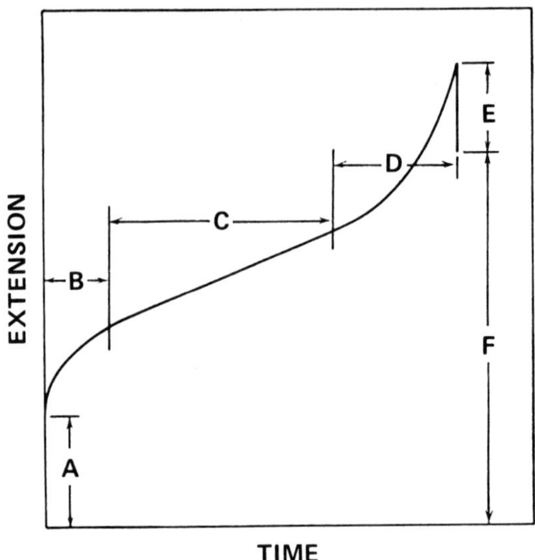

Figure 6.7. Creep curve–extension plotted against elapsed time: (A) elastic extension; (B) creep at decreasing rate; (C) creep at approximately constant rate; (D) creep at increasing rate; (E) elastic contraction; (F) permanent change of shape.

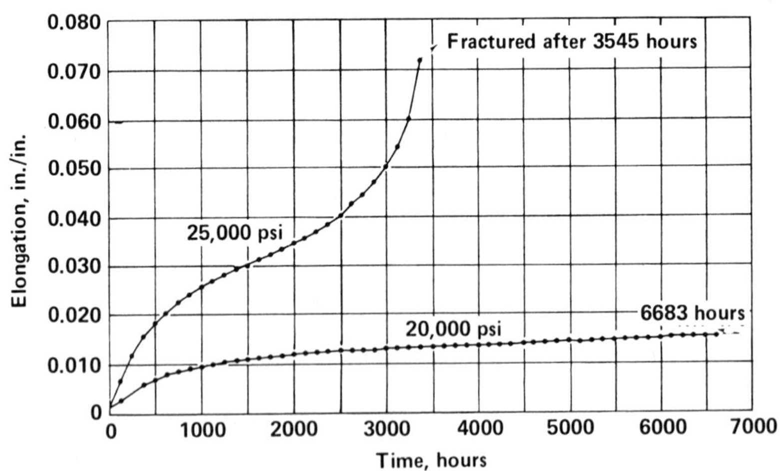

Figure 6.8. Typical time-elongation curves.

several plots similar to Figure 6.8, it is possible to construct a cross plot, which determines the slopes of the "second-stage creep." This is shown in Figure 6.9. For materials operating under creep conditions, the allowable stress is determined relative to the creep limit at the service temperature, σ_c^t:

$$\sigma_{allow} = \frac{\sigma_c^t}{\eta_c} \; ; \quad \eta_c = \frac{\sigma_c^t}{\sigma_{allow}} \tag{6.16}$$

Nevertheless, the creep limit is not a sufficient characteristic and could not serve as a criterion for strength design of equipment operating at high temperatures. Experimental data of creep limits are determined over a short period of time for insignificant rates of deformation. A more reliable approach is the stress-rupture or creep-rupture test [7], which is similar to the creep test except that high loads are used to produce greater creep rates and to rupture the specimen in a reasonable length of time. The strain in this test usually exceeds 50%. Figure 6.9 presents curves plotted from both creep test data and creep-rupture test data. The time of rupture versus stress is given in Figure 6.10. The ASME code [2] specifies that the allowable stresses at the high temperatures are based on 100% of the stress to produce a creep rate of 1/100% per 1000 hours (0.00001% hr), or on 60% of the average stress or 80% of the minimum stress to produce rupture at the end of 100,000 hours, whichever is the lower value.

Formulas for determining allowable stresses and safety factors on the basis of fatigue stresses may be expressed as follows:

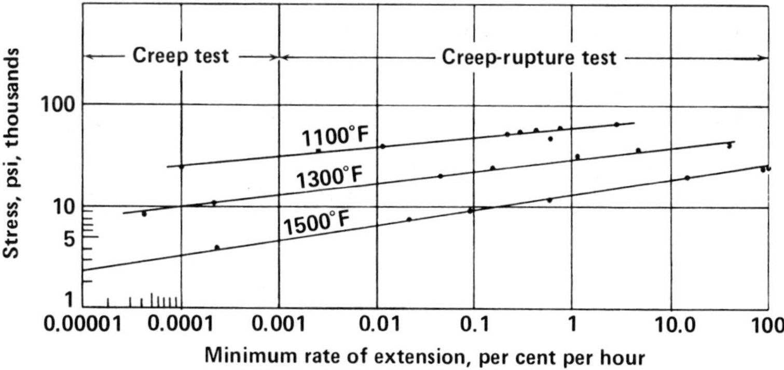

Figure 6.9. Typical correlation of creep-test and creep-rupture-test data.

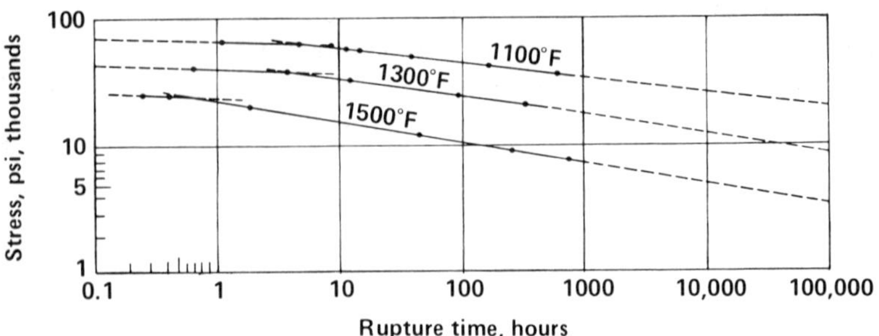

Figure 6.10. Typical stress vs rupture time curves.

$$\sigma_{allow} = \frac{\sigma_f^t}{\eta_f}\ ;\ \eta_f = \frac{\sigma_f^t}{\sigma_{allow}} \tag{6.17}$$

The values of nominal allowable stresses, σ_{allow}^*, of low-carbon and alloy steels depend on temperature and are given in the ASME code [2]. For other types of steel the nominal allowable stress, σ_{allow}^*, is taken to the least value of the three allowable stresses. The design on the basis of fatigue strength should be done if the service temperature of the wall exceeds ~470°C for carbon steels, and ~550°C for austenitic steels. The safety factors for tensile strength and yield point are taken: $\eta_T = 3.0$; $\eta_y = 1.65$; $\eta_f = 1.65\text{-}2.0$. The design ratings of σ_T, σ_y and σ_f should be assumed minimum values, which are obtained experimentally. Further, the actual allowable stress is determined by multiplying the nominal stress by a coefficient, η, which considers design features and influences of temperature that may weaken the shell. The following values of η may be used:

Shells welded without holes or with reinforced holes:

unheated $\eta = 1.0$
heated $\eta = 0.9$

Shells welded or seamless with holes without reinforcement:

unheated $\eta = 0.95$
heated $\eta = 0.85$

The actual allowable stress, σ_{allow}, should be decreased further by the coefficient ψ' in the following cases:

1. For equipment for which the volume of media or its pressure may be increased as result of action, $\psi' = 0.65$.
2. For equipment heated outside by open flame or electric coil, $\psi' = 0.85$.
3. For equipment cast from steel, $\psi' = 0.7$.
4. For all other cases and steam boilers, $\psi' = 1.0$.

The allowable stresses may be increased 1.5 times. The welded seam may not be as strong as the adjoining rolled-steel plate of the shell, and allowance should be made for such weakness by introducing a "joint efficiency factor," ψ, which has different values for the following shell joints.

1. For a butt joint made by manual welding with root undercut and subsequently welded on (Figure 6.11a), $\psi = 0.95$.
2. For a butt joint with backing strip made by automatic and manual arc-welding (Figure 6.11b), $\psi = 0.9$.
3. For a single-welded butt joint made by manual arc welding (Figure 6.11c), $\psi = 0.7$.
4. For a single-welded butt joint made by automatic arc-welding (Figure 6.11b), $\psi = 0.8$.
5. For a double-welded butt joint made by automatic welding (Figure 6.11d), $\psi = 1.0$.
6. For a butt-welded joint working for compression, $\psi = 1.0$.
7. For a joint welded by water gas, $\psi = 0.85$.
8. For a double-fillet T weld (Figure 6.11f), $\psi = 0.95$.
9. For a double-fillet T weld with intermittent weld, $\psi = 0.8$ (the design considers the total length of the seams).
10. For a double lap-weld (Figure 6.11e).

Parts of equipment are connected with butt-welds undercut and welded on; single and double-welded butt joints. The weld reinforcement of filler metal is not taken into consideration in determining the safety factor.

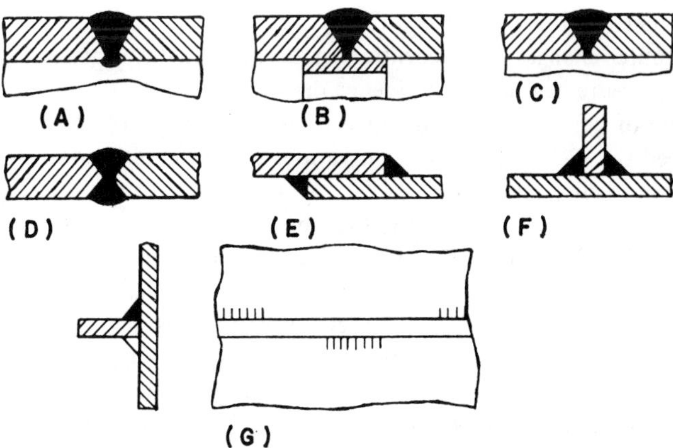

Figure 6.11. Types of welded joints.

For shells weakened by holes of the same diameter located along the shell axis, the joint efficiency factor is

$$\psi = \frac{t - d}{t} \tag{6.18}$$

where t = the spacing between holes
d = hole diameter

Some wall thickness, C, should be added to the designed shell thickness because the rolled up shell may continuously become thinner than the nominal thickness because of negative tolerances and corrosion. The allowance for negative tolerances of sheet thickness, C_1, is very small. It is taken as $C_1 = 0.1$ cm when (S-C) $\leqslant 2$ cm, and $C_1 = 0$ for (S-C) > 2 cm. For cast equipment, $C_1 \geqslant 0.2$ cm.

The corrosion allowance, C_2, depends on the aggressiveness of the processing fluid, the chemical resistance of the shell material and the life of the equipment. For apparatus of carbon steels operating under atmospheric corrosion, the allowance is $C_2 = 0.1$-0.3 cm. The selection of corrosion allowance for equipment operating in aggressive fluids is vital and complicated. The corrosion allowance is taken equal to the corrosion penetration rate, ν_p, cm/yr times the life of equipment, τ:

$$C_2 = \nu_p \times \tau, \text{ cm} \tag{6.19}$$

The corrosion penetration rate should be based on experimental data obtained under conditions similar, or very close, to actual operating conditions. Even minute changes in concentration of processed substances, temperature and pressure have a great influence on the corrosion rate.

The rate of corrosion of materials used for fabrication of apparatus should not exceed 0.1–0.5 mm/yr. Data on various corrosion-resistant materials can be found in the literature [8].

One always should account for the possibility of nonuniform and intergranular corrosion, which does not induce a visible thinning of a sheet. In designing shells, precautions should be taken to eliminate erosion. Erosion often promotes corrosion from a variety of conditions such as the formation of dead zones, cracks and local flow around a wall by aggressive fluids [9]. Design measures to avoid corrosion cannot be overstated.

In selecting a value for corrosion allowance, C_2, the possibility of double-side corrosion effects on the material should be considered. If the internal parts are surrounded by the aggressive fluid, the corrosion allowance, C_2, should be doubled. To prevent atmospheric corrosion, the equipment should be painted or covered with lacquer.

In the structural design of chemical and process equipment, it is widely accepted that the theory of strength based on the maximum tangential stresses describes the behavior of a greater number of structural materials more so than the theory of Huber-Mises-Hencky. According to this theory, the strength conditions are expressed as follows:

$$\sigma_{max} - \sigma_{min} \leqslant \sigma_{allow}$$

For a flat-stress state assumed for the shell material,

$$\sigma_{max} = \sigma \; ; \quad \sigma_{min} = \sigma_r = 0$$

Consequently,

$$\sigma \leqslant \sigma_{allow}$$

Thus, the equation for circumferential stress becomes the designing formula

$$\sigma \leqslant \frac{pd}{2t}$$

and the wall thickness of the cylindrical shell with regard to σ_{allow}, ψ, and C will take the form

$$t = \frac{pd}{2\sigma_{allow}\psi - p} + C, \text{ cm} \qquad (6.20)$$

For a given external diameter of the shell, d_0, the thickness of the wall is determined as

$$t = \frac{pd_0}{2\sigma_{allow}\psi + p} + C \qquad (6.21)$$

At $(\sigma_{allow}/p)\,\psi \geqslant 50$, the value of p in the denominators of Equations 6.20 and 6.21 may be omitted.

The design pressure is taken as follows: For vessels in which the working media is gas (vapor),

$$p = p_f \ , \quad N/m^2 \tag{6.22}$$

p_w = working pressure for vessels where the working media is a liquid-gas mixture (vapor).

$$p = p_w + p_e \ , \quad N/m^2 \tag{6.23}$$

Hydrostatic pressure, p_e, is determined by the following formula:

$$p_e \approx 10\, \rho_e H_e \ , \quad N/m^2 \tag{6.24}$$

where ρ_e = liquid density (kg/m^3)
H_e = height of liquid column (m).

The hydrostatic pressure may not be considered in the case in which it does not exceed 2.5% of the working pressure of the gas (vapor) in the vessel.

The nominal allowable stress is taken to be the least of the three following values:

$$\sigma^*_{allow} \leqslant \frac{\sigma^t_T}{\eta_T} \ ; \quad \sigma^*_{allow} \leqslant \frac{\sigma^t_y}{\eta_y} \ ; \quad \sigma^*_{allow} \leqslant \frac{\sigma^t_f}{\eta_f}$$

Because the coefficient in the denominator of Equation 6.12 is equal to 2.3, the values of the safety factors are decreased by the ratio 2/2.3 and, consequently, should be equal to

$$\eta_T = 2.6 \ ; \quad \eta_y = 1.45 \ ; \quad \eta_f = 1.45 - 1.75 \ .$$

Apparatuses are classified according to operating conditions:

1. Apparatuses for processing and storage of explosive and flammable substances at temperatures higher than the flash point, which include apparatuses handling corrosive and poisonous materials to which personnel are exposed.
2. Apparatuses for processing and storage of inert, flammable material including apparatuses for processing and storage of toxic, explosive and

flammable materials at temperatures below their flash points. In this class, the presence of personnel is not required.

3. Apparatuses for processing and storage of nonpoisonous and flammable substances operating under pressure, near which the constant presence of personnel is not required.

Apparatuses are divided further according to heated and unheated. Heated apparatus are those subjected from outside to flammable gases or radiation from electrical heaters. The apparatuses are considered to be unheated if they are insulated from radiation sources and are in contact with heat carriers.

Recommended correction factors, η, that depend on the service condition and heat effect are given in Table 6.1. Most apparatuses are related to Classes 1 and 2.

The design shell temperature, t_{sh}, is taken according to the temperature of the fluid, t_f, inside the vessel, shell insulation and conditions of heating (or cooling). For unheated shells insulated from the outside,

$$t_{sh} = t_f \quad {}^\circ C \tag{6.25}$$

$$t_{sh} = 32 + t_f(1.8) \quad {}^\circ F \tag{6.26}$$

For unheated shells having inside insulation or cooled from the outside,

$$t_{sh} = 0.5(t_{sh}^{out} + t_{sh}^{in}) \quad {}^\circ C \tag{6.27}$$

For gas or electric outside heating, Equation 6.27 may be used, but for temperatures not less than

$$t_w = t_f + 100^\circ C \tag{6.28}$$

For a shell heated by steam,

Table 6.1. Values of Correction Factor, η

	Class of Apparatus		
Apparatus	**1**	**2**	**3**
Heated	0.65	0.75	0.90
Unheated	0.70	0.9	1.00

$$t_{sh} = t_{st} \quad {}^{\circ}C \tag{6.29}$$

where t_{st} = temperature of saturated steam at pressure, p

When heating or cooling shells with a liquid that flows in, the design temperature of a wall, t_w, is taken to be the highest temperature of the heating fluid or the lowest temperature of the cooling liquid, t_f.

The design thickness of a cylindrical shell determined from Equations 6.20 and 6.21 should be rounded off to the nearest millimeter and taken from the sheet grades. If the shell is subjected to axial bending (in the center plane) and torsional moment (in the plane perpendicular to the axis) from loads (from gravity forces, etc.), then the equivalent stress from these loads, σ_{eq}, is calculated from the following formula:

$$\sigma_{eq} = \sqrt{(\sigma + 0.8\,\sigma_1)^2 + 3\,\tau^2}\ , \quad N/m^2 \tag{6.30}$$

where σ = axial tensile or compression stress (N/m^2)
σ_1 = bending stress (N/m^2)
τ = torsional stress (N/m^2)

The axial stress, σ, is determined from

$$\sigma_1 = \frac{P}{F}\ , \quad N/m^2 \tag{6.31}$$

where P = axial external load (N)
$F = \pi(d + t)(t - C)$ = shell cross-sectional area (m^2)

The bending stress is

$$\sigma_1 = \frac{M}{W}\ , \quad N/m^2 \tag{6.32}$$

where M = external bending moment acting in center plane ($N \cdot m$)
W = axial moment of resistance (m^3)

The torsional stress is

$$\tau = \frac{M_\tau}{2W}\ , \quad N/m^2 \tag{6.33}$$

where M_τ = external torsional moment acting in plane perpendicular to the axis, N-m

The equivalent stress in the shell from external loads should conform to the condition

$$\sigma_{eq} \leqslant \sqrt{1.2\,\sigma_{allow}^2 - \frac{d + (t - c)^2 p^2}{4(t - c)^2 \psi^2}}, \quad N/m^2 \tag{6.34}$$

In calculating σ_{eq} from the formula in Equation 6.30, the stresses σ, σ_1 and τ are taken for one section. If the greatest stress values are in different sections, then the section that has σ_{eq} at a maximum should be used. For example, to decrease the distance between supports (in horizontal vessels), insert reinforcing ribs, etc.

Shells will undergo axial and bending loads when

$$\frac{\ell'}{d} \geqslant 1 \tag{6.35}$$

These should be checked for longitudinal stability from the following formula:

$$(t - c) \geqslant \sqrt{\frac{1}{E^t}\left[K_1\left(1 - \frac{K}{2}\right) + 0.125\,KK_2\right]P + \left(4\,\frac{M}{d}\right)}, \quad m \tag{6.36}$$

where ℓ' = the design length (height) of the shell (m)
K_1, K_2 = coefficients given in Figure 6.12
E^t = modulus of elasticity at a given temperature

and

$$K = \frac{2M}{0.25\,Pd + M} \tag{6.37}$$

In all cases, the determined wall thickness of the cylindrical shell should be checked for stress with a hydrostatic pressure test and the following formula:

$$\sigma = \frac{[d + (t - c)]P'}{2(t - c)\psi} \leqslant \frac{\sigma_y}{1.2}, \quad N/m^2 \tag{6.38}$$

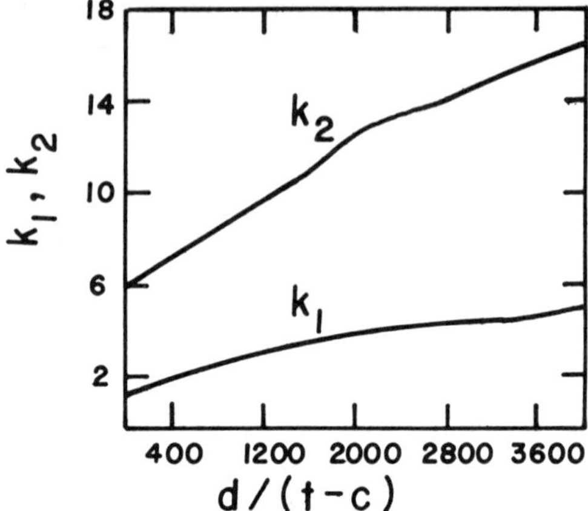

Figure 6.12. Diagram for determination of coefficients K_1 and K_2 for Equation 6.36.

The design test pressure is

$$p' = p_h + p_e \,, \quad N/m^2 \tag{6.39}$$

where p_h = hydrostatic test pressure (N/m^2) (from Table 6.2)
p_e = hydrostatic pressure of water (N/m^2) (from Equation 6.24)

If condition 6.38 is not fulfilled, the designed wall thickness, t, should be increased correspondingly.

The following examples illustrate the principles outlined above.

Example 6.1

Determine the wall thickness of a cylindrical welded shell for a vertical vessel operating under internal pressure. The vessel is shown in Figure 6.13.

Data

The material is carbon steel ($\sigma_T = 380 \times 10^6$ N/m^2; $\sigma_y = 240 \times 10^6$ N/m^2); permeability of shell material in the medium is 0.06 mm/yr; $C_1 = 1 \times 10^{-3}$ m; $C_2 = 0$; the medium-fluid ($\rho_e = 1.2 \times 10^3$ kg/m^3) is gas; p = 1.0 MN/m^2; t = +20°C; d = 2 m, H = 5 m; the shell has no holes; the longitudinal welded seam is a manual double butt joint ($\psi = 0.95$); the vessel is the second (unheated) operating class group 3 ($\eta = 1.0$).

Table 6.2. Hydrostatic Tests

Description	ASME Code Reference
Test must be at least $1\frac{1}{2}$ times the maximum design pressure multiplied by the lowest ratio of the stress value for the test temperature to that for the design temperature.	Par. UG-99 Par. UG-21 Par. UA-60(e)
If the allowable stress at design temperature is less than the allowable stress at test temperature, the hydrostatic test pressure must be increased proportionally. Thus,	
Test pressure $= 1.5 \times \dfrac{\text{allowable stress at test temperature} \times \text{design pressure}}{\text{allowable stress at design temperature}}$	
Combination units must be so tested that hydrostatic pressure is on one chamber without pressure in the other parts.	Par. UG-21 and UG-98 Par. UA-60(b)
Corrosion allowance must be included in calculating test pressure.	
Inspection must be made at a pressure not less than $\frac{2}{3}$ of test pressure.	
Test cast-iron vessels at twice the design working pressure; for design pressure under 30 psi, test at $2\frac{1}{2}$ times design pressure, but not to exceed 60 psi.	Par. UCI-99
Test for clad-plate vessels.	Par. UCL-52

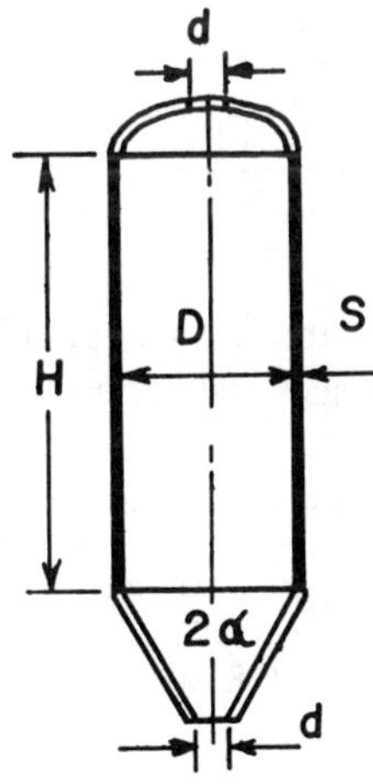

Figure 6.13. Sketch for Example 6.1.

Solution

The hydrostatic pressure in the lower part of the shell is determined from Equation 6.24:

$$P_e \approx 10\,\rho_e H_e = 10 \times 1.2 \times 10^3 \times 5 = 0.06 \times 10^6 \ \text{N/m}^2$$

The design internal pressure in the shell is determined from Equation 6.23:

$$P = P_w + P_e = 1 \times 10^6 + 0.06 \times 10^6 = 1.06 \times 10^6 \ \text{N/m}^2$$

The allowable tensile stress for steel is determined from

$$\sigma_{allow} = \frac{\sigma_T}{\eta_T}\,\eta = \frac{380 \times 10^6}{2.6} \times 1.0 = 146 \times 10^6 \ \text{N/m}^2$$

The allowable yield stress for steel is determined from

$$\sigma_{allow} = \frac{\sigma_y}{\eta_y} = \frac{240 \times 10^6}{1.5} \times 1.0 = 160 \times 10^6 \ \text{N/m}^2$$

Since the first stress is less than the second one, it is taken as the design stress. Because

$$\frac{\sigma_{allow}}{P}\,\psi = \frac{146 \times 10^6}{1.06 \times 10^6} \times 0.95 = 131 > 50$$

the value of p in the denominator of Equation 6.20 is omitted.

The design thickness of the shell is determined from Equation 6.20:

$$t = \frac{dP}{2\sigma_{allow}\psi} + C = \frac{2 \times 1.06 \times 10^6}{2 \times 146 \times 10^6 \times 0.95} + C = 7.65 \times 10^{-3} + C \quad \text{m}$$

$$C = C_1 + C_2 + C_3 = (1 + 0 + 0.8) \times 10^{-3} = 1.8 \times 10^{-3} \ \text{m}$$

where C_1 = corrosion allowance
C_2 = erosion allowance
C_3 = negative allowance for sheet thickness

Hence,

$$t = (7.65 + 1.8)10^{-3} = 9.45 \times 10^{-3} \text{ m}$$

We assume t = 10 mm.

The stress in the shell is checked for the hydrostatic test of the vessel with water ($p_e = 0.06 \times 10^6$ N/m^2). The design test pressure is determined from Equation 6.39:

$$p' = (p_h + p_e) = (1 + 0.3 + 0.06) \times 10^6 = 1.36 \times 10^6 \text{ N/m}^2$$

The stress in the shell wall at hydrostatic test is determined from Equation 6.38:

$$\sigma = \frac{[d + (t - C)]P'}{2(t - C)\psi} = \frac{[2 + (10 - 1.8) \times 10^{-3}] \times 1.36 \times 10^6}{2(10 - 1.8) \times 10^{-3} \times 0.95}$$

$$= 175 \times 10^6 \text{ N/m}^2 < \frac{\sigma_y}{1.2} = \frac{240 \times 10^6}{1.2} = 200 \times 10^6 \text{ N/m}^2$$

Example 6.2

Determine the wall thickness of a soldered cylindrical shell of a horizontal vessel operating under internal pressure. (The system is illustrated in Figure 6.14.)

Data

The material is copper ($E = 115 \times 10^9$ N/m^2; $\sigma_T = 200 \times 10^6$ N/m^2; $\sigma_y = 50 \times 10^6$ N/m^2; corrosion permeability 0.04 mm/yr ($C_1 = 0$; $C_2 = 0$)); the fluid is gas ($p = 0.4$ MN/m^2; $t = -120$°C; $d = 0.9$ m; $L = 1.5$ m; $\ell = 1.0$ m; $P = 10{,}000$ N; gravity force of the vessel together with water $G \simeq 20{,}000$ N). The shell has holes for tubes without reinforcement (2 holes with $d_1 = 0.05$ m and 4 holes with $d_2 = 0.025$ m). All holes are located on the same generatrix; the longitudinal lap-soldered seam with hard solder ($\psi = 1.0$). The vessel is to conform to the second operating class and third group ($\eta = 1.0$).

The internal pressure in the shell is determined from Equation 6.22:

$$p = p_w = 0.4 \times 10^6 \text{ N/m}^2$$

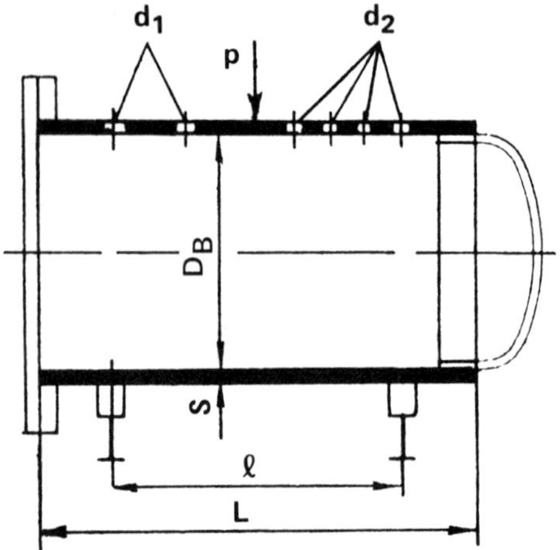

Figure 6.14. Sketch for Example 6.2.

The allowable tensile stress for copper is

$$\sigma_{allow} = \frac{\sigma}{\eta_T}\eta = \frac{200 \times 10^6}{2.6} \times 1.0 = 77 \times 10^6 \text{ N/m}^2$$

The allowable yield stress for copper is

$$\sigma_{allow} = \frac{\sigma_y}{\eta_y}\eta = \frac{50 \times 10^6}{1.5} \times 1.0 = 33.3 \times 10^6 \text{ N/m}^2$$

The last stress is the least one and therefore is taken as the design stress. The design stress factor for the shell in longitudinal direction is

$$\psi = \frac{L - (2d_1 + 4d_2)}{L} = \frac{1.5 - (2 \times 0.05 + 4 \times 0.025)}{1.5} = 0.867$$

because

$$\frac{\sigma_{allow}}{P}\psi = \frac{33.3 \times 10^6}{0.4 \times 10^6} \times 0.867 = 72 > 50$$

The value of p in the denominator of Equation 6.20 is omitted.

The design wall thickness of the shell is determined from Equation 6.20:

$$t = \frac{d_p}{2\sigma_{allow} \times \psi} + C = \frac{0.9 \times 0.4 \times 10^6}{2 \times 2.33 \times 10^6 \times 0.867} + C = 6.24 \times 10^3 + C \quad \text{m}$$

The allowance C at $C_3 = 0.5 \times 10^{-3}$ m is determined from the following formula:

$$C = C_1 + C_2 + C_3 = (0 + 0 + 0.5)10^{-3} = 0.5 \times 10^{-3} \text{ m}$$

Hence,

$$t = (6.24 + 0.5) \times 10^{-3} \text{ m}$$

We assume the wall thickness t = 8 mm. The stress in the wall is checked with a hydrostatic test of the vessel using water. The hydrostatic pressure in the upper part of the shell weakened by holes is equal to zero and, therefore, is not considered.

The design test pressure is determined from Equation 6.39:

$$p' = p_h + p_e = p_h = 1.5 \times 0.4 \times 10^6 = 0.6 \times 10^6 \text{ N/m}^2$$

The stress in the wall of the shell at hydrostatic test is determined from Equation 6.38:

$$\sigma = \frac{[d + (t - C)]P'}{2(t - C)\psi} = \frac{[0.9 + (8 - 0.5) \times 10^{-3}]0.6 \times 10^{-6}}{2(8 - 0.5)10^{-3} \times 0.867} = 41.9 \times 10^6 \text{ N/m}^2$$

$$41.9 \times 10^6 \approx \frac{\sigma_y}{1.2} = \frac{50 \times 10^6}{1.2} = 41.7 \times 10^6$$

The shell is checked for the stress in its wall from the external bending load (gravity force of the vessel filled with water G = 20,000 N and load p = 10,000 N). The maximum bending moment for the dangerous section of the shell (in the middle between supports), assuming that the load, G, is distributed uniformly along the shell, is

$$M = \frac{G}{8}(2\ell - L) + \frac{p\ell}{4} = \frac{2 \times 10^4}{4}(2.1 - 1.5) + \frac{1 \times 10^4 \times 1}{4} = 3.75 \times 10^3 \text{ N-m}$$

The section modulus of the dangerous section of the shell is

$$W \approx \frac{\pi d^2}{4}(t - C) = \frac{\pi(0.9)^2}{4}(8 - 0.5) \times 10^{-3}$$

$$= 4.77 \times 10^{-3} \text{ m}^3$$

The equivalent stress in the shell from external loads (the axial and torsional loads are absent) is determined from Equations 6.30 and 6.32:

$$\sigma_{eq} = 0.8\frac{M}{W} = 0.8\frac{3.75 \times 10^3}{4.77 \times 10^{-3}} = 0.63 \times 10^6 \text{ N/m}^2$$

Now we check condition 6.34:

$$\sigma_{eq} \leqslant \sqrt{1.2\,\sigma_{allow}^2 - \frac{[d + (t - C)]^2 P^2}{4(t - C)^2 \psi^2}}$$

or

$$0.63 \times 10^6 \text{ N/m}^2 < \sqrt{1.2(33.3 \times 10^6)^2 - \frac{[0.9 + (8 - 0.5)10^{-3}]^2 (0.4 \times 10^6)^2}{4[(8 - 0.5)10^{-3}]^2 (0.867)^2}}$$

$$= 23.4 \times 10^6 \text{ N/m}^2$$

We now check the shell for stability under external loads (the axial load is absent): Condition 6.35 is first fulfilled, because

$$\frac{l}{d} = \frac{1}{0.9} = 1.11 > 1$$

Coefficient K, in Equation 6.36 is determined from Equation 6.37:

$$K = \frac{2M}{0.25\ Pd + M} = \frac{2 \times 3.75 \times 10^3}{0.25 \times 0 \times 0.9 + 3.75 \times 10^3} = 2$$

For

$$\frac{d}{t - C} = \frac{0.9}{(8 - 0.5)10^{-3}} = 120$$

the coefficients K_1 and K_2 are taken from Figure 6.12: $K_1 = 1.6$ and $K_2 = 6.2$.

The stability of the shell determined from Equation 6.36 is kept because

$$\sqrt{\frac{1}{E^t}\left[K_1\left(1 - \frac{K}{2}\right) + 0.125\ KK_2\right]\left(P + \frac{4M}{d}\right)}$$

$$= \sqrt{\frac{1}{115 \times 10^9}\left[1.6\left(1 - \frac{2}{2}\right) + 0.125 \times 2 \times 6.2\right] \times \left(0 + \frac{4 \times 3.75 \times 10^3}{0.9}\right)}$$

$$= 0.472 \times 10^{-3}\ m < (t - C) = (8 - 0.5) \times 10^{-3}$$

$$= 7.5 \times 10^{-3}\ m$$

6.2.2 Design of Steel Rolled-Up Welded Cylindrical Shells Operating Under External Pressure

A variety of cylindrical vessels operate under partial vacuum, for example, vacuum condensers, distillation columns, crystallizers, etc. Such vessels operate under external pressure from the atmosphere. Jacketed vessels that are heated by condensing vapor under pressure in the jacket also produce an external pressure on the vessel.

Substitution of internal pressure by external pressure induces a change in the stress sign, and in the shells operating under external pressure, compression stresses arise instead of tensile.

A cylindrical shell under external pressure has an induced circumferential compressive stress equal to twice the longitudinal compressive stress. Under such a condition the shell is apt to collapse because of elastic instability caused by circumferential compressive stress.

If a thin-walled shell [at $(t - C)/d \leqslant 0.04$] is under external pressure, it is possible for the wall to collapse into the shell. The shell thickness needed to prevent this is determined from the condition of strength in compression:

$$t = \frac{Pd}{2\sigma'_{allow}\,\psi} \tag{6.40}$$

where $\quad d$ = external diameter
σ'_{allow} = allowable stress in compression

which can be calculated as

$$\sigma'_{allow} = \frac{\sigma'_y}{4}, \quad kg/cm^2 \tag{6.41}$$

and the joint efficiency, ψ, when designing a shell for compression, is taken equal to unity.

Long, thin shells will buckle at stresses below the yield point of the material. The corresponding critical pressure at which buckling occurs is a function only of the t:d ratio and the modulus of elasticity, E.

Based on the strength consideration, the design thickness of a cylindrical shell, under external pressure, p_e, satisfying the following condition,

$$0.4 \geqslant \frac{P_e}{\sigma_{eT}} \geqslant 0.2 \tag{6.42}$$

is determined from the following formula [2]:

$$t' = 0.5\,d\left\{\sqrt{\frac{\sigma_{eT}\psi}{\sigma_{eT}\psi - 1.73\,P_e}} - 1\right\} + C \tag{6.43}$$

where $\quad \psi$ = shell joint efficiency in the longitudinal direction

$$C = C_1 + C_2 + C_3$$

The allowance, C_1, is taken depending on the internal and external media with regard to corrosion from both sides of the wall. For usual cases in shell design under external pressure, p_e, assuming that

$$1 \leqslant \frac{\ell'}{d} \leqslant 8 \tag{6.44}$$

and

$$\left(\frac{P_e}{E^t} \cdot \frac{\ell'}{d}\right)^{0.4} \leqslant 0.523 \tag{6.45}$$

the design thickness of cylindrical shell, t', from materials with Poisson's ratio, $\tilde{\mu} \approx 0.3$, is determined on the basis of stability consideration:

$$t' = 1.25\,d\left(\frac{P_e}{E^t} \cdot \frac{\ell'}{d}\right)^{0.4} + C \tag{6.46}$$

where ℓ' = design length (height) of the cylindrical shell

The design length of a cylindrical shell is:

1. the distance between flanges, for flanged shells;
2. the length of cylindrical shell and head plus one-third of elliptical (spherical) part, when the shell is ended by an elliptical (or spherical) head;
3. the length of cylindrical shell down to the bottom, when shell is ended by a flat bottom; and
4. the distance between their centers, when the shell has stiffening rings.

For shells of the same materials ($\tilde{\mu} \approx 0.3$) in cases in which condition 6.45 is not fulfilled, as well as for shells of any plastic material,

$$0.1 \leqslant \frac{\ell'}{d} \leqslant 10 \tag{6.47}$$

the value of nominal (without allowance C) design wall thickness, t_0, is determined from Figure 6.15. This chart is given for the design of cylindrical shells of carbon steel at wall temperatures t_w = +20°C as a function of ℓ'/d and critical pressure, p_{cr}, which is expressed as

$$p_{cr} = p_e n_s \tag{6.48}$$

where n_s = stability factor
n_s = 4 for vertical shells
n_s = 5 for horizontal shells

Figure 6.15 may be used for shell design at $t_w > 20$°C for other materials. The thickness, t_0, is determined from the chart by using the value p'_{cr}, which,

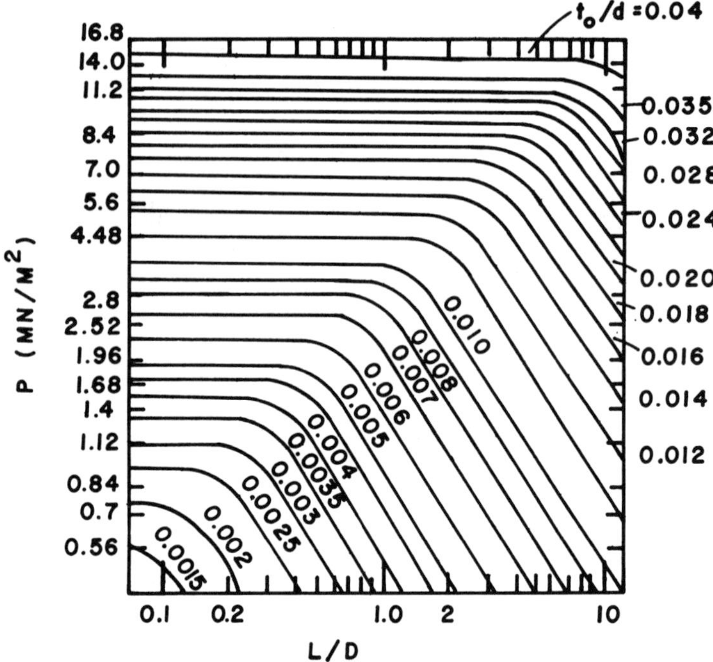

Figure 6.15. Chart for determining shell thickness of cylindrical vessels subjected to external pressure–carbon steel.

for sloped parts of the curve, is equal to

$$P'_{cr} = P_{cr} \frac{210 \times 10^9}{E^t}, \quad N/m^2 \tag{6.49}$$

For horizontal parts of the curve it is equal to

$$P'_{cr} = P_{cr} \frac{182 \times 10^6}{\sigma^t_{y,c}} \tag{6.50}$$

where $\sigma^t_{y,c}$ = yield point for compression

The determined value, t_0, should be checked for allowable external pressure ($p_{e,allow} \geqslant p_e$) from

$$P_{e,allow} = \left(\frac{t_0}{d}\right) \frac{2E^t}{n_s(z^2 - 1)[1 + 0.4\, z^2(\ell'/d)^2]^2}$$

$$+ \left(\frac{t_0}{d}\right)^3 \frac{0.67\, E^t}{n_s(1 - \tilde{\mu}^2)} \times \left[(z^2 - 1) + \frac{2z^2 - 1 - \tilde{\mu}}{1 + 0.4\, z^2(\ell'/d)^2}\right], \quad N/m^2 \quad (6.51)$$

where z = the number of waves in the cylinder at collapsing

The number of waves corresponding to the minimum value of allowable external pressure, $p_{e,allow}$, can be obtained from Figure 6.16. If the intersection of parameters t_0/d and ℓ'/d falls between the curves of the chart, $p_{e,allow}$ should be determined with the use of Equation 6.51 for a greater value of z, where $p_{e,allow}$ is always less.

In designing cylindrical shells under external pressure, the ratio ℓ':d selected should be as small as possible. The use of solid shells with a ratio ℓ':d > 5 is not recommended.

For shells with a nominal design wall thickness, t_0 can be obtained from Figure 6.16 and checked by Equation 6.51; an expression for the total wall thickness is

$$t' = t_0 + C \quad (6.52)$$

For final wall thickness selection, the directions given for selecting a shell under internal pressure should be followed. The chosen thickness of a

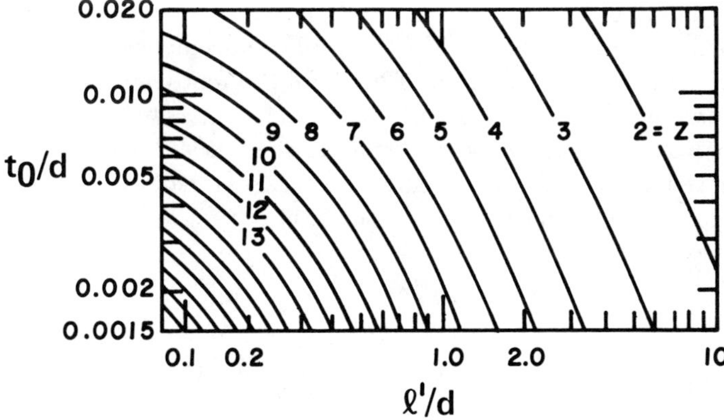

Figure 6.16. Chart for determination of the number of waves, z, in Equation 6.51.

cylindrical wall under the condition

$$\ell > 2.2\sqrt{dt} \ , \quad m \tag{6.53}$$

should be checked for combined stress, σ from bending and compression (or tensile) from the following formulas [10,11]:
For vertical vessels,

$$\sigma = \frac{(d + t)P_e}{1.1(t - c)} + \frac{M}{W} + \left| \frac{(d + 2t)P_e}{4(t - c)} \pm \frac{P \pm G}{\pi(d + t)(t - c)} \right| \leqslant \sigma_{allow} \ , \quad N/m^2 \tag{6.54}$$

For horizontal vessels,

$$\sigma = \frac{(d + t)P_e}{1.1(t - c)} + \frac{M}{W} + \frac{(d + 2t)P_e}{4(t - c)} \leqslant \sigma_{allow} \ , \quad N/m^2 \tag{6.55}$$

where p = external axial load (N)
G = gravity force of vessel and liquid (N)
M = external bending moment (N-m)
W = section modulus of the shell (m^3)

In Equations 6.54 and 6.55, the third term is considered only with external pressure against the shell front; otherwise it is equal to zero.

The fourth term in Equation 6.54 is assigned a plus sign when the front pressure and algebraic sum of axial load and gravity force act in the same direction, and a negative sign when the mentioned components act in different directions. The algebraic sum of the third and fourth terms is taken with plus signs. The shells with external jacket under excessive pressure, besides the abovementioned checking for combined stress from Equations 6.54 and 6.55 at condition 6.53, also should be checked for the wall stress, σ, at the hydrostatic external test with pressure [11] p_{et}:
For vertical vessels,

$$\sigma = \frac{(d + t)P_{e.t.}}{1.1(t - c)} + \frac{M}{W} + \left| \frac{(d + 2t)}{4(t - c)} \pm \frac{P + G}{\pi(d + t)(t - c)} \right| \leqslant \frac{\sigma_y}{1.2} \ , \quad N/m^2 \tag{6.56}$$

For horizontal vessels,

$$\sigma = \frac{(d + t)P_{et}}{1.1(t - c)} + \frac{M}{W} + \frac{(d + 2t)P_{e.t.}}{4(t - c)} \leqslant \frac{\sigma_y}{1.2}, \quad N/m^2 \tag{6.57}$$

The remarks made of the terms in Equations 6.54 and 6.55 are valid for Equations 6.56 and 6.57 as well.

If conditions 6.54–6.57 are not fulfilled, the wall thickness, t, should be increased correspondingly. The shells with stiffening rings also should be checked for local stability of the cylindrical wall between the rings:

$$\ell \leqslant \frac{0.65\, E^t(t - c)^2}{d\, P_{e.t.}} \sqrt{\frac{t - c}{d}}, \quad m \tag{6.58}$$

The holes in the cylindrical shell operating under external pressure should be reinforced. The stiffening rings are used to decrease wall thickness and increase the rigidity of thin-walled cylindrical shells. They are spaced at equal distances from each other along the shell generatrix either from outside or inside, as determined by design considerations.

The stiffening rings are welded to the shell with two intermittent seams (if continuous seams are not demanded), with the total length of each seam equal to no less than a half length of ring circle at the place of its connection with the shell. The spacing between the welds of intermittent seams should not be less than 8t. Figure 6.17 presents typical cross sections of stiffening rings and their connections with a shell. There are several types that shall be described briefly:

Type I

The ring is of a rectangular section, used for shells of any diameter and fabricated of sheet or strip metal. Such rings may be in a single piece or sectional. The connection of separate sectors is made mostly by a double-welded butt joint (Figure 6.18), with subsequent machining of the internal cylindrical surface of a ring.

Type II

A stiffening ring of an angle section is used for vessels with diameters greater than 1000 mm.

Types III and IV

Stiffening rings of other rolled stock are used for vessels with diameters greater than 2000 mm and with an inertia moment of the ring $J'_E > 200 \times 10^{-8}\ m^4$.

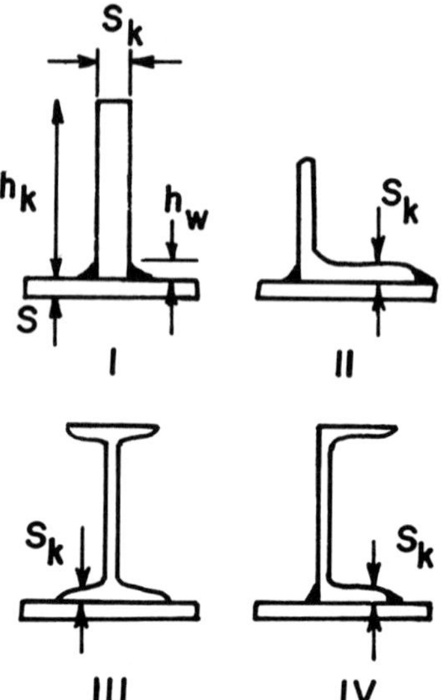

Figure 6.17. Typical cross sections of stiffening rings: at $S \leqslant S_K$, $h_W \approx S$; and at $S > S_K$, $h_W \approx S_K$.

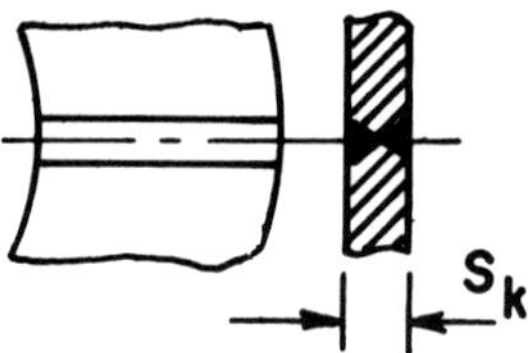

Figure 6.18. Welded joint of stiffening ring sectors.

The spacing between rings is determined by design and economic considerations. The inertia moment of the nominal design cross section of a stiffening ring is determined on the basis of the ring stability from Equation 6.61, after the following two conditions are met:

$$\frac{d}{t - c} \geqslant 60 \tag{6.59}$$

$$1 \leqslant \frac{\ell'}{d} \leqslant 8 \tag{6.60}$$

$$J'_r \approx \frac{P_e \ell' d^3}{9.2\, E^t}, \quad m^4 \tag{6.61}$$

where J'_r = inertia moment of nominal design ring cross section relating to the axis located on the nearest to the ring shell generatrix

For a ring of rectangular cross section and a recommended ring width:thickness ratio equal to five times the nominal ring design thickness, use the following formula:

$$t_{n.r.} = 0.42 \sqrt[4]{J'_r}, \quad m \tag{6.62}$$

If condition 6.60 is not fulfilled and $\ell'/d < 1$, the inertia moment of nominal ring design cross section of the stiffening ring, $J'_{r.c.}$, is determined from the following formula [11]:

$$J'_{r.c.} \approx \frac{0.23\, P_e d \sqrt[3]{dt}}{E^t}, \quad m^4 \tag{6.63}$$

where $J'_{r.c.}$ = inertia moment of nominal ring cross section relating to the axis passing through its gravity center parallel to the shell generatrix.

The nominal ring design thickness, $t_{n.r.}$, of a rectangular cross section with the abovementioned width:thickness ratio is obtained from the following formula:

$$t_{n.r.} = 0.555 \sqrt[4]{J'_{r.c.}}, \quad m \tag{6.64}$$

For the final section of the ring cross section sizes, allowance should be made for corrosion and negative tolerance of rolled stock by introducing a value, C, to the nominal design ring sizes and rounding them to the larger recommended sizes.

The approved sizes of the ring stiffening cross section should be checked for combined stress in the ring, σ, from bending and compression:

$$\sigma = \frac{0.61\, P_e \sqrt{dt}\, d}{F'_r + F'_w} + \frac{0.0038\, P_e \sqrt{ds}\, d^2 h}{J'_{r.w.}} \leqslant \sigma_{allow}, \quad N/m^2 \tag{6.65}$$

where F'_r = design area of cross section minus area corresponding to C (m^2)
F'_w = design area of cross section of the shell element (m^2)
J'_{rw} = inertia moment of the area $F'_r + F'_w$ relating to the axis passing through the gravity center parallel to shell generatrix (m^4)
h = distance from gravity center of the area $F'_r + F'_w$ to its more distant edge in the direction perpendicular to shell axis (m).

In Equation 6.65, F'_w, h and $J'_{r.w.}$ are determined as follows:

$$F'_w = [b_2 + 1.1\sqrt{d(t-c)}](t-c)\ , \quad m^2 \tag{6.66}$$

The expression in square brackets should not exceed the space between stiffening ring axes:

$$h = h_2 - \frac{0.5[F'_r h_r - F_w(t-c)]}{F'_r + F'_w}\ , \quad m \tag{6.67}$$

$$J'_{r.w.} = J'_r + J'_w + (h - 0.5\,h_2)^2 F'_r + [0.5(t-c) + h_2 - h]^2 F'_w\ , \quad m^4 \tag{6.68}$$

where b_2 = width of the ring contacting shell surface (m)
h_2 = height of ring cross section (m)
J'_r = inertia moment of area F'_2 relating to the axis passing through its gravity center parallel to shell generatrix (m^4)
J'_w = inertia moment of area F'_w relating to the axis passing through its gravity center parallel to shell generatrix (m^4)

The shell stiffening rings of vessels with jackets under excessive pressure also should be checked for stress, σ, at hydrostatic external test under pressure $p_{e.t.}$ from the following formula:

$$\sigma = \frac{0.61\,P_{e.t.}\sqrt{dt}\,d}{F'_r + F'_w} + \frac{0.0038\,P_{e.t.}\sqrt{dt}\,d^2 h}{J'_{r.w.}} \leqslant \sigma_y/1.2\ , \quad N/m^2 \tag{6.69}$$

If conditions 6.65 and 6.69 are not fulfilled, the designed sizes of ring cross section should be increased correspondingly. The following examples illustrate these principles.

Example 6.3

Determine the wall thickness of a welded cylindrical shell of a vessel operating under vacuum and external pressure for the following data (the system is shown in Figure 6.19).

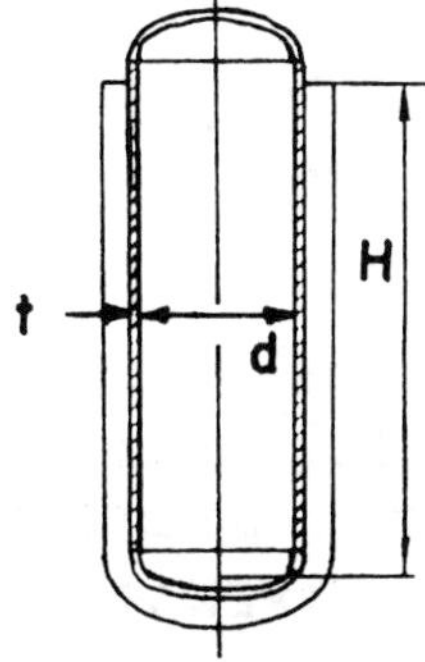

Figure 6.19. Sketch for Example 6.3.

Data

The material of the shell is 18-10 austenitic steel (E^{150} = 185 X 10^9 N/m^2; σ_y^{150} = 250 X 10^6 N/m^2, σ_y = 220 X 10^6 N/m^2—see Figures 6.20 (A) and (B)—$\tilde{\mu}$ = 0.3.

The material is corrosion resistant in the medium: There are no data on penetration (C_1 = 1 X 10^{-3} m, C_2 = 0); the external medium is a nonaggressive liquid; the internal medium is a gas under pressure, p = vacuum 10 N/m^2; external pressure, p_e = 0.6 MN/m^2; temperature of the wall, t_w = 150°C; d = 0.8 m; H = 2.4 m, vessel's gravity force, G = 10,000 N; vessel's gravity force filled with water in the jacket, G_1 = 20,000 N; the shell has no holes; longitudinal automatic double butt-welded seam, ψ = 1.0; the vessel is in the second operating class, and third group η = 1.0.

Solution

The design external pressure in the shell with regard to the vacuum in the vessel is

$$p_e = p_e + 0.1 \times 10^6 = (0.6 + 0.1)10^6 = 0.7 \times 10^6 \text{ N/m}^2$$

Checking condition 6.44,

$$1 < \frac{H}{d} = \frac{2.4}{0.8} = 3 < 8$$

Checking condition 6.45,

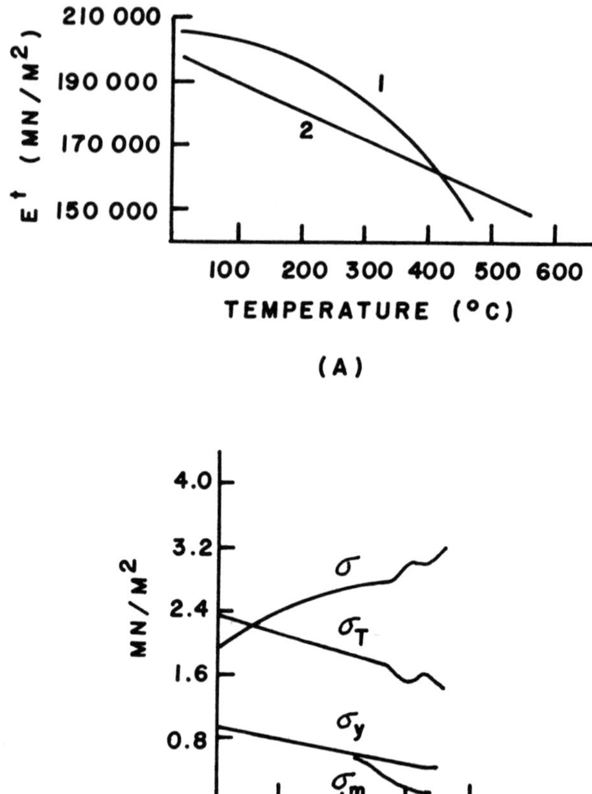

Figure 6.20. (A) Modulus of elasticity for steels at elevated temperatures: 1 = carbon steel; 2 = alloy austenitic steels; (B) mechanical properties of austenitic 18-10 steel at elevated temperatures.

$$\left(\frac{P_e}{E^t} \times \frac{H}{d}\right)^{0.4} = \left(\frac{0.7 \times 10^6}{185 \times 10^9} \times 3\right)^{0.4} = 0.0105 < 0.523$$

Because conditions 6.44 and 6.45 are fulfilled, the design wall thickness of the shell is determined from Equation 6.45:

$$t' = 1.25\, d\left(\frac{P_e}{E^t} \times \frac{H}{d}\right)^{0.4} + C = 1.25 \times 0.8 \times 0.0105 + C = 10.5 \times 10^{-3} + C$$

C at $C_3 = 0.8 \times 10^{-3}$ m is:

$$C = C_1 + C_2 + C_3 = (1.0 + 0 + 0.8)10^{-3} = 1.8 \times 10^{-3} \text{ m}$$

Hence,

$$t' = (10.5 + 1.8) \times 10^{-3} = 12.3 \times 10^{-3} \text{ m}$$

We take $t' = 12$ mm, which is 2.5% less than the design thickness, which is permissible.

Checking the shell for combined stress in its wall from bending and compression, according to condition 6.53,

$$\ell = H = 2.4 \text{ m} > 2.2\sqrt{dt} = 2.2\sqrt{0.8 \times 12 \times 10^{-3}} = 0.216 \text{ m}$$

The vessel is not subjected to external axial loads and bending moments. The load from the vessel gravity force ($G \approx 10{,}000$ N) acts on the shell in the opposite direction of the external pressure from the end plane. The combined stress in the shell is determined from formula 6.54:

$$\sigma = \frac{(d + t)P_e}{1.1(t - c)} + \left|\frac{(d + 2t)P_e}{4(t - c)} - \frac{G}{\pi(d + t)(t - c)}\right|$$

$$= \frac{(0.8 + 12 \times 10^{-3})0.7 \times 10^6}{1.1(12 - 1.8)10^{-3}} + \left|\frac{(0.8 + 2 \times 12 \times 10^{-3})0.7 \times 10^6}{4(12 - 1.8)10^{-3}}\right.$$

$$\left. - \frac{10{,}000}{\pi(0.8 + 12 \times 10^{-3})(12 - 1.8)10^{-3}}\right|$$

$$= 50.5 \times 10^6 + |14.1 \times 10^6 - 0.4 \times 10^6| = 64.2 \times 10^6 \quad \text{N/m}^2$$

$$< \sigma_{allow} = \frac{\sigma_y^t}{\eta_y} = \frac{2.10 \times 10^6}{1.5} = 140 \times 10^6 \quad \text{N/m}^2$$

Then the shell is checked for combined stress in the wall from bending and compression under external hydrostatic test of the vessel with water.

The hydrostatic pressure in the lower part of the shell is determined from Equation 6.24:

$$P_{ef} \approx 10\,\rho_f H_f = 10 \times 1 \times 10^3 \times 2.4 = 0.024 \times 10^6 \quad N/m^2$$

The design test pressure is determined from Equation 6.39:

$$P' = P_h + P_e = (0.7 + 0.3 + 0.024)10^6 = 1.024 \times 10^6 \quad N/m^2$$

The combined stress in the shell under hydrostatic test is determined from Equation 6.54:

$$\sigma = \frac{(d+t)P'_e}{1.1(t-c)} + \left|\frac{(d+2t)P'_e}{4(t-c)} - \frac{G_1}{\pi(d+t)(t-c)}\right|$$

$$= \frac{(0.8 + 12 \times 10^{-3})1.024 \times 10^6}{1.1(12 - 1.8) \times 10^{-3}} + \left|\frac{(0.8 + 2 \times 12 \times 10^{-3}) \times 1.024 \times 10^6}{4(12 - 1.8)10^{-3}}\right.$$

$$\left. - \frac{20\,000}{\pi(0.8 + 12 \times 10^{-3})(12 - 1.8) \times 10^{-3}}\right|$$

$$= 74 \times 10^6 + |20.6 \times 10^6 - 0.8 \times 10^6| = 93.8 \times 10^6 \quad N/m^2$$

$$< \frac{\sigma_y}{1.2} = \frac{220 \times 10^6}{1.2} = 183 \times 10^6 \quad N/m^2$$

Example 6.4

Determine the wall thickness of a welded cylindrical shell and the cross section of the stiffening rings for a horizontal vessel operating under external and internal pressure for the following data (see Figure 6.21).

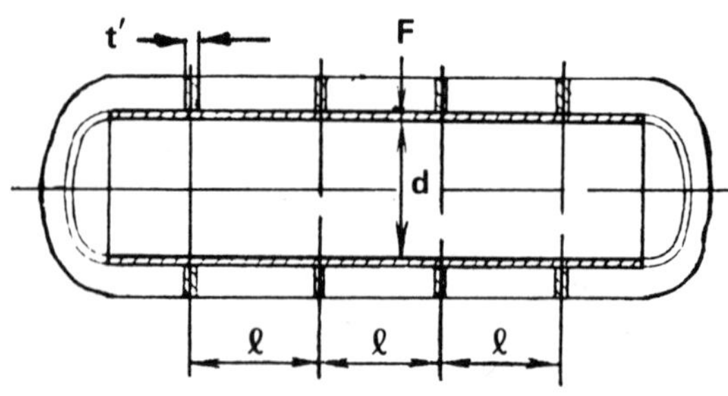

Figure 6.21. Sketch for Example 6.4.

Data

The material of the shell is aluminum alloy ($\sigma_T = 130 \times 10^6$ N/m^2; $\sigma_{o.r.} = 60 \times 10^6$ N/m^2; E = 72×10^9 N/m^2; $\tilde{\mu} = 0.33$); the material is resistant to the internal medium; there are no data for permeability ($C_1 = 1 \times 10^{-3}$ m; $C_2 = 0$); the internal medium is fluid ($\rho_e = 1.1 \times 10^3$ kg/m^3); the external medium–liquid ($\rho_{ef} = 0.8 \times 10^3$ kg/m^3)–the gas is not aggressive; internal pressure, p = 0.3 MN/m^2; wall temperature, $t_w = -190$°C d = 1.3 m; $\ell = 0.9$ m; the shell has reinforced holes; longitudinal automatic double butt-welded seam ($\psi = 0.85$). The vessel is in the first (heated) operating class and third group ($\eta = 0.9$). The shell is designed for internal pressure.

Solution

The hydrostatic pressure in the lower part of the shell at H_e = d is determined from Equation 6.24:

$$P_e \approx 10\, \rho_e H_e = 10 \times 1.1 \times 10^3 \times 1.3 \approx 0.014 \times 10^6 \quad \text{N/m}^2$$

The design internal pressure in the shell is determined from Equation 6.23:

$$P = P + P_e = 0.3 \times 10^6 + 0.014 \times 10^6 = 0.314 \times 10^6 \quad \text{N/m}^2$$

The allowable stress on the basis of tensile strength is determined as follows:

$$\sigma_{allow} = \frac{\sigma_T}{\eta_T}\,\eta = \frac{130 \times 10^6}{2.6} \times 0.9 = 45 \times 10^6 \quad \text{N/m}^2$$

The allowable stress on the basis of yield point is

$$\sigma_{allow} = \frac{\sigma_y}{\eta_y}\,\eta = \frac{60 \times 10^6}{1.5} \times 0.9 = 36 \times 10^6 \quad \text{N/m}^2$$

Because the second value is less than the first, it is taken as a design stress:

$$\frac{\sigma_{allow}}{P}\,\psi = \frac{36 \times 10^6}{0.314 \times 10^6} \times 0.85 = 97.5 > 50$$

The value of p in the denominator of Equation 6.20 can be neglected.

The design wall thickness of the shell is determined from Equation 6.20:

$$t = \frac{dP}{2\sigma_{allow}\psi} + C = \frac{1.3 \times 0.314 \times 10^6}{2 \times 36 \times 10^6 \times 0.85} + C = 6.75 \times 10^{-3} + C \ , \quad m$$

Next we design the shell for external pressure.

The hydrostatic pressure in lower part of the shell (assuming that the liquid level covers the shell) is determined from Equation 6.24:

$$P_e \approx 10\, \rho_{ef} H_{ef} = 10 \times 0.8 \times 10^3 \times 1.3$$
$$= 0.0104 \times 10^6 \quad N/m^2$$

The design external pressure on the shell is determined from Equation 6.23:

$$P_e = P + P_e = 0.16 \times 10^6 + 0.0104 \times 10^6 \approx 0.17 \times 10^6 \quad N/m^2$$

Because condition 6.60 is not fulfilled since

$$\frac{\rho}{d} = \frac{0.9}{1.3} = 0.69 < 1$$

and there is condition 6.47, $0.1 < \ell/d$, the value of the nominal design thickness of the wall, t_0, is determined from Figure 6.15.

The critical pressure at safety against buckling for $n_s = 5$ is determined from Equation 6.48:

$$P_{cr} = P_e \times n_s = 0.17 \times 10^6 \times 5 = 0.85 \times 10^6 \quad N/m^2$$

The values of critical pressures can be obtained from the appropriate charts assuming values for E^t and σ_y^t for a the wall temperature of $t_w = 20°C$:

For sloping parts of curves, this value is determined from Equation 6.49:

$$P'_{cr} = P_{cr} \frac{210 \times 10^9}{E^t} = 0.85 \times 10^6 \frac{210 \times 10^9}{72 \times 10^9}$$
$$= 2.48 \times 10^6 \quad N/m^2$$

For horizontal parts of curves we use Equation 6.50:

$$p'_{cr} = p_{cr} \frac{182 \times 10^6}{\sigma_y^t} = 0.85 \times 10^6 \frac{182 \times 10^6}{60 \times 10^6} = 2.58 \times 10^6 \quad N/m^2$$

$\ell/d = 0.692$, so in the chart p'_{cr} lies in the region of sloping parts of the curves and $t_0/d = 0.0065$. Hence,

$$t_0 = 0.0065d = 0.0065 \times 1.3 = 8.45 \times 10^{-3} \text{ m}$$

We then check the external allowable pressure, $p_{e,allow}$, from formula 6.51. The number of waves from the chart (Figure 6.16) is z = 7:

$$p_{e,allow} = \left(\frac{t_0}{d}\right) \frac{2E}{n_s(Z^2 - 1)[1 + 0.4 Z^2(\ell'/d)^2]^2}$$

$$+ \left(\frac{t_0}{d}\right)^3 \frac{0.67E}{n_s(1 - \tilde{\mu}^2)} \left[(Z^2 - 1) + \frac{2Z^2 - 1 - \tilde{\mu}}{1 + 0.4 Z^2(\ell'/d)^2}\right]$$

Hence, the following values are computed:

Z	$Z^2 - 1$	$1 + 0.4Z^2 \left(\frac{\ell'}{d}\right)^2$	$2Z^2 - 1 - \tilde{\mu}$	$1 - \tilde{\mu}^2$
7	$7^2 - 1 = 48$	$1 + 0.4 \times (7)^2 \times (0.692)^2 = 10.4$	$2(7)^2 - 1 - 0.33 = 96.67$	$1 - (0.33)^2 = 0.891$

$$p_{e,allow} = \frac{0.0065 \times 2 \times 72 \times 10^9}{5 \times 48 \times 10.4^2} + \frac{0.0065^3 \times 0.67 \times 72 \times 10^9}{5 \times 0.891} \times \left(48 + \frac{96.67}{10.4}\right)$$

$$= 0.036 \times 10^6 + 0.17 \times 10^6$$

$$= 0.206 \times 10^6 \text{ N/m}^2 > p_e = 0.17 \times 10^6 \quad \text{N/m}^2$$

For design wall thickness, t′, we select the greater value on the basis of shell design for external pressure.

The allowance C at $C_3 = 0.5 \times 10^{-3}$ is $C = C_1 + C_2 + C_3 = (1 + 0 + 0.5)10^{-3}$
$= 1.5 \times 10^{-3}$ m

Hence,

$$t' = t_0 + C = (8.45 + 1.5) \times 10^{-3} = 9.55 \times 10^{-3} \text{ m}$$

We take t = 10 mm. We now check the shell for combined stress in the wall from bending and compression: Condition 6.53 is fulfilled:

$$\ell = 0.9 \text{ m} > 2.2\sqrt{dt} = 2.2\sqrt{1.3 \times 10 \times 10^{-3}} = 0.251 \text{ m}$$

The shell is not subjected to external axial loads and bending moments; the bending moment from gravity force may be neglected. The combined stress in the shell is determined from formula 6.55:

$$\sigma = \frac{(d + t)P_e}{1.1(t - c)} + \frac{(d + 2t)P_e}{4(t - c)} = \frac{(1.3 + 10 \times 10^{-3})0.17 \times 10^6}{1.1(10 - 1.5)10^{-3}}$$

$$+ \frac{(1.3 + 2 \times 10 \times 10^{-3})0.17 \times 10^6}{4(10 - 1.5) \times 10^{-3}}$$

$$= 23.8 \times 10^6 + 6.6 \times 10^6$$

$$= 30.4 \times 10^6 \text{ N/m}^2 < \sigma_{allow} = 36 \times 10^6 \text{ N/m}^2$$

We now check the shell for combined stresses in the wall from bending and compression at external hydrostatic test of the vessel with water:

The hydrostatic pressure in the lower part of the shell is determined from formula 6.24:

$$P_e \approx 10\, \rho_{ef} H_{ef} = 10 \times 1 \times 10^3 \times 1.3 = 0.013 \times 10^6 \text{ N/m}^2$$

The design test pressure is determined from formula 6.23:

$$p_e' = p' + p_e = (1.5 \times 0.17 + 0.013)10^6 = 0.268 \times 10^6 \text{ N/m}^2$$

The total stress in the wall at hydrostatic test is determined from formula 6.55:

$$\sigma = \frac{(d + t)P'_e}{1.1(t - c)} + \frac{(d + 2t)P'_e}{4(t - c)}$$

$$= \frac{(1.3 + 0.01)0.268 \times 10^6}{1.1(10 - 1.5) \times 10^{-3}} + \frac{(1.3 + 2 \times 0.01)0.268 \times 10^6}{4(10 - 1.5)10^{-3}}$$

$$= 37.6 \times 10^6 + 10.4 \times 10^6 = 48 \times 10^6 \text{ N/m}^2 < \frac{\sigma_y}{1.2}$$

$$= \frac{60 \times 10^6}{1.2} = 50 \times 10^6 \text{ N/m}^2$$

To design the stiffening rings, condition 6.59 must be checked:

$$\frac{d}{t - c} = \frac{1.3}{(10 - 1.5) \times 10^{-3}} = 153 > 60$$

Hence, condition 6.59 is fulfilled. As condition 6.60 ($\ell'/d = 0.692 < 1$) is not fulfilled, we should determine the inertia moment of the nominal design cross section of the stiffening ring from formula 6.63:

$$J'_r = \frac{0.23 \times P_e d^3 \sqrt{dt}}{E} = \frac{0.23 \times 0.17 \times 10^6 \times (1.3)^3 \sqrt{1.3 \times 0.01}}{72 \times 10^9}$$

$$= 13.6 \times 10^{-8} \text{ m}^4$$

We select a stiffening ring with rectangular cross section (the ratio of height/thickness = 5).

The nominal design ring thickness is determined from formula 6.64:

$$t_r = 0.555 \sqrt[4]{J'_r} = 0.555 \sqrt[4]{13.6 \times 10^{-8}}$$

$$= 10.7 \times 10^{-3} \text{ m}$$

The sizes of the ring cross section are taken (at $C_1 = 0$; $C_2 = 0.8 \times 10^{-3}$ m): $t_2 = 12 - 0.8 = 11.2$ mm, $h_2 = 60$ mm.

The design area of the assumed ring cross section is

$$F'_r = t_r h_r = (11.2 \times 60)10^{-6} = 672 \times 10^{-6}\ m^2$$

The design length of a shell element at $b_2 = t_2$ (Equation 6.66) is

$$\ell'_w = b_r + 1.1\sqrt{d(t-c)} = 11.2 \times 10^{-3} + 1.1\sqrt{1.3(10-1.5) \times 10^{-3}} = 130 \times 10^{-3}\ m < \ell = 0.9\ m$$

The design area of the shell element is determined from formula 6.66:

$$F'_w = \ell'_w(t-c) = 0.13(10-1.5)10^{-3} = 1105 \times 10^{-6}\ m^2$$

The distance from the gravity center of cross section $F'_r + F'_w$ to its more distant edge in the direction perpendicular to shell axis is determined from Equation 6.67:

$$h = h_r - \frac{0.5[F'_r h_r - F'_w(t-c)]}{F'_r + F'_w} = 60 \times 10^{-3} - \frac{0.5[672 \times 60 - 1105 \times 8.5]10^9}{(672+1105)10^{-6}} = 51.3 \times 10^{-3}\ m$$

The inertia moment of cross section F'_r with respect to the axis passing through its gravity center parallel to the shell generatrix is

$$J'_r = \frac{t_r h_r^3}{12} = \frac{11.2 \times (60)^3 \times 10^{-12}}{12} = 20.2 \times 10^{-8}\ m^4$$

The inertia moment of cross section area, F'_w, with respect to the axis passing through its gravity center parallel to the shell generatrix, is

$$J'_w = \frac{F'_w(t-c)^2}{12} = \frac{1105(10-1.5)^2 10^{-12}}{12} = 0.665 \times 10^{-8}\ m^4$$

The inertia moment of cross-sectional area $F'_r + F'_w$ with respect to the axis passing through its gravity center parallel to the shell generatrix is determined from Equation 6.68:

$$J'_{r.w.} = J'_r + J'_w + (h - 0.5h_r)^2 F'_r + [0.5(t - c) + h_r - h]^2 F'_w$$
$$= 20.2 \times 10^{-8} + 0.665 \times 10^{-8} + (51.3 - 0.5 \times 60)^2 \times 10^{-6} \times 672 \times 10^{-6}$$
$$+ [0.5(10 - 1.5) + 60 - 51.3]^2 \times 10^{-6} \times 1105 \times 10^{-6}$$
$$= 69.9 \times 10^{-8}\ m^4$$

The total stress in the stiffening ring from bending and compression is determined from Equation 6.69:

$$\sigma = \frac{0.61 P_e \sqrt{dt}\, d}{F'_r + F'_w} + \frac{0.0038 P_e \sqrt{dt}\, d^2 h}{J'_{r.w.}}$$
$$= \frac{0.61 \times 0.17 \times 10^6 \sqrt{1.3 \times 10 \times 10^{-3}} \times 1.3}{(672 + 1105)10^{-6}}$$
$$+ \frac{0.0038 \times 0.17 \times 10^6 \sqrt{1.3 \times 10 \times 10^{-3}} \times (1.3)^2 \times 51.3 \times 10^{-3}}{69.9 \times 10^{-8}}$$
$$= 8.65 \times 10^6 + 9.13 \times 10^6$$
$$= 17.8 \times 10^6\ N/m^2 < \sigma_{allow} \approx 36 \times 10^6\ N/m^2$$

The total stress in the stiffening ring from bending and compression at external hydrostatic test of the vessel is determined from Equation 6.69:

$$\sigma = \frac{0.61 P'_e \sqrt{dt}\, d}{F'_r + F'_w} + \frac{0.0038 P'_e \sqrt{dt}\, d^2 h}{J'_{r.w.}}$$
$$= \frac{0.61 \times 0.268 \times 10^6 \sqrt{1.3 \times 10 \times 10^{-3}} \times 1.3}{(672 + 1105)10^{-6}}$$
$$+ \frac{0.0038 \times 0.268 \times 10^6 \sqrt{1.3 \times 10 \times 10^{-3}} \times (1.3)^2 \times 51.3 \times 10^{-3}}{69.9 \times 10^{-8}}$$
$$= 13.6 \times 10^6 + 14.4 \times 10^6 = 28 \times 10^6\ N/m^2$$
$$< \frac{\sigma_y}{12} = \frac{60 \times 10^6}{1.2} = 50 \times 10^6\ N/m^2$$

6.2.3 Design of Forged Cylindrical Shells Operating Under Internal Pressure

Forged cylindrical shells generally are used as thick-walled shells ($d_e/d \geqslant 1.1$) for fluid pressure in the vessel not less than 10 MN/m^2. Preferred materials for such shells are alloy steels of different brands.

Forged shells could be fabricated either from forged solid or sectional courses connected mostly by electroslag welding. Typical designs of forged steel cylindrical shells are given in Figure 6.22. The left-hand side of each design presents the solid version, and the right-hand side is a sectional one. Forged shells are used mostly for vertical vessels. It is recommended that holes be avoided in such cylindrical shells, but in the event they are used, they should be totally reinforced.

The design thickness of a forged cylindrical wall, t′, under internal pressure, p, is determined from Equations 6.20 and 6.21 if the following condition is fulfilled:

$$\frac{p}{\sigma_{allow}} < 0.4 \tag{6.70}$$

If, however, condition 6.71 holds,

$$0.4 \leqslant \frac{p}{\delta_{allow}} \leqslant 1.1 \tag{6.71}$$

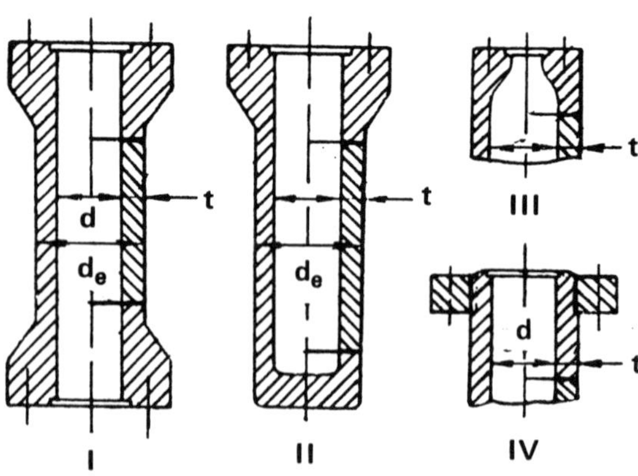

Figure 6.22. Typical designs of forged shells.

then the following formula should be used:

$$t' = 0.5d(10^{P/2.3\sigma_{allow}} - 1) + 10^{P/2.3\sigma_{allow}}C\ , \quad m \tag{6.72}$$

Equation 6.72 is not applicable for shells made from austenitic steels.

The value $10^{P/2.3\sigma_{allow}}$ in Equation 6.72 is determined from Figure 6.23.

Allowance $C = C_1 + C_2 + C_3$

$$C_3 = 0.5(\Delta_n + \Delta_p) + \Delta_a \tag{6.73}$$

where Δ_n = negative tolerance for d (m)
Δ_p = positive tolerance for d (m)
Δ_a = allowable eccentricity between the external and internal shell surfaces (m).

The design pressure is determined in all cases from Equation 6.22.

The design temperature of the wall is taken from the same consideration as for shells of welded rolled-up sheets.

The calculated wall thickness should be checked for its stress, σ, at the vessel hydrostatic test:

If calculations are made from either Equation 6.20 or 6.21, a check should be performed using Equation 6.38; if calculations are made from formula 6.72, then checking should be done according to the following:

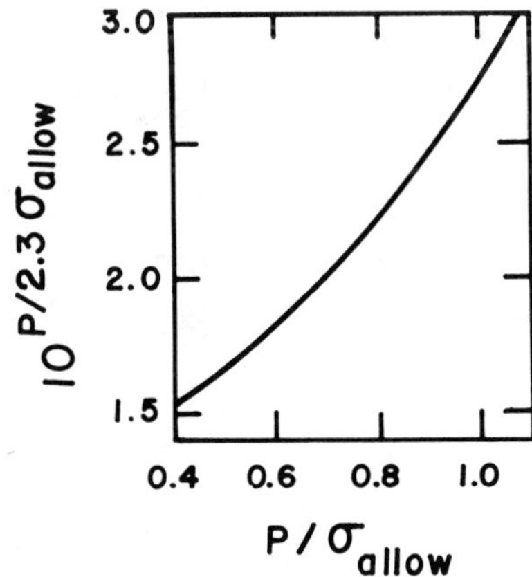

Figure 6.23. Diagram for determination of $P/2.3\ \sigma_{allow}$ in Equation 6.72.

$$\sigma = \frac{P_t}{\ln(\beta)} \leqslant \frac{\sigma_y}{1.2}, \quad N/m^2 \tag{6.74}$$

where

$$\beta = \frac{d + 2t}{d + 2(C_1 + C_2)} \tag{6.75}$$

The following example illustrates use of these formulas.

Example 6.5

Determine the wall thickness of a forged cylindrical shell of a vessel operating under internal pressure for the following data (Figure 6.24).

Data

The material of the shell is special alloy steel ($\sigma_T^{400} = 600 \times 10^6$ N/m^2; $\sigma_y^{400} = 350 \times 10^6$ N/m^2; $\sigma_y = 380 \times 10^6$ N/m^2); the permeability of the shell material in the medium 0.02 mm/yr ($C_1 = 0$; $C_2 = 0$); the medium is gas; $p = 40$ MN/m^2; wall temperature, $t_w = 300$°C, $d = 0.6$ m; the shell is solid, without holes ($\psi = 1$); electric-heated vessel of second class, group 1 ($\eta = 0.9$).

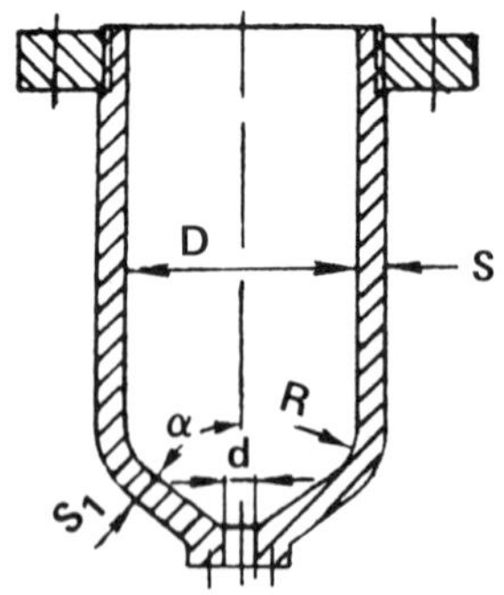

Figure 6.24. Sketch for Example 6.5.

Solution

The design temperature of the wall at electric heating is determined from Equation 6.28:

$$t_w = t_f + 100 = 300 + 100 = 400^\circ C$$

The design internal pressure is assumed according to Equation 6.22:

$$p = p_f = 40 \times 10^6 \ N/m^2$$

The allowable stress determined on the basis of tensile strength is

$$\sigma_{allow} = \frac{\sigma_T^t}{\eta_T}\eta = \frac{600 \times 10^6}{2.6} \times 0.9$$
$$= 208 \times 10^6 \ N/m^2$$

The allowable stress determined on the basis of yield point is

$$\sigma_{allow} = \frac{\sigma_y^t}{\eta_y}\eta = \frac{350 \times 10^6}{1.5} \times 0.9 = 210 \times 10^6 \ N/m^2$$

Because the first value of σ_{allow} is less than the second one, it is taken as a design stress.

Since condition 6.70 is fulfilled,

$$\frac{P}{\sigma_{allow}} = \frac{40 \times 10^6}{208 \times 10^6} = 0.192 < 0.4$$

The design wall thickness of the cylindrical shell is determined from Equation 6.20:

$$t' = \frac{dp}{2\sigma_{allow} - p} + C = \frac{0.6 \times 40 \times 10^6}{2 \times 208 \times 10^6 \times 1 - 40 \times 10^6} + C$$

Allowance $C = C_1 + C_2 + C_2$. The tolerance by the 7 class on $d + 2t'$, $\Delta_n = 2 \times 10^{-3}$ m; the tolerance by 5 class for $d\Delta_p = 0.9 \times 10^{-3}$ m; eccentricity is taken as $\Delta_a = 1.5 \times 10^{-3}$ m and C_3 from Equation 6.73:

$$C_3 = 0.5(\Delta_n + \Delta_p) + \Delta_a = 0.5(2 + 0.9)10^{-3}$$
$$+ 1.5 \times 10^{-3} = 2.95 \times 10^{-3} \text{ m}$$
$$C = C_1 + C_2 + C_3 = 0 + 0 + 2.95 \times 10^{-3} \text{ m}$$
$$= 2.95 \times 10^{-3} \text{ m}$$

Hence,

$$t' = (63.9 + 2.95)10^{-3} = 66.85 \times 10^{-3} \text{ m}$$

Assuming $t' = 67.5$ mm, then

$$d_e = d + 2t = 0.6 + 2 \times 0.0675 = 0.735 \text{ m}$$

We now check the stress in the wall at the hydrostatic vessel test with water: The design test pressure is determined first:

$$p_t \approx p_h = 1.25\, p_f = 1.25 \times 40 \times 10^6 = 50 \times 10^6 \text{ N/m}^2$$

The stress in the wall at the hydrostatic vessel test is determined from Equation 6.38:

$$\sigma = \frac{[d + (t - c)]P'}{2(t - c)\psi} = \frac{[0.6 + (6.75 - 2.95)10^{-3}]50 \times 10^6}{2(6.75 - 2.95)10^{-3} \times 1}$$

$$= 256 \times 10^6 \text{ N/m}^2 < \frac{\sigma_y}{1.2} = \frac{380 \times 10^6}{1.2} = 316 \times 10^6 \text{ N/m}^2$$

Example 6.6

Determine the wall thickness of a forged cylindrical shell of a vessel operating under internal pressure for the following data (Figure 6.24).

Data

The material of the shell is special alloy steel (σ_T = 1050 X 10^6 N/m^2; σ_y = 900 X 10^6 N/m^2); permeability of the material in the medium is less than 0.01 mm/yr (C_1 = 0, C_2 = 0); the medium is gas; p = 250 MN/m^2; wall temperature, t_w = 20°C; d = 0.2 m; it is a solid shell without holes, of the second class, group 2 vessel (η = 1).

Solution

The design internal pressure is taken according to Equation 6.22:

$$p = p_f = 250 \times 10^6 \text{ N/m}^2$$

The allowable stress determined on the basis of tensile strength is

$$\sigma_{allow} = \frac{\sigma_T}{\eta_T}\,\eta = \frac{1050 \times 10^6}{2.6} \times 1 = 403 \times 10^6 \text{ N/m}^2$$

The allowable stress determined on the basis of yield point is

$$\sigma_{allow} = \frac{\sigma_y}{\eta_y}\,\eta = \frac{900 \times 10^6}{1.5} \times 1 = 600 \times 10^6 \text{ N/m}$$

The first stress is less than the second one; therefore, it is taken as a design stress.

Because condition 6.71 is fulfilled,

$$0.4 < \frac{p}{\sigma_{allow}} = \frac{250 \times 10^6}{403 \times 10^6} = 0.62 < 1.1$$

The design thickness of the shell is determined from Equation 6.72, and the value $10^{P/2.3\sigma_{allow}}$ is taken from Figure 6.23:

$$\begin{aligned} t' &= 0.5d(10^{P/2.3\sigma_{allow}} - 1) + 10^{P/2.3\sigma_{allow}}C \\ &= 0.5 \times 0.2(1.85 - 1) + 1.85\ C \\ &= 85 \times 10^{-3} + 1.85\ C\ , \quad \text{m} \end{aligned}$$

Allowance $C = C_1 + C_2 + C_3$. The tolerance by 7 class on $d + 2t'$, $\Delta_n = 1.8 \times 10^{-3}$ m; the tolerance by class 5 is $d\Delta_p = 0.6 \times 10^{-3}$ m; eccentricity from design consideration $\Delta_a = 1.5 \times 10^{-3}$; thus, allowance C_3 is determined from Equation 6.73:

$$C_3 = 0.5(\Delta_n + \Delta_p) + \Delta_a = 0.5(1.8 + 0.6)10^{-3}$$
$$+ 1.5 \times 10^{-3} = 2.7 \times 10^{-3} \text{ m}$$

and

$$C = C_1 + C_2 + C_3 = 0 + 0 + 2.7 \times 10^{-3} \text{ m}$$

Hence,

$$t' = (85 + 1.85 \times 2.7) \times 10^{-3} = 90 \times 10^{-3} \text{ m}$$

We take $t' = 90$ mm and, consequently,

$$d_e = d + 2t = 0.2 + 2 \times 90 \times 10^{-3} = 0.38 \text{ m}$$

The stress in the wall is now checked for hydrostatic test of the vessel filled with water.

The design test pressure,

$$p_t = p_h = 1.25 p = 1.25 \times 250 \times 10^6 = 313 \times 10^6 \text{ N/m}^2$$

The value of β is determined from Equation 6.75:

$$\beta = \frac{d + 2t}{d + 2(C_1 + C_2)} = \frac{0.2 + 2 \times 90 \times 10^{-3}}{0.2 + 2(0 + 0)} = 1.9$$

The stress in the wall at hydrostatic test is determined from Equation 6.74:

$$\sigma = \frac{P'}{\ell n(\beta)} = \frac{313 \times 10^6}{\ell n(1.9)} = 488 \times 10^6 \text{ N/m}^2$$

$$< \frac{\sigma_y}{1.2} = \frac{900 \times 10^6}{1.2} = 750 \times 10^6 \text{ N/m}^2$$

If a thick-walled cylindrical shell ($d_e/d > 1.1$) is subjected to heat, throughput stresses are set up in axial and circumferential directions due to the temperature gradients. The stresses are distributed over the cross section of the wall similar to the stresses produced in a beam through a bending moment. They change from a maximum positive value on one surface to a maximum negative value on the other surface, the warmer side being compressed and the cooler side being in tension.

If the temperature gradient along the wall thickness exceeds 10°C, and on the condition that the wall design temperature eliminates creep, it is necessary to determine the additional temperature stresses in the wall σ_t^j and σ_t^e:

At the internal surface,

$$\sigma_t^i = \frac{E^t \alpha^t (t_w^i - t_w^e)}{2(1 - \widetilde{\mu})} \left(\frac{1}{\ell n(\beta)} - \frac{2\beta^2}{\beta^2 - 1} \right), \quad N/m^2 \tag{6.76}$$

At the external surface,

$$\sigma_t^e = \frac{E^t \alpha^t (t_w^i - t_w^e)}{2(1 - \widetilde{\mu})} \left(\frac{1}{\ell n(\beta)} - \frac{2}{\beta^2 - 1} \right), \quad N/m^2 \tag{6.77}$$

where α^t = coefficient of linear expansion
β = coefficient determined from Equation 6.75

The thermal stresses are neglected if the design temperature of the wall is

$t_w \geqslant +420°C$, for carbon steels

$t_w \geqslant +470°C$, for alloy steels

$t_w \geqslant +550°C$, for austenitic steels

In cases in which thermal stresses must be considered the assumed thickness of the shell, t, must be checked for its combined stresses from internal pressure and thermal gradient. As this occurs, the maximum combined tensile stress will be on the shell surface with the lower temperature.

The combined stress at the internal surface of the shell, σ^i, is determined from

$$\upsilon^i = \frac{1.73 P\beta^2}{(\beta^2 - 1)\psi} + o_t^i \leqslant \frac{\sigma_y^t}{1.2}, \quad N/m^2 \tag{6.78}$$

The combined stress at the external surface of the shell is determined from

$$\sigma^e = \frac{1.73P}{(\beta^2 - 1)\psi} + \sigma_t^e \leqslant \frac{\sigma_y^t}{1.2}, \quad N/m^2 \tag{6.79}$$

For any case, the following condition must be fulfilled:

$$0.5(\sigma^i + \sigma^e) \leqslant \sigma_{allow}, \quad N/m^2 \tag{6.80}$$

The values σ_y^t in Equations 6.78 and 6.79 should be taken at temperatures equal to t_w from Equation 6.27.

If conditions 6.78 or 6.79 are not fulfilled, then the assumed wall thickness, t, must be increased correspondingly, with allowance made for possible changes in design temperature of internal and external wall surfaces according to the data of checking thermotechnical designs.

6.2.4 Design of Forged Cylindrical Shells Operating Under External Pressure

The design thickness of a cylindrical shell operating under external pressure, p_e, is determined from Equation 6.43. Because forged cylindrical shells are used for vessels operating under internal pressure, $p > 10$ MN/m^2, their wall thickness is determined on the basis of internal pressure.

The presence of external pressure, p_e, on such vessels may arise through the pressure inside of heating or cooling jackets, where the pressure is considerably lower than the internal pressure of the shell. Consequently, it cannot be used to determine the shell thickness.

Therefore, in the presence of external pressure on such vessels it is usually necessary to check only the combined stress in the cylindrical shell from the external pressure (without consideration of internal pressure) and thermal gradient, if any exists.

It should be noted that the maximum combined compression stress will be on the shell surface having the higher temperature. The combined stress on the internal surface of the shell is determined from the formula

$$\sigma^i = \sigma_t^i - \frac{2P_e\beta^2}{(\beta^2 - 1)\psi} \leqslant \frac{\sigma_y^t}{1.2}, \quad N/m^2 \tag{6.81}$$

The combined stress on the external surface of the shell, σ^e, is determined from the formula

$$\sigma^{e} = \sigma_{t}^{e} - \frac{P_e(\beta^2 - 1)}{(\beta^2 - 1)\psi} \leqslant \frac{\sigma_y^t}{1.2}, \quad N/m^2 \tag{6.82}$$

where β is obtained from Equation 6.75
σ_t^i is obtained from Equation 6.76
σ_t^e is obtained from Equation 6.77
ψ is obtained from Equation 6.20 and 6.21

The values of σ_y^t in Equations 6.81 and 6.82 should be taken at the temperature of the wall, t_w, from Equation 6.27. Further, condition 6.80 must be fulfilled in any case and checked by design. The assumed shell thickness, t, also should be checked for the stress σ at the external hydrostatic test of the vessel from the following formula:

$$\sigma = \frac{2P_e'\beta^2}{(\beta^2 - 1)\psi} \leqslant \frac{\sigma_y}{1.2}, \quad N/m^2 \tag{6.83}$$

Example 6.7

Determine the wall thickness of a forged cylindrical shell of a vessel operating under internal and external pressures according to the following data (Figure 6.25): the material of the shell is steel 18-10 (E^{500} = 155×10^9 N/m^2; $\alpha^{500} = 18 \times 10^{-6}$ 1/°C; $\tilde{\mu} = 0.3$; $\sigma_T^{500} = 440 \times 10^6$ N/m^2; $\sigma_y^{500} = 140 \times 10^6$ N/m^2; $\sigma_y = 220 \times 10^6$ N/m^2); permeability of material from internal side, 0.2 mm/yr, from external side, less than 0.01 mm/yr ($C_1 = 0$, $C_2 = 0$); service life is 10 years; p = 12.5 MN/m^2; t_w = 450°C;

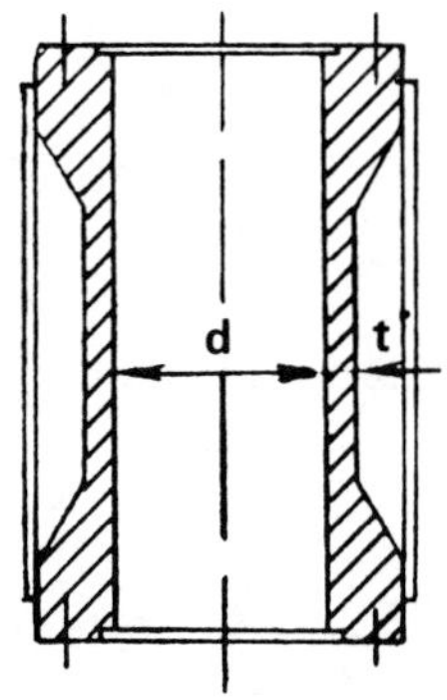

Figure 6.25. Sketch for Example 6.7.

$p_e = 5\ \text{MN/m}^2$; $T_{e.w.} = 500°\text{C}$, $d = 0.4$ m; solid shell without holes ($\psi = 1$) heated by liquid from outside; $t_w^e = 490°\text{C}$; $t_w^i = 470°\text{C}$; the vessel is Class 2, group 2; during operating, the pressure may be either only external or only internal.

Solution

The design wall temperature is assumed from Equation 6.25:

$$t_w = t_{e.w.} = 500°\text{C}$$

The design internal pressure is obtained from Equation 6.22:

$$p = p_f = 12.5 \times 10^6\ \text{N/m}^2$$

The design external pressure is obtained from Equation 6.22:

$$p_e = p_{e.f.} = 5 \times 10^6\ \text{N/m}^2$$

The allowable stress for steel 18-10 determined on the basis of tensile strength is

$$\sigma_{allow} = \frac{\sigma_T^t}{\eta_T}\eta = \frac{440 \times 10^6}{2.6} \times 1 = 16.9 \times 10^6\ \text{N/m}^2$$

The allowable stress determined on the basis of yield point is

$$\sigma_{allow} = \frac{\sigma_y^t}{\eta_y}\eta = \frac{140 \times 10^6}{1.5} \times 1 = 93.4 \times 10^6\ \text{N/m}^2$$

The second value of allowable stress (the lower one) is taken as an allowable design stress. Because condition 6.70 is fulfilled,

$$\frac{p}{\sigma_{allow}} = \frac{12.5 \times 10^6}{93.4 \times 10^6} = 0.134 < 0.4$$

The design thickness of the cylindrical shell is determined from Equation 6.20:

$$t' = \frac{d \times p}{2\sigma_{allow}\psi - p} + C = \frac{0.4 \times 12.5 \times 10^6}{2 \times 93.4 \times 10^6 \times 1 - 12.5 \times 10^6} + C$$

$$= 28.8 \times 10^{-3} + C \text{ m}$$

The corrosion allowance, C_1, from the internal side of the shell based on the vessel's service life, is

$$C_1 = 0.2 \times 10^{-3} \times 10 = 2 \times 10^{-3} \quad \text{m}$$

The allowance (Class 7) for d + 2t, $\Delta_e = 1.8 \times 10^{-3}$ m; the allowance (Class 5) for $d\Delta_i = 0.76 \times 10^{-3}$ m; the eccentricity (from design considerations) for $\Delta_a = 1.5 \times 10^{-3}$ m; the allowance for C_3 is from Equation 6.73:

$$C_3 = 0.5(\Delta_e + \Delta_i) + \Delta_a = 0.5(1.8 + 0.76) \times 10^{-3}$$

$$+ 1.5 \times 10^{-3} = 2.78 \times 10^{-3} \text{ m}$$

$$C = C_1 + C_2 + C_3 = (2 + 0 + 2.78) \times 10^{-3}$$

$$= 4.78 \times 10^{-3} \text{ m}$$

Hence, $t' = (28.8 + 4.78) \times 10^{-3} = 33.58 \times 10^{-3}$ m. We assume t = 35 mm, and, consequently,

$$d_e = d + 2t = 0.4 + 2 \times 35 \times 10^{-3} = 0.47 \text{ m}$$

The value β is determined from Equation 6.75:

$$\beta = \frac{d + 2t}{d + 2(C_1 + C_2)} = \frac{0.4 + 2 \times 35 \times 10^{-3}}{0.4 + 2(2 + 0)10^{-3}} = 1.17$$

The thermal stress on the internal shell surface is determined from Equation 6.76:

$$\sigma_t^i = \frac{\alpha^t E^t (t_w^i - t_w^e)}{2(1 - \widetilde{\mu})} = \left(\frac{1}{\ell n(\beta)} - \frac{2\beta^2}{\beta^2 - 1}\right)$$

$$= \frac{18 \times 10^{-6} \times 155 \times 10^9 (470 - 490)}{2(1 - 0.3)} \left(\frac{1}{\ell n(1.17)} - \frac{2 \times (1.17)^2}{(1.17)^2 - 1}\right)$$

$$= (-39.8 \times 10^6)(-1.03) = 41 \times 10^6 \text{ N/m}^2$$

The thermal stress on the external shell surface from Equation 6.77 is

$$\sigma_t^e = \frac{\alpha^t E^t (t_w^i - t_w^e)}{2(1 - \widetilde{\mu})} \left(\frac{1}{\ell n(\beta)} - \frac{2}{\beta^2 - 1}\right)$$

$$= (-39.8 \times 10^6) \left(\frac{1}{\ell n(1.17)} - \frac{2}{(1.17)^2 - 1}\right) = (-39.8 \times 10^6)(0.97)$$

$$= -38.6 \times 10^6 \text{ N/m}^2$$

The combined stress on the internal shell surface under internal pressure from Equation 6.78 is

$$\sigma^i = \frac{1.73 P \beta^2}{(\beta^2 - 1)\psi} + \sigma_t^i = \frac{1.73 \times 12.5 \times 10^6 \times (1.17)^2}{(1.17^2 - 1)} + 41 \times 10^6$$

$$= 121 \times 10^6 \text{ N/m}^2 > \frac{\sigma_y^t}{1.2} = \frac{140 \times 10^6}{1.2} = 116.5 \times 10^6 \text{ N/m}^2$$

These excesses $\sigma^i > \sigma_y^t/1.2$ for 4% may be admitted as an exception. The combined stress on the external surface of the shell under internal pressure from Equation 6.79 is

$$\sigma^e = \frac{1.73 P}{(\beta^2 - 1)\psi} + \sigma_t^e = \frac{1.73 \times 12.5 \times 10^6}{(1.17^2 - 1) \times 1} + (-38.6 \times 10^6)$$

$$= 20 \times 10^6 \text{ N/m}^2 < 116.5 \times 10^6 \text{ N/m}^2$$

Condition 6.80 is fulfilled:

$$0.5(\sigma^i + \sigma^e) = 0.5(121 + 20) \times 10^6 = 70.5 \times 10^6 \text{ N/m}^2$$

$$< \sigma_{allow} = 93.4 \times 10^6 \text{ N/m}^2$$

The combined stress at the internal surface of the shell under external pressure from Equation 6.81 is

$$\sigma_1^i = \sigma_t^i - \frac{2P_e\beta^2}{(\beta^2 - 1)\psi} = 41 \times 10^6 - \frac{2 \times 5 \times 10^6 \times (1.17)^2}{(1.17^2 - 1) \times 1}$$

$$= 4 \times 10^6 \text{ N/m}^2 < 116.5 \times 10^6 \text{ N/m}^2$$

The combined stress on the external shell surface under external pressure from Equation 6.82 is

$$\sigma_1^e = \sigma_t^e - \frac{P_e(\beta^2 + 1)}{(\beta^2 - 1)} = -38.6 \times 10^6 - \frac{5 \times 10^6(1.17^2 + 1)}{(1.17^2 - 1)}$$

$$= -70 \times 10^6 \text{N/m}^2 < 116.5 \times 10^6 \text{ N/m}^2$$

Condition 6.80 is fulfilled:

$$0.5(\sigma_1^i + \sigma_1^e) = 0.5(4 - 70) \times 10^6 = -33 \times 10^6 \text{ N/m}^2$$

$$< \sigma_{allow} = 93.4 \times 10^6 \text{ N/m}^2$$

We now check the stress in the shell under the internal hydrostatic test of the vessel with water.

The design test pressure is

$$P_t \approx P_h = 1.25P = 1.25 \times 12.5 \times 10^6$$

$$= 15.65 \times 10^6 \text{ N/m}^2$$

The stress in the shell under hydrostatic test from Equation 6.38 is

$$\sigma = \frac{[d + (t - c)]P'}{2(t - c)\psi} = \frac{[0.4 + (35 - 4.78) \times 10^{-3}]15.65 \times 10^6}{2(35 - 4.78) \times 10^{-3} \times 10^{-3} \times 1}$$

$$= 111.5 \times 10^6 \text{ N/m}^2 < \sigma_y/1.2 = \frac{220 \times 10^6}{1.2} = 183 \times 10^6 \text{ N/m}^2$$

Next, we determine the stress in the shell under the external hydrostatic test of the vessel with water.

The design test pressure is

$$P' \approx P_h = 1.25\ P_{e.f.} = 1.25 \times 5 \times 10^6$$
$$= 6.25 \times 10^6\ N/m^2$$

The stress in the shell under the hydrostatic test from Equation 6.83 is

$$\sigma = \frac{2P'_e\beta^2}{(\beta^2 - 1)\psi} = \frac{2 \times 6.25 \times 10^6 \times 1.17^2}{(1.17^2 - 1) \times 1} = 46.2 \times 10^6\ N/m^2$$

$$< \frac{\sigma_y}{1.2} = \frac{220 \times 10^6}{1.2} = 183 \times 10^6\ N/m^2$$

6.2.5 Design of Cast Cylindrical Shells

Cast cylindrical shells made from plastic materials are used in process equipment considerably less than are welded shells of rolled steel sheets. This can be explained by the fact that fabrication of chemical and process apparatuses has mostly an individual or small-scale character when casting is not economical.

High metal consumption of castings is a disadvantage because safety factors for the same materials are higher than for rolled steel. In addition, the design wall thickness should be increased at the expense of a considerable technological allowance, C_3.

However, casting is the only technology available for the fabrication of brittle construction materials. The most widespread among brittle materials for manufacturing chemical equipment are cast irons, which can be used for pressures up to 0.6 MN/m^2 and temperatures up to 250°C. Such apparatuses often are subjected to enameling, rubberizing, etc., from the inside.

Figure 6.26 shows designs of cast cylindrical shells (the shells are molded together with heads). If a shell has holes, they should be reinforced. The design thickness determination is recommended as follows:

1. For shells subjected to internal pressure, use Equations 6.20 or 6.21.
2. For shells subjected to external pressure or vacuum in the vessel, use Equations 6.43, 6.46 or Figure 6.15.
3. For shells operating without pressure or under pouring, use Equations 6.20 and 6.21 where p is assumed to be not less than 0.1 MN/m^2.

In all cases, $C = C_1 + C_2 + C_3$. C_3 is the technological allowance equal to the following: for steels, $C_3 > 3$ mm; for nonferrous metals and their alloys, $C_3 > 2$ mm; for cast irons, $C_3 > 6$ mm; for quartz glass, $C_3 > 3$ mm.

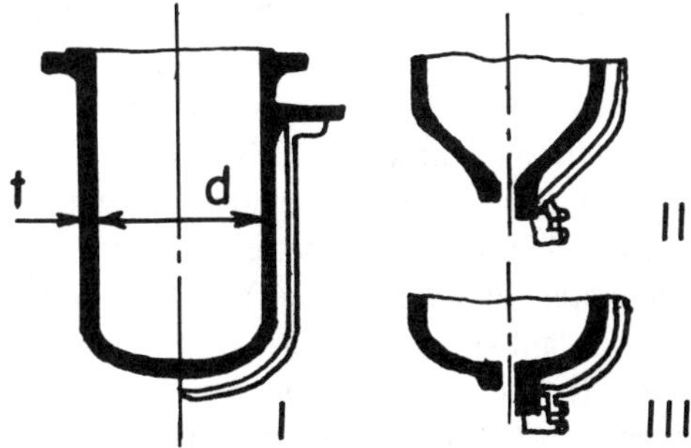

Figure 6.26. Principal designs of cast cylindrical shells.

For the final selection of shell thickness, t, it is necessary to follow directions given for forged shells. However, from design considerations it is recommended to take the following wall thickness: for steels, $t > 8$ mm; nonferrous metals and their alloys, $t > 6$ mm; cast iron, $t > 12$ mm; quartz glass, $t > 6$ mm.

If the cylindrical shell must operate under an internal pressure of more than 0.07 MN/m^2, the shell should be checked for stress under the hydrostatic test according to Equation 6.38. For brittle materials, one may assume, conditionally, $\sigma_y^t \simeq 0.5\,\sigma_T^t$.

6.3 BOX-TYPE SHELLS

Box-type shells are not used as widely in comparison to cylindrical designs. They use more material and are more complicated to fabricate. However, in some cases there is a need to use this design, for example, as casings for fabricating plunged coolers, condensers, dryers, casings for bag filters, etc. The rectangular form of shells is convenient for placing working devices inside the shell, as well as box-type shells themselves inside a building. Hence, the field of their application is limited mostly by pressure inside an apparatus up to 1.6 MN/m^2.

The application of box-type apparatuses for liquid storage is not a sound practice. The specific metal consumption for fabricating box-type storages ranges from 90 to 130 kg/m^3 of storage volume, whereas for cylindrical storages this ranges from 18 to 50 kg/m^3 of storage, that is, 3–5 times less than for box-type storages. The specific metal consumption decreases with increases in the storage volume in both cases. The maximum capacity of a box-type apparatus usually does not exceed 200 m^3 at pressure of a 2- to 2.5-meter water column. Therefore, their application may be justified only when increased metal consumption is counterbalanced by operating convenience or better utilization of the apparatus's volume.

Box-type shells are fabricated in three versions: (1) without reinforcement (small open tanks, as well as shells for apparatuses operating under pressure drops of several millimeters of water column); (2) with internal connections and tie rods (large open tanks); and (3) with reinforcing elements or tie rods from the outside. This last design is used because connecting rods inside the working space prevent assembly and disassembly of heat transfer surfaces or some other working devices.

Their fabrication may be done by welding of rolled steel sheets or by casting. Along with box-type shells, single flat rectangular walls are used often, which could be considered as elements forming the box-type shell. Therefore, the design considerations of these elements and a shell as a whole, independent of the material and fabrication method, are common.

To design box-type shells and their elements, the following directions should be followed:

1. The sizes of single flat rectangular walls should be small if possible.
2. Large flat rectangular walls should be reinforced by ribs.
3. The flat box-type shell should be butt-welded beyond the zone of smooth transition by radius.
4. The welded joints at the walls should be located at about one-fourth of the distance between opposite edges.
5. The holes for tubes and manholes, etc., should not have cross-welded seams; rather, they should be reinforced.

The design of angle joints for single flat rectangular walls is shown in Figure 6.27.

6.3.1 Design of Box-Type Shells Operating Under Internal Pressure

The design thickness of a flat rectangular wall with a fixed end operating under internal pressure is based on the following equation:

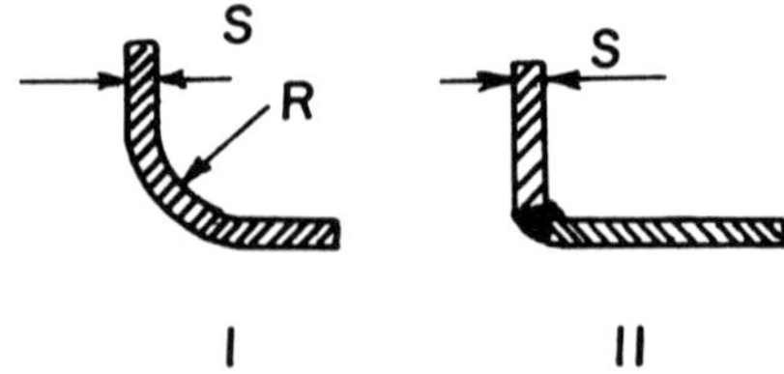

Figure 6.27. Angle joining of flat rectangular walls: I = walls of sheets, R > 5 s; for cast walls, R > 2s; II = for welded apparatuses that are open to the atmosphere.

$$t' = 0.5B\sqrt{\frac{P}{\sigma_{allow}}} + C \qquad (6.84)$$

where B = width (the smaller side) of the rectangular wall (mm)
C = allowance for corrosion, erosion and negative tolerance of a sheet (m)

The value of the design pressure is taken to be the same as in Equations 6.20 and 6.21. (The walls are considered to be fixed at edges when they have flanges and reinforcing rings, as well as wall angle joining.)

At the final selection of wall thickness, t, the following should be accounted for:

1. For walls of sheet-rolled metal, the directions given for the selection of wall thickness of cylindrical shells operating under internal pressure should be used.
2. For cast walls, the design wall thickness should be rounded from design considerations with regard to corresponding directions made above for cast cylindrical shells.

The assumed thickness of a flat rectangular wall, t, should be checked for stress, σ, under hydrostatic test from the following formula:

$$\sigma = \frac{0.25B^2P'}{(t-c)^2} \leqslant \frac{\sigma_y}{1.2}, \quad N/m^2 \qquad (6.85)$$

σ values for brittle materials can be assumed to be $\sigma_y \approx 0.5\sigma'$.

For decreasing the design thickness of a flat rectangular wall, as determined from Equation 6.84, reinforcement ribs should be considered. Figure 6.28 shows the basic types (I-III) of a flat rectangular wall reinforcement by ribs. Reinforcement ribs usually are located outside the wall.

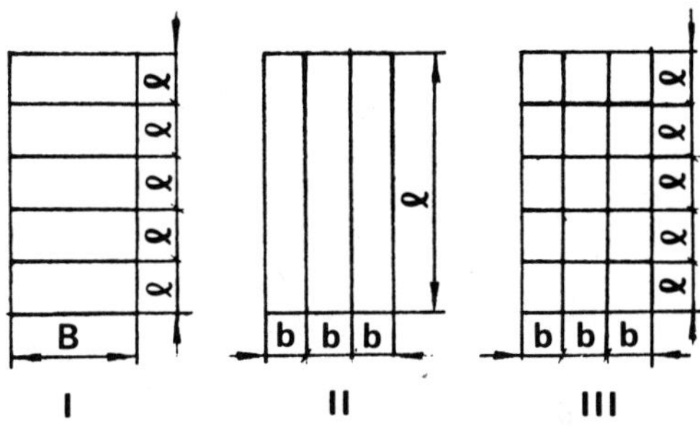

Figure 6.28. Basic schemes of flat rectangular wall reinforcements with ribs.

The design wall thickness when reinforced with ribs is determined from Equation 6.84 where B is replaced by the following: for type I use a value for ℓ; for type II, value b; for type III, the smaller value of ℓ or b.

The scheme of wall reinforcement with ribs and their number is determined from design considerations.

The preferred reinforcement scheme is type I. The typical cross sections of reinforcing ribs and their attachment to the flat wall are similar to those of Figure 6.17. The ribs of rectangular cross section usually are used for cast flat walls, as shown in Figure 6.29.

The approximate design of reinforcing ribs may be done by the following simplified method. It is taken conditionally that half the load on the flat rectangular wall is absorbed by the ribs (without regard to the flat wall). Then the design section modulus of each reinforcing rib, W_r', is determined from the following formulas:

At reinforcement type I,

$$W_r' = \frac{B^2 \ell P}{2\widetilde{K}\sigma_{b,allow}}, \quad m^3 \tag{6.86}$$

At reinforcement type II,

$$W_r' = \frac{L^2 b P}{4\widetilde{K}\sigma_{b,allow}} \tag{6.87}$$

At reinforcement type III in the transverse direction,

$$W_r' = \frac{B^2 \ell P}{4\widetilde{K}\sigma_{b,allow}} \tag{6.88}$$

And in the longitudinal direction,

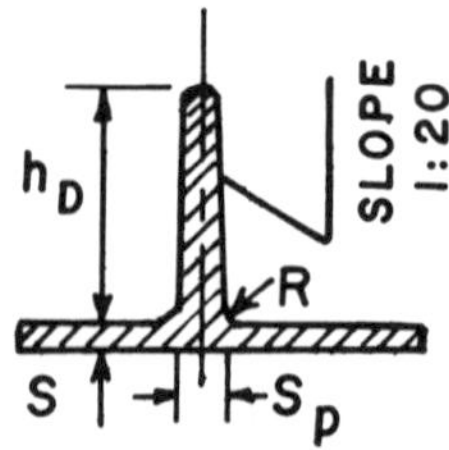

Figure 6.29. Cross section of reinforcing rib for a flat rectangular cast wall: $h_p = 5\, s_p$; $R \geqslant 0.5\, s_p$.

$$W'_{r,\ell} = \frac{L^2bP}{4\widetilde{K}\sigma_{b,allow}} \tag{6.89}$$

where B, L, b and ℓ are corresponding sizes shown in Figure 6.28.

$\widetilde{K}$ = coefficient depending on the method of rib fixing on the supports. at rigid fixing (with a flange, other ribs, etc.) $\widetilde{K} = 12$; for nonrigid fixing, $\widetilde{K} = 8$.

$$W'_r = W'_\ell \quad \text{at} \quad \frac{b}{t} = \frac{B^2}{L^2}$$

Equations 6.86–6.89 were derived based on the assumption that the rib acts as a beam on two supports uniformly loaded by pressure, p, exposed to a corresponding flat element of the wall.

The cross section of the rib is found from the section modulus, W'_r. In particular, for its rectangular cross section at the recommended thickness: height ratio of 1:5, the design rib thickness, S'_r, is determined as follows:

$$S'_r = 0.62\sqrt[3]{W'_r}\,, \quad m \tag{6.90}$$

The assumed sizes of the reinforced ribs cross section should be checked for their maximum stress (with regard to the flat wall) at operating conditions and at the hydrostatic test of the vessel.

For this purpose, the resistance moment of the element cross section with reinforcing rib $W_{r.w.}$ is first determined from the following formula:

$$W_{r.w.} = \frac{J_r + J_w + F_r(0.5h_r - y)^2 + F_w \times [y + 0.5(s - c)]^2}{h_r - y}\,, \quad m^3 \tag{6.91}$$

where F_r = assumed area of rib cross section (m^2)

F_w = assumed area of wall element cross section (m^2)

J_r = inertia moment of area, F_r with respect to the axis passing through its gravity center parallel to the wall (m^4)

J_w = inertia moment of area, F_w, with respect to the axis passing through its gravity center parallel to the wall (m^4)

y = distance from wall to gravity center of area normal to the wall (m)

Refer to Figure 6.30. In this figure, the values of F_w and y are

$$F_w = x(t - c) , \quad m^2 \tag{6.92}$$

$$y = \frac{F_r h_r - F_w(h - c)}{2(F_r + F_w)} , \quad m \tag{6.93}$$

The maximum stress in reinforcing ribs, σ, under operating conditions should correspond to the following conditions:

At reinforcement type I,

$$\sigma = \frac{B^2 \ell P}{\tilde{K} W_{r.w.}} \leqslant 1.2\sigma_{b,allow} , \quad N/m^2 \tag{6.94}$$

At reinforcement type II,

$$\sigma = \frac{L^2 b P}{\tilde{K} W_{r.w.}} \leqslant 1.2\sigma_{b,allow} , \quad N/m^2 \tag{6.95}$$

At reinforcement type III in the transverse direction,

$$\sigma = \frac{B^2 \ell p}{2\tilde{K} W_{r.w.}} \leqslant 1.2\sigma_{b,allow} , \quad N/m^2 \tag{6.96}$$

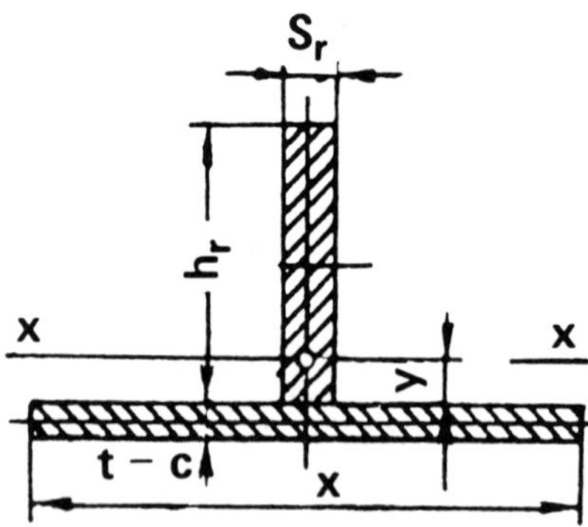

Figure 6.30. Cross section of an element of flat rectangular wall with a reinforcing rib. For transverse ribs, $x = \ell$; for longitudinal ribs, $x = b$ (see also Figure 6.28).

in the longitudinal direction,

$$\sigma = \frac{L^2bP}{2\widetilde{K}W_{r.w.}} \leqslant 1.2\sigma_{b,allow} \ , \quad N/m^2 \tag{6.97}$$

The maximum stress in reinforcing ribs, σ, at hydrostatic test of the vessel should correspond to the following conditions:

At reinforcement type I,

$$\sigma = \frac{B^2\ell P'}{\widetilde{K}W_{r.w.}} \leqslant \frac{\sigma_y}{1.2} \ , \quad N/m^2 \tag{6.98}$$

At reinforcement type II,

$$\sigma = \frac{B^2\ell P'}{\widetilde{K}W_{r.w.}} \leqslant \frac{\sigma_y}{1.2} \ , \quad N/m^2 \tag{6.99}$$

At reinforcement type III, in the transverse direction,

$$\sigma = \frac{B^2\ell P'}{2\widetilde{K}W_{r.w.}} \leqslant \frac{\sigma_y}{1.2} \ , \quad N/m^2 \tag{6.100}$$

In the longitudinal direction,

$$\sigma = \frac{L^2bP'}{2\widetilde{K}W_{r.w.}} \leqslant \sigma_y/1.2 \ , \quad N/m^2 \tag{6.101}$$

In cases in which conditions 6.95–6.101 are not fulfilled, the assumed sizes of ribs' cross sections (or their number) should be increased correspondingly. The four rectangular flat walls are connected among them at right angles from the box-type shell.

When determining the design thickness, each of the walls should be made based on Equation 6.84. If the walls have different widths and are not reinforced by ribs, then the thickness of all the walls of the shell usually is taken according to the calculation for the wider wall.

Figure 6.31 shows basic designs of box-type shells of process apparatuses in which the left-hand side of each type presents a version with flanges, and the right-hand side a version with bottoms.

The following example illustrates the design principles outlined above.

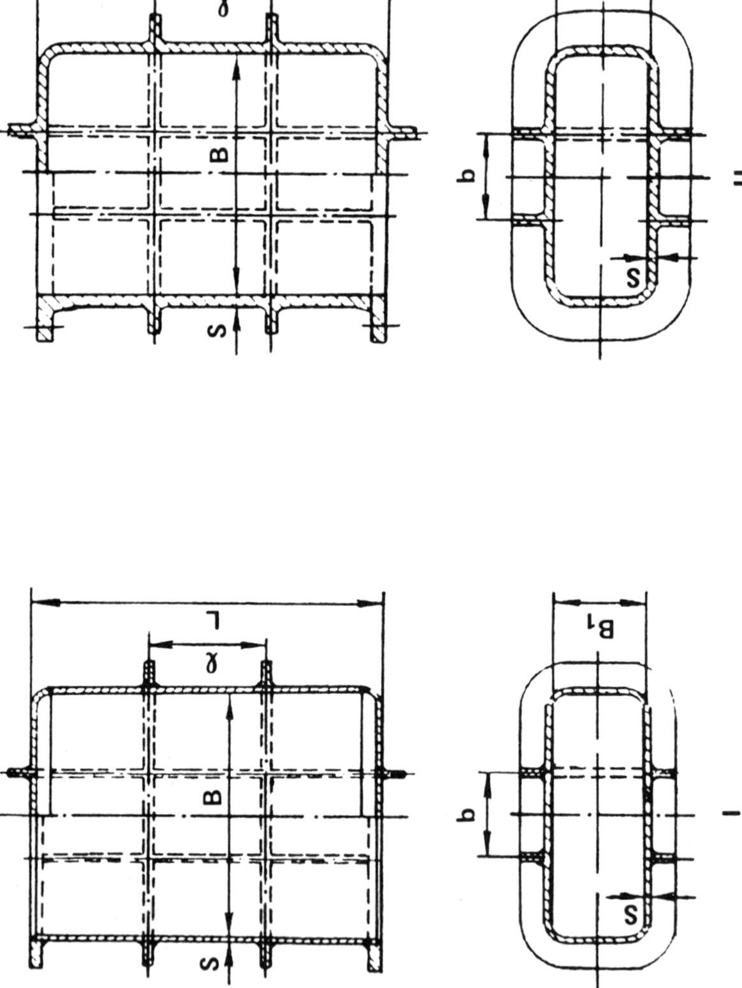

Figure 6.31. Basic designs of box-type shells: I = welded; II = cast.

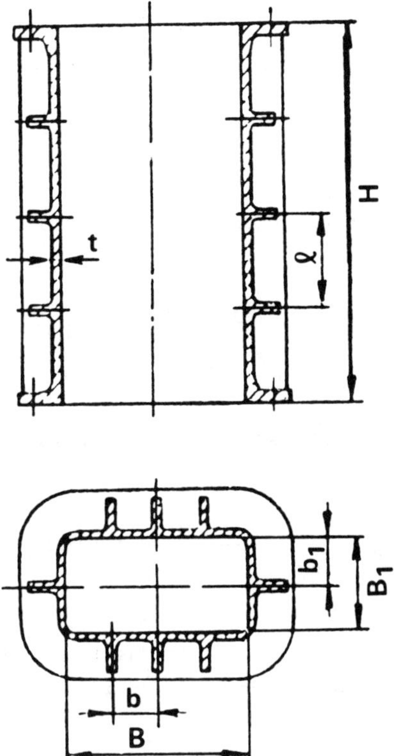

Figure 6.32. Sketch for Example 6.8.

Example 6.8

Determine the thickness of the walls of a cast box-type shell and cross section of reinforcing ribs of vertical apparatus operating under internal pressure according to the following data (Figure 6.32):

Data

The material of shell is steel ($\sigma_T^{150} = 400 \times 10^6$ N/m^2; $\sigma_y^{150} = 200 \times 10^6$ N/m^2; $\sigma_y = 220 \times 10^6$ N/m^2); the internal medium is gas, p = 7 MN/m^2; t = 150°C; B = 1.0 m; B_1 = 0.5 m; H = 2.0 m; ℓ = 0.5 m; b = b_1 = 0.25 m; the holes of the shell are reinforced; the apparatus is Class 2, group 2 ($\eta = 1.0$).

Solution

The design internal pressure is taken from Equation 6.22:

$$p = p_f = 1 \times 10^6 \text{ N/m}^2$$

The allowable stress for steel determined on the basis of tensile strength is

$$\sigma_{b,allow} = \frac{\sigma_T^t}{\eta_T} \eta = \frac{400 \times 10^6}{3.5} \times 1 = 114 \times 10^6 \text{ N/m}^2$$

The allowable stress for steel determined on the basis of yield point is

$$\sigma_{b,allow} = \frac{\sigma_y^t}{\eta_y} \eta = \frac{200 \times 10^6}{2} \times 1 = 100 \times 10^6 \text{ N/m}^2$$

The second value of stress (the smaller one) is taken for the design value.

The design wall thickness is determined from Equation 6.84:

$$t' = 0.5b \sqrt{\frac{P}{\sigma_{b,allow}}} + C = 0.5 \times 0.25 \sqrt{\frac{1 \times 10^6}{100 \times 10^6}} + 3 \times 10^{-3}$$

$$= 15.5 \times 10^{-3} \text{ m}$$

Checking the stress in the wall of the shell at the hydrostatic test of the apparatus from Equation 6.24,

$$P_e \approx 10 \rho_e H_e = 10 \times 1 \times 10^3 \times 2 = 0.02 \times 10^6 \text{ N/m}^2$$

The design test pressure is determined from Equation 6.39:

$$P' = P_h + P_e = 1.5 \times 1 \times 10^6 + 0.02 \times 10^6$$

$$= 1.52 \times 10^6 \text{ N/m}^2$$

The stress in the wall of the shell at the hydrostatic test of the vessel, from Equation 6.85, is

$$\sigma = \frac{0.25b^2P_b}{(t-c)^2} = \frac{0.25 \times 0.25^2 \times 1.52 \times 10^6}{[(16-3)10^{-3}]^2} = 140.5 \times 10^6 \text{ N/m}^2$$

$$< \sigma_y/1.2 = \frac{220 \times 10^6}{1.2} = 183 \times 10^6 \text{ N/m}^2$$

The design of reinforcing ribs: The design ribs section modulus (for wall with the width B) at $\widetilde{K} = 12$ (rigid fixing of rib) is determined from Equations 6.88 and 6.89:

$$W'_r = \frac{B^2 \ell P}{4\widetilde{K}\sigma_{b,allow}} = \frac{(1)^2 \times 0.5 \times 1 \times 10^6}{4 \times 12 \times 100 \times 10^6} = 104 \times 10^{-6} \text{ m}^3$$

$$W'_{r,\ell} = \frac{L^2 bP}{4\widetilde{K}\sigma_{b,allow}} = \frac{(2)^2 \times 0.25 \times 1 \times 10^6}{4 \times 12 \times 100 \times 10^6} = 208 \times 10^{-6} \text{ m}^3$$

Selection of ribs of rectangular cross section: The design of a rib thickness with a thickness:height ratio of 1:5 (Figure 6.29) is determined from Equation 6.90:

$$t'_r = 0.62 \sqrt[3]{W'_r} = 0.62 \sqrt[3]{104 \times 10^{-6}} = 29.1 \times 10^{-3} \text{ m}$$

$$t'_{r,\ell} = 0.62 \sqrt[3]{W'_{r\ell}} = 0.62 \sqrt[3]{208 \times 10^{-6}} = 36.7 \times 10^{-3} \text{ m}$$

As an example, we take the following sizes of ribs:

Transverse:

$$t'_r = 30 \text{ mm} \; ; \quad h_r = S_r \times 5 = 30 \times 5 = 150 \text{ mm}$$

Longitudinal:

$$t'_{r\ell} = 38 \text{ mm} \; ; \quad h_{r\ell} = S_{r\ell} \times 5 = 38 \times 5 = 190 \text{ mm}$$

Checking for maximum stress in ribs, first we find

$$F_r = t_r'h_r = 30 \times 150 \times 10^{-6} = 4500 \times 10^{-6}\ m^2$$

$$F_{r\ell} = t_{r\ell}'h_{r\ell} = 38 \times 190 \times 10^{-6} = 7200 \times 10^{-6}\ m^2$$

$$J_r = \frac{t_r'h_r^3}{12} = \frac{30 \times 10^{-3} \times 150^3 \times 10^{-9}}{12} = 844 \times 10^{-8}\ m^4$$

$$J_{re} = \frac{t_{r\ell}'h_{r\ell}^3}{12} = \frac{38 \times 10^{-3} \times 190^3 \times 10^{-9}}{12} = 2160 \times 10^{-8}\ m^4$$

$$J_w = \frac{\ell(t-c)^3}{12} = \frac{0.5\,[(16-3)10^{-3}]^2}{12} = 9.15 \times 10^{-8}\ m^4$$

$$J_{w\ell} = \frac{b(t-c)^3}{12} = \frac{0.25\,[(16-3)10^{-3}]^3}{12} = 4.58 \times 10^{-8}\ m^4$$

From Equation 6.92,

$$F_w = \ell(t-c) = 0.5(16-3)10^{-3} = 6500 \times 10^{-6}\ m^2$$

$$F_{w\ell} = b(t-c) = 0.25(16-3)10^{-3} = 3250 \times 10^{-6}\ m^2$$

From Equation 6.93,

$$y = \frac{F_r h_r - F_w(t-c)}{2(F_r + F_w)}$$

$$= \frac{4500 \times 10^{-6} \times 150 \times 10^{-3} - 6500 \times 10^{-6}\,(16-3)10^{-3}}{2(4500 + 6500) \times 10^{-6}}$$

$$= 26.8 \times 10^{-3}\ m$$

$$y_\ell = \frac{F_{r\ell}h_{r\ell} - F_{w\ell}(t - c)}{2(F_{r\ell} + F_{w\ell})}$$

$$= \frac{7200 \times 10^{-6} \times 190 \times 10^{-3} - 3250 \times 10^{-6}(16 - 3)10^{-3}}{2(2700 + 3250)10^{-6}}$$

$$= 63.5 \times 10^{-3}\ \mathrm{m}$$

From Equation 6.91,

$$W_r = \frac{J_r + J_w + F_r(0.5\,h_r - y)^2 + F_w[(y + 0.5(t - c)]^2}{h_r + y}$$

$$= \{844 \times 10^{-8} + 9.15 \times 10^{-8} + 4500 \times 10^{-6}$$

$$\times (0.5 \times 150 \times 10^{-3} - 26.8 \times 10^{-3})^2$$

$$+ 6500 \times 10^{-6}[26.8 \times 10^{-3} + 0.5(16 - 3) \times 10^{-3}]^2\}/$$

$$(150 - 26.8) \times 10^{-3} = 212 \times 10^{-6}\ \mathrm{m}^3$$

$$W_{rw\ell} = \frac{J_{r\ell} + J_{w\ell} + F_{r\ell}(0.5\,h_{r\ell} - y_\ell)^2 + F_{w\ell} \times [y_e + 0.5(t - c)]^2}{h_{r\ell} - y_e}$$

$$= 354 \times 10^{-6}\ \mathrm{m}^2$$

The maximum stress under operating conditions:

1. In transverse ribs, it is determined from Equation 6.96:

$$\sigma = \frac{B^2\ell p}{2\tilde{K}W_{rw}} = \frac{1^2 \times 0.5 \times 1 \times 10^6}{2 \times 12 \times 212 \times 10^{-6}} = 98 \times 10^6\ \mathrm{N/m^2} < 1.2\ \sigma_{b,allow}$$

$$= 120 \times 10^6\ \mathrm{N/m^2}$$

2. In longitudinal ribs, it is determined from Equation 6.97:

$$\sigma = \frac{L^2 bp}{2\tilde{K}W_{rw\ell}} = \frac{2^2 \times 0.25 \times 1 \times 10^6}{2 \times 12 \times 354 \times 10^{-6}} = 118 \times 10^6 \text{ N/m}^2$$

For the maximum stress at the hydrostatic test:

1. In transverse ribs, they are determined from Equation 6.100:

$$\sigma = \frac{B^2 \ell p'}{2\tilde{K}W_{rw}} = \frac{1^2 \times 0.5 \times 1.52 \times 10^6}{2 \times 12 \times 212 \times 10^{-6}} = 149 \times 10^6 \text{ N/m}^2 < \sigma_y/1.2$$

$$= \frac{220 \times 10^6}{1.2} = 183 \times 10^6 \text{ N/m}^2$$

2. In longitudinal ribs, they are determined from Equation 6.101.

$$\sigma = \frac{L^2 bp'}{2\tilde{K}W_{rw\ell}} = \frac{2^2 \times 0.25 \times 1.52 \times 10^6}{2 \times 12 \times 354 \times 10^{-6}} = 179 \times 10^6 \text{ N/m}^2$$

For a wall with a width $B_1 < B$, cross sections of longitudinal and transverse ribs are taken as the same as for a wall with a width B.

6.3.2 Design of Box-Type Shells Operating During Filling

The design thickness of a flat rectangular wall subjected to hydrostatic pressure is determined from Equation 6.84, where the design pressure, p, is taken to be equal to p_e from Equation 6.24 for a more loaded lower part of the wall.

The thickness of a flat rectangular wall made from bulk and lump materials in the absence of internal pressure usually is taken from design considerations with respect to rigidity and operating conditions.

For final selection of wall thickness, the directions given for cylindrical shells operating without pressure or open should be followed. In most cases (especially in welded shells), a wall thickness of more than 6 mm is not recommended.

In the case of reinforced flat walls of box-type shells, it is advisable to locate ribs parallel to the smaller side of the wall (Figure 6.17). By reinforcing the walls of the shell with vertical ribs (a common case) it is assumed that all the load from the hydrostatic pressure is accepted entirely by the ribs. Then the design section modulus, W'_r, of each rib considered as a beam, uniformly loaded by hydrostatic pressure and freely supported at the ends, is

$$W'_r = \frac{0.64\,\rho_\ell H_\ell^3 \ell}{\sigma_{b,allow}} \tag{6.102}$$

where H_ℓ = the height (maximum) of liquid column.
ℓ = distance between the rib axis (0.64—keep in mind the value of g)

The supports of ribs serve as bottoms for the shell. If the shell is open at the top, the upper edges of rectangular walls should be reinforced accordingly or provided with tie rods between opposite walls absorbing the load of the upper rib supports. Horizontal ribs usually are applied as an additional reinforcement whenever vertical ribs are used. As this takes place, no more than two horizontal ribs are used as a rule (depending on the size of the wall). Figure 6.33 shows the recommended spacing for horizontal reinforcing ribs on a vertical flat rectangular wall proceeding from the condition that ribs are loaded uniformly.

The design section modulus of horizontal reinforcing rib is determined as follows:

For one horizontal rib,

$$W'_r = \frac{0.156\rho_e H_e^2 \ell^2}{\sigma_{b,allow}}, \quad m^3 \tag{6.103}$$

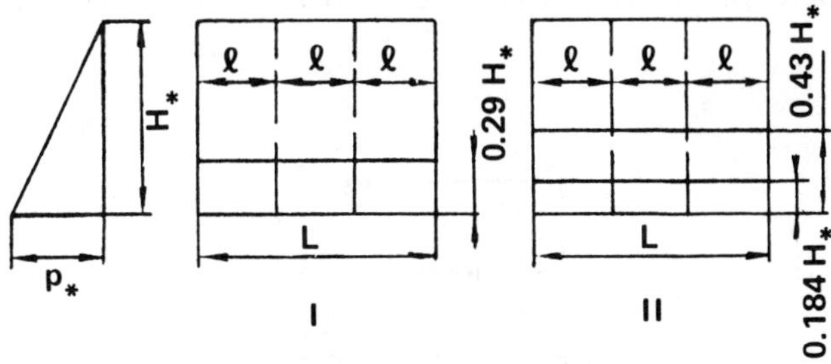

Figure 6.33. Spacing of the horizontal reinforcing ribs on the vertical rectangular wall subjected to hydrostatic pressure ($p_e = \rho_e g\, H_e$, N/m^2): I = one rib; II = two ribs.

For two horizontal ribs,

$$W_r' = \frac{0.104\,\rho_e H_e^2 \ell^2}{\sigma_{b,allow}}, \quad m^3 \tag{6.104}$$

In deriving formulas 6.103 and 6.104, the rib was considered as a beam freely supported at the ends and uniformly loaded with hydrostatic pressure on the corresponding flat element of the wall, taking into account the acceptance of this pressure also by vertical ribs (or perpendicular walls), bottoms or flanges of the shell. The section modulus is determined as follows:

For one horizontal rib,

$$W_r' = \frac{0.156\,\rho_e H_e^2 \ell^2}{\sigma'_{allow}}, \quad m^3 \tag{6.105}$$

For two horizontal ribs,

$$W_r' = \frac{0.104\,\rho_e H_e^2 \ell^2}{\sigma'_{allow}}, \quad m^3 \tag{6.106}$$

Example 6.9

Determine the thickness of the walls of the welded box-type shell, the cross section of the reinforcing ribs and their location for an apparatus operating during filling conditions according to the following data (Figure 6.34):

Data

The material of the shell is steel ($\sigma_{b.allow} = 146 \times 10^6$ N/m^2); the shell is rubberized on the inside ($C_1 = 0$; $C_2 = 0$); the medium is liquid $\rho_\ell = 1.3 \times$

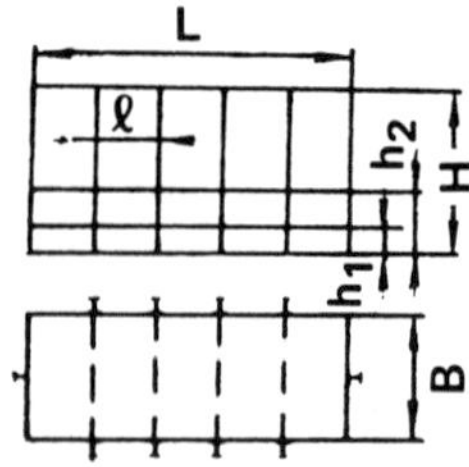

Figure 6.34. Sketch for Example 6.9.

10^3 kg/m^3, $t_w = 20°C$; L = 6 m; B = 2.4 m; H = 3 m; $\ell = 1.2$ m; the apparatus is of the second operating class, group 2 ($\eta = 1.0$).

Solution

The design hydrostatic pressure is determined first:

$$P = p_e \approx 10\rho_\ell H_\ell = 10 \times 1.3 \times 10^3 \times 3 = 0.039 \times 10^6 \text{ N/m}^2$$

Location of horizontal ribs along the vertical plane is determined from Figure 6.33:

$$h_1 = 0.184\ H_e = 0.184 \times 3 = 0.55 \text{ m}$$

$$h_r = 0.43\ H_e = 0.43 \times 3 = 1.29 \text{ m}$$

The design thickness of flat wall is determined from Equation 6.84 where $B = h_1 = 0.55$ mm.

$$t' = 0.5B\sqrt{\frac{P}{\sigma_{b,allow}}} + C = 0.5 \times 0.55 \times \sqrt{\frac{0.039 \times 10^6}{146 \times 10^6}} + C$$

$$= 4.5 \times 10^{-3} + C\ , \quad \text{m}$$

At $C_1 = 0$ and $C_2 = 0$, $C = C_3 = 0.5 \times 10^{-3}$. Hence,

$$t' = (4.5 + 0.5)10^{-3} = 5 \times 10^{-3} \text{ m}$$

We take t = 5 mm.

The design section modulus of a vertical reinforcing rib is obtained from Equation 6.102:

$$W_r' = \frac{0.64\rho_\ell H_\ell^3 \ell}{\sigma_{b,allow}} = \frac{0.64 \times 1.3 \times 10^3 \times 3^3 \times 1.2}{146 \times 10^6} = 185 \times 10^{-6} \text{ m}^3$$

We select the beam B1(24 X 7) from the American Standard ($W_r' = 12.2$ in.^{3}). The design resistance moment of the horizontal reinforcing ring is obtained from Equation 6.106:

$$W'_r = \frac{0.104\rho_\ell H_\ell^2 \ell^2}{\sigma_{b,allow}} = \frac{0.104 \times 1.3 \times 10^3 \times 3^2 \times 1.2^2}{146 \times 10^6} = 12 \times 10^{-6}\ m^3$$

We select the equal angle A4(4 X 4 X ¼ ") from the American Standard:

$$W_x = \frac{J_x}{b_0 - Z_0} = \frac{3}{4 - 1} = 1\ \text{in.}^3 \simeq 12\ \text{cm}^3$$

6.4 SPHERICAL SHELLS

From the point of view of material consumption, the most economical shape of a shell is spherical. However, as its fabrication is complicated, it has limited application. Spherical shells are used mostly for storage of large volumes of fluids under moderate pressures (up to 1.6 MN/m^2) and under vacuum.

The fabrication of spherical shells consists mostly of stamping petal-shaped plates and their subsequent butt-welding.

6.4.1 Design of Spherical Shells Operating Under Internal Pressure

A spherical shell is loaded uniformly by internal pressure, p (Figure 6.35).

The longitudinal force, U, acting on unity of length of its application line and equal to the circumferential force, T, is

$$U = T = \frac{p}{2R} \tag{6.107}$$

The longitudinal stress, σ_U, is equal to the circumferential stress, σ_T:

$$\sigma_U = \sigma_T = \frac{pR}{2S} \tag{6.108}$$

The radial displacement along the main radius is

$$\Delta = \frac{PR}{2ES}(1 - \tilde{\mu}) \tag{6.109}$$

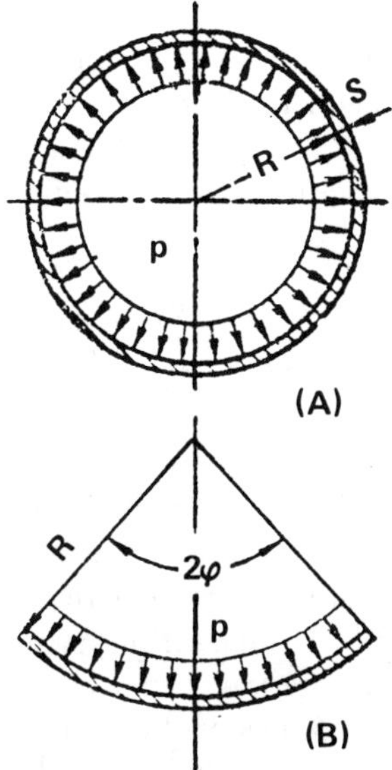

Figure 6.35. Scheme of spherical shell operating under internal pressure.

The displacement along the radius of a parallel circle for a central angle $2\phi^*$ (Figure 6.35(B)) is

$$\Delta' = \frac{PR}{2SE}(1 - \tilde{\mu}) \sin\phi^* \qquad (6.110)$$

Angular displacement is

$$\nu = 0 \qquad (6.111)$$

The design thickness of a spherical wall, t, under internal pressure is

$$t = \frac{dp}{4\sigma_{allow}\psi - p} + c \qquad (6.112)$$

If $(\sigma_{allow}/p)\ \psi \geqslant 25$, the value, p, in the denominator of Equation 6.112 can be omitted. The procedure of design, determination of values ρ, ψ, C, and the selection of allowable stress, σ_{allow}, are the same as those of the design of welded cylindrical shells made from rolled up sheet metal operating under internal pressure. The same method used for the selection of wall thickness is applied here. The assumed wall thickness of a spherical shell, t, should be checked for stress at hydrostatic pressure according to the following formula:

$$\sigma = \frac{[d + (t - c)]p'}{4(t - c)\psi} \leqslant \frac{\sigma_y}{1.2}, \quad N/m^2 \tag{6.113}$$

6.4.2 Design of Shells Operating Under External Pressure

The spherical shell operating under external pressure or vacuum may collapse if $(t - C)/d \leqslant 0.04$. The danger of pressing in increases as the shell deviates from an accurate spherical shape. The permissible deviation from roundness is no more than 0.5% in any direction.

The design thickness of a spherical shell operating under external pressure is determined from the following formula by using stability considerations [12]–the safety stability factor is equal to 5:

$$t = 0.8 \left\{ \frac{P\sqrt{3(1 - \bar{\mu}^2)}}{E^t} \right\}^{1/2} + c \tag{6.114}$$

At final selection of the thickness of the spherical shell, the directions for cylindrical shells operating under internal pressure should be followed.

6.4.3 Design of Shells Operating Under Low Operating Pressure (no more than 0.07 MN/m²) or During Filling

The design thickness of a spherical shell operating under open conditions with its lower part subjected to internal hydrostatic pressure is determined from Equation 6.112, where the design pressure, p, should be equal to p_e and calculated from Equation 6.24. The pressure should be no less than 0.1 MN/m^2 to provide rigidity. For vessels operating under low pressure (up to 0.07 MN/m^2) with gases, the design pressure to provide rigidity should be determined from the same formula where the design pressure is taken as $p = 0.1\ MN/m^2$.

For final selection of the shell thickness, the above directions should be followed for the selection of wall thicknesses of cylindrical shells operating under low pressure or open.

6.5 CONICAL SHELLS

Conical shapes are used for some apparatuses, as well as connecting parts in different cylindrical apparatus with changeable diameters.

The fabrication of conical shells is similar to that of cylindrical ones (rolling up and welding metal sheets, forging, casting). Design generally depends on the shell material, the operating pressure in an apparatus and similar considerations discussed previously for cylindrical shells. Figure 6.36 presents typical designs of conical shells used in process equipment. Three types are shown:

Type I is a shell (without beading) used for apparatuses operating either under internal or external pressure up to 1.6 MN/m^2 at $\gamma < 10°$, as well as without pressure or under filling up conditions at $\gamma < 45°$.

Type II is a shell of sheet metal, forged piece or casting used for apparatuses operating either under internal or external pressure at $\gamma < 45°$.

Type III is a shell of sheet metal (left-hand side), forged or cast from brittle materials (right-hand side) edged by flanges used for apparatuses operating under internal pressure, without pressure and open at $\gamma < 45°$. Reinforced holes for tubes, manholes, etc., should be avoided at seams and flanged areas.

6.5.1 Design of Shells Operating Under Internal Pressure

The design shell thickness (Figure 6.36) of types I and III operating under internal pressure is determined from the following formula:

$$t = \frac{p}{2d \cos\gamma(\sigma_{allow}\psi - p)} + C \qquad (6.115)$$

where γ = angle from Figure 6.36, grades

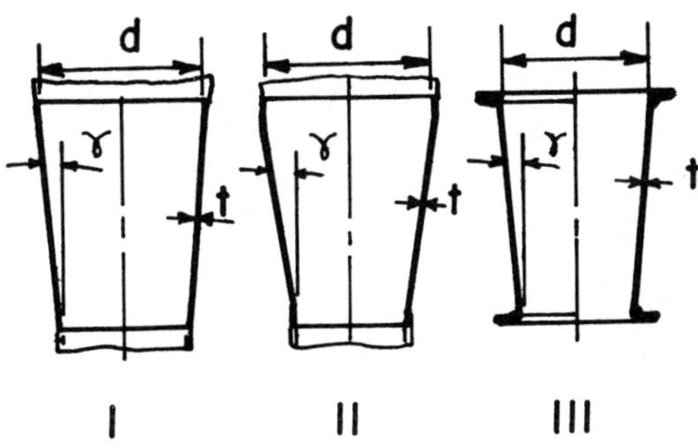

Figure 6.36. Typical designs of conical shells.

The design procedure, determination of p, ψ, and C and the selection of allowable stress, σ_{allow}, are similar to those of cylindrical shells operating under internal pressure. The same is related to the selection of shell thickness, t. The assumed wall thickness of a conical shell, t, should be checked for stress at hydrostatic test of an apparatus from the following formula:

$$\sigma = \left[\frac{dp'}{2\cos\gamma(t - c)} + p'\right]\frac{1}{\psi} \leqslant \frac{\sigma_y}{1.2}, \quad N/m^2 \qquad (6.116)$$

6.5.2 Design of Shells Operating Under External Pressure

At loading of thin-walled conical shells with external pressure (or if they are operating under vacuum at $(t - c)/d \leqslant 0.04$), pressing in of the shell can occur. The danger of pressing in increases with the deviation of the shell configuration from the correct cone shape. The permissible out-of-roundness is no more than 0.5% in any cross section.

For a reasonable approximation of the design of conical shells at $\gamma < 30°$ (Figure 6.36) for type I (if the larger and lower diameters are reinforced) and type III the formulas for cylindrical shells operating under external pressure may be used. In so doing, the value of the internal diameter of the shell in all formulas should be taken as the larger diameter of the conical shell.

The design of similar cast shells should be done with regard to the data given in the section for the design of cast cylindrical shells.

The design of conical shells according to type II (Figure 6.36) and all other types where $\gamma > 30°$, will be given in Chapter 7.

6.5.3 Design of Shells Operating Under Low Internal Pressure (up to 0.07 MN/m²) or During Filling Conditions

The design thickness of a conical shell, operating under open conditions with a more loaded lower part subjected to internal hydrostatic pressure, is determined from Equation 6.115, where the design pressure, p, is taken from Equation 6.24, but no less than 0.1 MN/m^2 (based on considerations of rigidity).

For apparatuses operating under low pressure (up to 0.07 MN/m^2) and subjected to gaseous medium, the design thickness of a conical shell to provide rigidity should be determined from the same formula in which the design pressure, p, is taken to be equal to 0.1 MN/m^2. For the final selection of wall thickness, the directions should be followed that are given for cylindrical shells operating under low internal pressure or open conditions.

NOMENCLATURE

a	area resisting longitudinal rupture (m^2)
B	width of rectangular box (m)
b_2	width of ring (m)
C	allowance factor
C_2	corrosion allowance (cm)
D_{opt}	optimum tank diameter (m)
d	inside diameter (m)
d_e	outer diameter (m)
E^t	modulus of elasticity at specified temperature t (N/m^2)
F	cross-sectional area (m^2)
G	gravity force of vessel and fluid (N)
H	shell height (m)
H_{opt}	optimum tank height (m)
J'	inertia moment of ring (m^4)
J'_r	inertial moment of nominal design ring cross section (m^4)
K_1, K_2	coefficients in Equation 6.36 (refer to Figure 6.12)
K	coefficient in Equations 6.86–6.89, which depends on the method of rib fixing on supports
L	length (m)
x'	design length or height (m)
M	bending moment (N-m)
M	external torsional moment (N-m)
n_s	stability factor (see Equation 6.48)
P	pressure (kg/cm^2)
p	internal pressure (kg/cm^2)
p'	longitudinal force (N)
P_e	hydrostatic pressure (N/m^2)
P_h	hydrostatic test pressure (N/m^2)
p'	shell design pressure (N/m^2)
R_2	radius of central surface (m)
S_k, S_∂	parameters defined in Figure 6.2
S'_r	rib thickness (m)
T	circumferential force (N)
t	temperature (°C) or thickness (mm)
t_{nr}	nominal ring thickness (m)
V	volume (m^3)
W	section modulus or axial moment of resistance (m^3)
y	distance normal from surface (m)
z	number of waves in cylinder at collapse
α	coefficient of linear expansion (in./(in.-°F))

β parameter defined by Equation 6.75
Δ displacement (m)
Δ_a allowable eccentricity (m)
Δ_n negative tolerance (m)
Δ_p positive tolerance (m)
η_τ safety factor (see Equation 6.14)
$\tilde{\mu}$ Poisson's ratio
ν angular displacement (rad)
ν_p corrosion penetration rate (cm/yr)
ρ density (kg/cm^3)
σ_{allow} allowable stress (kg/cm^2)
σ'_{allow} allowable stress in compression (kg/cm^2)
σ^*_{allow} nominal allowable stress (N/m^2)
σ_c^t creep stress at temperature t (N/m^2)
σ_f^t fatigue stress at temperature, t (N/m^2)
σ_T circumferential stress (N/m^2)
σ_t tensile strength (kg/cm^2)
σ_ν longitudinal stress (kg/cm^2)
σ_R radial stress (kg/cm^2)
σ_y yield stress (kg/cm^2)
τ time (yr) or torsional stress (N/m^2)
ϕ^* angle (°)
ψ joint efficiency factor

REFERENCES

1. Timoshenko, S., and G. H. MacCullough. *Elements of Strength of Materials* (New York: Van Nostrand Reinhold Company, 1938).
2. *ASME Boiler and Pressure Vessel Code*, Section 8, Division 1 (New York: American Society of Mechanical Engineers, 1977).
3. Brownell, L. E., and E. H. Young. *Process Equipment Design* (New York: John Wiley & Sons, Inc., 1959).
4. Huber, M. T. "Probleme der Static Technisch Wichtiger Orthotroper Platten," Warszawa (1929).
5. von Mises, R. "Göttinger Nachrichten," *Math. Phys.* (1913), p. 582.
6. Hencky, H. "Uber Langsame Stationäre Strömungen in Plastischen Massen mit Rucksicht auf Vorgänge Bei Walzen, Pressen und Ziehen von Metallen," *Z. Angew. Math. U. Mech.* 5:115 (1925).
7. "Ferrous Metals," Pt. 1, *Am. Soc. Testing Materials Standards* (1955).
8. *Corrosion Handbook*, H. H. Uhlig, Ed. (New York: John Wiley & Sons, Inc., 1947).
9. Domashnev, A. D., A. I. Rychkov, and O. Vlianii. "Konstruktivnykh form Apparatury na ee Korroziiy," *Kchim. Prom.* N7 (1948).

10. Timoshenko, S. *Theory of Elastic Stability* (New York: McGraw-Hill Book Co., 1936).
11. Vikhman Iu. L., I. F. Babitzky, and S. J. Volfson. "Raschet i Konstruirovanie Neftezavodskoi Apparatury," *GNTI, M.* (1953).
12. Kontozovich, Z. B. "Osnovy Rascheta Khimicheskikh Mashim i Apparatov," *Mashgiz, M.* (1960).

CHAPTER 7

DESIGN OF BOTTOMS AND HEADS

7.1 INTRODUCTION

The term "head" refers to the part of an apparatus that confines a shell from below, above and the sides. The bottoms and heads are integral parts of shells of process apparatuses. As a rule, they are nondetachable from the shells and fabricated of the same material. In welded or soldered equipment, heads usually are welded or soldered to a shell; in forged and cast equipment, they are either integral with a shell and/or welded to it; in cast equipment of brittle materials, heads are always fabricated together with the shell.

The shape of the head depends on several factors: shape of the shell to be connected with the head, process requirements made to the apparatus, pressure and temperature of the fluid to be processed, and various design considerations. Different head configurations normally fabricated are spherical, box-type, conical, elliptical and flat (round and rectangular).

Cylindrical apparatuses with formed heads are used for a wide variety of applications. These applications can be grouped into three categories: (1) functional use, (2) pressure consideration, and (3) size limitations. Among these are distillation columns, evaporators, crystallizers, heat exchangers, etc. If the working pressure of an apparatus is to be other than atmospheric pressure, formed heads are used to close the apparatus.

In general, all cylindrical apparatuses requiring a working pressure in the vapor space of about 5 lb/in. gauge or more, as well as equipment designed under vacuum, and small horizontal storage vessels supported off the ground, usually are fabricated with formed heads.

The formed flanged head (Figure 7.1(A)) is the most economical to fabricate and is produced by forming a flange with a radius on the flat plate. The radius of a flanged head decreases the abruptness in change of shape at the junction of the flat head and cylindrical shell, resulting in a

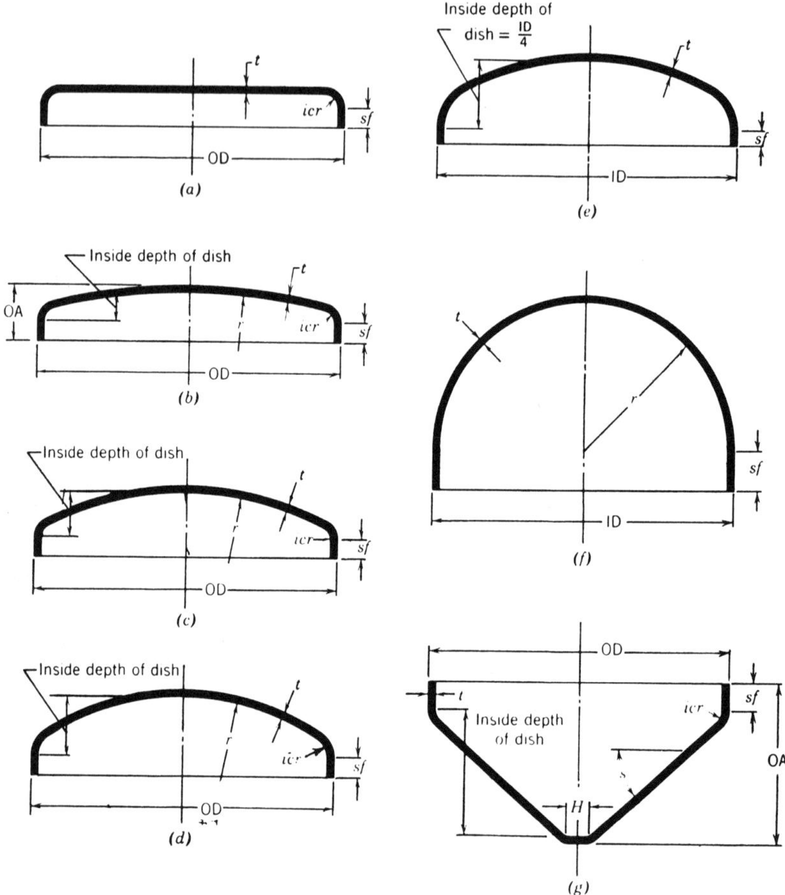

Figure 7.1. Various types of common formed heads [1].

reduction in local stresses. This type head is used widely in closing the ends of horizontal cylindrical storage tanks operating at atmospheric pressure. In addition, flanged heads may be used for the bottoms of vertical cylindrical apparatuses up to 20 ft in diameter resting on concrete slabs.

The pressure rating of a flanged head (type "b" and "c") can be increased if the heads are formed from a flat plate into dished-shape configurations in two radii (the "crown" radius and inside-corner radius). If the radius of a dish is equal to, or less than, the outside diameter, the head is referred to as a "flanged and standard dished head" (type "b"–flanged and shallow head, and type "c"–flanged and standard dished head). Neither type of head is permitted in vessels that must meet pressure-code requirements because of the high localized stresses that result from the small inside-corner radius. They are applied in construction of vertical process vessels for low pressures and for horizontal cylindrical storage tanks.

By forming the heads so that the inside-corner radius is made equal to at least three times the metal sheet thickness (for code requirements, the radius should never be less than 6% of the inside diameter and the radius of the dish may be made equal to, or less than, the diameter of the head), the pressure rating of flanged and dished heads can be increased. With this arrangement, the local stresses occurring in the inside corner of the head will be reduced. Figure 7.1(D) shows a cross section of a flanged and dished head meeting the ASME Code [2] for unfired pressure vessels in the general range of 15 to about 200 lb/in.^{2} gauge. Figure 7.1(E) shows a cross section of an elliptical dished head, which is preferred over code flanged and dished heads formed with two radii for many pressurized vessels in the range of 100 psi and for most vessels for pressures over 200 psi.

Figure 7.1(F) shows a cross section of a hemispherical head, which is the strongest of the formed heads for a given thickness. These heads can be used to resist approximately twice the pressure rating of an elliptical dished head or cylindrical shell of the same thickness and diameter.

Figure 7.1(G) shows a cross section of conical and toriconical head, which is used widely as bottom heads for a variety a process equipment such as evaporators, spray dryers, crystallizers, settling tanks, etc. The conical bottoms are very convenient for accumulation and removal of solids in equipment. Another common application of conical sections is for changing the diameter of cylindrical shells, which is often necessary in some distillation column designs.

The toriconical heads differing from simple conical heads are better suited for pressure-vessel applications because the localized stresses near the junction of the cone and shell are distributed more uniformly and are greatly reduced.

Figure 7.2(A) shows a cross section of a flanged and reversed dished head. This design often is used on the ends of vessels that face against a plane surface (e.g., the bottom of steel vessels resting on concrete slabs). Such a head configuration has the advantage of a bearing surface in one plane and the additional strength resulting from dishing.

Figure 7.2(B) shows a dished-only head, which can be used as a center piece for a large size built-up hemispherical head and also may be used in specially designed equipment where flanges are undesirable. Figure 7.2(C) shows a cross section of a flared and dished head, which may be used for cover plates of kettles, hoppers and specially designed equipment.

When internal pressure is applied to elliptical dished heads, deformation often occurs inward, as illustrated in Figure 7.3 [3], while the cylindrical

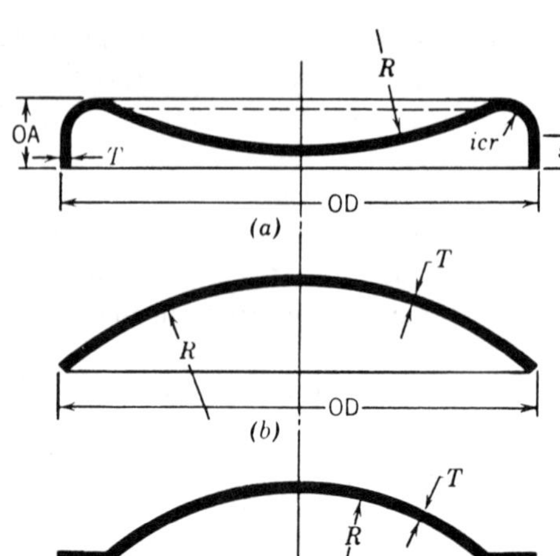

Figure 7.2. Other types of formed heads: (a) flanged and reverse dished; (b) dished only; (c) flared and dished [1].

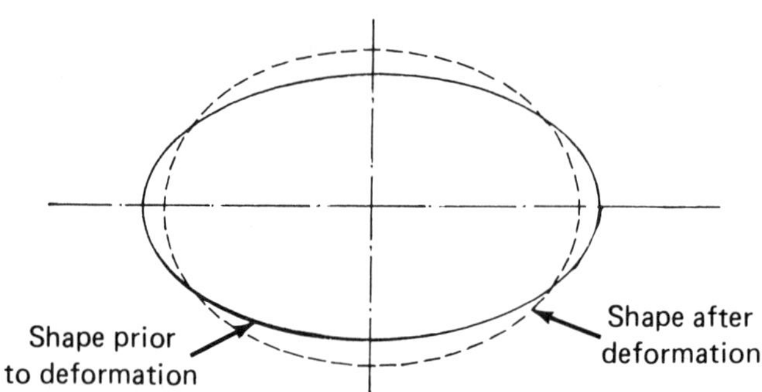

Figure 7.3. Deformation of an ellipsoid of revolution under internal pressure.

shell will tend to deform radially outward. However, the joining of the head to the shell restrains both deformations, which gives rise to additional bending moments and considerable bending stresses in the head and in the shell. There is generally a transition zone in which the stresses are greatest and continue to increase as the ratio of the inside-corner radius (icr) to radius (r) decreases (Figure 7.1).

Following the transition zone, a straight flange (sf) no less than 50 mm long is provided to carry out the welded joint beyond this zone and to prevent additional bending moments. Formed heads have been analyzed theoretically and experimentally [4]. Figures 7.4, 7.5 and 7.6 give results of some investigations of widely used shapes of flanged and elliptical heads.

The external diameter of all these heads is 1300 mm; wall thickness t = 20 mm; internal pressure p = 10 kg/cm^2. The radius of flanged and dished head is R = 910 mm, the inside corner radius is r = 1300 mm (Figure 7.4), height of the head is H = 335 mm, and the ratio $d_e/2H \approx 2$. The elliptical head (Figure 7.5) had the same ratio, $d_e/2H = 2$.

The third head is a flanged dished head of an older design formed with the radius R = 1700 mm and with very little inside-corner radius r = 40 mm; the head height was 175 mm and, consequently, the ratio $d_e/2H = 3.7$.

Analysis of experimental data shows that the stress distribution in formed heads is very complicated, and peak stresses existing in transition zones are generated as a result of considerable bending and tensile stresses. The greatest peaks of longitudinal and circumferential stresses arise in flanged heads at the transition points from the torus to the flanged

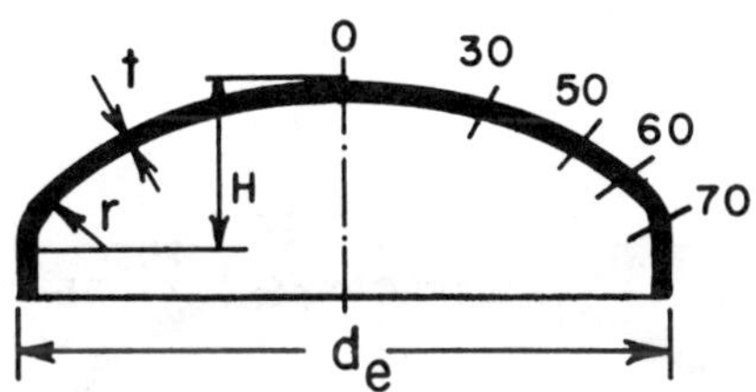

Measure Point	Longitudinal Stress (kg/cm^2) On the Surface External	Longitudinal Stress (kg/cm^2) On the Surface Internal	Circumferential Stress (kg/cm^2) On the Surface External	Circumferential Stress (kg/cm^2) On the Surface Internal
0	+230	+230	+230	+230
30	+225	+225	+270	+210
50	+315	+115	+210	+130
60	+195	+170	+40	−35
70	−125	+445	−210	−40

Figure 7.4. Distribution of stresses in flanged and dished heads: d_e = 1300 mm, t = 20 mm, $H/d_e = 0.25$, p = 10 kg/cm^2 [4].

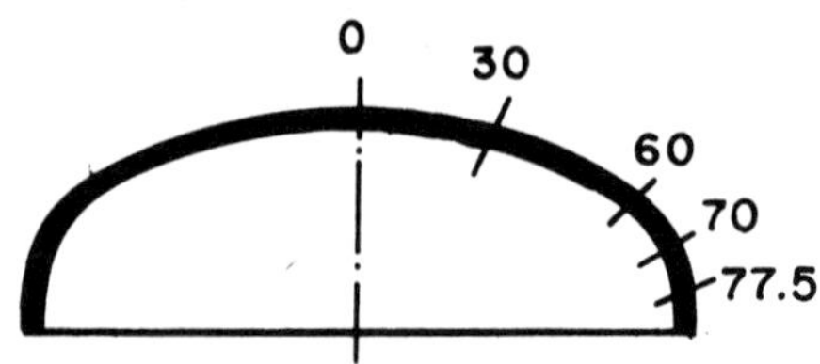

Measure Point	Longitudinal Stress (kg/cm²)		Circumferential Stress (kg/cm²)	
	On the Surface			
	External	Internal	External	Internal
0	+400	+245	+400	+245
30	+300	+230	+180	+155
60	+95	+300	−80	−40
70	−40	+340	−140	−30
77.5	+200	+110	+60	+30

Figure 7.5. Distribution of stresses in an elliptical head: d_e = 1300 mm, t = 20 mm, H/d_e = 0.25, p = 10 kg/cm [4].

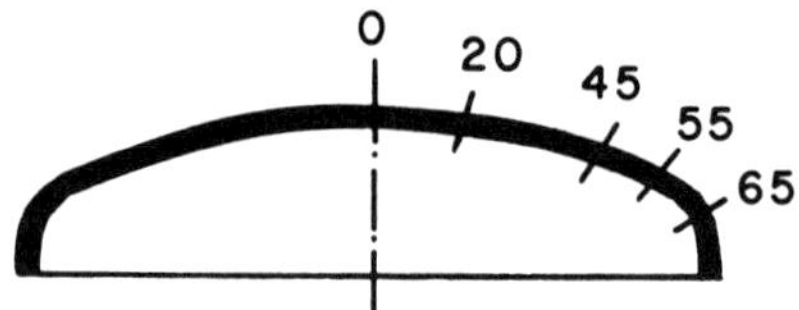

Measure Point	Longitudinal Stress (kg/cm²)		Circumferential Stress (kg/cm²)	
	On the Surface			
	External	Internal	External	Internal
0	+470	+350	+465	−340
20	+525	+295	+525	+324
45	+825	−5	+465	0
55	+285	+535	−110	−290
65	−950	+1400	−70	−140

Figure 7.6. Distribution of stresses in a flanged head: d_e = 1300 mm, t = 20 mm, H/d_e = 0.135, p = 10 kg/cm^2 [4].

part. In elliptical heads, the values of peak stresses are much less than in flanged heads and arise only at joining points of the ellipse with a cylinder.

The behavior of stresses and their values depend on the shape of the head and its proportioning. Elliptical heads work better than flanged and dished ones because their continuous changing of curvature causes bending stresses to become smaller. The value of the shape factor $d_e/2H$ greatly influences the degree of work of a head. The smaller $d_e/2H$, the more uniform the distribution of stresses. The spherical head providing optimal work of material has the shape factor $d_e/2H = 1$.

The openings in heads (without reinforcement) considerably complicate the stress distribution in the head material. Figures 7.7, 7.8 and 7.9 give results of experiments of three heads with openings of 320 X 450 mm. The first two heads were of the same size as the "blind" head in Figure 7.4, that is, d_e = 1300 mm, R = 910 mm, r = 130 mm, H = 335 mm. The third head was similar to the flanged head in Figure 7.6, with the following parameters: d_e = 1300 mm, R = 1700 mm, r = 40 mm, H = 175 mm.

All three heads with openings were fabricated of 20-mm-thick metal and were tested under the same pressure, p = 10 kg/cm^2, including blind (without holes) heads (see Figures 7.4, 7.5 and 7.6).

If we compare the test results of the dished and flanged head with the central and off-center openings, it is shown that considerable circumferential

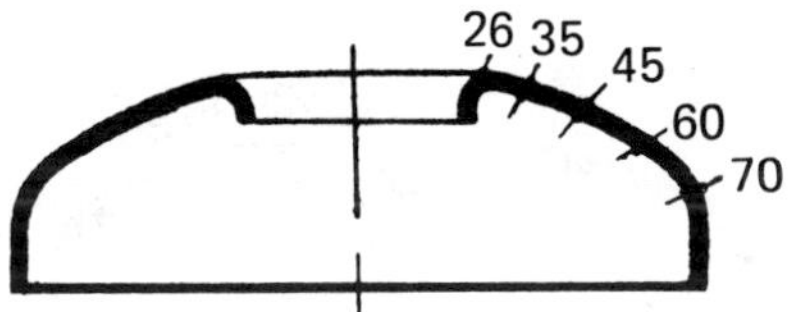

Measure Point	Longitudinal Stress (kg/cm^2)		Circumferential Stress (kg/cm^2)	
	On the Surface			
	External	Internal	External	Internal
26	+240	+100	+1410	+800
35	−135	+725	+320	+465
45	+130	+365	+85	+130
60	+130	−280	−70	−70
70	−70	+450	−240	−80

Figure 7.7. Stress distribution in a flanged and dished head with a central opening: d_e = 1300 mm, t = 20 mm, H/d_e = 0.25, p = 10 kg/cm^2.

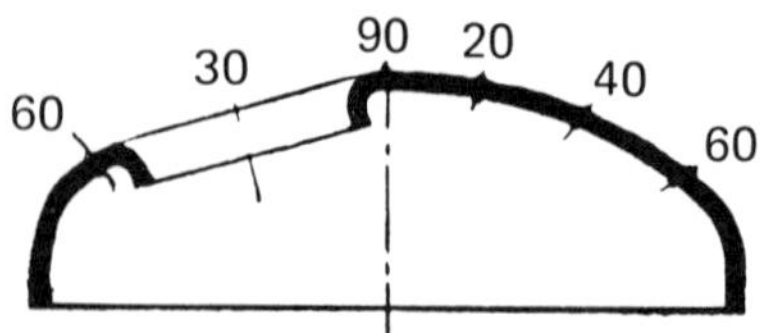

Measure Point	Longitudinal Stress (kg/cm^2)		Circumferential Stress (kg/cm^2)	
	On the Surface			
	External	Internal	External	Internal
0	−230	+585	+460	+510
9	+270	+290	+1270	+820
20	+160	+330	+180	+230
40	+385	+155	+310	+200
60 right side	+155	+430	+15	+50
60 left side	−210	+710	+185	+245
30	+270	+370	+1600	+1080

Figure 7.8. Stress distribution in a flanged and dished head with an eccentrically located opening: d_e = 1300 mm, t = 20 mm, H/d_e = 0.25, p = 10 kg/cm^2.

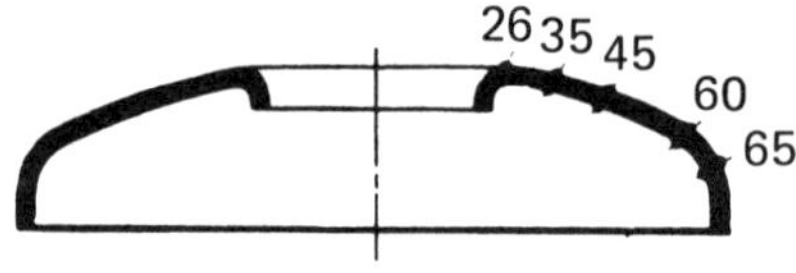

Measure Point	Longitudinal Stress (kg/cm^2)		Circumferential Stress (kg/cm^2)	
	On the Surface			
	External	Internal	External	Internal
26	+785	−400	+2480	+830
35	+305	+790	+1080	+540
45	+365	+580	+280	0
60	−310	+930	−525	−280
65	−685	+1210	−765	−200

Figure 7.9. Stress distribution in a flanged head with a central opening (old design): d_e = 1300 mm, t = 20 mm, H/d_e = 0.135, p = 10 kg/cm^2.

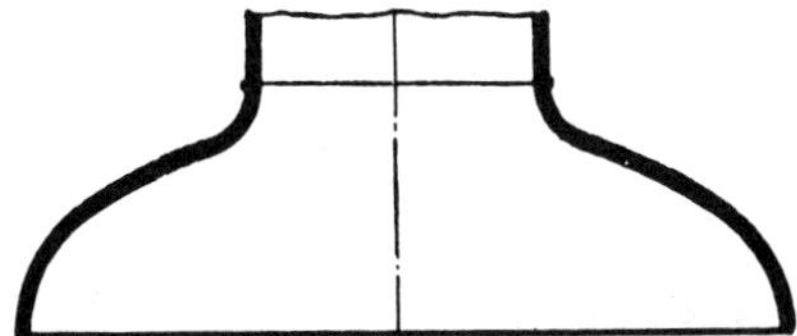

Figure 7.10. Head with an opening flanged outside.

stresses develop at the edge of the opening. These stresses extend to the edge of the eccentrically located opening at about 1600 kg/cm^2. The blind head having the same parameters developed maximum longitudinal stress in the transition zone at about +445 kg/cm^2.

The stress at the edge of the head opening of the older design (see Figure 7.9) was even higher than in the head reaching 2480 kg/cm^2 and greatly exceeded the maximum allowable stress of 1400 kg/cm^2 in the transition zone of the blind head. Stresses in the head also are influenced by the ratio of opening diameter to head diameter. The greater the diameter of the opening in the same head, the greater is the stress peak. Flanging the head outside (Figure 7.10) produces a more favorable stress distribution.

7.2 ELLIPTICAL HEADS

An elliptical-shaped head is advantageous in services in which pressure absorbing is important. Elliptical heads are recommended for use in the fabrication of cylindrical welded, soldered and cast horizontal and vertical apparatuses working under any pressure. Recommended sizes for elliptical dishes are given in Figure 7.11.

The heads of sizes larger than that of standard sheets are fabricated from two or three sheets welded before stamping or from stamped sections with subsequent welding (Figure 7.12).

7.2.1 Design of Heads Operating Under Internal Pressure

Head thickness specification is based on ultimate loads. The principal design formula was derived from analysis of the stressed membrane (momentless) state of an elliptical head fabricated from plastics [5-8].

$$t = \frac{dp}{3.8K\sigma_{allow}\,\psi - p} \cdot \frac{d}{2h_b} + c, \text{ m} \tag{7.1}$$

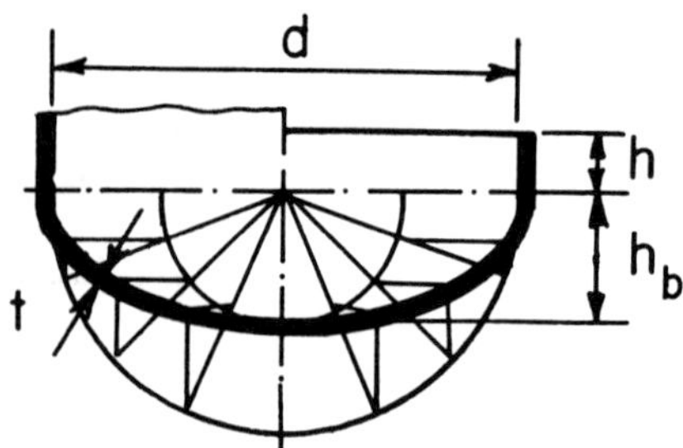

Figure 7.11. Sizes of elliptical heads, h = 0.25 d, but no less than 0.2 d; for heads of steel sheets, h ⩾ 2t, but no less than 25 mm.

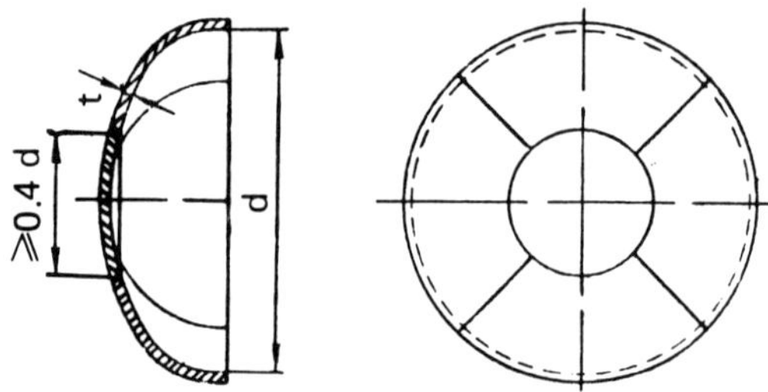

Figure 7.12. Elliptical head fabricated of stamped parts.

where h_b = the height of the domed part of the head (see Figure 7.11) (m)
K = dimensionless coefficient

Values of design pressure, p, and allowable stress, σ_{allow}, are taken from the design basis for cylindrical shells [2]. The maximum allowable stress values in tension for carbon and low-alloy steels, as assigned by ASME, are given in Table 7.1. The coefficient K in Equation 7.1 evaluates a weakening introduced by an opening without regard for its location on the head and is determined from the following [6]:

$$K = 1 - \frac{d_o}{d} \tag{7.2}$$

where d_o = the largest diameter (or size for noncircular openings) of a nonreinforced opening in the head (m).

For "blind" heads or completely reinforced openings, K = 1. For partial reinforcement of an opening, the value of K is assumed to be the same as a nonreinforced opening. Generally, it is better to locate the opening in the center of the head. If several openings in the head are required, their location should conform to Figure 7.13, where the larger opening is located closer to the center of the head. Locating openings in the transition zone should be avoided so as not to weaken it (although it is possible to make small holes on flanged parts of the head, such as for manometers, vents, etc.). The flanged part of the head should not be loaded with additional forces, such as those produced by supports, etc.

If one must provide an opening that has a diameter larger than half that of the head, then an inverse curved reducer without reinforcing is recommended. Final selection of head thickness, t, should be based on the directions given for selecting corresponding cylindrical shells operating under internal pressure with regard to the thickness of stamped heads. The thickness of a head for welded apparatuses should not be less than that of the jointed shell.

In cast apparatuses, a head thickness of 2–5 mm more than the thickness of the jointed shell usually is taken, and its connection should be provided with a smooth reducer.

The assumed head thickness, t, should be checked for stress with a hydrostatic test through the following formula:

$$\sigma = \frac{[d^2 + 2h_b(t - c)]p'}{7.6K\psi h_b(t - c)} \leqslant \frac{\sigma_y}{1.2}, \text{ N/m}^2 \tag{7.3}$$

The following example illustrates these principles.

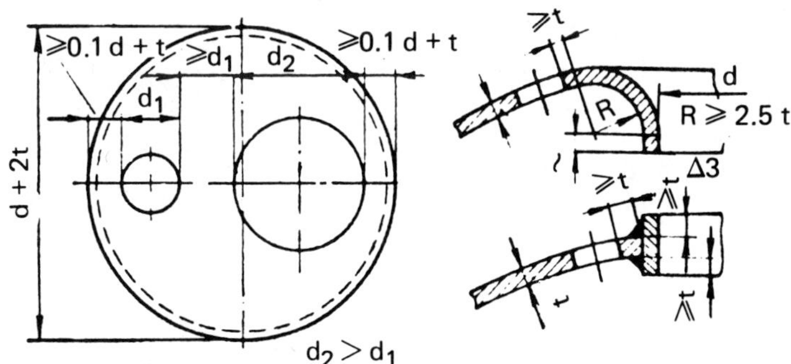

Figure 7.13. Location and design of openings in elliptical heads.

Table 7.1. Maximum Allowable Stress Values
(from the 1956 ASME Unfired-Pressure-Vessel Code, with

Material and Specification Number	Grade	Nominal Composition	P-Number	Spec Min Tensile	Notes	-20 to 650	700
Plate Steels							
Carbon steels							
SA-7	–	–	1	60,000	a,b	12,650	–
SA-30	Flange	–	1	55,000	–	13,750	13,250
SA-30	Firebox A	–	1	55,000	c	13,750	13,250
SA-30	Firebox B	–	1	48,000	c	12,000	11,650
SA-113	C	–	1	48,000	a,b	11,050	–
SA-129	A	–	1	40,000	–	10,000	–
SA-129	B	–	1	44,000	–	11,000	–
SA-129	C	–	1	42,000	–	10,500	–
SA-201	A	C-Si	1	55,000	–	13,750	13,250
SA-201	B	C-Si	1	60,000	–	15,000	14,350
SA-212	A	C-Si	1	65,000	–	16,250	15,500
SA-212	B	C-Si	1	70,000	–	17,500	16,600
SA-283	A	–	1	45,000	a,b	10,350	–
SA-283	B	–	1	50,000	a,b	11,500	–
SA-283	C	–	1	55,000	a,b	12,650	–
SA-283	D	–	1	60,000	a,b	12,650	–
SA-285	A	–	1	45,000	c,d	11,250	11,000
SA-285	B	–	1	50,000	c,d	12,500	12,100
SA-285	C	–	1	55,000	c,d	13,750	13,250
SA-299	–	C-Mn-Si	3	75,000	–	18,750	17,700
SA-300	–	–	–	–	e	–	–
Low-alloy steels							
SA-202	A	Cr-Mn-Si	4	75,000	–	18.750	17,700
SA-202	B	Cr-Mn-Si	4	85,000	–	21,250	19,800
SA-203	A,D	2½ and 3½ Ni	4	65,000	–	16,250	15,500
SA-203	B,E	2½ and 3½ Ni	4	70,000	–	17,500	16,600
SA-203	C	2½ Ni	4	75,000	–	18,750	17,700
SA-204	A	C–½ Mo	3	65,000	–	16,250	16,250
SA-204	B	C–½ Mo	3	70,000	–	17,500	17,500
SA-204	C	C–½ Mo	3	75,000	–	18,750	18,750
SA-225	A	Mn-V	3	70,000	f	17,500	17,500
SA-225	B	Mn-V	3	75,000	f	18,750	18,750
SA-301	A	½ Cr–½ Mo	3	65,000	–	16,250	16,250
SA-301	B	1 Cr–½ Mo	4	60,000	–	15,000	15,000
SA-302	A	Mn–½ Mo	3	75,000	–	18,750	18,750
SA-302	B	Mn–½ Mo	3	80,000	–	20,000	20,000
SA-353	–	9 Ni	10	90,000	g	22,500	–
SA-357	–	5 Cr–½ Mo	5	60,000	h	–	13,400

in Tension for Carbon and Low-Alloy Steels (lb/in.^{2}) [2]

permission of the American Society of Mechanical Engineers)

For metal temperatures not exceeding, °F									
750	800	850	900	950	1,000	1,050	1,100	1,150	1,200
–	–	–	–	–	–	–	–	–	–
12,050	10,200	8,350	–	–	–	–	–	–	–
12,050	10,200	8,350	6,500	–	–	–	–	–	–
10,700	9,300	7,900	6,500	–	–	–	–	–	–
–	–	–	–	–	–	–	–	–	–
–	–	–	–	–	–	–	–	–	–
–	–	–	–	–	–	–	–	–	–
–	–	–	–	–	–	–	–	–	–
12,050	10,200	8,350	6,500	4,500	2,500	–	–	–	–
12,950	10,800	8,650	6,500	4,500	2,500	–	–	–	–
13,850	11,400	8,950	6,500	4,500	2,500	–	–	–	–
14,750	12,000	9,250	6,500	4,500	2,500	–	–	–	–
–	–	–	–	–	–	–	–	–	–
–	–	–	–	–	–	–	–	–	–
–	–	–	–	–	–	–	–	–	–
–	–	–	–	–	–	–	–	–	–
10,250	9,000	7,750	6,500	–	–	–	–	–	–
11,150	9,600	8,050	6,500	–	–	–	–	–	–
12,050	10,200	8,350	6,500	–	–	–	–	–	–
15,650	12,600	9,550	6,500	4,500	2,500	–	–	–	–
–	–	–	–	–	–	–	–	–	–
15,650	12,600	9,550	6,500	4,500	2,500	–	–	–	–
17,700	12,800	9,900	6,500	4,500	2,500	–	–	–	–
13,850	11,400	8,950	6,500	4,500	2,500	–	–	–	–
14,750	12,000	9,250	6,500	4,500	2,500	–	–	–	–
15,650	12,600	9,550	6,500	4,500	2,500	–	–	–	–
16,250	15,650	14,400	12,500	10,000	6,250	–	–	–	–
17,500	16,900	15,000	12,750	10,000	6,250	–	–	–	–
18,750	18,000	15,900	13,000	10,000	6,250	–	–	–	–
17,500	–	–	–	–	–	–	–	–	–
18,750	–	–	–	–	–	–	–	–	–
16,250	15,650	14,400	12,500	10,000	6,250	–	–	–	–
15,000	14,750	14,200	13,100	11,000	7,500	5,000	2,800	1,550	1,000
18,750	18,000	15,900	13,000	10,000	6,250	–	–	–	–
20,000	19,100	16,800	13,250	10,000	6,250	–	–	–	–
–	–	–	–	–	–	–	–	–	–
13,100	12,800	12,400	11,500	10,000	7,300	5,200	3,300	2,200	1,500

Table 7.1,

Material and Specification Number	Grade	Nominal Composition	P-Number	Spec Min Tensile	Notes	-20 to 650	700
Pipes and Tubes							
Seamless carbon steels							
SA-53	A	–	1	18,000	c,i	12,000	11,650
SA-53	B	–	1	60,000	c,i	15,000	14,350
SA-83	A	–	1	–	c,i	11,750	11,450
SA-106	A	–	1	48,000	–	12,000	11,650
SA-106	B	–	1	60,000	–	15,000	14,350
SA-192	–	–	1	–	–	11,750	11,450
SA-210	–	–	1	60,000	–	15,000	14,350
SA-333	C	–	1	55,000	c,i	13,750	–
SA-334	C	–	1	55,000	c,i	13,750	–
Seamless low-alloy steels							
SA-209	T1	C–1/2 Mo	3	55,000	–	13,750	13,750
SA-209	T1a	C–1/2 Mo	3	60,000	–	15,000	15,000
SA-209	T1b	C–1/2 Mo	3	53,000	–	13,250	13,250
SA-213	T3	1 3/4 Cr–0.7 Mo	4	60,000	–	15,000	15,000
SA-213	T5	5 Cr–1/2 Mo	5	60,000	h	–	13,400
SA-213	T7	7 Cr–1/2 Mo	5	60,000	h	–	13,400
SA-213	T9	9 Cr–1 Mo	5	60,000	h	–	13,400
SA-213	T11	1 1/4 Cr–1/2 Mo-Si	4	60,000	–	15,000	15,000
SA-213	T12	1 Cr–1/2 Mo	4	60,000	–	15,000	15,000
SA-213	T5b	5 Cr–1/2 Mo-Si	5	60,000	h	–	13,400
SA-213	T3b	2 Cr–1/2 Mo	4	60,000	–	15,000	15,000
SA-213	T5c	5 Cr–1/2 Mo-Ti	5	60,000	h	–	13,400
SA-213	T17	1 Cr–V	4	60,000	j	15,000	–
SA-213	T21	3 Cr–0.9 Mo	5	60,000	–	15,000	14,800
SA-213	T22	2 1/4 Cr–1 Mo	5	60,000	–	15,000	15,000
SA-333	3	3 1/2 Ni	9	65,000	–	16,250	–
SA-333	5	5 Ni	9	65,000	–	16,250	–
SA-334	3	3 1/2 Ni	9	65,000	–	16,250	–
SA-334	5	5 Ni	9	65,000	–	16,250	–
SA-335	P1	C–1/2 Mo	3	55,000	–	13,750	13,750
SA-335	P2	1/2 Cr–1/2 Mo	3	55,000	–	13,750	13,750
SA-335	P3	1 3/4 Cr–0.7 Mo	4	60,000	–	15,000	15,000
SA-335	P3b	2 Cr–1/2 Mo	4	60,000	–	15,000	15,000
SA-335	P5	5 Cr–1/2 Mo	5	60,000	h	–	13,400
SA-335	P5b	5 Cr–1/2 Mo-Si	5	60,000	h	–	13,400
SA-335	P5c	5 Cr–1/2 Mo-Ti or Cb	5	60,000	h	–	13,400
SA-335	P11	1 1/4 Cr–1/2 Mo-Si	4	60,000	–	15,000	15,000
SA-335	P12	1 Cr–1/2 Mo	4	60,000	–	15,000	15,000
SA-335	P15	1 1/2 Si–1/2 Mo	3	60,000	–	15,000	15,000
SA-335	P7	7 Cr–1/2 Mo	5	60,000	h	–	13,400

continued

For metal temperatures not exceeding, °F									
750	800	850	900	950	1,000	1,050	1,100	1,150	1,200
10,700	9,300	7,900	6,500	–	–	–	–	–	–
12,950	10,800	8,650	6,500	–	–	–	–	–	–
10,550	9,200	7,850	6,500	–	–	–	–	–	–
10,700	9,300	7,900	6,500	4,500	2,500	–	–	–	–
12,950	10,800	8,650	6,500	4,500	2,500	–	–	–	–
10,550	9,200	7,850	6,500	4,500	2,500	–	–	–	–
12,950	10,800	8,650	6,500	4,500	2,500	–	–	–	–
–	–	–	–	–	–	–	–	–	–
–	–	–	–	–	–	–	–	–	–
13,750	13,450	13,150	12,500	10,000	6,250	–	–	–	–
15,000	14,400	13,750	12,500	10,000	6,250	–	–	–	–
13,250	13,000	12,750	12,500	10,000	6,250	–	–	–	–
15,000	15,000	14,400	13,100	11,000	7,800	5,500	4,000	2,500	1,200
13,100	12,800	12,400	11,500	10,000	7,300	5,200	3,300	2,200	1,500
13,100	12,500	11,500	9,500	7,000	5,000	3,500	2,500	1,800	1,200
13,100	12,800	12,500	12,000	10,800	8,500	5,500	3,300	2,200	1,500
15,000	15,000	14,400	13,100	11,000	7,800	5,500	4,000	2,500	1,200
15,000	14,750	14,200	13,100	11,000	7,500	5,000	2,800	1,550	1,000
13,100	12,800	12,400	10,900	9,000	5,500	3,500	2,500	1,800	1,200
15,000	14,700	14,000	12,500	10,000	6,200	4,200	2,750	1,750	1,200
13,100	12,800	12,400	11,500	10,000	7,300	4,800	2,800	1,800	1,200
–	–	–	–	–	–	–	–	–	–
14,500	13,900	13,200	12,000	9,000	7,000	5,500	4,000	2,700	1,500
15,000	15,000	14,400	13,100	11,000	7,800	5,800	4,200	3,000	2,000
–	–	–	–	–	–	–	–	–	–
–	–	–	–	–	–	–	–	–	–
–	–	–	–	–	–	–	–	–	–
–	–	–	–	–	–	–	–	–	–
13,750	13,450	13,150	12,500	10,000	6,250	–	–	–	–
13,750	13,450	13,150	12,500	10,000	6,250	–	–	–	–
15,000	15,000	14,400	13,100	11,000	7,800	5,500	4,000	2,500	1,200
15,000	14,700	14,000	12,500	10,000	6,200	4,200	2,750	1,750	1,200
13,100	12,800	12,400	11,500	10,000	7,300	5,200	3,300	2,200	1,500
13,100	12,800	12,400	10,900	9,000	5,500	3,500	2,500	1,800	1,200
13,100	12,800	12,400	11,500	10,000	7,300	4,800	2,800	1,800	1,200
15,000	15,000	14,400	13,100	11,000	7,800	5,500	4,000	2,500	1,200
15,000	14,750	14,200	13,100	11,000	7,500	5,000	2,800	1,550	1,000
15,000	14,400	13,750	12,500	10,000	6,250	–	–	–	–
13,100	12,500	11,500	9,500	7,000	5,000	3,500	2,500	1,800	1,200

Table 7.1,

Material and Specification Number	Grade	Nominal Composition	P-Number	Spec Min Tensile	Notes	−20 to 650	700
SA-335	P9	9 Cr–1 Mo	5	60,000	h	–	13,400
SA-335	P21	3 Cr–0.9 Mo	5	60,000	–	15,000	14,800
SA-335	P22	$2^1/_4$ Cr–1 Mo	5	60,000	–	15,000	15,000
SA-369	FP1	C–$^1/_2$ Mo	3	55,000	–	13,750	13,750
SA-369	FP2	$^1/_2$ Cr–$^1/_2$ Mo	3	55,000	–	13,750	13,750
SA-369	FP3b	2 Cr–$^1/_2$ Mo	4	60,000	–	15,000	15,000
SA-369	FP11	$1^1/_4$ Cr–$^1/_2$ Mo-Si	4	60,000	–	15,000	15,000
SA-369	FP12	1 Cr–$^1/_2$ Mo	4	60,000	–	15,000	15,000
SA-369	FP21	3 Cr–1 Mo	5	60,000	–	15,000	14,800
SA-369	FP22	$2^1/_4$ Cr–1 Mo	5	60,000	–	15,000	15,000
SA-369	FP5	5 Cr–$^1/_2$ Mo	5	60,000	h	–	13,400
SA-369	FP7	7 Cr–$^1/_2$ Mo	5	60,000	h	–	13,400
SA-369	FP9	9 Cr–1 Mo	5	60,000	h	–	13,400
Forgings							
Carbon steels							
SA-105	I	–	1	60,000	c	15,000	14,350
SA-105	II	–	1	70,000	c	17,500	16,600
SA-181	I	–	1	60,000	c	15,000	14,350
SA-181	II	–	1	70,000	c	17,500	16,600
SA-266	I	–	1	60,000	–	15,000	14,350
SA-266	II	–	1	70,000	–	17,500	16,600
SA-266	III	–	–	75,000	–	18,750	17,700
SA-350	LF1	–	1	60,000	–	15,000	–
Low-alloy steels							
SA-182	F1	C–$^1/_2$ Mo	3	70,000	–	17,500	17,500
SA-182	F12	1 Cr–$^1/_2$ Mo	4	70,000	h	–	16,150
SA-182	F5	5 Cr–$^1/_2$ Mo	5	90,000	h	–	17,500
SA-182	F11	$1^1/_4$ Cr–$^1/_2$ Mo	4	70,000	h	–	16,150
SA-182	F9	9 Cr–1 Mo	5	100,000	h	–	21,200
SA-182	F22	$2^1/_4$ Cr–1 Mo	5	70,000	–	17,500	17,500
SA-336	F1	C–$^1/_2$ Mo	3	70,000	–	17.500	17,500
SA-336	F2	1 Cr–$^1/_2$ Mo	4	70,000	h	–	16,150
SA-336	F5a	5 Cr–$^1/_2$ Mo	5	80,000	h	–	16,500
SA-336	F22	$2^1/_4$ Cr–1 Mo	5	80,000	–	20,000	20,000
SA-350	LF3	$3^1/_2$ Ni	9	70,000	–	17,500	–
Castings							
Carbon steels							
SA-95	–	–	–	70,000	k,l	17,500	16,600
SA-216	WCA	–	1	60,000	k	15,000	14,350
SA-216	WCB	–	1	70,000	k	17,500	16,600
SA-352	LCB	–	–	65,000	k	16,250	–

continued

For metal temperatures not exceeding, °F									
750	800	850	900	950	1,000	1,050	1,100	1,150	1,200
13,100	12,800	12,500	12,000	10,800	8,500	5,500	3,300	2,200	1,500
14,500	13,900	13,200	12,000	9,000	7,000	5,500	4,000	2,700	1,500
15,000	15,000	14,400	13,100	11,000	7,800	5,800	1,200	3,000	2,000
13,750	13,450	13,150	12,500	10,000	6,250	–	–	–	–
13,750	13,450	13,150	12,500	10,000	6,250	–	–	–	–
15,000	14,700	14,000	12,500	10,000	6,200	1,200	2,750	1,750	1,200
15,000	15,000	14,400	13,100	11,000	7,800	5,500	4,000	2,500	1,200
15,000	14,750	14,200	13,100	11,000	7,500	5,000	2,800	1,550	1,000
14,500	13,900	13,200	12,000	9,000	7,000	5,500	4 000	2,700	1,500
15,000	15,000	14,400	13,100	11,000	7,800	5,800	4,200	3,000	2,000
13,100	12,800	12,400	11,500	10,000	7,300	5,200	3,300	2,200	1,500
13,100	12,500	11,500	9,500	7,000	5,000	3,500	2,500	1,800	1,200
13,100	12,800	12,500	12,000	10,800	8,500	5,500	3,300	2,200	1,500
12,950	10,800	8,650	6,500	4,500	2,500	–	–	–	–
14,750	12,000	9,250	6,500	4,500	2,500	–	–	–	–
12,950	10,800	8,650	6,500	4,500	2,500	–	–	–	–
14,750	12,000	9,250	6,500	4,500	2,500	–	–	–	–
12,950	10,800	8,650	6,500	4,500	2,500	–	–	–	–
14,750	12,000	9,250	6,500	4,500	2,500	–	–	–	–
15,650	12,600	9,550	6,500	4,500	2,500	–	–	–	–
–	–	–	–	–	–	–	–	–	–
17,500	16,900	15,000	12,750	10,000	6,250	–	–	–	–
15,500	14,850	14,200	13,100	11,000	7,500	5,000	2,800	1,550	1,000
16,000	14,500	13,000	11,500	10,000	7,300	5,200	3,300	2,200	1,500
15,500	15,000	14,400	13,100	11,000	7,800	5,500	4,000	2,500	1,200
20,000	17,700	15,400	13,100	10,800	8,500	5,500	3,300	2,200	1,500
17,500	17,500	16,000	14,000	11,000	7,800	5,800	4,200	3,000	2,000
17,500	16,900	15,000	12,750	10,000	6,250	–	–	–	–
15,500	14,850	14,200	13,100	11,000	7,500	5,000	2,800	1,550	1,000
15,500	14,500	13,000	11,500	10,000	7,300	5,200	3,300	2,200	1,500
20,000	18,000	16,000	14,000	11,000	7,800	5,800	4,200	3,000	2,000
–	–	–	–	–	–	–	–	–	–
14,750	12,000	9,250	6,500	4,500	2,500	–	–	–	–
12,950	10,800	8,650	6,500	4,500	2,500	–	–	–	–
14,750	12,000	9,250	6,500	4,500	2,500	–	–	–	–
–	–	–	–	–	–	–	–	–	–

Table 7.1,

Material and Specification Number	Grade	Nominal Composition	P-Number	Spec Min Tensile	Notes	−20 to 650	700	
Low-alloy steels								
SA-217	WC1	C–1/2 Mo	3	65,000	k,m	16,250	16,250	
SA-217	WC4	Ni-Cr–1/2 Mo	4	70,000	k,m	17,500	17,500	
SA-217	WC5	Ni-Cr–1 Mo	4	70,000	k,m	17,500	17,500	
SA-217	WC6	1 1/4 Cr–1/2 Mo	4	70,000	k,m	17,500	17,500	
SA-217	WC9	2 1/4 Cr–1 Mo	5	70,000	k,m	17,500	17,500	
SA-217	C5	5 Cr–1/2 Mo	5	90,000	k,m	22,500	21,600	
SA-217	C12	9 Cr–1 Mo	5	90,000	k,m	22,500	22,000	
SA-352	LC1	–	3	65,000	k	16,250	–	
SA-352	LC2	–	4	65,000	k	16,250	–	
SA-352	LC3	–	9	65,000	k	16,250	–	
Bolting								
Carbon steels								
SA-261	BO	–	–	100,000	n,o	16,250	14,950	
SA-307	B	–	–	55,000	p	–	–	
SA-325	–	–	–	–	n,o	18,750	17,200	
Low-alloy steels								
SA-193	B5	5 Cr–1/2 Mo	–	–	n,o	20,000	20,000	
SA-193	B7	1 Cr–0.2 Mo	–	–	n,o	20,000	20,000	
SA-193	B7a	1 Cr–0.6 Mo	–	–	n,o	20,000	20,000	
SA-193	B14	1 Cr–0.3 Mo-V	–	–	n,o	20,000	20,000	
SA-193	B16	1 Cr–1/2 Mo-V	–	–	n,o	20,000	20,000	
SA-320	L7, L9,	–	–	–				
	L10, L43	–	–	–	m,q	–	–	–
SA-354	BB	–	–	–	n,o	18,750	17,200	
SA-354	BC	–	–	–	n,o	20,000	18,400	
SA-354	BD	–	–	–	n,o	20,000	18,400	
Bars								
Carbon steels								
SA-306	50	–	1	50,000	–	12,500	–	
SA-306	55	–	1	55,000	–	13,750	–	
SA-306	60	–	1	60,000	–	15,000	–	

Note: The stress values in this table may be interpolated to determine values for inter-
table. All stress values in bearing are 1.60 times the values in the above table.

[a]See section 13.5, first paragraph of the ASME code.

[b]These stress values are one-fourth the specified minimum tensile strength multiplied

[c]For service temperatures above 850°F, it is recommended that killed steels containing
with large amounts of aluminum and rimmed steels may have creep and stress-rupture
on which the values in the above table are based.

[d]Flange quality in this specification not permitted over 850°F.

[e]The stress values to be used for temperatures below −20°F when steels are made to

[f]Between temperatures of 750 and 1000°F inclusive, the stress values for specifications

continued

For metal temperatures not exceeding, °F									
750	800	850	900	950	1,000	1,050	1,100	1,150	1,200
16,250	15,650	14,400	12,500	10,000	6,250	–	–	–	–
17,500	17,000	15,800	14,000	10,000	6,250	–	–	–	–
17,500	17,000	15,800	14,000	11,000	7,800	5,500	4,000	–	–
17,500	17,000	15,800	14,000	11,000	7,800	5,500	4,000	–	–
17,500	17,000	15,800	14,000	11,000	7,800	5,800	4,200	3,000	2,000
20,400	19,000	17,000	13,600	10,000	7,300	5,200	3,300	2,200	1,500
21,000	19,400	17,300	15,000	11,750	8,500	5,500	3,300	2,200	1,500
–	–	–	–	–	–	–	–	–	–
–	–	–	–	–	–	–	–	–	–
–	–	–	–	–	–	–	–	–	–
12,500	10,000	6,900	–	–	–	–	–	–	–
–	–	–	–	–	–	–	–	–	–
15,650	–	–	–	–	–	–	–	–	–
20,000	20,000	17,250	13,750	10,300	7,300	4,800	2,750	–	–
20,000	20,000	16,250	12,500	8,500	4,500	–	–	–	–
20,000	20,000	17,250	13,750	10,300	7,300	4,800	2,750	–	–
20,000	20,000	18,750	16,650	14,250	11,000	6,250	2,750	–	–
20,000	20,000	18,750	16,650	14,250	11,000	6,250	2,750	–	–
–	–	–	–	–	–	–	–	–	–
15,650	–	–	–	–	–	–	–	–	–
16,750	–	–	–	–	–	–	–	–	–
16,750	–	–	–	–	–	–	–	–	–
–	–	–	–	–	–	–	–	–	–
–	–	–	–	–	–	–	–	–	–
–	–	–	–	–	–	–	–	–	–

mediate temperatures. All stress values in shear are 0.80 times the values in the above

by a quality factor of 0.92, except for SA-283, Grade D, and SA-7.
not less than 0.10% residual silicon be used. Killed steels that have been deoxidized properties in the temperature range above 850°F, which are somewhat less than those

conform with specification SA-300 shall be those given in the column for −20 to 650°F.
SA-212, Grade B may be used until high-temperature test data become available.

Table 7.1,

[g]See special ruling in regard to welding with electrodes having a tensile strength less
[h]Maximum allowable stress values for temperatures below 700°F are given in the follow-

Specification Number	Grade	P-Number	For metal temperatures not exceeding °F: −20 to 400	500	600	650
SA-213	T5	5	15,000	14,500	14,000	13,700
SA-213	T7	5	15,000	14,500	14,000	13,700
SA-213	T9	5	15,000	14,500	14,000	13,700
SA-213	T5b	5	15,000	14,500	14,000	13,700
SA-213	T5c	5	15,000	14,500	14,000	13,700
SA-335	P5	5	15,000	14,500	14,000	13,700
SA-335	P5b	5	15,000	14,500	14,000	13,700
SA-335	P5c	5	15,000	14,500	14,000	13,700
SA-336	F2	4	17,500	17,500	17,500	16,800
SA-336	F5	5	20,000	19,200	18,000	17,300

[i]Only killed steel (silicon) shall be used above 900°F.
[j]Between temperatures of 650 and 1000°F inclusive, the stress values for specification
[k]To these stress values a quality factor shall be applied, as specified in the code.
[l]This material not suitable for welding.
[m]These stress values apply to normalized and drawn material only.
[n]These stress values are established from a consideration of strength only and will be long period of time without retightening is required, lower stress values may be neces- relaxation properties.
[o]Between temperatures of −20 and 400°F, stress values equal to the lower of the follow- yield strength.
[p]Not permitted above 150°F; allowable stress value, 7000 psi.
[q]For temperatures below 400°F, stress values equal to 20% of the specified minimum

Example 7.1

Determine the head thickness of the upper elliptical head for the shell of a vertical apparatus designed in Example 6.1 (see Figure 6.13). The unit has the following parameters: the material of the head is steel; d = 2.0 m; h = 0.5 m; the head has one nonreinforced opening, d_o = 0.15 m; the head is manual welded in two sections; the seam is a double butt-weld ($\psi = 0.95$); $\sigma_{allow} = 146 \times 10^6$ N/m²; $\sigma_y = 240 \times 10^6$ N/m²; p = 1 MN/m²; P′ = 1.3 MN/m².

Solution

The coefficient K from Equation 7.2 is

$$K = 1 - \frac{d_o}{d} = 1 - \frac{0.15}{2.0} = 0.925$$

continued

than the base metal.
ing table:

Specification Number	Grade	P-Number	For metal temperatures not exceeding °F: −20 to 100	500	600	650
SA-357	–	–	15,000	14,500	14,000	13,700
SA-369	FP5	5	15,000	14,500	14,000	13,700
SA-369	FP7	5	15,000	14,500	14,000	13,700
SA-369	FP9	5	15,000	14,500	14,000	13,700
SA-335	P7	5	15,000	14,500	14,000	13,700
SA-335	P9	5	15,000	14,500	14,000	13,700
SA-182	F12	4	17,500	17,500	17,500	16,800
SA-182	F5a	5	22,500	21,600	20,100	19,000
SA-182	F11	4	17,500	17,500	17,500	16,800
SA-182	F9	5	25,000	24,000	22,700	22,000

SA-201, Grade B may be used until high-temperature test data become available.

satisfactory for average service. For bolted joints, where freedom from leakage over a
sary, determined from the relative flexibility of the flange and bolts and corresponding

ing will be permitted: 20% of the specified tensile strength, or 25% of the specified

tensile strength will be permitted.

Then,

$$\frac{\sigma_{allow}}{P} K\psi = \frac{146 \times 10^6}{1 \times 10^6} \times 0.925 \times 0.95 = 128 > 30$$

The value, P, in the denominator of Equation 7.1 is omitted and the design head thickness determined as follows:

$$t = \frac{dp}{3.8\sigma_{allow} K\psi} \cdot \frac{d}{2h_b} + C = \frac{2.0 \times 1.0 \times 10^6}{3.8 \times 146 \times 10^6 \times 0.925 \times 0.95}$$

$$\times \frac{2.0}{2 \times 0.5} + C = 8.2 \times 10^{-3} + C \ , \quad m$$

Allowable C at $t - c = 8.2 < 1$ mm is taken from Example 6.1, with an increase of 2.0 mm:

$$C = (1.8 + 2) \times 10^{-3} = 3.8 \times 10^{-3} \text{ m}$$

Hence,

$$t = (8.2 + 3.8) \times 10^{-3} = 12 \times 10^{-3} \text{ m}$$

Assume t = 12 mm. The stress in the head at the hydrostatic vessel test is computed from Equation 7.3:

$$\sigma = \frac{[d^2 + 2h_b(t - c)]p'}{7.6K\psi h_b(t - c)}$$

$$= \frac{[4 + 2 \times 0.5(12 - 3.8)10^{-3}] \times 1.3 \times 10^6}{7.6 \times 0.925 \times 0.95 \times 0.5(12 - 3.8) \times 10^{-3}}$$

$$= 190 \times 10^6 \text{ N/m}^2 < \sigma_y/12 = \frac{240 \times 10^6}{1.2}$$

$$= 200 \times 10^6 \text{ N/m}^2$$

7.2.2 Design of Heads Operating Under External Pressure

A thin-walled elliptical head (at $(t - c)/d \leqslant 0.04$) under external pressure or vacuum operation tends to deform inward. Any head deviation from the exact elliptical shape will reduce the strength of the head. This results in increased stress concentration, causing the external pressure to aggravate the condition [2,9,10].

The radius of curvature of an elliptical head reaches the center of the head and is equal to twice the radius of the shell (the major-to-minor axis is usually equal to 2). Specifying the head wall thickness on the basis of the maximum radius curvature would result in a considerable overdesign because the radius of curvature decreases as the point under consideration is moved away from the center toward the junction with the shell, resulting in increased rigidity and greater elastic stability.

The design thickness of an elliptical head, t, operating under external pressure [2,11] is

$$t = \frac{dP_e}{3.8\sigma_{c,allow}KK_1\psi - P_e} \frac{d}{2h_b} + C, \text{ m} \tag{7.4}$$

where K_1 = coefficient; for blind heads and heads with reinforced openings $K_1 = 0.74$; for heads with no reinforced openings, $K_1 = 0.64$

At $(\sigma_{c,allow}/P)KK_1\psi \geqslant 30$, the value p in the denominator of Equation 7.4 may be omitted. Final thickness specification should be based on the same criteria used for vessels operating under internal pressures.

For heads with jackets operating under excessive pressure, the wall thickness of the head, t, should be checked for stress, σ, under a hydrostatic test of the jacket from the following formula:

$$\sigma = \frac{(d^2 + 2h_b(t - c))P_e'}{7.6KK_1\psi h_b(t - c)} \leqslant \sigma_y/1.2, \ N/m^2 \tag{7.5}$$

The following example illustrates these principles.

Example 7.2

An elliptical head is to be designed for the horizontal vessel described in Example 6.4 (Figure 6.21). The required head thickness is to be determined for the following conditions: the material of the head is aluminum alloy; $d = 1.3$ m; $h_b = 0.325$ m; the welded two parts of the head are without openings; the seam is automatic double butt-weld ($\psi = 0.85$) $\sigma_{allow} = 36 \times 10^6$ N/m^2; $p = 0.314$ MN/m^2; $p_e = 0.17$ MN/m^2; $p' = 0.45$ MN/m^2; $K = 1$; $K_1 = 0.74$.

Solution

$$\frac{\sigma_{allow}}{P} K\psi = \frac{36 \times 10^6}{0.314 \times 10^6} \times 1 \times 0.85 = 97.5 > 30$$

Hence, the value, p, in denominator of Equation 7.1 is omitted. The design thickness, assuming internal pressure, is determined from Equation 7.1:

$$t = \frac{dp}{3.8\sigma_{allow}K\psi} \frac{d}{2h_b} + C$$

$$= \frac{1.3 \times 0.314 \times 10^6}{3.8 \times 36 \times 10^6 \times 1 \times 0.85} \times \frac{1.3}{2 \times 0.325} + C = 7 \times 10^{-3} + C \ , \quad m$$

The design head thickness, assuming external pressure, is determined from Equation 7.4:

$$t = \frac{d_e P_e}{3.8\sigma_{c,allow} K K_1 \psi} \frac{d_e}{2h_b} + C$$

$$= \frac{1.3 \times 0.17 \times 10^6}{3.8 \times 36 \times 10^6 \times 1 \times 0.74 \times 0.85} \times \frac{1.3}{2 \times 0.325} + C = 5.1 \times 10^{-3} + C \ , \quad m$$

The final design value is the greater one, as determined from Equation 7.1. The allowance C at $t - c = 7 < 10$ mm is assumed. From Example 6.4, we use a 2 mm increase:

$$C = (1.8 + 2) \times 10^{-3} = 3.8 \times 10^{-3} \text{ m}$$

Hence,

$$t = (7 + 3.8) \times 10^{-3} = 10.8 \times 10^{-3} \text{ m}$$

Assume t = 12 mm (the closest from the gauge). The stress in the head from the internal hydrostatic test is determined from Equation 7.6:

$$\sigma = \frac{[d^2 + 2h_b(t - c)]p'}{7.6K\psi h_b(t - c)} = \frac{[1.3^2 + 2 \times 0.325(12 - 3.8)10^{-3}]0.45 \times 10^6}{7.6 \times 1 \times 0.85 \times 0.325(12 - 3.8) \times 10^{-3}}$$

$$= 44 \times 10^6 \text{ N/m}^2 < \frac{\sigma_y}{1.2} = \frac{60 \times 10^6}{1.2} = 50 \times 10^6 \text{ N/m}^2$$

7.3. HEMISPHERICAL HEADS [12-16]

Hemispherical heads are used in the fabrication of welded horizontal cylindrical apparatuses of large diameters (at d > 4000 mm and, in some cases, at lower values of d), regardless of pressure, and for vertical apparatuses operating under pressure. The fabrication of sectional hemispherical heads of large diameters is easier than that of torispherical heads.

A typical design of a hemispherical head welded of stamped sections (spherical segment and several spherical sectors) is shown in Figure 7.14. The number of sectors and segment diameters are selected using rational sheet cutting, stamping, minimum length of welded seams and other design considerations.

The thickness usually is assumed to be equal to that of a cylindrical shell. Because of its greater strength, the hemispherical head requires less reinforcement if large access openings for manways or nozzles are needed. The directions given in Sections 6.4 and 7.2 should be followed for the design of these shells.

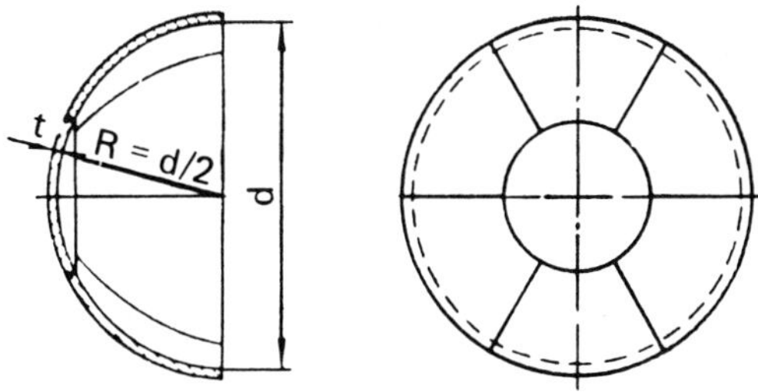

Figure 7.14. Typical design of a hemispherical head welded of stamped sections.

7.4 TORISPHERICAL HEADS [15]

A torispherical (or flanged and dished) head longitudinal section, shown in Figure 7.15, consists of a central dished section b-c with radius, R, and transition section a-b with inside-corner radius, r, and flange with height, h.

The radius of the dished section is $R \leqslant d$, where d is the diameter of the head; also $r \geqslant 0.1R$ and $h \geqslant 50$ mm.

The height of the head is

$$H = R - A \tag{7.6a}$$

where

$$A = \sqrt{(R - r)^2 - (0.5d - r)^2} \tag{7.6b}$$

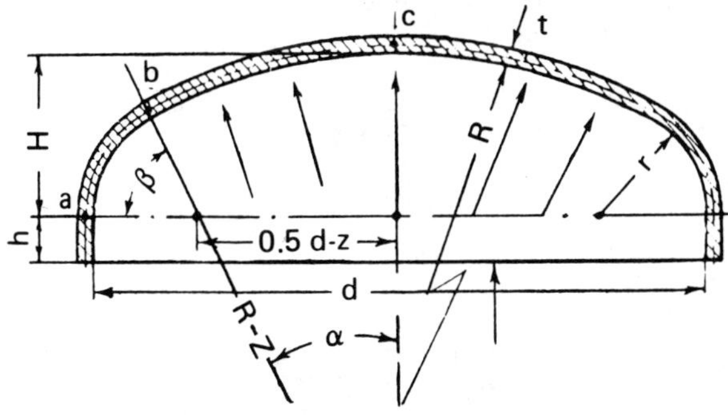

Figure 7.15. Sketch of a torispherical head.

The stress $\sigma = PR/2t$ (Equation 6.106) is related only to the central section of flanged and dished head (b-c); in the transition zone (a-b) of the inside-corner radius, r, the stress will be higher than calculated from Equation 6.106. The internal excessive pressure tends to decrease R and increase r, causing the vessel's shape to become more ellipsoid. This results in the generation of bending stresses in the transition zone.

The stress increase in the transition zone is evaluated by an overstress coefficient, y. The coefficient, y, has a value greater than unity and depends on ratio r/R, that is, y = f(r/R). The larger r/R, the more favorable a stress distribution in the head. The coefficient, y, is introduced into Equation 6.106:

$$t = \frac{PRy}{2\sigma_{allow}} \tag{7.7a}$$

And taking into account allowance C,

$$t = \frac{PRy}{2\sigma_{b,allow}\psi} + C \tag{7.7b}$$

The allowable internal pressure for a given torispherical head is, thus,

$$P = \frac{2\sigma_{b,allow}\psi(t - c)}{Ry} \tag{7.8}$$

The highest resistance to internal pressure observed experimentally has an elliptical head, where d is the large axis and 2H is the small axis of an ellipse [17]. Therefore, the flanged and dished heads made by a basket-handle curve should have a shape closely approximating an ellipse.

The condition mentioned is fulfilled when the ratio r/R approaches a maximum, that is, at

$$\frac{r}{R} = \frac{\sqrt{K^2 + 1} - K}{\sqrt{K^2 + 1} - 1} \tag{7.9}$$

and

$$r = \frac{d}{2}\,\frac{K^2 + 1 - K\sqrt{K^2 + 1}}{K^2 + K - K\sqrt{K^2 + 1}} \tag{7.10}$$

and

$$R = \frac{d}{2}\,\frac{K^2 + 1 - \sqrt{K^2 + 1}}{K^2 + K - K\sqrt{K^2 + 1}}$$

where

$$K = \frac{d}{2H} \tag{7.11}$$

According to experiments [17], the best conditions of head work with respect to stress distribution are obtained for K = 2 – 3. At K = 2, the heads are the thinnest.

If the seamless head thickness t' is obtained from Equation 7.8,

$$t_1 = \frac{PRy}{2\sigma_{allow}} + C$$

and the welded shell thickness is

$$t = \frac{pd}{2\sigma_{allow}\psi} + C$$

Then, for the case of $t_1 = t$, we obtain

$$\frac{PRy}{2\sigma_{b,allow}} + C = \frac{pd}{2\sigma_{t,allow}\psi}$$

As the average value of $\psi = 0.9$, $\sigma_{b,allow} = 1.05\,\sigma_{t,allow}$, and $R = 0.9d$. Then,

$$\frac{p \times 0.9\,dy}{2\sigma_{b,allow}} = \frac{pd}{2\sigma_{t,allow}}$$

or

$$0.9y = \frac{1.05}{0.9} \text{ and } y = 1.3$$

where $\sigma_{b,allow}$ = allowable bending stress
$\sigma_{t,allow}$ = allowable tensile stress

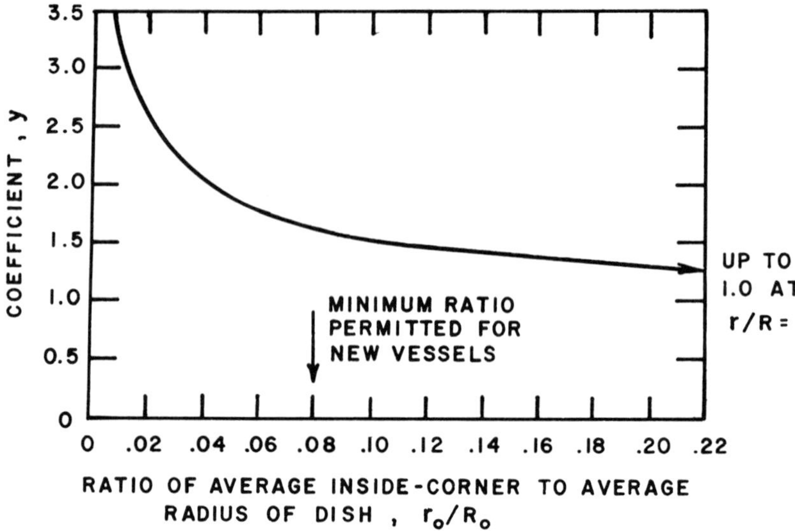

Figure 7.16. Plot of the ratio of average inside-corner radius to the average radius of the dish (r_0/R_0) vs the overstress coefficient, y.

The overstress coefficient y = 1.3, according to the plot given in Figure 7.16, corresponds to the ratio r/R = 0.19. From Equation 7.9, the ratio r/R = 0.19 corresponds to K = 2:

$$\frac{r}{R} = \frac{\sqrt{4+1} - 2}{\sqrt{4+1} - 1} = 0.19$$

From Equation 7.11,

$$R = \frac{d}{2}\,\frac{2^2 + 1 - \sqrt{4+1}}{4 + 2 - 2\sqrt{4+1}} \simeq 0.9d$$

Thus, at r/R = 0.19 and R = 0.9 d, the overstress coefficient y = 1.3, and the thickness of seamless heads are equal to those of a cylindrical shell.

Example 7.3

Determine the thickness of a flanged and dished steel welded head approximating an ellipse for an apparatus with d = 2000 mm and operating under internal pressure p = 16 kg/cm^2; at t = 250°C, ψ = 0.85. The head thickness is determined from Equation 7.7.

Solution

First the r/R ratio is determined:

$$\frac{r}{R} = \frac{340}{1800} \approx 0.19$$

At r/R = 0.19, the overstress coefficient is y = 1.3 (from Figure 7.16). Assuming

$$\sigma_{t,allow} = \frac{\sigma_t}{4} = \frac{3800}{4} = 950 \text{ kg/cm}^2 \; ,$$

we find $\sigma_{b,allow} = 1.05\ \sigma_{t,allow} = 1.05 \times 950 = 1000$ kg/cm^2. Then,

$$t = \frac{PRy}{2\sigma_{b,allow}\,\psi} + C = \frac{16 \times 180 \times 1.3}{2 \times 1000 \times 0.85} + 0.2$$

$$= 2.4 \text{ cm or } 24 \text{ mm}$$

7.5 SPHERICAL DISHED HEADS [15,16]

Because of difficulties in fabrication and the relatively high costs of flanged and dished heads, spherical dished heads sometimes are used for cylindrical vessels operating under pressures up to 0.07 MN/m^2 and temperatures up to 200°C. They are welded directly to shells without reinforcement, as shown in Figure 7.17 (point I), or with a stiffening ring from the outside (point II).

The nominal design thickness of a spherical dished head is determined by the use of the condition that the longitudinal bending stress, σ, at the point of junction with a shell should not exceed an allowable tensile stress, that is, $\sigma = \sigma_{t,allow}$. Figures 7.18 and 7.19 provide plots for determining the limits of application of spherical dished heads as a function of diameter and pressure in a cylindrical vessel [17].

Example 7.4.

We wish to design heads for the following three vessels: (1) d = 1500 mm, p = 1 kg/cm^2; (2) d = 2000 mm, p = 2 kg/cm^2; and (3) d = 3000 mm, p = 3 kg/cm^2.

According to Figure 7.18, for vessel 1 the design of the head from Figure 7.17 (point I) is acceptable; for vessel 2, from Figure 7.17 (point II); and for vessel 3, only the flanged and dished heads are acceptable.

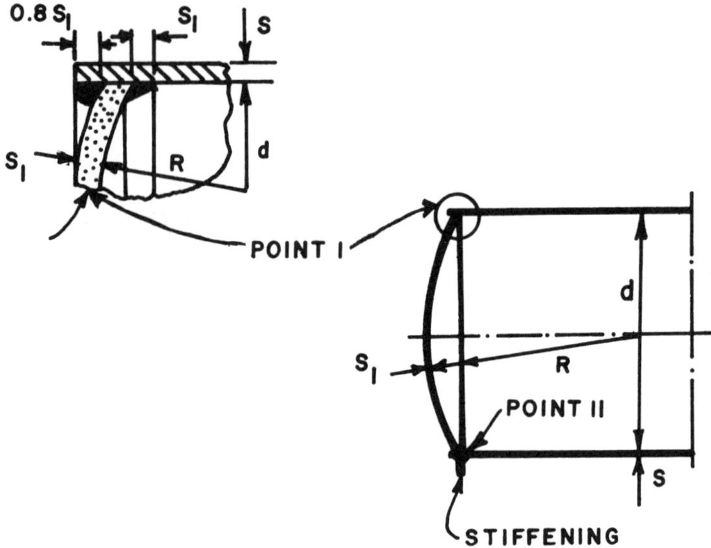

Figure 7.17. Design of a spherical dished head.

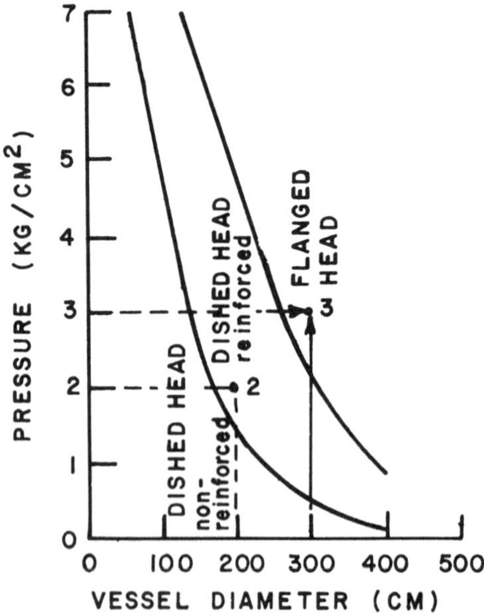

Figure 7.18. Diagram for determination of application limit of dished heads.

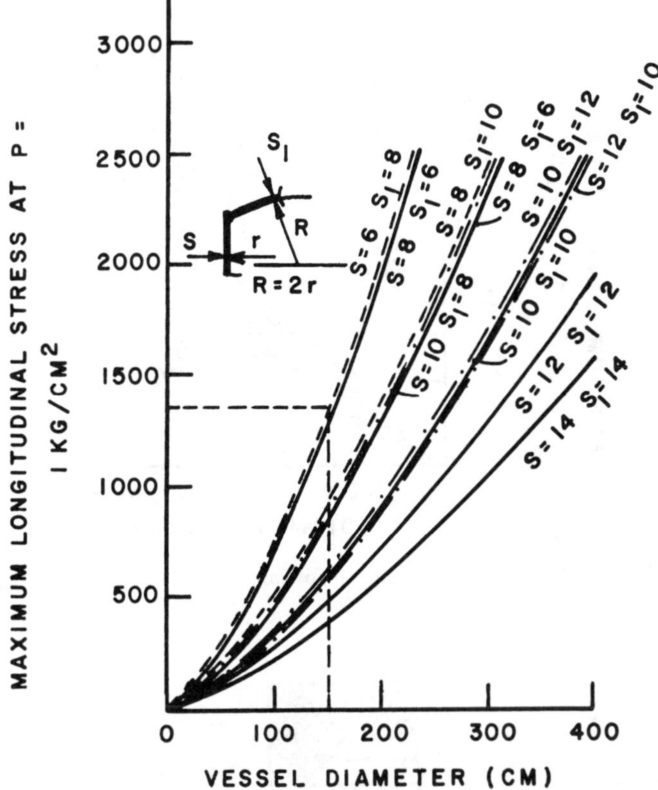

Figure 7.19. Diagram for determination of application limit of dished heads.

Solution

According to Figure 7.19, where stresses are given for pressure p = 1 kg/cm^2, the longitudinal stress in the region joining the head to the shell is equal to 1350 kg/cm^2 for a thickness of the shell s = 6 mm, and for the head, s_1 = 8 mm.

For vessel 2, according to Figure 7.20, where stresses also are given for vessel pressure, p = 1 kg/cm^2 at s = 10 mm, and the section of stiffening ring 60 × 20 mm, the longitudinal stress in the region joining the head to the shell is 700 kg/cm^2 if the operating pressure is p = 1 kg/cm^2. Because in our case the pressure p = 2 kg/cm^2 and the stress is 700 × 2 = 1400 kg/cm^2. A better connection than is given in Figure 7.17 (point I) from the viewpoint of a weld seam is the design given in Figure 7.21.

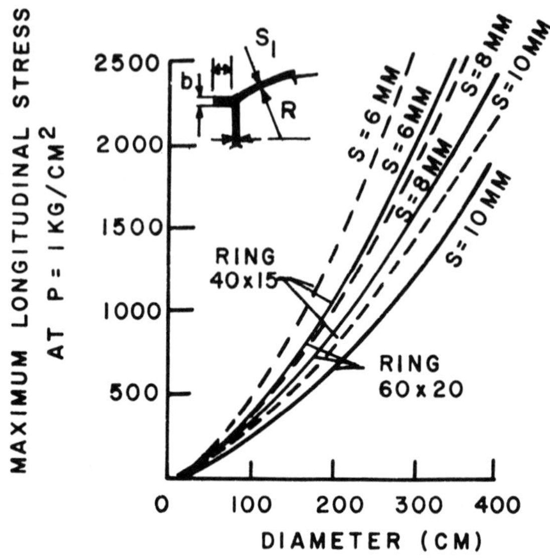

Figure 7.20. Diagram for determination of application limit of dished heads.

Figure 7.21. Welded joint of a dished head.

7.5.1 Design of Dished Heads Operating Under External Pressure

Dished heads operating under external pressure tend to be pressed inward. When loaded from the outside, dished heads are more stable at a smaller radius of spherical section of the head with respect the vessel diameter, d, and greater inside-corner radius, r. The out-of-roundness results in increased stress concentrations, and external pressure tends to aggravate this condition.

Heads operating under external pressure should be designed based on Equations 6.106 and 7.7. The design thickness should be multiplied by a coefficient of 1.7. The maximum allowable pressures calculated for these heads should be divided by 1.7.

When operating under external pressure, elastic stability of the heads should be checked for compression stress, which should not exceed half the yield point.

Example 7.5.

Determine the thickness of the head for Example 7.3 if the vessel operation is under external pressure.

Solution

According to the above method, we obtain the desired head thickness as follows:

$$(2.4 - 0.2) \times 1.7 + 0.2 \approx 4.0 \text{ cm}$$

Example 7.6.

Determine the head thickness of a low-carbon steel vacuum column of 6400 mm inside diameter. The tensile strength is $\sigma_t = 3800$ kg/cm^2. We assume $\sigma_{t,allow} = 820$ kg/cm^2 at t = 375°C. $\sigma_{b,allow} = 1.05 \times \sigma_{t,allow} = 860$ kg/cm^2.

Solution

The shape of the head (see Figure 7.17) is determined by the following parameters: radius of a central section of a dished spherical head R = 0.95 d = 0.95 X 6400 = 6080 mm; we assume R = 6100 mm; inside-corner radius r = 0.1R = 610 mm; ratio r/R = 0.1. According to Figure 7.16, the overstress coefficient is y = 1.54. Hence, the head thickness is

$$s_1 = 1.7 \frac{PRy}{2\sigma_{b,allow}\psi} + C = \frac{1.7 \times 610 \times 1.54}{2 \times 860 \times 0.85} + 0.4 = 1.10 + 0.4 = 1.5 \text{ cm}$$

We assume a head thickness equal to the column shell thickness:

$$s_1 = 18 \text{ mm}$$

If the head is to be hemispherical,

$$s_1 = 1.7 \frac{pR}{2\sigma_{allow}\psi} + C = \frac{1.7 \times 320}{2 \times 820 \times 0.85} + 0.4 = 0.8 \text{ cm}$$

Taking into account the convenience of joining the head with the shell, we assume the head thickness (in case of a hemispherical design) to be equal to that of the shell, that is,

$$s = 18 \text{ mm}$$

7.6 TORICONICAL HEADS [18,19]

Conical and toriconical heads are used preferentially for the fabrication of cylindrical (welded, forged and cast) vertical vessels. They are employed mostly for the lower part in the cases where application of elliptical, spherical and flat heads is desirable. Cone angles at the apex of 60° or 90° are commonly used in the heads and, in some cases, angles are up to 150°. Cones having an angle of 60° are usually applied for the removal of viscous liquids and suspensions, and for the disposal of adhesion-powdered and lump materials.

If conical head designs are employed for the removal of bulk materials, an additional angle equal to half the central angle, α, must exceed the angle of repose of the bulk material, β (Figure 7.22). For most materials, the angle of repose is 35-50°; therefore, the apex (2α) is usually equal to 60-80°.

If the head is used to serve as a diffuser for decreasing pressure drop, the apex (from inlet of the flow) should not exceed 38-40°. Diffusers with 2α = 50-90° not only decrease the pressure drop, but can increase it by 15–20% over an abrupt expansion. This is shown from the resistance coefficient:

$$m = \frac{\zeta}{[1 - (D/d)^2]^2}$$

Figure 7.23 is a plot of the resistance coefficient versus the apex angle, 2α. Minimum resistance, which provides the best fluid distribution over the diffuser cross section, is achieved at 2α = 5-20°. However, the use of diffusers with such angles is not recommended. Diffusers are usually fabricated with 2α not less than 15-20°.

Conical heads for thin-walled apparatuses at small central angles are capable of withstanding considerable pressures. When the hole diameter at the cone apex is more than half of its base, a stamped inverse concave conical reducer is recommended. This is shown in Figure 7.24.

For an angle $\alpha \leqslant 15°$, a welded (not flanged) conical reducer is recommended, as shown in Figure 7.25. A flanged shallow conical head with $\alpha > 15°$ (Figure 7.26) is used for apparatuses operating under internal pressure. This head has openings at the cone apex that are relatively small

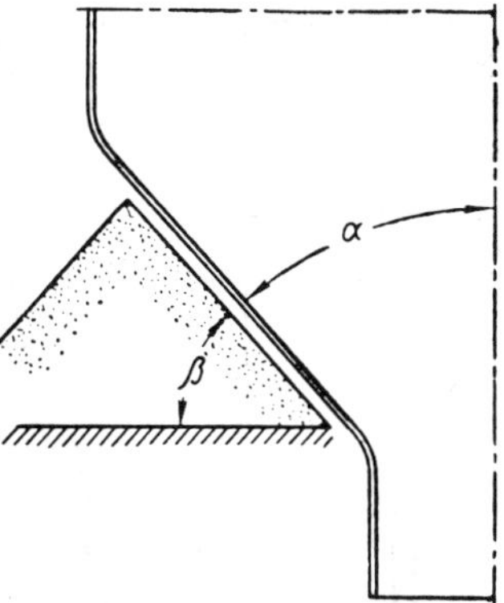

Figure 7.22. Sketch for determination of the shape of a cone.

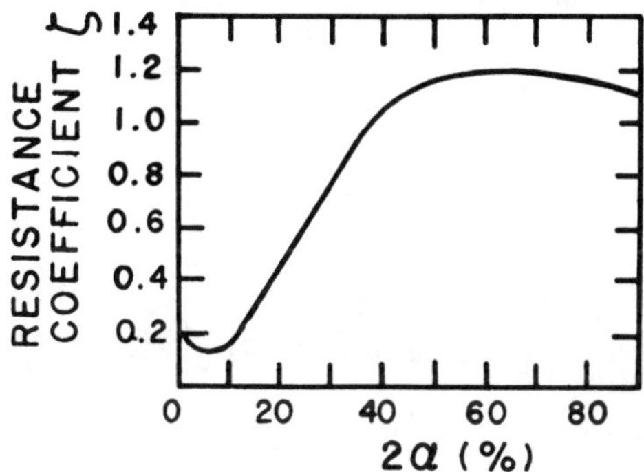

Figure 7.23. Resistance coefficient of a cone vs angle 2α.

compared to the diameter. The reinforcement of the cylindrical shell by stiffening rings should be provided where it is joined with the head, if the angle α is more than 15°. The most stressed place in conical heads is the large cone base. At the cone apex the stresses are theoretically equal to zero; therefore, the holes should be located in the apex of the cone.

7.6.1 Design of Heads Operating Under Internal Pressure [11,18,19]

The design thickness of a conical wall of the head (at $2\alpha \leqslant 150°$) operating under internal pressure, p, is determined from the following formulas:

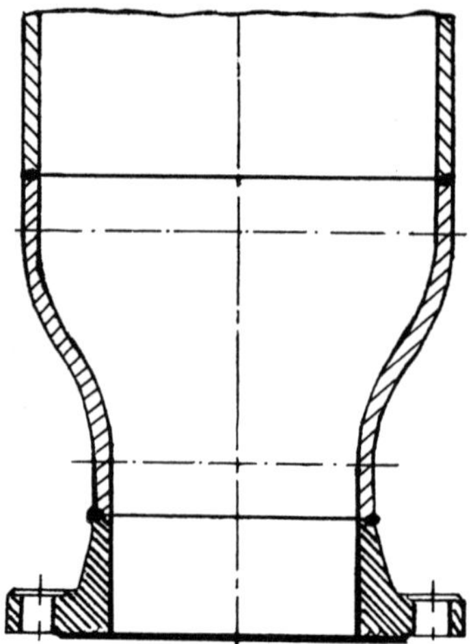

Figure 7.24. Inverse concave conical reducer.

$$t = \frac{dpy}{2\sigma_{b,allow}\psi} + C, \text{ m} \tag{7.12}$$

$$t = \frac{D'P}{2\cos\alpha(\sigma_{allow} - P)} + C, \text{ m} \tag{7.13}$$

where
- y = shape factor of head, determined from Figures 7.27 and 7.28
- D' = design diameter of the head, m (Figure 7.29); for flanged heads, $D' = d - 2(r(1 - \cos\alpha) + 10\ t\sin\alpha)$, but not less than $0.5(d - 2r(1 - \cos\alpha) + d_0)$; for unflanged heads, $D' = d$
- ψ = joint efficiency of circumferential welded seam if the head is provided with it; if there are no circumferential seams on the cone, then the distance between the center line of the mentioned seam and transition $\geqslant 0.5\sqrt{dt/\cos\alpha}$, $\psi = 1$
- ψ' = strength factor of conical wall in the longitudinal direction

At $(\sigma_{allow}/P)\psi \geqslant 50$, the value, p, in the denominator of Equations 7.12 and 7.13 may be omitted.

Determination of values p, ψ and C is made in the same manner as for cylindrical shells operating under internal pressure. The same is related to the final selection of wall thickness, t.

The assumed value of wall thickness, t, should be checked for stress at the hydrostatic test of the apparatus from the following formula:

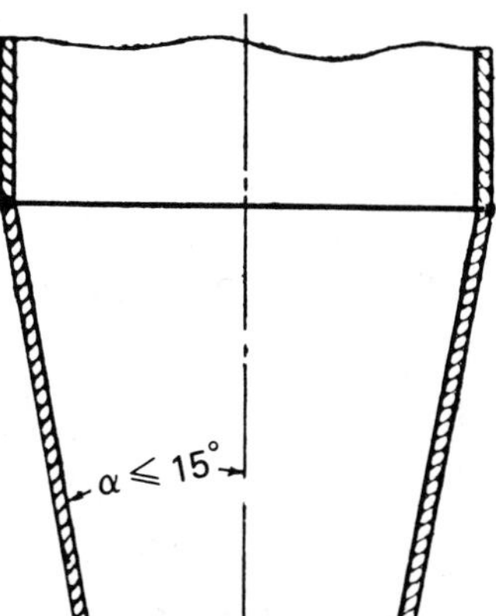

Figure 7.25. Welded (not flanged) conical reducer.

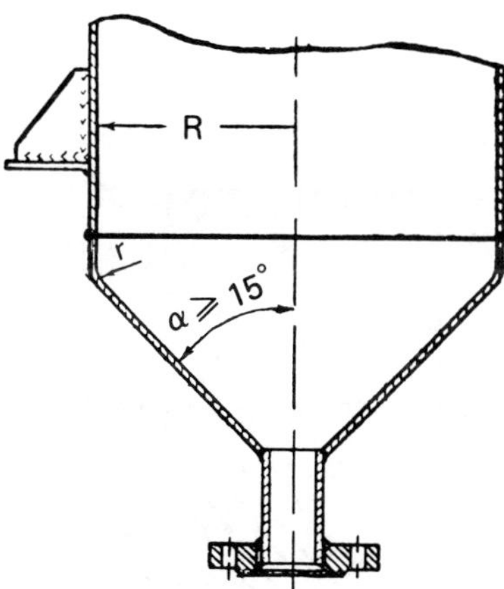

Figure 7.26. Flanged shallow conical head.

$$\sigma = \frac{dp'y}{2(t - c)\psi} \leqslant \frac{\sigma_y}{1.2}\ , \ N/m^2 \qquad (7.14)$$

Equation 7.14 is based on the design thickness of the wall as determined from Equations 7.12 and 7.15.

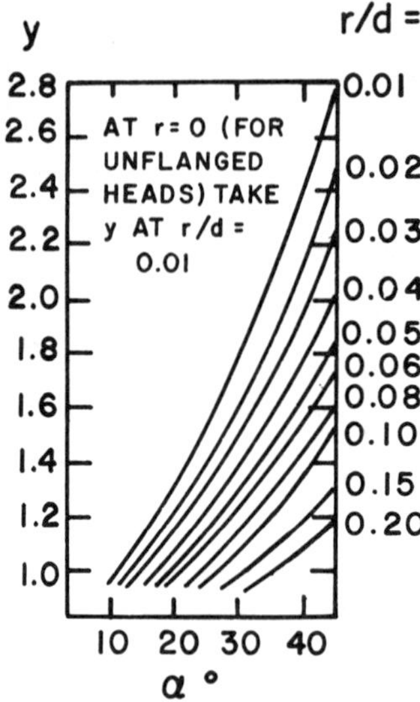

Figure 7.27. Diagram for determination of shape factor, y, in design of conical heads ($2\alpha \leqslant 90°$) from Equation 7.12.

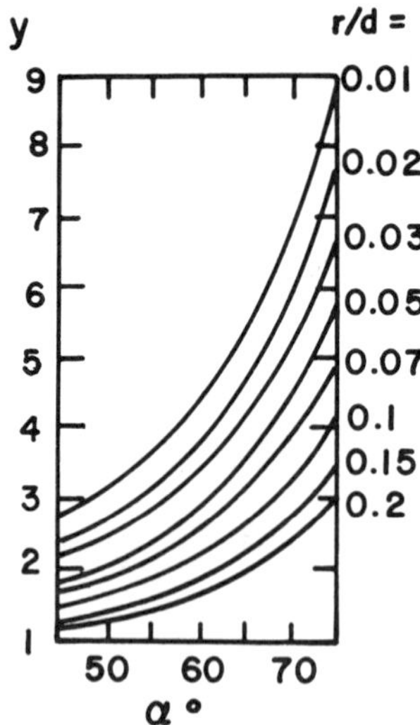

Figure 7.28. Diagram for determination of shape factor, y, in design of conical heads ($2\alpha > 90°$ up to 150°) from Equation 7.12.

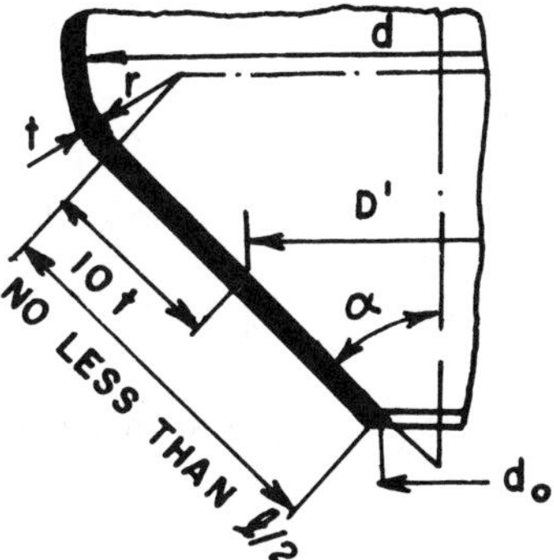

Figure 7.29. Sketch for design of conical heads.

$$\sigma = \left[\frac{D'P'}{2\cos\alpha(t - c)} + P'\right]\frac{1}{\psi} \leqslant \frac{\sigma_y}{1.2}, \quad N/m^2 \qquad (7.15)$$

if the design thickness of the wall was determined from Equation 7.13.

Example 7.7.

Determine the wall thickness of the lower flanged head for a welded shell of a vertical vessel designed in Example 6.1 (refer to Figure 6.13) for the following data: the material of the head is carbon steel; d = 2.0 m; r/d = 0.15; $\alpha = 30°$; there is a central hole underneath the head $d_0 = 0.15$ m; the head is welded ($\psi = 0.95$); $\sigma_{allow} = \sigma_{b,allow} = 146 \times 10^6$ N/m^2; p = 1.06 MN/m^2; p′ = 1.35 MN/m^2.

Solution

The shape factor of the head is determined from Figure 7.27: y = 0.98.
The design wall thickness of the head is determined from Equation 7.12:

$$t = \frac{dpy}{2\sigma_{b,allow}\psi} + C = \frac{2 \times 1.06 \times 10^6 \times 0.98}{2 \times 146 \times 10^6 \times 0.95} + C = 7.45 \times 10^{-3} + C, \text{ m}$$

Assuming t = 8 mm, we determine the design diameter, D′:

$$D' = d - 2[r(1 - \cos\alpha) + 10t\sin\alpha]$$

$$= 2 - 2[0.3(1 - \cos 30^\circ) + 10 \times 8 \times 10^{-3} \sin 30^\circ]$$

$$= 1.84 \text{ m} > 0.5[d - 2r(1 - \cos\alpha) + d_0]$$

$$= 0.5[2 - 2 \times 0.3(1 - \cos 30^\circ) + 0.15] = 1.035 \text{ m}$$

The strength factor of the conical head in the longitudinal direction is $\psi' = \psi = 0.95$. Because

$$\frac{\sigma_{allow}}{P}\psi = \frac{146 \times 10^6}{1.06 \times 10^6} \times 0.95 = 131 > 50$$

the value of p in the denominator of Equation 7.13 is omitted.

The design wall thickness of the head is determined from Equation 7.13:

$$t = \frac{D'P}{2\cos\alpha\,\sigma_{allow}\,\psi} + C = \frac{1.84 \times 1.06 \times 10^6}{2\cos 30^\circ \times 146 \times 10^6 \times 0.95} + C$$

$$= 8.1 \times 10^{-3} + C, \text{ m}$$

The final design wall thickness of the head is assumed to be greater than the wall thickness as determined from Equation 7.13.

"C" allowance is taken from Example 6.1.

$$C = C_1 + C_3 = (1 + 0.8) \times 10^{-3} = 1.8 \times 10^{-3} \text{ m}$$

Hence,

$$t = (8.1 + 1.8) \times 10^{-3} = 9.9 \times 10^{-3} \text{ m}$$

Assume t = 10 mm. The stress in the wall of the head at the hydrostatic test of the vessel with water is checked from Equation 7.15,

$$\sigma = \left[\frac{D'P'}{2\cos\alpha(t - c)} + P'\right]\frac{1}{\psi}$$

$$= \left[\frac{1.84 \times 1.35 \times 10^6}{2\cos 30^\circ(10 - 1.8) \times 10^{-3}} + 1.35 \times 10^6\right]\frac{1}{0.95}$$

$$= 186 \times 10^6 \text{ N/m}^2 < \sigma_y/1.2 = 240 \times 10^6/1.2 = 200 \times 10^6 \text{ N/m}^2$$

Example 7.8.

Determine the wall thickness of a forged conical head of the vertical vessel designed in Example 6.5 (Figure 6.24) using the following data: the head material is alloy steel, d = 0.6 m; r/d = 0.15; $\alpha = 60^\circ$; underneath the head there is a central hole d_0 = 0.05 m. The head is forged together with the shell (ψ = 1); $\sigma_{allow} = \sigma_{b.allow}$ = 208 X 10^6 N/m^2; p = 40 MN/m^2; p′ = 50 MN/m^2.

Solution

The shape factor of the head at x = 60° and r/d = 0.15 is determined from Figure 7.28: y = 1.9.

The design wall thickness of the head is determined from Equation 7.12:

$$t = \frac{dpy}{2\sigma_{b,allow}\psi} + C = \frac{0.6 \times 40 \times 10^6 \times 1.9}{2 \times 208 \times 10^6 \times 1} + C = 110 \times 10^{-3} + C, \text{ m}$$

The design diameter is

$$D' = 0.5[d - 2r(1 - \cos\alpha) + d_0]$$

$$= 0.5[0.6 - 2 \times 0.09(1 - \cos 60^\circ) + 0.05]$$

$$= 0.28 \text{ m}$$

The design wall thickness of the head is determined from Equation 7.13:

$$t = \frac{D'P}{2\cos\alpha(\sigma_{allow}\psi - P)} + C$$

$$= \frac{0.28 \times 40 \times 10^6}{2\cos 60^\circ(208 \times 10^6 \times 1 - 40 \times 10^6)} + C = 66.7 \times 10^{-3} \text{ m}$$

The final design wall thickness of the head is assumed to be the greater one determined from Equation 7.12.

The allowance C at C_2 = 0, taking into account the value t − c = 110 mm > 20 mm, may be taken as C = 0. Assume t = 110 mm. The stress in the wall at the hydrostatic test of the vessel with water is checked from Equation 7.14:

$$\sigma = \frac{dP'y}{2(t-c)\psi} = \frac{0.6 \times 50 \times 10^6 \times 1.9}{2 \times 110 \times 10^{-3} \times 1}$$

$$= 259 \times 10^6 \text{ N/m}^2 < \sigma_y/1.2 = \frac{380 \times 10^6}{1.2} = 316 \times 10^6 \text{ N/m}^2$$

7.6.2 Design of Heads Operating Under External Pressure

If thin-walled conical heads (at $(t - c)/d \leqslant 0.04$) are loaded with external pressure or vacuum in the vessel, walls may tend to press inward. The permissible out-of-roundness for such heads should not exceed 1.0%.

The design wall thickness of a conical head (at $2\alpha \leqslant 150°$) operating under external pressure, p_e, is determined from the following formulas, and the final design thickness is assumed to be the greater value obtained:

$$t = \frac{dP_e y}{2\sigma_{b,allow} K_1 \psi} + C \tag{7.16}$$

and

$$t = \frac{D'P_e}{2 \cos\alpha K_1 (\sigma_{allow}\psi - Pe)} + C \tag{7.17}$$

where K_1 = coefficient (see Equation 7.4).

At $(\sigma_{allow}/p)\psi \geqslant 50$, the value, p, in the denominator of Equation 7.17 may be omitted. In other respects, the design is similar to that for conical heads operating under internal pressure. If the head has a jacket operating under an excessive pressure, the assumed wall thickness of the head, t, should be checked for stress at hydrostatic conditions as follows:

If the design wall thickness is obtained from Equation 7.16,

$$\sigma = \frac{dP'_e y}{2K_1(t-c)\psi} \leqslant \sigma_y/1.2, \text{ N/m}^2 \tag{7.18}$$

and if the design wall thickness is obtained from Equation 7.17,

$$\sigma = \frac{D'P'_e}{2 \cos\alpha K_1(t-c)} + P'_u \frac{1}{\psi} \leqslant \frac{\sigma_y}{1.2}, \text{ N/m}^2 \tag{7.19}$$

7.6.3 Design of Heads Operating Under Low Internal Pressure (up to 0.07 MN/m^2) and During Filling Conditions

Dished and unflanged heads are used for such conditions. The design pressure is taken from Equation 6.24, and for providing head rigidity, a value no less than 0.1 MN/m^2 is assumed.

For vessels operating under low pressure (no more than 0.07 MN/m^2) and subjected to gas medium, the design thickness of the wall is determined from Equation 7.12. The design pressure in Equation 7.12 is assumed to be equal to 0.1 MN/m^2.

For the head thickness of an apparatus operating under atmospheric pressure and used for handling bulk materials, the wall thickness should be selected on the basis of design considerations.

For final head thickness selection, the directions should be followed that apply to cylindrical vessels operating under similar conditions. Use of unflanged conical heads for welded apparatuses produces additional radial compression stresses in the region of the cylindrical shell where it is joined with the head. Therefore, such shells (at $\alpha > 15°$) should be reinforced by a stiffening ring, as shown in Figure 7.30.

The cross-sectional area of the stiffening ring is determined as follows:

$$F = \frac{d^2 p \tan\alpha}{8\sigma_{allow}}, \quad m^2 \tag{7.20}$$

Example 7.9.

Determine the thickness of the lower head of a vertical welded vessel operating under internal pressure from the following data: the material is carbon steel; material permeability in the medium is ~0.1 mm/yr (C_1 = 1 mm;

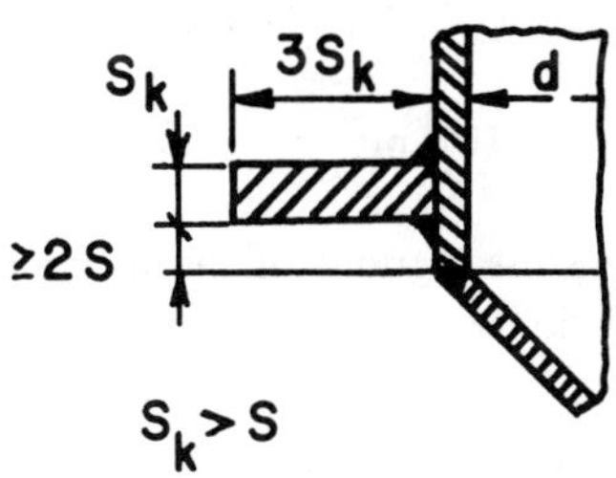

Figure 7.30. Design of a stiffening ring for welded vessels with conical unflanged heads.

$C_2 = 0$); the medium is liquid ($\rho_e = 1.2 \times 10^3$ kg/m^3); gas: p = 0.114 MN/m^2, t = 20°C; d = 2.4 m; H = 4.5 m; t = 6 mm; the head is reinforced by a stiffening ring. The head is conical, unflanged with a central hole $d_0 = 0.05$ m; $\psi = 0.95$;

$$\sigma_{b,allow} = \sigma_{allow} = 146 \times 10^6 \text{ N/m}^2;\ 2\alpha = 90°$$

Solution

The head shape factor at $\alpha = 45°$ and r/d = 0.01 is determined from Equation 7.22, y = 2.75. The design head thickness is obtained from Equation 7.12:

$$t = \frac{dpy}{2\sigma_{b,allow}\psi} + C = \frac{2.4 \times 0.114 \times 10^6 \times 2.75}{2 \times 146 \times 10^6 \times 0.95} + C$$

$$= 2.71 \times 10^{-3} + C, \text{ m}$$

Allowance C at $C_3 = 0.6$ mm is computed from

$$C = C_1 + C_2 + C_3 = (1 + 0 + 0.6) \times 10^{-3} = 1.6 \times 10^{-3} \text{ m}$$

Hence,

$$t = (2.71 + 1.6) \times 10^{-3} = 4.31 \times 10^{-3} \text{ m}$$

Assume s = 6 mm.

The cross-sectional area of the stiffening ring placed in the lower part of the shell is obtained from Equation 7.20:

$$F = \frac{d^2p \tan\alpha}{8\sigma_{allow}} = \frac{(2.4)^2 \times 0.114 \times 10^6 \times \tan 45°}{8 \times 146 \times 10^6}$$

$$= 5.61 \times 10^{-4} \text{ m}^2$$

Taking the sizes of the cross sections of the stiffening ring from Figure 7.30, we obtain the design thickness:

$$S_k' = \sqrt{\frac{F}{3}} = \sqrt{\frac{5.61 \times 10^{-4}}{3}} = 13.7 \times 10^{-3} \text{ m}$$

Assuming $S_k = 14$ mm, the ring width will be $b = 3S_k = 3 \times 14 = 42$ mm.

In designing conical heads, it was assumed that the internal pressure in a vessel is constant. We shall now consider the case in which the internal pressure changes along the height of the conical head. The design of a conical head subjected to internal pressure, as well as any shell of revolution, follows the following expression [7]:

$$\frac{\sigma_1}{R_1} + \frac{\sigma_2}{R_2} = \frac{P}{t} \tag{7.21}$$

where σ_1 = longitudinal stress (kg/cm^2)
σ_2 = circumferential stress (kg/cm^2)
R_1 = radius of curvature of the central surface of the longitudinal curve (cm), in this case the longitudinal curve is the cone generatrix
R_2 = radius of curvature of the central surface in the direction perpendicular to the cone generatrix (cm)
t = thickness of the conical head (cm)
p = internal pressure (kg/cm^2)

This formula considers the action of tensile forces without the influence of bending moments on the shell [3].

The radius, R_1, for a cone at any of its points is equal to ∞; at $\sigma_1/R_1 = 0$,

$$\sigma_2 = \frac{PR_2}{t} \tag{7.22}$$

The value of the longitudinal tensile stress, σ_1, is determined from the condition that vertical components of longitudinal stresses developing in any section "nn" (Figure 7.31) balance the load, Q, acting on the part of the shell below the section "nn":

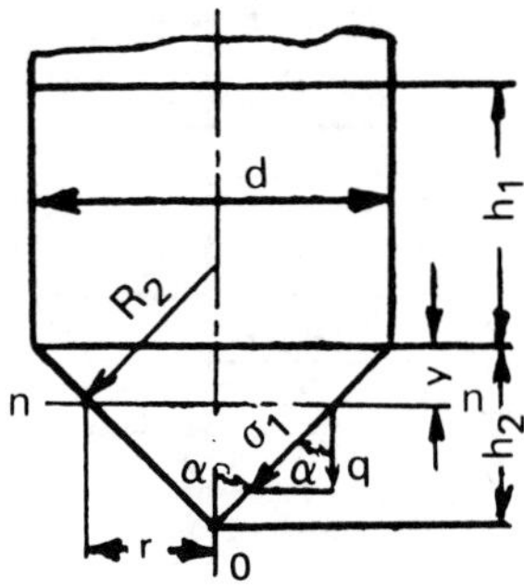

Figure 7.31. Sketch for deriving Equation 7.23.

$$Q = \Sigma q = \Sigma t\sigma_1 \cos\alpha$$

or

$$Q = 2\pi r\sigma_1 \cos\alpha$$

Hence,

$$\sigma_1 = \frac{Q}{2\pi rt \cos\alpha} \tag{7.23}$$

The circumferential tensile stress is determined from Equation 7.22. As the side $n\ell$ in the triangle $n\ell o$ (Figure 7.32) is equal to

$$n\ell = r = (h_2 - y) \tan\alpha \tag{7.24}$$

so, from the triangle $n\ell z$,

$$R_2 = \frac{n\ell}{\cos\alpha} = \frac{(h_2 - y) \tan\alpha}{\cos\alpha} \tag{7.25}$$

Substituting R_2 into Equation 7.22, we obtain

$$\sigma_2 = \frac{P(h_2 - y) \tan\alpha}{t \cos\alpha}$$

The maximum value of σ_2 is obtained at the point at which $y = 0$, that is, at the point of joining the conical head with the cylindrical shell:

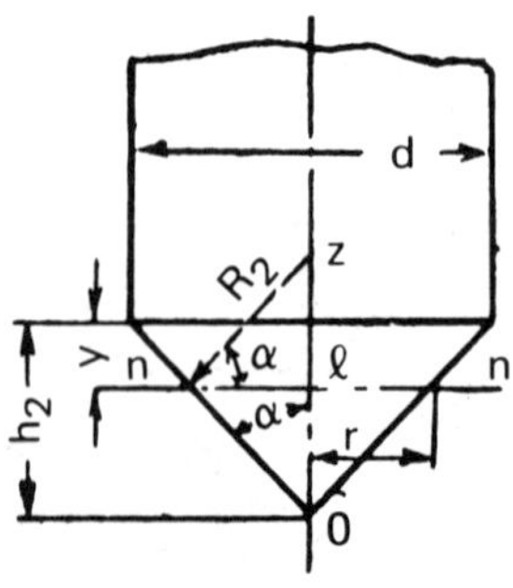

Figure 7.32. Sketch for deriving Equation 7.24.

$$\sigma_{2,max} = \frac{ph_2 \tan\alpha}{t \cos\alpha}$$

Let us consider now a vessel filled with liquid to a definite maximum height, as shown in Figure 7.31. The intensity of the liquid pressure on the wall of the cone at any of its points is

$$p = \rho(h_1 + y) \tag{7.26}$$

where y = the distance of a given point from the cone base (cm). Substituting into Equation 7.22 the value of p from Equation 7.26 and the value of R_2 from Equation 7.25, we obtain

$$\sigma_2 = \frac{\rho(h_1 + y)(h_2 - y) \tan\alpha}{t \cos\alpha} \tag{7.27}$$

The maximum value of σ_2 will correspond to a value of y determined from the first derivative of the right-hand side of Equation 7.27, which is equated to zero:

$$\frac{d}{dy}\left(\frac{\rho(h_1 + y)(h_2 - y)}{t \cos\alpha}\right) = 0$$

Hence,

$$y = \frac{h_2 - h_1}{2} \tag{7.28}$$

Substituting y from Equation 7.28 into Equation 7.27, we obtain the maximum value of σ_2:

$$\sigma_{2,max} = \frac{\rho \tan\alpha}{4t \cos\alpha}(h_1 + h_2)^2 \tag{7.29}$$

Taking into account joint efficiency coefficient ψ and allowance C, and changing $\tan\alpha$ for $d/2h_2$, we obtain

$$\sigma_{2,max} = \frac{\rho d}{8\psi(t - c) \cos\alpha} \frac{(h_1 + h_2)^2}{h_2} \tag{7.30}$$

Substituting the value σ_2 for the allowable stress, σ_{allow}, we obtain the final formula for determining the wall thickness:

$$t = \frac{\rho d}{8\psi\sigma_{allow}\cos\alpha} \frac{(h_1 + h_2)^2}{h_2} + C \qquad (7.31)$$

Equation 7.30 is valid if $h_1 < h_2$. For this case, y values determined from Equation 7.28 will be positive. For the case in which $h_1 = h_2$, $y = 0$.

If $h_1 > h_2$, then according to Equation 7.28 y becomes negative. This shows that the maximum stress in the conical part is beyond the limit of a given vessel, and by substituting a negative value for y in Equation 7.27 an impractical answer is obtained.

Thus, if $h_1 = h_2$ or $h_1 > h_2$, the maximum value of σ_2 will be at $y = 0$, that is, at the cone base; substituting this value in Equation 7.27, we obtain

$$\sigma_{2,max} = \frac{\rho \tan\alpha}{t \cos\alpha} h_1 h_2 \qquad (7.32)$$

Introducing joint efficiency coefficient ψ and allowance C, and changing $\tan\alpha$ for $d/2h_2$, we obtain

$$\sigma_{2,max} = \frac{\rho h_1 d}{2\psi(t - c)\cos\alpha} \qquad (7.33)$$

And substituting σ_2 for allowable stress, σ_{allow}, we obtain the formula for determining wall thickness:

$$t = \frac{\rho h_1 d}{2\psi\sigma_{allow}\cos\alpha} + C \qquad (7.34)$$

If the liquid level is $h_1 = 0$, the value of y from Equation 7.28 will be

$$y = \frac{h_2}{2}$$

That is, the maximum stress that will be in the middle of the cone and its value as determined from Equation 7.27 will become

$$\sigma_{2,max} = \frac{\rho h_2 d}{8t \cos\alpha}$$

At $y = h_2$ at the cone apex, the stresses are equal to zero.

Now we will determine the longitudinal stress, σ_1, expressed by Equation 7.23.

Here, Q is the liquid pressure on the part of the cone section below

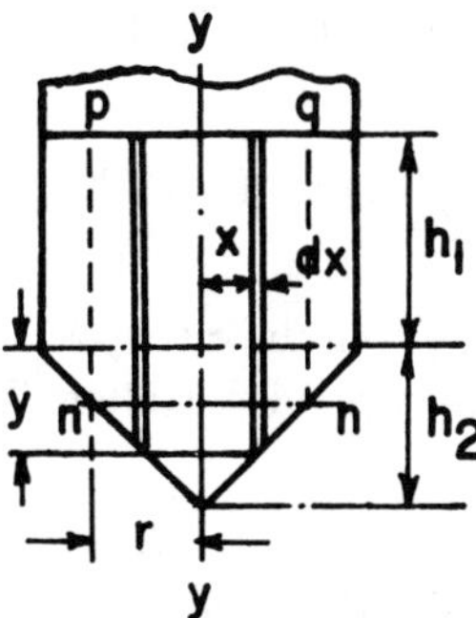

Figure 7.33. Sketch for deriving Equation 7.35.

"nn" (Figure 7.33). This pressure is equal to the weight of the liquid confined in the volume, V, of the cylindrical part, npqn, and conical part non. Thus,

$$\rho V = Q$$

but

$$V = \int_0^r 2\pi x dx(h_1 + y')$$

and

$$y' = h_2 - \cot n\alpha$$

Then

$$V = 2\pi \int_0^r x(h_1 + h_2 - \cot n\alpha)dx$$

$$= \frac{\pi r^2}{3}(3h_1 + 3h_2 - 2r \cot n\alpha)$$

and, according to Equation 7.23,

$$\sigma_1 = \frac{\pi r^2 \rho(3h_1 + 3h_2 - 2 \cot n\alpha)}{3 \times 2\pi r t \cos\alpha} = \frac{\rho r(3h_1 + 3h_2 - 2 \cot n\alpha)}{6t \cos\alpha}$$

substituting the value, r, from Equation 7.24, we obtain

$$\sigma_1 = \frac{\rho \tan\alpha}{6t \cos\alpha}(h_2 - y)(3h_1 + h_2 + 2y) \tag{7.35}$$

The maximum value of σ_1 will be at y, as determined from the first derivative of the right-hand side of Equation 7.35 equated to zero. That is,

$$\frac{d}{dy}\left[\frac{\rho \tan\alpha}{6t \cos\alpha}(h_2 - y)(3h_1 + h_2 + 2y)\right] = 0$$

Hence,

$$y = \frac{h_2 - 3h_1}{4} \tag{7.36}$$

Substituting y from Equation 7.36 into Equation 7.35, we obtain

$$\sigma_{1,max} = \frac{3\rho \tan\alpha}{16t \cos\alpha}(h_1 + h_2)^2 \tag{7.37}$$

Equation 7.37 is valid for $h_1 < h_2/3$. In the case in which $h_1 \geqslant h_2/3$, the maximum value $\sigma_{1,max}$ will be at $y = 0$, and from Equation 7.35 we obtain

$$\sigma_{1,max} = \frac{\rho \tan\alpha}{6t \cos\alpha} h_2(3h_1 + h_2) \tag{7.38}$$

Introducing ψ and C, and substituting into Equations 7.37 and 7.38 $\tan\alpha$ for $d/2h_2$, we obtain the final formulas:

At $h_1 < h_2/3$,

$$\sigma_{1,max} = \frac{3}{32}\,\frac{\rho d}{\psi(t - c)\cos\alpha}\,\frac{(h_1 + h_2)^2}{h_2} \tag{7.39}$$

At $h_1 \geqslant h_2/3$,

$$\sigma_{1,max} = \frac{\rho d}{12\psi(t - c)\cos\alpha}(3h_1 + h_2) \tag{7.40}$$

Let us now consider the case in which a gas or vapor exists under low pressure over the liquid surface in the vessel. Changing pressure, p, by an equivalent liquid column with a weight of unity volume liquid, ρ, the height of an equivalent liquid column is

$$h_{eq} = P/\rho \tag{7.41}$$

The design for this case is based on Equations 7.30, 7.31 or 7.33, 7.34 and 7.39, or 7.40, where h_1 is changed by

$$H = h_1 + h_{eq} \tag{7.42}$$

If the gas (vapor) pressure over the liquid interface is well above that of the liquid, the liquid pressure should be recalculated from Equation 7.41 in kg/cm^2 and added to the gas pressure. The design procedure is similar to that of a vessel operating under constant pressure.

Connection of conical unflanged heads to cylindrical shells is allowable only for shells operating under low pressures and if the angle, α, is small ($\alpha \leqslant 30°$).

The thickness of a conical unflanged head at $\alpha \leqslant 30°$ is determined from the following formula:

$$t = \frac{pr}{(\psi\sigma_{allow} - 0.6p)\cos\alpha}, \text{ cm} \tag{7.43}$$

Furthermore, it should be noted that t is temperature dependent. The following data show the general dependency of $p/\psi\sigma_{allow}$ with temperature difference, ΔT.

$\frac{p}{\psi\sigma_{allow}}$	0.001	0.002	0.003	0.004	0.005	0.006
ΔT°C	13	18	22	25	28	31

If $\Delta T° < \alpha°$, a stiffening ring should be inserted where the conical head is joined to the cylindrical shell, as shown in Figure 7.34. The stiffening rings are noted by "a."

The cross section of a stiffening ring is determined from

$$F = \frac{q_1 d}{2\sigma_{c,allow}}, \text{ cm}^2 \tag{7.44}$$

where $\sigma_{c,allow}$ = the allowable stress for compression (kg/cm^2), which is the same as the allowable tensile stress:

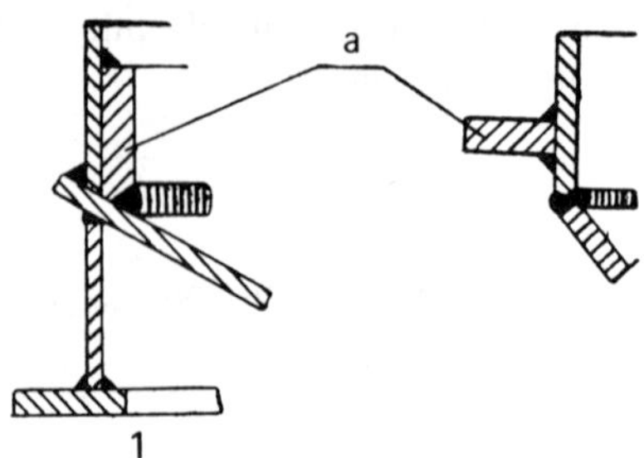

Figure 7.34. Reinforcing elements.

$$q_1 = q \tan\alpha = \frac{pd \tan\alpha}{4} \tag{7.45}$$

q is the force per 1 m of the stiffening ring circumferential length (Figure 7.35).

The internal gas pressure in the vessel is

$$Q = \frac{p\pi d^2}{4} \tag{7.46}$$

Consequently,

$$q = \frac{Q}{\pi d} = \frac{pd}{4} \tag{7.47}$$

In some cases, at small values of the central angle ($\alpha \leqslant 15^\circ$) the application of an additional reinforcing ring is not necessary.

Example 7.10.

Determine the stress in the conical head of an apparatus where d = 4500 mm inside diameter and height H = 6000 mm (Figure 7.36). The head thickness t = 8 mm; allowance C = 3 mm; there is additional pressure above the liquid level p = 0.2 kg/cm^2; $\alpha = 60^\circ$. Temperature of fluid, t = 70°C. $\rho = 0.0009$ kg/cm^3; $\psi = 0.95$.

Solution

The height of the conical head is

$$h_2 = 2250 \times \tan 30^\circ = 2250 \times 0.577 = 1300 \text{ mm}$$

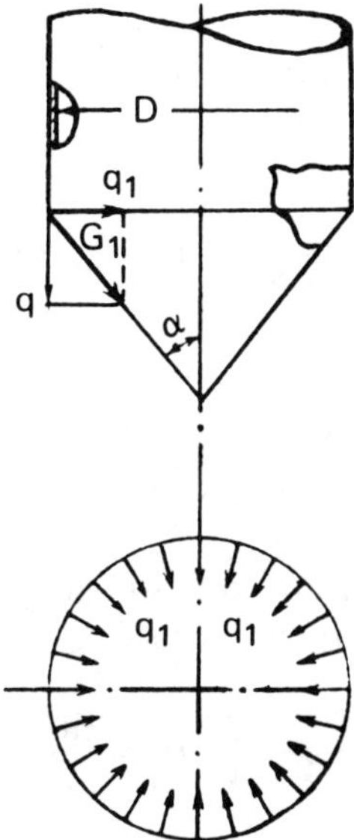

Figure 7.35. Sketch for the determination of the cross section of a reinforcing element.

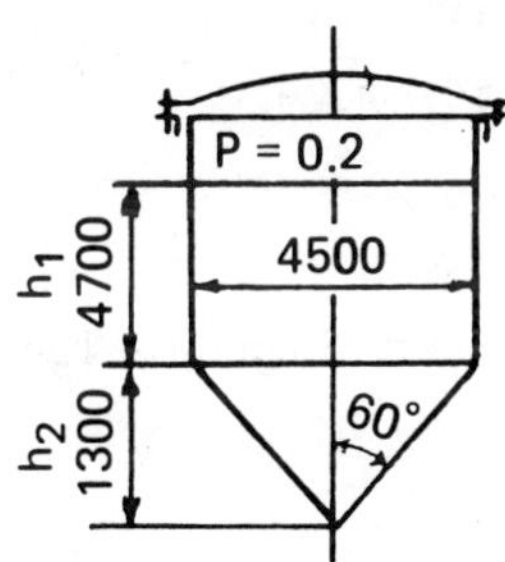

Figure 7.36. Sketch for Example 7.10.

The height of the liquid column equivalent to the pressure p = 0.2 kg/cm^2 is

$$h_{eq} = P/\rho = \frac{0.2}{0.0009} = 222 \text{ cm}$$

Since $H = h_1 + h_{eq} > h_2$, the design of the head for a maximum tensile stress $\sigma_{2,max}$ along the generatrix is computed from Equation 7.33:

$$\sigma_{2,max} = \frac{\rho h_1 d}{2\psi(t - c)\cos\alpha} = \frac{\rho H d}{2\psi(t - c)\cos\alpha}$$

Hence, from Equation 7.42,

$$H = h_1 + h_{eq} = 470 + 222 = 692 \text{ cm}$$

and

$$\sigma_{2,max} = \frac{0.0009 \times 692 \times 450}{2 \times 0.85 \times (0.8 - 0.3)\cos 60^\circ} = 66 \text{ kg/cm}^2$$

The cross-sectional area F of the stiffening ring is determined from Equation 7.44. First we determine the load, Q, on the conical head:

$$Q = \frac{\pi d^2}{4}\frac{h_2}{3}\rho + \frac{\pi d^2}{4}H\rho = \frac{\pi d^2}{4}\rho\left(\frac{h_2}{3} + H\right)$$

$$= \frac{\pi(450)^2}{4} \times 0.0009\left(\frac{130}{3} + 692\right) \approx 105\,000 \text{ kg}$$

From Equation 7.47,

$$q = \frac{Q}{\pi d} = \frac{105\,000}{\pi \times 450} = 75 \text{ kg/cm}$$

From Equation 7.45,

$$q_1 = q\tan\alpha = 75 \tan 60^\circ = 124 \text{ kg/cm}$$

Hence,

$$F = \frac{q_1 d}{2\sigma_{c,allow}} = \frac{124 \times 450}{2 \times 1200} = 23 \text{ cm}^2$$

We can assume a stiffening ring 3 cm thick by 8 cm high.

Figure 7.37 shows a conical flange head. The flange radius should not be less than 0.12 internal head radius, and no less than three times of transition part thickness, S_1. The thickness of the conical part of the flanged head is determined from the formula for an unflanged cone head, but instead of using r, we use r_1, the radius of the part where the cone passes into the torous section:

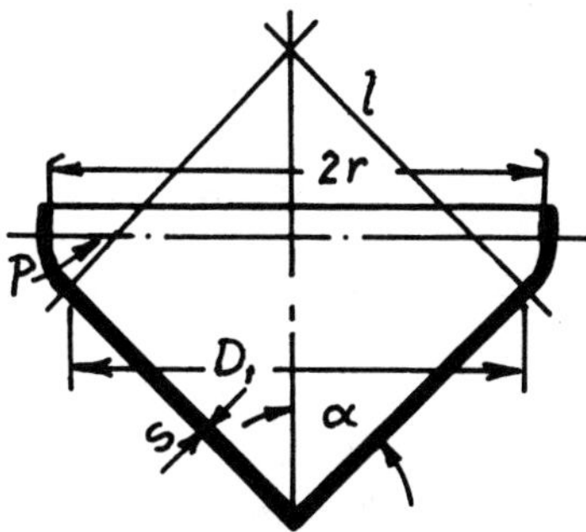

Figure 7.37. Conical flanged head.

$$r_1 = r - \dot{\xi}(1 - \cos\alpha) \qquad (7.48)$$

where $\dot{\xi}$ is the flange radius and, for proper design, $\dot{\xi} \geqslant 0.12$ r. The thickness of the flanged part is

$$S_1 = K' \frac{Pr_1}{(\psi\sigma_{allow} - 0.1P)\cos\alpha} \qquad (7.49)$$

where K′ is a coefficient taken from Table 7.2 and depends on the ratio $r_1/\dot{\xi}\cos\alpha$. The maximum ratio of $r_1/\dot{\xi}\cos\alpha = 16.66$. If the transition part is formed by flanging the head, then the head thickness is assumed to be the greater value between S and S_1.

The design of flanged conical heads at $\alpha = 45\text{-}75°$ may be done using the following experimental formula [7]:

Table 7.2. Values of Coefficient K′

$r_1/\dot{\xi}\cos\alpha$	1.00	1.25	1.50	1.75	2.00	2.25	2.50	2.75	3.00	3.25	3.50
K′	1.00	1.03	1.06	1.08	1.10	1.13	1.15	1.17	1.18	1.20	1.22
$r_1/\dot{\xi}\cos\alpha$	4.0	4.5	5.0	5.5	6.0	6.5	7.0	7.5	8.0	8.5	9.0
K′	1.25	1.28	1.31	1.34	1.36	1.39	1.41	1.44	1.46	1.48	1.50
$r_1/\dot{\xi}\cos\alpha$	9.5	10.0	10.5	11.0	11.5	12.0	13.0	14.0	15.0	16.0	16.66
K′	1.52	1.54	1.56	1.58	1.60	1.62	1.65	1.69	1.72	1.75	1.77

$$\sigma_{allow} = \frac{ypd}{2(t - c)\psi} \tag{7.50}$$

where

$$y = 0.5 + n \tan\alpha \tag{7.51}$$

The value of n depends on the ratio of r/d:

r/d	0.1	0.05	0.033
n	1	1.4	1.7

Figure 7.38 shows three different designs of reducers for cones to the cylindrical shell.

Figure 7.38(A) presents a standard design in which the conical head is flanged. As the transition part is the most loaded, it is sound practice to increase the cross section of metal, as shown in Figure 7.38(B). This connection is made of a forged reducing ring of increased wall thickness. To avoid using an expensive forged ring, it is possible to apply a welded reducer ring, as in Figure 7.38(C).

Rolling of conical heads is a difficult operation, especially if the opening at the cone apex is small and the wall of the head is thick. At $\alpha > 60°$ conical heads are very thick, heavy and expensive. They tend to increase the size of the vessel, and the ratio of surface volume is unfavorable. Therefore, they should not be used without sufficient justification.

For large-diameter vessels, conical heads may be ended by spherical parts, as in Figure 7.39 as these are easier to fabricate.

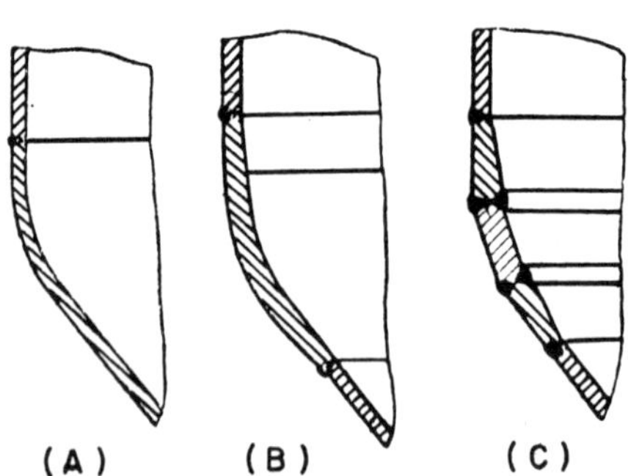

Figure 7.38. Designs of reducers from cone to cylinders.

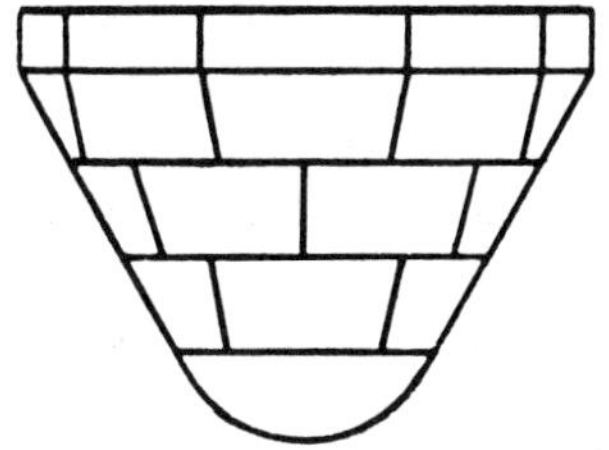

Figure 7.39. Combined conical head.

7.7 FLAT HEADS

Flat heads, flanged and unflanged, are used widely as closures for cylindrical vessels, hand holes, manholes, etc. They are simple to fabricate and inexpensive. However, their application is limited to low-pressure service or to closures for small-diameter vessels (up to 400 mm); and in cases of high pressure drop, they become heavy and bulky. Figure 7.40 shows both types.

The use of flat plates as closures for vessels is recommended only in cases in which they provide easy assembly of internal devices and parts or are easily connected to other parts, such as tube plates. The sharp discontinuity in shape at the junction of a cylindrical vessel and a flat plate induces localized stress concentrations at the junction zone.

The nature of these stress concentrations is complex in that bending moments, shear and stress reversals must be considered in addition to the membrane stresses resulting from internal pressure. Under the influence of internal pressure, the difference between the flexibility of the flat head

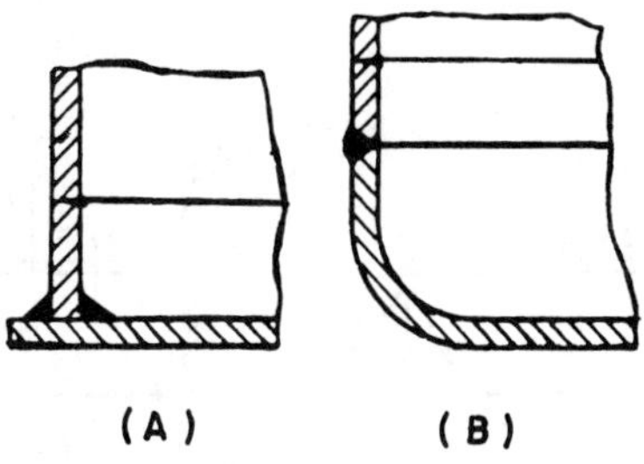

Figure 7.40. Flat heads: (A) unflanged; (B) flanged.

and of the cylindrical shell induces both elements of the rigid juncture in attempting to deform radially and angularly. This results in shear and flexural stresses, which may cause severe concentration [7] picks. Flat heads may be designed using either theoretical formulas or semiempirical formulas.

Circular flat plates described below are used mostly in manufacturing chemical equipment.

1. **A uniformly loaded flat circular plate simply supported at the edge (Figure 7.41(A)):**
The stress at the center of the plate is

$$\sigma_r = \sigma_\tau = 1.24p(r/S)^2 \tag{7.52}$$

where σ_r = radial stress (kg/cm^2)
σ_τ = circumferential stress (kg/cm^2)
r = radius of a plate (cm)
S = thickness of a plate (cm)

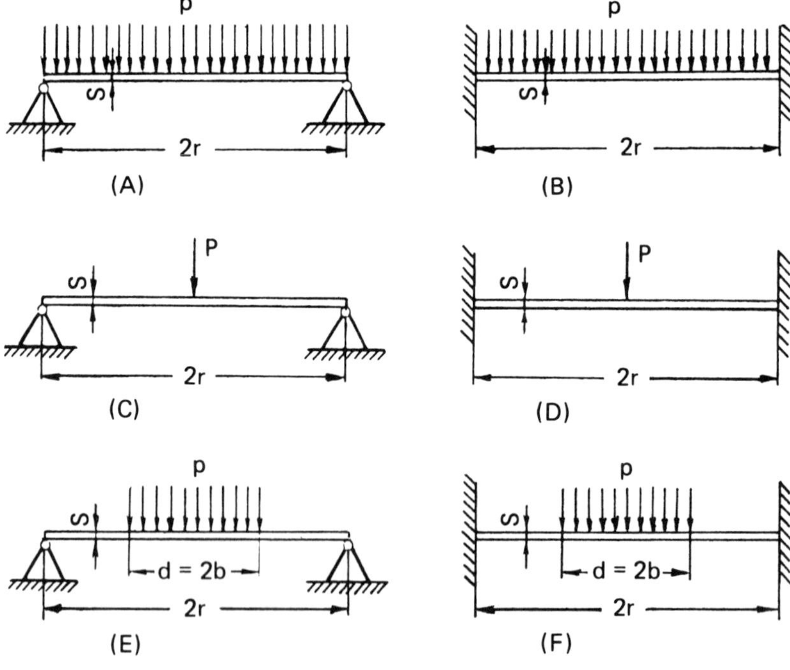

Figure 7.41. Design schemes of circular plates.

The stress at the edge of the plate is

$$\sigma_r = 0 \; ; \quad \sigma_\tau = 0.52p(r/s)^2 \tag{7.53}$$

and the deflection at the center is

$$f = \frac{0.7pr^4}{ES^2} \tag{7.54}$$

2. **Uniformly loaded flat circular plate with the edges clamped (Figure 7.41(B)):**
The stress at the center of the plate is

$$\sigma_r = \sigma_\tau = 0.49p(r/s)^2 \tag{7.55}$$

The stress at the edge of the plate is

$$\sigma = -0.75p(r/s)^2 \tag{7.56}$$

$$\sigma_\tau = \tilde{\mu}\sigma_r \tag{7.57}$$

At the periphery the lower fibers are compressed. Deflection at the center is

$$f = 0.17\,\frac{pr^4}{ES^3} \tag{7.58}$$

3. **Circular plate simply supported at the edge with a single load, P, at the center (Figure 7.41(C)):**
Maximum tensile stress is

$$\sigma_{max} = \frac{P}{S^2}\,(0.63\,\ln(r/s) + 1.16) \tag{7.59}$$

Deflection at the center of the plate is

$$f = 0.55\,\frac{Pr^2}{ES^3} \tag{7.60}$$

4. **Circular flat plate with the edges clamped with a single load, P, at the center (Figure 7.37D):**
Maximum tensile stress is

$$\sigma_{max} = \frac{P}{S^2}\,(0.63\ \ln(r/s) + 0.68) \tag{7.61}$$

Deflection at the center of the plate is

$$f = 0.218\,\frac{Pr^2}{ES^2} \tag{7.62}$$

5. **Uniformly loaded flat circular plate simply supported over the circumference: the load is applied on the circle diameter d = 2b, and with the ratio b/r = β (Figure 7.41(E)):**
The stress at the center of the plate is

$$\sigma_r = \sigma_\tau(1.5 - 0.26\beta^2 - 1.95\ \ln\beta)p(b/s)^2 \tag{7.63}$$

The stress at the edge of the plate is

$$\sigma_r = 0\ ;\quad \sigma_\tau = 0.525(2 - \beta^2)p(b/s)^2 \tag{7.64}$$

Deflection at the center of the plate is

$$f = (1.73 - 1.03\beta^2 + 0.68\beta^2\ \ln\beta)p\,\frac{r^2b^2}{ES^3} \tag{7.65}$$

6. **Circular flat plate clamped at the edges, uniformly loaded by pressure subjected on the circle radius, b; the ratio b/r = β (Figure 7.41(F)):**
The stress at the center of the plate is

$$\sigma_r = \sigma_\tau = 0.49(\beta^2 - 4\ \ln\beta)p(b/s)^2 \tag{7.66}$$

The stress at the edge of the plate is

$$\sigma_r = -0.75(2 - \beta^2)p(b/s)^2 \tag{7.67}$$

$$\sigma_\tau = \tilde{\mu}\sigma_r$$

Deflection at the center of the plate is

$$f = (0.68 - 0.51\beta^2 + 0.68\beta^2\ \ln\beta)\,\frac{pr^2b^2}{ES^3} \tag{7.68}$$

7. **Circular ring-shaped plate clamped at the internal contour; uniformly distributed moment load M_0 kg-cm/cm at the perimeter of the plate (Figure 7.42) is as follows:**
The stresses at internal contour are

Radial stress:

$$\sigma_r = \frac{12}{1.3 + \dfrac{0.7}{a^2}} \frac{M_0}{S^2}, \quad kg/cm^2 \tag{7.69}$$

Circumferential stress:

$$\sigma_T = \tilde{\mu}\sigma_r, \quad kg/cm^2 \tag{7.70}$$

Stresses at external contour:

$$\sigma_r = \frac{\sigma M_0}{S^2}, \quad kg/cm^2 \tag{7.71}$$

$$\sigma_T = \frac{6\left(1.3 - \dfrac{0.7}{a^2}\right)}{1.3 + \dfrac{0.7}{a^2}} \frac{M_0}{S^2} \tag{7.72}$$

Deflection at r = a is

$$f = 5.46 \frac{1 - \dfrac{2 \ln a}{a^2} - \dfrac{1}{a^2}}{1.3 + 0.7/a^2} \frac{M_0 a^2}{ES^3} \tag{7.73}$$

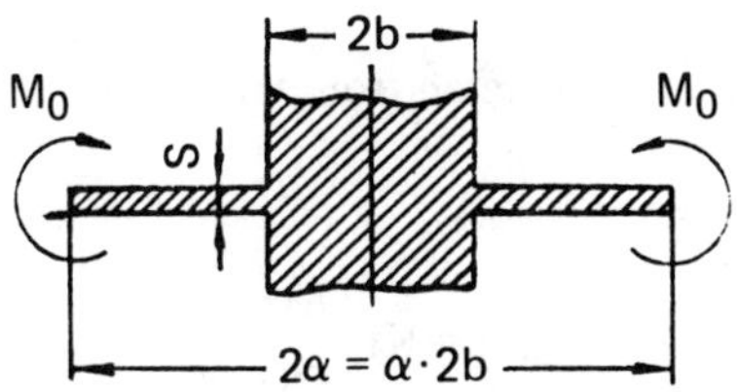

Figure 7.42. Ring-shaped plate loaded with end moments.

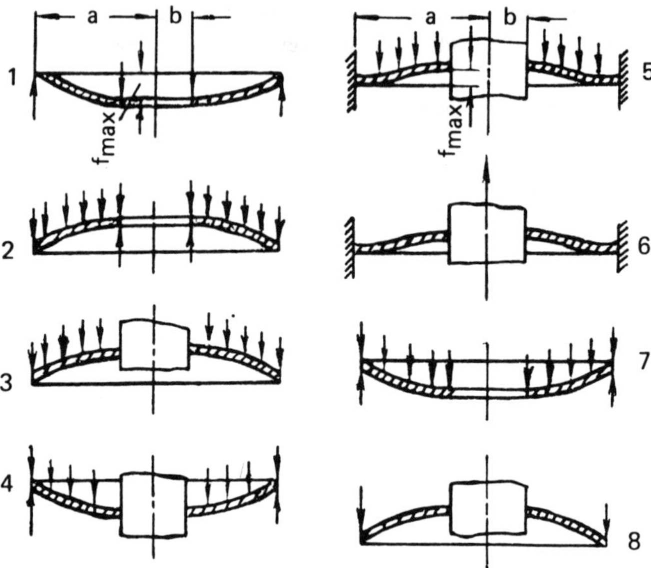

Figure 7.43. Schemes for design of ring-shaped plates.

Figure 7.43 gives various schemes for steel ring-shaped plates where $\widetilde{\mu} = 0.3$. Maximum stresses are determined from the following formulas:

For a uniformly distributed load,

$$\sigma_{max} = K \frac{qa^2}{S^2} \tag{7.74}$$

for a load distributed at the contour,

$$\sigma_{max} = K \frac{P}{S^2} \tag{7.75}$$

Maximum deflections are calculated from the following formulas:

For a uniformly distributed load,

$$f_{max} = K_1 \frac{qa^4}{ES^3} \tag{7.76}$$

For a load distributed at the contour,

$$f_{max} = K_1 \frac{Pa^3}{ES^3} \tag{7.77}$$

where p = resultant load distributed at the contour (kg)
q = load per unit surface area (kg/cm)
S = plate thickness (cm)

The values of coefficients K and K_1 for different ratios of a/ℓ, for the different types of loads illustrated in Figure 7.43 are described in Table 7.3.

Flat heads and circular plates used in the fabrication of process equipment are mostly uniformly loaded with the pressure of a fluid. In this case, the theoretical equation for the design of flat plates may be presented in the following form [20]:

$$\sigma = KP\left(\frac{d}{S}\right)^2 \tag{7.78}$$

from which the thickness of the flat head is

$$S = d\sqrt{\frac{PK}{\sigma_{allow}}} \tag{7.79}$$

where K = coefficient considering the rate and character of edge clamping of a circular head. Introducing the allowance constant, C, in Equation 7.79, we can write

$$S = d\sqrt{\frac{PK}{\sigma_{allow}}} + C \tag{7.80}$$

Depending on the design of the plate attachment to the vessel (Figure 7.44), the coefficient K can be obtained from the *ASME Boiler and Pressure Vessel Code, Unfired Pressure Vessels* [2].

Figure 7.45 shows three more widely used flat welded head designs used for shell connections. Of the three, Figure 7.45(A) is the most favorable. The inside-corner radius should be of sufficient size so that r = 1.0–1.5 s. With this size criterion, the stress concentration in the transition place is reduced, and the weld itself is carried out of the most stressed zone. Figure 7.45(B) is a good arrangement also. Its circular grooves decrease the stress concentration, although not as effectively as the smooth reducer

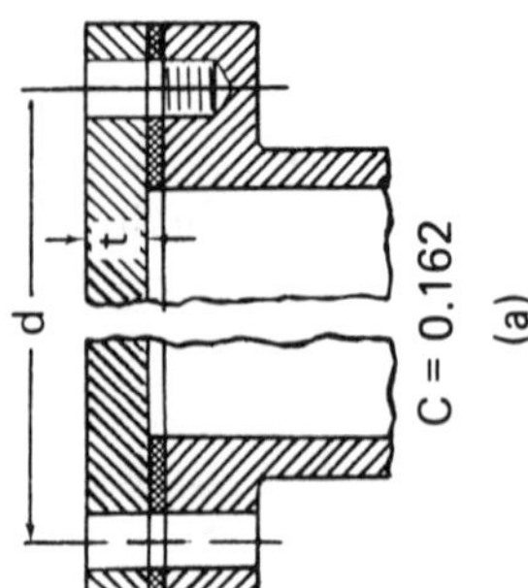
d
t
C = 0.162
(a)

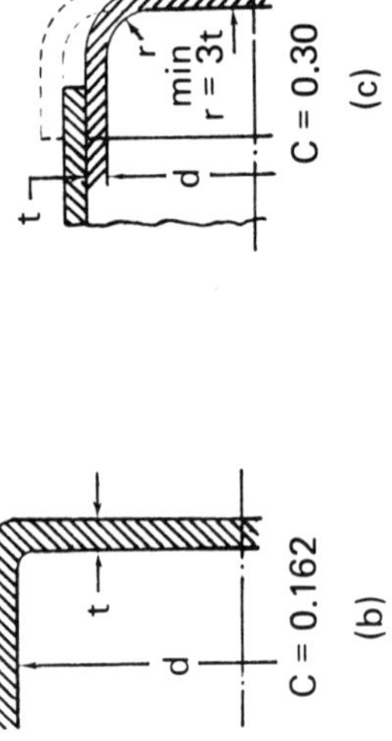
d
t
C = 0.162
(b)
t
d
min
r = 3t
r
t
C = 0.30
(c)

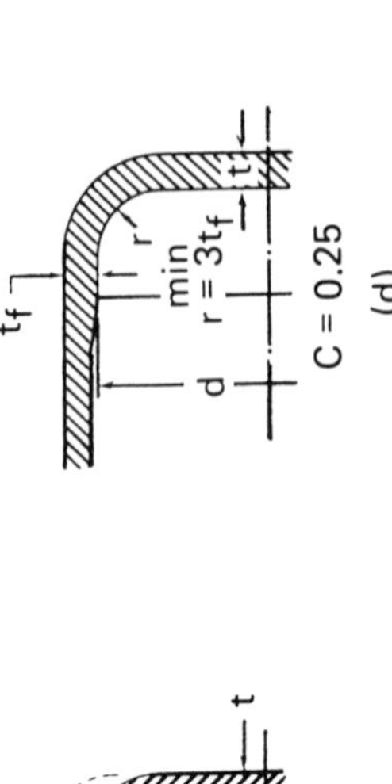
t_f
d
min
r = $3t_f$
r
t
C = 0.25
(d)

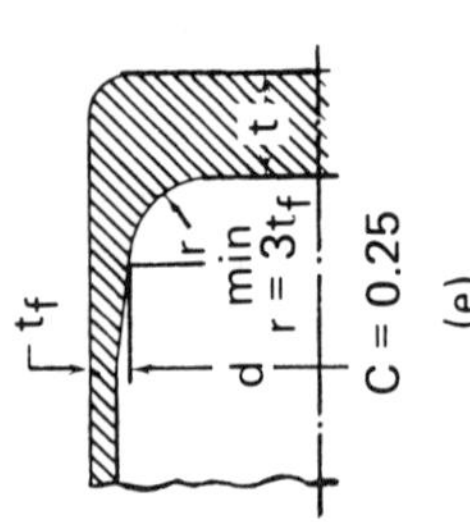
t_f
d
min
r = $3t_f$
r
t
C = 0.25
(e)

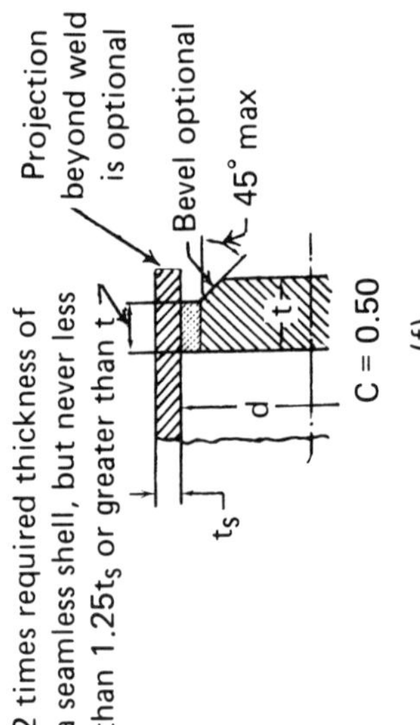
t_1 = 2 times required thickness of
a seamless shell, but never less
than 1.25t_s or greater than t
t_s
d
Projection
beyond weld
is optional
Bevel optional
45° max
t
C = 0.50
(f)

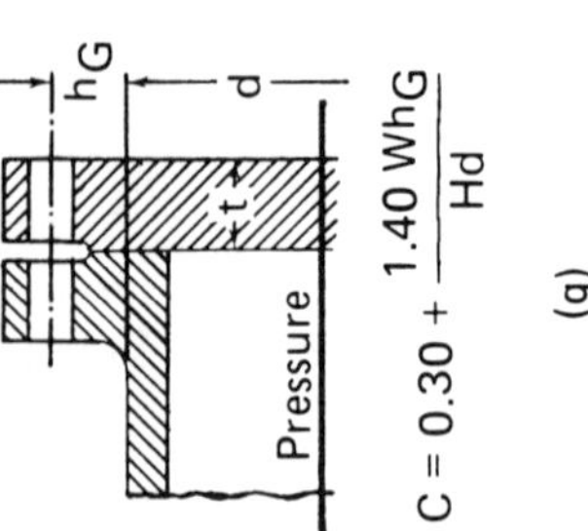
h_G
d
t
Pressure
C = 0.30 + $\frac{1.40\,Wh_G}{Hd}$
(g)

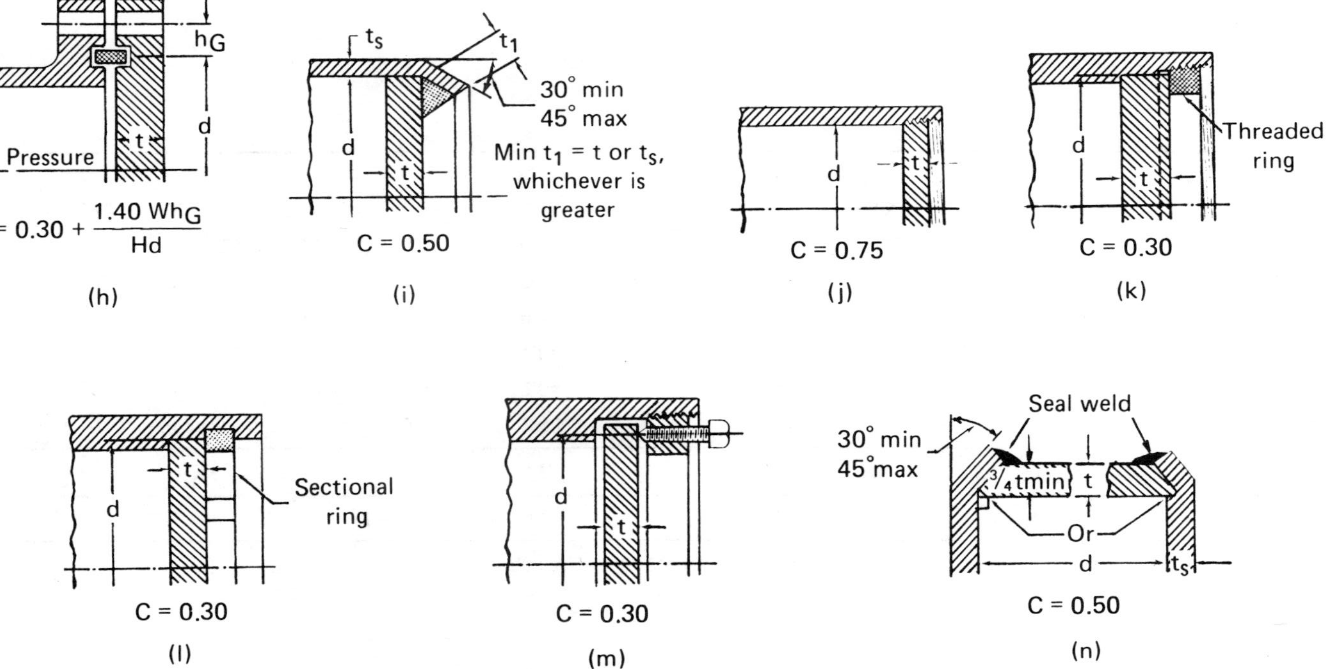

Figure 7.44. Different types of flat heads and covers as recommended in the 1956 edition of the ASME *Boiler and Pressure Vessel Code*, *Unfired Pressure Vessels* [21] (courtesy of the American Society of Mechanical Engineers).

Table 7.3. Values of Coefficients K and K_1

	1.25		1.50		2		3		4		5	
a/b	K	K_1	K	K_1	K	K_1	K	K_1	K	K_1	K	K_1
1	1.10	0.341	1.26	0.519	1.48	0.672	1.88	0.734	2.17	0.724	2.34	0.704
2	0.66	0.202	1.19	0.491	2.04	0.902	2.34	1.22	4.30	1.300	5.10	1.310
3	1.135	0.00231	0.41	0.0183	1.04	0.0938	2.15	0.293	2.99	0.448	3.69	0.546
4	0.122	0.00343	0.336	0.0313	0.74	0.125	1.21	0.291	1.45	0.417	1.59	0.492
5	0.090	0.00077	0.273	0.0062	0.71	0.329	1.54	0.110	2.23	0.179	2.80	0.234
6	0.115	0.00129	0.220	0.0064	0.405	0.0237	0.703	0.062	0.933	0.092	1.13	0.114
7	0.592	0.184	0.976	0.414	1.44	0.664	1.88	0.824	2.08	0.830	2.19	0.813
8	0.227	0.0051	0.427	0.0249	0.753	0.0877	1.205	0.209	1.514	0.293	1.745	0.350

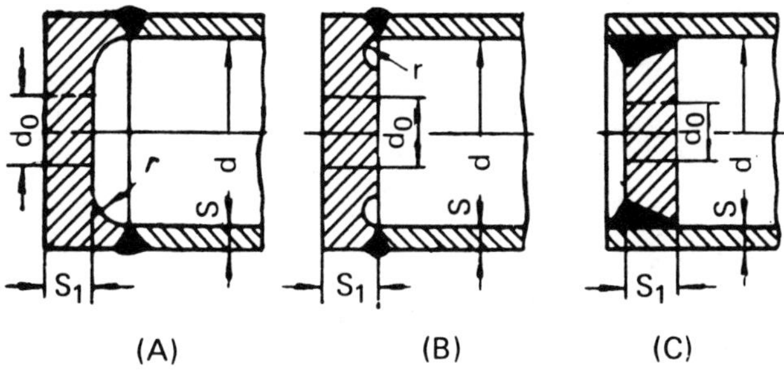

Figure 7.45. Examples of welded connection of heads to shells.

used in design (A). In designing such grooves, the minimum inside-corner radius should be

$$r \geqslant \frac{S}{3} \quad \text{and} \quad r \geqslant 0.5 \text{ cm}$$

and the minimum head thickness at the circular groove should be

$$S_2 \geqslant \frac{2}{3} S_1 \quad \text{and} \quad S_2 \geqslant S$$

The most unfavorable design is version (C). In this case, the weld is located in the zone of action of bending stresses, where the concentration of stresses is very pronounced. Joint efficiency should be considered in determining the head thickness for this version. Design criteria are given in Table 7.4. Designs such as version (C) generally are not recommended, and certainly not for pressures in excess of 35 kg/cm^2.

For apparatuses in which the shell and head are fabricated of different materials, design calculations should be based on the material having the poorer mechanical properties.

The values of allowable stresses and safety factors for flat heads are assumed to be less than those of cylindrical shells. The safety factors based on tensile strength, yield point and fatigue strength have the following values:

$$n_T = 4.7 \ , \quad n_y = 2.07 \ \text{and} \ n_f = 2.07$$

For determination of the thickness of flat heads connected to the vessel with bolts, the following formula derived from the method of ultimate loads is recommended [7].

Table 7.4. Maximum Allowable Efficiencies for Arc- and Gas-Welded Joints

(from the 1956 ASME Unfired-Pressure-Vessel Code, with permission of the American Society of Mechanical Engineers)

Type of Joint	Limitations	Basic Joint Efficiency (%)	Radiographed	Thermally Stress Relieved	Maximum Joint Efficiency (%)
Double-welded butt joint	None		No	No	80
Single-welded butt joint with backing strip	Longitudinal joints not over $1\frac{1}{4}$ in. No thickness limitation on circumferential joints.	80	No Yes Yes	Yes No Yes	85 90 95
Single-welded butt joint without backing strip	Circumferential joints only, not over $\frac{5}{8}$ in. thick	70	No No	No Yes	70 75
Double full-fillet lap joint	Longitudinal joints not over $\frac{3}{8}$ in. thick. Circumferential joints not over $\frac{5}{8}$ in. thick.	65	No No	No Yes	65 70
Single full-fillet lap joint with plug welds	Circumferential joints only, not over $\frac{5}{8}$ in. thick; attachment of heads not over 24 in. in outside diameter to shells not over $\frac{5}{8}$ in. thick.	60	No No	No Yes	60 65
Single full-fillet lap joint without plug welds	Only for attachment of heads convex to pressure to shells not over $\frac{5}{8}$ in. thick and for attachment of heads concave to pressure not over 24 in. in outside diameter to shells not over $\frac{1}{4}$ in. thick.	50	No No	No Yes	50 55

$$S_1 = 0.73d \sqrt{\frac{\sigma_y^b}{\sigma_y^h}\left[\frac{\ell z}{d} - \frac{2.2P}{\sigma_y^b}\left(\frac{D_g}{D_b}\right)\left(\frac{\ell}{D_b} - 0.14\right)\right]} \tag{7.81}$$

where σ_y^h = yield point of head material at operating temperature (kg/cm^2)
σ_y^b = yield point of bolt material at operating temperature
D_g = outside diameter of a gasket (cm)
D_b = diameter of bolting circle (cm)
ℓ = arm of a moment bending the head, $\ell = (D_b - D_g)/2$ (cm)
d_b = external bolt diameter (cm)
z = number of bolts

The design is shown in Figure 7.46.

The thickness of heads calculated from Equation 7.81 is less than that obtained from Equation 7.79; however, flat heads for cast vessels or those fabricated of brittle materials should be designed from Equation 7.79.

7.7.1 Design of Flat Circular Heads Operating Under Internal Pressure

Flat circular plates are recommended for use in the following cases:

1. for forged and cast apparatus when there is no need for conical heads and when elliptical or spherical heads are not desired;
2. for welded apparatuses operating under pressure at $d < 400$ mm;
3. for welded apparatuses when there is no need for conical heads and when elliptical heads are not desired; in these cases reinforcement with ribs are recommended; and
4. for welded vertical apparatuses resting on a solid foundation and operating under atmospheric pressure or under open conditions.

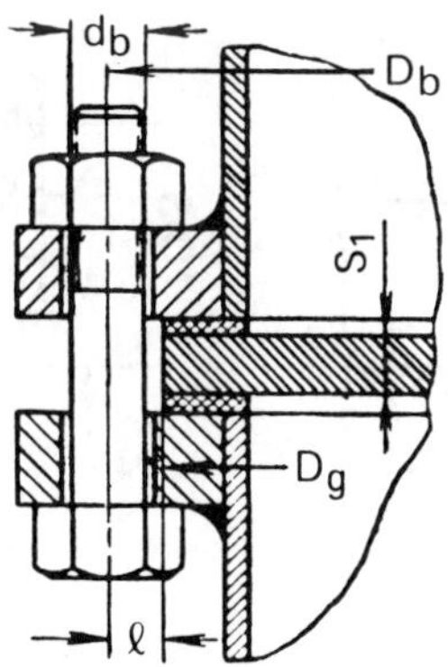

Figure 7.46. Sketch of a design of heads and flanges.

Basic designs of flat circular heads operating under pressure are presented in Figure 7.47. As shown, there are three types:

- **Type I.** The left-hand side is a design of a welded head, and the right-hand side is a design of forged or cast head.
- **Type II.** This is a more rational design of a welded head; the thickness of its forging blank is less than that of type I.
- **Type III.** This is a design of a welded head that can be used for operating at pressures up to 1.6 MN/m^2.

The basic designs of welded flat circular unflanged heads for apparatuses operating at atmospheric pressure or under filling up conditions are presented in Figure 7.48.

The basic designs of welded circular flat heads reinforced with radial ribs are given in Figure 7.49.

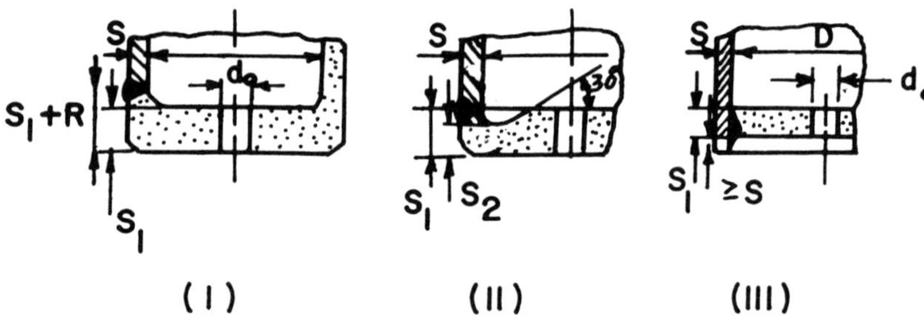

Figure 7.47. Typical designs of flat circular heads operating under pressure. $R \geqslant S/3$ but not less than 5 mm; $S_2 \geqslant {}^2/_3\, S^3$ but not less than S; $d_0 \leqslant 0.6$ d.

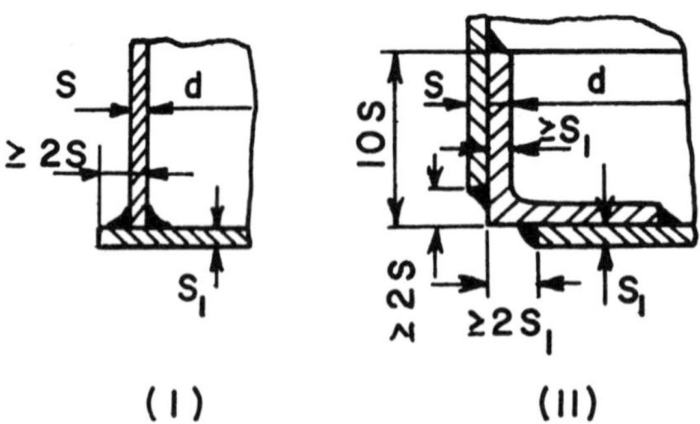

Figure 7.48. Typical designs of flat circular heads operating under atmospheric pressure or under filling up conditions.

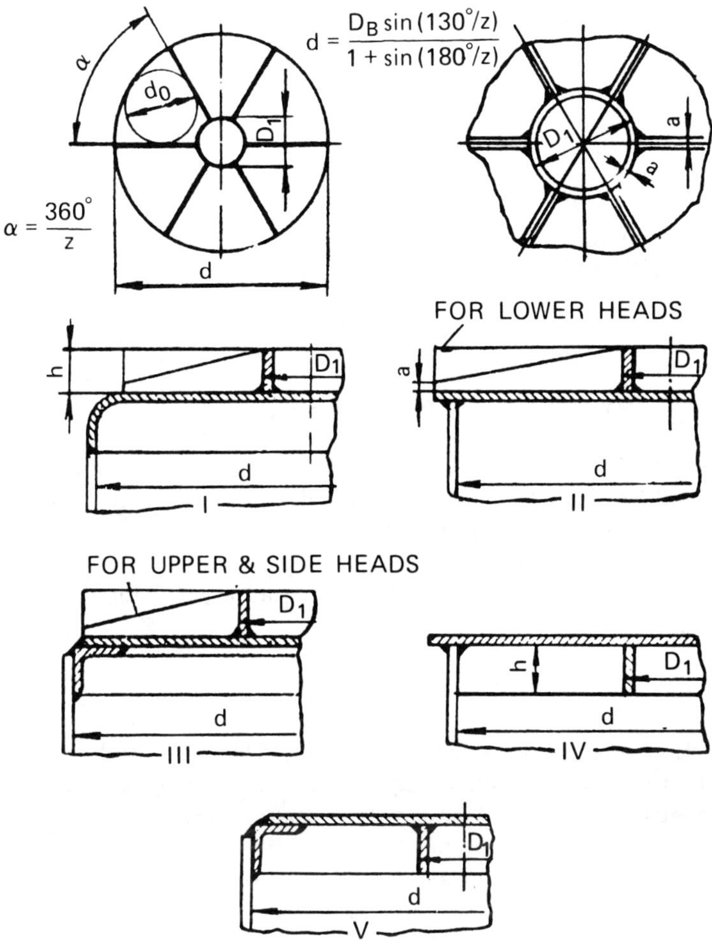

Figure 7.49. Typical designs of welded flat circular heads reinforced with ribs: I–III = head reinforced from outside; IV, V = heads reinforced from inside.

The design thickness of a flat head without reinforcement with ribs S_1 (Figure 7.47) operating under internal pressure is determined from the following formula [6]:

$$S_1' = \frac{0.93S}{1 - 0.43\dfrac{d_0}{d}} \sqrt{\frac{0.19}{\sigma_{b,allow}}\left(\frac{d}{S}\right)^2 - 1} + C, \text{ m} \tag{7.82}$$

If the following conditions are fulfilled,

$$\frac{P}{\sigma_{allow}} \geqslant 11.2(S/d)^2 \tag{7.83}$$

$$\frac{P}{\sigma_{allow}} \leqslant \frac{S}{d+S} \tag{7.84}$$

If condition 7.83 is not fulfilled, the following formula may be used:

$$S_1' = \frac{S}{1 - 0.43\dfrac{d_0}{d}} + C, \text{ m} \tag{7.85}$$

If condition 7.84 is not fulfilled, the thickness of the cylinder wall, S, joining with the head should be increased to fulfill this condition. Then calculate the head thickness from either Equation 7.82 or 7.85, depending on fulfillment of condition 7.83.

Coefficient k = 0.85 for types I and II; and k = 0.5 for type III. For allowance,

$$C = C_1 + C_2 + C_3, \; C_3 = 2 - 5 \text{ mm} \tag{7.86}$$

The determination of design pressure, allowable stress and temperature is similar to the determination of cylindrical shells joining with heads. If the heads contain no holes, then let $d_0 = 0$ in Equations 7.82 and 7.85.

The assumed thickness of the head should be checked for stress at the hydrostatic test in the vessel from the following formula:

$$\sigma = \frac{d^2P'}{5.25(S_1 - C)^2\left[1.15\left(1 - 0.86\dfrac{d_0}{d}\right) + \left(\dfrac{S-C}{S_1-C}\right)^2\right]} \leqslant \frac{\sigma'}{n_T} \tag{7.87}$$

If the following conditions are fulfilled,

$$\frac{S_1}{S} \geqslant \frac{1}{1 - 0.43\dfrac{d_0}{d}} \tag{7.88}$$

$$\frac{S_1}{S} \leqslant \frac{0.93}{1 - 0.43\dfrac{d_0}{d}}\sqrt{\frac{0.55}{\left(1 + \dfrac{S}{d}\right)\dfrac{S}{d}} - 1} \tag{7.89}$$

And from the formula

$$\sigma = \frac{d^2P'}{11.2\left(1 - 0.86\,\dfrac{d_0}{d}\right)(S_1 - C)^2} \leqslant \frac{\sigma'}{n_T} \tag{7.90}$$

if condition 7.88 is not fulfilled.

The value of C in (s − C) is taken to be the same as for a cylindrical wall and, in (s_1 − C), as for a head. If condition 7.88 is not fulfilled, the stress in the head at the hydrostatic test may be determined from Equation 6.37.

Example 7.11.

Determine the thickness of a flat circular head (Figure 7.47, type I, right-hand side) for a shell of the vessel designed in Example 6.5 (Figure 6.24) for the following data: the material of the head is alloy steel; $d = 0.6$ m; $d_0 = 0.05$ m; the head is forged integral with the shell; $\sigma_{allow} = \sigma_{b,allow} = 208 \times 10^6$ N/m^2; $p = 40$ MN/m^2; $p' = 50$ MN/m^2; $S = 67.5 \times 10^{-3}$m.

Solution

Checking conditions 7.83 and 7.84,

$$\frac{P}{\sigma_{allow}} = \frac{40 \times 10^6}{208 \times 10^6} = 0.192 > 11.2\left(\frac{S}{d}\right)^2$$

$$= 11.2\left(\frac{67.5 \times 10^{-3}}{0.6}\right)^2 = 0.142$$

$$\frac{P}{\sigma_{allow}} = 0.192 < 2.9\,\frac{S}{d + S} = 2.9\,\frac{67.5 \times 10^{-3}}{0.6 + 67.5 \times 10^{-3}} = 0.293$$

The design head thickness (at K = 0.85) is determined from Equation 7.82:

$$S_1' = \frac{0.93\,S}{1 - 0.43\,\dfrac{d_0}{d}}\sqrt{\frac{0.19P}{\sigma_{b,allow}K}\left(\frac{d}{S}\right)^2 - 1} + C$$

$$= \frac{0.93 \times 67.5 \times 10^{-3}}{1 - 0.43\,\dfrac{0.05}{0.6}}\sqrt{\frac{0.19 \times 40 \times 10^6}{208 \times 10^6 \times 0.85}\left(\frac{0.6}{0.0675}\right)^2 - 1} + C$$

$$= 101 \times 10^{-3} + C, \text{ m}$$

Allowance C at $C_1 = 0$ and $C_2 = 0$ from Equation 7.86; $C = C_3 = 4$ mm. Hence,

$$S = S_1 + C = (101 + 4) \times 10^{-3} = 105 \times 10^{-3}\ \text{m}$$

We next check the stress in the head at the hydrostatic test with water. As a preliminary check, conditions 7.88 and 7.89 are examined:

$$\frac{S_1}{S} = \frac{105}{67.5} = 1.56 > \frac{1}{1 - 0.43\dfrac{d_0}{d}} = \frac{1}{1 - 0.43(0.05/0.6)} = 1.04$$

$$\frac{S_1}{S} = 1.56 < \frac{0.93}{1 - 0.43\left(\dfrac{d_0}{d}\right)}\sqrt{\frac{0.55}{\left(1 + \dfrac{S}{d}\right)\dfrac{S}{d}} - 1}$$

$$= \frac{0.93}{1 - 0.43(0.05/0.6)}\sqrt{\frac{0.55}{\left(1 + \dfrac{0.0675}{0.6}\right)\left(\dfrac{0.0675}{0.6}\right)} - 1} = 1.78$$

The stress in the head is determined from Equation 7.87:

$$\sigma = \frac{d^2P'}{5.25(S_1 - C)^2\left[1.15\left(1 - 0.86\dfrac{d_0}{d}\right) + \dfrac{(S - C)^2}{(S_1 - C)^2}\right]}$$

$$= \frac{(0.6)^2 50 \times 10^6}{5.25(105 - 4)^2 10^{-6}\left[1.15\left(1 - 0.86\dfrac{0.05}{0.6}\right) + \dfrac{(67.5 - 2.95)^2 10^{-6}}{(105 - 4)^2 10^{-6}}\right]}$$

$$= 230 \times 10^6\ \text{N/m}^2 < \sigma'/n_T = 700 \times 10^6/2.6 = 269 \times 10^6\ \text{N/m}^2$$

The design thickness of a flat head reinforced with ribs is determined from Equation 6.84, in which the value, B, is assumed to be equal to the diameter of the inscribed circle between ribs and the contour of the head (Figure 7.49). The assumed thickness of the flat head should be checked for stress at the hydrostatic test of the vessel from Equation 6.85.

An approximate design of radial reinforcing ribs (Figure 7.49) may be done by the following simplified method. It is assumed that half the load on the head (from internal pressure) is absorbed only by the ribs (this does not account for the head). Then the design section modulus of the rib, W_r', assuming it to be a beam simply supported at the edges of the head, is

$$W'_r = \frac{0.065d^3P}{z\sigma_{b,allow}}, \text{ m}^3 \tag{7.91}$$

where z = number of ribs

After determining W'_r, the design cross section of the reinforcing rib can be determined. In particular, for a rectangular section with a thickness: height ratio of 1:5, the design thickness of the rib can be calculated from Equation 6.89.

The assumed values of the cross sections of the ribs should be checked for maximum stresses, accounting for the presence of the head under operating conditions and at hydrostatic test of an apparatus similar to the design of rectangular flat walls (see section 6.3). We first determine the section modulus of the cross section of the head element with a reinforcing rib, $W_{r.w}$ (refer to Figure 6.30) from Equation 6.90. The value is

$$x = \frac{\pi D_1}{z}$$

The maximum stress in the reinforcing ribs under operating conditions should correspond to the following condition:

$$\sigma = \frac{0.13d^3P}{zW_{r,w}} \leqslant 1.2\sigma'_T \tag{7.92}$$

The maximum stress in the reinforcing ribs at hydrostatic test should correspond to the following condition:

$$\sigma = \frac{0.13d^3P'}{zW_{r,w}} \leqslant \frac{\sigma_y}{1.1}, \text{ N/m}^2 \tag{7.93}$$

If the conditions 7.92 and 7.93 are not fulfilled, the assumed total cross section should be increased accordingly.

Example 7.12.

Determine the thickness of the upper flat circular flanged head and cross section of reinforcing ribs (Figure 7.49, type I) for a vertical welded vessel of the following data: the material is carbon steel; $p = 0.3$ MN/m²; $d = 1$ m; $z = 6$; $\sigma_{b,allow} = 146 \times 10^6$ N/m²; $p' = 0.45$ MN/m²; $C_1 = 1$ mm; $C_2 = 0$.

Solution

The design diameter of the head between the ribs (Figure 7.49) is

$$d_0 = \frac{d \sin(180°/z)}{1 + \sin(180°/z)} = \frac{1 \times 0.5}{1 + 0.5} = 0.333 \text{ m}$$

We assume $D_1 = 0.33$ m.

The design wall thickness of the head from Equation 6.84 is

$$S' = 0.5d'_0 \sqrt{\frac{P}{\sigma_{b,allow}}} + C = 0.5 \times 0.33 \sqrt{\frac{0.3 \times 10^6}{146 \times 10^6}} + C$$

$$= 7.54 \times 10^{-3} + C, \text{ m}$$

For allowance C (at $C_3 = 0.8$ mm),

$$C = C_1 + C_2 + C_3 = (1 + 0 + 0.8) \times 10^{-3} = 1.8 \times 10^{-3} \text{ m}$$

Hence,

$$s = (7.54 + 1.8) \times 10^{-3} = 9.34 \times 10^{-3} \text{ m}$$

We assume s = 10 mm.

Checking the stress in the wall of the head at hydrostatic test of the apparatus from Equation 6.85,

$$\sigma = \frac{0.25d^2P'}{(S - C)^2} = \frac{0.25(0.333)^2 0.45 \times 10^6}{(10 - 1.8)^2 \times 10^{-6}}$$

$$= 185 \times 10^6 \text{ N/m}^2 < \sigma_y/1.2 = \frac{240 \times 10^6}{1.2} = 200 \times 10^6 \text{ N/m}^2$$

The design section modulus of cross section of reinforcing ribs from Equation 7.91 is

$$W'_r = \frac{0.065d^3P}{z\sigma_{b,allow}} = \frac{0.065 \times 1^3 \times 0.3 \times 10^6}{6 \times 146 \times 10^6} = 22.3 \times 10^{-6} \text{ m}^3$$

We assume the reinforcing ribs to be of rectangular cross section. With the thickness to height ratio equal to 1:5, we determine the rib's design thickness from Equation 6.89:

$$S_r' = 0.62 \sqrt[3]{W_r'} = 0.62 \sqrt[3]{22.3 \times 10^{-6}} = 17.5 \times 10^{-3}\ \text{m}$$

We assume the rib sizes are $S_r = 18$ mm; $h_2 = S \times S_2 = 5 \times 18 = 90$ mm. Checking for maximum stress in ribs, preliminary determination,

$$F_r = S_r h_r = 18 \times 90 \times 10^{-6} = 16.2 \times 10^{-4}\ \text{m}^2$$

$$J_r = \frac{S_r h_r^3}{12} = \frac{18(90)^3}{12} 10^{-12} = 109 \times 10^{-8}\ \text{m}^4$$

$$F_w = \frac{\pi D_1}{z}(S - C) = \frac{\pi \times 0.33}{6}(10 - 1.8)10^{-3} = 14.2 \times 10^{-4}\ \text{m}^2$$

$$J_w = \frac{\pi D_1 (S - C)^2}{12z} = \frac{\pi \times 0.33(10 - 1.8)10^{-9}}{6 \times 12} \approx 0.8 \times 10^{-8}\ \text{m}^4$$

From formula 6.93,

$$y = \frac{F_r h_r - F_w(S - C)}{2(F_r + F_w)} = \frac{16.2 \times 10^{-4} \times 0.09 - 14.2 \times 10^{-4}(10 - 1.8)10^{-3}}{2(16.2 + 14.2)10^{-4}}$$

$$= 22.1 \times 10^{-3}\ \text{m}$$

From Equation 6.91,

$$W_{rw} = \frac{J_r + J_w + F_r(0.5h_r - y)^2 + F_w[y + 0.5(S - C)]^2}{h_r - y}$$

$$= \{109 \times 10^{-8} + 0.8 \times 10^{-8} + 16.2 \times 10^{-4}(0.5 \times 0.09 - 22.1 \times 10^{-3})^2$$

$$+ 14.2 \times 10^{-4} \times [22.1 \times 10^{-3} + 0.5(10 - 1.8) \times 10^{-3}]^2\}$$

$$\div (0.09 - 22.1 \times 10^{-3}) = 43.2 \times 10^{-6}\ \text{m}^3$$

The maximum stress in the ribs under operating conditions from Equation 7.92 is

$$\sigma = \frac{0.13 d^3 P}{z W_{r,w}} = \frac{0.13 \times 1^3 \times 0.3 \times 10^6}{6 \times 43.2 \times 10^{-6}} = 151 \times 10^6\ \text{N/m}^2$$

$$< 1.2\sigma_{b,allow} = 1.2 \times 146 \times 10^6 - 175 \times 10^6\ \text{N/m}^2$$

The maximum stress in ribs at hydrostatic test from Equation 7.93 is

$$\sigma = \frac{0.13d^3P'}{zW_{r,w}} = \frac{0.13 \times 1^3 \times 0.45 \times 10^6}{6 \times 43.2 \times 10^{-6}} = 226 \times 10^6 \text{ N/m}^2$$

$$> \sigma_y/1.1 = 240 \times 10^6/1.1 = 218 \times 10^6 \text{ N/m}^2$$

that is, 3.5% more.

7.8 FLAT ANNULAR HEADS

Flat annular heads supported at the internal and external edges operating under internal pressure, p, are used in the fabrication of process equipment in the form of heads proper, partitions, trays for distillation columns, etc.

Annular heads may be either unreinforced or reinforced (usually with radial ribs). The design thickness of an unreinforced flat annular plate may be calculated from the following approximate formula:

$$S' = 0.25(d_1 - d_2)\sqrt{\frac{P}{\sigma_{b,allow}\phi}} + C, \text{ m} \tag{7.94}$$

where d_1 = the greater head diameter (m)
d_2 = the smaller head diameter (m)
$\phi = (t - d_0)/t$ = coefficient of head weakening with holes (only for perforated heads)

The design pressure, p, is assumed to correspond to that of cylindrical shells. For a perforated head such as sieve plates, the pressure is the sum of the hydrostatic pressure of the liquid column plus the specific load due to gravity.

The design thickness of a reinforced circular head with radial ribs, S′, may be determined from the following approximate formula:

$$S' = 0.5\frac{d \sin\left(\frac{180°}{z}\right)}{1 + \sin\left(\frac{180°}{z}\right)}\sqrt{\frac{P}{\sigma_{b,allow}\phi}} + C, \text{ m} \tag{7.95}$$

The design number of radial ribs is

$$z > \frac{180°}{\sin^{-1}\left(\frac{d - d_2}{d + d_2}\right)} \tag{7.96}$$

Radial ribs may be calculated approximately as beams simply supported at the two edges of the head absorbing the load on the corresponding element of the head.

The assumed head thickness should be checked for stress in the head if it is subjected to the hydrostatic test as follows:

For unreinforced heads,

$$\sigma = \frac{0.0625(d - d_2)^2 P'}{(S - C)^2} \leqslant \frac{\sigma_y}{1.2}\ \text{N/m}^2 \tag{7.97}$$

For reinforced heads,

$$\sigma = \frac{0.25 d^2 \sin^2\left(\frac{180^\circ}{z}\right) P'}{(S - C)^2 \left[1 + \sin\left(\frac{180^\circ}{z}\right)\right]^2} \leqslant \frac{\sigma_y}{1.2}\ \text{N/m}^2 \tag{7.98}$$

Example 7.13.

Determine the thickness of a reinforced circular sieve plate with radial ribs simply supported at the internal and external edges, as well as the number and size of ribs from the following data: the material of the plate is copper; $\sigma_{b,allow} = 33.4 \times 10^6$ N/m^2; d = 1 m; $d_2 = 0.2$ m; liquid column on the plate $H_e = 20$ mm, liquid density $\rho_e = 1 \times 10^3$ kg/m^3; C = 0.25 mm; perforation $d_0 = 1$ mm; t = 4 mm.

Solution

The design pressure from the mass of the plate is 10 kg/m^2:

$$P \simeq 10(\rho_e H_e + m) = 10(1 \times 10^3 \times 0.02 + 10) = 300\ \text{N/m}^2$$

The design number of ribs is determined from Equation 7.96:

$$z > \frac{180^\circ}{\sin^{-1}\left(\frac{d - d_2}{d + d_2}\right)} = \frac{180^\circ}{\sin^{-1}\left(\frac{1 - 0.2}{1 + 0.2}\right)} = \frac{180^\circ}{41.8^\circ} = 4.3$$

We assume z = 8.

The weakening coefficient of the plate is

$$\phi = \frac{t - d_0}{t} = \frac{4 - 1}{4} = 0.75$$

The design plate thickness from Equation 7.95 is

$$S = 0.5\frac{d\sin(180^\circ/z)}{1+\sin(180^\circ/z)}\sqrt{\frac{P}{\sigma_{b,allow}\phi}} + C$$

$$= 0.5\frac{1 \times \sin 22.5^\circ}{1+\sin 22.5}\sqrt{\frac{300}{33.4 \times 10^6 \times 0.75}} + 0.25 \times 10^{-3}$$

$$= 0.73 \times 10^{-3}\ \text{m}$$

We assume s = 1 mm. The load absorbed by one rib is

$$P = \frac{0.785(d^2 - d_2^2)p}{z} = \frac{0.785(1^2 - 0.2^2)300}{8} = 28.2\ \text{N}$$

Section modulus of the rib is

$$W' = \frac{P(d - d_2)}{2 \times 8\sigma_{b,allow}} = \frac{28.2(1 - 0.2)}{2 \times 8 \times 33.4 \times 10^6} = 0.0423 \times 10^{-6}\ \text{m}^3$$

Assuming rib thickness s_2 = 2 mm, we calculate its design height as follows:

$$h_r' = \sqrt{\frac{6W_r'}{S_r}} = \sqrt{\frac{6 \times 0.0423 \times 10^{-6}}{2 \times 10^{-3}}} = 11.3 \times 10^{-3}\ \text{m}$$

Assume h_r = 12 mm.

7.9 FLAT RECTANGULAR HEADS

For the design of flat rectangular heads, the data given in Section 6.3 should be used. Flat rectangular heads for apparatuses operating under pressure should have profiles similar to flat flanged circular heads. This is shown in Figure 7.50.

The design of flat rectangular heads for welded apparatuses operating under atmospheric pressure or at filling up conditions is similar to Figure 7.48. The rectangular flat plate uniformly loaded and simply supported at the edges has the maximum stress in its center (Figure 7.51).

$$\sigma = \frac{0.75Pb^2}{S^2(1 + 1.61\epsilon^3)} \tag{7.99}$$

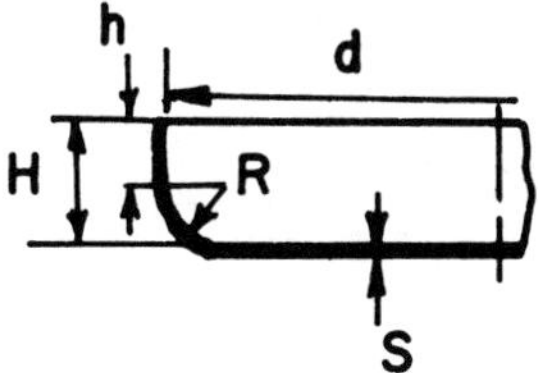

Figure 7.50. Flat rectangular head (where $R \geqslant 5s$).

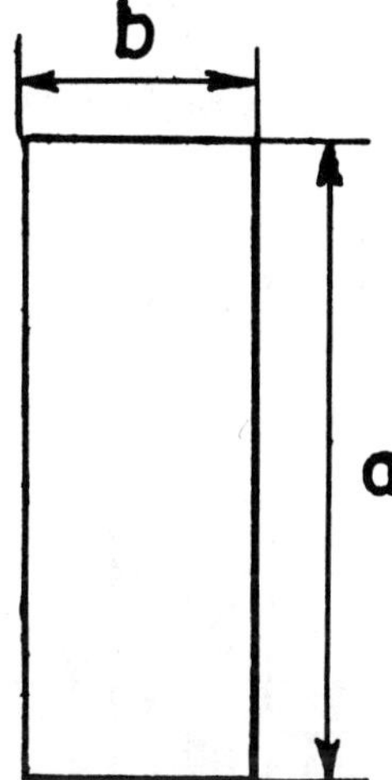

Figure 7.51. Rectangular plate uniformly loaded and simply supported at the edges.

where

$$\frac{b}{a} = \epsilon \leqslant 1 \quad \text{and} \quad P = pab$$

and the maximum vertical deflection is

$$f_{max} = \frac{0.1422Pb^4}{ES^3(1 + 2.21\epsilon^3)} \tag{7.100}$$

For a rectangular plate uniformly loaded and simply supported at the edges, the maximum stress (in the center) is

$$\sigma = \beta \frac{P}{S^2} \tag{7.101}$$

where β is determined from Table 7.5 and Figure 7.52. For a rectangular plate uniformly loaded with a clamped edge, refer to Figure 7.53. The maximum stress at the center of the plate is

$$\sigma_{max} = \frac{0.5Pb^2}{S^2(1 + 0.623a^6)} \tag{7.102}$$

The deflection of the rectangular clamped plate with width b (at $b/a \leqslant 0.5$) is

$$f_{max} = \frac{Pb^4}{384\ N} \tag{7.103}$$

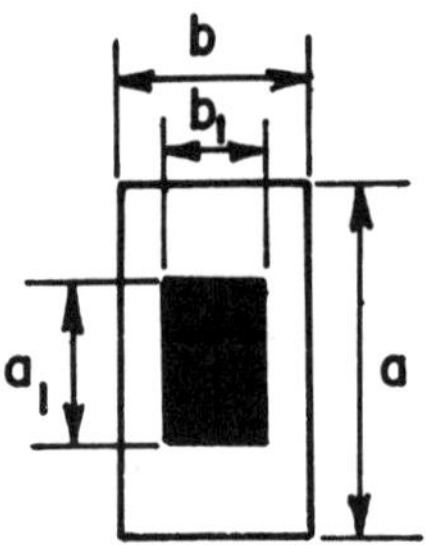

Figure 7.52. Sketch for use with Table 7.5.

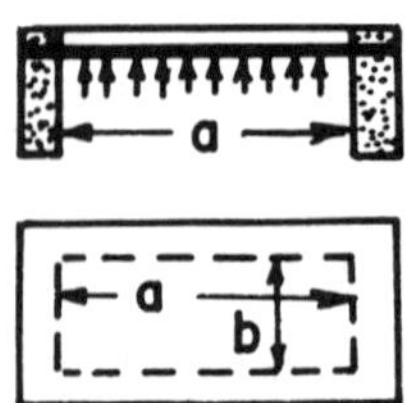

Figure 7.53. Rectangular plate uniformly loaded with a clamped edge.

Table 7.5 Values of Parameter β in Equation 7.101[a]

$\frac{a_1}{a}$	$\frac{b_1}{a}$: b = a						b = 1.4a						b = 2a					
	0	0.2	0.4	0.6	0.8	1.0	0	0.2	0.4	0.8	1.2	1.4	0	0.4	0.8	1.2	1.6	2.0
0	–	1.82	1.38	1.12	0.93	0.76	–	2.00	1.55	1.12	0.84	0.75	–	1.64	1.20	0.97	0.78	0.64
0.2	1.82	1.28	1.08	0.90	0.76	0.63	1.78	1.43	1.23	0.95	0.74	0.64	1.73	1.31	1.03	0.84	0.68	0.57
0.4	1.39	1.07	0.84	0.72	0.62	0.52	1.39	1.13	1.00	0.80	0.62	0.55	1.32	1.08	0.88	0.74	0.60	0.50
0.6	1.12	0.90	0.72	0.60	0.52	0.43	1.10	0.91	0.82	0.68	0.53	0.47	1.04	0.90	0.76	0.64	0.54	0.44
0.8	0.92	0.76	0.62	0.51	0.42	0.36	0.90	0.76	0.68	0.57	0.45	0.40	0.87	0.76	0.63	0.54	0.44	0.38
1.0	0.76	0.63	0.52	0.42	0.35	0.30	0.75	0.62	0.57	0.47	0.38	0.33	0.71	0.61	0.53	0.45	0.38	0.30

[a] At $\tilde{\mu} = 0.3$.

At $b/a > 0.5$,

$$f_{max} = \frac{Pb^4}{N\pi^4} \times \frac{1}{3 + 2a^2 + 3a^4} \quad (7.104)$$

where

$$N = \frac{ES^3}{12(1 - \tilde{\mu}^2)} \approx \frac{ES^3}{11} \quad (7.105)$$

E is the modulus of elongation.

NOMENCLATURE

A	= parameter defined by Equation 7.6
b	= ring width (mm)
D'	= design diameter of head (m)
d	= diameter (mm)
F	= area (m^2)
f	= deflection (m)
H	= head height (mm)
h_b	= height of domed portion of head (m)
K	= dimensionless coefficient (see Equation 7.2)
K_1	= coefficient for blinds and heads (see Equation 7.4)
K'	= coefficient (see Table 7.2)
M_0	= moment load (kg-cm/cm)
m	= resistance coefficient (see Figure 7.23)
n	= safety factor
P	= pressure (N/m^2)
Q	= total load (kg)
q	= point load per unit length or per unit surface area (N/m or N/m^2)
R,r	= radius (m)
S	= thickness (mm)
V	= volume (m^3)
W'_r	= design section modulus (m^3)
X	= section modulus of cross section of head element (m^3)
y	= overstress coefficient
z	= number of bolts
α	= apex angle (°)
β	= angle of repose (°)
ϵ	= parameter defined in Equation 7.99
ξ	= flange radius (m)

σ_r = radial stress (kg/cm^2)
$\sigma_{b,allow}$ = allowable bending stress (kg/m^2)
$\sigma_{t,allow}$ = allowable tensile stress (kg/m^2)
σ_T = circumferential stress (kg/cm^2)
ψ = joint efficiency
ψ' = strength factor of conical wall

REFERENCES

1. "Lukens Spun Heads, Pricing and Engineering Data," Lukens Steel Company, Coatesville, PA (1955).
2. *ASME Boiler and Pressure Vessel Code, Section VIII, Rules for Construction of Unfired Pressure Vessels* (New York: American Society of Mechanical Engineers, 1956).
3. Brownell, L. E., and E. H. Young. *Process Equipment Design* (New York: John Wiley & Sons, Inc., 1959).
4. Shwartzman, M. "Rastchet na prochnost kotelnykh agregatov," *Gosenergoizdat, M.* (1957).
5. Troyansky, E. A. "Metally kotlostroenia i rastchet prochnosti kotlov," *Gosenergoizdat, M.* (1956).
6. "Normy rastcheta elementov parovykh na prochnost," *Gosenergoizdat, M.* (1957).
7. Timoshenko, S. *Theory of Plates and Shells* (New York: McGraw-Hill Book Co., 1940).
8. Watts, G. W., and H. A. Lang. "The Stresses in a Pressure Vessel with a Flat Head," *Trans. Am. Soc. Mech. Eng.* 74:1083 (1952).
9. Strum, R. G., and H. I. O'Brien. "Computing Strength of Vessels Subjected to External Pressure," *Trans. Am. Soc. Mech. Eng.* 69:353 (1947).
10. Bergman, E. O. "The New-type Code Chart for Design of Vessels under External Pressure," *Trans. Am. Soc. Mech. Eng.* 74(10):1217 (1952).
11. "Sosudy, I., apparaty. Edinye Vsesousaye normy i metody rascheta ha prochnost uslov i detalei," RTM 42-62 M, *Niikhimmash* (1962).
12. Hohn, E. "Strain in Dished Heads," *Mech. Eng.* 49(6):632 (1927).
13. Coates, W. M. "State of Stress in Full Heads of Pressure Vessels," *Trans. Am. Soc. Mech. Eng.* 52:APM-52-12 (1930).
14. Greene, T. W. "Design of Ellipsoidal Heads for Pressure Vessels," *Trans. Am. Soc. Mech. Eng.* 51:APM-51-13 (1929).
15. Boardman, H. C. "Optimum Torispherical Pressure Vessel Heads," *The Water Tower* 35(3):8 (1949).
16. Watts, G. W., and H. A. Lang. "The Stresses in a Pressure Vessel with a Hemispherical Head," *Trans. Am. Soc. Mech. Eng.* 75(1):83 (1953).
17. Vikhman, Iu. L., I. F. Babitzki and S. I. Volvson. "Raschet i konstruirovanie neftezavodskoi apparatury, *GNTI, M.* (1953).
18. Watts, G. W., and H. A. Lang. "Stresses in a Pressure Vessel with a Conical Head," *Trans. Am. Soc. Mech. Eng.* 74:315 (1952).
19. O'Brien, H. L., E. Witterstrom, M. G. Dykhuizen and R. G. Strum. "Design Correlations for Cylindrical Pressure Vessels with Conical or Toriconical Heads," *Welding J. (NY), Res. Suppl.* 29:336-s (1950).

CHAPTER 8

REINFORCEMENT OF OPENINGS

8.1 INTRODUCTION AND DESIGN BASICS

Shells and heads of process apparatuses very often are furnished with different kinds of openings, such as nozzles, manholes, loading hatches, inspection windows, stuffing boxes, etc. Such openings may be unreinforced or reinforced. Unreinforced openings are generally those prepared for rolled-out tubes for threading and for openings tightened with locks and nozzles.

The diameter of unreinforced openings in cylindrical and conical shells should not exceed 0.6 of the internal diameter of a cylinder or cone in the place of the opening (however, this size should not exceed 200 mm).

The elliptical and rectangular openings in cylindrical and conical shells should be so oriented that their smaller size is directed along the generatrix of the cylinder (cone). Reinforcement metal may include any one or combination of the four metal parts listed below:

1. the metal in the attachment flange of an attached fitting,
2. the metal of a reinforcing pad,
3. the metal of any excess shell-plate thickness beyond that required from a calculation of the minimum plate thickness,
4. the metal of a nozzle.

The basic sizes of reinforcing elements (the thickness of sleeves, diameters of pads and their height) are determined on the basis that the metal removed from the opening must be replaced by the metal of reinforcement. If this condition is not fulfilled, even at the given reinforcement, the opening is assumed to be only partially reinforced.

It is not necessary to reinforce all openings. Many vessels are designed to operate under low pressure; therefore, all openings do not require reinforcement. Small-sized openings welded or brazed to a vessel do not require reinforcement. The specific sizes not requiring reinforcement are: (1) 3-in. nominal pipe size in a shell or head with a thickness of

⅜ in. or less, and (2) 2-in. nominal pipe size in a shell or head with a thickness greater than ⅜ in. Threaded, studded or expanded connections having a nominal pipe size of 2 in. or less also do not require reinforcement.

The design of the opening, including the knuckle radii and the angle, must meet the requirements for conical or turiconical closures. The determination of the limiting hole diameter, which may not be reinforced, is calculated from the following formula [1-3]:

$$d_{eim} = 3.7 \sqrt[3]{D(S - C)(1 - K_s)} \tag{8.1}$$

where D = internal diameter of cylindrical shell (cm)
S = wall thickness (cm)
C = corrosion allowance
K_s = safety factor of the vessel, which is determined from the formula

$$K_s = \frac{PD}{(2.3\sigma_{allow} - P)(S - C)} \tag{8.2}$$

For unreinforced threaded holes, the external diameter of the thread is assumed to be a design diameter. The elliptical and box-type heads work under more favorable conditions than cylindrical shells because the material is loaded more uniformly in parts near the axis. Therefore, it is possible to use unreinforced holes of greater diameter than in cylindrical shells. The largest permissible hole diameter in the heads is

$$d_{eim} = 0.95D(1 - K_i) \tag{8.3}$$

where

$$K_1 = \frac{P}{4\sigma_{allow}}\left[\frac{D}{(S - C)}\frac{D}{2h} - 1\right] \tag{8.4}$$

For an oval opening that can be formed at an angular connection of a nozzle to the shell, for example, the large axis of the ellipse is taken as the design diameter. Obviously removing metal from the walls to form an opening causes certain weakening effects, which are intensified by high stress concentrations near the metal around the cavity. The maximum stress [1-9] may be three to four times as large as the stresses in the normal or plain portion of the shell. This is illustrated in Figure 8.1.

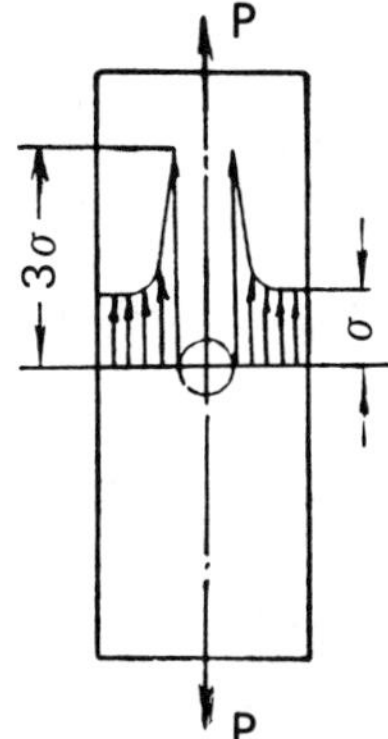

Figure 8.1. Stress concentration at the edge of a hole.

As shown in Figure 8.1, the stress concentration has a local character that spreads over the part of the shell in the immediate vicinity of the hole. Concentration usually is expressed by the stress concentration factor, which is the ratio of the high stress, σ_h, to the stress in any section of the solid portion of the shell. The stresses are highest at the edge of the hole and decrease very rapidly as the distance from the hole increases. The stress concentration near the hole in the thin-walled cylindrical shell operating under internal pressure may reach very high values. In this case, the concentration factor [2,3,5,6] is

$$\dot{\alpha} = 3 + 3.45 \frac{d^2}{Dt} \tag{8.5}$$

where d = diameter of the hole (cm)
D = diameter of the shell (cm)
t = shell thickness (cm)

The following formula was derived for the maximum circumferential tension at the edge of an elliptical opening assuming that a uniform stress acted at the edge of the flat plate in one direction only, and that one axis of the ellipse is parallel to the direction of stress:

$$\sigma_h = \sigma_a \left(1 + \frac{2a}{b}\right) \tag{8.6}$$

where a = length of semi axis perpendicular to the direction of the stress on edge of plate

b = length of semi axis parallel to the direction of stress on edge of plate

Equation 8.5 gives the following concentration factors:

11 in. X 15 in. manhole: $\dot{\alpha} = 3.73$
10 in. X 16 in. manhole: $\dot{\alpha} = 4.20$
circular openings: $\dot{\alpha} = 3.00$

In the case of a small elliptical hole located in the center of a head operating under internal pressure, the maximum stress concentration is equal to (2a/b) times the uniform stress.

The force breaking a cylindrical shell along the generatrix and resulting from the action of internal pressure is

$$P = DLp \tag{8.7}$$

where the parameters are defined in Figure 8.2. The wall cross-sectional area resisting the breaking force is

$$F = 2SL \tag{8.8}$$

According to Figure 8.2, by weakening the shell with the hole, d_0, the cross-sectional area of the wall is decreased by Sd_0. The weakening of the shell in the most dangerous longitudinal direction depends on the diameter of the opening. Reinforcement must be provided to restore the strength of the shell. A more economical approach would be to attach a reinforcing pad by welding it to the shell and to the nozzle, as shown in Figure 8.3. The weld

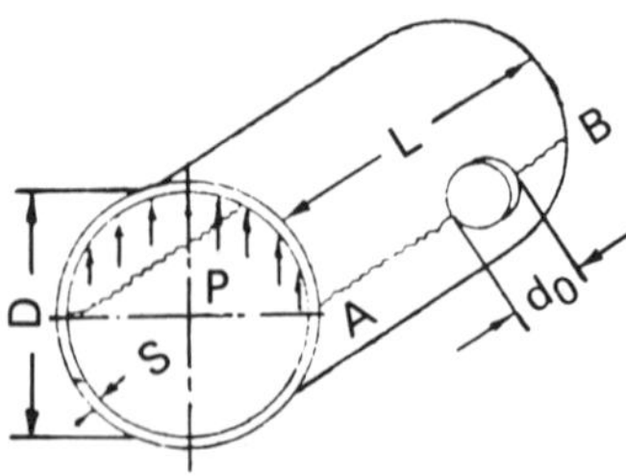

Figure 8.2. Forces acting on a cylindrical shell.

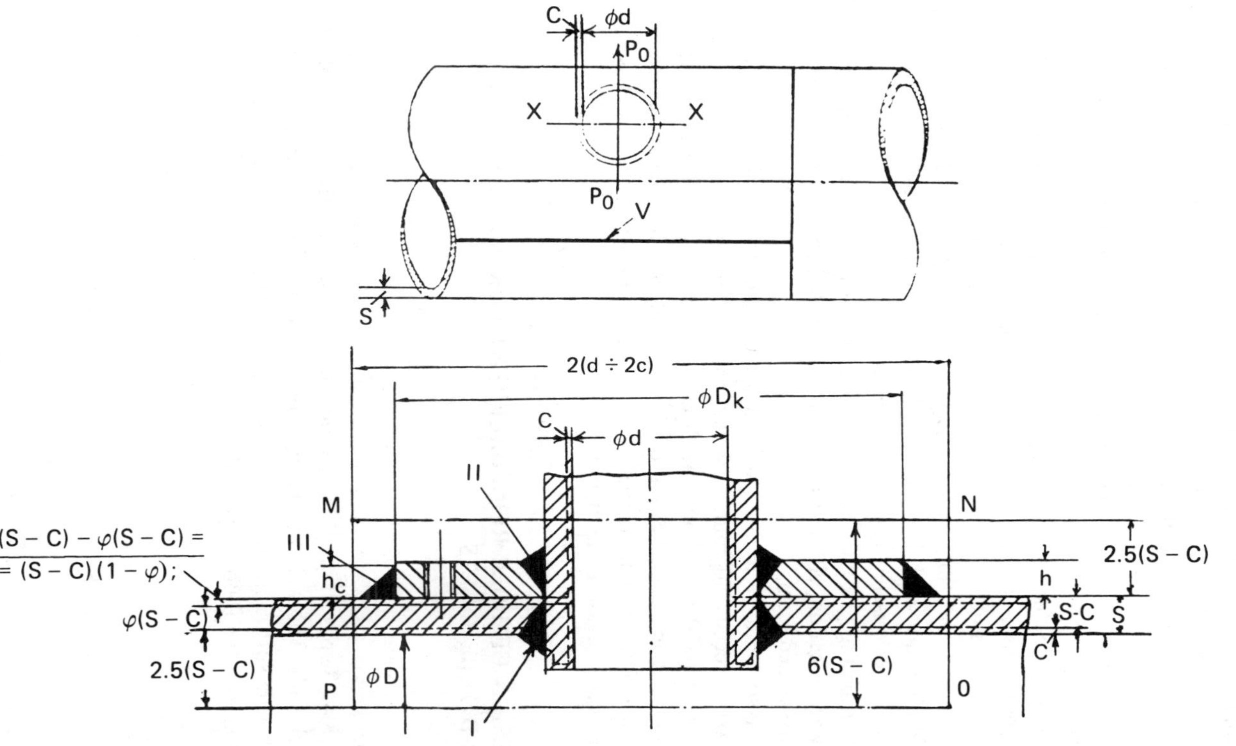

Figure 8.3. Reinforcement of single opening.

is placed at the point where it is weakened, and its cross-sectional area is made approximately equal to that of the metal removed. Note that the additional metal used in constructing the reinforcement should not exceed the limit of the rectangle MNOP in Figure 8.3 with the length of $2(d + 2C)$ and height of $2.5(S - C) + (S - C) + 2.5(S - C) = 6(S - C)$, where d is the internal diameter of the nozzle, S is the shell thickness and C is the corrosion allowance.

The neck of the nozzle extending both inward and outward of the MNOP zone is not considered part of the reinforcing element. This assumption has practically no influence on the size of the pad if the nozzle is thin-walled. In some cases, it is desirable to decrease the size of a pad, in which case thick-walled nozzles are used. In this case, all the metal in the limit of the MNOP zone should be considered part of the reinforcing element.

The reinforcing pad may be located both inward and outward of the shell. For inward location, allowance should be provided for corrosion in the pad and at its welded seams.

Openings should be reinforced in all cases when the nominal diameter of the nozzle exceeds 50 mm. However, the reinforcing of holes with diameters less than 50 mm should be done if the spacing between the centers of the holes is less than the sum of their diameters, when the holes cross the longitudinal and circumferential welded seams, or when the vessel is subjected to sharp pressure fluctuations.

8.2 DESIGN OF HOLE REINFORCEMENT FOR NOZZLES AND TUBE INLETS

Typical designs [4] of shell openings requiring reinforcement in vessels that have nozzles and tubes under internal pressure are presented in Figure 8.4. Types II and IX are applicable only for flat, elliptical and spherical walls. The weld on the shell or head-nozzle, tube and sleeve (types I-IV) provide sufficient reinforcement of an opening if their wall thickness, S, corresponds to the following conditions:

At $S_1 < S$,

$$S_1 \geqslant 0.5\left[S_1' + \sqrt{(S_1')^2 + 0.8(2S_0' - S + C)d}\right], \text{ m} \qquad (8.9)$$

At $S_1 \geqslant S$,

$$S_1 \geqslant 0.2\left(2\frac{S_0'}{S} + \frac{C}{S} - 1\right)d + S_1', \text{ m} \qquad (8.10)$$

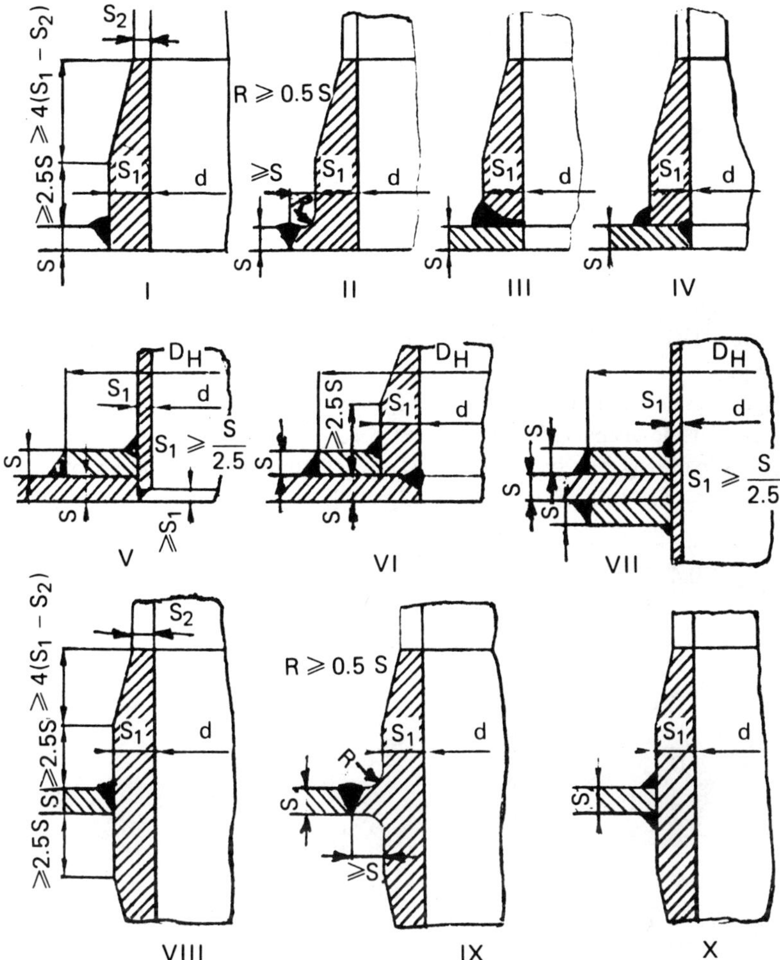

Figure 8.4. Typical design of reinforced shell opening in vessels under internal pressure. Reinforcement-joined tubes: I–IV = with sleeves; V–VI = with pads. Reinforcement of tube inlet: VII = with a pad; VIII–X = with a sleeve.

where S_0' = nominal design thickness of the wall to be reinforced, as determined from the corresponding formulas for internal or external pressures at $\psi = 1$; $K = 1$ and $C = 0$, m

S_1' = design thickness of the wall of a nozzle, tube or sleeve from the calculation for internal or external pressure

C = allowance to the thickness of the wall to be reinforced for corrosion, erosion and negative tolerance of sheet thickness

If conditions 8.9 and 8.10 are not fulfilled, the welding nozzles, tubes or sleeves should be reinforced further.

If the condition

$$d \leqslant \frac{2.5S - S_1'}{0.2\left(2\dfrac{S_0'}{S} + \dfrac{C}{S} - 1\right)} \tag{8.11}$$

is fulfilled, it is recommended to apply reinforcement with a sleeve (types I-IV); otherwise a pad should be used—type V at $S_1 \leqslant S$ or type VI at $S_1 > S$.

The design thickness of the reinforcing sleeve, S_1, for types I-IV (at $S_1 \geqslant S$) is determined from Equation 8.10. The design diameter of the reinforcing pad, D_p' (for types V and VI), whose thickness should equal that of the reinforcing walls, S, is determined as follows:

At $S_1 \leqslant 0.4S$,

$$D_p' \geqslant (2S_0' + C)\frac{d}{S}\ , \quad m \tag{8.12}$$

At $S > S_1 > 0.4S$,

$$D_p' \geqslant (2S_0' + C)\frac{d}{S} - 2\left(2.5\frac{S_1}{S} - 1\right)(S_1 - S_1')\ , \quad m \tag{8.13}$$

At $S_1 \geqslant S$,

$$D_p' \geqslant (2S_0' + C)\frac{d}{S} - 3(S_1 - S_1')\ , \quad m \tag{8.14}$$

The inlets of the tubes welded to the shell are themselves sufficient reinforcements for the openings if their wall thickness, S, corresponds to the following conditions:

At $S_1 < S$,

$$S_1 \geqslant 0.5[S_1' + 0.5C_1 + \sqrt{(S_1' + 0.5C_1)^2 + 0.4(2S_0' - S + C)d}\,]\ , \quad m \tag{8.15}$$

At $S_1 \geqslant S$,

$$S_1 \geqslant 0.1\left(2\frac{S_0'}{S} + \frac{C}{S} - 1\right)d + S_1' + 0.5C_1\ , \quad m \tag{8.16}$$

where C_1 = corrosion allowance to the thickness of the tube from the outside for its part inside the vessel (m).

If conditions 8.15 and 8.16 are not fulfilled, the welded tube inlets should be reinforced further. If the condition

$$d \leqslant \frac{2.5S - S_1 - 0.5C_1}{0.1\left(2\frac{S_0'}{S} + \frac{C}{S} - 1\right)}, \quad m \tag{8.17}$$

is fulfilled, it is recommended that a sleeve reinforcement of types VIII–X be applied; otherwise, a pad reinforcement (type VII) should be used. The thickness of the reinforcing sleeve, S_1, for types VIII–X at $S_1 \geqslant S$ can be determined from Equation 8.16.

The design diameter, D_p', for a reinforcing pad (type VII) at its recommended thickness is determined as follows:

At $S_1 \leqslant 0.4S$

$$D_p' \geqslant \frac{2(S_0' + S)}{2S - C}\frac{d}{S}, \quad m \tag{8.18}$$

At $S > S_1 > 0.4S$

$$D_p' \geqslant \frac{(2S_0' + S)d - (2.5S_1 - S)(2S_1 - 2S_1' - C_1)}{2S - C}, \quad m \tag{8.19}$$

At $S_1 \geqslant S$,

$$D_p' \geqslant \frac{(2S_0' + S)d - 3S(2S_1 - 2S_1' - C_1)}{2S - C}, \quad m \tag{8.20}$$

For one reinforcing pad, the design diameter, D_p', may be approximated from Equations 8.18–8.20, depending on the ratio S_1/S. The assumed diameter of the pad should be

$$D_p \geqslant d + 2S_1 + 0.05, \quad m$$

To locate an inspection hole, set d = 10 mm and check the tightness of welded seams. In cylindrical and conical shells, this hole should be located in the plane of the shell's cross section. The design diameter of the pad for elliptical, spherical and flat walls should be increased by 0.02 m.

Example 8.1

Check whether there is need for an opening reinforcement for a nozzle 273 X 9 mm in the cylindrical shell described in Example 6.1. If the hole must be reinforced, determine the size of the reinforcing element.

Solution

From Example 6.1, d = 2 m; p = 1.06 MN/m^2; σ_{allow} = 146 X 10^6 N/m^2; C = 1.8 mm; C_1 = 1 mm t = 10 mm.

The nominal design shell thickness is determined from Equation 6.20, assuming $\psi = 1$ and C = 0:

$$S_0' = \frac{dp}{2\sigma_{allow}} = \frac{2 \times 1.06 \times 10^6}{2 \times 146 \times 10^6} = 7.26 \times 10^{-3}$$

The design thickness of the nozzle wall, assuming it operates under internal pressure, is determined from Equation 6.20:

$$S_1' = \frac{dp}{2\sigma_{allow}} + C = \frac{0.255 \times 1.06 \times 10^6}{2 \times 146 \times 10^6} + C = 0.9 \times 10^{-3} + C$$

Allowance C for the nozzle (at $C_3 = 0.5$) is

$$C = C_1 + C_2 + C_3 = (1 + 0 + 0.5)10^{-3}\ 1.5 \times 10^{-3}$$

Hence,

$$S_1' = (0.9 + 1.5) \times 10^{-3} = 2.4 \times 10^{-3}\ \text{m}$$

We now check to see whether the reinforcement by the nozzle itself is sufficient; at S_1 = 9 mm < S = 10 mm, from Equation 8.9,

$$0.5\,[(S_1') + \sqrt{(S_1')^2 + 0.8\,(2S_0' - S + C)d'}]$$

$$= 0.5[2.4 \times 10^{-3} + \sqrt{(2.4 \times 10^{-3})^2 + 0.8(2 \times 7.26 - 10 + 1.8)10^{-3} \times 0.255}]$$

$$= 19.2 \times 10^{-3}\ \text{m}$$

Because 19.2×10^{-3} m $> S_1$, condition 8.9 is not fulfilled, that is, it is necessary to use an additional reinforcement.

To select the proper type of reinforcement, we will check condition 8.11:

$$\frac{2.5S - S_1'}{0.2\left(2\frac{S_0'}{S} + \frac{C}{S} - 1\right)} = \frac{(2.5 \times 10 - 2.4)\,10^{-3}}{0.2\left(2 \times \frac{7.26}{10} + \frac{1.8}{10} - 1\right)}$$

$$= 0.179 \text{ m} < d$$

Since condition 8.11 is not fulfilled, we select the pad reinforcement type V (Figure 8.4).

The design diameter of the reinforcing pad at $S = 10 > S_1 = 9 > 0.4 \times 10 =$ 4 mm is determined from Equation 8.13:

$$D_p' = (2S_0' + C)\frac{d}{S} - 2\left(2.5\frac{S_1}{S} - 1\right)(S_1 - S_1')$$

$$= (2 \times 7.26 + 1.8)\,10^{-3} \times \frac{0.255}{10 \times 10^{-3}} - 2\left(2.5 \times \frac{9}{10} - 1\right)(9 - 2.4)\,10^{-3}$$

$$= 0.4 \text{ m}$$

Assume $D_p = 400$ mm.

Example 8.2

Determine the size of a reinforcing element for a tube inlet 159×4.5 mm ($C = 1.5$ mm) in the cylindrical shell apparatus described in Example 6.1.

Solution

From Example 8.1, $S_0' = 7.26 \times 10^{-3}$.

The design thickness of the inlet tube wall operating under internal pressure is determined from Equation 6.20:

$$S_1' = \frac{dp}{2\sigma_{allow}} + C = \frac{0.15 \times 1.06 \times 10^6}{2 \times 146 \times 10^6} + 1.5 \times 10^{-3}$$

$$= 2.05 \times 10^{-3} \text{ m}$$

Checking if the reinforcement of the hole by the tube itself is sufficient at $S_1 = 4.5$ mm < 10 mm from condition 8.15:

$$0.5\,[S_1' + 0.5C_1 + \sqrt{(S_1' + 0.5C_1)^2 + 0.4(2S_0' - S + C)\,d}]$$

$$= 0.5\,[2.05 \times 10^{-3} + 0.5 \times 10^{-3}$$

$$+ \sqrt{(2.55 \times 10^{-3})^2 + 0.4(2 \times 7.26 - 10 + 1.8)10^{-3} \times 0.15}]$$

$$= 11.08 \times 10^{-3}\ \text{m}$$

Because 11.08×10^{-3} m $> S_1$, condition 8.15 is not fulfilled and it is necessary to provide an additional reinforcement. To select the reinforcement type, condition 8.17 must be checked:

$$\frac{2.5S - S_1' - 0.5C_1}{0.1\left(2\dfrac{S_0'}{S} + \dfrac{C}{S} - 1\right)} = \frac{(2.5 \times 10 - 2.05 - 0.5 \times 1) \times 10^{-3}}{0.1\left(2 \times \dfrac{7.26}{10} + \dfrac{1.8}{10} - 1\right)}$$

$$= 0.357\ \text{m} > d = 0.15\ \text{m}$$

Condition 8.17 is fulfilled, thus we select the sleeve reinforcement type VIII (Figure 8.4).

The design thickness of the reinforcing sleeve is determined from Equation 8.16:

$$S_1 = 0.1\left(2\frac{S_0'}{S} + \frac{C}{S} - 1\right) d + S_1' + 0.5C_1$$

$$= 0.1\left(2 \times \frac{7.26}{10} + \frac{1.8}{10} - 1\right) \times 0.15 + (2.05 + 0.5 \times 1)\,10^{-3}$$

$$= 12.03 \times 10^{-3}\ \text{m}$$

Assume $S_1 = 14$ mm.

8.3 DESIGN OF REINFORCEMENT OF OPENINGS COVERED WITH HEADS AND FOR DETACHABLE INLETS OF TUBES

Typical designs of opening reinforcements in walls closed with flanged or threaded covers, stoppers and for detachable tube inlets are shown in

Figure 8.5. Type II is applied only for flat, elliptical and spherical walls. The design height of a reinforced boss, h', for all types at diameter D_b, which is chosen from design considerations, is determined from:

$$h' \geqslant \frac{(2S_0' - S + C)\,d}{D_b - d - 2d_0} + 2d_0\ , \quad m \tag{8.21}$$

In this formula, D_b must be no more than 2d, independent of the chosen D_b. The design diameter of reinforced D_b' for all types at the chosen height is

$$D_b' \geqslant d + \frac{(2S_0' - S + C)\,d}{h} + 2d_0\ , \quad m \tag{8.22}$$

where $d_0 = d_b$, the thread diameter in the boss for a stud

Example 8.3

Determine the height of a reinforced boss, h (type IV) in the flat wall of the box-type shell of the apparatus described in Example 6.8 for a flanged tube inlet D = 125 mm at D_e = 0.245 m and d_b = 16 mm.

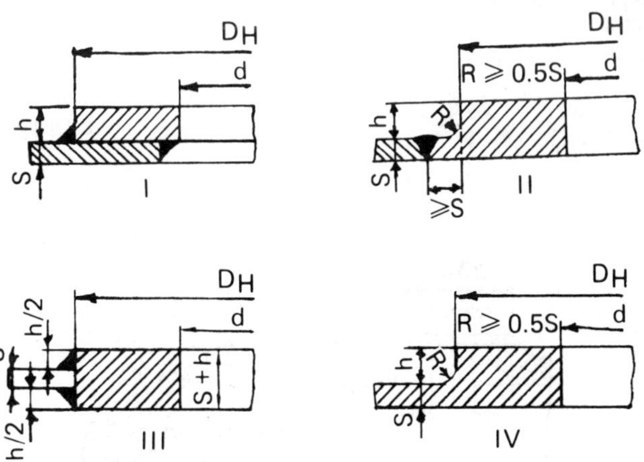

Figure 8.5. Examples of typical reinforced openings closed with covers, stoppers and for detachable tube inlets for apparatuses operating under pressure: I–III = reinforcements for welded apparatuses; IV = reinforcement for cast apparatuses.

Solution

From Example 6.8 we have $p = 1\ MN/m^2$; $\sigma_{b,allow} = 100 \times 10^6\ N/m^2$; $C_1 = 0$; $S = 16$ mm; $B_1 = 0.25$ m. The nominal design wall thickness is determined from Equation 6.84, assuming

$$C = 0:\ S_0' = 0.5B_1 \sqrt{\frac{P}{\sigma_{b,allow}}} = 0.5 \times 0.25 \sqrt{\frac{1 \times 10^6}{100 \times 10^6}} = 12.5 \times 10^{-3}\ m$$

The design height of the reinforced boss is determined from Equation 8.21:

$$h' = \frac{(2S_0' - S + C)}{D_H - d - 2d_0} = \frac{(2 \times 12.5 - 16 + 3) \times 10^{-3} \times 0.125}{0.245 - 0.125 - 2 \times 0.016}$$

$$= 17.1 \times 10^{-3}\ m$$

Assume h = 18 mm.

8.4 DESIGN OF OPENING REINFORCEMENTS IN FORGED CYLINDRICAL SHELLS

A typical example of reinforcement is shown in Figure 8.6. It is recommended that $d \leqslant D_b$.

The design thickness of the circular reinforced part, S_k', is determined from the following:

$$S_k' \geqslant 2S_0' + C\ ,\quad m \tag{8.23}$$

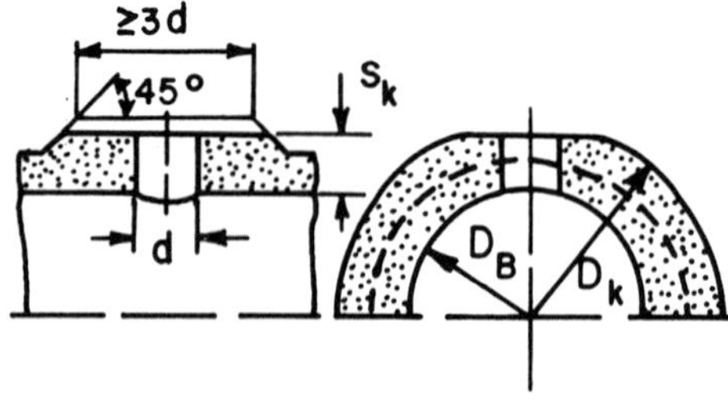

Figure 8.6. Example of hole reinforcement in a cylindrical forged shell.

8.5 CONDITIONS OF REINFORCEMENT OF HOLE SERIES IN CYLINDRICAL SHELLS

If the cylindrical shell has a series of holes of the same diameter located along the longitudinal direction, the maximum allowable diameter of the unreinforced holes is determined from the following formulas:

For an unstaggered order of hole locations (Figure 8.7),

$$d_{max} \leqslant \left(1 - \frac{S_0'}{S - C}\right) t \qquad (8.24)$$

For a staggered order of hole locations (Figure 8.8),

$$d_{max} \leqslant \left(1 - \frac{S_0' \sqrt{1 - 0.75(1 - K)^2}}{S - C}\right) 0.5 t_1 \sqrt{1 + K^2}\;, \quad m \qquad (8.25)$$

where $K = t/t_1$

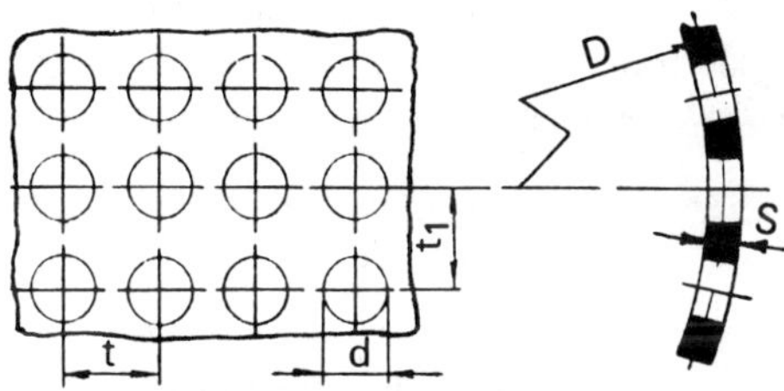

Figure 8.7. Unstaggered location of holes in a cylindrical shell.

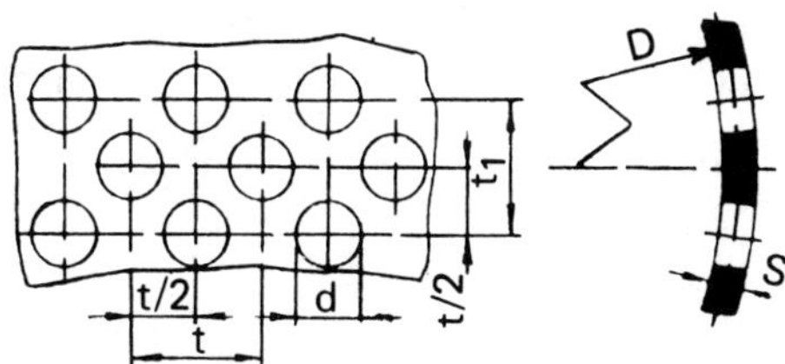

Figure 8.8. Staggered location of holes in a cylindrical shell.

If the diameter of the unreinforced hole is greater than d_{max}, it is necessary to increase either t, t_1 or S to maintain conditions 8.24 and 8.25.

If the tubes are welded or soldered with a hard solder (at $t \geqslant 2d_i$), the holes are assumed to be reinforced when the following condition is fulfilled:

$$SS_1(S - S_1') \geqslant (2S_0' - S + C)\, d_i \tag{8.26}$$

where d_i = tube internal diameter (m)

8.6 DESIGN OF WELDED REINFORCEMENT OF OPENINGS

The design of reinforcements for openings involves the determination of the cross-sectional area, F_1, removed from an opening and a cross-sectional area, F_2, of the reinforcement. For the reinforcement of a weakened wall, the following condition should be fulfilled:

$$F_1 \leqslant F_2$$

Several cases are considered below:

1. The opening is located beyond longitudinal welded seams of a shell and reinforced with a pad (refer to Figure 8.3):

To determine F_1, we assume

$$d_p = (d + 2C) \tag{8.27}$$

According to this condition, the opening must be located beyond the seams on the solid portion of the shell, where the design wall thickness is

$$S' = \psi\,(S - C) \tag{8.28}$$

where S = wall thickness (cm) determined from the following:

$$S = \frac{pd_e}{2\sigma_{allow}\,\psi + p} + C$$

or

$$S = \frac{pd}{2.3\sigma_{allow}\psi - p} + C$$

Then

$$F_1 = (d + 2C)S' = (d + 2C)\psi(S - C) \quad (8.29)$$

To determine F_2, we introduce D_k as the outside diameter of the reinforcing pad (Figure 8.3) and h as its thickness. The following assumptions are applied:

- The metal of the pad and the nozzle wall at the height, h, with total reinforcement area is defined as follows:

$$F_2' = [D_k - (d + 2C)]h \quad (8.30)$$

- Assume that excessive metal in the solid portion of the wall of the vessel exists.

The design thickness in the solid portion of the shell is S', and the total wall thickness minus corrosion allowance C is $(S - C)$; therefore, the excessive thickness of the wall in the solid place of the shell is

$$(S - C) - S' \quad (8.31)$$

Substituting S' from Equation 8.28, we obtain

$$(S' - C) - S' = (S - C) - (S - C)\psi = (S - C)(1 - \psi) \quad (8.32)$$

Thus, the area corresponding to the excessive metal thickness in the wall of the shell considered as a reinforcement is equal to

$$F_2'' = [D_k - (d + 2C)][(S - C)(1 - \psi)] \quad (8.33)$$

Finally, the total reinforcement area is

$$F_2 = F_2' + F_2'' = [D_k - (d + 2C)]h + [D_k - (d + 2C)](S - C)(1 - \psi)$$

or

$$F_2 = [D_k - (d + 2C)]\,[h + (S - C)(1 - \psi)] \tag{8.34}$$

Assuming $F_1 = F_2$ and substituting their values from Equations 8.29 and 8.34, we obtain

$$D_k = \frac{(d + 2C)(S - C)\psi}{h + (S - C)(1 - \psi)} + (d + 2C) \tag{8.35}$$

The thickness of the pad, h, usually is assumed to be equal to the shell thickness. The outside diameter of the reinforcing pad, D_k, should not exceed twice the diameter of the nozzle, $D_k \leqslant 2d$.

The increase in pad diameter beyond the zone MNOP (Figure 8.3) has no efficient reinforcing action.

Assuming that $C = 0$, $\psi = 0.85$ and $h = S$ in Equation 8.35, we obtain

$$D_k = 1.74d \tag{8.36}$$

and at $\psi = 1.0$,

$$D_k = 2.0d \tag{8.37}$$

Thus, when the opening is located beyond the welded seam there is no need to make the diameter of the pad, D_k, more than 2.0d, if the thickness of the pad is $h = S$.

2. The opening is located beyond the welded seams of a welded spherical dished head and reinforced with a pad (refer to Figure 8.3):

The opening itself and its reinforced elements should not be located in the transition part of the head characterized by the greatest stresses.

To determine F_1, the length of the opening is $d_p = (d + 2C)$. The opening is located in the solid portion of the head beyond the seams, and the transition part of the design wall thickness is

$$S' = \frac{pR}{2\sigma_{allow}} \tag{8.38}$$

and

$$F_1 = (d + 2C)S' \tag{8.39}$$

Comparing S' with the actual head thickness, it is shown that the last one is greater than S' because it was determined from the following formula:

$$S = \frac{pRy}{2\sigma_{allow}\,\psi} + C \tag{8.40}$$

Equation 8.40 includes the joint efficiency ψ, overstressing coefficient y in the head transition zone, and a corrosion allowance C. From Equation 8.40, $pR/2\sigma_{allow} = S'$, and substituting into Equation 8.29, we obtain

$$F_1 = (d + 2C)(S - C)\frac{\psi}{y} \tag{8.41}$$

or

$$F_1 = (d + 2C)\left(\frac{S - C}{y}\right)\psi \tag{8.42}$$

The determination of F_2 involves the use of Equation 8.30:

$$F_2' = [D_k - (d + 2C)]\,h \tag{8.43}$$

The excess head thickness in the solid portion beyond the transition curve is similar to Equation 8.31:

$$\begin{aligned}(S - C) - S' &= (S - C) - (S - C)\frac{\psi}{y} \\ &= S - C\left(1 - \frac{\psi}{y}\right)\end{aligned} \tag{8.44}$$

Also, an expression similar to Equation 8.33 is obtained:

$$F_2'' = [D_k - (d + 2C)]\left[(S - C)\left(1 - \frac{\psi}{y}\right)\right] \tag{8.45}$$

From Equations 8.43 and 8.45, the total reinforcement area is

$$F_2 = F_2' + F_2'' = [D_k - (d + 2C)]h + [D_k - (d + 2C)]\left[(S - C)\left(1 - \frac{\psi}{y}\right)\right] = [D_k - (d + 2C)]\left[h + (S - C)\left(1 - \frac{\psi}{y}\right)\right] \tag{8.46}$$

Because $F_1 = F_2$,

$$D_k = \frac{(d + 2C)(S - C)\psi}{y\left[h + (S - C)\left(1 - \frac{\psi}{y}\right)\right]} + (d + 2C) \tag{8.47}$$

For $C = 0$, $\psi = 0.85$, $h = S$, $y = 1.3$, we obtain from Equation 8.47

$$D_k = \frac{0.85d}{1.3 + 0.45} + d \approx 1.5d \tag{8.48}$$

3. The opening reinforced with the pad has a welded seam that crosses in the spherical dished head:

Using the same reasoning, we obtain

$$D_k = \frac{(d + 2C)(S - C)}{hy + (S - C)(y - 1)} + (d + 2C) \tag{8.49}$$

At $C = 0$, $h = S$ and $y = 1.3$,

$$D_k = \frac{d}{1.3 + 0.3} + d \simeq 1.63d \tag{8.50}$$

4. The opening reinforced with the pad is located in the solid flanged spherical head:

Reasoning as before, we obtain

$$D_k = \frac{(d + 2C)(S - C)}{hy + (S - C)(y - 1)} + (d + 2C)$$

5. The opening is located where there is no possibility to put a standard pad, as, for example, in a solid flanged spherical head near the transition zone (Figure 8.9):

The reinforcement of such openings may be done by different methods, for example, by additional increases in the thickness of the part itself where the opening is increased, as well as the wall thickness of the nozzle.

The reinforcing element in all these cases is the metal of the thickened walls of the parts. Only those reinforced parts of a wall located in the limits MNOP of Figure 8.3 should be accounted for. From Figure 8.9, for reinforcing openings, the design values of welded head thicknesses should be increased by S_1. Also, thicknesses should be increased by S_2. It is necessary to check whether these increases provide the required reinforcement of the opening.

To determine F_1, the length of the opening is assumed to be $d_p = d + 2C$. If the opening is made beyond both the welded seams of the head and the transition zone, but immediately adjacent to it, then the head thickness, S', must be determined as follows:

$$S' = \frac{pRy}{2\sigma_{allow}} \tag{8.51}$$

and the desired quantity is

$$F_1 = d_p S' = (d + 2C)S' \tag{8.52}$$

To determine F_2, assume the following to comprise the reinforcing elements:

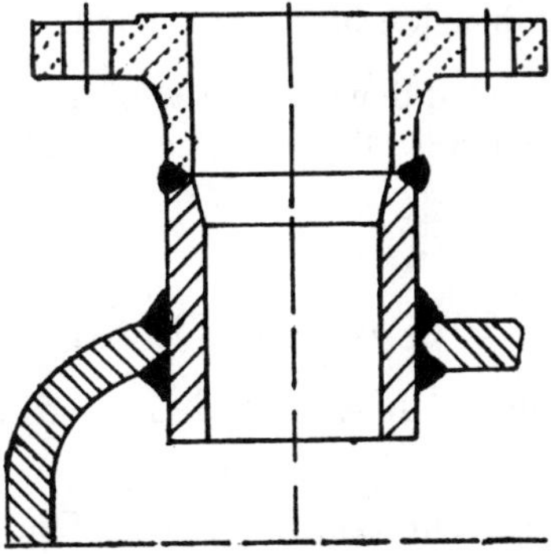

Figure 8.9. Solid flanged spherical head near the transition zone.

- additional metal added to the head thickness
- additional metal added to the nozzle thickness

The reinforced head thickness is

$$S_1 = (S - C) - S' \tag{8.53}$$

Assuming that the metal of the head within the thickness limits from Equation 8.53 serves as a reinforcement in the zone MNOP of Figure 8.3 along the length 2(d + 2C) minus the opening inside the nozzle, we can write

$$\begin{aligned} F_2' &= [2(d + 2C) - (d + 2C)]\,[(S - C) - S'] \\ &= (d + 2C)[(S - C) - S'] \end{aligned} \tag{8.54}$$

We then determine the area of the nozzle wall, F_2'', which may be considered as a reinforcement.

The design wall thickness of the nozzle is

$$S_n' = \frac{pd}{2\sigma_{allow}\psi - p} + C \tag{8.55}$$

The wall thickness of the nozzle, which may be considered as a reinforcement of the opening, is

$$S_2 = [(S_n - C) - S_n'] \tag{8.56}$$

where S_n = the total thickness of the nozzle wall

Assuming the metal of additional thickness of the nozzle in the zone limits MNOP (Figure 8.3), comprise the reinforcement of the opening, the reinforcement area will be

$$\begin{aligned} F_2'' &= 2 \times 2.5(S - C)[(S_n - C) - S_n'] \times 2 \\ &= 10(S - C)[(S_n - C) - S_n'] \end{aligned} \tag{8.57}$$

Hence, the total opening reinforcement area from Equations 8.54 and 8.57 is

$$F_2 = F_2' + F_2'' = (d + 2C)[(S - C) - S'] + 10(S - C)[(S_n - C) - S_n'] \quad (8.58)$$

The condition $F_1 \leqslant F_2$, must be fulfilled. That is,

$$(d + 2C)S' \leqslant (d + 2C)[(S - C) - S'] + 10(S - C)[(S_n - C) - S_n']$$

Taking into account Equations 8.53 and 8.56, we obtain

$$(d + 2C)S' \leqslant (d + 2C)S_1 + 10(S - C)S_2 \quad (8.59)$$

By selecting proper values for S_1 and S_2, condition 8.59 will be met.

If the opening is located near the shell seam, the welded seam joining the reinforcing pad to the shell should not approach the seam closer than twice the thickness of the shell and, as a minimum, no less than 30 mm.

Note that the openings in heads for manholes, nozzles, etc., should be located so that the edge of the reinforcing pad or the nozzles' external circumference does not approach the matching line of the head and the shell to within one-fourth of external diameter of the shell. If the vessel diameter $D \geqslant 1500$ mm, the maximum size of the circular or elliptical opening in the shell should not exceed $\frac{1}{3}D$. If the size of the vessel is $D < 1500$ mm, the maximum opening is taken to be 500 mm (but no more than half the vessel diameter).

If more than one opening is needed, the larger one should be located close to the head's center. If the opening in a closure of a cylindrical vessel exceeds one-half the inside diameter of the shell, the opening and the closure should be fabricated by one of the methods given in Figure 8.10.

If the distance between two or more adjacent holes is such that their reinforcement pads overlap, then this group of holes is reinforced with one reinforcement pad, the strength of which should be equal to the total strength of a single reinforcement, which would be necessary for each hole separately.

Example 8.4

Two heat exchangers are inserted in a spherical head of a reboiler 3000 mm in inside diameter operating under an internal pressure of 6 atm. The openings in the head are located as shown in Figure 8.11.

Check the opening reinforcement by use of additional wall thickening of the head and nozzles. The outside diameter of each nozzle is 720 mm, the

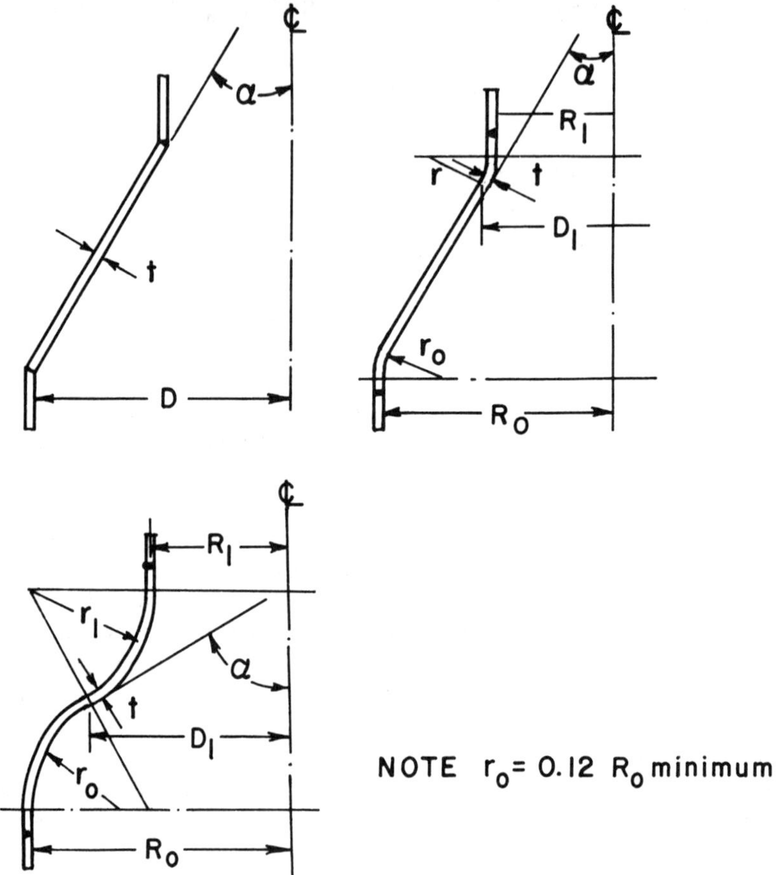

Figure 8.10. Large head openings–reverse-curve and conical shell-reducer sections [4].

wall thickness is 36 mm and the head thickness is 18 mm. The openings are made beyond the welded seams of the head. The inside diameter of the nozzle d = 720 − 2 X 3.6 = 64.8 cm.

Solution

Assume that the reinforcement zone is beyond the transition curve of the head. The design length of the opening is determined from Equation 8.27:

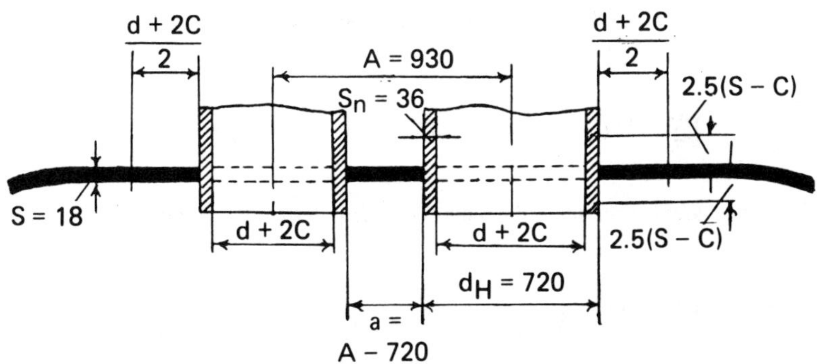

Figure 8.11. Sketch for Example 8.4.

$$d_p = 64.8 + 2 \times 0.4 = 65.6 \text{ cm}$$

From Equation 8.38

$$S' = \frac{pR}{2\sigma_{allow}} = \frac{6 \times 270}{2 \times 1000} \approx 0.84 \text{ cm}$$

where R = 2700 for a normalized head for
D = 3000 mm

From Equation 8.52,

$$F_1 = (d + 2C)S' = 65.6 \times 0.84 = 55 \text{ cm}^2$$

Because there are two openings, their total design area is

$$2F_1 = 55 \times 2 = 110 \text{ cm}^2$$

Since the area for the head reinforcing is adequate (Figure 8.11), we can state the following:

1. The space between two openings is the following:
   $a = A - 72 = 93 - 72 = 21$ cm with area

$$F_2'' = 21(S - C - S') = 21(1.8 - 0.4 - 0.84) = 11.8 \text{ cm}^2$$

2. Two areas at the sides of openings with length $(d + 2C)/2$ and total area:

$$F_2''' = 2\left(\frac{d + 2C}{2}\right)[(S - C) - S']$$
$$= (64.8 + 2 \times 0.4)[(1.8 - 0.4) - 0.84] = 36.6 \text{ cm}^2$$

The total area of the cross section in the wall thickness of the metal of the head itself, which may be assumed as reinforcement, is

$$F_2' = F_2'' + F_2''' = 11.8 + 36.6 = 48.4 \text{ cm}^2$$

The theoretical wall thickness of the nozzle is

$$S_n' = \frac{pd}{2\sigma_{allow}\psi - p} = \frac{6 \times 64.8}{2 \times 950 \times 0.85 - 6} = 0.32 \text{ cm}$$

From Equation 8.57,

$$F_2'' = 10(S - C)[(S_n - C) - S_n']$$
$$= 10(1.8 - 0.4)[(3.6 - 0.4) - 0.32] \approx 40.5 \text{ cm}^2$$

The cross-sectional area of the two nozzles is

$$2F_2'' = 2 \times 40.5 = 81.0 \text{ cm}^2$$

The total area of reinforcement is

$$F_2 = F_2' + F_2'' = 48.4 + 81.0 = 129.4 \text{ cm}^2$$

Thus, $2\,F_1 = 2 \times 55 = 110 \text{ cm}^2 < F_2 = 129.4 \text{ cm}^2$

Consequently, the reinforcement additions to the wall thicknesses of the head and nozzle openings were chosen correctly.

8.7 DESIGN OF WELDED SEAMS OF REINFORCING PADS

Reinforcing pads should be welded around the shell's internal and external perimeters. Note that the internal circular seam of the pad does not increase the safety factor.

The design of a welded seam of a reinforcing pad involves checking the size of a leg, h_c (see Figure 8.3). The seam should be designed with sufficient strength to withstand stresses both from above and below. That is, the dangerous section xx in Figure 8.3 should be able to withstand the load, P_0, delivered by the reinforcing ring.

To determine the load, P_0, consider the shell element 1 cm long along the line AB in Figure 8.2. At wall thickness, S, the element's cross-sectional area is equal to $1 \times S'$, where $S' = S - C$.

The shell element with the area $1 \times S'$ at stress σ_{allow} is capable of withstanding the load from the internal pressure $q = 1 \times S' \sigma_{allow} \psi$.

At the opening length (d + 2C), the total load, P_0, is due to the cross-sectional area of the removed metal:

$$P_0 = 1 \times S'(d + 2C)\, \sigma_{allow}\, \psi \qquad (8.60)$$

The strength equation for a seam along each half of the pad is

$$P_0 = \frac{\pi D_k}{2}\, 0.7 h_c \psi_c \sigma_{s, allow} \qquad (8.61)$$

where ψ_c = the seam efficiency of the pad's welded seam
$\sigma_{s, allow}$ = shear stress

Equating the right-hand sides of Equations 8.60 and 8.61 and assuming that $\sigma_{s, allow} = 0.8\, \sigma_{allow}$,

$$h_c = \frac{2S'(d + 2C)\, \sigma_{allow}\, \psi}{\pi D_k\, 0.56 \sigma_{allow}\, \psi_c} \qquad (8.62)$$

Noting that.

$$\frac{D_k}{d + 2C} = \psi$$

$$h_c = \frac{(S - C)\,2\psi}{\pi 0.56\,\psi\psi_c} \qquad (8.63)$$

At $\psi = 0.95$, $\psi_c = 0.8$ and minimum value of $\psi = 1.6$:

$$h_c = 0.85\,(S - C)$$

h_c of the fillet seam usually is equal to the pad thickness: $h_c = h$.

NOMENCLATURE

a	length along semiaxis perpendicular to direct of stress (mm)
b	length of semiaxis parallel to direct of stress (mm)
C	corrosion allowance (cm)
D	diameter (cm)
D'_p	design diameter of reinforcing pad (m)
F	area (cm^2)
h	height (m)
K_s	safety factor (see Equation 8.2)
k_1	parameter defined in Equation 8.4
L	length (m)
p	local pressure (N/m^2)
S	thickness (m)
S_0	nominal design thickness (m)
t	shell thickness (mm)
y	overstressing coefficient
$\dot{\alpha}$	stress concentration factor (see Equation 8.5)
Δ_{eim}	design hole diameter (cm)
ψ	joint efficiency factor

REFERENCES

1. Taylor, J. H., and E. O. Waters. "Effect of Openings in Pressure Vessels," *Trans. Am. Soc. Mech. Eng*. APM 56-3 (March 1934).
2. Timoshenko, S. *Theory of Elasticity* (New York: McGraw-Hill Book Co., 1934).

3. Inglis, C. E. "Stresses in Plates Due to the Presence of Cracks and Sharp Corners," *Trans. Inst. Naval Architects (Great Britain),* LX (1): 219 (1913).
4. *ASME Boiler and Pressure Vessel Code, Section VIII, Rules for Construction of Unfired Pressure Vessels* (New York: American Society of Mechanical Engineers, 1956).
5. Timoshenko, S. "On Stresses in a Plate with a Circular Hole," *J. Franklin Inst.* (April, 1924).
6. Timoshenko, S., and A. Dietz. "Stress Concentration Produced by Holes and Fillets," *Trans. Am. Soc. Mech. Eng.* (May 1925), pp. 199-237.
7. Siebert, E., F. Koerber and A. Pomp. "Versuche neber die Anstrengung und Formmenderung gewölbter Kesselboeden mit and ohne Mannloch bei der Beanspruchung durch inneren Druck," Verlag Stahl und Eisen, GmbH, Dusseldorf (1926).
8. Jacobus, D. S. "Openings in Cylindrical Drums," *Mech. Eng.* 368 (May, 1932).
9. Nadai, A., R. V. Band and A. Wahl. "Stress Distribution and Plastic Flow in an Elastic Plate with Circular Hole," *Mech. Eng.* (3) 187-192 (1930).

CHAPTER 9

PIPES, COMPENSATORS AND VALVES

9.1 PIPES

If pumps represent the heart of a chemical processing plant, then pipes and conduits are the veins and arteries. Piping and its components provide the means for transporting materials between and from various pieces of processing equipment. In many cases, they are an integral part of the equipment itself, such as in tubular heat exchangers, for example. In addition, many internal and external assemblies and parts of apparatuses are constructed from piping.

Among the major types of piping are piling pipe, line pipe, mechanical tubing, pressure tubing and standard pipe. Piling pipe is a round-welded or seamless pipe, employed as foundation pipe. The pipe cylinder acts as a permanent load-carrying unit and often is filled with concrete to form cast-in-place concrete piles. It is employed by the construction industry in buried installations for foundation work. Piling pipe is designed according to American Society for Testing and Materials (ASTM) specifications.

Line pipe is used for transporting gas, oil and water. Pipe diameters range from $^1/_8$ in. to 42 in. This pipe is fabricated to American Petroleum Institute (API) and American Water Works Association (AWWA) specifications.

Mechanical tubing includes welded or seamless tubing fabricated in sizes ranging from $^3/_{16}$ in. to $10^3/_4$ in. o.d. inclusive for carbon and alloys. Specifications are not tight, and only exact outside diameter and decimal wall thickness must be met.

Pressure tubing is used for conveying fluids at elevated temperatures and/or pressures. It is suitable for head applications and is fabricated to meet exact o.d. and decimal wall thicknesses. Sizes range from 0.5 in. to 0.6 in. o.d. inclusive, at ASTM specifications. Examples of pressure tubing are air heater tubes, boiler tubes, header tubes, heat exchanger and condenser tubes, oil-still tubes, superheater tubes and pressure tubing couplings and coupling stock.

Standard pipe is used in low pressure applications, for example, in transporting air, steam, gas, water, oil, etc. Applications are in machinery, building, sprinkling and irrigation systems, and water wells. This type pipe can transport fluids at elevated temperatures and pressures that are not subjected to external heat. Diameters range from 1/8 in. to 42 in. o.d. and are designed according to ASTM specifications. Examples of applications are machine piping, pressure piping, water and gas service pipe, nipple pipe and pipe for plating or enameling.

Structural pipe and tubing includes welded or seamless pipe and tubing that is often used in structural or aboveground load-bearing operations. The construction industry is the largest user of this type of pipe. It is manufactured in nominal wall thicknesses and sizes according to ASTM specifications in round, square, rectangular and odd cross sections.

Fabricating assemblies and parts from pipes often requires that the piping be bent. Bending pipes fabricated from plastic materials with outside diameters less than 50 mm is most often accomplished without heating. For diameters greater than 50 mm, heat must be applied to prevent cracking or breakage. The average bending radius, R_{av}, is recommended to be greater than or equal to three times the pipe's nominal diameter [1]. Pipes constructed from nonferrous metals and their alloys may be bent at $R_{av} = 2$ D.

For steep bending of carbon steel pipes, stamped elbows with a small bend radius (R = D) are recommended. If there is a need to provide a pipe bend with the radius $R < 2D$ (at $p \leqslant 5 Mn/m^2$) as well as for bends of large diameters, welded sectoral bends should be employed. The basic design for 90° bends are shown in Figure 9.1.

Bends with either larger or smaller radii may be fabricated in a similar

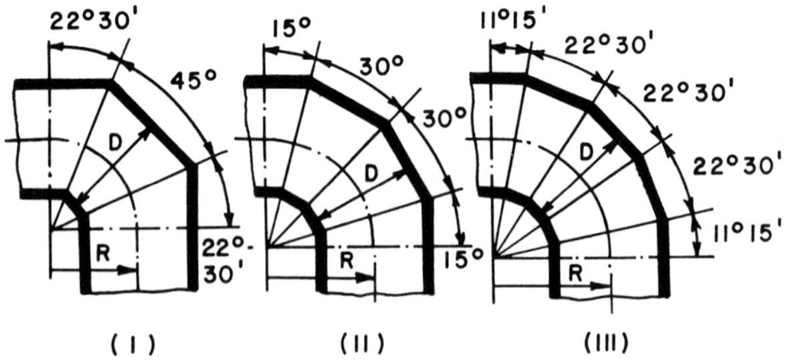

Figure 9.1. Examples of welded sectoral bends with a 90° angle: I = $R \geqslant 0.75D$; II = $R \geqslant 0.9D$; III = $R \geqslant D$.

fashion. The design of a return bend for $p \leqslant 5\text{Mn/m}^2$ with an angle of 180° and small distance between its parallel branches is shown in Figure 9.2.

Examples of nondetachable pipe connections between the pipes themselves are shown in Figure 9.3. The length of the socket, ℓ, in connections with soft soldering according to types VI and VII, is determined from the following formula:

$$\ell = \frac{pD}{4\tau_{sh}} \tag{9.1}$$

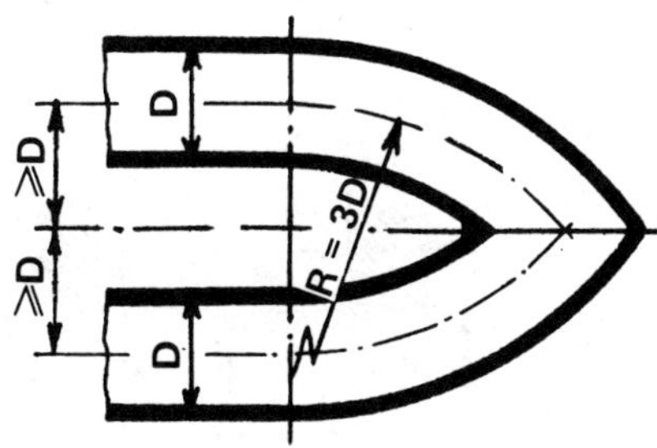

Figure 9.2. Example of welded return bend (α = 180°) with a small distance between its parallel branches.

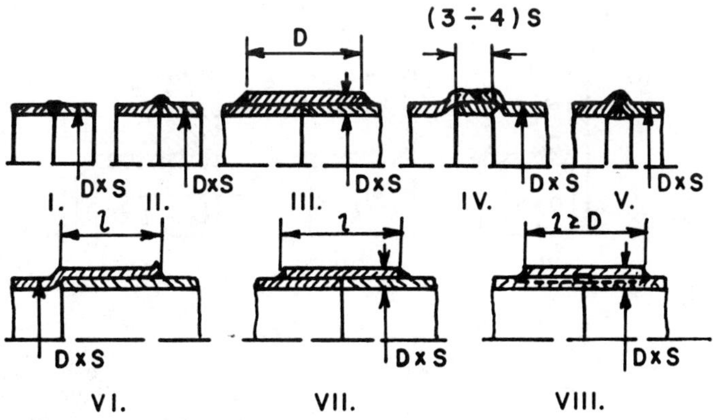

Figure 9.3. Examples of nondetachable pipe connections between the pipes themselves: I–V = welded connections (I–II, $p < 20\ \text{MN/m}^2$; III, $p \leqslant 10\ \text{MN/m}^2$; IV–V, $p \leqslant 20\ \text{MN/m}^2$); VI–VII = soldered connections, $p \leqslant 5\ \text{MN/m}^2$; VIII = threaded connections with soldering or welding, $p \geqslant 5\ \text{MN/m}^2$.

Examples of nondetachable pipe connections to different parts of apparatuses without reinforcement of their walls are presented in Figure 9.4.

9.1.1 Design of Pipes Operating Under Internal Pressure

The design thickness of the wall of the pipe S′, straight or bent (with $R \geqslant 2D$) and operating under internal pressure, P, is determined from the following formulas:

At $P/\sigma_{allow} < 0.4$ from the following formulas [1],

$$S' = \frac{DP}{2\sigma_{allow}\psi + P} + C \ , \quad m \tag{6.21}$$

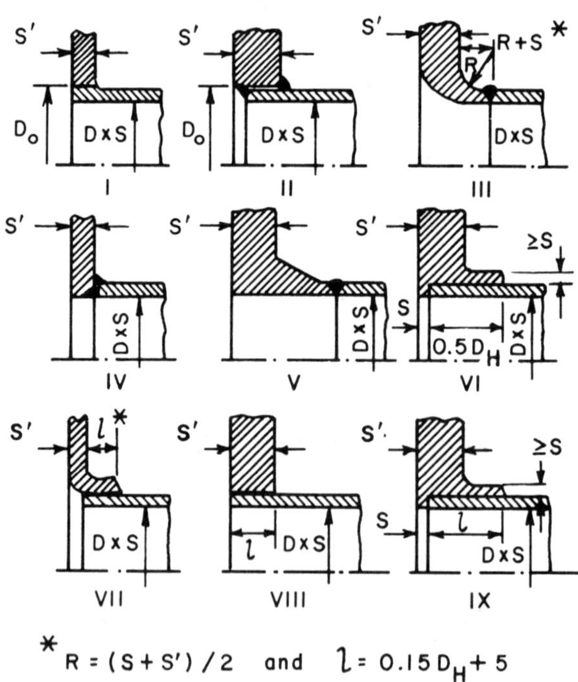

Figure 9.4. Examples of nondetachable pipe connections to the different parts of apparatuses. I–V = welded connections (I, $s \geqslant 0.6\ s_1$)–to walls for $p \leqslant 5$ MN/m^2 and to flanges at $p \leqslant 2.5$ MN/m^2; III–to walls for $p \leqslant 10$ MN/m^2; IV–at $s \geqslant 0.4\ s_1$–to walls and flanges for $p \leqslant 5$ MN/m^2; V–at $s < 0.6\ s_1$–to flanges for $p > 2.5$ MN/m^2; VI–threaded connection on tinning–to walls and flanges for $p \geqslant 5$ MN/m^2; VII–IX–soldered connections (VII–to walls for $p \leqslant 1.6$ MN/m^2); VIII–to walls and flanges for $p \leqslant 5$ MN/m^2; IX–to walls and flanges for $p \leqslant 1.6$ MN/m^2.

At $P/\sigma_{allow} \geqslant 0.4$,

$$S' = 0.5D \frac{10^{(P/2.3\sigma_{allow})} - 1}{10^{(P/2.3\sigma_{allow})}} + C \ , \quad m \tag{9.2}$$

Allowance $C = C_1 + C_2 + C_3$, where

$$C_3 = S_0 K \geqslant 0.5 \times 10^{-3} \ m \tag{9.3}$$

where K = coefficient is taken from Table 9.1

The assumed pipe thickness S should be checked for stress at the hydrostatic test from the following formula:

$$\sigma = \frac{d + (S - C)P'}{2(S - C)\psi} \leqslant \frac{\sigma_y}{1.2} \ , \quad N/m^2 \ . \tag{6.38}$$

if the design was based on formula 6.21; otherwise, use the following formula:

$$\sigma = \frac{P'}{\ell n \beta} \leqslant \sigma_y/1.2 \ , \quad N/m^2 \tag{6.74}$$

where

$$\beta = \frac{d + 2S}{d + 2(C_1 + C_2)} \tag{6.75}$$

Example 9.1

Determine the wall thickness of seamless pipe in an apparatus operating under internal pressure for the following data: the pipe material is carbon steel (σ_y = 250 X 10^6 N/m^2; σ_T = 120 X 10^6 N/m^2); D = 0.57 m;

Table 9.1. Coefficient k Values for Equation 9.3

Tolerance on Wall Thickness (%)	−15	−12.5	−10	−5
K	0.2	0.18	0.15	0.1

p = 20 Mn/m^2; tolerance by wall thickness is 15%; p′ = 20 Mn/m^2; the fluid is inert.

Solution

The ratio

$$\frac{p}{\sigma_{allow}} = \frac{20 \times 10^6}{120 \times 10^6} = 0.167 < 0.4$$

Therefore, the design thickness of the pipe is determined from Equation 6.21:

$$S' = \frac{DP}{2\sigma_{allow}\,\psi + P} + C = \frac{0.057 \times 20 \times 10^6}{2 \times 120 \times 10^6 + 20 \times 10^6} + C$$

$$= 4.38 \times 10^{-3} + C, \text{m}$$

Allowance C (at $C_1 = 0$ and $C_2 = 0$) is determined from Equation 9.3; K = 0.2 (from Table 9.1):

$$C = C_3 = S_0K = 4.38 \times 10^{-3} \times 0.2 = 0.876 \times 10^{-3} \text{ m}$$

Hence,

$$S' = (4.38 + 0.876) \times 10^{-3} = 5.256 \times 10^{-3} \text{ m}$$

We assume S = 5.5 mm.

The stress in the wall of the pipe at hydrostatic test from Equation 6.38 is

$$\sigma = \frac{[d + (S - C)]P'}{2(S - C)\psi} = \frac{[0.046 + (5.5 - 0.876)10^{-3}]25 \times 10^6}{2(5.5 - 0.876)10^{-3}}$$

$$= 137 \times 10^6 \text{ N/m}^2 < \sigma_y/1.2 = \frac{250 \times 10^6}{1.2}$$

$$= 208 \times 10^6 \text{ N/m}^2$$

9.1.2 Design of Pipes Operating Under External Pressure

Pipes operating under external pressure should be constructed to almost exact cylindrical shapes. The permissible out-of-roundness in any cross section of such pipe should not be more than 1% of the pipe diameter. The design thickness of the pipe, S′, straight or bent (with R ⩾ 2 D), operating under external pressure is calculated from the following formulas (Figure 9.5):

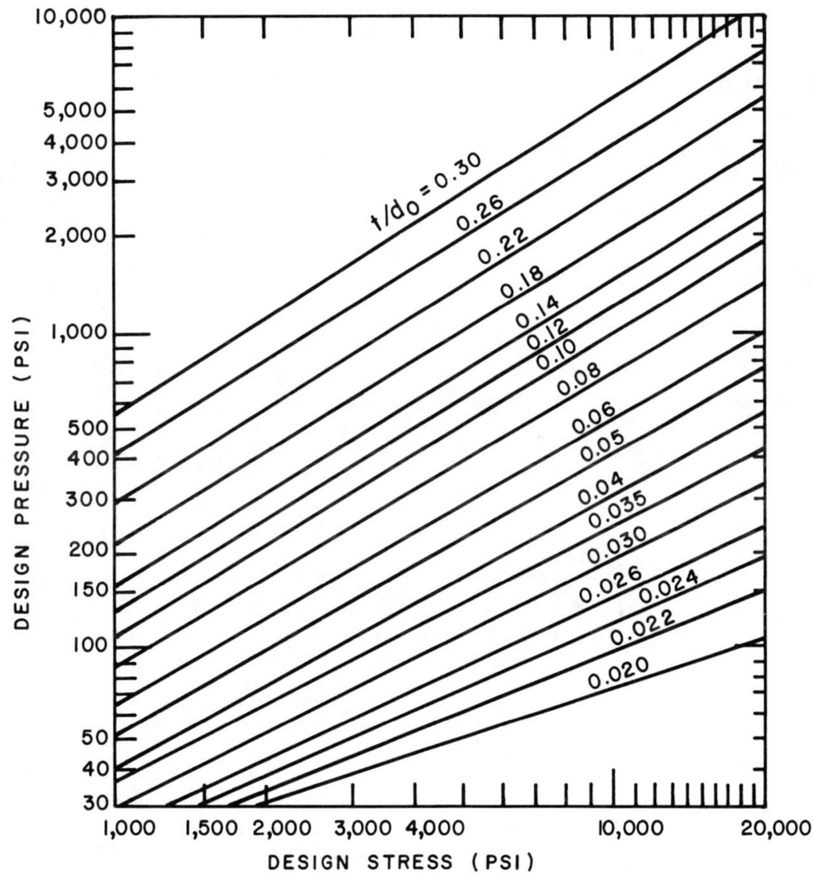

Figure 9.5. Chart for determination of wall thickness of tubes under external pressure [1] (courtesy of the American Society of Mechanical Engineers).

At $P_e/\sigma_{allow} \geqslant 0.2$,

$$S' = 0.5\, D\left(1 - \sqrt{\frac{\sigma_{allow}\,\psi - 1.73\, P_e}{\sigma_{allow}\,\psi}}\right) + C\ , \quad m \tag{9.4}$$

At $P_e/\sigma_{allow} < 0.2$,

$$S' = \frac{DP_e}{2\sigma_{allow}\,\psi} + C\ , \quad m \tag{9.5}$$

where C is taken from $C = C_1 + C_2 + C_3$. C_1 considers external and internal corrosion and C_3 is obtained from Equation 9.3.

The assumed wall thickness of the pipe, S, if the calculation was made from Equation 9.5, should be checked for stability at a safety factor $\eta = 4$ [2]. For pipe lengths $L \leqslant 8\,D$, use Equation 6.51. At $L > 8\,D$, if the condition

$$\frac{E^t}{1 - \tilde{\mu}^2}\left(\frac{S - C}{D - S}\right)^2 < \sigma_y^t\ , \quad N/m^2 \tag{9.6}$$

is fulfilled, use the following:

$$P_e = \frac{E^t(S - C)^3}{2(1 - \tilde{\mu}^2)(D - S)} \geqslant P_\ell\ , \quad N/m^2 \tag{9.7}$$

If condition 9.6 is not fulfilled, then

$$P_e = \frac{S - C}{D - S}\left\{\frac{0.5\,\sigma_y^t}{1 + \dfrac{\sigma_y^t}{E^t}\left(\dfrac{D - S}{S - C}\right)^2}\right\} \geqslant P_\ell\ , \quad N/m^2 \tag{9.8}$$

Example 9.2

Determine the wall thickness of a seamless pipe operating under external pressure for the following data: the material of the pipe is copper ($E^t = 115 \times 10^9$ N/m^2; $\sigma_y^t = 50 \times 10^6$ N/m^2; $\sigma_{allow} = 33 \times 10^6$ N/m^2; $\tilde{\mu} = 0.33$); $D = 0.02$ m; $p_e = 2.5$ Mn/m^2; tolerance by wall thickness is 10% of wall thickness; $C_1 = 1$ mm; $L = 2$ m.

Solution

Ratio

$$\frac{P_e}{\sigma_{allow}} = \frac{2.5 \times 10^6}{33 \times 10^6} = 0.076 < 0.2$$

therefore, the design thickness of the pipe wall is determined from Equation 9.5:

$$S' = \frac{DP_e}{2\sigma_{allow}\,\psi} + C = \frac{0.02 \times 2.5 \times 10^6}{2 \times 33 \times 10^6 \times 1}$$

$$= 0.76 \times 10^{-3} + C$$

Allowance C (at $C_2 = 0$) from Equation 9.3 is

$$C = C_1 + S_0 K = (1 + 0.76 \times 0.15) \times 10^{-3} = 1.11 \times 10^{-3}\ \text{m}$$

Assume S = 3 mm. Because L > 8 D, we must check condition 9.6:

$$\frac{E^t}{1 - \tilde{\mu}^2}\left(\frac{S - C}{D - S}\right)^2 = \frac{115 \times 10^9}{1 - 0.33^2}\left(\frac{3.0 - 1.11}{20 - 3.0}\right)^2$$

$$= 1610 \times 10^6\ \text{N/m}^2 > \sigma_y^t = 50 \times 10^6\ \text{N/m}^2$$

Since condition 9.6 is not fulfilled, the allowable external pressure is determined from Equation 9.8:

$$P_{allow} = \frac{S - C}{D - S}\left\{\frac{0.5\,\sigma_y^t}{1 + \frac{\sigma_y^t}{E^t}\left(\frac{20 - 3.0}{3.0 - 1.11}\right)^2}\right\}$$

$$= \frac{3.0 - 1.11}{20 - 3.0} \times \frac{0.5 \times 50 \times 10^6}{1 + \frac{50 \times 10^6}{115 \times 10^9}\left(\frac{20 - 3.0}{3.0 - 1.11}\right)^2}$$

$$= 2.68 \times 10^6\ \text{N/m}^2 > P_{allow} = 2.5 \times 10^6$$

Consequently, the pipe is rigid.

Figure 9.5 (from the 1956 ASME code) gives the allowable design pressure for pipes subjected to external pressure as a function of the allowable stress of the material and the ratio of t/d_0 (pipe thickness, in./pipe outside diameter, in.). In case of corrosion or erosion, additional metal must be supplied. If the pipe is threaded, additional metal equal to (0.8/n) inches, where n is the number of threads per inch, must be provided.

9.1.3 Reinforced Plastic Piping

Steel, cast iron and various other conventional piping materials have been replaced in recent years by special-grade plastics and fiberglass-reinforced plastics (FRP). Plastic piping offers the advantages of lightweight, durability and strength. In addition, FRP is especially resistant to corrosion and chemical attack, provided the proper bonding resin is employed. Corrosion is a key consideration since the steel industry alone devotes roughly 40% of its production toward the replacement of corroded materials [3]. If FRP pipe and components are prepared properly, significantly fewer replacements are needed. This, in turn, results in significant savings in energy, and manufacturing and maintenance costs. Specific mechanical properties depend on the plastic's and resin's resistance to heat and chemicals being handled.

Plastic and FRP pipe are available in a wide range of diameters and lengths. In the U.S., inside diameters range from less than 1 in. to more than 12 ft. FRP pipe diameters of 60 ft to more than 100 ft have been proposed for ocean thermal applications in which enormous volumes of water must be circulated between the surface and ocean floor.

The most widely used FRP pipes are fabricated from polyesters and vinyl esters. Maximum operating temperatures are as high as 300°F, depending on the resin. Operating pressures vary with pipe diameter, wall thickness and the method of fabrication.

Filament-wound pipe and ductwork can be employed in a wide number of high-pressure applications and are used extensively in chemical processing plants, pulp and paper mills, auto manufacturing plants, food processing facilities, bulk storage and loading complexes, and various conventional and nuclear power generating plants. Table 9.2 lists some properties of large-diameter, filament-wound pipe.

FRP is well suited for many fluid flow applications, with a weight that is roughly one-sixth that of steel and one-twelfth that of concrete. It has the strength of steel with the additional properties of inertness (i.e., FRP pipe will not decay, corrode or support combustion), dimensional stability and flexibility.

FRP piping can be made corrosion resistant to almost every kind of

Table 9.2. Nominal Properties of Vinyl Ester, Filament-Wound Pipe (supplied by Fibercast Co./Div. of Youngstown Sheet and Tube Co.) (pipe diameter = 14-72 in.; specific gravity = 1.86; coefficient thermal expansion = 10.5×10^{-6} in./in./°F)

Temperature (°F)	Hoop Stress on Reinforced Thickness (psi)	Axial Stress (psi)	Axial Tensile (psi)	Coefficient of Elasticity	Beam Bending Modulus of Elasticity (psi $\times 10^6$)	Beam Bending (psi)	Pipe-Fitting Layup Joint Shear (psi)
77	50,000	25,000	9,000	2.0	1.0	9,000	600–1,000
150	50,000	25,000	7,000	1.5	0.4	7,000	600–1,000
180	50,000	25,000	5,000	1.2	0.15	5,000	600–1,000
200	40,000	20,000	4,000	0.8	0.04	2,000	600–1,000

chemical or compound, which explains why it is used widely in the Chemical Process Industry.

Pipe and fittings can absorb more than 40% diametrical deflection without undergoing structural change, and components will return to their original configuration once the overload has been removed.

Filament-wound fiberglass-reinforced pipe has excellent flow properties. The interior surfaces of piping and components are generally smooth and glass-like. This smooth finish greatly reduces and, in many cases, eliminates, material buildup. This, in turn, significantly reduces pumping costs. Figure 9.6 is a flowchart that can be used for rough estimates for large-diameter pipes.

There are several types of wall construction used with FRP filament-wound pipe. Standard pipe generally has no circumferential reinforcement ribs. These pipes are designed in this manner to meet the pressure/bending stress

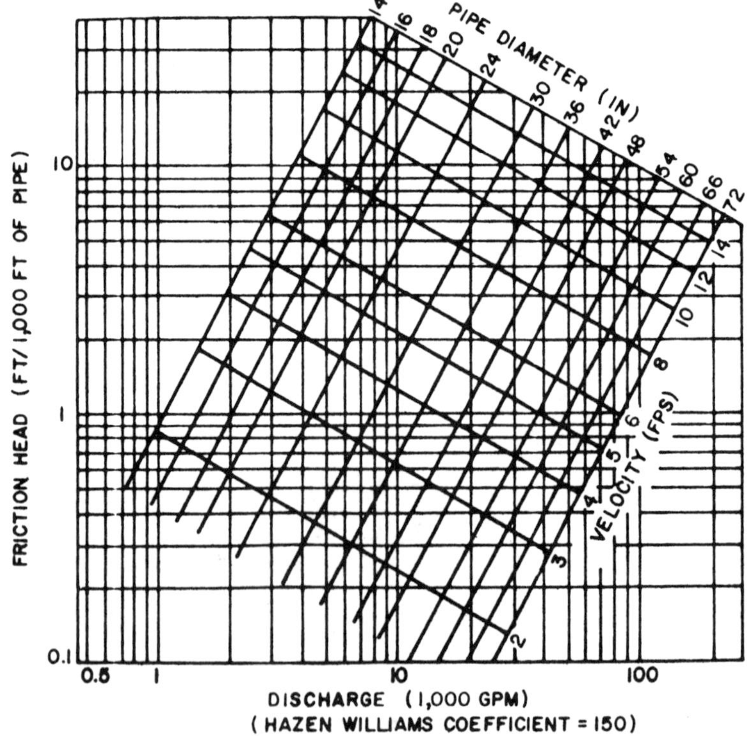

Figure 9.6. Flowchart for large-diameter, filament-wound pipe. (Courtesy of Fibercast Co./Div. of Youngstown Sheet and Tube Co.)

criteria of various applications and are suitable for buried pipe installations or for aboveground or hanger-suspended pipe systems. Standard wall designs also are often used as tunnel liners.

Rib-wall construction is used on large-diameter pipes to provide additional strength for buried and subaqueous installations. Usually, soil compaction is difficult to achieve in these situations. Prestressed ribs are circumferentially wound over cured pipe after the specified overall wall thickness has been reached. The rib reinforcement must be positioned based on closely calculated spacing requirements. This addition greatly increases the pipe wall stiffness. The design reduces the nominal pipe wall thickness while maintaining the same strength:weight ratio. Two basic types of rib reinforcement are the half-elliptical and the trapezoidal designs [3].

In the former design, a solid filament-wound, half-elliptical rib is provided to give additional pipe wall stiffness (Figure 9.7). These are employed in substandard burial conditions or excessive burial depths. The half-elliptical rib is constructed from continuous strands of tensioned high-strength glass

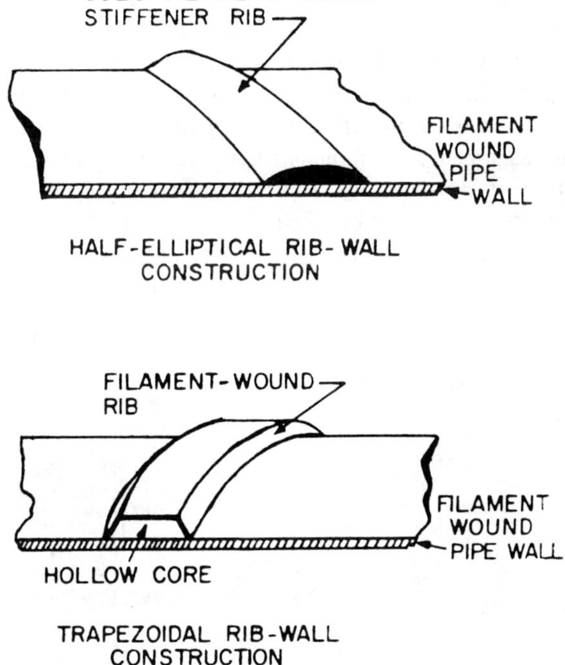

Figure 9.7. Two types of rib reinforcements used on filament-wound pipe.

reinforcement, which are helically wrapped at high helix angles over the completed outer wall. The rib generally is wider than it is high.

The trapezoidal rib design (Figure 9.7(B)) is employed under conditions in which the ultimate degree of pipe stiffness is required. The rib also is constructed from highly tensioned continuous strands of impregnated fiberglass. A helix angle is employed that gives a much thicker rib profile. This type of rib configuration enables the pipe to withstand excessive burial loads and extremely high impact strengths. As such, it provides the most rugged construction, capable of withstanding shock that normally would crumble or damage concrete or steel piping.

There are a variety of procedures for connecting FRP filament-wound pipe, in addition to the numerous structural components often encountered in large, complex construction jobs. If the pipe is to pass through a wall, for example, a filament-wound thrust ring is recommended. This is wound onto the pipe during fabrication. The thrust ring ensures anchoring and serves as a medium to transfer hydrodynamic and thermal stresses to the wall. When subsidence is expected, the pipe can be adapted to accommodate the anticipated settlement. Usually, a neoprene rubber compression wrap can be employed at the pipe-to-concrete termination encasement to dampen or minimize shear possibilities at this termination point.

When expansion joints are used, a flex-coupling can be employed. Expansion joints usually are used less frequently in filament-wound FRP piping networks than with such systems as steel. They are important, however, when severe expansion thrusts are anticipated at connection points, for adjusting misalignment at equipment connections, or where provisions for value replacements are necessary. This is illustrated in Figure 9.8.

Because of its great versatility FRP pipe can be readily incorporated into existing systems constructed from other materials, provided proper connec-

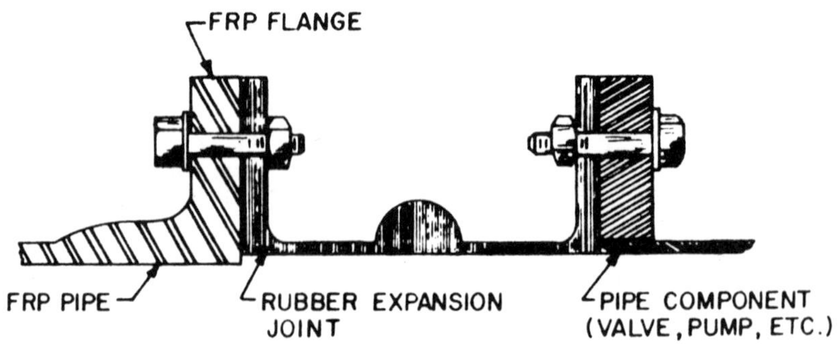

Figure 9.8. Typical expansion joint tie-in.

tions are used. Two methods are employed for making connections to concrete pipe. One encloses the mated sections of pipe in a concrete encasement; the other approach employs a specially designed spigot joint to fit a concrete bell.

Mating FRP pipe with steel pipe can be done in a number of different ways. One way is to use a concrete-encased adhesive joint connection, as shown in Figure 9.9(A). Here, the steel pipe is sealed adhesively within the bell end of the plastic pipe and the entire joint encased in concrete. A second method, shown in Figure 9.9(B), employs flange ends on both pieces of pipe. An appropriate size flange must be fabricated onto the end of the FRP pipe and drilled to prespecified orientation. FRP pipe also may be joined to steel by means of a flexcoupling, similar to that shown in Figure 9.8.

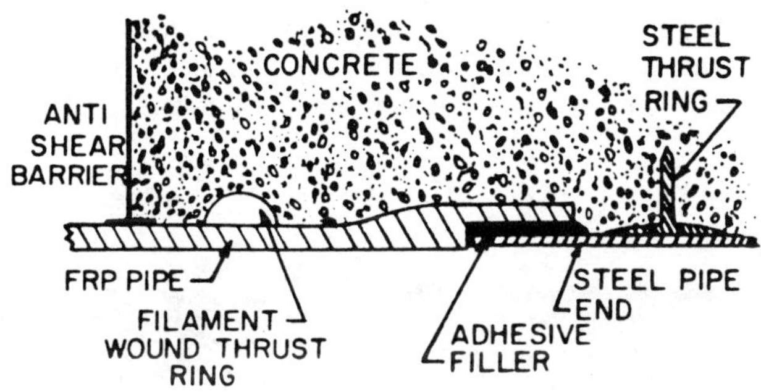

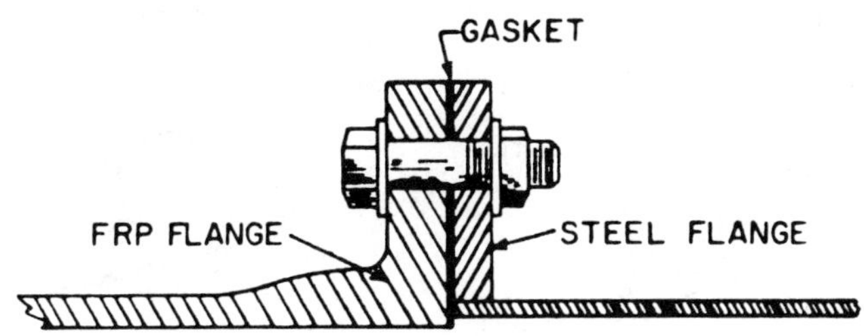

Figure 9.9. Two methods for connecting FRP pipe to steel pipe. (A) FRP pipe to steel pipe connection, (B) FRP flange to steel flange.

Pipe can be supported on hangers or saddles, which may be constructed from either plastic or steel. FRP saddles can be cut from actual portions of the pipe to be supported. Recommended saddle lengths should be at least one-half the pipe diameter. Note that the pipe and saddle will not have exactly the same radius; therefore, the gap must be filled with a suitable filler material to provide full bearing support. Clamps and "U" bolts should be used to provide snug, but not excess, clamp pressure on the pipe. Some typical support designs are shown in Figures 9.10 and 9.11.

The recommended saddle angle should be 180° for any one of the following conditions:

1. when pipe diameters exceed 18 in.,
2. when the pipe wall is rated at less than 50 psig, or
3. when the pipe is designed for vacuum.

Valves, flowmeters or other flow equipment that cause weight concentrations to occur in the line should be supported rigidly. This additional support should be independent of the remainder of the piping system.

Support spacings should provide for maximum tolerable deflection, as well as maximum stress at the points of maximum deflection and maximum stress at the hangers. Table 9.3 gives recommended support spacings. (Note that when complete line drainage requirements exist, lines should be

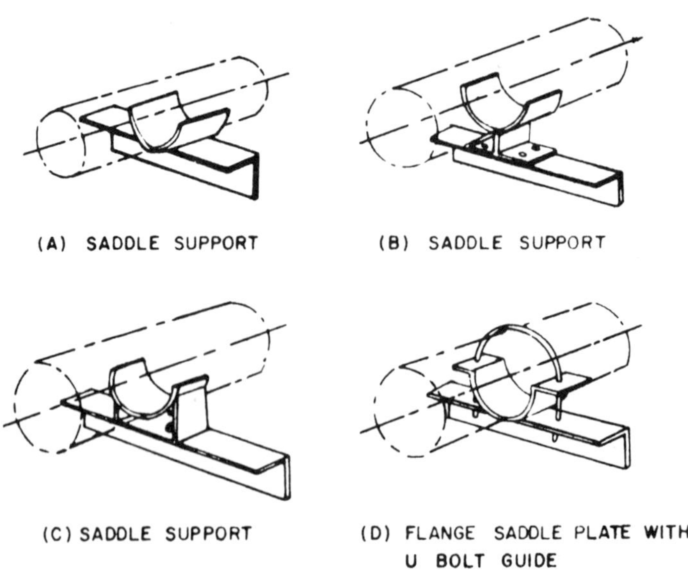

Figure 9.10. Typical support arrangements.

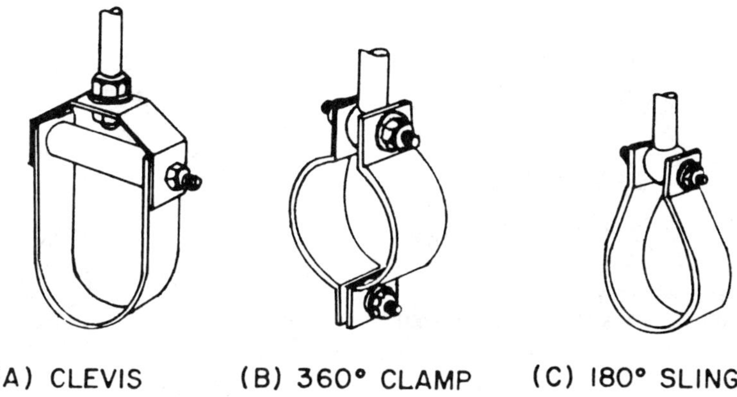

Figure 9.11. Typical hanger support arrangements.

Table 9.3. Recommended Support Spacing Span (ft) for Fluids with Specific Gravity 1.2 (Ceilcote, Div. of General Signal Corp.)

Inside Diameter (in.)	Internal Pressure Rating, psig					
	25	50	75	100	125	150
2	8.0	8.0	8.0	8.0	8.0	8.0
4	10.0	10.0	10.0	10.5	10.5	10.5
6	10.5	10.5	11.5	11.5	12.0	12.5
8	11.5	12.5	13.0	13.0	13.5	14.0
10	12.0	13.0	14.0	14.5	15.0	15.5
12	12.5	14.0	15.0	15.5	16.0	17.0
14	14.5	15.0	15.5	16.5	17.5	18.5
16	15.0	15.5	16.5	17.5	18.5	
18	15.5	17.0	18.0	19.0	20.0	
20	15.5	17.5	18.5	20.0		
24	12.5	19.0	20.5	22.0		

sloped at a minimum of 0.5 in. for every 10 ft of pipe length. The system should be provided with periodic low spots with flanged drain connections for positive drainage during a shutdown or cleanout.)

The most effective way to restrain pipe movement by applied forces is through the use of an anchor. Figure 9.12 illustrates anchor arrangements. The design can be used to restrain thermal expansion; as such, it must meet maximum anticipated end forces. It also must prevent the pipe from sliding within the arrangement.

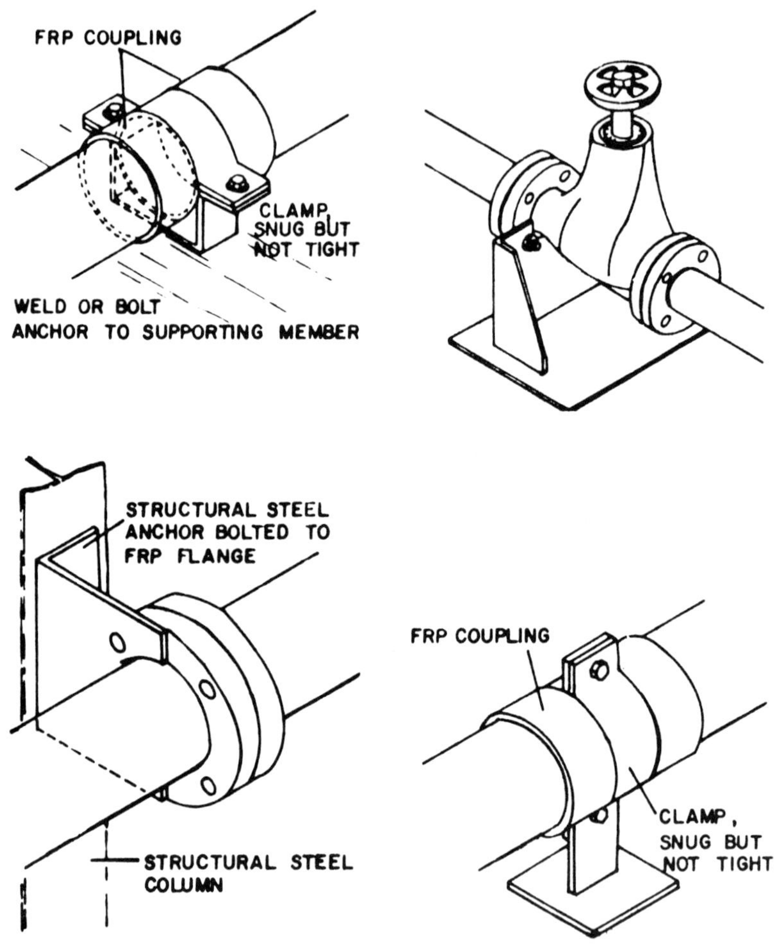

Figure 9.12. Various anchor arrangements that restrain pipe movements in all directions.

In situations in which a full anchor is not necessary but movement must be restricted, a tie-down can be used. Tie-downs are employed when long lengths of pipe are left to expand. It is generally recommended that the pipe be tied down in the middle to ensure equal expansion in both directions. The expansion may be directed to one end by placing the tie-down away from the center. Another situation in which tie-downs can be used is when the velocity and surge forces promote pipe movement.

When expansion must be directed away from equipment to prevent the expanding line from overstressing a tank shell or vessel nozzle, tie-downs also should be used. Tie-downs basically consist of a snug "U" bolt and a 6-in.-wide strap joint, which is placed over the U-bolt to maintain the pipe in a fixed condition. The support must be fastened securely to a rigid foundation. This arrangement should not be used for supporting vertical runs.

Guides should be used to minimize snaking when the pipe has been restrained from expanding. A guide is essentially any type of device or support that prevents the pipe from transverse movement (but not axial movement). Under conditions in which axial movement is excessive, sliding guides should be employed. These will protect the pipe from abrasion. Figure 9.13 illustrates this type of guide.

Most FRP piping systems can be designed without expansion joints because although plastic piping systems have roughly twice the rate of thermal expansion of steel, they have a comparatively low modulus of elasticity. Usually, a slip-type joint or a bellows-type expansion joint is recommended for added protection. Anchors and supporting structures must be designed properly for the end-force from operating line pressures when expansion joints are employed. Between any two anchors only one expansion joint should be installed. The pipe should be guided close to the expansion joint to prevent shear and bending loads and to eliminate any misalignment during axial movement. Guides should be positioned on both sides of the expansion joint at 4 and 12 pipe diameters away from the joint.

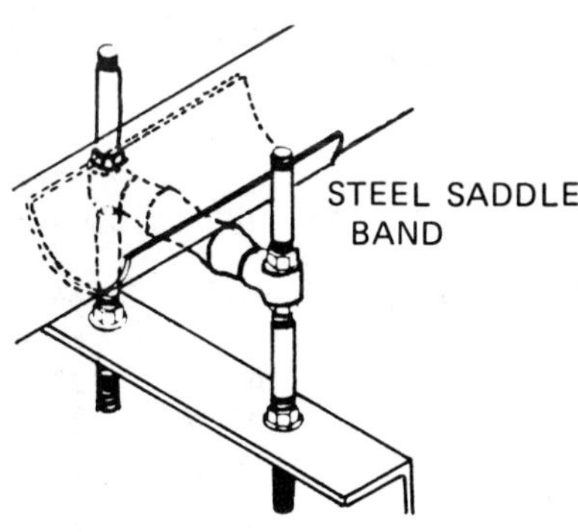

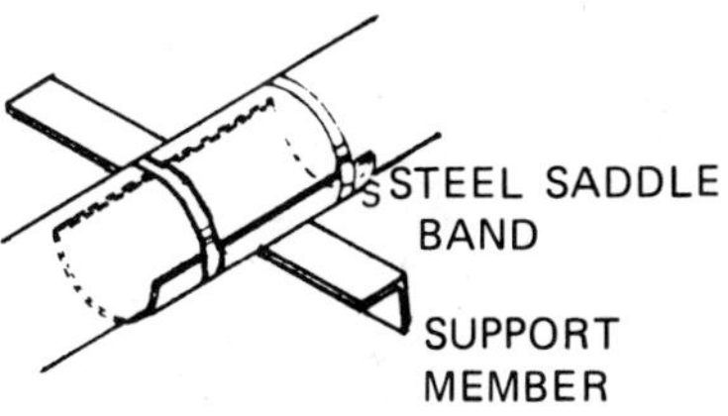

Figure 9.13. Typical sliding guide arrangements: (A) adjustable pipe support (pipe can roll or move axially); (B) pipe support (pipe can move sideways or axially).

FRP pipe is commercially available in a wide range of diameters, wall thicknesses and pressure ratings. Wall thickness specifications are based on an allowable strain, which can yield structural safety factors, in some cases in excess of 20:1. The strain level is defined as the maximum strain that can be tolerated to ensure long-term integrity of the inner liner. Note that when considering FRP pipe versus concrete pipe, corrugated metal or other types for buried installations, the choice should be based on the safety factor. Table 9.4 gives wall thicknesses of filament-wound FRP pipe at various pressures.

In addition to having high pressure ratings, filament-wound pipe can be

Table 9.4. Recommended Wall Thickness for

	Internal Pressure					
	25		50		75	
Inside Diameter (in.)	Wall (nominal) (in.)	wt/ft (lb/ft)	Wall (nominal) (in.)	wt/ft (lb/ft)	Wall (nominal) (in.)	wt/ft (lb/ft)
2	0.188	0.945	0.188	0.945	0.188	0.945
2.5	0.188	1.18	0.188	1.18	0.188	1.18
3	0.188	1.42	0.188	1.42	0.188	1.42
4	0.188	1.89	0.188	1.89	0.188	1.89
5	0.188	2.36	0.188	2.36	0.188	2.36
6	0.188	2.63	0.188	2.83	0.188	2.83
8	0.188	3.78	0.188	3.78	0.188	3.78
10	0.188	4.72	0.188	4.72	0.188	4.72
12	0.188	5.67	0.188	5.67	0.188	5.67
14	0.188	6.61	0.188	6.61	0.188	6.61
16	0.188	7.56	0.188	7.56	0.188	7.56
18	0.188	8.50	0.188	8.50	0.188	8.50
20	0.188	9.45	0.188	9.45	0.188	9.45
24	0.188	11.3	0.188	11.3	0.214	12.9
30	0.188	14.2	0.188	14.2	0.268	20.2
36	0.188	17.0	0.214	19.4	0.321	29.0
42	0.188	19.8	0.250	26.4	0.375	39.6
48	0.188	22.7	0.286	34.5	0.429	51.8
54	0.188	25.5	0.321	43.6	0.482	65.4
60	0.188	28.3	0.357	53.8	0.536	80.8
72	0.214	38.7	0.429	77.6	0.643	116.4
84	0.250	52.8	0.500	108.1	0.750	158.3
96	0.286	69.0	0.571	137.8	0.857	206.8
108	0.321	87.1	0.642	174.3	0.964	261.7
120	0.357	107.7	0.714	215.3	1.07	322.7
144	0.429	155.3	0.857	310.2	1.29	467.0

fabricated with the proper resins to be highly resistant to corrosion and chemical attack. It is used by chlorine manufacturers in transporting hot chlorinated acid brine; in the paper industry when handling corrosive wastes at high temperatures; in transporting caustic, chlorine and sodium hypochlorite from bleach plants to waste treatment facilities; and in the food processing industry when handling both acetic and citrus acids in addition to process treatment chemicals.

Unlike most metals that undergo electrochemical attack, plastic pipe generally will be affected by chemical reaction or by solvation, which is the penetration of the plastic by a corrosive element. This causes softening,

Filament-Wound FRP Pipe at Various Pressures [3]

Rating, psi					
100		125		150	
Wall (nominal) (in.)	wt/ft (lb/ft)	Wall (nominal) (in.)	wt/ft (lb/ft)	Wall (nominal) (in.)	wt/ft (lb/ft)
0.188	0.945	0.188	0.945	0.188	0.945
0.188	1.18	0.188	1.18	0.188	1.18
0.188	1.42	0.188	1.42	0.188	1.42
0.188	1.89	0.188	1.89	0.188	1.89
0.188	2.36	0.188	2.36	0.188	2.36
0.188	2.83	0.188	2.83	0.188	2.83
0.188	3.78	0.188	3.78	0.188	3.78
0.188	4.72	0.188	4.72	0.188	4.72
0.188	5.67	0.188	5.67	0.214	6.45
0.188	6.61	0.208	7.32	0.250	8.80
0.190	7.64	0.238	9.57	0.286	11.5
0.214	9.68	0.268	12.1	0.321	14.5
0.238	12.0	0.298	15.0	0.357	17.9
0.286	17.3	0.357	21.5	0.429	25.9
0.357	27.0	0.446	33.6	0.536	40.4
0.429	38.8	0.536	48.5	0.643	58.2
0.500	52.8	0.625	66.0	0.750	79.2
0.571	68.9	0.714	86.1	0.857	103.4
0.643	87.3	0.804	109.1	0.964	130.8
0.714	107.7	0.893	134.7	1.07	161.4
0.857	155.1	1.07	193.6	1.29	233.4
1.00	211.1	1.25	263.9	1.50	316.7
1.14	275.1	1.43	345.0	1.71	412.6
1.29	350.0	1.61	437.0	1.93	524.0
1.43	431.0	1.79	540.0	2.14	645.0
1.71	619.0	2.14	775.0	2.57	930.0

swelling in the walls and, eventually, failure. Because of the fundamental differences between electrochemical and solvation reactions, conventional corrosion rates should not be used in evaluating or comparing the chemical resistance of plastic pipe to other materials. Chemical resistance for FRP usually is determined by immersing the laminates in a bath of the corrosive media and simulating operating conditions. Tests of strength retention, surface hardness and visual inspection for chemical attack provide detailed qualitative information on plastic and FRP pipe characteristics. Table 9.5 gives the chemical resistance of filament-wound pipe to various fluids and the range of maximum operating temperatures for which FRP pipe is commercially available.

The specific corrosion-resistant service for a reinforced plastic pipe is determined from the resin:glass ratio used in the laminate construction. The nature of the resin plays the dominant role in corrosion resistance to specific chemicals. The interior layer of the pipe generally has a low glass content (on the order of 20–30% glass content by weight). This ratio provides high corrosion resistance. To meet desired wall thicknesses and strengths, a higher glass content is used in fabricating laminates for the remaining thickness. A typical overall ratio is roughly 70% resin to 30% glass reinforcement by weight. The outer wall is often covered with a glass surfacing mat that is resin rich. Such a surface provides excellent resistance to weathering (sunlight, heat, cold and other weathering conditions), fumes and spillage.

Contact-molded or hand layup piping often is used where maximum chemical resistance is required throughout a laminate and where pressure requirements are not severe. For many applications, however, filament-wound FRP pipe is preferred over contact-molded because of the former's higher structural properties. These properties in turn possess economic advantages since thinner laminate thicknesses can be employed to meet physical requirements. Table 9.6 gives recommended wall thicknesses of contact-molded pipe at various pressures.

Contact molding is a process in which layers of resin-impregnated fabrics are built up, one layer at a time, on a mold-forming pipe. Little or no pressure is applied in forming the resulting laminate during curing. The structural wall of the pipe usually consists of chopped glass strand mat and/or woven roving. Normally, a 60% resin, 40% glass content by weight is used. An inner wall with a very high resin content is made for protection against chemical attack and corrosion. The resin content at the inner wall is on the order of 80–90%, and the reinforcement consists of chopped strand mat (roughly 0.10-in.-thick mat). An intermediate layer usually is formed, with resin contents in the range of 70–80%.

Properties of contact-molded FRP pipe of various wall thicknesses are given in Tables 9.7 and 9.8.

Table 9.5. Chemical Resistance of Filament-Wound FRP Pipe (polyester resin systems) (Fibercast Co./Div. of Youngstown Sheet and Tube Co.)

Chemical	Strength (%)	Maximum Operating Temperature (°F)
Acetic Acid	0-10	100-150
	10-50	100
Acetic Acid, glacial	–	100
Aceto	0-5	150
Acrylic Acid	–	75
Adipic Acid, air	Solution	200-250
	–	300
Alcohol, ethyl	0-10	150
Alcohol, isopropyl	10	150
Alcohol, methyl	10	150
Alcohol, methyl isobutyl	10	150
Alcohol, secondary butyl	10	150
Allyl Chloride	–	100
Aluminum Chloride	–	100-300
Aluminum Fluoride	–	150
Aluminum Hydroxide	–	150
Aluminum Nitrate	–	150-250
Aluminum Sulfate	–	200-300
Alums	–	200-300
Ammonia Gas, dry	–	150
Ammonia, wet	–	100
Ammonium Carbonate	–	100-200
Ammonium Chloride	–	200
Ammonium Fluoride	25	150-200
Ammonium Hydroxide	0-10	150-200
	10-20	150
	20-30	100
Ammonium Nitrate	–	200-250
Ammonium Persulfate	–	200
Ammonium Phosphate	–	150
Ammonium Sulfate	–	200-250
Amyl Acetate	–	75
Amyl Chloride	–	75
Aniline	–	75
Anitmony Trichloride	–	150-200
Barium Carbonate	–	200-250
Barium Chloride	–	200-250
Barium Hydroxide	0-10	150-200
Barium Sulfate	–	200-250
Barium Sulfide	–	150-300
Beer	–	200
Benzene	–	100
Benzene Sulfonic Acid	–	200
Benzoic Acid	–	200
Black Liquor	–	200
Borax	–	200-250
Boric Acid	–	200
Bromic Acid	–	150

Table 9.5, continued

Chemical	Strength (%)	Maximum Operating Temperature (°F)
Bromine, liquid	–	NR[a]
Bromine Water	–	100
Butadiene	–	100
Butane	–	100
Butyl Acetate	–	75
Butyl Cellusolve	–	150
Butyric Acid	0-15	150-200
	25-50	150
Calcium Bisulfide	–	200
Calcium Carbonate	–	150-300
Calcium Chlorate	–	200
Calcium Chloride	–	200-300
Calcium Hydroxide	0-50	150-200
Calcium Hypochlorite	0-20	200
Calcium Nitrate	–	200-250
Calcium Sulfate	–	200-250
Carbon Dioxide	–	200-250
Carbon Disulfide	–	NR
Carbon Tetrachloride	–	100
Carbonic Acid	–	150
Chloroacetic Acid	0-25	100-150
Chlorine, dry	–	200
Chlorine, wet	–	200
Chlorine Dioxide	15	150
Chlorine Water	–	200
Chlorobenzene	100	100
Chloroform	100	100
Chromic Acid	5	150
	10	150
	20	75-150
	100	NR
Chromic Fluoride	–	75
Citric Acid	–	200-250
Copper Chloride	–	200-250
Copper Fluoride	–	200-250
Copper Nitrate	–	200-250
Copper Sulfate	–	200
Crude Oil, sour	–	200-300
Crude Oil, sweet	–	200-300
Deionized Water	–	200-300
Diacetone Alcohol	–	150
Dichlorobenzene	100	150
Dichloroethylene	100	75
Diethylene Triamine	100	NR
Dimenthylamine	–	NR
Ethyl Acetate	–	150
Ethyl Ether	–	75

Table 9.5, continued

Chemical	Strength (%)	Maximum Operating Temperature (°F)
Ethylene Chlorohydrin	–	NR
Ethylene Diamine	–	NR
Ethylene Glycol	–	200
Ethylene Oxide	–	NR
Fatty Acids	–	200
Ferric Chloride	–	200-300
Ferric Nitrate	–	200-250
Ferric Sulfate	–	200
Ferrous Chloride	–	200-250
Ferrous Sulfate	–	200
Fluorine Gas, wet	–	75
Fluorosilicic Acid	10	200
Fluoroboric Acid	–	200
Formaldehyde	40	150-200
Formic Acid	25	100
Freon®	–	150
Gas, natural	–	200
Gasoline, sour	–	200-300
Gasoline, 108 octane	–	150
Glucose	–	200-300
Glycerine	–	150-300
Glycol, ethylene	–	200
Glycol, propylene	–	200-250
Heptane	–	150
Hexane	–	75
Hydraulic Fluid	–	150-200
Hydrobromic Acid	50	150-200
Hydrochloric Acid	0-37	200
Hydrocyanic Acid	10	150
Hydrofluoric Acid	10	150
Hydrogen	–	150
Hydrogen Peroxide	10	150
	20-30	75-150
Hydrogen Sulfide		
Aqueous	–	200-250
Dry	–	200-250
Hypochlorous Acid	–	200
Kerosene	–	200-250
Lactic Acid	–	200
Lauric Acid	–	200
Lead Acetate	–	200-250
Levulinic Acid	25	200-250
Magnesium Carbonate	–	200-250
Magnesium Chloride	–	200-300

Table 9.5, continued

Chemical	Strength (%)	Maximum Operating Temperature (°F)
Magnesium Hydroxide	–	150-250
Magnesium Nitrate	–	200-250
Magnesium Sulfate	–	200-250
Maleic Acid	100	150-200
Mercury	–	200-300
Mineral Oils	–	200-300
Naphtha	–	200
Naphthalene	–	100-150
Nickel Chloride	–	200-300
Nickel Nitrate	–	200
Nitric Acid	0-15	100
	20	75
Oleic Acid	–	200
Perchloric Acid	70	75
Phenol	1	150
Phosphoric Acid	0-75	200-250
	75-85	200
	85-110	200
Phosphorus Pentoxide	0-54	200
Picric Acid	–	75
Potassium Bicarbonate	–	150-250
Potassium Bromide	–	200
Potassium Carbonate	–	150-250
Potassium Chloride	–	200-300
Potassium Dichromate	–	200-250
Potassium Hydroxide	–	150-200
Potassium Permanganate	10-25	150
Potassium Sulfate	–	150-200
Propane	–	100
Silver Nitrate	–	200-250
Soaps	–	200-250
Sodium Acetate	–	150
Sodium Bicarbonate	–	150-250
Sodium Bromide	–	200
Sodium Carbonate	–	100-250
Sodium Chlorate	–	150
Sodium Chloride	–	200-300
Sodium Cyanide	–	200-250
Sodium Fluoride	–	200-250
Sodium Hydroxide	0-5	150-200
	>5	150-200
Sodium Hypochlorite	0-10	75-100
Sodium Methoxide	40	150
Sodium Nitrate	–	200-300
Sodium Peroxide	–	75
Sodium Sulfate	–	200-300

Table 9.5, continued

Chemical	Strength (%)	Maximum Operating Temperature (°F)
Sodium Sulfite	–	200
Sodium Thiosulfate	–	150-200
Stearic Acid	–	150
Sulfamic Acid	25	200
Sulfite Liquors	–	200
Sulfur Dioxide, (wet and dry)	–	200
Sulfuric Acid	10	200-250
	25	150-200
	50	100-200
	60	75-200
	70	75
Sulfurous Acid	0-7	200
Tartaric Acid	–	200-250
Tetraethyl Lead	–	100
Toluol	100	150
Trichloroacetic Acid	–	200
Trichloroethylene	–	150
Triethylamine	–	75
Tung Oil	–	200
Turpentine	–	75
Urea	–	150
Vinegar	–	150
Vinyl Acetate	–	150
Water, deionized	–	200-250
Water, distilled	–	200-300
Water, fresh	–	200-300
Water, salt	–	200-300
Xylene	–	150
Zinc Chloride	–	200-250
Zinc Sulfate	–	200-250

[a]NR = not recommended.

Properties, particularly those of layup pipe, vary with temperature. Consequently, the operating temperature must be considered in design and materials selection. For example, tensile strength will tend to increase at elevated temperatures, whereas flexural properties (i.e., flexural modulus, flexural strength) decrease as temperature goes up. This must be considered when the modulus is used in computing deflections, vacuum capability and any elastic stability conditions.

Table 9.6. Recommended Wall Thickness for

	Internal Pressure					
	25		50		75	
Inside Diameter (in.)	Wall (nominal) (in.)	wt/ft (lb/ft)	Wall (nominal) (in.)	wt/ft (lb/ft)	Wall (nominal) (in.)	wt/ft (lb/ft)
2	0.187	0.863	0.187	0.863	0.187	0.863
2.5	0.187	1.08	0.187	1.08	0.187	1.08
3	0.187	1.30	0.187	1.30	0.187	1.30
4	0.187	1.73	0.187	1.73	0.187	1.73
5	0.187	2.17	0.187	2.17	0.250	2.89
6	0.187	2.59	0.187	2.59	0.250	3.45
8	0.187	3.45	0.250	4.60	0.250	4.60
10	0.187	4.32	0.250	5.76	0.312	7.20
12	0.187	5.18	0.250	6.91	0.375	10.4
14	0.250	8.05	0.312	10.7	0.375	12.1
16	0.250	9.21	0.312	11.5	0.437	16.1
18	0.250	10.36	0.375	15.5	0.500	20.7
20	0.250	11.51	0.375	17.3	0.500	23.0
24	0.250	13.82	0.437	24.2	0.625	34.5
30	0.312	21.59	0.500	34.5	0.750	51.8
36	0.375	31.08	0.625	51.8	0.937	77.7
42	0.375	36.3	0.750	72.5	1.06	102.8
48	0.437	48.4	0.812	89.8	1.25	138.2
54	0.500	62.2	0.937	116.6	1.37	171.0
60	0.500	69.1	10.00	138.2	1.50	207.3
72	0.625	103.6	1.25	207.3	1.81	300.5
84	0.750	145.1	1.44	278.1	2.12	411.0
96	0.812	179.6	1.62	359.2	2.44	540.0

FRP pipe also can be manufactured by centrifugal casting and pressure molding. Wall thicknesses vary, and the use of fillers in both methods is optional. The manufacturing processes produce pipe that is greatly different in construction and physical properties; however, the general benefits each has to offer are approximately the same. Table 9.9 compares various physical properties of the three principal manufacturing techniques.

9.1.4 Service Considerations for Plastic Pipe

Physical properties and strengths vary with the resin system (for FRP) and fabrication method. The coefficient of thermal expansion is approximately 10.5×10^{-6} in./in./°F for centrifugally cast FRP pipe and 13×10^{-6} in./in./°F for filament-wound FRP pipe. Figure 9.14 gives typical data on the

Contact-Molded FRP Pipe at Various Pressures [3]

Rating, psi					
100		125		150	
Wall (nominal) (in.)	wt/ft (lb/ft)	Wall (nominal) (in.)	wt/ft (lb/ft)	Wall (nominal) (in.)	wt/ft (lb/ft)
0.187	0.863	0.187	0.863	0.187	0.863
0.187	1.08	0.187	1.08	0.250	1.44
0.187	1.29	0.250	1.73	0.250	1.73
0.250	2.30	0.250	2.30	0.250	2.30
0.250	2.89	0.312	3.61	0.375	4.33
0.250	3.45	0.312	4.32	0.375	5.18
0.312	5.76	0.375	6.91	0.437	8.06
0.375	8.64	0.437	10.07	0.500	11.9
0.437	12.1	0.500	13.82	0.625	17.3
0.500	16.1	0.625	20.15	0.750	24.2
0.562	20.7	0.687	25.3	0.812	30.0
0.625	25.9	0.750	31.1	0.937	38.9
0.687	31.7	0.875	40.3	1.00	46.1
0.812	44.9	1.00	55.3	1.12	62.2
1.00	69.1	1.25	86.4	1.50	103.6
1.25	103.6	1.50	124.4	1.81	150.3
1.44	139.0	1.75	169.3	2.12	205.9
1.62	179.6	2.00	221.1	2.44	269.4
1.81	225.4	2.25	279.8	2.75	342.6
2.00	276.3	2.50	345.4	3.00	415.0
2.44	404.0	3.00	497.0	3.62	601.8
2.81	545.0	3.50	677.0	4.25	820.8
3.25	720.0	4.00	885.0	4.81	1065.0

total expansion per 100 ft of pipe as a function of temperature change. In general, FRP pipe acts as a thermal insulator (i.e., it conducts heat more slowly than metal, which expands and contracts more rapidly). This must be considered carefully to prevent any excess stress on FRP pipe or components when connections are made to metal piping systems. Tanks, steel piping and other structures can transmit significant stresses to plastic pipe; hence, these deflections created by the nonplastic system must be compensated for by independent means. In a piping system, nonplastic lines should be anchored immediately before the plastic connection to ensure that any thermal strain is not transmitted to the plastic system.

Usually, directional changes in piping systems can compensate for small expansions due to temperature changes (the same is often true for contractions due to a drop in temperatures). Expansion joints or expansion

Table 9.7. Physical Properties of Contact-Molded FRP Pipe (Ceilcote Co./Div. of General Signal Corp.)

Wall Thickness (in.)	Ultimate Tensile Strength (psi)	Flexural Strength (psi)	Flexural Modulus of Elasticity (psi)	Compressive Strength (psi)
0.187	9,000	16,000	700,000	18,000
0.250	12,000	19,000	800,000	20,000
0.312	13,500	20,000	900,000	21,000
0.375	15,000	22,000	1,000,000	22,000

Table 9.8. General Properties of Contact-Molded FRP Pipe (Ceilcote Co./Div. of General Signal Corp.)

Specific Gravity	1.3-1.7
Impact Strength (ft-lb)	30-40
Barcol Hardness	35-45
Thermal Conductivity @ 212°F (Btu/ft^2-hr-°F/in.)	1.5
Linear Coefficient of Thermal Expansion (32-212°F) (in./in.-°F)	15×10^{-6}
Heat Distortion Temperature at 264 psi (°F)	250-350
Electrical Resistivity (ohm-cm)	10^{14}
Dielectric Constant	3.16
Machinability	Good
Light Transmission	Translucent

loops should be designed into the system wherever possible to prevent overstressing.

When expansion joints are used they should be preset to properly allow for both expansion and contraction. The amount of precompression necessary for an expansion joint can be computed from the following formula:

$$\widetilde{C}_p = \frac{R_M(T_i - T_m)}{\Delta T} \tag{9.9}$$

where $\widetilde{C}_p$ = amount of precompression (in.)
R_M = expansion joint rated movement (in.)
T_i = temperature at which system was installed (°F)
T_m = minimum operational temperature anticipated (°F)
ΔT = operational temperature difference anticipated (°F)

Table 9.9. Average Properties of Various Types of FRP Pipe [3]

	Filament-Wound, Epoxy or Polyester Resins	Centrifugally Cast Epoxy or Polyester Resin	Contact-Molded Polyester Resin
Modulus of Elasticity in Axial Tension @ 77°F (psi)	$1.0\text{-}2.7 \times 10^6$	$1.3\text{-}1.5 \times 10^6$	$0.8\text{-}1.8 \times 10^6$
Ultimate Axial Tensile Strength @ 77°F (psi)	8,000-10,000	25,000	9,000-18,000
Ultimate Hoop Tensile Strength @ 77°F (psi)	24,000-50,000	35,000	9,000-10,000
Modulus of Elasticity in Beam Flexure @ 77°F (psi)	$1\text{-}2 \times 10^6$	$1.3\text{-}1.5 \times 10^6$	$1.0\text{-}1.2 \times 10$
Coefficient of Thermal Expansion (in./in./°F)	$8.5\text{-}12.7 \times 10^{-6}$	13×10^{-6}	15×10^{-6}
Heat Deflection Temperature @ 264 psi (°F)	200-300	200-300	200-250
Thermal Conductivity (Btu/ft^2-hr-°F/in.)	1.3-2.0	0.9	1.5
Specific Gravity	1.8-1.9	1.58	1.3-1.7
Corrosive Resistance	E[a]	E	NR[b]

[a]E = excellent, will resist most corrosive chemicals.
[b]NR = not recommended for highly alkaline or solvent applications.

If expansion joints are not used, then the piping arrangement must be designed to absorb the expansion or contraction. End loads are caused when the pipe is restrained and no proper means of expansion is provided. Such situations require guiding to keep the pipe straight. Guides direct forces in the axial direction and prevent buckling.

When the piping system is allowed to move freely with temperature change, care must be taken to ensure that all fixed points, anchors, laterals, elbows, etc., are within the recommended safe stress level. High stress levels can be minimized by employing one or more 90° bends with adequate pipe leg between anchors or rigid mounting stations.

The amount of expansion that can be handled by right-angle pipe legs can be estimated from the plot shown in Figure 9.15, which is a plot of the allowable deflection versus pipe length with pipe diameter as a parameter. For pure thermal stress, the amount of expansion at the point of leg attachment is expressed by

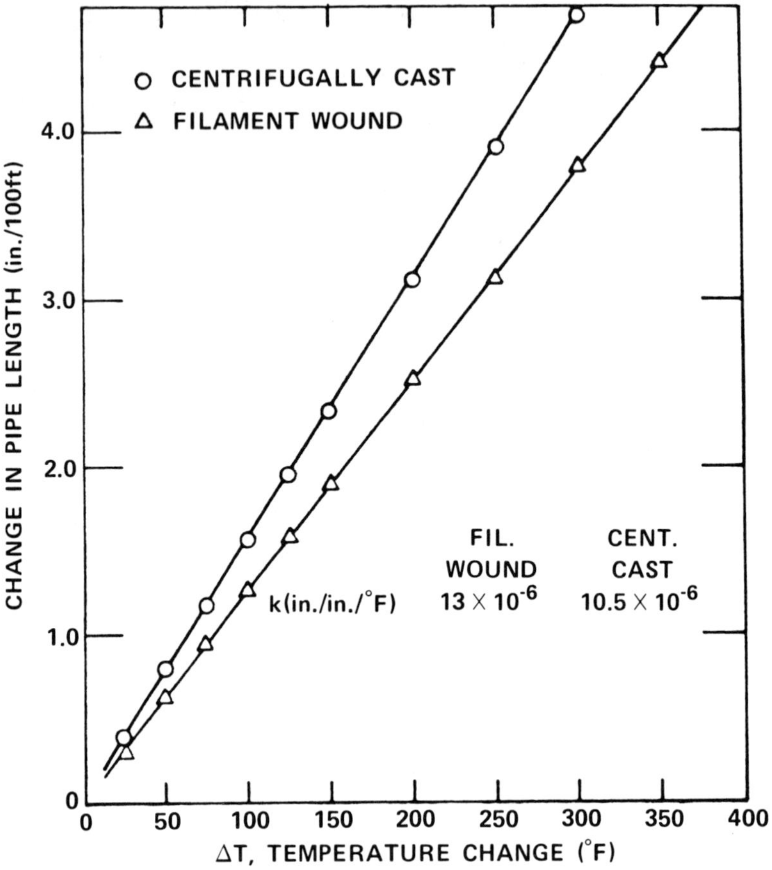

Figure 9.14. Expansion as a function of temperature change [4].

$$\Delta x = \alpha L \Delta T \tag{9.10}$$

where Δx = expansion (in.)
α = coefficient of thermal expansion (in./in./°F)
L = pipe length (in.)
ΔT = temperature change (°F)

From Equation 9.10 and Figure 9.15, the necessary pipe leg length can be obtained to take up the anticipated expansion.

Another approach to this problem is to anchor the pipe at some intermediate point and direct the expansion to another portion of the

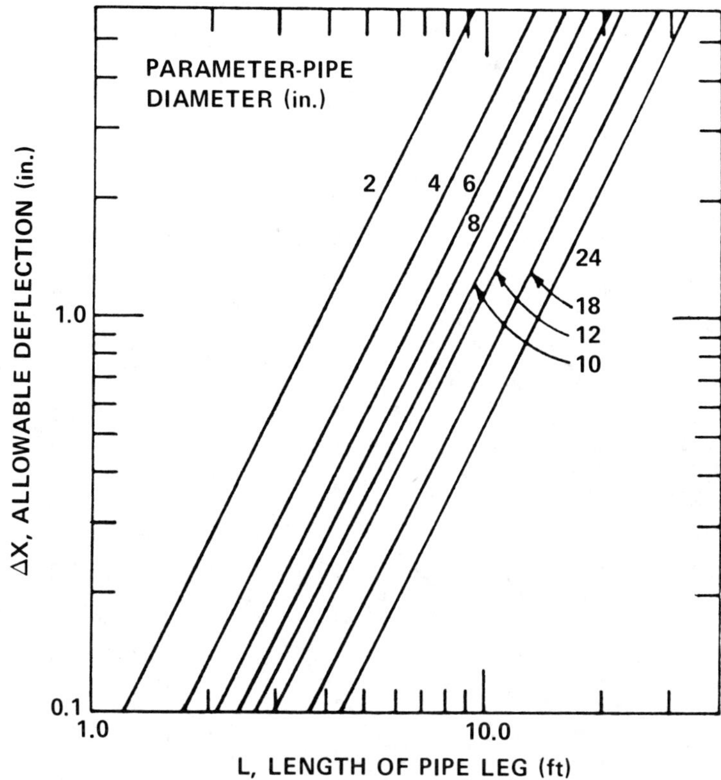

Figure 9.15. Allowable deflection for right-angle pipe legs [5].

system. To eliminate expansion completely, long, straight runs can be anchored at both ends. If the latter arrangement is selected, anchors must be designed to meet the end force, which is the total force resulting from both the internal operating pressure and the thermal expansion stress. The end force resulting from the internal operating pressure of the system can be estimated from the plot shown in Figure 9.16, which is a plot of end force, F, versus operating pressure, P, with pipe diameter as a parameter.

The amount of end force exerted on anchors due to thermal stress from expansion and contraction can be estimated from the plots shown in Figures 9.17 and 9.18, respectively.

Note that when U bolts are used as anchoring devices, the pipe should be protected from excessive pressure. In general, supports that make only a point contact or have narrow supporting areas should not be used unless additional protection is supplied.

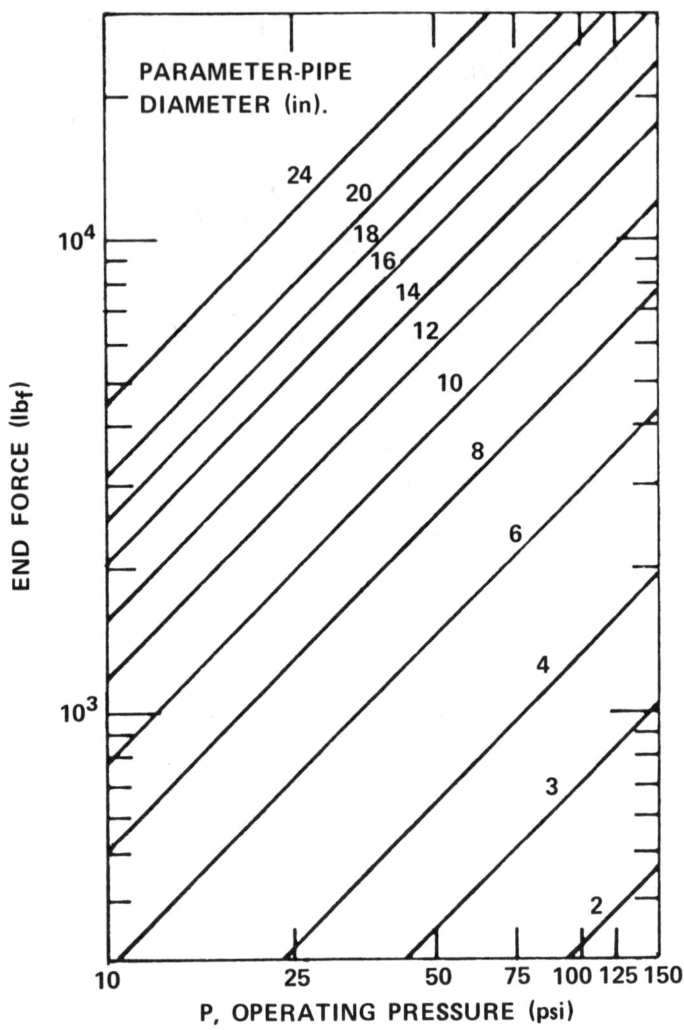

Figure 9.16. End force exerted on anchors due to internal pressure [5].

9.1.5 Pressure and Vacuum Service Considerations

The wall thickness for FRP pipe that is to be fabricated for pressure service can be computed from Barlow's formula:

$$\delta = \frac{PD}{2\sigma} \tag{9.11}$$

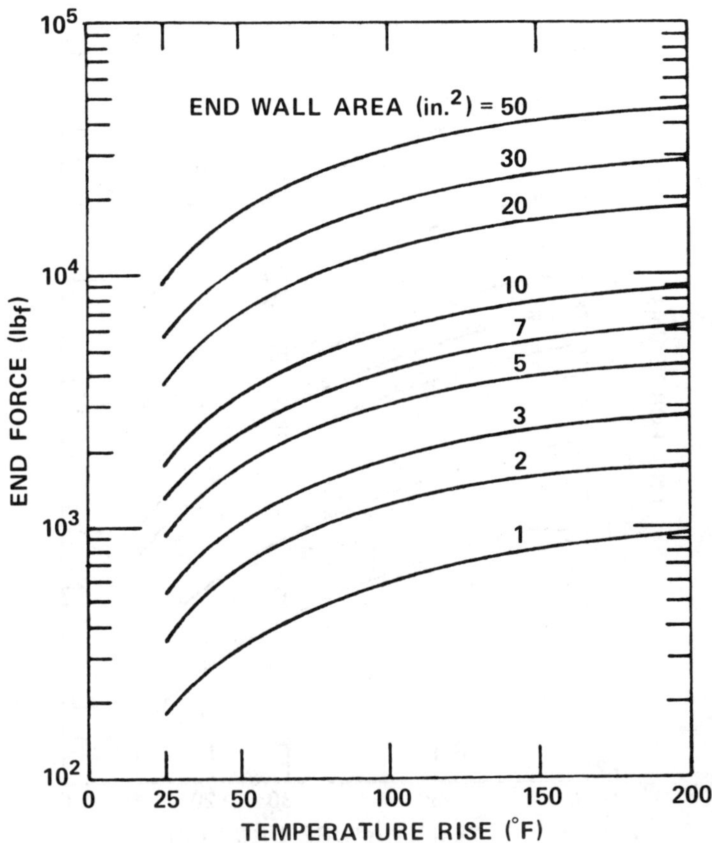

Figure 9.17. End force exerted on anchors due to expansion [5].

where δ = laminate wall thickness required (in.)
P = operating pressure (psi)
D = inside diameter of pipe (in.)
σ = design stress (psi)

After the system has been installed, it should be tested for leaks by filling the lines with water and pressurizing to one and half times the rated pressure of the pipe. Testing with compressed air or gases is dangerous and is not recommended.

Consideration should be given to protection against the occurrence of high vacuums during operation, as reinforced plastics have relatively low moduli of elasticity. Adequate vacuum protection can be accomplished by employing vacuum-release valves or stand pipes. Vacuum lines should be

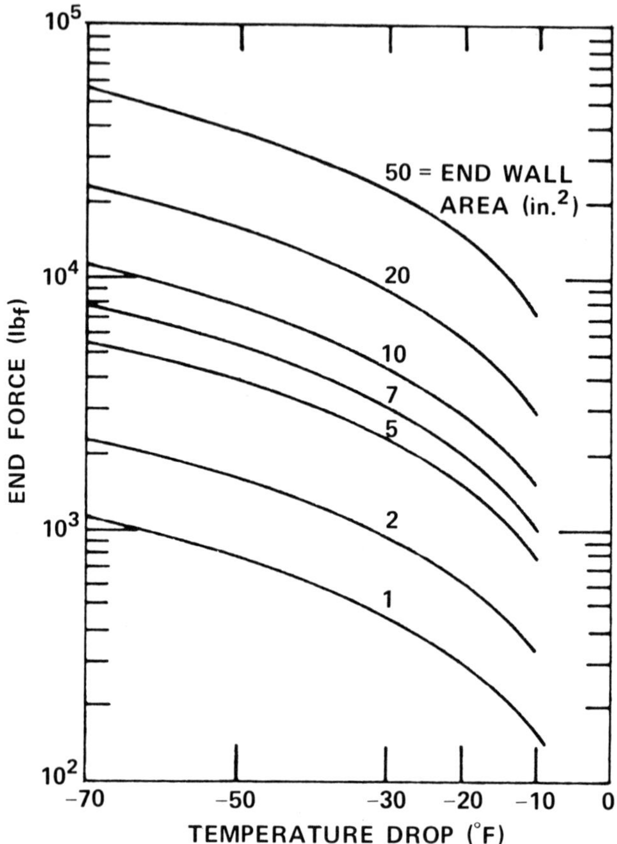

Figure 9.18. End force exerted on anchors due to contraction [5].

supported by 180° saddles and should not be exposed to any external loads. Whenever flow velocity is disrupted suddenly (e.g., pump shutoff or valve closure), vacuum conditions may arise.

Disturbances or changes in a flowline can cause excessive or instantaneous rises in line pressure. This is caused most often by closing a valve or by sharp expansions or contractions in the line or other constrictions, such as sharp bends, tees, etc. Water hammer can occur in any size pipeline, and it is important to note that the normal working pressure does not play a role in the intensity of the hammer. The impact of the flow stoppage is proportional to the mass of liquid holdup being stopped and the square of the flow velocity. The potential for these disturbances should be accounted for in the design.

Air chambers are recommended for cushioning the shock; however, frequent maintenance and spot checking is required as they often become inoperable due to waterlogging. Water hammer does not always create a noise problem. It is most often detected by violent vibration of the pipeline. Furthermore, the point of region where water hammer is detected may not be the source. Shock waves can travel over a considerable distance through the line and be the cause of damage or noise problems far from the wave source. Under certain conditions, shock forces caused by water hammer can reach levels of sufficient magnitude to rupture the piping system. The buildup of shock waves is illustrated in Figure 9.19.

Proper design practices can eliminate, or at least minimize, water hammer. Check valves on pumps should be designed to close as quickly as possible to maintain at a minimum the liquid velocity flowing back through the check valve. Shock waves from quick-closing valves are most difficult to eliminate in short lines. Reliable estimates should be made of the impact of water hammer on pressure rise to ensure that the system will function properly.

The velocity of the shock wave through the line can be estimated from the following expression:

$$\omega = \frac{12}{\left[\frac{\rho}{g_c}\left(\frac{1}{\Upsilon} + \frac{D}{\epsilon\chi}\right)\right]^{1/2}} \tag{9.12}$$

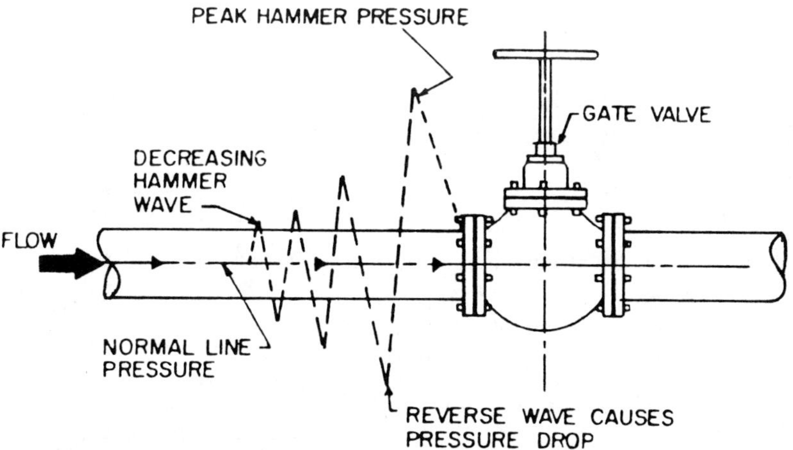

Figure 9.19. Buildup of shock waves by quickly closing a valve at the end of a line.

where ω = shock wave velocity (fps)
Υ = bulk modulus of compressibility of the fluid (psi)
ρ = fluid density (lb/ft^3)
D = pipe inside diameter (ft)
ϵ = modulus of elasticity of the pipe (psi)
χ = reinforced wall thickness of the pipe (ft)

For an estimate of the instantaneous rise in pressure that would be experienced from a fast valve closure at the end of a line, the following expression can be used:

$$\Delta P' = \frac{\rho \omega \bar{U}}{144\, g_c} \tag{9.13}$$

where $\Delta P'$ is the pressure rise above normal (psi). Equation 9.13 is subject to the following constraint:

$$\dot{\theta} \leqslant \frac{2L}{\omega} \tag{9.14}$$

where $\dot{\theta}$ = elapsed time for closing the valve completely (sec)
L = length of straight pipe into the valve (ft)

For a situation in which the valve is closed slowly, the following set of expressions are recommended for use along with Figure 9.20:

$$\hat{K} = \frac{\omega \bar{U}}{2gh_0} \tag{9.15}$$

$$N = \frac{\omega t}{2L} \tag{9.16}$$

where g = acceleration due to gravity (32.2 ft/sec^2)
$\hat{K}$ = top axis of plot in Figure 9.20
N = left axis of plot in Figure 9.20
h_0 = normal head for steady flow (ft)

Equations 9.15 and 9.16 are subject to the following condition:

$$t > \frac{2L}{\omega} \tag{9.17}$$

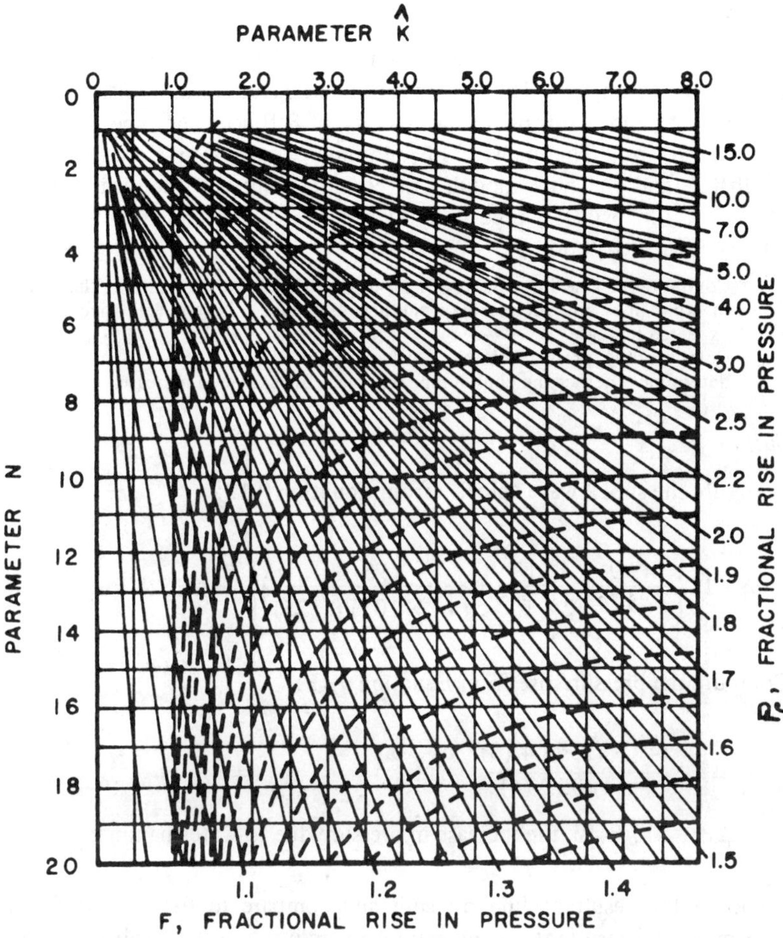

Figure 9.20. Pressure rise prediction plot for slow valve closure [6].

Once parameters $\hat{K}$ and N are known, Figure 9.20 can be used to estimate the fractional rise in pressure, P_f, defined by

$$P_f = \frac{h_0 + \Delta P'}{h_0} \tag{9.18}$$

The following example problem illustrates the use of Equations 9.12 9.18 and Figure 9.20.

Example 9.3

A 5-in. i.d. FRP pipeline, 2000 ft long, is to be used to transport oil at near ambient conditions. The system is to be designed for a normal operating head of 70 ft and a full pipe flow velocity of 8 ft/sec. The pipe selected is rated at 170 psi. The density of oil under the system's conditions is 49.9 lb/ft^3, and the bulk modulus of compressibility of the oil is approximately 2×10^5 psi. The pipe wall thickness is ⅛ in., and the modulus of elasticity is 4.2×10^6 psi.

1. Compute the rise in pressure for an instantaneous valve closure at the end of the line and determine whether the pipe chosen will meet the expected increase.
2. Compute the pressure rise if a valve closure time of 15 sec is used and determine whether the chosen pipe will meet the expected increase.

Solution

Compute the shock wave velocity from Equation 9.12:

$$\omega = 12\left\{\frac{49.9}{32.174}\left(\frac{1}{2 \times 10^5} + \frac{5}{(4.2 \times 10^6)(0.125)}\right)\right\}^{1/2} = 2528 \text{ fps}$$

Compute the pressure rise from Equation 9.13:

$$\Delta P = \frac{(49.9)(2528.4)(8)}{(144)(32.174)}$$

$\Delta P' = 217.8$ psi, the increase in pressure due to water hammer

Compute the resultant line pressure and compare to the pressure rating of the pipe. The resultant line pressure from an instantaneous valve closure is the sum of the pressure rise and normal head, h_0:

$$h_0 = (70 \text{ ft } H_2O)(0.433) = 30.3 \text{ psi}$$

Therefore, line pressure = 217.8 + 30.3 = 248.1 psi. This is above the recommended operating pressure of the pipe selected.

For Part 2, check which boundary condition, t, matches Equations 9.14 and 9.17:

$$\frac{2L}{\omega} = \frac{2(2{,}000)}{2528.4} = 1.6 \text{ sec}$$

Since $t > 1.6$ sec, Equations 9.15 and 9.16 are applicable.

Compute parameters $\hat{K}$ and N from Equations 9.15 and 9.16, respectively:

$$\hat{K} = \frac{(2528.4)(8)}{(2)(32.2)(70)} = 4.49$$

$$N = \frac{(2528.4)(15)}{2(2000)} = 9.48$$

Locate the point designated by the values for $\hat{K}$ and N on the plot in Figure 9.20 and interpolate between diagonal lines to the P_f axis to obtain $P_f = 1.6$.

Compute the total line pressure at the peak:

$$P_f = \frac{h_0 + \Delta P'}{h_0} = 1.6$$

$$h_0 + \Delta P' = (1.6)(70) = 112 \text{ ft}$$

or the total line pressure in psi would be

$$(112)(0.433) = 48.5 \text{ psi}$$

which gives a 71% safety margin over the recommended operating pressure of the pipe.

Whenever possible, the piping system should be tested hydrostatically before initiating service. As in actual operation, care should be taken to avoid water hammer. As a further precaution against damage from water hammer, as well as external vibrations or shock loads, connections to pumps and other equipment should consist of flexible connections. Flexible connections compensate for vibrations by absorbing shock, thus eliminating the placing of undue strain on pipe and fittings. Expansion joints such as a bellows design are recommended. These are often constructed from Teflon(R)*; however, a wire-reinforced rubber hose can serve the purpose in many low- to moderate-pressure applications. Note that the latter requires much higher forces to compress and elongate and, as such, is not considered the better of the two options.

---

*Registered trademark of E. I. du Pont de Nemours and Co., Inc., Wilmington, Delaware.

9.1.6 Guidelines for Buried Pipe Installations

The depth, grade and general layout or contour of a trench will depend on the project specifications. However, there are some general guidelines.

The width of the trench floor should be roughly 1- to 2-ft wider than the diameter of the pipe to be installed. This will allow sufficient clearance for working crews and is optimum for compacting the backfill. If unstable soil conditions exist, the trench width should be widened to remove unwanted matter adjacent to the pipe. Stable backfill material should be used to replace the additional removal. As a standard safety practice, trenches should be shored and drained.

Bell and spigot joints allow the pipe to be deflected at each point for alignment adjustments and for curves, as shown in Figure 9.21. Table 9.10 gives rough estimates of the amount of joint deflection possible and the resulting radius of curve for two different lengths of pressure and non-pressure pipe. The table should be used in conjunction with Figure 9.21.

If the design calls for curves of shorter radius than can be obtained with deflected joints, then fittings will be required. Extra trench space should then be required for working.

The trench bedding should be uniform and continuous. The floor should be free of rocks greater than 1 in., lumps or debris. It is often recommended that excavation be at least 1 in. below grade and then the bottom can be built up with rock-free granular soil. For rougher terrain, where the soil is saturated with stones or the trench has a rock base, overexcavation is recommended to at least 6 in. below the grade. This can then be replaced with granular soil or crushed stone.

High spots in the trench floor promote uneven bearing on the pipe and may cause damage due to stress from backfill. They may also cause excessive wear, particularly when pulsation occurs in lines due to pumping.

Bells should be provided with depressions on the trench bottom to provide clearance and provide the proper support along the pipe length. For assembly, pipe should be strung along the top of the trench, with the bells facing the direction in which they are to be installed. For small-diameter pipes, the following procedure is recommended:

1. If the trench wall is straight, having firm edges, lower the pipe by hand.
2. If the trench is not too deep, a crew can be stationed on the bottom to reach the pipe as it is lowered.

For larger-diameter pipes, or for relatively deep trenches that do not fit the criteria in steps 1 and 2 above, the following is recommended:

1. Wooden skids should be placed along the side of the trench. (This prevents dirt and rocks from entering the bell and protects against any damage from rocks extending out of the trench wall.)

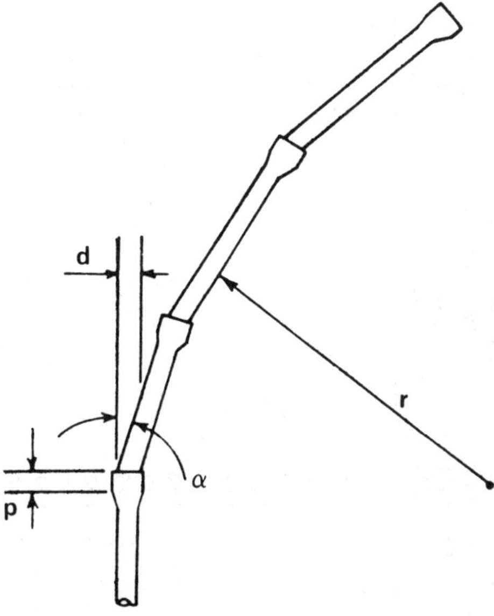

Figure 9.21. Important dimensions in bending lines in a trench: α = maximum angular deflection per joint; p = maximum pull per joint; d = maximum linear deflection per joint; and r = minimum curve radius.

2. Using rope or cable, slowly lower the pipe by hand. Each man should stand firmly on the other rope end.
3. Stop-knots should be used on the ropes under the workmen's feet to prevent heavy pipe lengths from slipping. For very large pipe, cranes or other types of heavy-duty hoists may be necessary.

Once the pipe has been positioned on the trench floor, the joints can be cleaned. The bell inside surface, spigot groove and any rubber gaskets should be cleaned thoroughly and solvents given sufficient time to evaporate. The bell sealing surface and the gasket should be lubricated. The gasket should be stretched over the spigot and snapped into the groove, making sure that it is stretched evenly around the groove and not twisted. Care should be taken not to recontaminate cleaned surfaces.

There are a number of techniques used for making the final connections. For pipe diameters up to 18 in., one or more leverage bars and a block of wood can be used as shown in Figure 9.22(A). The wood block protects the pipe from point loads. The block should be positioned horizontally against the bell, and the bar should be embedded firmly in the trench floor behind the block. The spigot should be pushed firmly into the mouth of the bell until it reaches the shoulder.

Table 9.10. Permissible Curve Radius and Joint Deflection for Pressure and Nonpressure Pipe [7]

Inside Diameter (in.)	α, Maximum Angular Deflection per Joint (°)		p, Maximum Pull per Joint (in.)	Length = 10 ft: d, Maximum Linear Deflection per Joint (in.)	Length = 10 ft: r, Minimum Curve Radius (ft)	Length = 20 ft: d, Maximum Linear Deflection per Joint (in.)	Length = 20 ft: r, Minimum Curve Radius (ft)
Nonpressure Pipe							
8	5	0	$\frac{3}{4}$	$10\frac{1}{2}$	115	21	230
10	5	0	1	$10\frac{1}{2}$	115	21	230
12	5	0	$1\frac{1}{4}$	$10\frac{1}{2}$	115	21	230
14	5	0	$1\frac{3}{8}$	$10\frac{1}{2}$	115	21	230
15	5	0	$1\frac{3}{8}$	$10\frac{1}{2}$	115	21	230
16	4	0	$1\frac{1}{4}$	$8\frac{3}{8}$	145	$16\frac{3}{4}$	290
18	4	0	$1\frac{3}{8}$	$8\frac{3}{8}$	145	$16\frac{3}{4}$	290
20	4	0	$1\frac{1}{2}$	$8\frac{3}{8}$	145	$16\frac{3}{4}$	290
21	4	0	$1\frac{5}{8}$	$8\frac{3}{8}$	145	$16\frac{3}{4}$	290
24	4	0	2	$8\frac{3}{8}$	145	$16\frac{3}{4}$	290
27	4	0	2	$8\frac{3}{8}$	145	$16\frac{3}{4}$	290
30	3	50	$2\frac{1}{8}$	$7\frac{5}{16}$	177	$14\frac{5}{8}$	354
33	2	45	$1\frac{3}{4}$	$5\frac{3}{16}$	250	$11\frac{1}{2}$	500
36	2	35	$1\frac{3}{4}$	$5\frac{7}{16}$	266	$10\frac{7}{8}$	532
39	2	35	$1\frac{3}{4}$	5	286	10	572
42	2	15	$1\frac{3}{4}$	$4\frac{11}{16}$	308	$9\frac{3}{8}$	616
45	2	0	$1\frac{3}{4}$	$4\frac{1}{4}$	344	$8\frac{1}{2}$	688
48	2	0	$1\frac{3}{4}$	$4\frac{1}{4}$	344	8	688
54	2	0	$1\frac{7}{8}$	$4\frac{1}{4}$	290	$8\frac{1}{2}$	580
60	2	0	2	$4\frac{1}{4}$	290	$8\frac{1}{2}$	580

Pressure Pipe							
8	3	0	1/2	6 1/4	190	12 1/2	380
10	3	0	5/8	6 1/4	190	12 1/2	380
12	3	0	3/4	6 1/4	190	12 1/2	380
14	3	0	3/4	6 1/4	190	12 1/2	380
15	3	0	7/8	6 1/4	190	12 1/2	380
16	3	0	3/4	6 1/4	190	12 1/2	380
18	3	0	1	6 1/4	190	12 1/2	380
20	3	0	1 1/8	6 1/4	190	12 1/2	380
21	3	0	1 1/4	6 1/4	190	12 1/2	380
24	3	0	1 3/8	6 1/4	190	12 1/2	380
27	3	0	1 1/2	6 1/4	190	12 1/2	380
30	3	0	1 5/8	6 1/4	190	12 1/2	380
33	2	0	1 1/4	4 3/16	285	8 3/8	570
36	1	50	1 1/4	3 3/4	310	7 1/2	620
39	1	40	1 1/4	3 1/2	345	7	690
42	1	36	1 1/4	3 11/32	356	6 11/16	712
45	1	30	1 1/4	3 1/8	380	6 1/4	760
48	1	24	1 1/4	2 15/16	410	5 7/8	820
54	2	0	1 7/8	4 1/4	290	8 1/2	580
60	2	0	2	4 1/4	290	8 1/2	580

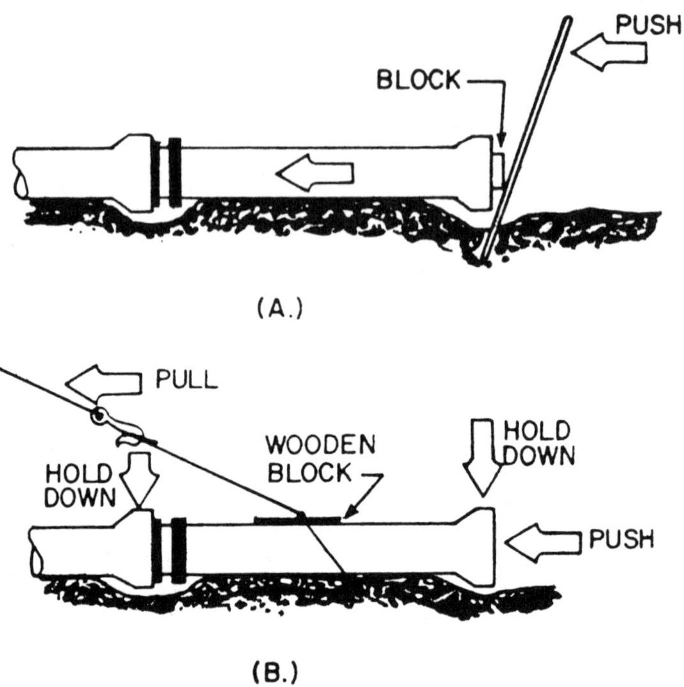

Figure 9.22. Two methods of making joints in buried installations.

Another technique is shown in Figure 9.22(B), in which a choker cable can be attached to a crane boom. This requires an adequate downward load onto the bell ends of the pipe. Care must be taken to ensure that the cable angle is low to prevent buckling or misalignment. Choker cables and come-along jacks also may be used.

If fittings are used to make changes in direction, thrust blocking must be provided against solid trench walls. These fitting areas should not be machine dug as the excavator could well dig too deeply and may damage the bearing surface of the trench wall.

The primary function of thrust blocking is to prevent movement of the piping component, that is, it serves as an anchor. A major part of the system design is to determine where thrust blocks are needed and how large they should be. Figure 9.23 shows some typical cases in which concrete thrust blocks can be employed. As a general rule when pouring concrete, the widest practical support should be provided without covering piping components.

Closures are also a significant part of the system design. Closures are the arrangement at the last length of pipe that closes the line, and the assembly usually requires a special length of pipe with a coupling. The most common way to do this is to use a long bell-shaped pipe. Another approach is to use a flexible coupling that is to be installed over the pipe. The latter method requires a pipe with either a resin outer wall, smooth outside diameter or some specially prepared standard pipe from the supplier. A third type of closure makes use of a plain end section of pipe where a full circle repair clamp can be employed as the coupling.

If pipe is to be laid under road crossings it is advisable to use conduits through which the pipe can pass. This will protect pipe from damage due to excessive loads or burial depths. Care should be taken to ensure that the pipe is properly bedded at the entrance and exit points of the conduit so that stress or wear do not occur to the pipe. Conduits also are useful in preventing pipe movement caused by pressure surges or flotation.

When installing pipe through casings or conduit, skids must be used. The use of skids prevents the pipe from snagging on the inside wall of the casing during installation. It also prevents the pipe from resting on the bells. Skids should be 2–2½ in. long or continuous and should be sufficiently high to provide clearance between the pipe bells and the casing. The leading ends of the skids should be rounded so that the pipe may be readily pushed through the casing. Skids must be strapped tightly to the pipe by metal strapping.

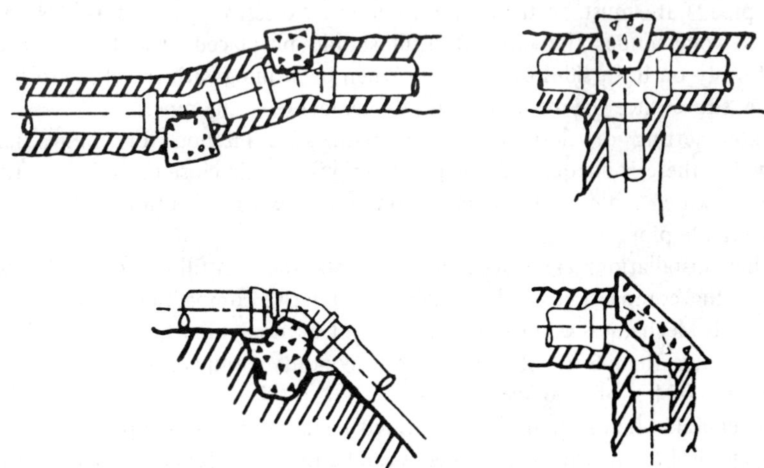

Figure 9.23. Typical examples of the use of thrust blocks.

Care should be taken to install bottom skids parallel to the centerline of the pipe, as this will prevent the pipe from rolling in the casing. Normally, six skids are recommended, mounted at 60° intervals about the periphery of the pipe and at selected axial intervals.

Pressure grouting also may be used to restrict pipe movement in the conduit; however, it may be necessary to reposition skids to accommodate the grouting hose. The recommended grout mix ratio is four parts sand to one part cement. Sufficient water should be added to obtain a pea soup consistency. The grouting machine should be equipped with a pressure gauge at its discharge. Once the delivery pressure has built up, the grout will discharge from the hose and fill the void between the pipe and conduit. Delivery pressure should be monitored carefully as just a 2–3 psi excess over the initial required delivery pressure might be enough to collapse some pipe.

Once the pipe has been laid out and final connections are made, the installation should be backfilled with sufficient fill to hold the line in place. All fittings and joints should be left open for inspection during the test period. Primary backfill serves the function of a continuous and firm support for the line. Primary backfill, sometimes called the pipe zone, must be placed and compacted so that it provides sufficient and even support about the periphery of the pipe.

Damp, stone-free, nonlumpy soil should be used for primary backfill. The soil or filler material should be relatively clean of vegetation, debris and stones over 1 in. in size. Filler materials that can be used in place of soil include sand, graded gravel or crushed rock.

Compaction may be carried out by shoveling the backfill thoroughly under the pipe. Care must be taken not to shovel excessive amounts as this will create voids under the pipe. Backfill should be placed in 6- to 8-in. lifts uniformly on both sides of the pipe. When backfilling with gravel or crushed stone, shovel slice the material under the haunches of the pipe.

Water jetting or flooding can be used as a method for compaction provided there is adequate drainage. If the pipe is laid close to, or below, the ground water table, water compaction must be used cautiously to avoid floating the pipe.

When installations are placed under streets, the backfill compaction must match the compaction of the roadbed. A relative compaction of 85-90% is required. Mechanical compactors should not be used directly over the pipe if there is less than 6 in. of cover. For heavy-duty compaction equipment, a minimum of 1 ft of coverage is required.

In general, when installing underground installations, only a portion of the line should be covered so that checks can be made for deflections against the 5% limit. A good installation should average around 2%, with maximum vertical deflections peaking at 3%.

Once primary backfilling is completed, the entire trench may be filled in with secondary backfill material. Secondary backfill is most often the native soil. It should be graded adequately so that vegetation and large stones are removed. In situations with live loads, the secondary backfill should be placed and compacted in short lifts. Figure 9.24 illustrates recommended trench configurations and composition. Note that bed depth and recommended compaction densities will vary with the specific pipe used.

Line Testing

During installation, it is often advisable to pressure test segments of the assembly before it is completed. Portions to be tested should be backfilled and compacted to prevent movement under pressure. The following factors should be considered for testing:

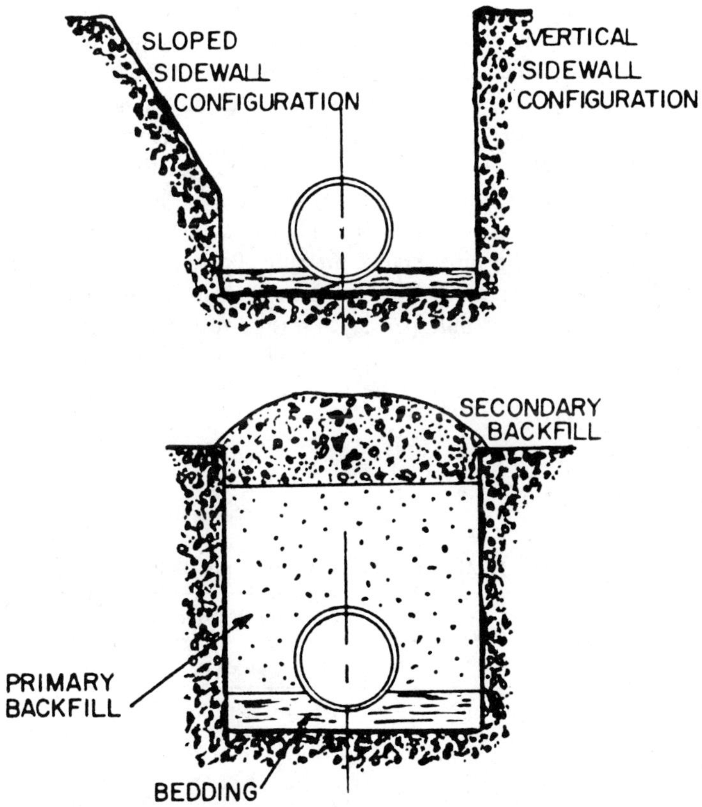

Figure 9.24. Recommended trench configurations and compositions.

1. Sufficient backfill must be provided to restrict all movement under test pressures.
2. Thrust blocks at fittings must be permanent and capable of withstanding test pressures.
3. Test ends must be capped and anchored to withstand thrusts generated under test pressures.

Lines can be tested with either air or water, although the latter is preferred. When the line is being filled with water, air becomes entrapped in the water. The air will begin to escape and becomes entrapped in the line once the flowing has stopped and the water settles down. An air vent or air relief valve should be supplied at the highest point in the line so that entrapped air can be forced out of the pipe, that is, water should displace the air so that the entire length of pipe undergoing testing is filled. This must be done before pressure is applied to the system.

It is generally recommended to fill the line 24 hours in advance at a nominal pressure prior to initiating tests. Once the line air has been expelled, the water can be pumped at the specified test pressure with a pressure gauge located at a low point in the line, where pressures are usually highest. Possible sources of leakage are in joints at saddles, transition fittings, valves and adaptors. Pipe joints normally are not trouble areas.

The project engineer should establish allowable leakage standards. As a general rule, a pipeline is acceptable if losses are less than 100 gallons per inch of diameter per mile of pipe per day for tests conducted on nonpressure lines with up to 8.7 psi of head. Usually pressure lines are not tested in excess of 1.5 times the pipe's rated pressure. Pressure lines are considered acceptable if the water needed to maintain pressure is less than 30 gallons per inch of diameter per mile per day. Pressure tests at 1.5 times the maximum operating pressure should be conducted for a minimum of 2 hours. As a rough guide, Table 9.11 can be used for allowable leakage criteria at different test pressures.

Table 9.11. Allowable Leakage Guide (gal/1000 ft/hr)

Pipe Inside Diameter (in.)	Test Pressure, psi				
	50	100	150	200	250
4	0.38	0.54	0.66	0.77	0.85
6	0.57	0.81	0.99	1.15	1.28
8	0.76	1.08	1.32	1.53	1.71
10	0.96	1.35	1.66	1.91	2.14
12	1.15	1.62	1.99	2.30	2.56

For critical operations, a more severe cyclic test is recommended. This involves subjecting the system to 5–10 pressurization cycles at 1½ times the maximum rating of the lowest-rated component in the arrangement. Sudden pressure surges or water hammer should be avoided during testing. Water hammer can cause pressures several times higher than the system's rating.

9.2 COMPENSATORS [8-11]

During operation, different parts of process equipment reach different temperatures, which results in additional thermal stresses to rigid structures. The rigid construction is not applicable if the total stresses (from both the pressure of the processing fluid and thermal stresses) become greater than is permissible. Compensators should be provided for apparatuses operating under such conditions. Compensators are widely used for a multitude of chemical and process equipment. There are two major types:

1. flexible compensators (lens, bellows, membranes) set between parts of equipment having different temperatures, and
2. stuffing-box compensators, which permit free movement of one part of equipment against another.

The interaction force due to thermal stresses induced between rigidly connected parts of an apparatus, such as a shell and tubes in a heat exchanger, is determined by the following formula:

$$P = \frac{|\alpha_s^t t_s - \alpha_p^t t_p|}{\dfrac{1}{E_s^t F_s} + \dfrac{1}{E_p^t F_p}} \qquad (9.19)$$

where t_s and t_p = the design temperatures of the shell and tubes proceeding from the maximum difference between them during operation, starting-up and shutdown of equipment (°C)

α_s^t and α_p^t = the coefficients of linear expansion for the shell and tubes material at t_s and t_p (1/°C)

E_s^t and E_p^t = the modulus of elongation for materials of the shell and tube at corresponding temperatures t_s and t_p (N/m^2)

F_s and F_p = the cross-sectional areas of the shell and tubes, (m^2), respectively

Apart from the force, P^t, there may be a pressure force, P, acting on rigidly connected parts of an apparatus and induced by the pressure of the fluid, for example, in single-pass heat exchangers. This pressure force from inside and outside the tubes causes elongating in the axial direction. It can be determined from the following formula:

$$P = 0.785\ [(D^2 - d_e^2 Z)P_e + d^2 ZP] \quad , \quad N \tag{9.20}$$

where D = the design diameter of tube sheet (m)
d_e and d = the outside and inside diameters of tubes, respectively (m)
P_e and P = the outside and inside pressures of the tubes, respectively (N/m²)
Z = the number of tubes

The rigid connection created between a shell with tubes in a heat exchanger is allowable if the following conditions are met:

1. At $t_s > t_p$ the shell is compressed in the axial direction, and the tubes are elongated:

$$\sigma_{max}^{s} = \frac{P^t}{F_s} - \frac{PE_s^t}{E_s^t F_s + E_p^t F_p} \leqslant \frac{\sigma_{y,s}^t}{1.2} \quad , \quad N/m^2 \tag{9.21}$$

$$\sigma_{max}^{p} = \frac{P^t}{F_p} + \frac{PE_p^t}{E_s^t F_s + E_p^t F_p} \leqslant \frac{\sigma_{y,p}^t}{1.2} \quad , \quad N/m^2 \tag{9.22}$$

2. At $t_s < t_p$ the shell is elongated in the axial direction, and the tubes are compressed:

$$\sigma_{max}^{s} = \frac{P^t}{F_s} + \frac{PE_s^t}{E_s^t F_s + E_p^t F_p} \leqslant \frac{\sigma_{y,s}^t}{1.2} \quad , \quad N/m^2 \tag{9.23}$$

$$\sigma_{max}^{P} = \frac{P^t}{F_p} - \frac{PE_p^t}{E_s^t F_s + E_p^t F_p} \leqslant \frac{\sigma_{y,p}^t}{1.2} \quad , \quad N/m^2 \tag{9.24}$$

The design axial force, P_p, acting in the place of tube fixing in the sheet under operating conditions is determined from the following formula:

$$P_p = 0.785\ (d_e^2 - d^2)\sigma_{max}^{P} \quad , \quad N \tag{9.25}$$

If conditions 9.21–9.24 are not fulfilled, it becomes necessary to insert a compensator in the equipment.

9.2.1 Lens Compensators

Lens compensators with one or more lenses are used widely in process equipment (such as heat exchangers, piping, etc.).

Examples of circular metal lens compensators are given in Figure 9.25. Lens compensators may be used in either vertical or horizontal apparatuses, as well as in piping systems for operation under pressures up to 0.6 MN/m^2 and sometimes up to 1.6 MN/m^2.

The compensating capability of a compensator is approximately proportional to the number of lenses. However, the use of more than four lenses in one compensator is not recommended. To decrease the resistance to the motion of the fluid, a sleeve is provided inside the shell or tube, which is welded from one side (in vertical apparatuses and piping from above; in horizontal ones from the inside of the liquid motion).

To increase the compensating capacity, lens compensators should be compressed before insertion if they are intended to supplement elongation. If they are intended to supplement compression, they should be stretched. Compressing and extending a compensator is accomplished by its full deformation to double its compensating capability.

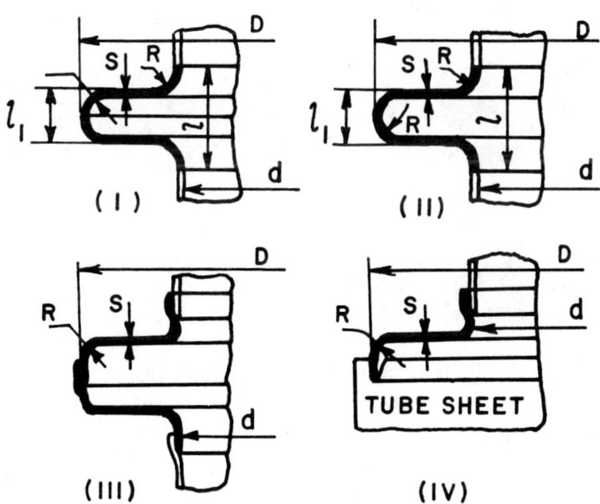

Figure 9.25. Examples of circular metal lens compensators used in process equipment: I = welded from two stamped semilenses; II = welded from several sectorial parts connected along radial planes; III = from copper and brass of two semilenses connected with soft solder; IV = from copper and brass of one semilens inserted between the shell and tube sheet with soft soldering. In all cases, $R \geqslant S$.

The design diameter of a lens compensator for a steel apparatus operating under pressure in the intertubular space $p_{in} > 0.6\ MN/m^2$ as well as for an apparatus of nonferrous metals and their alloys (independent of pressure) is determined from the following formula:

$$D' \geqslant \frac{0.5d}{0.5 - \sqrt{\dfrac{P_{in}}{\sigma_{allow}}}} \tag{9.26}$$

The deformation (tension or compression) of one lens [8, 9] is

$$\Delta_e = \frac{\alpha_1}{2} \frac{(1 - \beta)P_c d^2}{\pi E S^3} \tag{9.27}$$

where d = inside diameter of a lens (cm)
P_c = force to induce the deformation of one lens by $\Delta\ell$ (kg)
S = lens wall thickness (cm)

The coefficients α_1 and β are

$$\alpha_1 = 6.9 \left(\frac{1 - \beta^2}{\beta^2} - \frac{4\ \ell n^2 \beta}{1 - \beta^2} \right) \frac{1}{1 - \beta}$$

$$\beta = d/D \tag{9.28}$$

where D = outside diameter of a lens (cm)
d = inside diameter of a lens (cm)

The values of α_1 may be determined from Figure 9.26.

Because of a known rigidity of the system, the tubes become compressed (or elongated) by the value

$$\Delta_P = \frac{P_p L}{E F_p} \tag{9.29}$$

where P_p is the compression (elongation) force, N, of the tubes. The shell and compensator become elongated (compressed) by the value

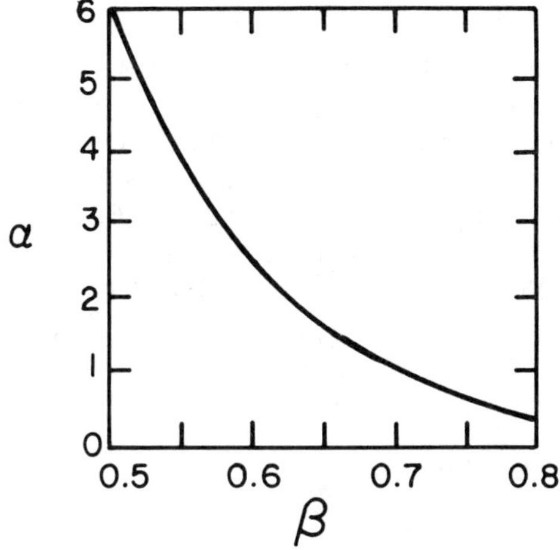

Figure 9.26. Diagram for design of lens compensators [9].

$$\Delta_s = \frac{P_s L}{EF_s} + \frac{\alpha_1(1-\beta)P_s d^2}{2\pi ES^3} = \frac{P_s}{E}\left[\frac{L}{F_s} + \frac{\alpha_1(1-\beta)d^2}{2\pi S^3}\right] \tag{9.30}$$

Taking into account that $P_p = P_S$, as well as $\Delta^s = \Delta - \Delta^p$, where $\Delta = \alpha\Delta tL$, we obtain from Equations 9.29 and 9.30 the compensator reaction, P_c, along its axis at given wall thickness, S, and ratio $\beta = d/D$:

$$P_c = \frac{L|\alpha_c^t t_c - \alpha_p^t t_p|}{\dfrac{L}{E_c^t F_c} + \dfrac{L}{E_p^t F_p} + \dfrac{0.159\,\alpha_1(1-\beta)d^2}{E_c^t S^3}} \tag{9.31}$$

where α_1 = coefficient determined from Figure 9.26
L = length of the shell and tubes (m)

For compensator type IV in Figure 9.25 (of one semilens), the deformation $\Delta\ell$ computed from formula 9.27 should be decreased by a factor of two. The design number of lenses at a given length of tubes, L, is determined from the following formula:

$$Z_e' = \frac{L}{\Delta_e}\left[|\alpha_c^t t_c - \alpha_p^t t_p| - P_c\left(\frac{1}{E_c^t F_c} + \frac{1}{E_p^t F_p}\right)\right] \tag{9.32}$$

Example 9.4

Determine whether it is necessary to insert a compensator for a heat exchanger and whether there is a need to calculate basic sizes of the compensator type I (Figure 9.25) for the following data: $d = 0.5$ m; $L = 3.0$ m; $P_p = 2.5 \times 10^6$ N/m^2; $P_e = 1 \times 10^6$ N/m^2; $t_p = 250$°C; $t_s = 100$°C; $S_s = 4$ mm; $C_s = 15$ mm; tubes 25 X 2; $C_p = 1.2$ mm; $Z = 187$; the material of the shell, tubes and compensator is carbon steel ($\sigma_{y,s}^{100} = 240 \times 10^6$ N/m^2; $\sigma_{y,p}^{250} = 200 \times 10^6$ N/m^2); $E_s^{100} = 190 \times 10^9$ N/m^2; $E_p^{250} = 176 \times 10^9$ N/m^2; $\alpha_s^t = \alpha_p^t = 0.11 \times 10^{-4}$ 1/°C; $\sigma_{allow} = 150 \times 10^6$ N/m^2.

Solution

Cross-sectional area of the shell is

$$F_s = \pi(d + S_k)S_k = \pi(0.5 + 0.004) \times 0.004$$
$$= 63.2 \times 10^{-4}\ m^2$$

The cross-sectional area of tubes is

$$F_p = \pi(d_e - S_p)S_pZ = \pi(0.025 - 0.002) \times 0.002 \times 187$$
$$= 271 \times 10^{-4}\ m^2$$

Interaction force between the shell and tubes due to the thermal stresses, from Equation 9.19, is determined from

$$P^t = \frac{\alpha_s^t|t_s - t_p|}{\dfrac{1}{E_s^tF_s} + \dfrac{1}{E_p^tF_p}} = \frac{(0.11 \times 10^{-4})|100 - 250|}{\dfrac{1}{190 \times 10^9 \times 63.2 \times 10^4} + \dfrac{1}{176 \times 10^9 \times 271 \times 10^{-4}}}$$
$$= 158 \times 10^4\ N$$

The total force elongating the shell and tubes due to fluid pressure in the tubular and intertubular spaces at the design diameter of tubular sheet, D, is determined from Equation 9.20:

$$P = 0.785\ [(D^2 - d_e^2Z)P_e + d^2ZP]$$
$$= 0.785\ [(0.5^2 - 0.025^2 \times 187) \times 10^6 + 0.021^2 \times 187 \times 2.5 \times 10^6]$$
$$= 26.6 \times 10^4\ N$$

The maximum stress in the shell from Equation 9.23 is

$$\sigma^s_{max} = \frac{P^t}{F_s} + \frac{PE^t_s}{E^t_s F_s + E^t_p F_p}$$

$$= \frac{158 \times 10^4}{63.2 \times 10^{-4}} + \frac{26.6 \times 10^4 \times 190 \times 10^9}{190 \times 10^9 \times 63.2 \times 10^{-4} + 176 \times 10^9 \times 271 \times 10^{-4}}$$

$$= 250 \times 10^6 + 8.46 \times 10^6 = 258.5 \times 10^6 \text{ N/m}^2$$

$$> \frac{\sigma^{100}_{y,s}}{1.2} = \frac{240 \times 10^6}{1.2} = 200 \times 10^6 \text{ N/m}^2$$

that is, condition 9.13 is not fulfilled. Maximum stress in the tubes, from Equation 9.24 is

$$\sigma^p_{max} = \frac{P^t}{F_p} - \frac{PE^t_p}{E^t_s F_s + E^t_p F_p}$$

$$= \frac{158 \times 10^4}{271 \times 10^{-4}} - \frac{26.6 \times 10^4 \times 176 \times 10^9}{190 \times 10^9 \times 63.2 \times 10^{-4} + 176 \times 10^9 \times 271 \times 10^{-4}}$$

$$= 58.3 \times 10^6 - 7.85 \times 10^6 = 50.5 \times 10^6$$

$$< \frac{\sigma^{250}_{y,p}}{1.2} = \frac{200 \times 10^6}{1.2} = 167 \times 10^6 \text{ N/m}^2$$

Because condition 9.23 is not fulfilled, the heat exchanger should be provided with a compensator.

The design diameter of a lens compensator is determined from Equation 9.26:

$$D' = \frac{0.5d}{0.5 - \sqrt{\frac{P_{in}}{\sigma_{allow}}}} = \frac{0.5 \times 0.5}{0.5 - \sqrt{\frac{1 \times 10^6}{150 \times 10^6}}} = 0.598 \text{ m}$$

Assume D = 0.6 m.

The design thickness of the lens wall at C = 1.5 mm, from Equation 6.21, is:

$$S' = \frac{dp}{2\sigma_{allow}} + C = \frac{0.6 \times 1 \times 10^6}{2 \times 150 \times 10^6} + 0.0015 = 0.0035 \text{ m}$$

Assume S = 4 mm. At $\beta = d/D = 0.5/0.6 = 0.834$, from Figure 9.26, $\alpha_1 \cong 0.2$. The reaction of the compensator is determined from Equation 9.31:

$$P_c = \frac{L\alpha_c^t|t_c - t_p|}{\dfrac{L}{E_c^t F_c} + \dfrac{L}{E_p^t F_p} + \dfrac{0.159\,\alpha_1(1-\beta)d^2}{E_c^t S^3}}$$

$$= \frac{3 \times 0.11 \times 10^{-4}|100 - 250|}{\dfrac{3}{190 \times 10^9 \times 63.2 \times 10^{-4}} + \dfrac{3}{176 \times 10^9 \times 271 \times 10^{-4}} + \dfrac{0.159 \times 0.2\,(1 - 0.834)\,0.5^2}{190 \times 10^9\,(0.004)^3}}$$

$$= 4.45 \times 10^4 \text{ N}$$

Deformation of one lens from Equation 9.27 is

$$\Delta_e = \frac{0.159\,\alpha_1(1-\beta)P_c d^2}{E_c^t S^3} = \frac{0.159 \times 0.2\,(1 - 0.834)\,4.45 \times 10^4 \times 0.5^2}{190 \times 10^9 \times 0.004^3}$$

$$= 4.84 \times 10^{-3} \text{ m}$$

The design number of lenses from Equation 9.32 is

$$Z_e' = \frac{L}{\Delta_e}\left[|\alpha_c^t t_c - \alpha_p^t t_p| - P_c\left(\frac{1}{E_c^t F_c} + \frac{1}{E_p^t F_p}\right)\right]$$

$$= \frac{3}{4.84 \times 10^{-3}}\left[0.11 \times 10^{-4}|100 - 250| - 4.45 \times 10^4\left(\frac{1}{190 \times 10^9 \times 63.2 \times 10^{-4}} + \frac{1}{176 \times 10^9 \times 271 \times 10^{-4}}\right)\right] = 0.995$$

Assume $Z_e = 1$.

Example 9.5

Design the compensator of a heat exchanger with a heating surface of 50 m^2; diameter of the shell, d_s = 478 mm; wall thickness of the shell, s = 6 mm; lens inside diameter, d = 466 mm; lens outside diameter, D = 750 mm; tubes 25 X 3; number of tubes, Z = 120; length of the tubes, L = 600 mm; the temperature of the shell is 120°C higher than that of the tubes; the operating pressure in the intertubular space, p = 10 kg/cm^2.

Solution

The wall thickness of the compensator is determined from Equation 6.21:

$$S' = \frac{PD}{2\sigma_{allow}} + C = \frac{10 \times 75}{2 \times 900} + 0.1 = 0.52 \text{ cm}$$

Assume S = 0.6 cm = 6 mm.

The force originating in the tubes (shell) due to the temperature difference Δt is determined from Equation 9.19:

$$P_s = P_p = \frac{2\pi \alpha_1 \Delta t L E S^3}{(1 - \beta)\alpha_1 d^2}$$

$$= \frac{2\pi \times 0.00012 \times 120 \times 600 \times 2.1 \times 10^6 \times 0.6^3}{(1 - 466/750)\ 2.1 \times 46.6^2} \simeq 1100 \text{ kg}$$

The stress in the compensator resulting from the force, P_c, is determined from the following formula:

$$\sigma_c = \frac{P_c(1 - \beta)}{\pi(S - C)^2} = \frac{1100\ (1 - 0.62)}{\pi(0.5)^2} \simeq 540 \text{ kg/cm}^2$$

if $S < 1$ cm and $\beta \leqslant 0.7$.

The load on the tubes is computed from Equation 9.20:

$$P_p = 0.785 \times p \times (D^2 - d_e^2 Z) + P_{ther} = 0.785 \times 10\ (73.8^2 - 120 \times 2.5^2)$$
$$+ 1100 \simeq 3800 \text{ kg}$$

The load on one tube is

$$P'_p = \frac{P_p}{Z} = \frac{38{,}000}{120} \simeq 316 \text{ kg}$$

The load on 1 cm of tube circumference length is

$$\sigma = \frac{P'_p}{\pi de} = \frac{316}{\pi \times 2.5} \simeq 40 \text{ kg/cm}$$

This value is within an allowable limit. This calculation was made for steady-state behavior of heat transfer. The other condition that may arise is when the heat carrier is either in the intertubular or tubular space. In the first case the tubes and shell are heated approximately equally, but in the second case, the high heating of the tubes may result in high stresses, causing failure of the rolled connections.

9.2.2 Stuffing-Box Compensators

Along with lens compensators, stuffing-box compensators are used widely in the fabrication of process equipment, especially for round parts of apparatuses with inside diameters of less than 150 mm.

The important advantage of stuffing-box compensators over lens designs is their ability to be applied under high pressures. At the same time, stuffing-box compensators have a major disadvantage; they may leak and, therefore, there is a need to tighten them periodically. Examples of stuffing-box compensator designs with soft packing are given in Figure 9.27. The packing material is selected according to the fluid and its design

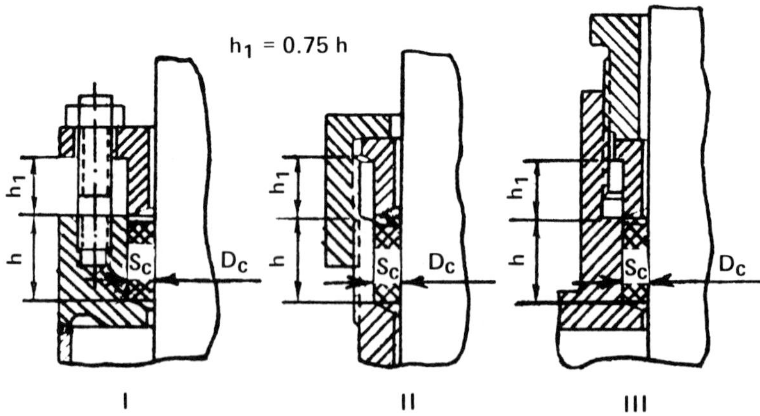

Figure 9.27. Examples of stuffing-box compensators: I = $D_c \geqslant 25$ mm; II, III = $D_c <$ 25 mm.

temperature. At high fluid temperatures, water cooling of the packing through a jacket is required. In designing stuffing-box compensators with soft packing, the design thickness of the soft packing, S'_c, is determined from the following formula [12]:

$$S'_c = 0.044\sqrt{D_c}\ , \quad m \tag{9.33}$$

The value of S'_c should be $3 < S'_c < 25$ mm. Values for the height of the packing are given in Table 9.12.

The design compression force of the stuffing-box, P'_c, is determined approximately from the following formula [12]:

$$P'_c = \pi(D_c + S_c)S_c q\ , \quad N \tag{9.34}$$

where q = specific load on the stuffing-box sleeve

Values are given in Table 9.13.

The structural design of basic elements of stuffing-box compensators, such as studs, thread, flange, sleeve, etc., are made according to the design compression force, P'_c, from the corresponding formulas given in Chapters 11 and 12.

Example 9.6

Determine the basic dimensions of a stuffing-box compensator with soft packing. The design is type I in Figure 9.27, and the design compression force in setting the compensator between a heat exchanger shell and its tube sheet is given in Example 9.1 (instead of a lens compensator).

Table 9.12. Height of the Soft Packing for Stuffing-Box Compensators

p, MN/m^2	$\leqslant 0.6$	>0.6 to 1.6	>1.6 to 2.5	>2.5
h_1 mm	$3S_c$	$4S_c$	$5S_c$	$6S_c$

Table 9.13. Specific Load on the Stuffing-Box Sleeve for Soft Packing

p, MN/m^2	0.6	1	1.6	2.5	4	6.5	10	>10
q, MN/m^2	1.8	2.5	3	5	7.5	10	12.5	P

Solution

From Example 9.1, we have $D_c = d = 0.5$ m:

$$P_c = 1 \times 10^6 \text{ N/m}^2$$

The design packing thickness from Equation 9.33 is

$$S'_c = 0.044\sqrt{D_c} = 0.044\sqrt{0.5} = 0.0311 \text{ m}$$

Assume S_c = 25 mm (the maximum recommended size). The packing height from Table 9.12 is

$$h = 4S_c = 4 \times 25 = 100 \text{ mm}$$

The height of the compression sleeve from Figure 9.27 is

$$h_1 = 0.75 \times h = 0.75 \times 100 = 75 \text{ mm}$$

The specific load of the sleeve on the packing from Table 9.13 is $q = 2.5 \times 10^6$ N/m^2.

The design compression force for the packing from Equation 9.34 is, hence,

$$P'_c = \pi(D_c + S_c)S_c q = \pi(0.5 + 0.025) \times 0.025 \times 2.5 \times 10^6 = 10.3 \times 10^4 \text{ N}$$

9.3 VALVES AND THEIR SELECTION

Valves provide the means to adjust inputs of process control systems. Proper valve selection is essential for a well-designed piping system. For systems requiring relatively short time lags and having large capacities, fairly simple control systems are adequate, normally permitting the use of high-sensitivity controllers with quick-opening valves. For this type service, flow characteristics become secondary considerations, and valve characteristics such as size and material become primary.

As time lag increases and capacity decreases, required control schemes become more complex. In general, the larger the time lag, the smaller and more precise must be the variations in flowrate of the controlling medium for given changes in process conditions.

Common control valves are illustrated in Figures 9.28 through 9.32. Table 9.14 provides brief descriptions of various control valves.

Knowledge of where the various parts are located in different valve types ensures proper installation and helps to maintain them. Preventive maintenance, such as regular lubrication, tightening or replacing packing, or replacing composition discs or resurfacing seats and discs on a regular basis will greatly extend the life of a valve. Figures 9.33 and 9.34 show the basic parts of common valves.

A brief description of the major valve types and guidelines for their installation and maintenance are given below.

9.3.1 Gate Valves

As implied by the name, when the wheel is turned to lift the wedge of a gate valve it is literally like opening a flood gate. Fluid flows freely through a full pilot with little pressure drop. As with all standard valves, turning the handwheel counterclockwise opens it, while turning it clockwise closes it. Gate valves are neither meant to be throttling valves, nor are they designed for frequent operating cycles. Operation of a gate valve with the wedge partially open can cause vibration and rapid wear of the wedge and seat surfaces, leading to possible valve failure.

Nonrising stem gate valves are particularly useful when there are space limitations. The gate itself is threaded and rides up on the stem when the stem is rotated. Other variations include inside screw, resisting stem and outside screw and yoke (see Figure 9.33). "U" bolt gate valves were designed for quick disassembly and are recommended for use with viscous fluids where the fluid tends to collect in the bonnet structure.

Unless specially constructed, gate valves can be installed with either face to the inlet side. Normally, discoloration or residue on the bonnet indicates a packing leak. The packing nut should be tightened or the packing replaced. Packing material must be compatible with the fluid in service. The packing nut must not be overtightened, and gland nuts must be tightened evenly to prevent a cocked gland flange.

If valves are to be installed in a dirty or dusty environment, leather or metal stem jackets are recommended. A clean handwheel provides a firm grip when the valve is to be opened or closed.

Valves that require frequent use should be lubricated regularly. Most outside screw (OS) and yoke (Y) valves have a special grease fitting for this purpose.

Table 9.14. Descriptions of Common Control Valve Types

Valve Type	Description	Applications	Limitations/Disadvantages
Single-Seated	Quick-opening disc type.	Used on simple, high-sensitivity, small time-lag applications/ open-and-shut service.	Impractical for throttling control if line pressures fluctuate widely, unless controller sensitivity is very high. Not recommended for throttling control above 2-in. size.
Double-Seated	Upstream pressure enters body between the seats, tending to create equal upward and downward forces. Balanced valve eliminates line pressure fluctuations.	Used for close throttling control, low-sensitivity applications.	Seating surfaces subject to wear. Cannot supply tight shutoff.
Throttle-Plug	Double-seated quick-opening inner valve with guided discs.	For large load changes. Handles large pressure drops well.	Limited to larger sizes.
V-Port	See Figure 9.31.	Used where widely varying flow-rates encountered and where full throttling control over entire flow range is needed. Has greatest controllable range.	
Single-Seat Throttling	In sizes smaller than 1 in., they are generally preferred to double-seat type. Provide tight shutoff of controlling medium.	Same as single-seated.	Impractical for throttling control for lines with large pressure fluctuations.

Needle-Type	See Figure 9.28. Practically any required flow characteristic is possible.	For control of very small flows.	Limited to small sizes.
Three-Way Type	See Figure 9.28. Provide highly stable flows.	Service requirements include: • mixing fluids (e.g., mixing hot and cold water); • feeding two fluids to a vessel or pipeline, used as a throttling controller; and • to divert fluid from one vessel or pipeline to two load demands.	
Angle Valves	Basically single-port valves.	Used to facilitate piping or where a self-draining piping system is needed. Used for handling fluids containing solids, slurries, flashing fluids.	Subject to shamming.
Butterfly Valves	Can be fitted with either manual or automatic valve positions. Main body may be diaphragm, cylinder, float, electric motor or solenoid operated.	For control of low-pressure, low-velocity fluids.	

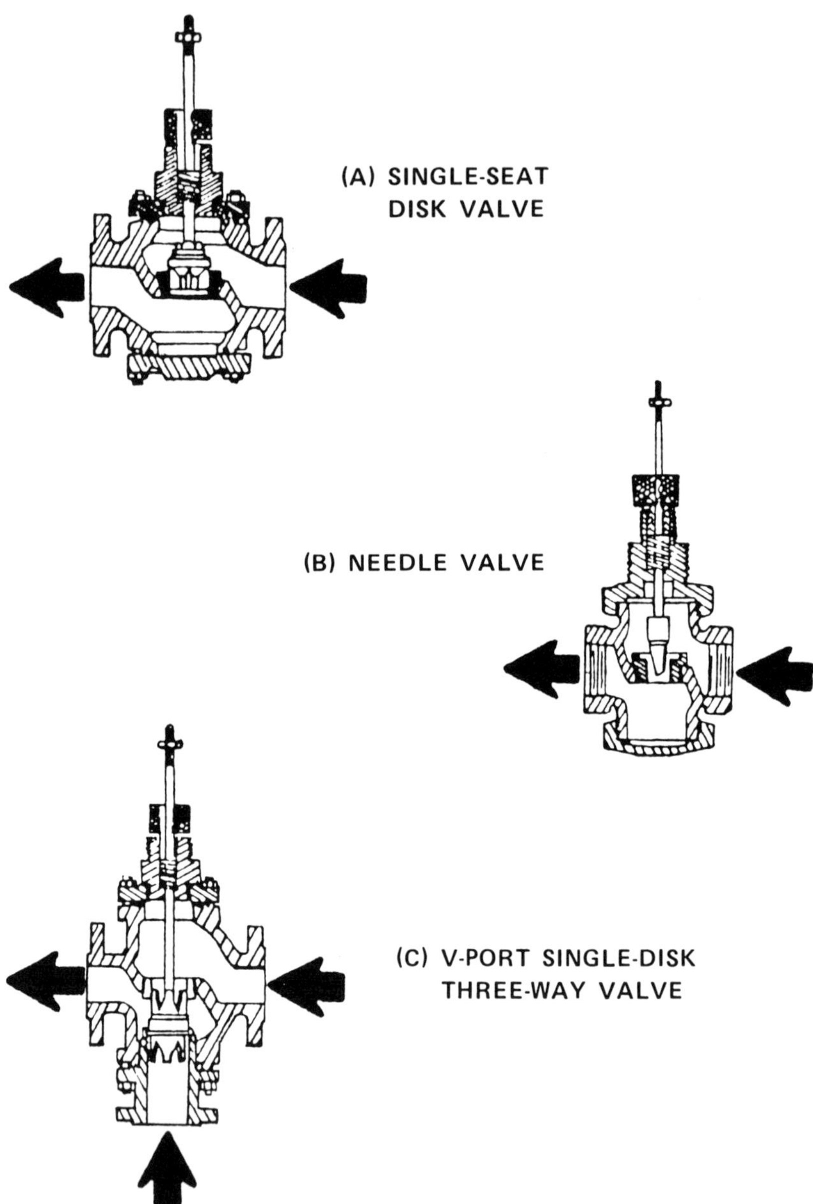

Figure 9.28. Various valve bodies: (A) single-seat disc valve, used mainly on simple, high-sensitivity, small time-lag applications and for open-and-shut service; also used for stable load conditions and constant line pressures; (B) needle valve, used for control of very small flows; employed generally where requirements are for less than ½-in. valve size; (C) V-port, single-disc three-way valve, suitable for mixing service.

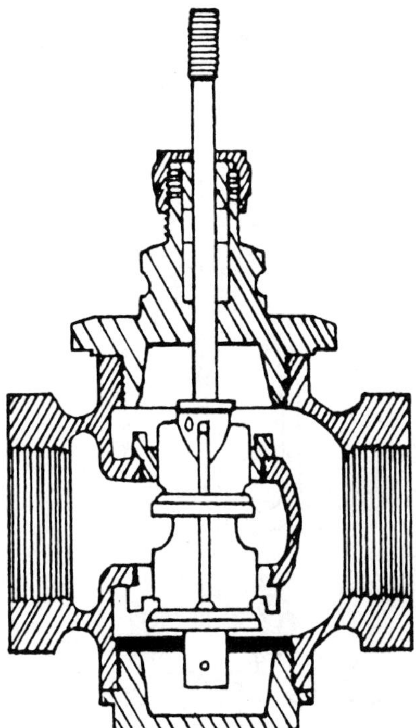

Figure 9.29. Double-seated, quick-opening inner valve with guided discs. These are used where close throttling flow is needed. Often used on low-sensitivity applications.

Gate valve wedges and seats usually wear together; therefore, they should be replaced together. Most wear occurs on the outside of the wedge. A temporary repair can be made by reversing the wedge.

The interior of the valve body should be inspected periodically for corrosion or scale, which may impede the flow. Scale deposits on smooth interior walls can promote eddies, which accelerate corrosion.

Check the stem for signs of dezincification (a form of corrosion peculiar to bronze gate valves, which leaves the metal soft and spongy). Replace the stem immediately if dezincification is detected.

9.3.2 Globe Valves

Globe valves are used most often for throttling or modulation. The choice of the disc and seat material must be based on its compatibility with the fluid in service. Regrind/renewable-type globes have removable bronze or

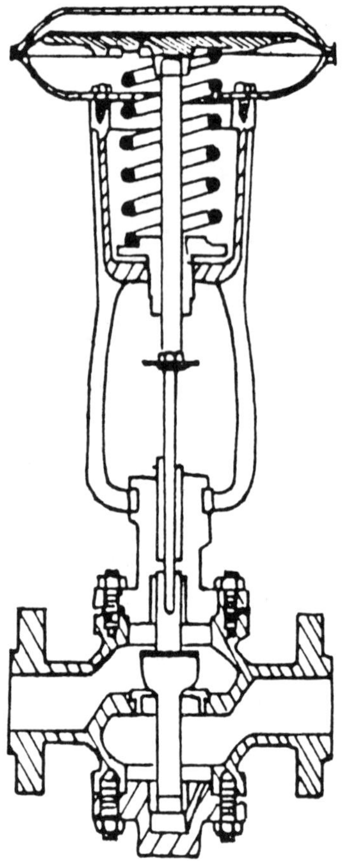

Figure 9.30. Single-seated, globe-type body with solid-plug inner valve. The throttle plug controls flowrate by means of variable port openings, as determined by curvature of the plug proper.

corrosion-resistant stainless steel seats, which are threaded into the valve body and may be reground or replaced.

Regrindable globes have permanent or integral seats, which can be reground by using a regrinding tool or by using the metal valve disc pinned to the stem.

Renewable-composition disc globe valves are used when normal wear results from frequent throttling. The disc is changed by removing the old disc from the disc holder and slipping on a new disc, which is held in place by the disc nut (see Figure 9.33).

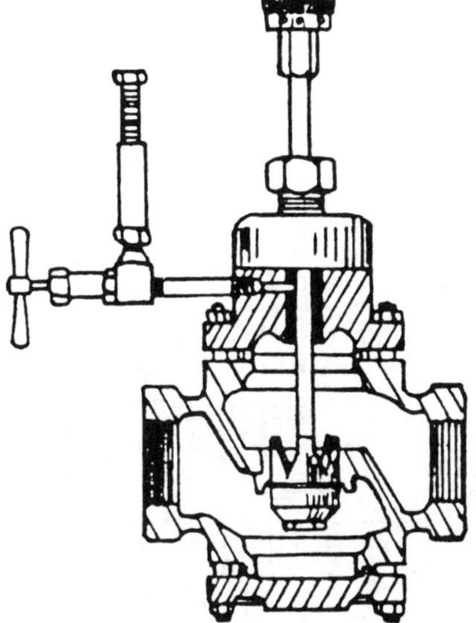

Figure 9.31. Single-seat, V-port inner valve.

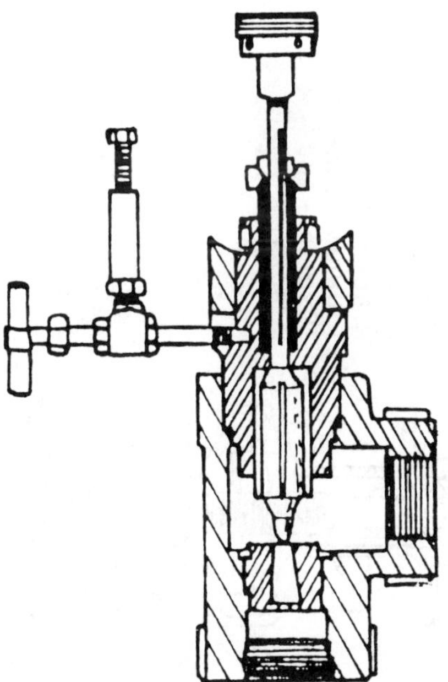

Figure 9.32. Angle valve.

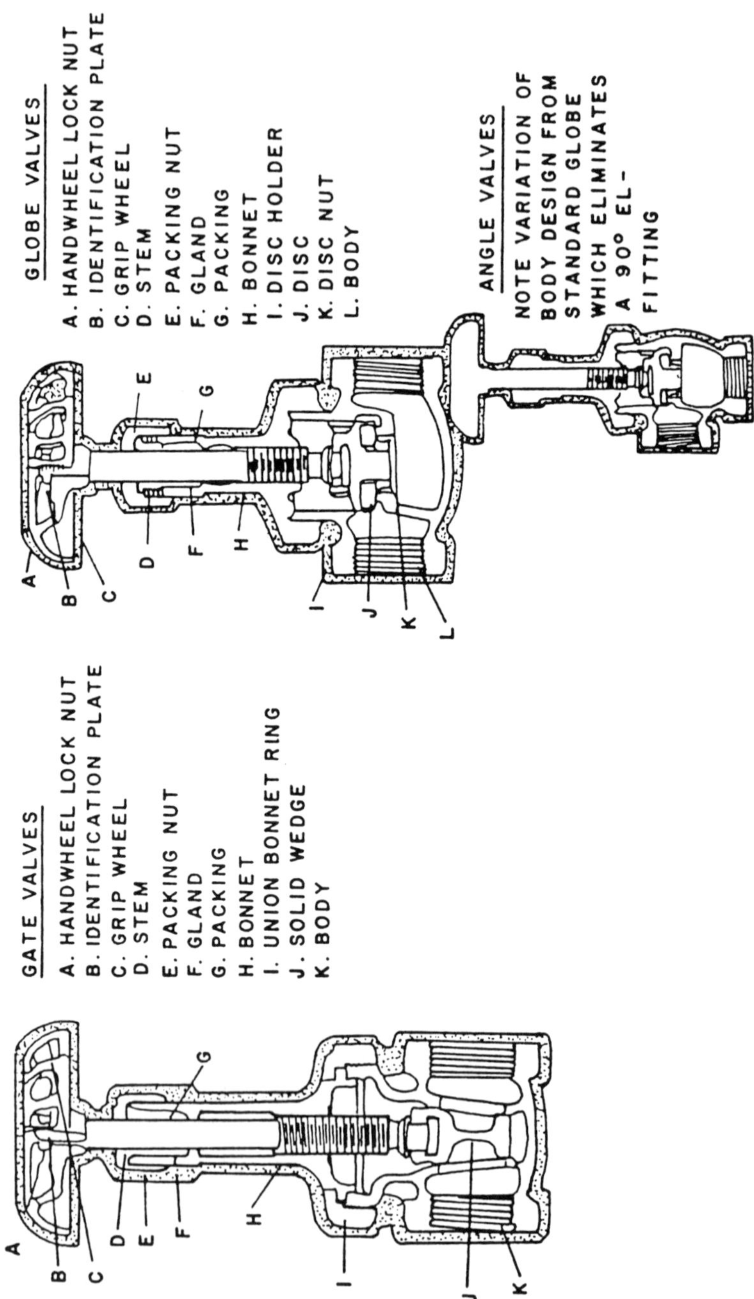
GATE VALVES
A. HANDWHEEL LOCK NUT
B. IDENTIFICATION PLATE
C. GRIP WHEEL
D. STEM
E. PACKING NUT
F. GLAND
G. PACKING
H. BONNET
I. UNION BONNET RING
J. SOLID WEDGE
K. BODY
GLOBE VALVES
A. HANDWHEEL LOCK NUT
B. IDENTIFICATION PLATE
C. GRIP WHEEL
D. STEM
E. PACKING NUT
F. GLAND
G. PACKING
H. BONNET
I. DISC HOLDER
J. DISC
K. DISC NUT
L. BODY
ANGLE VALVES
NOTE VARIATION OF
BODY DESIGN FROM
STANDARD GLOBE
WHICH ELIMINATES
A 90° EL-
FITTING

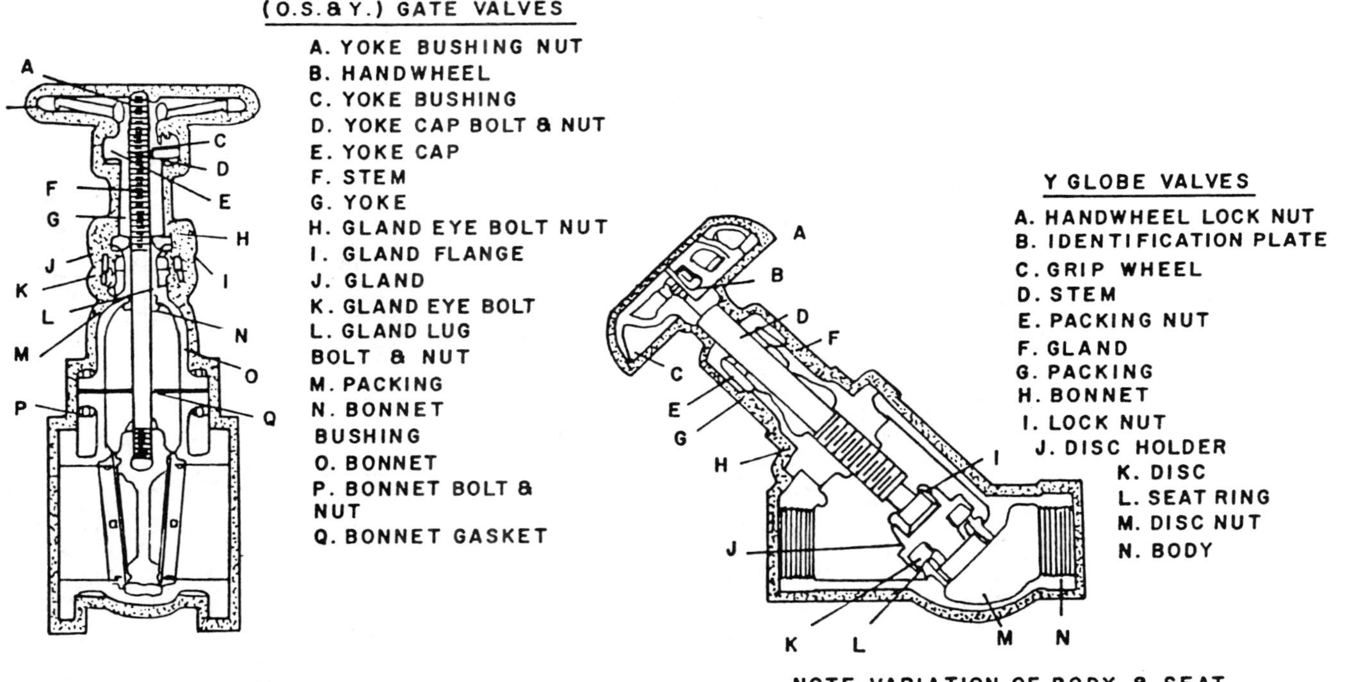

Figure 9.33. Basic parts of common valves.

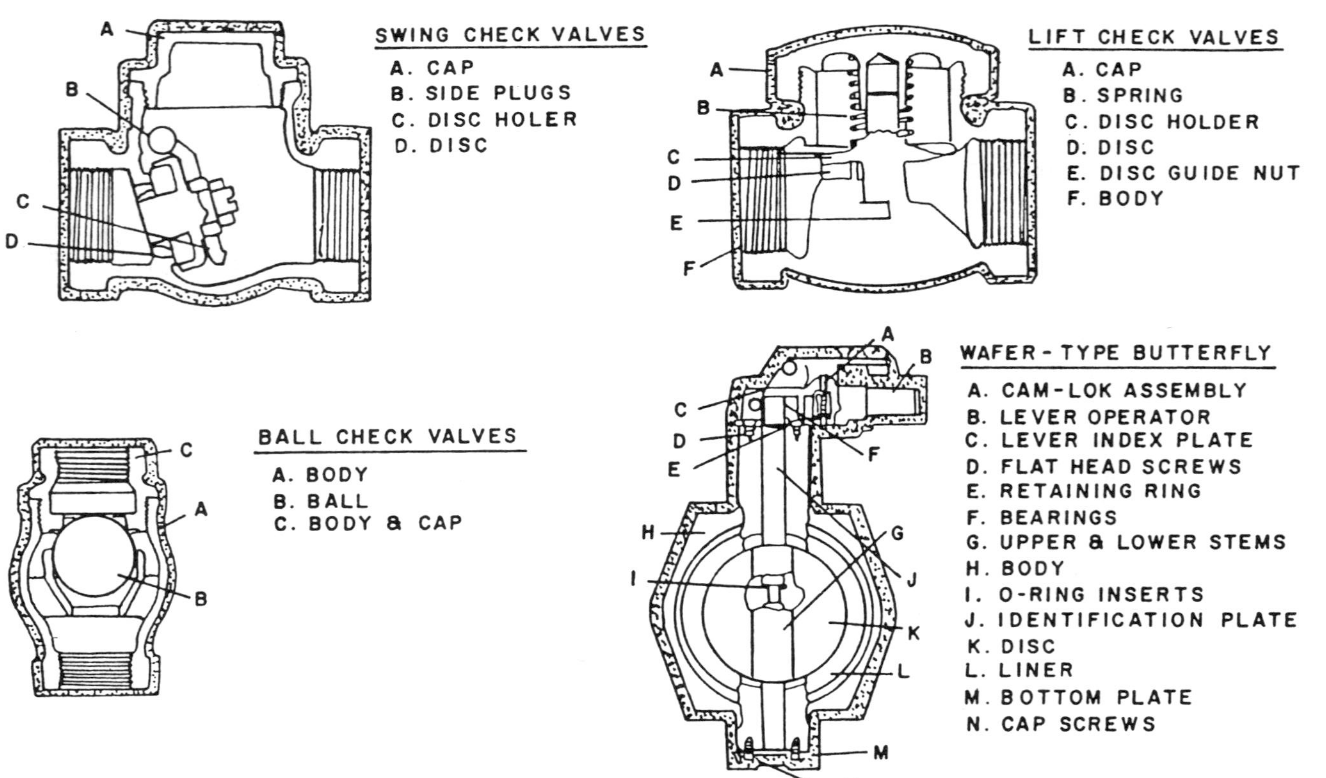
SWING CHECK VALVES
A. CAP
B. SIDE PLUGS
C. DISC HOLER
D. DISC
LIFT CHECK VALVES
A. CAP
B. SPRING
C. DISC HOLDER
D. DISC
E. DISC GUIDE NUT
F. BODY
BALL CHECK VALVES
A. BODY
B. BALL
C. BODY & CAP
WAFER - TYPE BUTTERFLY
A. CAM-LOK ASSEMBLY
B. LEVER OPERATOR
C. LEVER INDEX PLATE
D. FLAT HEAD SCREWS
E. RETAINING RING
F. BEARINGS
G. UPPER & LOWER STEMS
H. BODY
I. O-RING INSERTS
J. IDENTIFICATION PLATE
K. DISC
L. LINER
M. BOTTOM PLATE
N. CAP SCREWS

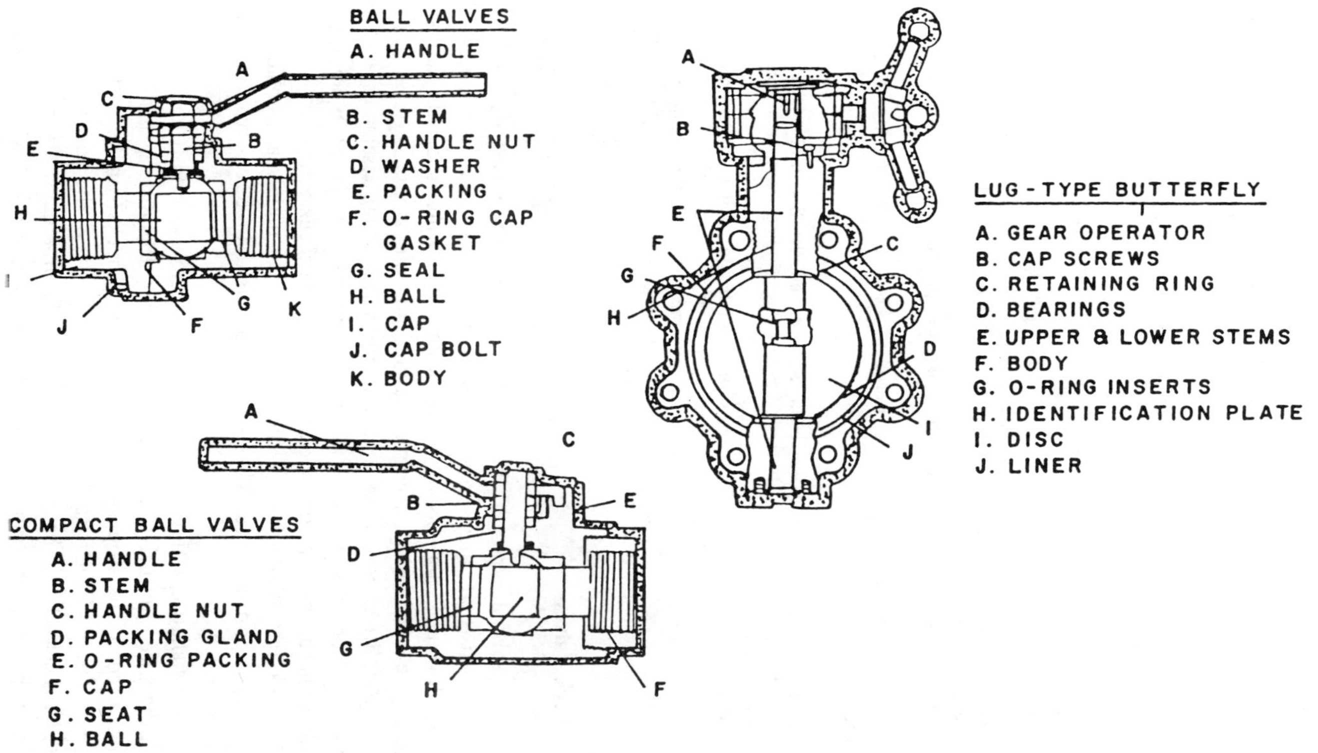

Figure 9.34. Basic parts of common valves.

About 90% of all globe valves are installed with the inlet pressure beneath the seat. This allows the valve to be opened somewhat more easily. Under severe operating conditions (e.g., in high-pressure steam service), some mechanical advantage may be obtained in closing the valve by installing it with pressure flow above the seat.

The following procedure should be followed in replacing a composition disc in a globe valve.

- Step 1. Isolate the valve to be repaired by closing the main valve and draining the line.
- Step 2. Open the valve slightly to separate the disc from the seat and remove the bonnet using a flat-jawed wrench.
- Step 3. Slip the disc holder off the stem and replace the disc by removing the disc nut.
- Step 4. Reassemble the valve by reversing the operation.

The following procedure should be used in repairing regrind/renewable discs and seats:

- Step 1. After the valve has been isolated and the line drained, remove the bonnet and place a small amount of grinding compound on the metal disc.
- Step 2. Position a pin in the disc nut notch and stem hole, reinsert the trim and screw down the union ring. Then back off one full turn.
- Step 3. Employ the stem as your grinding tool, exercising care not to overgrind. When finished, clean away the grinding compound, remove the pin, lubricate the threads and reassemble the valve.

9.3.3 Check Valves

There are several types of check valves. Swing check valves are similar to gate valves in that they ensure almost free flow. Fluid velocities are generally low and nonpulsating; otherwise, the swing disc will flap and pound on the seat causing excessive wear and possible failure.

Horizontal lift checks have a diaphragm setting that resembles that of globe valves. Generally these are used to prevent backflow in steam, air, gas or water lines having higher flow velocities than swing check valves.

Ball checks are similar to lift checks; however, a metal ball replaces the metal or composition disc. These valves are used in heavy, viscous fluid service.

For all installations, flow through a check valve must be from under the seat; otherwise, normal line pressure will force the disc against the seat causing a complete stoppage of flow. Swing checks can be installed in horizontal lines or in vertical lines where the flow is in the upward direction only. Lift checks should be installed only in horizontal lines.

Check valves usually trap fluid downstream of the valve, which must be drained when maintaining the line. A drain valve should be installed downstream or the valve body should be drilled, tapped and boss plugged.

Check valves should be installed as far from pumps as possible. Pulses will cause the disc to flap or pound on the seat, causing possible damage to the valve.

Many lift checks have replaceable seat rings. Discs are composition or metal and may be replaced or reground. The regrinding technique used is similar to that of globe valves, except that a screwdriver is used in a milled slot on top of the disc holder.

Most swing checks are provided with an access opening in line with the seat. This allows regrinding of the seat and disc with a screwdriver.

9.3.4 Ball Valves

Ball valves are used widely because of their small pressure drop due to the straight-through hole. A quarter turn will fully open or close such a valve.

Elastomeric seats are used to provide a tight closure against a polished metal ball. The elastomeric material also serves as a lubricant, promoting easy operation.

Note that elastomeric seats, as well as the body and ball material, must be compatible with the fluid. In selecting a ball valve, features such as a blowout-proof stem and ample packing box should be included.

When installing ball valves, pipe alignment is critical since strain on the valve itself may pinch or bind the ball. When soldering bronze ball valves, some manufacturers recommend complete disassembly to avoid heat damage to the seats. A constant check on valve body temperature should be maintained during soldering operations.

Check for possible leaks at two locations: the packing or seats. If there is a packing nut under the handle, tighten or replace the packing. Care should be taken not to overtighten the packing or the seam may be ruined.

When installing threaded ball valves in a line, a pipe wrench should be placed on the pipe and the body of the valve screwed onto the pipe and tightened with a flat-jawed wrench on the body and hex flats. Then the other pipe should be screwed into the cap while holding the cap hex flats stationary with a flat-jawed wrench. The pipe cap then should be tightened with a pipe wrench. The cap should not be allowed to turn and cause excessive pressure on the ball and seats.

When reassembling a ball valve, one should torque up bolts or cap screws evenly. The ball should be rotated during the final tightening to avoid binding. One should then check that the stem is squarely in the milled slot in the ball for proper alignment.

9.3.5 Butterfly Valves

Butterfly valves are best suited to relatively low-pressure service. Generally, the fluid friction coefficient is low. Because of the narrow design of the body, sediment buildup is usually minimal.

There are two basic types of butterfly valves:

1. the wafer design, which fits between pipe flanges and inside the circle of bolts; and
2. the lug type, which can be connected directly to a flange or flanges.

Butterfly valves operate with a quarter turn. Manual operators may be lever type or geared. The position of the disc is noted by an indicator on the gearbox or from the position of the handle.

During installation, care should be taken not to damage the elastomeric liner. To install the wafer type, fit the bottom lug bolts loosely in the pipe flanges. Then drop the valve in place on top of the bolts, insert the remaining bolts and tighten (using the alternate, cross-over method to ensure even distribution of torque).

The valve always should be closed during installation or maintenance to maintain the disc within the body and to avoid possible damage. Gaskets generally are not necessary since the valve liner maintains a tight seal between the valve and pipe flanges.

9.3.6 Valve Selection

Each valve type has its specific uses. Even the highest-quality valve installed in the wrong service will require frequent attention or may result in failure, causing unnecessary downtime and lost production. Therefore, selection must be based not only on materials, size and pressure rating, but on intended application as well.

The simplest fluid control application is on/off service. This type of control is required on the main water supply to a building or feedwater control to a boiler system. In both cases, it is necessary to provide the fullest possible flow with a minimal pressure drop.

Gate valves are used most often for this type service. Gate valves provide almost full flow, minimal pressure drop, minimum turbulence and fluid trapped in the line. These valves often are used where operation is infrequent and should never be used for throttling.

For modulation or throttling, the amount of fluid or gas in the service must be controlled or varied. Whenever the flow is reduced there is a corresponding drop in pressure across the valve. Globe, "Y," angle and needle valves are best suited for this type service. The globe family of valves is designed specifically for throttling; hence, particular attention must be given to the several types of seating materials available to avoid unnecessary wear. Whether they consist of composition or metal, they must be compatible with the fluid in service.

There are many services in which the reversal of flow could have disastrous results. One example is the failure in a boiler system, which could force live, superheated steam back into a city water system. Check valves are designed to prevent backflow by automatically seating when the direction of flow is reversed. The three basic types are swing, check, lift check and ball check.

Quarter-turn valves display features such as versatility and speed of operation, economy and lightweight service. Such valves are fully opened or closed by turning the hand 90° (hence, the name "quarter-turn valves"). Ball valves probably are the most versatile of all quarter-turn valves because of the variety of metals and different services in which they are used. In these designs, a highly polished ball seats against elastomeric seals. They are easier to operate than their older counterpart, the plug valve. Butterfly valves often are used to save weight and, thus, cost. They are used in larger, low-pressure services. The body is usually iron, with a choice of disc metals to meet required service conditions. The tight closure is provided by a one-piece elastomeric liner seal.

Valves are manufactured primarily from bronze, iron and steel. Proper body material must be combined with the right seating and packing materials to resist wear and corrosion throughout the life of the valve.

Bronze is the oldest material used to make valves. It is used most often in valves for hot and cold water, and low-pressure steam systems, and for other noncorrosive services. Bronze is durable, noncorrosive and easily fabricated. It is often used as seating services in larger iron body valves to ensure tight closure by eliminating rust in this critical area.

Standard iron valves generally are used in medium-to-large pipelines to control noncorrosive fluids and gases where pressures do not exceed 250 psi at 450°F, or 500 psi in cold water, oil or gas. Variations in trim combinations and seating elements make them suitable for long service in many mildly corrosive services.

Carbon steel is a strong, ductile material. Valves made from this metal are used in higher-pressure services, such as steam lines up to 600 psi at 850°F. Many steel valves are available with butt-weld ends for economy. They are used generally in high-pressure steam service, as well as other higher-pressure, noncorrosive services.

There are hundreds of stainless steel alloys developed for specific uses. Generally, 316SS can be used for more than 90% of corrosive services.

Valves forged from carbon steel are used in services up to 2000 psi and temperatures up to 1000°F in gate, globe and check valves. Forged steel ball valves may be used in high-pressure services, but are limited in terms of temperature because of their elastomeric seals.

9.3.7 Valve Ratings

Valves generally have several identifying marks that provide information on the valve class, its type service, pressure and temperature ratings. Information identifying a valve and its class may be found in two places on a valve: cast into the body or stamped onto an identification plate, which is found under the handwheel nut on bronze, iron and forged steel valves or fastened to the yoke or yoke bonnet on steel valves.

Information cast on the body usually includes the valve size and pressure class (primary and secondary ratings).

Identification plates include the name of the manufacturer, figure number and other important data, such as size, body, stem, disc and seat material, pressure class, and secondary pressure rating and temperature rating, if it is different from the standard.

The primary rating indicates the safe operating pressure for a valve at some elevated temperature. This temperature is dependent on the materials used and the fabrication of the valve. When specific data are not available, a rule of thumb to follow is the temperature of saturated steam or the primary rating indicated on the valve body. For example, if a valve has a primary rating of 150 psi, as indicated by 150S," then it has a maximum operating temperature of 367°F (temperature of saturated steam at 150 psi). The safe operating pressure of a valve is always relative to the working temperature. Basically, as the pressure increases there must be a proportional decrease in temperature to remain within the safe operating limits of the valve.

For example, a cast carbon steel valve rated at 150 psi may be operated safely up to 100°F at 275 psi. This same valve may sustain a working temperature of 800°F if the working pressure does not rise above 92 psi.

Manufacturers publish charts for a broad range of working pressures and temperatures in their comprehensive valve catalogs. Note that these are the maximum allowable nonshock pressures at the tabulated temperatures.

9.3.8 Valve Installation

Threaded and Flanged Joints

Long pipe threads permit the pipe to enter the valve too deeply and press against the seats, preventing proper seating of the disc or wedge and damaging the valve. Pipe-sealing compounds should be applied to the pipe-end only and not inside the valve body, where it can cause failure of the valve to the seat. The following steps should be followed when installing valves with threaded and flanged joints.

- Step 1. Ream out all burrs that can impede the flow. Then blow out welding scale and other foreign matter before making up the joints. Wipe away all chips and dirt from the pipe threads.
- Step 2. Line up the flanged joints carefully. Misaligned flanges can place unnecessary strain on a valve. Use a thread lubricant on the flange bolts and pull up the joint snugly. Pipe hangers are necessary to support valves and piping and to minimize pipeline stress, which can adversely affect the proper operation of valves.
- Step 3. Use a pipe wrench on the pipe near the valve and a flat-jawed wrench on the valve itself on the end nearest the pipe. Keep the valve closed tightly during installation to prevent twisting and distorting. Place the pipe in a pipe vise (never the valve).
- Step 4. On systems with flanged valves and piping, tighten bolts with the alternate-side cross-over method, which gives a much better bearing over the flange faces and reduces stress on the valve parts.
- Step 5. Check for proper flow direction. In many cases a valve can be mounted either way. But particularly for globe and check valves, the fluid flow direction must be correct.
- Step 6. Whenever possible, install valves with the bonnet up to prevent dirt, condensate and sediment from collecting in the bonnet structure and causing unnecessary wear on the stem threads.
- Step 7. Locate valves where they can be reached conveniently. If this is not possible, install chain-wheel operators extension stems or powered actuators. Be sure to allow clearance for rising-stem valves to open completely.
- Step 8. Insulate piping systems containing hot fluids to prevent injury from contact and to prevent freezing of lines exposed to subfreezing temperatures.

Soldered Joints

The following steps should be followed when installing valves with soldered joints.

- Step 1. Carefully measure the tubing run between fittings or valves and cut the tubing with a sharp hacksaw or round-wheel tubing cutter. The cut should be at a perfect right angle to the tubing.
- Step 2. Place the end of the tubing into the valve or fitting to doublecheck your measurement. If the joint is to be made in an existing line, make sure it is drained completely and dry. Heat from the torch will turn any water present into steam, preventing a perfect solder joint.
- Step 3. Remove all burrs from the outside and inside edges of the cut and clean it with a piece of fine sandpaper, emery cloth or steel wool. Then wipe clean both tubing and the inside of the valve or fitting.
- Step 4. Coat the outside end of the tubing and inside end of the valve or fitting lightly with flux and twist the tubing or fitting at least 180° once or twice to distribute the flux evenly.
- Step 5. Apply the flame to the fitting end and allow it to reach the proper temperature. The flame should not touch the solder or tubing directly. Fill the joint completely and wipe off the excess with a damp cloth to make a tight, good-looking, professional joint.

NOMENCLATURE

c	speed of sound (fps)
C_1, C_2, C_3	allowance coefficients (m)
C_p	amount of compression (in.)
D	diameter (m)
E^t	modulus of elongation at temperature t (N/m^2)
F	cross-sectional area (m^2)
ℓ	length (m)
n	number threads per inch
P	pressure (N/m^2)
q	specified load (MN/m^2)
R_{av}	average bending radius (m)
R_m	expansion joint rated moment (in.)
S'	thickness (m)
S'_c	design thickness of soft packing (mm)
T	temperature (°F)
t	temperature (°C)
W	shock wave velocity (fps)
Z	number of tubes
α	coefficient of thermal expansion (in./in./°F)
Δ	displacement (m)
Δ_x	expansion (in.)
δ	laminate wall thickness (in.)
ϵ	modulus of elasticity (psi)
η	safety factor
$\dot{\theta}$	elapsed time (sec)
$\tilde{\mu}$	Poisson's ratio
σ_{allow}	allowable stress (kg/m^2)
χ	reinforced wall thickness (ft)

REFERENCES

1. *ASME Boiler and Pressure Vessel Code*, Section 8, Division 1 (New York: American Society of Mechanical Engineers, 1977).
2. Timoshenko, S. *Theory of Elastic Stability* (New York: McGraw-Hill Book Co., 1936).
3. Cheremisinoff, N. P., and P. N. Cheremisinoff, *Fiberglass-Reinforced Plastics Deskbook* (Ann Arbor, MI: Ann Arbor Science Publishers, Inc., 1978).
4. Fibercast Co. *Fibercast Piping Design Manual*, (1977).
5. Ceilcote Co. "Corrosion-Resistant Fiberglass-Reinforced Plastic Pipe," *Bulletin 1010* (1977).

6. A. O. Smith Inland, Inc. "General Installation Instructions," *Manual No. 9474-3* (April 15, 1977).
7. Cheremisinoff, P. N. "Plastic Panel Coverings: Handle Tough Plant Conditions," *Chem. Eng.* (1958).
8. Vikhman, Iu.L., I. F. Babitzki and S. I. Volfson. "Raschet i Konstruirovanie neftezavodskoi apparatury," *Gosnauchtekhizdat, M.* (1963).
9. Kuzmak, E. "Tekhnologia neftenogo mashinostroenia," *Gostopizdat, M.* (1947).
10. Lossievski, V. A. "Gibkost gladkikh ivoenistykh kompensatorov," *Teplo i sila*, N6 (1933).
11. Palatnikov, E. A. "Raschet ocevykh kompen satorov vvodimykh v truboprovody, oborongiz," *M. Moscow* (1957).
12. "Mashinostroenie Enziklopedicheski spravochnik," *Mashgiz, M.* I(2) (1948).

CHAPTER 10

OBTURATION

10.1 INTRODUCTION

In the design of process equipment, we are concerned with numerous connections that are very important to chemical and process technology. Apertures in an apparatus may have varied diameters and purposes, placing different requirements on the sealing closures used. The sealing connections—flanged, threaded, bayonet and bow—must be tough, tight, detachable and immobile. These connections must be strongest for equipment requiring frequent assembly, for those working with great fluctuations of temperature and pressure, and especially in corrosive environments.

Obturation is defined as the tightening of immovable detachable connections and may be classified in the following manner.

I. By the tightening force:
   1. Forced sealing at the expense of stress developed by bolts; and
   2. By "unsupported-area" seal (a self-energized seal), which develops a force reaction on an area smaller than the area exposed to the pressure medium.

II. By the kind of tightening material:
   1. sealing with soft metal gaskets working for crumpling; and
   2. sealing with rigid metal rings or lens working in the region of elastic deformation.

III. By the design:
   1. sealing with soft metal gasket;
   2. cone sealing;
   3. double-cone sealing; and
   4. sealing with elastic metal ring, lens sealing.

Obturation is achieved by pressing down with a certain force, providing the hermetic nature of surfaces to be tightened directly with each other or through gaskets from softer material located between them. This pressing is accomplished by the use of bolts or studs (in flanged connections), with threads (in threaded connections), with bows (in bow connections), etc. Depending on the action of pressure in the apparatus, there are two types of obturation:

1. "Supported-area" obturation (not self-energizing), in which the pressing force of tightened surfaces decreases under operating conditions.
2. "Unsupported-area" obturation (not self-energizing), in which the pressing force of the tightened surfaces is increased or, in some cases, does not change under the conditions of operating pressure.

Therefore, the initial compression at the assembly of the connection for a supported-area obturation should differ from that of unsupported-area obturation. In the first case, the preliminary compression must be maximum; in the second case it must be minimum.

The unsupported-area obturation is more reliable for providing tightness. Besides, it is more effective because it demands less force at initial compression. However, the obturation used in process equipment is mostly of the unsupported-area type because it is often impossible to accomplish such obturation from design considerations.

The unsupported-area type is used mainly at high pressures (for connecting covers, different inlets, etc.) for some connections in apparatuses of medium and low pressures, such as connections at the top tube sheet with shells in coil heat exchangers, as well as in vacuum connections.

The most commonly used types of obturations in process equipment are presented in Table 10.1, where three groups are noted according to the method of sealing: with gaskets, without gaskets, and special type.

The gasket-type obturation is most common. It is used to make connections of low, medium and high pressures, as well as at vacuum.

Obturation without gaskets is applied basically for small diameters of apertures at high pressures, and in cases in which it is impossible to use gaskets because of temperature limitations.

Special obturation has limited applications, mostly to high pressure operations. Selection of the proper type of obturation requires information on equipment operation and maintenance (O&M) needs. Gasket-type obturation is preferred when frequent equipment disassembling is needed without changing gaskets. This requirement demands the use of highly elastic materials. Disassembling can be done many times without changing gaskets with such materials as paronite, fiber, fluoroplastics, corrugated metal, asbestos insert, solid flat metal, rings of iron or soft steel. Obturation without gaskets and obturation with lens gaskets demand additional polishing after each disassembling. The shape of sealing in all types of obturation is almost always circular. However, obturation of types I, II, III and IV, along with circular shapes, also may have rectangular and profile configurations.

10.2 OBTURATION WITH GASKETS

Wherever permitted by temperature, corrosion and other conditions, obturation with gaskets is used to provide better leak-proof connections and

Table 10.1. Principal Types of Obturation Used in Process Equipment for Detachable, Immovable Tough-Tight Connections

		Operating Conditions			
Type	Sketch	Items	Fluid	P MN/m^2	t, °C
		With Gaskets			
I	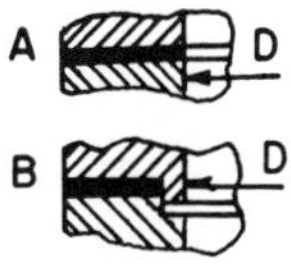 A D B D	With a gasket between flat surfaces	Inert, low aggressive	$\leqslant 2.5$	$\leqslant 500$
II	 D	With a gasket in male-and-female flange		$\geqslant 2.5$ up to 20	
III	 A D B D	With a gasket in tongue-and-groove flange	aggressive, toxic flame and explosion	$\leqslant 100$ vacuum	$\leqslant 800$ $\leqslant 250$
IV	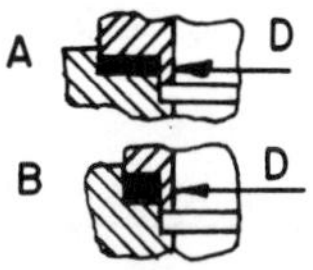 A D B D	With a gasket in the lock	any	>2.5 up to 100 vacuum	$\leqslant 800$ $\leqslant 250$
V	 D	With a gasket in the groove	Inert, low aggressive		$\leqslant 100$
VI	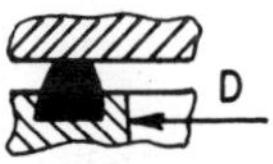 D	With a gasket of dovetail shape		<0.1 and vacuum	
VII	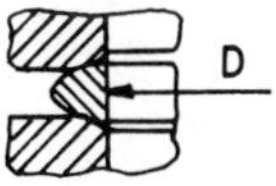 D	With a gasket between two conical surfaces	any	$\geqslant 20$ up to 150	$\geqslant -50$ up to 800
VIII	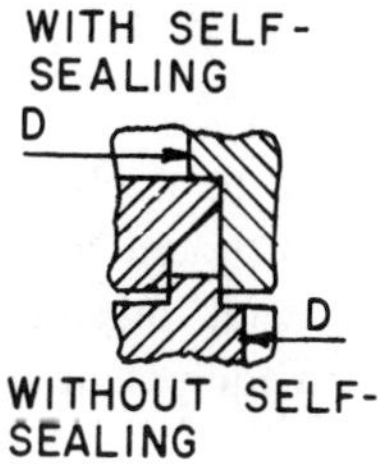 WITH SELF-SEALING D D WITHOUT SELF-SEALING	With a gasket in the lock	dangerously explosive and fire hazard	$\geqslant 10$ up to 100	$\leqslant 500$

Table 10.1, continued

Type	Sketch	Operating Conditions			
		Items	Fluid	P MN/m^2	t, °C
Without Gaskets					
IX	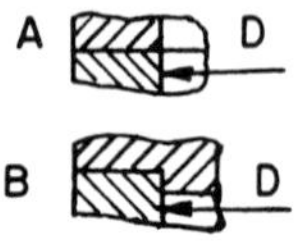	With flat sealing surfaces		⩽150	
X	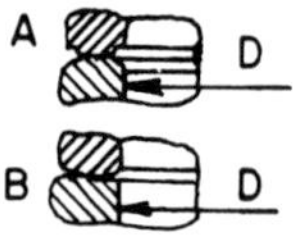	With flat and spherical or two spherical sealing surfaces	any	>10 up to 100	in material resistance limits
XI	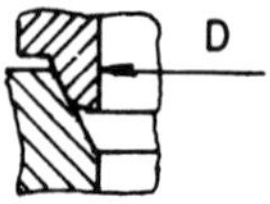	With conical and spherical sealing surfaces		⩽20	
XII	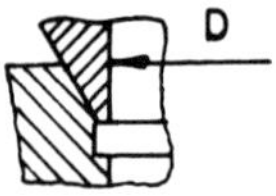	With two conical sealing surfaces		⩾10	
Special					
XIII	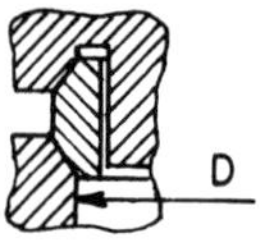	With obturator having double cone	any	⩾30 up to 100	⩾−50 up to 500
XIV	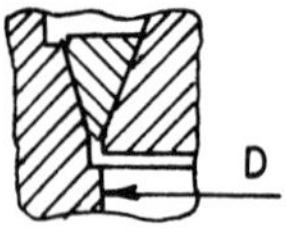	With obturator having outside and inside cone		⩾30	in material resistance limits

decrease the compression force, which is necessary to apply for sealing surfaces. Gaskets should be fabricated of material that should be milder than the sealed surfaces. Gaskets may be fabricated of metal, organic and non-organic materials. The principal types used in process equipment are given in Table 10.2.

Table 10.2. Principal Types of Gaskets Used in Process Equipment

Type	Sketches	Item	Material	Operating Conditions	
				Type of Obturation	p, Mn/m^2
1		Rectangular section	Rubber, rubber with cotton fabric insertion fiber, asbestos, fluoroplastic paronite.	I: II: III-A; IV-A; V	⩽10 >10
			Aluminum, copper, lead, nickel, steel, monel.	III-A and IV-A	⩾10
2		Round section	Aluminum, brass, lead, steel, monel, vacuum rubber.	III-B and IV-B	Vacuum ⩾2.5
3		Serrated section	Steel.	II and III-A in connection of pipes and fittings.	⩾10 at t ⩾ 200°C
4		Corrugated section	Asbestos, jacketed with steel, aluminum, nickel, copper, brass.	I and II	>0.3 to 10 at t ⩾ 400°C up to 500°C
5		Profile section	Rubber, plastic.	VI	⩽0.1 and vacuum
6		Lens section	Carbon and alloy steels.	VII	⩾20 up to 150
7		Trapezoid section	Aluminum, copper, nickel, steel, monel.	VIII	>10 up to 100

The sizes of gaskets of rectangular cross section (type I, except the rubber gaskets for obturation type V), serrated (type III) and corrugated (type IV) are given in Table 10.3.

Obturation types I-A are less by 1–2 mm than the width of sealed surfaces (the outside sizes of a gasket and surface are equal). In obturation types I-B, III-A and IV-A, the width of the gasket is equal to that of the sealed surface or a groove. The sizes of gaskets with round cross sections are given in Table 10.4.

After selecting the obturation type and constructive elements, it is necessary to determine the design seating force to provide a tight seal. The design seating force for gaskets types 1–5 and 7 is determined as follows:

Table 10.3. Sizes of Gaskets of Rectangular Cross Section, Serrated and Corrugated Shapes

Type of Gasket[a]	1			2	3	4
Material of Gasket	Rubber[b], Fiber, Asbestos, Vegetable Fiber, Plastic, Fluoroplastic Rubber with Cotton Fiber	Paronite	Cardboard	Metals		Asbestos Board Jacketed with Metal
at D, thickness (mm)						
<100	2	2	1.5	2.5	2.5	3
100-300	2.5	2	2	3	3	3
300-1000	3	3	2	4-5	4	3
>1000	4	4	–	–	–	–

[a]From Table 10.2.
[b]Except obturation type V.

Table 10.4. Sizes of Gaskets with Round Cross Sections

D, mm	Diameter of Gasket Cross Section: Metals	Rubber
⩽6	2	–
>6 up to 50	3	
>50 up to 300	4-5	
>300 up to 1000	5-8	6-8
>1000	–	10-12

The gasket mean diameter is equal to that of the sealed surface.

$$P'_n \geqslant 3D_n b_{ef} q \ , \quad N \tag{10.1}$$

where D_n = central diameter of the sealing, m
b_{ef} = effective width of the sealing, m
q = specific load on the sealing area, N/m^2

The force, P'_n, for rectangular or a profile shape of the sealing is calculated from formula 10.1 if the average length of the sealing surface is assumed in place of $3D_n$.

The values of design-effective width of the sealing, b_{ef}, for different types of specific loads, q, depending on gasket material, are given in Tables 10.5, 10.6 and 10.7.

Table 10.5. Design-Effective Width of Sealing, b_{ef}, for Different Types of Gaskets

Design Effective Width of Sealing	Type of Gasket (from Table 10.2)						
	1; 2; 4; 7		3 (see Figure 10.1)		See Table 10.7		
	Width of Gasket						
	$b \leqslant 10$	$b > 0$	$b \leqslant 10$	$b > 10$	A	B	C
b_{ef}	b	$3.16\sqrt{b}$	$b_0 Z$	$\frac{3.16 b_0 Z}{\sqrt{b}}$	3.5	5.5	8

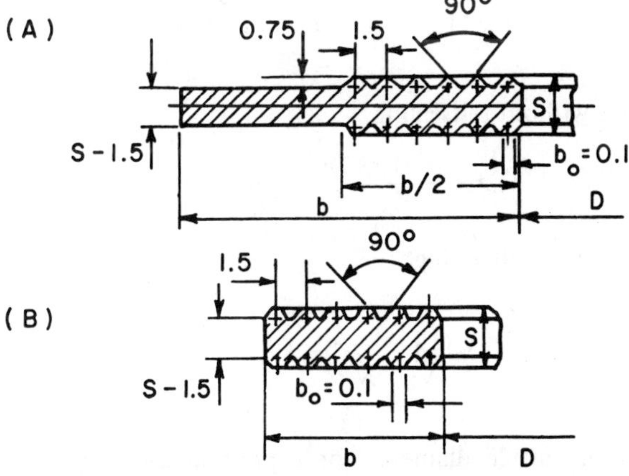

Figure 10.1. Constructive elements of serrated gaskets (type III): (A) obturation type II; (B) obturation type III. The number of teeth depends on value b.

Table 10.6. Values of Specific Loads for Gaskets of Different Materials

	Rubber, at p, MN/m^2		Fluoro-plastic	Rubber with Cotton Fiber	Asbestos, Plastics	Paronite	Fiber Poly-ethylene	Metals
	$\leqslant 3$	>3						
q, MN/m^2	1.7p+ 1.5	0.7p+ 4.5	p + 4	10	20	30	40	$1.2\sigma_y$ MN/m^2

Table 10.7. Design Elements of Profile Gaskets (Type V)

	D_n	b	h	R
		mm		
A	$\leqslant 200$	7.2	6.5	2.7
B	>200 to 500	11	10	4
C	>500	16.5	15	6.5

The design width of the seal, b′, in obturation types III and IV with metal gaskets types 1 and 2, and for obturation type VIII, is based on the condition that there are no residual bearing deformations on the sealing surfaces and is determined from the formulas outlined below:

In connections without selfsealing,

$$b' \geqslant \frac{0.33\, Dp'}{\sigma_y - 0.33p' - 1.1q}, \quad m \tag{10.2}$$

In connections with selfsealing,

$$b' \geqslant 0.5D\left(1 - \sqrt{1 - 1.1\frac{p'}{\sigma_y}}\right), \quad m \tag{10.3}$$

where D = the inside diameter for supported-area connection seals and outside diameter for unsupported-area seals

σ_y = yield point of sealed surfaces (N/m^2)

The unsupported-area seal conditions with gaskets type IV and VII are provided if the following inequality is fulfilled:

$$b_{ef} < 0.5D\left(1 - \sqrt{1 - \frac{p}{q}}\right) \tag{10.4}$$

The design seating force for gasket type VI is determined from the following formula [1]:

$$P'_n \geqslant 3D_k q_{line} \frac{\sin(\alpha + \rho_{f_2})}{\cos \rho_{f_2}} \tag{10.5}$$

where D_k = diameter of sealing (contact) of the lens (m)
q_{line} = specific load along the line of sealing (from Figure 10.2)
α = slope angle of conical surface for a lens, in grades
ρ_{f_2} = 8-10°, friction angle of the lens along the conical surface, in grades

The relationship between D_k and q_{line} is given in Figure 10.2. The lens gasket should be checked for tensile strength with the following formulas:

At operating conditions,

$$\frac{Dph}{F} \leqslant \frac{\sigma_y^t}{1.2}, \quad N/m^2 \tag{10.6}$$

At hydrostatic test,

$$\frac{Dp'h}{F} \leqslant \frac{\sigma_y}{1.2}, \quad N/m^2 \tag{10.7}$$

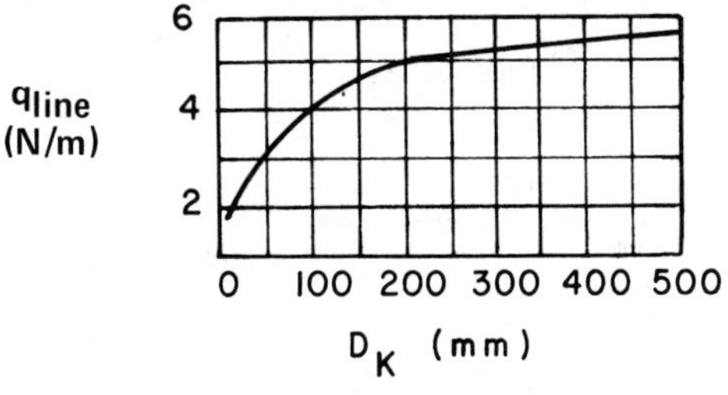

Figure 10.2. Diagram for determination of specific load along the sealing lino (Equation 10.5).

where σ_y = yield point for sealed surfaces (N/m^2)
F = diametral cross-sectional area of the lens (m^2)

Values for D and h are given in Tables 10.8 and 10.9.

Example 10.1

Determine the sizes of the construction elements and the design seating force of obturator type I-B for a flanged connection for a rectangular cover on the basis of the following data: $L_b \times B_b = 0.6 \times 0.4$ m; b = 20 mm; p = 0.1 MN/m^2; t = 20°C; the material of sealing surfaces is steel; the gasket is rubber.

Solution

From Tables 10.10 and 10.3 we have $b_1 = b + c = 20 + 1 = 21$ mm; h = 6 mm; s = 3 mm. The average length of the sealing surface is

$$L_{av} = 2(L_b + b) + 2(B_b + 2) = 2(0.6 + 0.02) + 2(0.4 + 0.02) = 2.08 \text{ m}$$

The effective width of the sealing is determined from Table 10.5:

$$b_{ef} = 3.16\sqrt{b} = 3.16\sqrt{20} = 14.1 \text{ mm}$$

The specific load on the gasket is determined from Table 10.6:

$$q = (1.7p + 1.5) \times 10^6 = (1.7 \times 0.1 + 1.5) \times 10^6 = 1.67 \times 10^6 \text{ N/m}^2$$

Table 10.8. Diameters of the Sealing Conical Surfaces for Lens Gaskets (mm) (see Table 10.9)

D_n	6	10	15	25	32	40	60	70	90	100
D_1	10	18	28	37* 40	43* 48 55	55* 65	72* 82	90* 95	100* 105 115 120	115* 120 125 130 132

D_n	125	150	200	250 ÷ 300	350 ÷ 500	600 ÷ 1000
D_1	145* 150 162 165	175* 185 195	225* 235 245	$D_n + 50$	$D_n + 55$	$D_n + 60$

Table 10.9. Construction Elements for Lens Gaskets (mm)

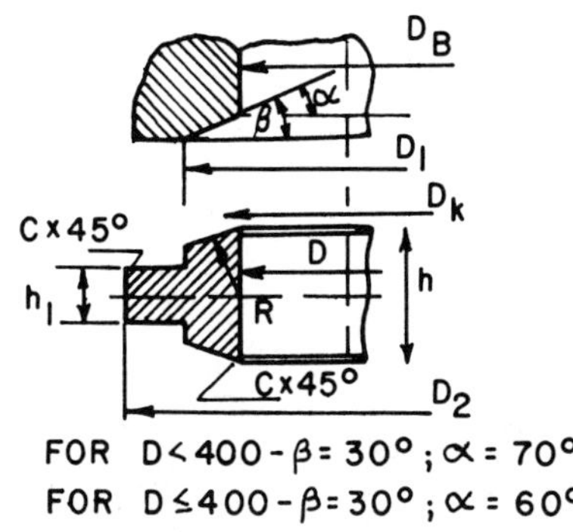

D_n	<25	⩾25 to 50	>50 to 100	>100 to 200	>200 to 400	>400
D	D_6	$D_6 + 1$	$D_6 + 2$		$D_6 + 3$	
D_k	D + 4	D + 5	D + 7	D + 10	D + 14	D + 20
D_1	see Table 10.8					
D_2	$\geqslant D_1 + 2(r + c + 1)$					
R	$\dfrac{D_k}{2 \sin\beta}$					
h	$\geqslant 2\left(\sqrt{R^2 - \dfrac{D^2}{4}} - \sqrt{R^2 - \dfrac{D_1^2}{4}}\right) + h_1 + 2r + 2$					
h_1	4	$\dfrac{D_n}{5}$	$\dfrac{D_n}{6}$	$\dfrac{D_n}{7}$	30	10
r	1	2	3	4	5	6
c	0.25	0.5		1.0	1.5	2.0

The design seating force from Table 10.1 is

$$p_n' = L_{av} \times b_{ef} \times q = 2.08 \times 0.0141 \times 1.67 \times 10^6 = 4.9 \times 10^4 \text{ N}$$

Example 10.2

Determine the size of construction elements and the design seating force of obturator type III for a flange connection on the basis of the following data: $D_b = 1.0$ m; $p = 3$ MN/m^2; $t = 190$°C; the gasket is paronite; the material of the metal surfaces is alloy steel 18-10.

Table 10.10. Construction Elements for Obturation Type I (mm)

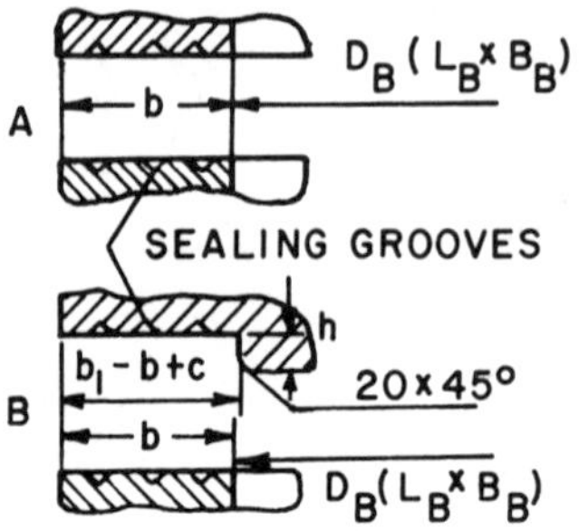

D_y	h	c
≤300	5	0.5
>300 to 1000	6 ÷ 8	1.0
>1000	10	1.5

[a]The selection of $D_b \approx D_n$ is made from design considerations.

Solution

From Tables 10.3, 10.6 and 10.11, we have a = 10 mm; b = 15 mm; b_1 = 16 mm; h = 6 mm; s = 3 mm; q = 30 × 10^6 N/m^2.

The average diameter of the seal is

$$D_n = D_b + 2a + b_1 = 1.0 + (2 \times 10 + 16)10^{-3}$$
$$= 1.036 \text{ m}$$

The value b_{ef} is determined from Table 10.5:

$$b_{ef} = 3.6\sqrt{b} = 3.16\sqrt{15} = 12.2 \text{ mm}$$

The design seating force from Table 10.1 is

$$P_n' = 3D_n b_{ef} q = 3 \times 1.036 \times 12.2 \times 10^{-3} \times 30 \times 10^6$$
$$= 114 \times 10^4 \text{ N}$$

Example 10.3

Determine the size of construction elements and the design seating force of obturator type IV-A for connecting the cover of an apparatus from the following data: D_b = 0.4 m; p′ = 25 MN/m^2; t = 300°C; the gasket is copper (σ_y = 50 × 10^6 N/m^2); the material of sealing surfaces is steel (σ_y = 350 × 10^6 N/m^2).

Table 10.11. Construction Elements of Sealing Surfaces for Obturator Type III[a]

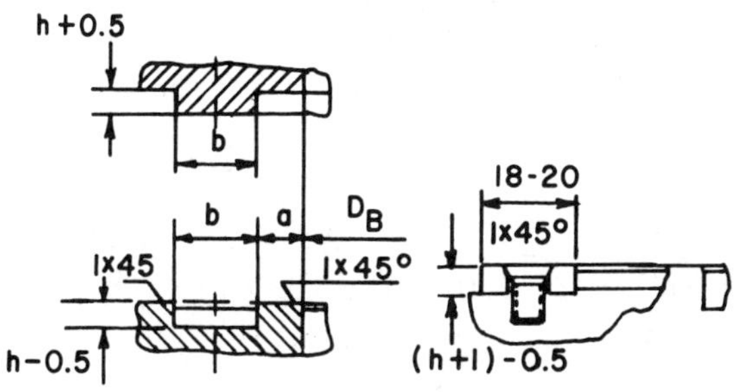

P, MN/m²	D (mm)	a (mm)	b (mm)	b_1 (mm)	h (mm)
	For nonmetal gaskets of rectangular cross section				
>1.6 to 10	⩾400 to 500	8	13	14	5
	>500 to 800	9			
	>800 to 1000	10	15	16	6
	>1000 to 1200	12	16	18	7
	>1200	14	18	20	
	For metal gaskets of rectangular cross section				
⩾10	<50	5	3 ÷ 8*		5
	⩾50 to 100	6			
	>100 to 200	7	4,5 ÷ 12*		6
	>200 to 400	8	6 ÷ 18		8
	>400 to 600	9	8 ÷ 24		
	>600	10	10 ÷ 30		10

P, MN/m²	D_n (mm)	a (mm)	b (mm)	b_1 (mm)		h (mm)
		For gaskets of round cross section				
Non-metal						
Vacuum	⩽1500	Depending on D_n, see above	d see Table 10.4	at d ⩽ 4	at d > 1	d + 2
Metal						
⩾2.5 to 10	⩽1000			d + 0.5	d + 1	

[a]The selection $D_b \approx D_n$ is made from design considerations. Value $b = b_1$ is determined either from formula 10.1 or 10.2.

Solution

From Tables 10.3, 10.6 and 10.12, we have $D = D_b + 10 = 400 + 10 = 410$ mm; $s = 4$ mm; $h = 2s + 1 = 2 \times 4 + 1 = 9$ mm; $h_1 = 2.5 \times s = 2.5 \times 4 = 10$ mm; $q = 1.2\ \sigma_y = 1.2 \times 50 \times 10^6 = 60 \times 10^6$ N/m^2. The design width of sealing is determined from Equation 10.2:

$$b' = \frac{0.33Dp'}{\sigma_y - 0.33p' - 1.1q} = \frac{0.33 \times 0.41 \times 25 \times 10^6}{350 \times 10^6 - 0.33 \times 25 \times 10^6 - 1.1 \times 60 \times 10^6}$$

$$= 12.3 \times 10^{-3} \text{ m}$$

Assume $b = 13$ mm; $b_{ef} = 3.16\sqrt{b} = 3.16\sqrt{13} = 11.4$ mm (from Table 10.5):

Table 10.12. Construction Elements of Sealing Surfaces for Obturator Type IV[a]

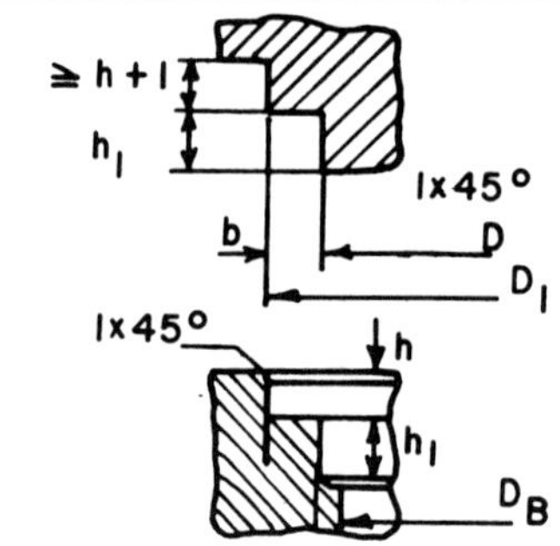

D_n	D	b*, mm, at p: ≥10 MN/m^2	≤10 MN/m^2	Vacuum	≤10 MN/m^2	h	h_1
		for gaskets: rectangular section		round section			
mm		Metal	Nonmetal		Metal	mm	
≤50	$D_b + 4$	3–8[b]		4			
>550 to 100	$D_b + 6$			5			
>100 to 200	$D_b + 8$	4,5–12	By Table 10.11	6	d, cm Table 10.4	2s + 1	2.5s
>200 to 400	$D_b + 10$	6–18		7		s–thickness of gasket	
>400 to 600	$D_b + 12$	6–21		8–10			
>600		10–30		10–12			

[a]Selection of $D_b \simeq D_n$ is made from design considerations.
[b]The value b is determined either from formula 10.2 or 10.3.

$$D_n = D + b = 410 + 13 = 423 \text{ mm}$$

The design seating force from Equation 10.1 is

$$p'_n = 3D_n \times b_{ef} \times q = 3 \times 0.423 \times 11.4 \times 10^{-3} \times 60 \times 10^6 = 86.6 \times 10^4 \text{ N}$$

Example 10.4

Determine the size of construction elements and the design seating force of obturator type VII for a flange connection from the following data: $D_b = 0.3$ m; $p = 25$ MN/m^2; $t = 100$°C; the lens material is alloy steel ($\sigma_y = 320 \times 10^6$ N/m^2); $p' = 31.3 \times 10^6$ N/m^2.

Solution

From Tables 10.4, 10.8 and 10.9 we have $D_1 = D_b + 50 = 300 + 50 = 350$ mm; $D = D_b + 3 = 303$ mm. $D_k = D + 14 = 317$ mm; $r = 5$ mm; $C = 1.5$ mm; $h_1 = 30$ mm; $D_2 = D_1 + 2(r + C + 1) = 350 + 2(5 + 1.5 + 1) = 365$ mm; $\beta = 20°$; $\alpha = 70°$.

$$R = \frac{D_k}{2 \sin \beta} = \frac{317}{2 \sin 20°} \simeq 460 \text{ mm}$$

$$h = 2\left(\sqrt{R^2 - \frac{D^2}{4}} - \sqrt{R^2 - \frac{D_1^2}{4}}\right) + h_1 + 2r + 2$$

$$= 2\left(\sqrt{460^2 + \frac{303^2}{4}} - \sqrt{460^2 - \frac{350^2}{4}}\right) + 30 + 2 \times 5 + 2$$

$$= 60 \text{ mm} \; ; \quad q_e = 5.3 \times 10^5 \text{ N/m}^2$$

$$F \approx 28.5 \times 10^{-4} \text{ m}^2$$

The axial compression force of the lens at $\rho_{f2} = 8°$ is determined from Equation 10.7:

$$\frac{DP'h}{F} = \frac{0.303 \times 31.3 \times 10^6 \times 0.06}{28.5 \times 10^{-4}} = 200 \times 10^6 \text{ N/m}^2$$

$$< \frac{\sigma_y}{1.2} = \frac{320 \times 10^6}{1.2} = 267 \times 10^6 \text{ N/m}^2$$

Hence, condition 10.7 is fulfilled.

10.3 OBTURATION WITHOUT GASKETS

Gasketless obturation is used for high-pressure connections and in cases where it is not possible to apply gaskets because of temperature, corrosion or other limitations. The recommended construction elements of basic types of gasketless obturations are presented in Tables 10.13, 10.14, 10.15 and 10.16.

Obturation type IX with flat polished sealing surfaces may be used for all flanged connections at $D_n \geqslant 50$ mm and $p \leqslant 150$ MN/m^2.

Obturation type X is used for threaded connections at $D_n < 50$ mm at $p > 10$ up to 100 MN/m^2.

Table 10.13. Construction Elements of Sealing Surfaces for Obturator Type IX[a,b]

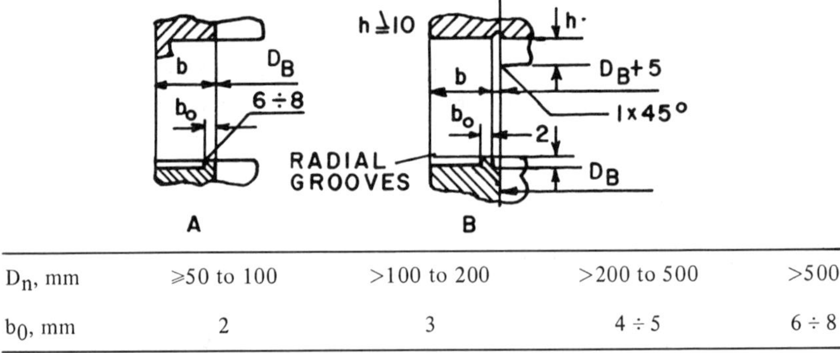

D_n, mm	⩾50 to 100	>100 to 200	>200 to 500	>500
b_0, mm	2	3	4 ÷ 5	6 ÷ 8

[a]The selection $D_b \simeq D_n$ is made on the basis of design considerations.
[b]b′ is determined from formula 10.8.

Table 10.14. Construction Elements of Sealing Surfaces for Obturator Type X (mm)[a,b,c]

D_n	⩽10	>10 to 20	>20 to 50
D_k	D_b + 4	D_b + 6	D_b + 8
R	4	5	6
b	4	6	8

[a]The sealing surfaces of pipe connections are fabricated according to A.
[b]For connecting pipe to apparatus, the surface from the pipe side is made according to A and, from apparatus side, according to B.
[c]The selection $D_b \simeq D_n$ is made on the basis of design considerations.

Table 10.15. Construction Elements of Sealing Surfaces for Obturator Type XI (mm)[a]

D_k	$D_b + 1.5(0.05\,D_b + 5)$
R	$0.578\,D_k$
D_1	$D_b + 2(0.05\,D_b + 5)$
h	$0.05\,D_b + 8$
h_1	$0.5\,R + 0.4(0.05\,D_b + 5)$
h_2	$0.167(0.05\,D_b + 5) + 3$

[a]The selection $D_b \simeq D_n$ is made from design considerations.

Table 10.16. Construction Elements of Sealing Surfaces for Obturator Type XII[a]

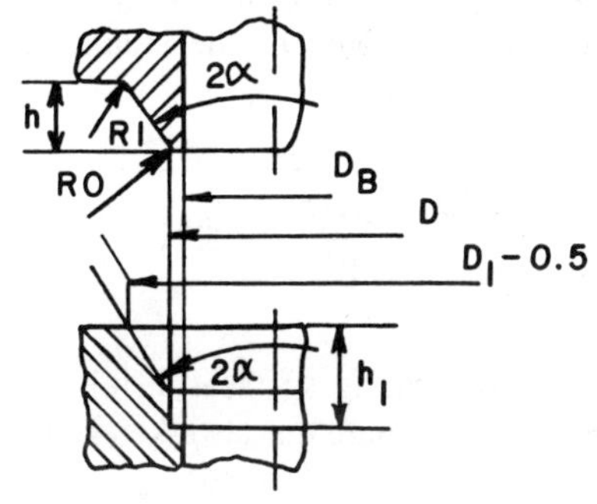

Values		For connection cups, cover etc.	For connection pipes, nipples etc.
α	Grades	$15 \div 30$	
α_1		For connection without self-sealing–$\alpha - 1$. For connection with self-sealing–$\alpha + 1$.	
D		D_b	$D_b + (0.05\,D_b + 4)$
D_1		For connection without self-sealing–from (10.9). For connection with self-sealing–from (10.10) but no less D + 5.	
h	mm	$\geqslant \dfrac{D_1 - D}{2\tan\alpha} + 5$	
h_1		–	$\geqslant \dfrac{(D_1 - D) + (0.05\,D_b + 4)}{2\tan\alpha}$

[a]The selection of $D_b \simeq D_n$ is made on the basis of design considerations.

Obturation type XI unsupported-area and supported-area seals are used in flanged and threaded connections at $D_n \leqslant 500$ mm and $p \leqslant 20$ MN/m^2. The recommended taper of sealed surfaces is $1/0.866(2\alpha = 60°)$.

Obturation type XII unsupported-area and supported-area seals are used for flanged and threaded connections at $D_n \leqslant 100$ mm and $p \geqslant 10$ MN/m^2. The taper for sealed surfaces is taken within the limits $2\alpha = 30 - 60°$ for the female part. This provides better sealing in the supported-area seal connections, which should have taper of 2° in excess of the end. The unsupported-area seal connections should be 2° or less.

The hardness of the material of the connecting parts in all gasketless obturations should be different: for the ends, they should be less hard than the nests (females). In obturation type X for pipe connections, both sealed surfaces made according to sketch A (see Table 10.14) are chosen from the same material.

The design seating force of sealed surfaces in obturation type IX is determined from Equation 10.1, where $q = 4p$ (for $p \geqslant 10$ MN/m^2) and b_{ef} is taken from Table 10.5 for gasket type I.

The design width of sealing in this obturation is based on the assumption of absence of residual deformations on the sealing surfaces, and is determined from the following formula:

$$b' \geqslant \sqrt{\frac{0.33(D + b_0)^2 p'}{\sigma_y - 4.4p} + 0.25D^2} - 0.5D \tag{10.8}$$

The design seating force of sealing surfaces in obturation types X–XII is determined from formula 10.5, where q_{line} is taken from Figure 10.2.

In obturation type XII (Table 10.16), the diameter of sealing (contact) D_k is assumed: in connections without self-sealing, $D_k = D$; in connections with self-sealing, $D_k = D_1$.

The design diameter of the end of the next D_1' for supported-area seal obturation is determined from the following formula:

$$D_1' \geqslant D\sqrt{\frac{1.32p' + 4.4\dfrac{q_{line}}{D}}{\sigma_y} + 1}\,, \quad \text{m} \tag{10.9}$$

and for an unsupported-area seal connection, from the following formula:

$$D_1' \geqslant D + \frac{0.66p'D + 2.2q_{line}}{\sigma_y - 0.66p'} \tag{10.10}$$

The unsupported-area seal obturation types XI and XII provide a reliable performance when the following condition is fulfilled:

$$q_{line} < 0.25 D_k p \frac{\cos \rho_{fr}}{\sin(\alpha + \rho_{fr})}, \quad \text{N/m} \qquad (10.11)$$

Example 10.5

Determine the size of construction elements and the design seating force for obturation type IX-B of a flanged cover for the following data: D_b = 0.3 m; p = 70 MN/m^2; p′ = 87.5 MN/m^2; the material of sealed surfaces is steel ($\sigma_y = 650 \times 10$ N/m^2).

Solution

From Table 10.13 we have D = D_b + 5 = 300 + 5 = 305 mm; b_0 = 5 mm; h = 10 mm. The design width of the sealing is determined from formula 10.8:

$$b' = \sqrt{\frac{0.33(D + b_0)^2 p'}{\sigma_y - 4.4p} + 0.25D^2} - 0.5D$$

$$= \sqrt{\frac{0.33(0.305 + 0.005)^2 \times 87.5 \times 10^6}{650 \times 10^6 - 4.4 \times 70 \times 10^6}} - 0.5 \times 0.305$$

$$= 0.025 \text{ m}$$

Assume b = 25 mm. The average diameter of the sealing is D_n = D + b = 0.305 + 0.025 = 0.33 m. The effective width of the seal (from Table 10.5) is

$$b_{ef} = 3.16\sqrt{b} = 3.16\sqrt{25} = 15.8 \text{ mm}$$

The design seating force of sealing surfaces at q = 4p from formula 10.1 is

$$p_n' = 3D_n b_{ef} q = 3 \times 0.33 \times 15.8 \times 10^{-3} \times 4 \times 70 \times 10^6 = 438 \times 10^4 \text{ N}$$

Example 10.6

Determine the size of construction elements and the design seating force of obturator type XI for a pipe flange connection for the following data: D_b = 0.2 m; p = 20 MN/m^2; the material of the sealed surfaces is steel 18-10; $\alpha = 30°$.

Solution

From Table 10.15 we have:

$$D_k = D_b + 1.5(0.05 D_b + 5) = 200 + 1.5(0.005 \times 200 + 5) \simeq 223 \text{ mm}$$

$$R = 0.578 D_k = 0.578 \times 223 = 130 \text{ mm}$$

$$D_1 = D_b + 2(0.05 D_b + 5) = 200 + 2(0.05 \times 200 + 5) = 230 \text{ mm}$$

$$h = 0.05 D_b + 8 = 0.05 \times 200 + 8 = 18 \text{ mm}$$

$$h_1 = 0.5R + 0.4(0.05D_b + 5) = 0.5 \times 130 + 0.4(0.05 \times 200 + 5) = 71 \text{ mm}$$

$$h_2 = 0.167(0.05D_b + 5) + 3 = 0.167(0.05 \times 200 + 5) + 3 = 5.5 \text{ mm}$$

The design seating force at $q_{line} = 5 \times 10^5$ N/m (from Figure 10.2) is determined from formula 10.5:

$$P_n' = 3D_k q_{line} \frac{\sin(\alpha + \rho_{f_2})}{\cos \rho_{f_2}} = 3 \times 0.223 \times 5 \times 10^5 \times \frac{\sin(30^\circ + 8^\circ)}{\cos 8^\circ}$$

$$= 20.8 \times 10^4 \text{ N}$$

Example 10.7

Determine the size of construction elements and the design seating force for obturation type XII without self-sealing for a nipple thread connection with the following data: $D_b = 0.02$ m; $p = 100$ MN/m^2; $p' = 125$ MN/m^2; the material of the nipple is mild steel ($\sigma_y = 600 \times 10^6$ N/m^2).

Solution

From Table 10.16 and Figure 10.2 we have $\alpha = 15^\circ$; $\alpha_1 = 14^\circ$; $D = D_b + (0.05D_b + 4) = 20 + (0.05 \times 20 + 4) = 25$ mm; $q_{line} = 2.3 \times 10^5$ N/m; $\rho_{f_2} = 8^\circ$; $D_k = D = 0.025$ m. The design diameter of the cone nest is determined from formula 10.9:

$$D_1' = D \sqrt{\frac{1.32p' + 4.4 q_{line}/D}{\sigma_y} + 1}$$

$$= 0.025 \sqrt{\frac{1.32 \times 125 \times 10^6 + 4.4 \times \dfrac{2.3 \times 10^5}{0.025}}{600 \times 10^6} + 1} = 0.029 \text{ m}$$

Assume $D_1 = 0.03$ m. The design seating force of sealed surfaces is determined from formula 10.5:

$$P'_n = 3D_k q_{line} \frac{\sin(\alpha + \rho_{f_2})}{\cos \rho_{f_2}}$$

$$= 3 \times 0.025 \times 2.3 \times 10^5 \frac{\sin(15^\circ + 8^\circ)}{\cos 8^\circ} = 6800 \text{ N}$$

10.4 SPECIAL OBTURATION [1-5]

Special obturation is used mostly for high-pressure connections. Performance under high pressure requires careful assembly of the apparatus. The quality of this assembly depends completely on the quality of the manufactured parts of the installation. Particularly strict requirements are placed on threads and sealing surfaces. Threads should be rounded, i.e., without sharp crests. To reduce the danger of so-called "seizing," threads should not be dimensioned the same as the sealing surface. Usually, the separation distance left should be 3-5 mm.

Sealing surfaces require equally careful manufacturing. The slightest scratch, groove or crack will break the hermetic seal of the obturator. Hence, sealing surfaces usually are polished.

It is extremely important that the plane of the sealing element be strictly perpendicular to the axis of the thread, i.e., the tightening device should result in the sealing surface contacting the part simultaneously at all points on the circumference. This requirement must be met more strictly for the harder parts being sealed. The better this requirement is fulfilled, the less force that is required for sealing.

The internal surfaces of high-pressure apparatus sealing rings should also be made with extreme care. Some applications require polishing to within accuracies of thousandths of a millimeter. The higher the pressure, the greater the possibility for penetration of the pressure-transmitting medium into microfissures, gaps and irregularities on surfaces. Finally, stresses are concentrated at fissures, cracks, sharp angles, etc., which may lead to the rupture of parts.

Special attention should be given to concentricity or eccentricity of mating apertures. Strict concentricity will provide proper operation of the obturator. The cylindricity or conicity of holes or parts is no less important. If the tolerances of conicity are not observed, the operation of sealing lenses is impeded.

Inasmuch as the sealing will be compressed when it moves toward the smaller diameter, it will be ineffective in the opposite direction of movement.

First we will consider sealing using gaskets of soft metal. For example, in the closure presented in Figure 10.3 the gasket fabricated of annealed copper, aluminum or Armco-iron is crushed with main studs. This connection is simple in fabrication and operation, but may be used for cold apparatuses with diameters up to 600 mm, which are opened approximately once a year.

The closure with a machined copper gasket of triangular cross section, where a screw socket is used instead of studs, is shown in Figure 10.4. The upper part of the screw is milled (Figure 10.5), forming a bayonet lock. This permits rapid opening and closing of the head by turning it at the angle equal to that of the milled sector. The location of the socket and head after assembly is fixed with a slide block.

Crushing the gasket is accomplished by special studs that are screwed in the head and the gasket is pressed into the ring. This design, operating up to 850 atm, permits application of narrow gaskets 8–12 mm wide. Consequently, it is possible to decrease the cross section of studs that crush the gasket.

The closure operating up to 350 atm with a machined copper gasket for a column 800 mm in diameter is shown in Figure 10.6, which is somewhat simpler in design than the closure in Figure 10.4.

Figure 10.7 shows a closure in which the stress from gas pressure acting on the top head 4 is absorbed with socket 5 consisting of two halves, fixed with

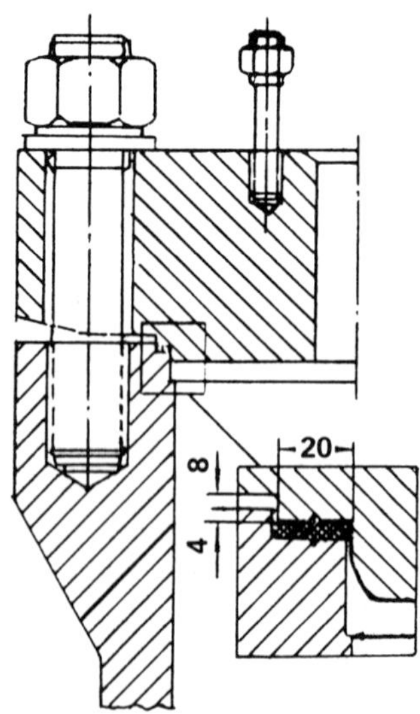

Figure 10.3. Closure with a flat metal gasket sealed with main studs.

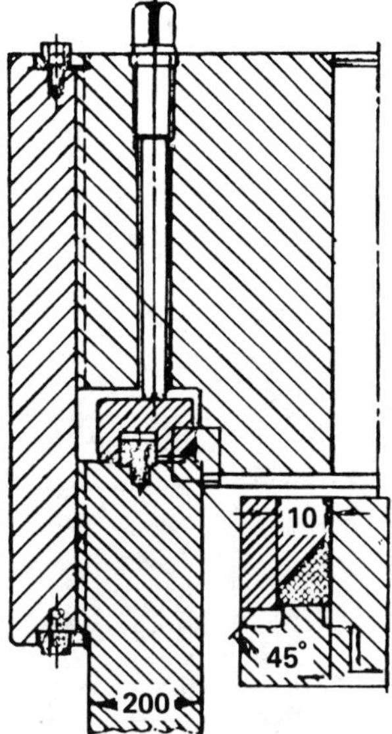

Figure 10.4. Closure with a metal gasket of triangular cross section sealed with special studs to the main fixture with a screw socket.

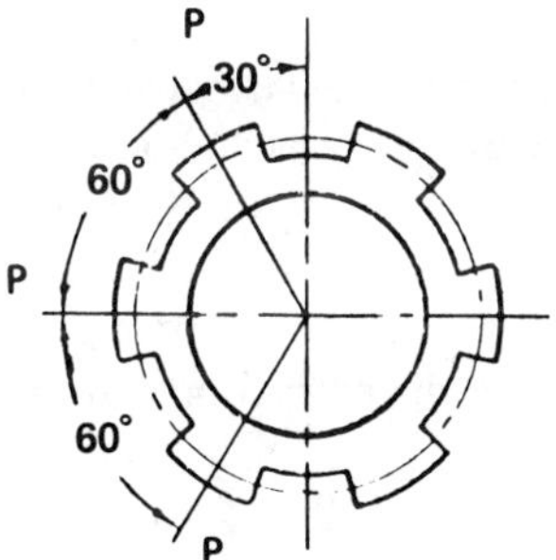

Figure 10.5. Milling of thread in the upper part of the socket.

bandages 6. The sealing is achieved owing to the crushing of copper gasket 2 with a pressing ring 3, which, in turn, is under action of a pressing bolt 7. The height of the socket in this design is lower than that of Figure 10.4 because in the bayonet lock, half the thread is cut away.

The unsupported-area seals (self-energizing seals have found a wide application in high-pressure closure designs) develop a force reaction on an area

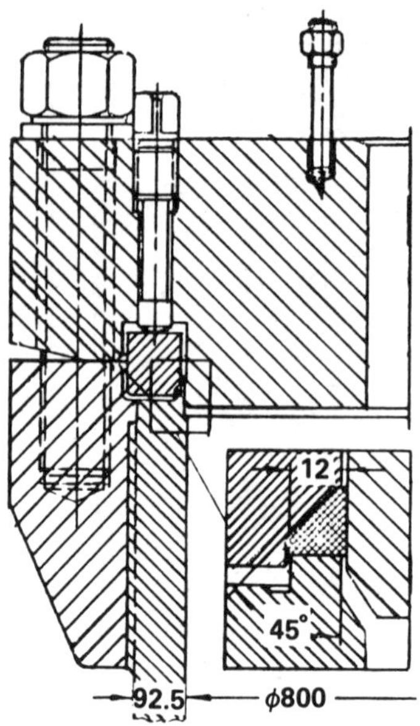

Figure 10.6. Closure with a metal gasket of triangular cross section sealed with studs of main fixing with short studs for columns up to 1200 mm in diameter and pressures up to 350 atm.

smaller than that exposed to the pressure medium (see Figure 10.7). Therefore, the pressure received by the projected area of the head is transmitted through the head to the area of sealing that is less than the area of the head; consequently, the stresses generated in the obturator are greater than the working pressure. In this case, the greater the pressure the greater the sealing of the connection. The force of internal pressure presses down the head of the apparatus to the sealing ring, "r," of the trapezoidal square section. The shape of the sealing ring determines the appearance of radial force, p_0, acting on the shell of the apparatus through the sealing ring and generating in the last tensile and crumpling stresses (shown in Figure 10.8).

The force $p_0 = p_1 \tan \alpha$ may reach large values at high values of angle α and, in the final analysis, must be accepted by the shell of the apparatus due to the ring deformation.

The force p_1 is equal to

$$p_1 = p \frac{\pi d^2}{4} \times \frac{1}{\pi d} = p \frac{d}{4}, \quad \text{kg/cm}$$

where p = internal pressure (kg/cm^2)

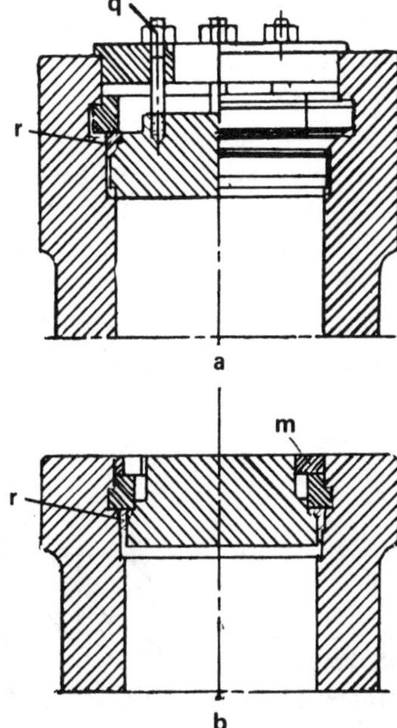

Figure 10.7. Closure of unsupported-area type.

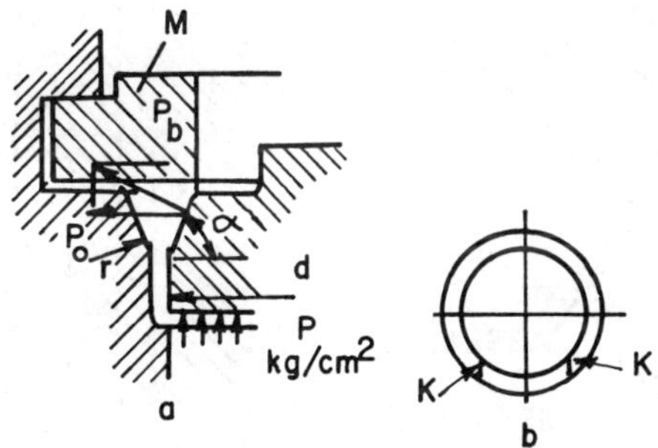

Figure 10.8. Diagram of forces acting on a trapezoidal ring.

The force, p_1, induces thrust collar M to work by bending and shear (Figure 10.8). The thrust collar is detachable and consists of three to four parts, depending on the size of an apparatus. This collar enters into the groove of the shell and disassembles easily. Besides the collar, for example, consisting of three parts, two planes of parting are made not in the radial direction but as is shown by "K" in Figure 10.8b. Before starting the unit for generating internal pressure, it is necessary to provide a preliminary tightening, which is made with studs, "q" (Figure 10.7a) or with a nut, "m" (Figure 10.7b) at small diameters of the apparatus.

In the generation of internal pressure, all the load from the nuts is removed. The unsupported-area seals can withstand especially high pressures because the seating force increases with pressure. Such an obturation provides reliability and hermetic nature at high temperatures, and lowers the price of the design.

The deterioration of tightness is possible along two cone surfaces of ring "2" (Figure 10.8a). However, the force, p_0, is usually so great that it provides the safety sealing.

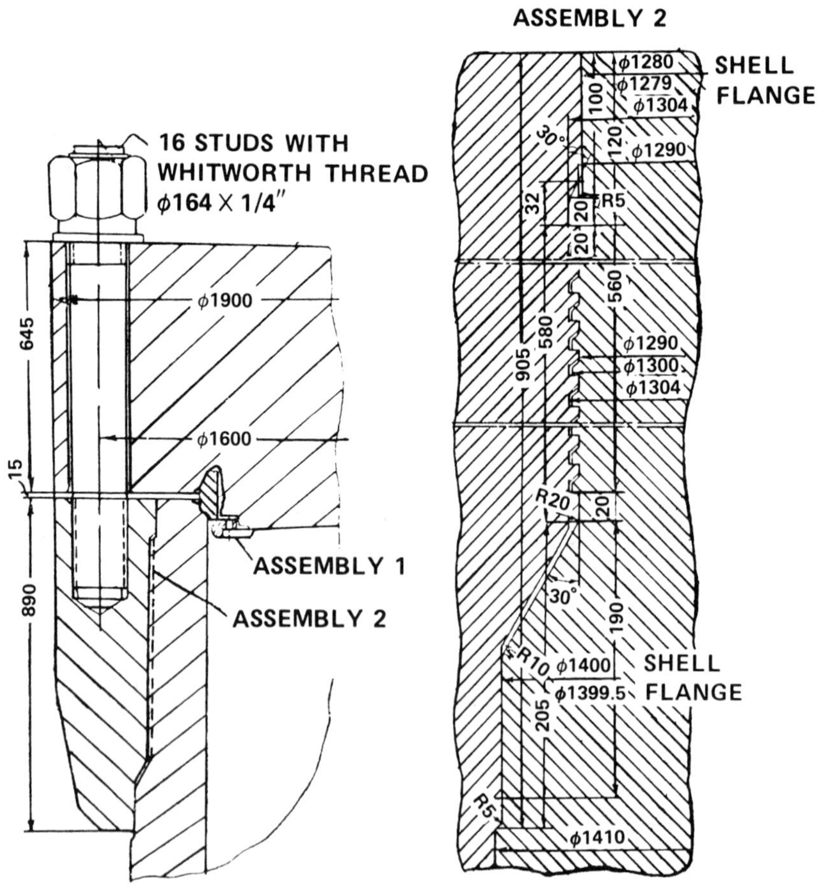

Figure 10.9. Obturation with a double-cone

Sometimes the contact of the sealing surfaces is made along the obturating band close to the contact line, or gasket spherical side surface and cone nest, or at two cone surfaces with different taper, or at two spherical surfaces with different radii of spheres. In all these cases there arises a considerable specific pressure at the seal.

The closures presented in Figures 10.9 and 10.10 are used widely in high-pressure equipment. These closures have obturators, "K" with two cone surfaces, 1, and two ring grooves, p, on each surface. Each obturator is furnished with aluminum gaskets 1 mm thick to provide tightness.

The stress in aluminum gaskets is higher than the yield point, and the aluminum filling microgap provides a tight connection. The obturators are

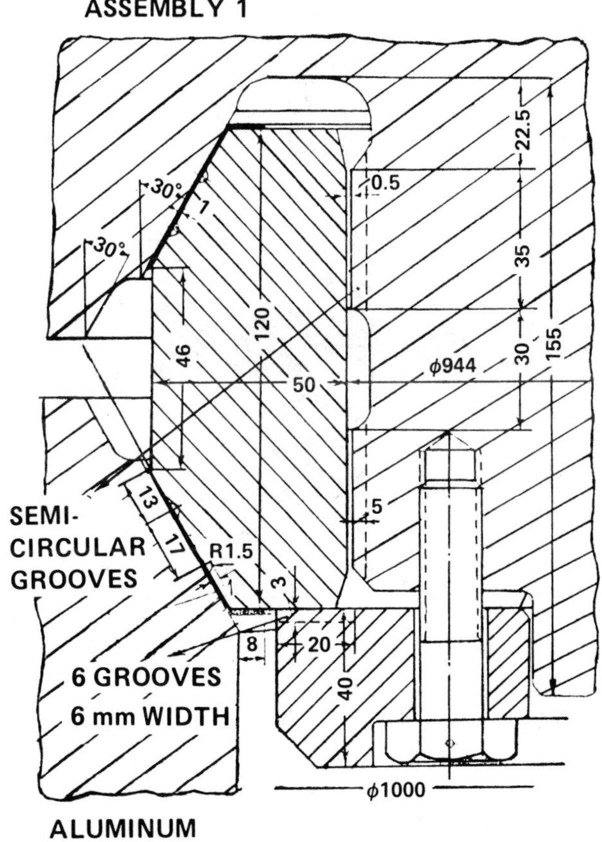

lens for the cover of a high-pressure vessel.

fixed to the covers of the apparatus with rings, "m," and bolts, "r," in Figure 10.10. There is a gap, "c," for providing the desired tightening between the cover and end of an apparatus. A small increase in an obturator diameter induced by the internal pressure provides a better tightening of all connections.

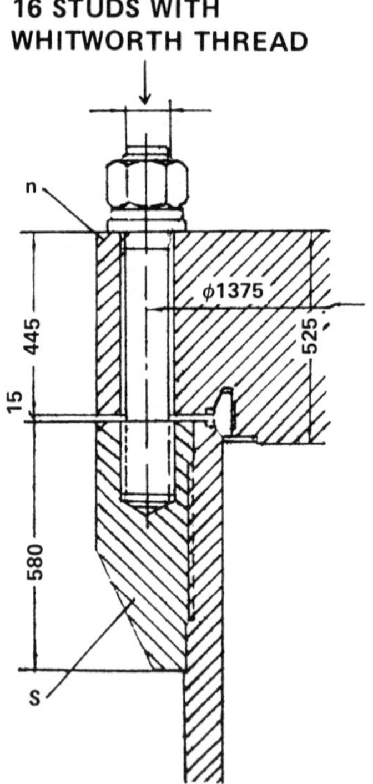

Figure 10.10. Obturation with a double-cone

Connection with double-cone obturators permits a reduction to a minimum in the bending moment acting on the flange because of a decrease in the lever arm between the lens and bolt circle and a decrease in the vertical component of the reactive lens force.

The application of threaded flanges eliminates the influence of flange

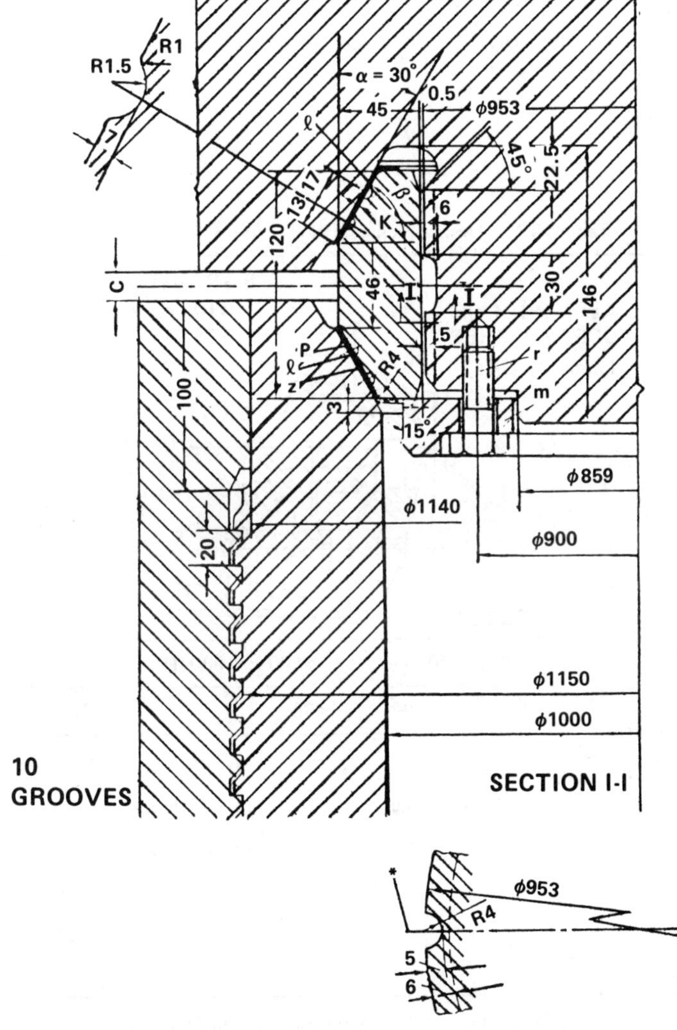

lens for the cover of a high-pressure vessel.

bending moments on the cylindrical shell of the vessel and makes possible the fabrication of cylindrical parts of the shell without reinforcement (i.e., it is then possible to make the shell of uniform wall thickness along the vessel).

The recommended construction elements typical for unsupported-area obturations with an obturator located between the sealed surfaces (which are mostly used for connection covers of apparatus) are given in Tables 10.17 and 10.18.

Obturation type XII is used at $D_b \geqslant 500$ mm up to 1500 mm for $p \geqslant 30$ up to 100 MN/m^2. The material of the sealing surfaces (shell and cover) and obturator may be the same for $\sigma_y \geqslant 6p$.

The design seating force for a given obturation is determined from the following formula [1]:

Table 10.17. Design Elements of Obturator Type XIII[a]

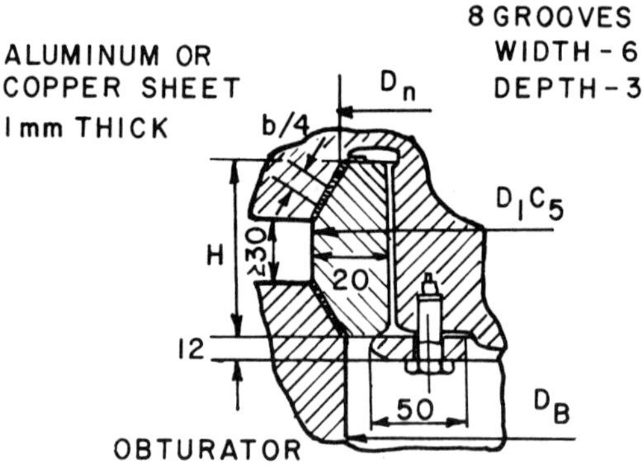

D_n, mm		500 to 800		>800 to 1000 D		>1000	
P, MN/m^2		<70	≥70	<70	≥70	<70	≥70
s		30	35	40	50	50	60
b	mm	From formula (10.13), but no less than 40					
D_1		D_b + b, but no less than 40					
D_n		$\dfrac{D_1 + D_b}{2}$					
H		b tang 60° + 30					

[a]The selection of $D_b \simeq D_n$ is made on the basis of design considerations.

Table 10.18. Construction Elements of Obturator Type XIV[a]

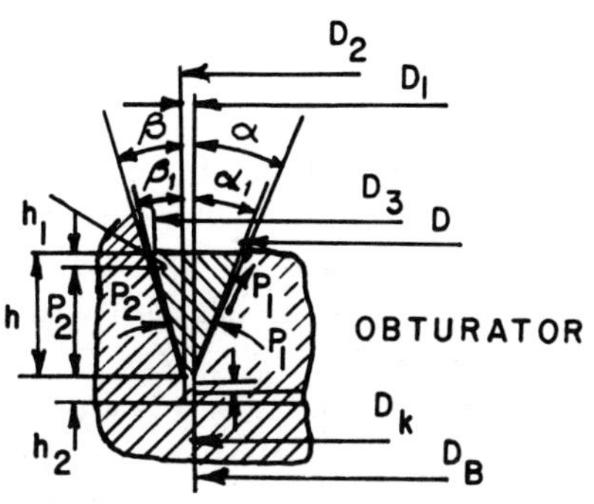

α	Grade	$18 \div 24$	D_2	mm	$D_k + (2 \div 3)$
α_1	Grade	$\alpha - 1$	D_3	mm	From formula (10.20)
β	Grade	$12 \div 15$	a	mm	$1 \div 2$
β_1	Grade	$\beta - 1$	h	mm	$\frac{D_3 - D_2}{2\ \text{tang}\alpha} + h_1$
D	mm	From formula (10.19)	h_1	mm	$3 \div 5$
D_k	mm	$D_b + (2 \div 5)$	h_2	mm	$5 \div 8$
D_1	mm	$D_2 - 2a$	F	m^2	$\approx \frac{(D_1 - D)^{2*}}{4\ \text{tang}\alpha}$

[a]The selection $D_b \simeq D_n$ is made on the basis of design considerations.

$$P_n' \geqslant 3D_n b_{eq} q \frac{\sin(\alpha + \rho_{f_2})}{\cos \rho_{f_2}}, \quad N \tag{10.12}$$

where D_n = mean diameter of sealing (m)
b_{eq} = 0.75b (m) (see Table 10.17)
$q = p(N/m^2)$
ρ_{f_2} = 8-10° friction angle of lens along the cone surface, grades
α = slope angle of conical surface for the lens

The design width of sealing, b', is determined from the condition of absence of residual deformations of crumpling on the sealed surfaces from the following formula:

$$b' \geqslant \frac{0.33 D'_b p'}{\sigma_y - p'\left[0.165 + 0.825 \dfrac{\sin(\alpha + \rho_{f_2})}{\cos \rho_{f_2}}\right]} \tag{10.13}$$

Obturation type XIV (see Table 10.18) is used for $D_b > 50$ mm and $p \geqslant 30$ MN/m^2. The material of the obturator with $\sigma_y \geqslant 2p$ should be softer than the sealed surfaces (shell and cover). It can be used in several assemblies, but a preliminary polishing of sealed surfaces is required before each assembly. The sequence of assembly is as follows: (1) a slight pressing down of the obturator from outside, then (2) a pressing of the obturator with a thrust applied on it at the same time. As a result, the obturator is expanded in the radial direction providing the preliminary sealing on both cone surfaces.

The pressure of the medium in the apparatus with the thrust on the obturator from the outside will evenly wedge up the obturator and thereby improve the tightness of the connection.

The design force of the preliminary pressing down of the cover to the obturator is determined from the following formula:

$$P'_0 \geqslant 3 D_1 b_0 \sigma_y \frac{\sin(\alpha + \rho_{f_2})}{\cos \rho_{f_2}} \tag{10.14}$$

where $b_0 = 0.002$ m, the width of the band of preliminary sealing
σ_y = yield point of obturator material (N/m^2)

Starting from equilibrium conditions for the wedge-shaped obturator at the hydrostatic test, the forces acting (see Table 10.18) are determined from the following formulas:

$$P_1 = \frac{0.78 D_1^2 p' \cos \rho_{f_2}}{\sin(\alpha + \rho_{f_2})}, \quad \text{N} \tag{10.15}$$

$$P_2 = \frac{\cos(\alpha + \rho_{f_2})}{\cos(\beta - \rho_{f_2})} - 6.3 F \sigma_y \frac{\cos \rho_{f_2}}{\cos(\beta - \rho_{f_2})}, \quad \text{N} \tag{10.16}$$

$$P_4 = P_1 \frac{\sin(\alpha + \beta) \cos \rho_{f_2}}{\cos(\beta - \rho_{f_2})} + P_3 - 6.3 \sigma_y \tan(\beta - \rho_{f_2}), \quad \text{N} \tag{10.18}$$

$$P_3 = 0.78 (D_2^2 - D_1^2) p', \quad \text{N} \tag{10.17}$$

The sizes of the obturator are determined from the following formulas:

$$D \leqslant \sqrt{D_1^2 - \frac{1.4\,P_1 \sin\alpha}{\sigma_y}}, \quad m \tag{10.19}$$

$$D_3 \geqslant \sqrt{D_2^2 + \frac{1.4\,P_2 \sin\beta}{\sigma_y}}, \quad m \tag{10.20}$$

Example 10.8

Determine the size of construction elements and the seating force of unsupported-area obturator type XIII for the flanged connection of a cover using the following data: D_b = 1 m; p = 60 MN/m^2; p′ = 75 MN/m^2; the material of sealed surfaces and obturator is steel ($\sigma_y = 650 \times 10^6$ N/m^2).

Solution

From Table 10.17, we have: $\alpha = 30^\circ$; s = 40 mm; $\rho_{f_2} = 8^\circ$.
The design width of sealing is determined from formula 10.13:

$$b' = \frac{0.33\,D_b p'}{\sigma_y - p'\left[0.165 + 0.825\,\dfrac{\sin(\alpha + \rho_{f_2})}{\cos\rho_{f_2}}\right]}$$

$$= \frac{0.33 \times 1.0 \times 75 \times 10^6}{650 \times 10^6 - 75 \times 10^6\left[0.165 + 0.825\,\dfrac{\sin(30^\circ + 8^\circ)}{\cos 8^\circ}\right]}$$

$$= 0.0414 \text{ m}$$

Assume b = 42 mm. From Table 10.17 we have $D_1 = D_b + b = 1.0 + 0.042 =$ 1.042 m; $D_n = 0.5(D_1 + D_b) = 0.5(1.042 + 1.0) = 1.021$ m; $b_{ef} = 0.75b =$ $0.75 \times 0.042 = 0.0315$ m; H = btan 50° + 0.03 = 0.042 tan 60° + 0.03 = 0.103 m; the gap $\delta = D_b/2000 = 1000/2000 = 0.5$ mm. The design seating force of sealed surfaces is determined from Equation 10.12:

$$P_n' = 3\,D_n b_{ef} q\,\frac{\sin(\alpha + \rho_{f_2})}{\cos\rho_{f_2}} = 3 \times 1.021 \times 0.0315 \times 60 \times 10^6 \times \frac{\sin(30^\circ + 8^\circ)}{\cos 8^\circ}$$

$$= 360 \times 10^4 \text{ N}$$

Example 10.9

Determine the size of construction elements and design forces acting on an unsupported-area obturation type XIV for connecting a cover, using the following data: $D_b = 0.3$ m; $p' = 187.5$ MN/m^2; the material of the obturator is steel ($\sigma_y = 500 \times 10^6$ N/m^2).

Solution

From Table 10.18 we have $a = 2$ mm; $h_1 = 4$ mm; $h_2 = 6$ mm; $\alpha = 20°$; $\beta = 12°$; $\alpha_1 = 20 - 1 = 19°$; $\beta_1 = 12 - 1 = 11°$; $D_k = D_b + 0.005 = 0.3 + 0.005 = 0.305$ m; $D_2 = D_k + 0.003 = 0.308$ m; $D_1 = D_2 - 2a = 0.308 - 2 \times 2 = 0.304$ m.

The design force of preliminary pressing in of the cover is determined from formula 10.14:

$$P_0 = 3D_1 b_0 \sigma_y \frac{\sin(\alpha + \rho_{f_2})}{\cos \rho_{f_2}}$$

$$= 3 \times 0.304 \times 0.002 \times 500 \times 10^6 \frac{\sin(20° + 8°)}{\cos 8°} = 43.1 \times 10^4 \text{ N}$$

To determine the force acting on the obturator at hydrostatic test and the sizes of the obturator, use Table 10.18.

The force P_1 is determined from formula 10.15:

$$P_1 = \frac{0.78 D_1^2 p' \cos \rho_{f_2}}{\sin(\alpha + \rho_{f_2})}$$

$$= \frac{0.78 \times 0.304^2 \times 187.5 \times 10^6 \cos 8°}{\sin(20° + 8°)} = 2860 \times 10^4 \text{ N}$$

The design inside diameter of the obturator, D', is determined from formula 10.19:

$$D' = \sqrt{D_1^2 - \frac{1.4 P_1 \sin \alpha}{\sigma_y}} = \sqrt{(0.304)^2 - \frac{1.4 \times 2860 \times 10^4 \sin 20°}{500 \times 10^6}}$$

$$= 0.255 \text{ m}$$

Assume $D = 0.25$ m

$$F = \frac{(D_1 - D)^2}{4 \tan \alpha} = \frac{(0.304 - 0.25)^2}{4 \tan 20°} = 20 \times 10^{-4} \text{ m}^2$$

The force P_2 is determined from formula 10.16:

$$P_2 = P_1 \frac{\cos(\alpha + \rho_{f_2})}{\cos(\beta - \rho_{f_2})} - 6.3F\sigma_y \frac{\cos \rho_{f_2}}{\cos(\beta - \rho_{f_2})}$$

$$= 2860 \times 10^4 \frac{\cos(20^\circ + 8^\circ)}{\cos(12^\circ - 8^\circ)} - 6.3 \times 20 \times 10^{-4} \times 500 \times 10^6 \frac{\cos 8^\circ}{\cos(12^\circ - 8^\circ)}$$

$$= 1904 \times 10^4 \text{ N}$$

The design outside diameter of obturator, D_3', is determined from formula 10.20:

$$D_3' = \sqrt{D_2^2 + \frac{1.4P_2 \sin\beta}{\sigma_y}}$$

$$= \sqrt{(0.308)^2 + \frac{1.4 \times 1904 \times 10^4 \sin 12^\circ}{500 \times 10^6}} = 0.326 \text{ m}$$

Assume $D_3 = 0.33$ m. The force, p_3, is determined from formula 10.17:

$$P_3 = 0.78(D_2^2 - D_1^2)p' = 0.78(0.308^2 - 0.304^2) \times 187.5 \times 10^6$$

$$= 36.5 \times 10^4 \text{ N}$$

The force, p_4, is determined from the formula 10.18:

$$P_4 = P_1 \frac{\sin(\alpha + \beta)\cos\rho_{f_2}}{\cos(\beta - \rho_{f_2})} + P_3 - 6.3F\sigma_y \tan(\beta - \rho_{f_2})$$

$$= 2860 \times 10^4 \frac{\sin(20^\circ + 12^\circ)\cos 8^\circ}{\cos(12^\circ - 8^\circ)} + 36.5 \times 10^4$$

$$- 6.3 \times 20 \times 10^{-4} \times 500 \times 10^6 \tan(12^\circ - 8^\circ) = 1500 \times 10^4 \text{ N}$$

$$h' = \frac{D_3 - D_2}{2\tan\beta} + h_1 = \frac{330 - 308}{2\tan 12^\circ} + 4 = 55.6 \text{ mm}$$

Assume h = 56 mm.

NOMENCLATURE

b_{ef} effective design width of sealing (m)
D_k diameter of sealing (m)
D_n central diameter of sealing (m)
F area (m^2)
h height (m)
p' design pressure (N/m^2)
q specific load (N/m^2)
Z number of bolts
σ stress (N/m^2)
σ_y yield point of sealed surfaces (N/m^2)
α slope angle of conical surface for a lens, grade
ρ_{f_2} friction angle of lens, grades

REFERENCES

1. Korndorf, B. A. "High Pressure Techniques in Chemistry," *Gostehimzdat* (1952).
2. Vollbrecht, H. *Dechema Monograph,* 34:205 (1959).
3. Ushakov, I. P. *Tr. Gos. Inst. Azo. Prom., Goshhimizdat,* N12 (1961).
4. Boon, E. F., and H. H. Lok. *Verlag Deut. Ing. Z.* 100:1613 (1958).
5. Semerchan, A. A., N. Z. Shishkov and V. K. Isaikov. *Pribory i Techhn. Eksperim.,* 4:152 (1963).

CHAPTER 11

DESIGN BASICS AND PRACTICES FOR FLANGES

11.1 REQUIREMENTS FOR FLANGE CONNECTIONS

Among the variety of attachments and accessories essential to vessels and piping, flanges are perhaps the most widely used items. They are employed on the shell of a vessel to provide for disassembly and removal or cleaning of internal parts. There are usually several flanged openings in any vessel: inlets and outlets for products and heat carriers, manholes for installing working devices and inspection, and so on. Flanges also are used for making connections for piping and for nozzle attachments of openings larger than 1½-in. nominal pipe size.

The need for detachable connections results from operating considerations, but also arises for reasons of equipment fabrication (such as the need in casting or forge-and-pressing shops) machine sizes, crane lifting capacity and others. Even when production possibilities permit fabrication of a vessel of a desired size in one piece, it is often made detachable for convenience of assembly and maintenance.

Flanges used for vessels and piping must conform to the following requirements:

1. They must provide tightness under given operating conditions.
2. They must provide sufficient strength.
3. They must permit quick and frequent assembling and disassembling.
4. They must provide mass production, with standard specifications at reasonable cost.

To avoid fabricating flanges for each service pressure, pipe and shell diameter, a rating system was developed. This rating is as follows: 150-, 300-, 400-, 600-, 900-, 1500- and 2500-lb flanges.

The ratings corresponding to service pressures and temperatures are given in Table 11.1. The table shows a decrease in the rating of flanges with temperature. For example, a 150 lb flange has a rating 230 psi at

Table 11.1. Carbon-Steel Flange Pressure Ratings in Water, Steam and Oil Service [1] (values are for flanges with standard facings; excluded are ring joints)

		Primary Service Pressure Ratings (psi)						
		150	300	400	600	900	1500	2500
		Maximum Hydrostatic-Shell-Test Pressures (temperature of test water should not exceed 125°F) (psi)						
		350	900	1200	1800	2700	4500	7500
Service Temperatures								
(°F)	(°C)	Maximum Nonshock Service-Pressure Ratings (psi)						
100	37.8	230	600	800	1200	1800	3000	5000
200	93.3	210	580	770	1160	1740	2900	1830
300	148.9	190	560	740	1120	1680	2800	4660
400	204.4	170	540	710	1075	1615	2690	4475
500	260.0	150[a]	500	665	1000	1500	2500	4465
600	315.6	130	445	590	890	1330	2220	3700
700	371.1	110	380	500	760	1140	1900	3160
800	426.7	92	300[a]	400[a]	600[a]	900[a]	1500[a]	2500[a]

[a]Primary service pressure ratings.

100°F and a rating of 150 psi service pressure at 500°F. A 2500-lb flange has a rating of 5000 psi at 100°F and 2500 psi at 800°F.

11.2 FLANGE SELECTION

A variety of types and sizes of flanges are available for various temperature and pressure services. Flanges differ in design, method of attachment to the pipe or shell and form (round, square, irregularly shaped), as well as in their facing surfaces.

The flanges from forged steel or cast iron (for low-pressure service) are fabricated in accordance with American Standard Association (ASA) B 16.5-1953. Forged-steel flanges are manufactured in the following standard types for all pressure ratings:

- Welding neck
- Slip-on
- Screwed
- Lap-joint
- Blind

11.2.1 Welding-Neck Flanges

Figure 11.1 and its accompanying table (Table 11.2) show a flange sectional view and list the dimensions of several standard 150-lb welding-neck flanges from ½- to 24-in. nominal pipe size.

Welding neck flanges have a long tapered hub, which provides a gradual transition from the flange ring to the pipe wall. This design decreases the discontinuity stresses and increases the strength of the flange. This type of flange works together with a shell (pipe). Because of the integral connection with a shell, it is more unloaded and may be fabricated thinner than slip-on flanges.

Welding-neck flanges are used for extreme service conditions, such as large mechanical loads, wide fluctuations in process parameters, high and low temperatures. These flanges are preferred for handling flammable or explosive fluids, where failure or leakage of a flange joint might bring disastrous consequences.

11.2.2 Slip-On Flanges

This type of a flange has no rigid connection with a pipe, therefore, it is more loaded and, consequently, somewhat thicker than a welding-neck flange. The slip-on type of flange, which is presented in Figure 11.2 and Table 11.3 (the sketch of a standard 150-lb slip-on flange and dimensions for such flanges of ½- to 24-in. nominal pipe sizes), is widely used because of its greater ease of alignment in welding assembly and because of its low initial cost.

The strength of a slip-on flange as calculated from internal pressure considerations is approximately two-thirds that of a corresponding welding-

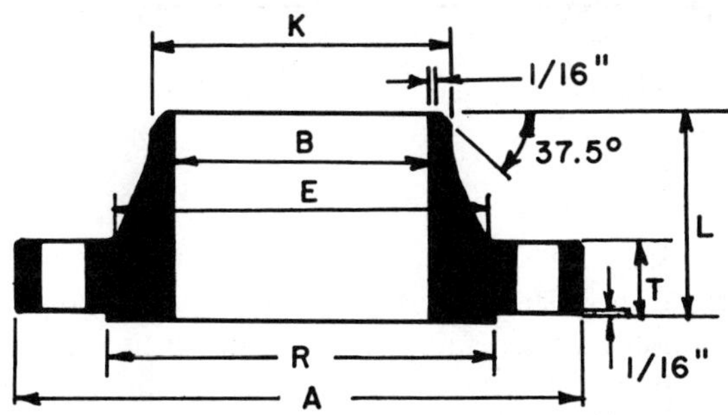

Figure 11.1. Dimensions of welding-neck flanges (refer to Table 11.2).

Table 11.2. Dimensions of Standard American Standard Forged and Rolled

Nominal Pipe Size (in.)	Outside Diameter of Flange A	Thickness of Flange (minimum) T	Outside Diameter of Raised Face R	Diameter of Hub at Base E	Diameter of Hub at Point of Welding K	Length through Hub L
1/2	3 1/2	7/16	1 3/8	1 3/16	0.81	1 7/8
1	4 1/4	9/16	2	1 15/16	1.32	2 3/16
1 1/2	5	11/16	2 7/8	2 9/16	1.90	2 7/16
2	6	3/4	3 5/8	3 1/16	2.38	2 1/2
2 1/2	7	7/8	4 1/8	3 9/16	2.88	2 3/4
3	7 1/2	15/16	5	4 1/4	3.50	2 3/4
4	9	15/16	6 3/16	5 5/16	4.50	3
6	11	1	8 1/2	7 9/16	6.63	3 1/2
8	13 1/2	1 1/8	10 5/8	9 11/16	8.63	4
10	16	1 3/16	12 3/4	12	10.75	4
12	19	1 1/4	15	14 3/8	12.75	4 1/2
14	21	1 3/8	16 1/4	15 3/4	14.00	5
16	23 1/2	1 7/16	18 1/2	18	16.00	5
20	27 1/2	1 14/16	23	22	20.00	5 11/16
24	32	1 7/8	27 1/4	26 1/8	24.00	6

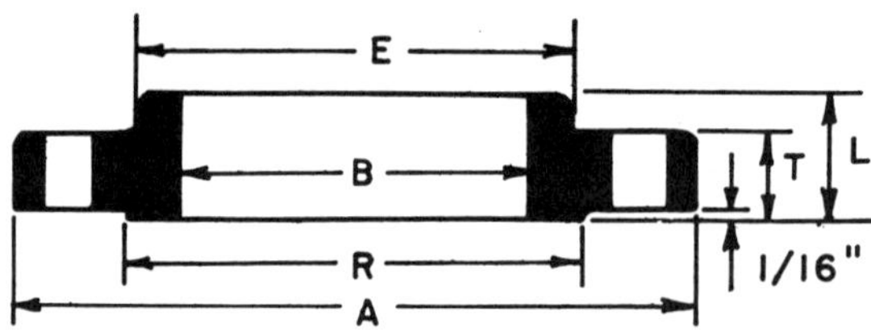

Figure 11.2. Dimensions of forged slip-on flanges.

neck type of flange. The use of this type flange should be limited to moderate services in which pressure and temperature fluctuations, vibrations and shock are not expected to be severe. It also is used for equipment and piping made from mild nonferrous metals, such as aluminum and copper, or for brittle material (ferrosilicon, ceramics), as well as in those cases in which it is necessary to save expensive construction materials as, for example, high-alloy steels and nonferrous alloys.

150-lb Steel Welding-Neck Flanges [1]
–ASA B16e-1939
Steel–ASTM A 181

Inside Diameter of Standard Wall Pipe		Drilling Template			
B	No. of Holes	Diameter of Holes	Diameter of Bolts	Bolt Circle	Approximate Weight Each (lb)
0.62	4	5/8	1/2	2 5/8	2
1.05	4	5/8	1/2	3 1/8	2
1.61	4	5/8	1/2	3 7/8	4
2.07	4	3/4	5/8	4 3/4	6
2.47	4	3/4	5/8	5 1/2	8
3.07	4	3/4	5/8	6	10
4.03	8	3/4	5/8	7 1/2	15
6.07	8	7/8	3/4	9 1/2	24
7.98	8	7/8	3/4	11 3/4	39
10.02	12	1	7/8	14 1/4	52
12.00	12	1	7/8	17	80
13.25	12	1 1/8	1	18 3/4	102
15.25	16	1 1/8	1	21 1/4	127
19.25	20	1 1/4	1 1/8	25	170
23.25	20	1 3/8	1 1/4	29 1/2	260

11.2.3 Lap-Joint Flanges

The lap-joint flange (Figure 11.3) is used for extreme service conditions, to high pressures of up to several hundred atmospheres and temperatures up to 530°C. The principal advantage of this flange is that the bolt holes are easily aligned. This simplifies the erection of large-diameter vessels and unusually stiff piping.

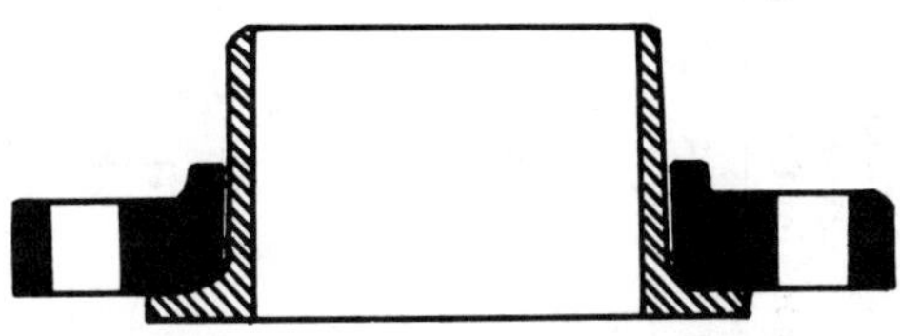

Figure 11.3. Sectional view of steel lap-joint flange and lap-joint stud.

Table 11.3. Dimensions of Standard 150-lb Forged Slip-on Flanges (refer to Figure 11.2) [1]
American Standard—ASA B16e-1939
Forged and Rolled Steel—ASTM A 181

Nominal Pipe Size (in.)	Outside Diameter of Flange	Thickness of Flange (minimum)	Outside Diameter of Raised Face	Diameter of Hub at Base	Length through Hub	Diameter of Bore	Drilling Template				Approximate Weight Each (lb)
	A	T	R	E	L	B	No. of Holes	Diameter of Holes	Diameter of Bolts	Bolt Circle	
1/2	3 1/2	7/16	1 3/8	1 3/16	5/8	0.88	4	5/8	1/2	2 3/8	2
3/4	3 7/8	1/2	1 11/16	1 1/2	5/8	1.09	4	5/8	1/2	2 3/4	2
1	4 1/4	9/16	2	1 15/16	11/16	1.38	4	5/8	1/2	3 1/8	2
1 1/2	5	11/16	2 7/8	2 9/16	7/8	1.97	4	5/8	1/2	3 7/8	3
2	6	3/4	3 5/8	3 1/16	1	2.44	4	3/4	5/8	4 3/4	5
2 1/2	7	7/8	4 1/8	3 9/16	1 1/8	2.94	4	3/4	5/8	5 1/2	7
3	7 1/2	15/16	5	4 1/4	1 3/16	3.56	4	3/4	5/8	6	8
4	9	15/16	6 3/16	5 5/16	1 5/16	4.56	8	3/4	5/8	7 1/2	13
6	11	1	8 1/2	7 9/16	1 9/16	6.72	8	7/8	3/4	9 1/2	19
8	13 1/2	1 1/8	10 5/8	9 11/16	1 3/4	8.72	8	7/8	3/4	11 3/4	30
10	16	1 3/16	12 3/4	12	1 15/16	10.88	12	1	7/8	14 1/4	43
12	19	1 1/4	15	14 3/8	2 3/16	12.88	12	1	7/8	17	64
14	21	1 3/8	16 1/4	15 3/4	2 1/4	14.19	12	1 1/8	1	18 3/4	85
16	23 1/2	1 7/16	18 1/2	18	2 1/2	16.19	16	1 1/8	1	21 1/4	93
20	27 1/2	1 11/16	23	22	2 7/8	20.19	20	1 1/4	1 1/8	25	155
24	32	1 7/8	27 1/4	26 1/8	3 1/4	24.19	20	1 3/8	1 1/4	29 1/2	210

However, the lap-joint flange costs about 30% more than a welding-neck flange of the same size and rating, but has only about 10% the fatigue life.

11.2.4 Screwed Flanges

Screwed flanges (Figure 11.4) are used in extremely high-pressure service with alloy steel, which is not easily welded. They are employed in applications involving no bending or thermal cycles. For these conditions, the flanges are susceptible to leakage.

11.2.5 Blind Flanges

This flange (Figure 11.5 and Table 11.4 give dimensions for such flanges of ½- to 24-in. nominal pipe size) is used for blanking off pressure-vessel openings. The blind flange also is used to block off the ends of piping and valves. A valve followed by a blind flange permits addition to the line while it is "on stream."

11.3 NONSTANDARD FLANGES

Nonstandard large-diameter welding-neck or slip-on flanges are fabricated by rolling a hot annular blank in sizes ranging from 26 in. to 96 in. in a variety of ratings. As an example, Figure 11.6 and Table 11.5 give a sectional view of a welding-neck flange and the dimensions for welding-neck and slip-on flanges, respectively.

Figure 11.4. Sectional view of a screwed flange.

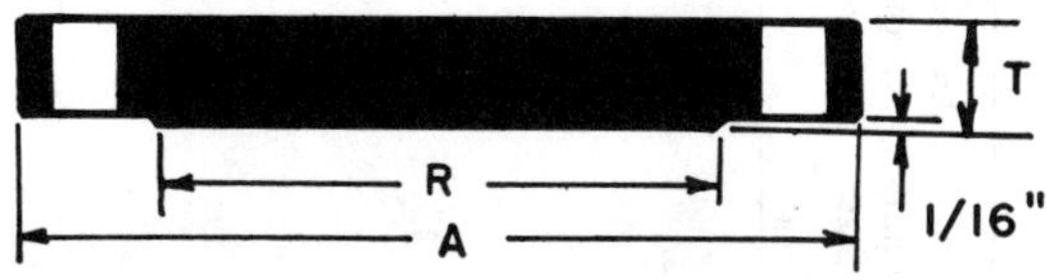

Figure 11.5. Dimensions of a blind flange (refer to Table 11.4).

Table 11.4. Dimensions of a Standard 150-lb Blind Flange
American Standard–ASA B163e-1939
Forged and Rolled Steel–ASTM A 181

Nominal Pipe Size (in.)	Outside Diameter of Flange	Thickness of Flange (minimum)	Outside Diameter of Raised Face	Drilling Template				Approximate Weight Each (lb)
	A	T	R	No. of Holes	Diameter of Holes	Diameter of Bolts	Bolt Circle	
1/2	3 1/2	7/16	1 3/8	4	5/8	1/2	2 3/8	2
3/4	3 7/8	1/2	1 11/16	4	5/8	1/2	2 3/4	2
1	4 1/4	9/16	2	4	5/8	1/2	3 1/8	2
1 1/2	5	11/16	2 7/8	4	5/8	1/2	3 7/8	3
2	6	3/4	3 5/8	4	3/4	5/8	4 3/4	4
2 1/2	7	7/8	4 1/8	4	3/4	5/8	5 1/2	7
3	7 1/2	15/16	5	4	3/4	5/8	6	9
4	9	15/16	6 3/16	8	3/4	5/8	7 1/2	17
6	11	1	8 1/2	8	7/8	3/4	9 1/2	26
8	13 1/2	1 1/8	10 5/8	8	7/8	3/4	11 3/4	45
10	16	1 3/16	12 3/4	12	1	7/8	14 1/4	70
12	19	1 1/4	15	12	1	7/8	17	110
14	21	1 3/8	16 1/4	12	1 7/8	1	18 3/4	311
16	23 1/2	1 7/16	18 1/2	16	1 1/8	1	21 1/4	170
20	27 1/2	1 11/16	23	20	1 1/4	1 1/8	25	272
24	32	1 7/8	27 1/4	20	1 3/8	1 1/4	29 1/2	411

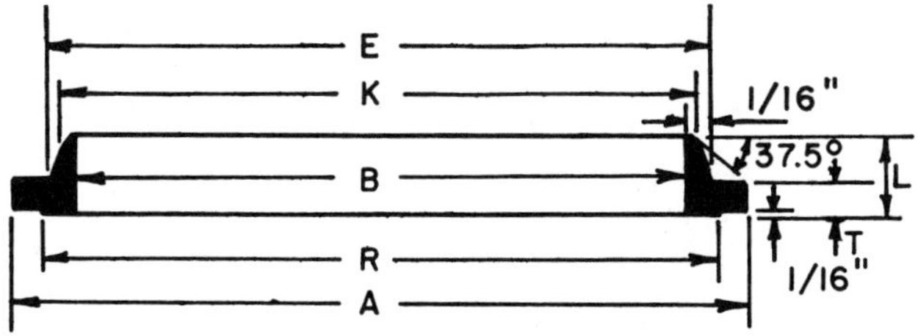

Figure 11.6. Nonstandard flange (refer to Table 11.5) [1].

11.4 FLANGES OF SPECIAL TYPES

Square, oval and irregularly shaped flanges sometimes are used for pipe connections and nozzle attachments of openings no larger than 1½-in. nominal pipe size. This type of flange is common in high-pressure and cryogenic equipment.

11.4.1 Welded Around Flange Connections

In vessels used for processing highly toxic substances any leakage is unacceptable. Therefore, it is advisable to design all-welded equipment without flanges and attach the piping to the equipment by welding. However, if in such equipment the flange connection is inevitable, such as for a manhole, for example, it is carried out without gaskets, but with the use of a thickening seam (Figure 11.7).

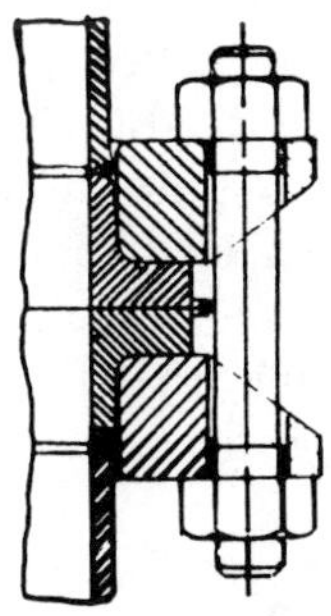

Figure 11.7. Flange with a thickening seam.

Table 11.5. Dimensions of Nonstandard Forged and

Size (in.)	Outside Diameter of Flange A	Thickness of Flange T	Outside Diameter of Raised Face R	Compressed-Asbestos-Gasket Size	Taper-Hub Diameters E	K
				50-lb Pressure at 100°F		
26	31 1/2	1 1/4	28 5/8	27 3/4 × 28 5/8	27 1/8	26 1/2
30	35 1/2	1 1/4	32 5/8	31 3/4 × 32 5/8	31 1/8	30 1/2
34	40 1/4	1 1/4	37	35 7/8 × 37	35 3/8	34 5/8
36	42 1/4	1 1/4	39	37 7/8 × 39	37 3/8	36 5/8
42	49	1 1/4	45 1/2	44 1/8 × 45 1/2	43 3/4	42 3/4
48	55	1 1/4	51 1/2	50 1/8 × 51 1/2	49 3/4	48 3/4
54	61 1/4	1 3/8	57 3/4	56 1/4 × 57 3/4	56	54 7/8
60	67 1/4	1 5/8	63 3/4	62 1/4 × 63 3/4	62	60 7/8
66	74	1 7/8	70 1/8	68 3/4 × 70 1/8	68	67
72	80	2 1/4	76 1/8	74 3/8 × 76 1/8	74	73
				50-lb Pressure at 100°F		
26	33	1 1/4	30	29 1/8 × 30	28 1/2	28
30	37	1 1/4	34	33 1/8 × 34	32 1/2	32
31	41 1/2	1 1/4	38 1/4	37 1/8 × 38 1/4	36 5/8	36 1/8
36	43 1/2	1 1/4	40 1/4	39 1/8 × 40 1/4	38 5/8	38 1/8
42	50	1 1/4	46 1/2	45 1/8 × 46 1/2	44 3/4	44 1/4
48	56	1 1/4	52 1/2	51 1/8 × 52 1/2	50 3/4	50 1/4
54	62 1/2	1 3/8	59	57 1/2 × 59	57 1/4	56 1/2
60	68 1/2	1 5/8	65	63 1/2 × 65	63 1/4	62 1/2
66	75 1/2	1 3/4	71 5/8	69 7/8 × 71 5/8	69 1/2	68 3/4
72	81 1/2	2	77 5/8	75 7/8 × 77 5/8	75 1/2	74 3/4

For disassembly, the seam is broken through then welded again for assembly. The thickening seam holds out 6-10 assemblies. It is then necessary to change the lap-joint studs. In designing such connections, the strength of a seam is not taken into consideration as it serves only for sealing. The trapezoidal form of the flange section (Figure 11.7) is made to decrease the flange weight. In this case, the flange strength is only slightly decreased because the outer flange fibers are less loaded.

11.4.2 Detachable Flanges

To disassemble a vessel, it is sometimes necessary to detach the flange from a pipe. Screwed flanges often are used in this case. However, it is generally

Large-Diameter Flanges [1]
Rolled Steel

Length through Hub	Diameter of Bore	Drilling Template			Approximate Weight Each (lb)
L	B	No. of Holes	Diameter of Holes	Bolt Circle	
–welding-neck type					
3	26	32	1	29 5/8	98
3	30	36	1	33 5/8	112
3 1/4	34	40	1 1/8	38 1/8	149
3 1/4	36	40	1 1/8	40 1/8	157
3 1/2	42	48	1 1/4	46 3/4	209
3 3/4	48	52	1 1/4	52 3/4	214
4	54	61	1 1/4	59	312
4 3/8	60	72	1 1/4	65	398
4 7/8	66	72	1 3/8	71 1/2	556
5 1/4	72	80	1 3/8	77 1/2	705
–slip-on type					
2 1/4	26 11/16	32	1	31	122
2 1/4	30 11/16	36	1	35	148
2 1/2	34 13/16	40	1 1/8	39 3/8	181
2 1/2	36 13/16	44	1 1/8	44 3/8	191
2 3/4	42 15/16	48	1 1/4	47 3/4	234
2 7/8	48 15/16	56	1 1/4	53 3/4	269
3 1/8	55 1/16	68	1 1/4	60 1/4	335
3 5/8	61 1/16	72	1 1/4	66 1/4	451
4	67 3/16	72	1 3/8	73	591
4 1/2	73 3/16	80	1 3/8	79	728

not the best solution to this problem: flanges become so rusted that they cannot be unscrewed. It is often better to use detachable flanges with a split ring set in a pipe groove, as shown in Figure 11.8. The detachable flanges for lap-joint pipes are fabricated sectionally, either in two- or in four-ring parts, as shown in Figure 11.9.

This flange design is simpler in fabrication and less susceptible to corrosion than are the screwed flanges.

11.4.3 Flanges for Pipes and Shells From Fragile Materials

It is not advisable that flanges made from ferrosilicon, ceramics, glass, porcelain, polyvinyl chloride (PVC) plastics and similar materials to be

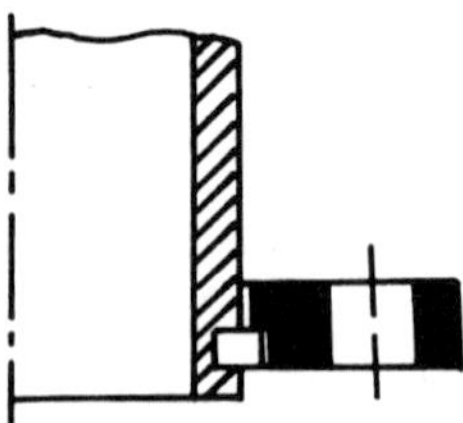

Figure 11.8. Detachable flange with a split ring.

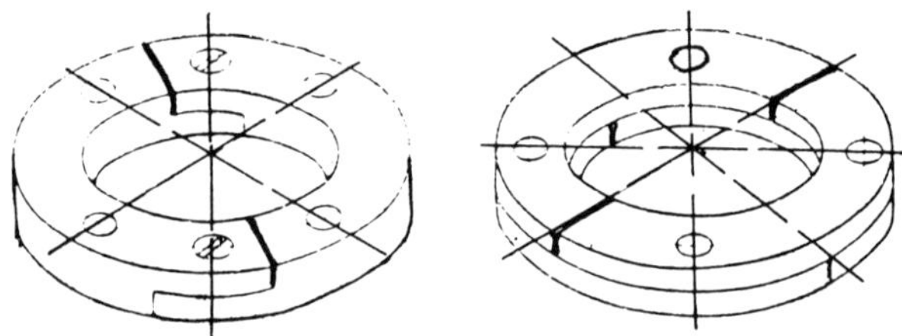

Figure 11.9. Detachable flange in two and four sections.

fabricated integrally with equipment because of their inadequate flexing resistance. The ends of pipes, shells and heads from such materials are made with thick conical parts on which special flanges are placed. They, as well as the detachable flanges, are made in two fashions: (1) detachable flanges with two-ring parts, and (2) flanges with split rings.

Figure 11.10 shows the detachable flange from malleable cast iron, where its two halves are fixed by bolts.

Figure 11.11 shows a sectional view of a steel flange with a bore somewhat larger than a thicker part of a pipe (shell), which leans against the two cast iron half rings. Flanges with split rings have higher strength and are cheaper than the detachable flanges. They are, however, less compact.

The detachable flanges from malleable cast iron are used in the case in which smaller sizes permit design of a more compact vessel, for example, a sprinkling cooler from siliceous cast iron pipes, and so on. A flange design with split rings for glass pipe connections is presented in Figure 11.12. The half rings are made from polyamide plastics; the gasket between the polished pipe sections and thick pipe ends is made from rubber. The

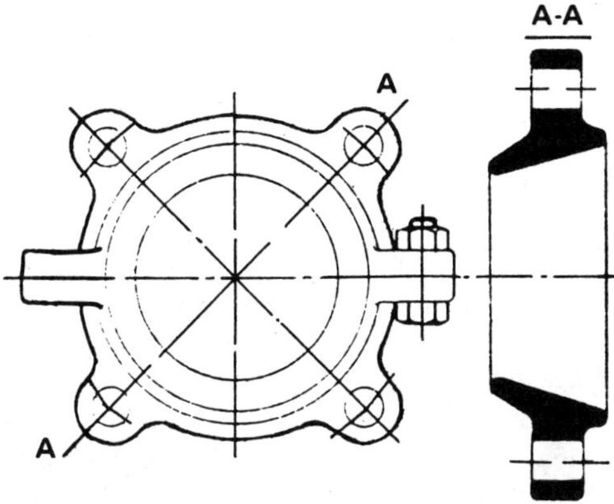

Figure 11.10. Detachable flange for ceramic pipe.

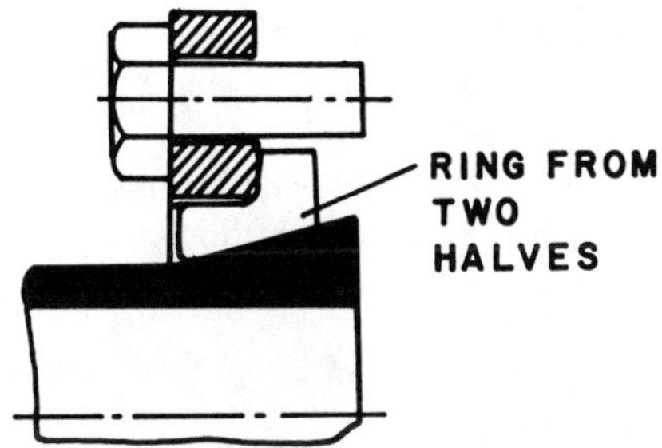

Figure 11.11. Flange with a split ring for pipes from siliceous cast iron.

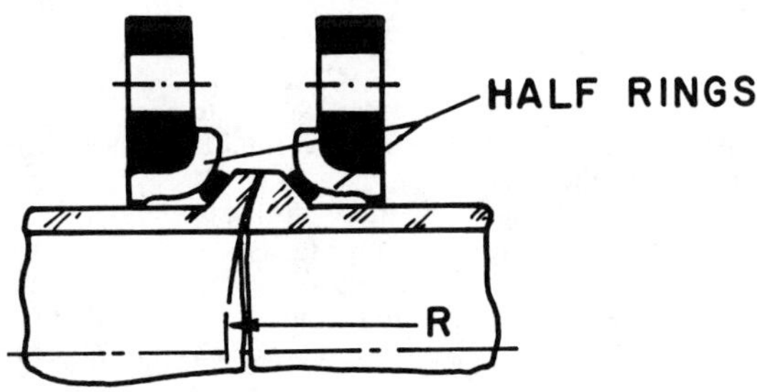

Figure 11.12. Connection for glass pipes.

spherical edge joints permit pipe connections to be made at an angle extending to 20°.

11.5 FLANGE FACINGS

Standard flanges are used with a variety of machined faces, as shown in Figure 11.13 (corresponding dimensions are given in Table 11.6). The raised face flanges are used extensively because of the simplicity of their design and adequacy for average service conditions. Flanges with ratings of 750 lb and 300 lb have 1/16-in.-high raised faces, and those for higher ratings have ¼-in.-high raised faces. For better yield and holding, a gasket is placed over the raised fall of a flange, which is machined with spiral or concentric grooves (approximately 1/64-in. deep).

The male and female facings (3/16-in. deep recess on the female face and ¼-in.-high raised face on the male part) confine the gaskets and minimize the possibility of blowout. However, this type of facing has no protection against forcing the gasket into the vessel.

The tongue-and-groove face gives protection against deforming soft gaskets into the interior of the vessel. This facing is less subject to errosive or corrosive contact with the working fluid.

The ring-joint type of facing offers the greatest protection under severe service conditions or with the use of hazardous fluids. While using this type of facing, the internal pressure in a vessel acts on the ring, increasing the sealing force of the joint.

11.6 GASKET SELECTION

A gasket is interposed between the flange facings to provide better tightness of a flange connection and to decrease the necessary compressive force ("seating" force). The gasket must conform to the following requirements:

1. The gasket must be sufficiently plastic to interpose its material through deformation under load and seal the minute surface irregularities to prevent leakage of the fluid.
2. The gasket must not change its elasticity during operation.
3. The gasket must not change the flange facing.

The decision as to which gasket material is to be selected depends on temperature, pressure and the aggresiveness of the sealed medium.

Gaskets usually made in thicknesses from 1/64 in. to 1/8 in. are commonly fabricated from such materials as paper, rubber, cloth, asbestos, plastics, copper, lead, aluminum, nickel, monel and soft iron. Paper, cloth and

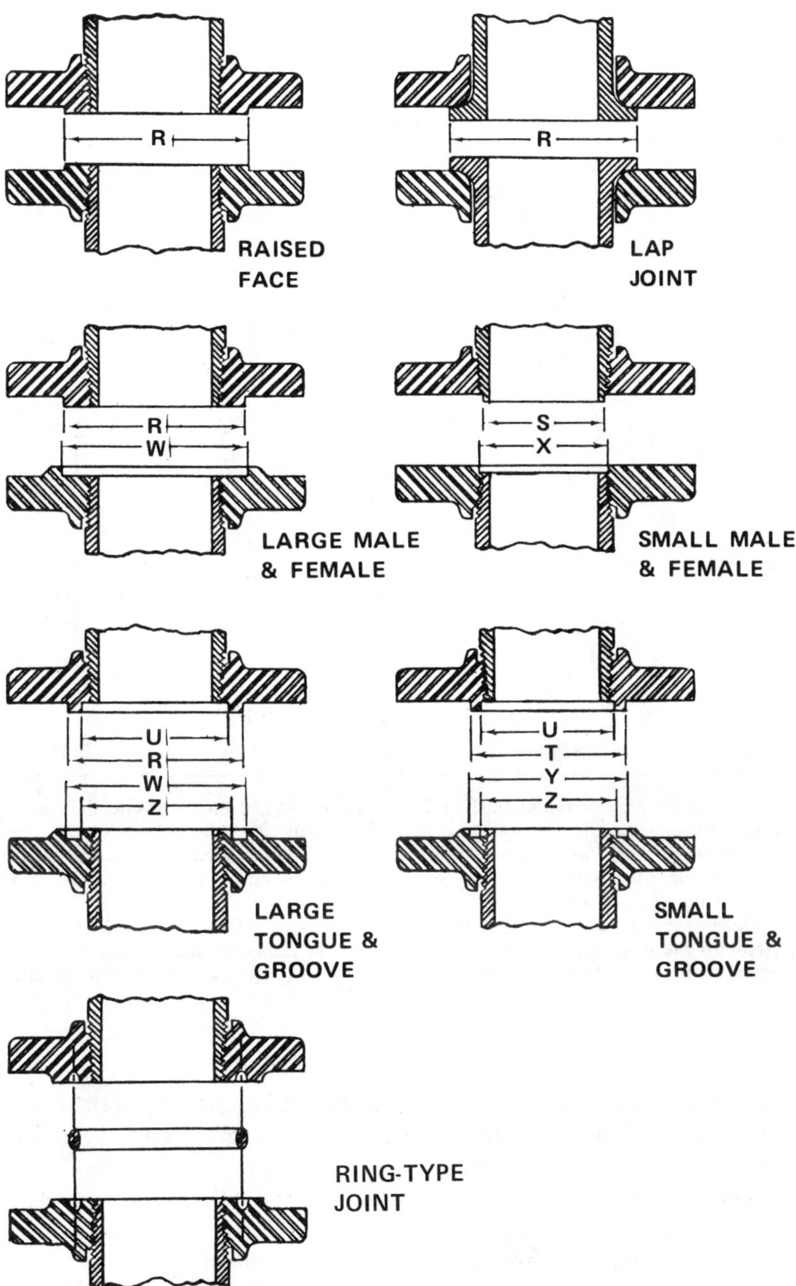

Figure 11.13. American Standard flange facings.

Table 11.6. American Standard Flange Facings for

Nominal Pipe Size (in.)	Outside Diameter[a]			Inside Diameter of Large and Small Tongue[a,b]
	Raised Face, Lap-Joint, Large Male, and Large Tongue[b]	Small Male[b,c]	Small Tongue[b]	
	R	S	T	U
1/2	1 3/8	23/32	1 3/8	1
3/4	1 11/16	15/16	1 11/16	1 5/16
1	2	1 3/16	1 7/8	1 1/2
1 1/2	2 7/8	1 3/4	2 1/2	2 1/8
2	3 5/8	2 1/4	3 1/4	2 7/8
2 1/2	4 1/8	2 11/16	3 3/4	3 3/8
3	5	3 5/16	4 5/8	4 1/4
4	6 3/16	4 5/16	5 11/16	5 3/16
6	8 1/2	6 3/8	8	7 1/2
8	10 5/8	8 3/8	10	9 3/8
10	12 3/4	10 1/2	12	11 1/4
12	15	12 1/2	14 1/4	13 1/2
14	16 1/4	13 3/4	15 1/2	14 3/4
16	18 1/2	15 3/4	17 5/8	16 3/4
20	23	19 3/4	22	21
24	27 1/4	23 3/4	26 1/4	25 1/4

[a]A tolerance of plus or minus 0.016 in. (1/64 in.) is allowed on the inside and outside
[b]Gaskets for male-female and tongue-groove joints shall cover the bottom of the recess
[c]Care should be taken in the use of joints of these dimensions (they apply particularly enough to permit sufficient bearing surface to prevent crushing the gasket. Threaded Standard Locknut Thread.
[d]Regular facing for 150- and 300-lb steel flanged fittings and flange standards is a 1/16-in. mitted on the 400-, 600-, 900-, 1500- and 2500-lb flange standards, but it must be
[e]Regular facing for 400-, 600-, 900-, 1500- and 2500-lb flange standards is a 1/4-in. raised

rubber gaskets are used below 250°F. Asbestos-composition gaskets may be used up to 650°F, and ferrous and nickel-alloy metal gaskets may be used up to the maximum temperature rating of the flanges.

Laminated gaskets (with a metal jacket and soft filler, usually asbestos) require less bolt load to seat and keep tight than solid-metal, flat-ring gaskets. They can be used for temperatures of up to about 850°F.

Serrated solid metal gaskets having concentric grooves machined into the faces greatly reduce the contact area on initial tightening, thereby reducing the bolt load. Serrated gaskets are used extensively in power

150, 300, 400, 600, 900, 1500 and 2500-lb Flanges

Outside Diameter[a]						
Large Female and Large Groove[b]	Small Female[b,c]	Small Groove[b]	Inside Diameter of Large and Small Groove[a,b]	Raised Face, 150- and 300-lb Standards[d]	Height Raised Face, Large and Small Male and Tongue, 400-, 600-, 900-, 1500-, and 2500-lb Standards[e]	Depth of Groove or Female
W	X	Y	Z			
1 7/16	25/32	1 7/16	15/16	1/16	1/4	3/16
1 3/4	1	1 3/4	1 1/4	1/16	1/4	3/16
2 1/16	1 1/4	1 15/16	1 7/16	1/16	1/4	3/16
2 15/16	1 13/16	2 9/16	2 1/16	1/16	1/4	3/16
3 11/16	2 5/16	3 5/16	2 13/16	1/16	1/4	3/16
4 3/16	2 3/4	3 13/16	3 5/16	1/16	1/4	3/16
5 1/16	3 3/8	4 11/16	4 3/16	1/16	1/4	3/16
6 1/4	4 3/8	5 3/4	5 1/8	1/16	1/4	3/16
8 9/16	6 7/16	8 1/16	7 7/16	1/16	1/4	3/16
10 11/16	8 7/16	10 1/16	9 5/16	1/16	1/4	3/16
12 13/16	10 9/16	12 1/16	11 3/16	1/16	1/4	3/16
15 1/16	12 9/16	14 5/16	13 7/16	1/16	1/4	3/16
16 5/16	13 13/16	15 9/16	14 11/16	1/16	1/4	3/16
18 9/16	15 13/16	17 11/16	16 11/16	1/16	1/4	3/16
23 1/16	19 13/16	22 1/16	20 15/16	1/16	1/4	3/16
27 5/16	23 13/16	26 5/16	25 3/16	1/16	1/4	3/16

diameters of all facings.
with minimum clearances taking into account the tolerances prescribed in footnote a.
on lines where the joint is made on the end of pipe) to ensure that the pipe used is thick
companion flanges are furnished with plain face and are threaded with American

raised face included in the minimum flange thickness. A 1/16-in. raised face also is per-
added to the minimum flange thickness.
face not included in minimum flange thickness dimensions.

stations in which soft or laminated gaskets are unsatisfactory and where the bolt load is excessive with a flat-ring metal gasket.

Corrugated gaskets with asbestos filling have a rigid surface with concentric rings. These gaskets require less seating force than laminated or serrated gaskets and are widely used in low-pressure fluid operations. The same gaskets, but without asbestos, are used extensively in sealing water, steam, gas, oil, acid and other chemicals at higher temperatures.

Ring-joint gaskets of oval and octagonal cross sections are used for high-pressure service. These gaskets are fabricated of solid metal-soft

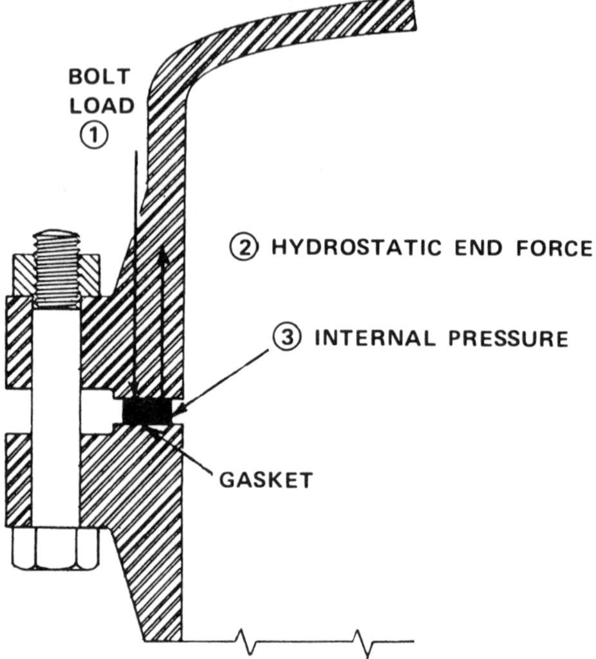

Figure 11.14. Major forces acting on a gasket.

iron, soft steel, Monel and stainless steels. Before using alloy-steel rings, they should be heat treated to soften them. The flange connection plastic rings may be used for corrosion resistance, and as a means of electrically insulating in the case of low-temperature services.

The amount of force ("seating" or "yield" force) that must be applied to the gasket to cause its material to flow and seal the surface irregularities depends on the gasket's width and thickness, roughness of the facings and on the type of the gasket itself. As a result of applying force to the gasket there arises a "yield" stress that represents the minimum load that ensures the tightness of connection for whatever pressures are used in the vessel.

For pressure vessels (Figure 11.14), an end force tends to separate the flanges and decrease the stress on the gasket. Leakage will occur in this case if the hydrostatic end force is great enough that the difference between it and the bolt-load force decreases the gasket load below a critical value.

The "gasket factor" is defined as the ratio of the gasket stress to the internal vessel pressure. It is a property of the gasket material and construction and is independent of the internal pressure. In selecting a gasket, it is necessary to determine the total amount of force required to make the gasket yield and to keep a tight seal under operating conditions. Some common types of gaskets, gasket factor, minimum design seating stress and recommended facings are given in Table 11.7. The effective width of the gasket, b, for various types of facings is presented in the Figure 11.15.

Because the yield force at a given value of yield stress is proportional to the gasket surface area, it is irrational to place the gasket over the total flange facing (Figure 11.16). Although such a layout makes it easier for the assembly of a flange connection and unloads the flanges from the bending moments, the bolt load will be increased unnecessarily; further, the diameter and number of bolts become so great that load fails to provide the proper distribution over the flange. The higher the pressure in the vessel, the narrower the gasket required. For high-pressure service, the sealing area of metal gaskets or rings is determined by the strip width of elastic deformation of compressed parts.

The type of facing greatly influences the flange performance and the amount of bolt load. Due to the turning of the flange section under the bolt load, the outer gasket part is more compressed than the inside gasket part. With this in mind, *the effective gasket width* (which is less than its actual width) is used in flange designing.

A preliminary estimate of gasket proportions may be obtained if we neglect the elastic deformations of bolts, gasket and flanges:

(Gasket seating force) – (Hydrostatic pressure force) = (Residual gasket force)

The residual gasket force cannot be less than that required to prevent leakage of the fluid from the vessel under operating pressure:

$$\frac{\pi}{4}(d_o^2 - d_i^2)y - \frac{\pi d_o^2}{4} p = \frac{\pi}{4}(d_o^2 - d_i^2)pm \qquad (11.1)$$

where y = yield stress (lb/in.^{2}) (see Table 11.7)
p = internal pressure (lb/in.^{2})
m = gasket factor (see Table 11.7)
d_o = outside diameter of gasket (in.)
d_i = inside diameter of gasket (in.)

Rewriting Equation 11.1, we obtain

$$\frac{d_o}{d_i} - \sqrt{\frac{y - pm}{y - p(m + 1)}} \qquad (11.2)$$

Table 11.7. Gasket Factors (m) for Operating Conditions and Minimum Design Seating Stress (y)

Gasket Material		Gasket Factor (m)	Minimum Design Seating Stress (y)	Sketches and Holes	Refer to Figure 11.5 Facing Limitations	
Rubber without fabric or a high percentage of asbestos fiber:						
below 75, Shore durometer		0.50	0			
75 or higher, Shore durometer		1.00	200		Use 1, 4, 6 only	
Asbestos with a suitable binder for the operating conditions	1/8 thick	2.00	1600			
	1/16 thick	2.75	3700			
	1/32 thick	3.50	6500			
Rubber with cotton-fabric insertion		1.25	400			
Rubber with asbestos-fabric insertion, with or without wire reinforcement	3-ply	2.25	2200			
	2-ply	2.50	2900		None	
	1-ply	2.75	3700			
Vegetable fiber		1.75	4400		1, 4, 6	
Spiral-wound metal, asbestos filled	Carbon	2.50	2900			
	Stainless	3.00	4500			
Serrated steel	Asbestos filled	2.75	3700			
Corrugated metal,	Soft aluminum	2.50	2900			
asbestos inserted	Soft copper or brass	2.75	3700			II
or	Iron or soft steel	3.00	4500			
Corrugated metal, jacketed	Monel or 4–6% chrome	3.25	5500		Use 1a only	
asbestos filled	Stainless steels	3.50	6500			

Gasket type	Material	m	y	Sketch	Facing
Corrugated metal	Soft aluminum	2.75	3700		
	Soft copper or brass	3.00	4500		
	Iron or soft steel	3.25	5500		
	Monel or 4–6% chrome	3.50	6500		
	Stainless steels	3.75	7600		
Flat metal, jacketed, asbestos filled	Soft aluminum	3.25	5500	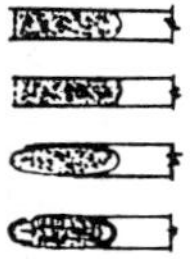	Use 1a, 2[b] only
	Soft copper or brass	3.50	6500		
	Iron or soft steel	3.75	7600		
	Monel	3.50	8000		
	4–6% chrome	3.75	9000		
	Stainless steels	3.75	9000		
Grooved iron or soft steel with or without metal jacket	Soft aluminum	3.25	5500	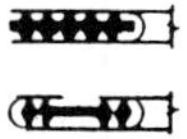	Use 1, 2, 3, only
	Soft copper or brass	3.50	6500		
	Iron or soft steel	3.75	7600		
	Monel or 4–6% chrome	4.00	8800		
	Stainless steels	4.25	10100		
Solid flat metal	Soft aluminum	4.00	8800		None
	Soft copper or brass	4.75	13000		
	Iron or soft steel	5.50	18000		
	Monel or 4–6% chrome	6.00	21800		
	Stainless steels	6.50	26000		
Ring joint	Iron or soft steel	5.50	18000	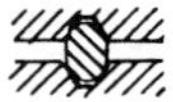	Use 8 only
	Monel or 4–6% chrome	6.00	21800		
	Stainless steels	6.50	26000		

[a]This table gives a list of many commonly used gasket materials and contact facings with suggested design values of m and y that have generally proved satisfactory in actual service when using effective gasket seating width, b, given in Figure 11.16.

[b]The surface of a gasket having a lap should be against the smooth surface of the facing and not against the nipple.

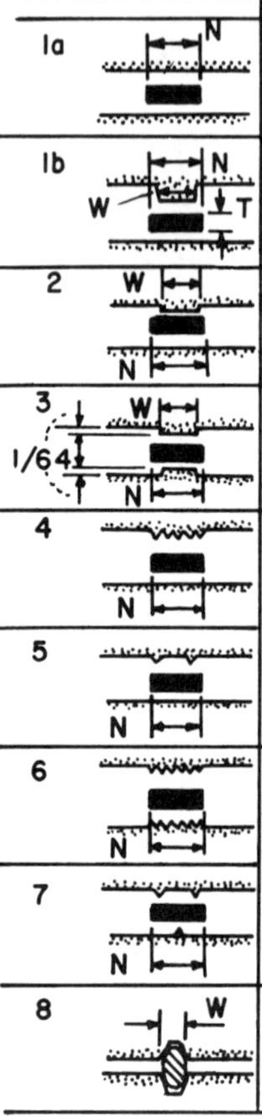

FACING SKETCH	BASIC GASKET SEATING WIDTH, b_o	
	Column I	Column II
1a	$\frac{N}{2}$	$\frac{N}{2}$
1b	$\frac{W+T}{2}$; $\left(\frac{W+N}{4}\text{ max}\right)$	Same as Column I.
2	$\frac{W+N}{4}$	$\frac{W+3N}{8}$
3 (1/64)	$\frac{W}{2}$; $\left(\frac{N}{4}\text{ min}\right)$	$\frac{W+N}{4}$; $\left(\frac{3N}{8}\text{ min}\right)$
4, 5	$\frac{3N}{8}$	$\frac{7N}{16}$
6, 7	$\frac{N}{4}$	$\frac{3N}{8}$
8	$\frac{W}{8}$	

Effective Gasket Seating Width, b

$b = b_o$ when $b_o \leq 1/4''$

$b = \sqrt{b_o}/2$ when $b > 1/4''$

LOCATION OF GASKET-LOAD REACTION

H_G h_G G b O.D. Contact Face For $b_o > 1/4''$

H_G h_G G Gasket Face For $b_o \leq 1/4''$

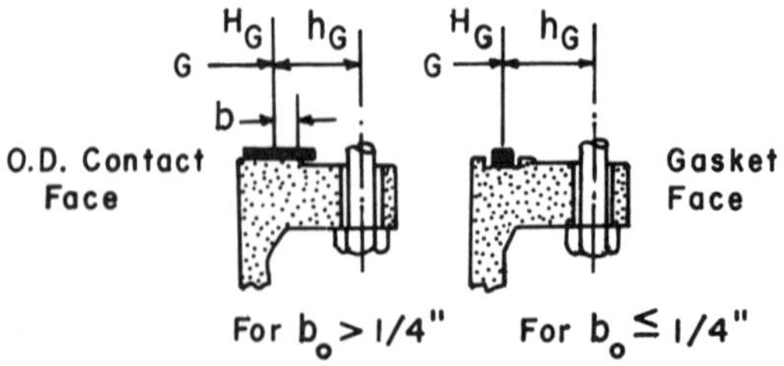

Figure 11.15. Effective gasket width and location of gasket load reaction. (Note that gasket factors listed apply only to flanged joints where the gasket is contained entirely within the inner edges of the bolt holes.)

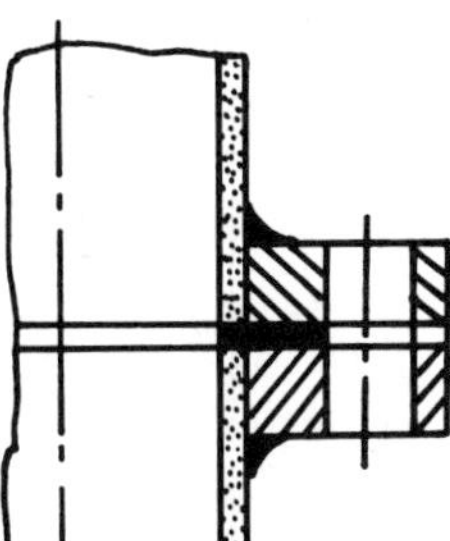

Figure 11.16. Gasket resting on total flange facing.

When it is desirable to reduce the gasket width, a gasket seating stress slightly in excess of y should be used (very large excesses of y can crush the gasket or cause it to be squeezed out between the flange faces).

11.7 OPTIMUM BOLT SELECTION

The following forces load the bolts: the force required to seat the gasket and the force required to withstand the internal pressure and to maintain the gasket-factor pressure (mp) at the same time. The greater of these two forces is taken as the designing force. The required bolting area is determined by dividing the designing force by the allowable bolting stress.

The minimum bolt spacing based on wrench clearances limits the number of bolts that can be placed in a given bolt circle. The maximum bolt spacing is limited by the flange deflection, providing a no-leak joint. The Taylor Forge Co. recommends the following empirical formula for maximum bolt spacing:

$$B_{s(max)} = 2d + \frac{6t}{m + 0.5} \tag{11.3}$$

where $B_{s(max)}$ = maximum bolt spacing for a tight joint (in.)
d = bolt diameter (in.)
t = flange thickness (in.)
m = gasket factor (see Table 11.7)

It is desirable to use a minimum diameter bolt circle, which may be determined from Table 11.8. This diameter either will be the diameter necessary

Table 11.8. Bolt Data [1]

Bolt Size (d)	Standard Thread		8-thread Series		Bolt Spacing[a]		Minimum Radial Distance (R)	Edge Distance (E)	Nut Dimension (across flats)	Maximum Fillet Radius (r)
	No. of Threads	Root Area	No. of Threads	Root Area	Minimum B_s	Preferred				
1/2"	13	0.126	No. 8		1 1/4"	3"	13/16"	5/8"	7/8"	1/4"
5/8"	11	0.202	thread		1 1/2"	3"	15/16"	3/4"	1 1/16"	5/16"
3/4"	10	0.302	series		1 3/4"	3"	1 1/8"	13/16"	1 1/4"	3/8"
7/8"	9	0.419	below 1"		2 1/16"	3"	1 1/4"	15/16"	1 7/16"	3/8"
1"	8	0.551	8	0.551	2 1/4"	3"	1 3/8"	1 1/16"	1 5/8"	7/16"
1 1/8"	7	0.693	8	0.728	2 1/2"	3"	1 1/2"	1 1/8"	1 13/16"	7/16"
1 1/4"	7	0.890	8	0.929	2 13/16"	3"	1 3/4"	1 1/4"	2"	9/16"
1 3/8"	6	1.054	8	1.155	3 1/16"		1 7/8"	1 3/8"	2 3/16"	9/16"
1 1/2"	6	1.294	8	1.405	3 1/4"		2"	1 1/2"	2 3/8"	5/8"

1⅝"	5½	1.515	8	1.680	3½"	2⅛"	1⅝"	2 9/16"	⅝"
1¾"	5	1.744	8	1.980	3¾"	2¼"	1¾"	2¾"	⅝"
1⅞"	5	2.049	8	2.304	4"	2⅜"	1⅞"	2 15/16"	⅝"
2"	4½	2.300	8	2.652	4¼"	2½"	2"	3⅛"	11/16"
2¼"	4½	3.020	8	3.423	4¾"	2¾"	2¼"	3½"	11/16"
2½"	4	3.715	8	4.292	5¼"	3 1/16"	2⅜"	3⅞"	13/16"
2¾"	4	4.618	8	5.259	5¾"	3⅜"	2⅝"	4¼"	⅞"
3"	4	5.621	8	6.324	6¼"	3⅝"	2⅞"	4⅝"	15/16"

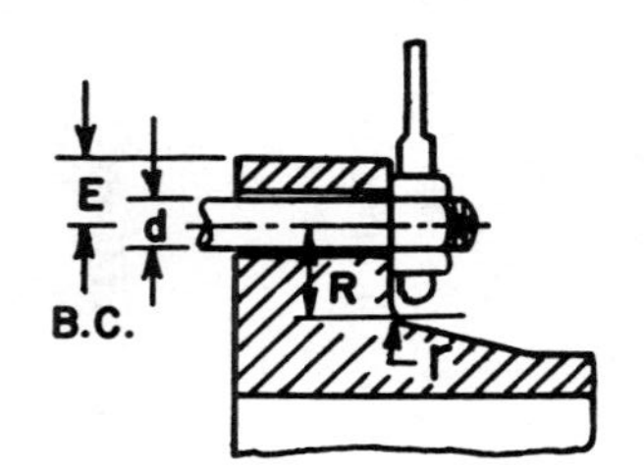

[a] B_s = center-to-center distance between bolts, inches.

to satisfy the radial clearances ($d = B + 2(g_1 + R)$) or the diameter necessary to satisfy the bolt-spacing requirement ($d = (NB_s/\pi)$), whichever is greater. The optimum design is achieved when these two controlling diameters are approximately equal. The following example illustrates the recommended procedure.

Example 11.1

A gasket has the following properties: inside diameter, B = 32 in.; hub thickness, g_1 = 1¼ in.; allowable bolt stress = 20,000 psi; maximum bolt load = 456,000 lb.

Solution

The minimum bolting area, A_B (min.), is

$$A_{B(min)} = \frac{W}{\sigma_{allow}} = \frac{446{,}000}{20{,}000} = 22.3 \text{ in.}^2$$

Root area, B_s, and radial distance, R, are taken from Table 11.9. Analysis of the data in Table 11.9 indicates that the minimum bolt circle will result from the use of 32 bolts 1⅛ in. in diameter. The size of the bolt circle is 37½ in.

11.8 PRINCIPLES OF FLANGE DESIGN

Despite the simplicity of a flange connection, there is no universally adopted method of design. Approximate methods that had been acceptable cannot be used for high-temperature, high-pressure designs involving large equipment. Flanges designed according to the old formulas cannot provide

Table 11.9. Selection of Optimum Bolt Size

Bolt Size	Root Area	Minimum No. of Bolts	Actual No. (N)	B_s	R	$\frac{NB_s}{\pi}$	$B + 2(g_1 + R)$
¾	0.302	73.7	76	3	1⅛	72.5	36¾
⅞	0.419	53.3	56	3	1¼	53.4	37
1	0.551	40.4	44	3	1⅜	42.0	37¼
1⅛	0.728	30.6	32	3	1½	30.5	37½
1¼	0.929	24.0	24	3	1¾	22.9	38

sufficient tightness even for large vessels working under standard conditions. At the same time, the sizes of most units demand economy in design without any compromise for safety. The result has been a steadily increasing need for deriving a number of different methods for flange design that are correct over a wide range of operating conditions.

Only the analysis of Timoshenko (2) is described here. This analysis applies to both ring flanges and hubbed flanges. Let us first analyze a loose flange. A loose flange is a ring of the rectangular section loaded by torque moments evenly distributed on the circumference (Figure 11.17). Suppose that the deformation consists of a rotation of the square section on the angle, ν, without any distortion of the rectangular flange cross section. R_i and R_o are inside and outside ring radii, and r is the fiber radius. For the fiber under investigation with radius r, its relative elongation as a result of rotation on the angle, ν, will be

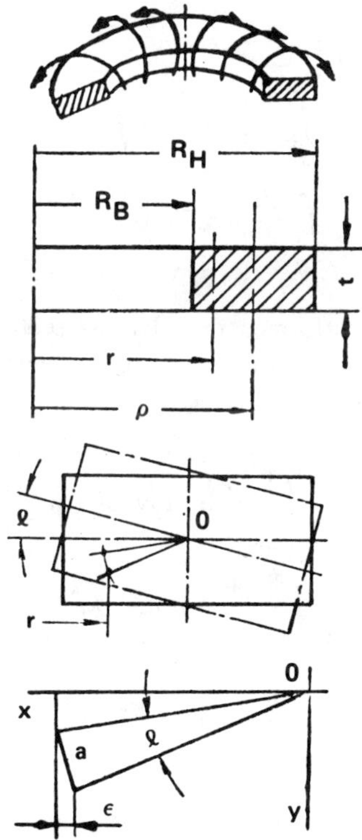

Figure 11.17. Sketch of flange design.

$$\epsilon = \nu y / r \tag{11.4}$$

and the stress corresponding to this elongation is

$$\sigma = \frac{E\nu y}{r} \tag{11.5}$$

According to equilibrium conditions for a half ring, the sum of all normal forces acting on its cross section must equal zero, and the moment of these forces with respect to the x-axis must be

$$M = M_t \rho \tag{11.6}$$

where M_t = the torque moment per unit length of the base line
ρ = radius of the base line

The equilibrium conditions for a ring of any cross section may be written as follows:

$$\left.\begin{aligned} \int_F \frac{E\nu y}{\rho}\, dF = 0 \\ \int_F \frac{E\nu y^2}{\rho}\, dF = 0 \end{aligned}\right\} \tag{11.7}$$

The integration is performed on all cross sections of the ring, F. The equilibrium conditions for the ring with a rectangular cross section are expressed by the following equations:

$$\left.\begin{aligned} \int_{-t/2}^{+t/2} \int_c^d \frac{E\nu y}{r}\, drdy = 0 \\ \int_{-t/2}^{t/2} \int_{R_i}^{R_o} \frac{E\nu y}{r}\, drdy = M \end{aligned}\right\} \tag{11.8}$$

Integrating the last equation, we obtain

$$\frac{E\nu t^3}{12} \ln \frac{R_o}{R_i} = M \tag{11.9}$$

Hence, we obtain the value of the rotation angle:

$$\nu = \frac{12M}{Et^3 \ \ell n R_o/R_i} = \frac{12M_t \rho}{Et^3 \ \ell n \ R_o/R_i} \tag{11.10}$$

Substituting ν in the equation for stress, we obtain

$$\sigma = \frac{E\nu y}{r} = \frac{12My}{rt^3 \ \ell n(R_o/R_i)} \tag{11.11}$$

The maximum stresses will arise at the inside angles of the ring cross sections. In this case,

$$r = R_i$$

and

$$y = \pm\frac{t}{2}$$

and maximum stress is

$$\sigma_{max} = \frac{6M}{t^2 R_i \ \ell n(R_o/R_i)} = \frac{6M_t \rho}{t^2 R_i \ \ell n(R_o/R_i)} \tag{11.12}$$

It is possible to directly determine the stress and the thickness of the loose flange:

$$\sigma = \frac{6W\ell}{\pi t^2 D_i 2.3 \ \ell og_{10}(D_o/D_i)} = 1.66 \frac{W\ell}{t^2 D_i \ \ell og_{10}(D_o/D_i)} \ , \quad kg/cm^2 \tag{11.13}$$

and

$$t = 1.29 \sqrt{\frac{W\ell}{\sigma_{allow} D_i \ \ell og(D_o/D_i)}} \tag{11.14}$$

where W = bolting load (kg)
ℓ = lever arm of moment

The design of integral flanges connected homogeneously with a pipe is more complicated. In this case, the flange under action of forces, P, not only rotates on the angle, ν, but bends the pipe itself (Figure 11.18). The force, P, is related to the length of the inside circle of the pipe. The force related to the outside pipe circle is PR_i/R_o. Because the pipe and the flange each have different rigidity at the point of the junction of the flange with the pipe, there will arise cutting forces, P_o, and bending moments, M_o.

We will find the values P_o and M_o related to the unity length of the pipe inside the circle from the condition of continuity at the point of junction between the flange and pipe. Considering that the flange is extremely rigid in the radial direction, the radial displacement resulting from forces P_o can be neglected. The rotation angle of the pipe rim is equal to the rotation angle of the flange itself, ν. Determining the values of the boundary factors for a flange connected to a pipe we obtain the following equations for M_o and P_o:

$$\frac{1}{2K^3D}(P_o - KM_o) = 0 \tag{11.15}$$

$$-\frac{1}{2K^2D}(P_o - 2KM_o) = \nu \tag{11.16}$$

The design of integral flanges connected homogeneously with a pipe is more complicated. In this case, for the flange under action of forces, P, the coefficient is

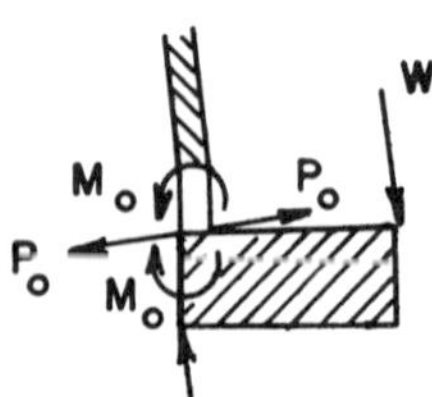

Figure 11.18. Analysis of forces and moments in an integral flange.

$$K = \sqrt{\frac{3(1-\bar{\mu}^2)}{R_i^2 S^2}} \tag{11.17}$$

where S = the pipe thickness at the connection point with the flange
R_i = inside pipe radius

From Equation 11.15 we obtain

$$P_o = KM_o \tag{11.18}$$

and from Equation 11.16 we have

$$M_o = 2KD\nu \quad \text{and} \quad P_o = 2K^2D\nu \tag{11.19}$$

The torque moment is related to the unity of the flange center line and arises under action of forces, W:

$$\begin{aligned} M_1 &= \frac{R_i}{\rho}\,[W(R_o - R_i) - M_o - P_o\,t/2] \\ &= \frac{R_i}{\rho}\,[W(R_o - R_i) - M_o - KM_o\,t/2] \end{aligned} \tag{11.20}$$

Substituting this value in our equation for determining the rotation angle, ν, we will find its value, and by substitution of ν we obtain

$$M_o = 2KD\nu \tag{11.21}$$

and, hence,

$$M_o = 2KD\frac{12R_i}{Et^2\,\ell n(R_o/R_i)}\left[W(R_o - R_i) - M_o - KM_o\,\frac{t}{2}\right] \tag{11.22}$$

Introducing the value of cylinder rigidity for a pipe,

$$D = \frac{ES^3}{12(1-\bar{\mu}^2)} \tag{11.23}$$

Finally, we obtain

$$M_o = W(R_o - R_i)\frac{1}{1 + \frac{Kt/2}{2} + \frac{1-\bar{\mu}^2}{2KR_i}(t/S)^3 \,\ell n(R_o/R_i)} \tag{11.24}$$

The maximum stress from bending the pipe is

$$\sigma = \frac{6M_o}{S^2} \tag{11.25}$$

This stress surpasses the stress from the inside pressure so much that the membrane stress may be neglected.

11.9 OVERVIEW OF FLANGE DESIGN

In designing flange connections, two different problems must be considered: the elements of the flange connection must be strong, and the connection itself must be tight. The problem of structural strength is fairly simple and definite. However, the tightness of flange connections is strongly dependent on the quality of flange fabrication and its assembly. In addition, the changes in specification of gasket materials are more pronounced than those in construction flange materials. Therefore, the tightness design of a flange is more approximate than its strength design. We shall review two practical methods of flange design.

The first one is based on the principle of critical loads developed by Zahkarov [3], and the second method is based on the principle of critical stresses, developed by Waters et al. [4]. Both methods are examined practically and proven to be completely reliable as long as they are based on the numerous theoretical analyses and on lasting operation.

11.9.1 Flange Design by Method of Critical Loads

This method takes into account the plastic properties of the material and reflects the actual conditions of flange performance. Because the design principle of critical loads permits the appearance of stresses exceeding local yield limits in some loaded parts, it may only be used for designing flanges fabricated from plastic materials.

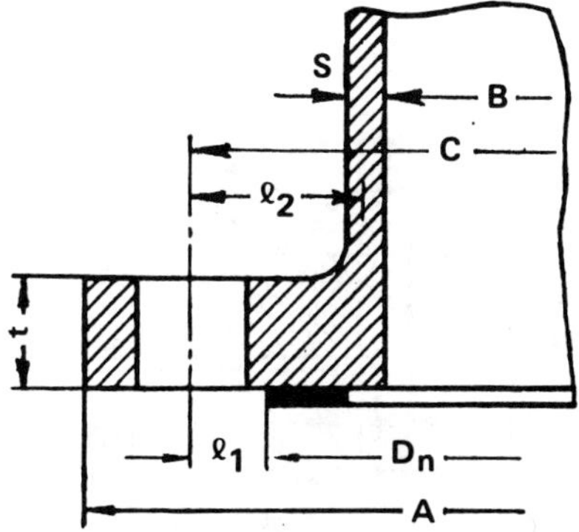

Figure 11.19. Sketch for designing an integral flange.

Design of integral flanges—The following formulas for lever arms are derived from Figure 11.19:

$$\ell_1 = \frac{C - D_n}{2} \; ; \; \ell_2 = \frac{C - B - S}{2} \quad , \quad \text{cm} \tag{11.26}$$

The thickness of the integral flange, t, is determined from the following formula:

$$t = 0.61 d_o \sqrt{\frac{\sigma_b}{K\sigma_f}\left[\frac{\ell_1}{\ell_2} Z + 0.7\psi \frac{\rho}{\sigma_b}\left(\frac{D_n}{d_o}\right)^2\right]} \quad , \quad \text{cm} \tag{11.26-A}$$

where d_o = outside bolt diameter (cm)
σ_b = allowable bolt stress (kg/cm^2)
σ_f = allowable flange stress (kg/cm^2)

and

$$K = 1 + \frac{B}{2\ell_2}\left[\frac{A}{B}\left(1 - 2\frac{\rho_2}{C}\right) + \frac{S^2}{t^2} - 1\right] \tag{11.27}$$

For a tapered-hub flange (Figure 11.20), the coefficient, K, is determined from

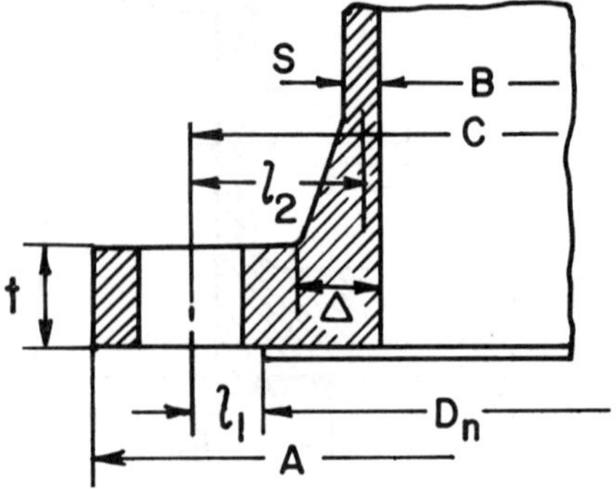

Figure 11.20. Sketch for designing a tapered-hub flange.

$$\gamma = \frac{2h}{\sqrt{B(\Delta - s)}} \tag{11.28}$$

where s = thickness of hub at small end
h = hub length
Δ = thickness of hub at back of flange

There are two cases:

1. Where $\gamma \geqslant 1.0$ and $t \geqslant 0$, and

$$K = 1 + \frac{B}{2\,\ell_2}\left[\frac{A}{B}\left(1 - 2\,\frac{\ell_2}{C}\right) + \frac{\Delta^2}{t^2} - 1\right] \tag{11.29}$$

2. $\gamma < 1.0$ and $h \geqslant 0$,

$$K = 1 + \frac{B}{2\,\ell_2}\left[\frac{A}{B}\left(1 - 2\,\frac{\ell_2}{C}\right) + \frac{s^2}{t^2} - 1\right] + \frac{B}{2\,\ell_2}\left(\frac{\Delta^2}{t^2} - \frac{s^2}{t^2}\right)\gamma \tag{11.30}$$

$$\psi = \left(1 - \frac{\ell_1}{\ell_2}\right)\left[\left(\frac{C - 2\,\ell_2}{D_n}\right)^2 + 1\right] + 0.2\,\frac{\ell_1}{\ell_2} \tag{11.31}$$

The flange design procedure is outlined in the following steps:

1. Sketch the flange connection.
2. Assign values A, B, C, D (refer to Figure 11.20).

3. Assume a preliminary value for the flange thickness, t.
4. Determine s/t and Δ/t.
5. Calculate ℓ_1 and ℓ_2.
6. Calculate γ, K and Ψ.
7. Determine the flange thickness.

If the deviation in the flange thickness, t, between the assumed and calculated values is more than 5%, the design computations must be repeated until satisfactory convergence is achieved. The flange thickness, t, is taken as an average value without deducting the groove depth, t_g, for a gasket (Figure 11.21).

The holes weaken the integral flanges insignificantly; therefore, their influence is not taken into account.

Design of Ring Flanges–The design of a ring flange is simpler because the pipe is not involved in the work of the flange. The thickness of a loose type ring is

$$t_{\ell t} = 0.58 d_b \sqrt{\frac{Z\sigma_b}{K\psi\sigma_f}}, \quad \text{cm} \tag{11.32}$$

The weakening of a loose flange by bolt holes has an unfavorable effect on the structural strength of the flange and is considered by the coefficient

$$\psi = 1 - \frac{2d}{A - B} \tag{11.33}$$

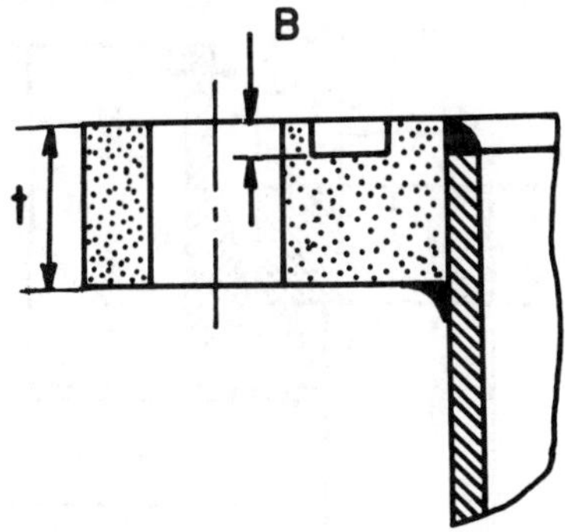

Figure 11.21. Sketch for establishing flange thickness.

where d is the hole diameter for a bolt. The coefficient, K, is obtained from the following formula:

$$K = 1 - \frac{A}{B} + \frac{A - B}{2\ell} \tag{11.34}$$

where the moment lever arm from Figure 11.22 is

$$\ell = \frac{C - D_1}{2} \tag{11.35}$$

where D_1 is hub diameter. The design of screwed flanges is made in the same manner as loose flanges.

Design of a Lap-Joint Stub—The shoulder thickness of a stub is determined by a formula similar to that used for integral flanges:

$$t_{sh} = 0.58 d_b \sqrt{\frac{\sigma_b}{K\sigma_f}\left[Z\frac{\ell_1}{\ell_2} + 0.7\psi\,\frac{P}{\sigma_b}\left(\frac{D_n}{d_b}\right)^2\right]} \tag{11.36}$$

In this case, the lever arm's length is shown in Figure 11.22.

The coefficients K, γ, ψ are the same as for integral flanges.

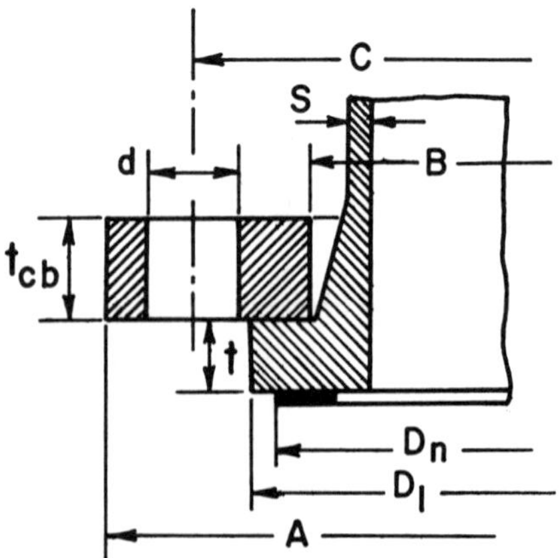

Figure 11.22. Sketch for design of a stub shoulder.

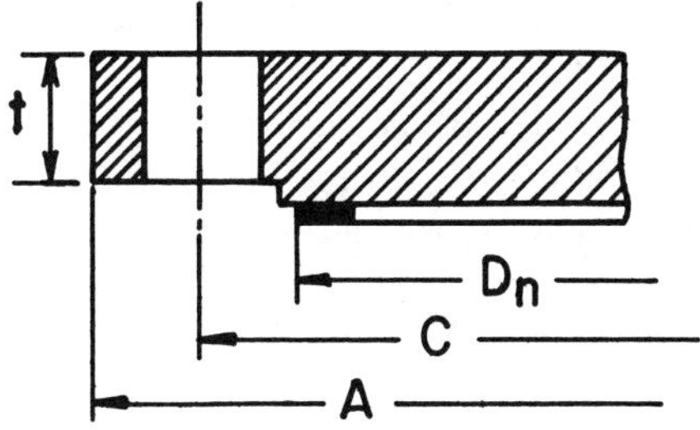

Figure 11.23. Sketch for design of a blind flange.

Design of a Blind Flange–The design of a blind flange (Figure 11.23) employs the method of critical loads, which gives considerably less flange thickness than by the method of critical stresses. This thickness is

$$t_{bf} = 0.41D_n \sqrt{\frac{P}{\sigma_f}\left\{1 + 7.3\,\frac{\ell}{C}\left[0.57Z\left(\frac{d_b}{D_n}\right)\frac{\sigma_b}{P} - 1\right]\right\}} \tag{11.37}$$

where ℓ = lever arm:

$$\ell = \frac{C - D_n}{2} \tag{11.38}$$

In designing structural members, a partial transition of material is allowed for from the elastic to plastic state (taking place mostly in the points of stress concentrations). Abrupt junctions and recesses and small transient fillets cease to be dangerous. This permits the design of slip-on flanges as welding-neck flanges. These features are characteristic of steady-state loaded parts manufactured from plastic materials under reversed loads. In parts from brittle materials, the presence of undercutting or conjunctions without a smooth passage (for example, from chilled steel or pig iron) will bring about destruction.

11.9.2 Flange Design by Method of Critical Stresses

Flange design by the method of critical stresses is based on the procedures developed by Waters et al. [4], Waters and Taylor [5], and Waters et al. [6]. The design equations are derived under the assumption that the flange is bent by circular moments and that the gasket or contacting flange surfaces are entirely within the bolt circle. The effects of each loading on the flange (longitudinal, radial and tangential) are analyzed independently and are assumed to be linearly related in such a way that the complete solution may be obtained by superposition, as shown in Figure 11.24.

From the designer's point of view, all flanges are divided into two groups:

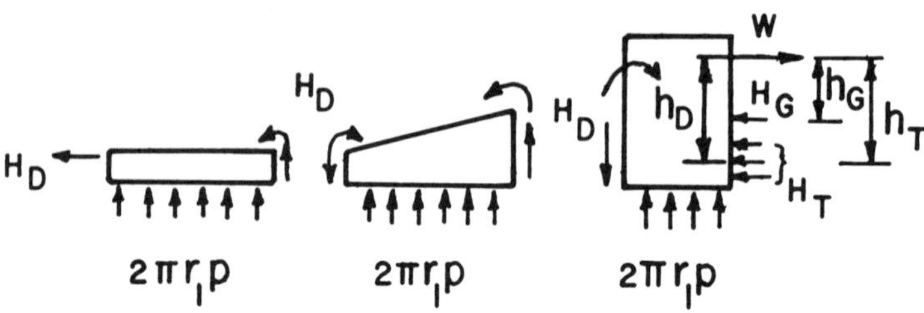

Figure 11.24. Combined loading of tapered-hub flange.

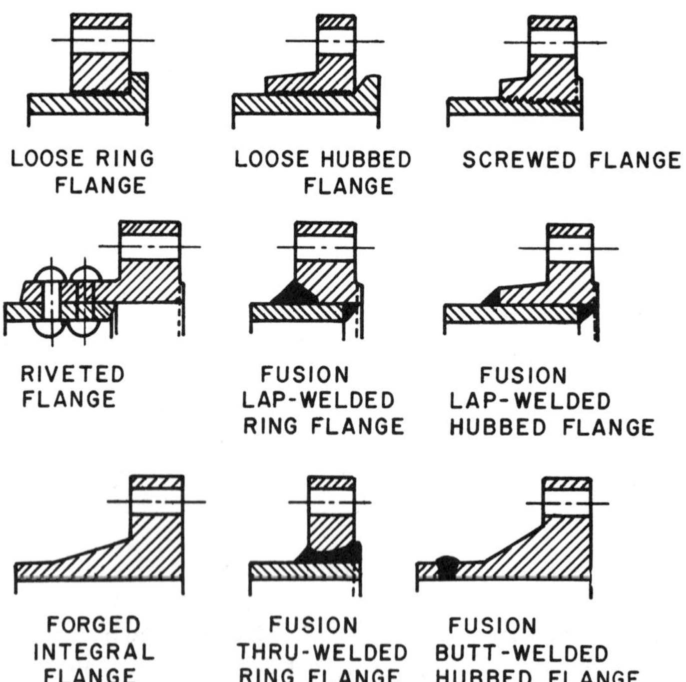

Figure 11.25. Various types of flanges subject to the analytical method of Waters et. al [4].

loose flanges and integral flanges (Figure 11.25). Among loose flanges are all design types in which the ring is set down over a shell (e.g., a pipe), either loose or coupled to it. Thus, screwed flanges, lap-joint flanges, as well as fusion lap-welded ring flanges belong to the loose flange group.

Among integral flanges are those that are fabricated together with vessels (cast or forged flanges), fused through welded ring flanges, and fused through butt-welded hubbed flanges, which guarantee that the shell will work together with the flange. To simplify the design, it is possible to consider any flange as a loose one and subsequently design it when neither one of the following values is exceeded:

$$g_0 \leqslant 1.6 \text{ cm} \ ; \quad \frac{B}{g_0} \leqslant 300 \ ; \quad p \leqslant 20 \text{ kg/cm}^2 \ ; \quad t^0 \leqslant 370^\circ\text{C}$$

where g_0 = thickness of hub at small end (Figure 11.26B)
B = inside diameter of flange (cm)
p = working pressure (N/m^2)
t^0 = working temperature (°C)

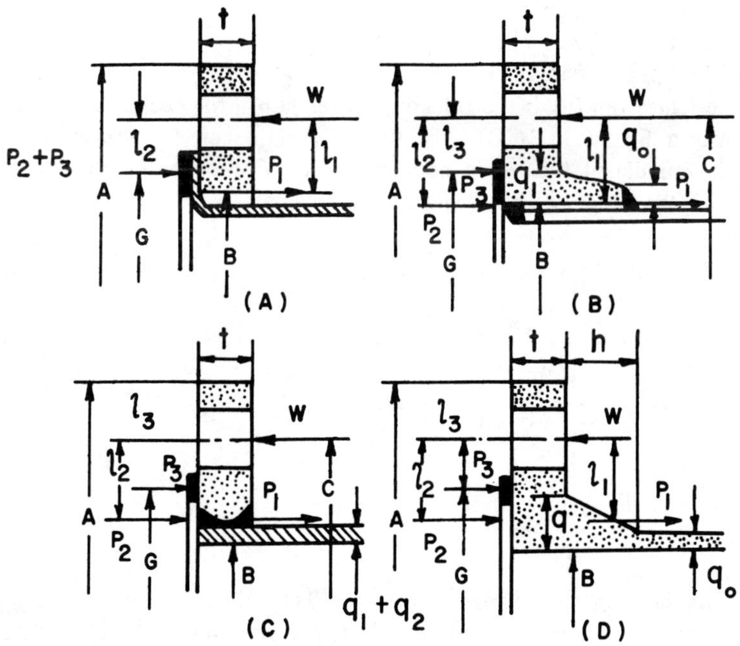

Figure 11.26. Schematic of flange loadings.

11.9.3 Determining Bending Moments

Flanges are loaded by the following forces, which give rise to the bending moments:

The force produced by the internal pressure on the area enclosed by the inside flange rim:

$$P_1 = 0.785B^2P \text{ (kg)} \tag{11.39}$$

The force equal to the difference of loads produced by total pressure and the force P_1:

$$P_2 = P - P_1 = 0.785B^2p \text{ (kg)} \tag{11.40}$$

The force acting on the gasket and equal to the load difference between the designing bolt load and the total load from the force P:

$$P_3 = W - P \text{ (kg)} \tag{11.41}$$

These forces are applied at the following distances forming the lever arms—Figure 11.26(C)(D).

1. In integral flanges, the force P_1 acts at the distance $0.5g_1$ from the inside flange rim (g_1 is the thickness of the hub at the back of the flange), as shown in Figure 11.26(D). The force P_3 acting on the gasket is applied in the middle of the gasket contact surface. Consequently, the lever arms bending the flange will be equal to the following:

$$\ell_1 = R + 0.5g_1 \tag{11.42}$$

$$\ell_2 = \frac{R + g_1 + \ell_3}{2} \tag{11.43}$$

$$\ell_3 = \frac{C - G}{2} \tag{11.44}$$

$$R = \frac{C - B}{2} - g_1 \tag{11.45}$$

2. In the loose-type flanges, where the flange directly contacts the gasket, the force P_1 is applied to the inside flange rim, and P_3 is applied to the middle of gasket contact surface (Figure 11.26(A)).

The bending lever arms are

$$\ell_1 = \frac{C - B}{2} \;;\quad \ell_3 = \frac{C - G}{2} \;;\quad \ell_2 = \frac{\rho_1 + \rho_2}{2} \tag{11.46}$$

These lever arms are taken for any flange type if they are designed as loose flanges.

3. In loose-type flanges, sitting on a lap-joint stud or on the flange ring (see Figure 11.26(A)) the force P_1 is applied to the inside flange rim, and the force acting on the gasket P_3 to the point in the middle of the contact strip between the flange and lap-joint. In this case, the moment lever arms are

$$\ell_1 = \frac{C - B}{2} \;;\quad \ell_2 = \ell_3 = \frac{C - G}{2} \tag{11.47}$$

and the bending moments are

$$M_1 = P_1 \ell_1 \;;\quad M_2 = P_2 \ell_2 \;;\quad M_3 = P_3 \ell_3 \tag{11.48}$$

and the total moment is

$$M_0 = M_1 + M_2 + M_3 \tag{11.49}$$

11.9.4 Determining Flange Stresses

It is first necessary to determine the values of stress correction factors f, d and e entering the flange-stress equations. All these parameters depend on the flange form and are determined by the ratios of the thickness of the hub at the back of the flange to the thickness of the hub at the small end g_1/g_0, and the hub length to factor $h_0 = \sqrt{Bg_0}$. The parameter f is found from Figure 11.27 for calculated values of g_1/g_0 and $h/\sqrt{Bg_0}$. If the obtained parameter f is less than unity, it is considered equal to unity. The values of stress correction factors ℓ and d for loose flanges are calculated by use of the following formulas:

$$\ell = \frac{F_\ell}{h_0} \;;\quad d = \frac{U}{V_\ell}\, h_0 g_0^2 \tag{11.50}$$

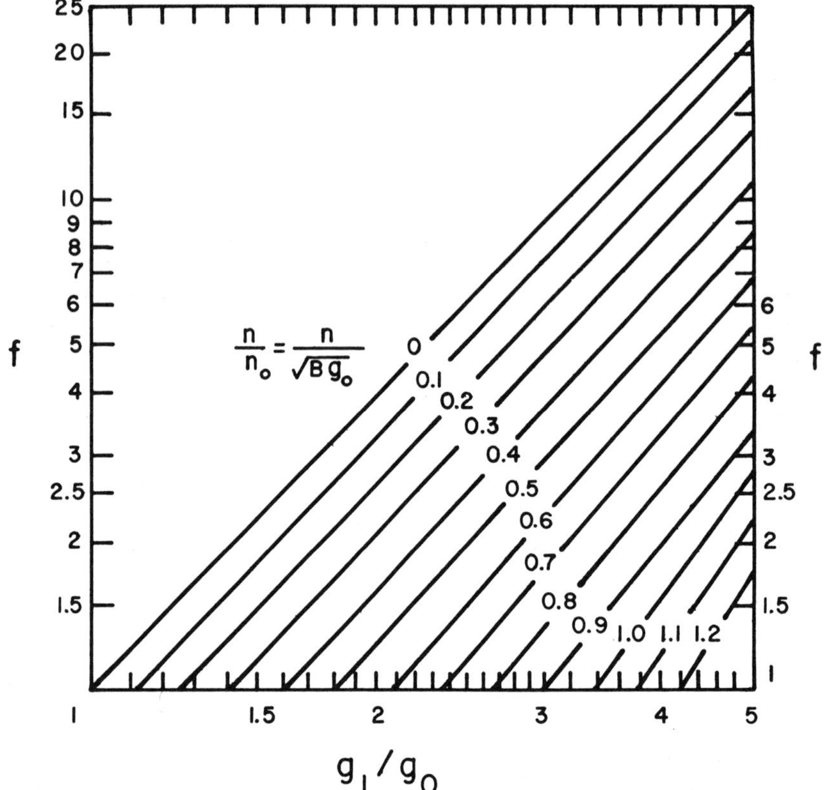

Figure 11.27. Values of stress correction factor, f: f = 1 (minimum); f = 1 for hubs of uniform thickness ($g_1/g_0 = 1$); and f = 1 for loose hubbed flanges.

and for integral flanges,

$$\ell = \frac{F_I}{h_0} \quad ; \quad d = \frac{U}{V_i} h_0 g_0^2 \tag{11.51}$$

where F_ℓ and F_I are determined from Figures 11.28 and 11.29; V_ℓ and V_I are determined from Figures 11.30 and 11.31, and V_1 and V_I are determined from Figure 11.30. The factors U, T, Y and Z are functions of the ratio of the outside flange diameter to inside flange diameter. They are determined from Figure 11.32.

By combining these factors, the quality, L, may be defined as

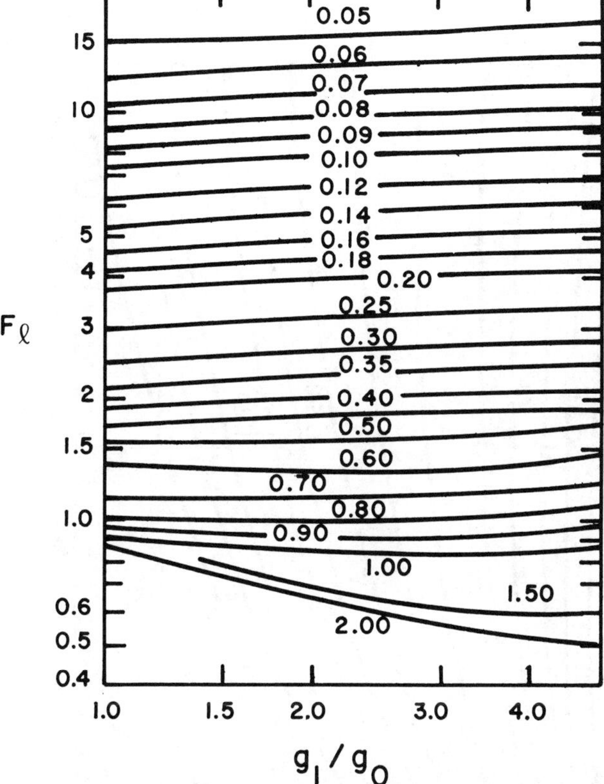

Figure 11.28. F_ℓ values for loose flanges.

$$L = \frac{t_\ell + 1}{T} + \frac{t^3}{d} \tag{11.52}$$

The following equations give the desired stresses:

For integral-type flanges and all hubbed flanges,

Longitudinal hub stress:

$$f_H = \frac{fM_0}{Lg_1^2 B} \tag{11.53}$$

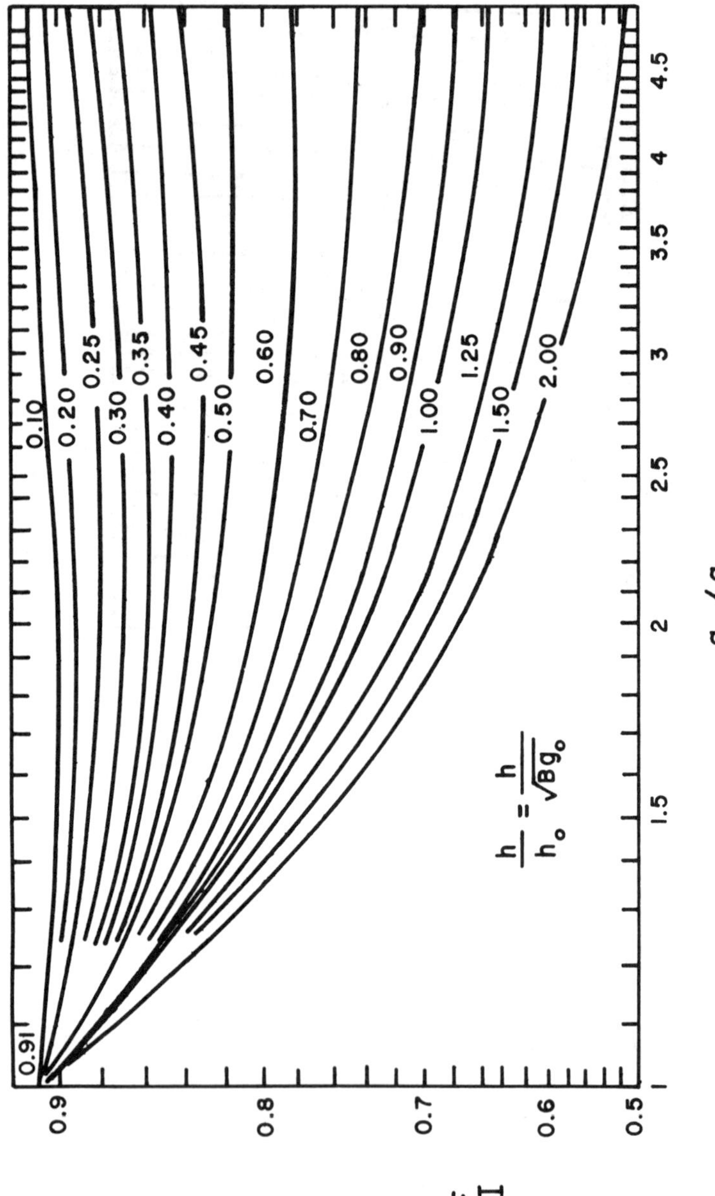

Figure 11.29. F_I values for integral flanges.

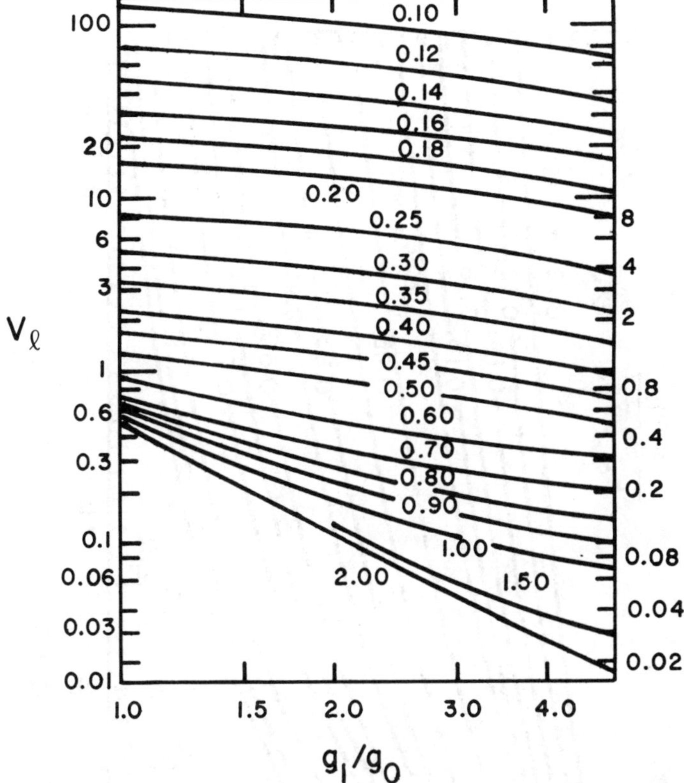

Figure 11.30. V_ℓ values for loose flanges.

Radial flange stress:

$$f_R = \frac{(0.75t_\ell + 1)}{Lt^2B} \tag{11.54}$$

Tangential flange stress:

$$f_T = \frac{YM_0}{t^2B} - Zf_R \tag{11.55}$$

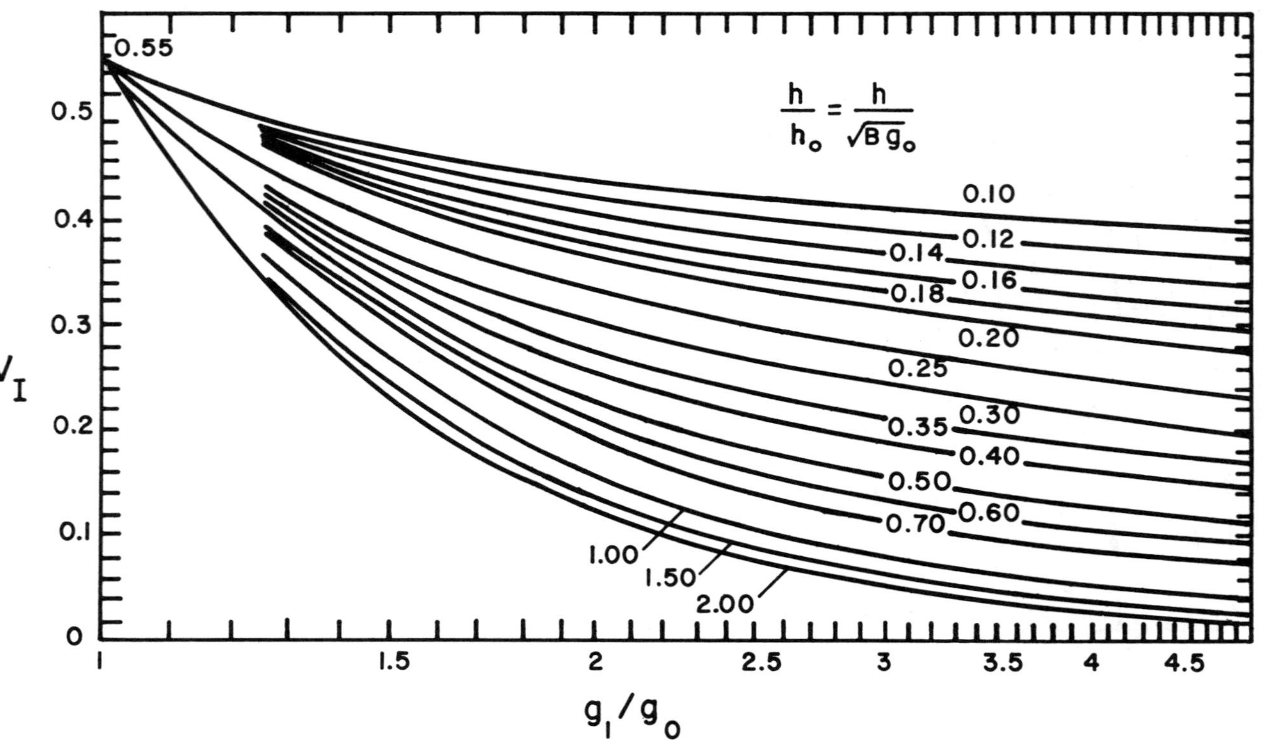

Figure 11.31. Values of V_I for integral flanges.

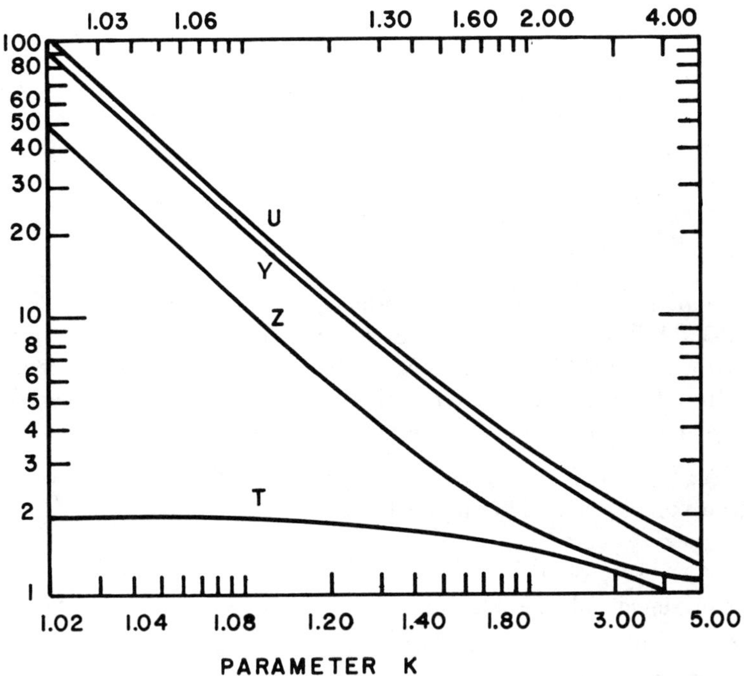

Figure 11.32. Values of T, U, Y and Z when $\tilde{\mu} = 0.3$ ($\tilde{\mu}$ = Poisson's ratio) Parameters are defined by the following equations:

$$T = \frac{K^2(1 + 8.55246 \log_{10}K) - 1}{(1.04720 + 1.9448K^2)(K - 1)}$$

$$U = \frac{K^2(1 + 8.55246 \log_{10}K) - 1}{1.36136(K^2 - 1)(K - 1)}$$

$$Y = \frac{1}{K - 1}\ 0.66845 + 5.71690\frac{K^2 \log_{10}K}{K^2 - 1}$$

$$Z = \frac{K^2 + 1}{K^1 - 1};\quad K = \frac{A}{B}$$

For ring flanges of the loose type,

$$f_T = \frac{YM_0}{t^2B} \tag{11.56}$$

$$f_R = 0 \tag{11.57}$$

$$f_H = 0 \tag{11.58}$$

11.9.5 Allowable Stresses

For purposes of design it is important that the stresses given in Equations 11.53, 11.54 and 11.55 do not exceed the corresponding allowable stress. This may be achieved by limiting the averages of the induced stresses, the average of $f_h + f_r$ and the average of $f_h + f_r$ to not more than the allowable stress in tension. These limitations may be summarized as follows:

The maximum induced longitudinal stress is

$$f_h \leqslant 1.5\, f_{allow} \tag{11.59}$$

The maximum induced radial stress is

$$f_R \leqslant f_{allow} \tag{11.60}$$

The maximum induced tangential stress is

$$f_T \leqslant f_{allow} \tag{11.61}$$

This last equation is valid for

$$\frac{f_H + f_T}{2} \leqslant f_{allow} \tag{11.62}$$

and

$$\frac{f_H + f_R}{2} \leqslant f_{allow} \tag{11.63}$$

The loose flanges seating on the lap-joint stub may induce shearing stresses, which may damage the stub, especially if it is made from mild metal. The shearing stress in the lap-joint must not exceed 0.8 f_{allow}.

11.10 CODE RULES FOR DESIGNING FLANGES

11.10.1 Bolt Loads

There are two bolt loads: (1) the bolt load, W_{m2}, for tightening-up conditions, i.e., exerting sufficient force, Hy, to seat the gasket to produce a tight joint. This load is equal to the effective area of the gasket times the gasket yield stress:

$$W_{m2} = Hy = \pi bGy \tag{11.64}$$

(2) the bolt load under the operating condition, W_{m1}, exerting the necessary force to resist the internal pressure and to keep the flange connection tight during operation. The internal pressure produces an end force, given by

$$H = \frac{\pi}{4} G^2 P \tag{11.65}$$

The force required to keep the gasket from leaking is given by

$$H_p = 2b\pi Gmp \tag{11.66}$$

Therefore,

$$W_{m1} = H + Hp \tag{11.67}$$

11.10.2 Bolt Areas

Because elevated temperatures considerably reduce the allowable stress of the bolting steel in comparison with its allowable stress at room temperature, different bolting areas will be required depending on the operating temperature. For the operating condition, the minimum bolting area is determined by

$$A_{m1} = \frac{W_{m1}}{f_b} \tag{11.68}$$

and for the bolting-up condition (no internal pressure),

$$A_{m2} = \frac{W_{m2}}{f_a} \tag{11.69}$$

For determining the actual bolt size, it is first necessary to calculate the maximum bolt spacing that limits the deflection between the flanges, causing leakage from the gasket joint. The following empirical equation obtained by the Taylor Forge Co. is used for maximum bolt spacing:

$$B_s(\max) = 2d + \frac{6t}{m + 0.5} \tag{11.70}$$

where $B_s(\max)$ = maximum bolt spacing for a tight joint (in.)
d = bolt diameter (in.)
t = flange thickness (in.)
m = gasket factor

A minimum diameter bolt circle is recommended for use that may be determined by setting up a table listing the number of bolts required, root area, bolt spacing, B_s, and radial spacing, R, as a function of bolt size. The minimum bolt-circle diameter will be either the diameter corresponding to the radial clearances $(d = B + 2(g_1 + R))$, or the diameter necessary to satisfy the bolt spacing requirement $(d + (NB_s/\pi))$, whichever is greater.

The best design is achieved when these two controlling diameters are approximately equal.

The integral bolt number used is usually a multiple of four. This means that the actual bolt area provided exceeds the minimum required bolt area which results in an overstressing of the flange in the bolting-up operation. The code specifies that the design load, W, for the bolting-up condition be based on the average of the minimum and the actual bolting areas:

$$W = \left(\frac{A_m + A_b}{2}\right) f_a \tag{11.71}$$

and for the following operating condition:

$$W = W_{m1} \tag{11.72}$$

as a margin of safety against such overstressing.

11.10.3 Flange Moments

The various axial forces on the flange produce bending moments. The moment summation is made about the bolting axis. For integral-type flanges, the total moment is equal to the sum of the moments acting on the flange. That is,

Flange loads x lever arms = moments:

$$H_d = 0.785B^2P \qquad h_D = R + \frac{g_1}{2} \qquad M_D = H_D \times h_D \tag{11.73}$$

$$H_T = H - H_D \qquad h_T = \frac{R + g_1 + h_G}{2} \qquad M_T = H_T \times h_T \tag{11.74}$$

$$H_G = W - H \qquad h_G = \frac{C - G}{2} \qquad M_G = H_G \times h_G \tag{11.75}$$

and the total moment is

$$M_0 = M_D + M_T + M_G \tag{11.76}$$

For loose-type flanges in which the flange bears down directly onto the gasket, the force H_D is assumed to act on the inside diameter of the flanges and on the gasket load at the center line of the gasket face. The lever arms for the moments are

$$h_D = \frac{C - B}{2} \tag{11.77}$$

$$h_G = \frac{C - G}{2} \tag{11.78}$$

$$h_T = \frac{H_D + H_G}{2} \tag{11.79}$$

The lever arms given in Equations 11.77, 11.78 and 11.79 apply to the flanges shown in sketches (B), (C) and (D) of Figure 11.33.

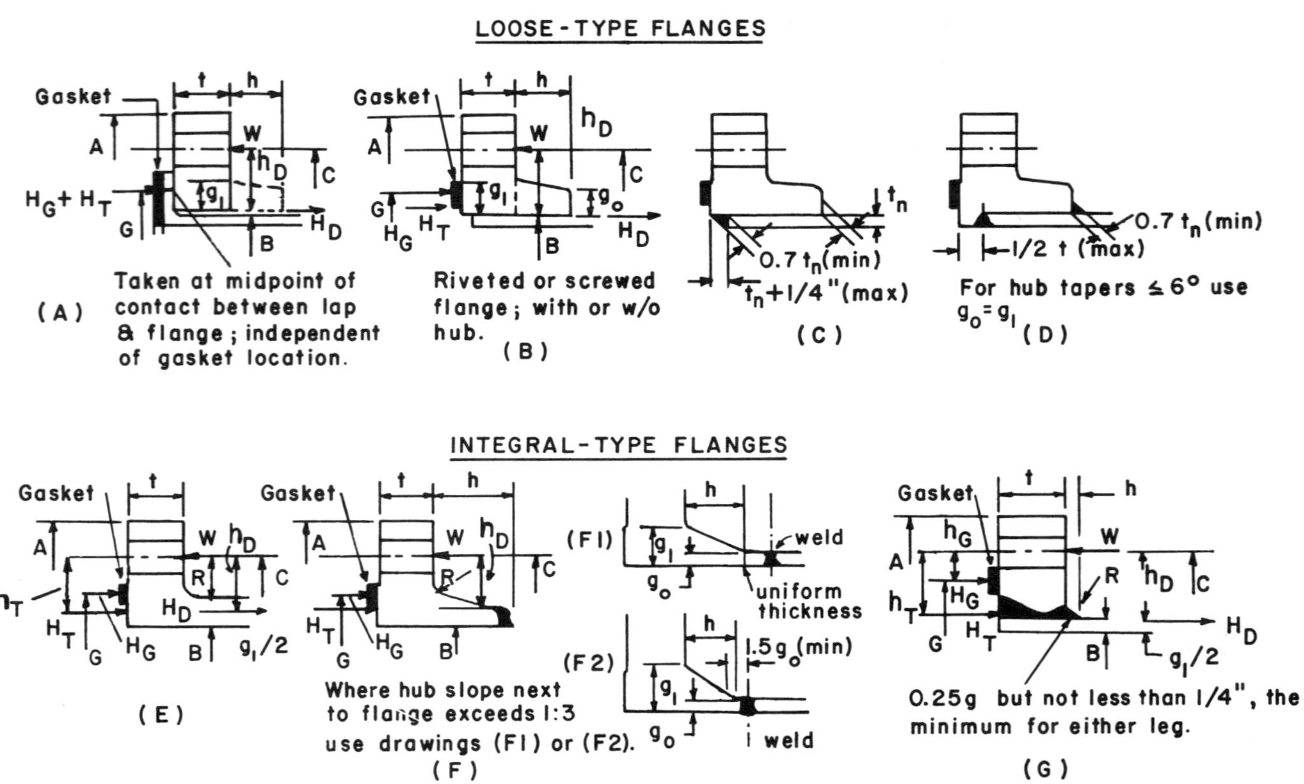
LOOSE-TYPE FLANGES
Gasket
A
W
h_D
C
$H_G + H_T$
G
g_1
B
H_D
t
h
Taken at midpoint of contact between lap & flange; independent of gasket location.
(A)
H_G
H_T
g_o
Riveted or screwed flange; with or w/o hub.
(B)
t_n
0.7 t_n(min)
t_n+1/4"(max)
(C)
0.7 t_n(min)
1/2 t (max)
For hub tapers ≤ 6° use $g_o = g_1$
(D)
INTEGRAL-TYPE FLANGES
h_T
R
$g_1/2$
(E)
Where hub slope next to flange exceeds 1:3 use drawings (F1) or (F2).
(F)
(F1)
weld
uniform thickness
g_o
(F2)
1.5 g_o(min)
weld
h_G
0.25g but not less than 1/4", the minimum for either leg.
(G)

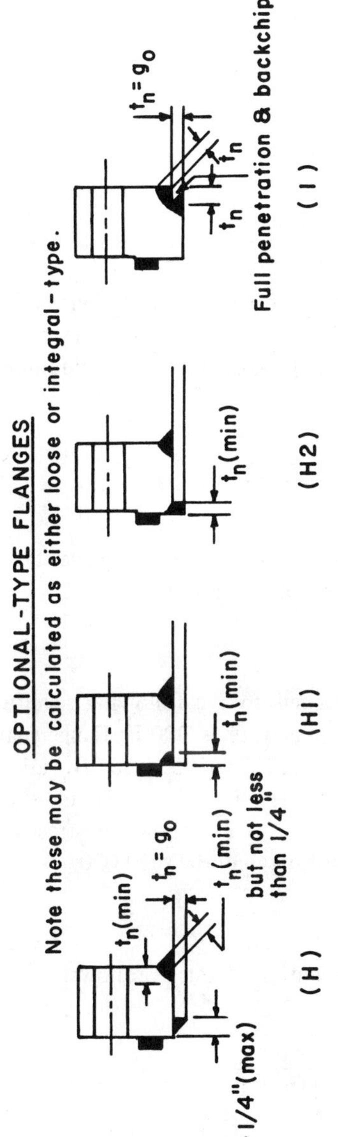

Figure 11.33. Types of flanges recommended by the ASME Code.

11.11 DESIGN PROCEDURE FOR RING FLANGES (LOOSE TYPE)

Flanges corresponding to the sketches (G), (H) and (I) in Figure 11.33 are designed as integral flanges if any of the following values are exceeded:

$$g_0 > 5/8 \text{ in.}$$

$$\frac{B}{g_0} > 300$$

$$P > 300 \text{ psi}$$

for operating temperatures >700°F. If the operating conditions and proportions do not exceed the above limits, flanges corresponding to sketches H through I may be designed as loose-type flanges. The minimum flange thickness for this case is determined from

$$t = \sqrt{(YM_{max})/(fb)} \tag{11.80}$$

The design procedure for a ring-type flange with a plane face for a heat-exchanger shell is illustrated by the following problem:

Example 11.2.

A loose type flange has to be designed to the following specifications: design pressure = 150 psi; design temperature = 300°F; flange material = ASTM A-201, grade B; bolting steel = ASTM A-193, grade B; gasket material = asbestos composition; shell outside diameter = 31 in. = B; shell thickness = 3/8 in.; shell inside diameter = 30 1/r; allowable stress of flange material = 15,000; allowable stress of bolting material = 20,000.

Solution

The calculation of gasket width by Equation 11.2 is as follows:

$$\frac{d_0}{d_i} = \sqrt{\frac{y - pm}{y - p(m + 1)}}$$

Assuming a gasket thickness of 1/16 in. from Figure 11.15, we obtain

$$y = 3700$$

$$m = 2.75$$

Therefore,

$$\frac{d_o}{d_i} = \sqrt{\frac{3700 - (150)(2.75)}{3700 - (150)(3.75)}} = 1.021$$

Assume that d_i of the gasket equals 32¾ in. Then,

$$d_o = (1.021)(32.75) = 33.5 \text{ in.}$$

The minimum gasket width is

$$\frac{33.5 - 32.75}{2} = \frac{3}{8} \text{ in.}$$

Therefore, use a ½-in. gasket width. The mean gasket diameter is G = 32.75 + 0.5 = 33.25 in.

11.11.1 Calculation of Bolt Load

To estimate bolt loads, the following formula may be used:

$$W_{m2} = Hy = b\pi Gy \tag{11.81}$$

From Figure 11.17,

$$b_0 = \frac{N}{2} = \frac{0.5}{2} = 0.25 \text{ in.}$$

$$b = b_0 \quad \text{if} \quad b_0 \leqslant 0.25 \text{ in.}$$

Therefore,

$$b = 0.25$$

Substituting, we obtain

$$Hy - 0.25(\pi)(33.25)(3700) = 93{,}600 \text{ lb}$$

The load to keep the joint tight under operation can be obtained from the following formula:

$$H_p = 2b\pi Gmp = 2(0.25)(\pi)(33.25)(2.75)(150) \tag{11.82}$$
$$= 21{,}600 \text{ lb}$$

For the load internal pressure,

$$H = \frac{\pi G^2}{4} P = \frac{\pi(33.25)^2}{4}(150) = 130{,}500 \text{ lb} \tag{11.83}$$

For the total operating load,

$$W_{m1} = H + H_p = 130{,}500 + 21{,}600 = 152{,}100 \text{ lb} \tag{11.84}$$

where W_{m1} is greater than W_{m2}. Therefore, the controlling load is 152,100 lb. Calculating the minimum bolting area, A_{m1}, as given by the following formula,

$$A_{m1} = \frac{W_{m1}}{f_b} = \frac{152{,}100 \text{ lb}}{20{,}000 \text{ psi}} = 7.6 \text{ in.}^2$$

Calculations of the optimum bolt size are given in Table 11.10. From Table 11.10 the minimum bolt circle is 34.3 in. when using ¾-in. bolts. For simplicity in dimensioning, specify 28 bolts of ¾-in. diameter on a 35-in. bolt circle:

Table 11.10. Optimum Bolt Sizes Computed

Bolt Size	Root Area	Minimum No. of Bolts	Actual No. of Bolts[a]	R	B_s	E	$\frac{NB_s}{\pi}$	ID + 2x $(1.415g_0 + R)$[b]
5/8	0.202	37.6	40	15/16	3	3/4	38.2	34.0
3/4	0.302	25.2	28	1 1/8	3	13/16	26.8	34.3
7/8	0.19	18.15	20	1 1/4	3	15/16	19.1	34.6

[a]Use multiples of 4.
[b]Allow (1/0.707) g_0 for weld.

Bolt circle diameter, $C = 35$ in.

Calculating the flange outside diameter,

$$\text{Flange OD} = \text{bolt} - \text{circle diameter} + 2E = 35 + 2\,(^{13}/_{16}) = 36\tfrac{5}{8} = A$$

And checking the gasket width,

$$A_b \text{ actual} = (28)\,(0.302) = 8.45 \text{ in.}^2$$

$$\text{Minimum gasket width} = \frac{A_{b\,\text{actual}}\, f_{\text{allow}}}{2y\pi G} = \frac{(8.45)\,(20{,}000)}{2(3700)\,(\pi)\,(33.25)} = 0.219 \text{ in.}$$

(compared with ½ in. specified)

11.11.2 Moment Computations

For a bolting-up condition (no internal pressure), the design load is given by the following formula:

$$W = \frac{1}{2}\,(A_b + A_m)\, f_a = \frac{1}{2}\,(7.6 + 8.45)\,(20{,}000)$$

$$= 160{,}500 \text{ lb}$$

The corresponding lever arm is given by Equation 11.71:

$$h_G = \frac{1}{2}\,(C - G) = \frac{1}{2}\,(35 - 32.25) = 1.375 \text{ in.}$$

The flange moment is as follows:

$$\text{Flange moment, } M_a = Wh_G = (160{,}500)\,(1.375)$$

$$= 220{,}500 \text{ in.-lb}$$

where h_G is the radial distance from the gasket-load reaction to the bolt circle. For operating condition ($W = W_{m1}$), using Equation 11.73 to calculate the hydrostatic end force on the area inside the flange,

$$H_D = 0.785B^2\,p = (0.785)\,(31)^2\,(150) = 113{,}000 \text{ lb}$$

The lever arm, h_D, from Equation 11.77 is

$$h_D = \frac{C - B}{2} = \frac{35 - 31}{2} = 2.0 \text{ in.}$$

The moment, M_D, from Equation 11.73 is

$$M_D = H_D \times h_D = (113{,}000)\,(2.0) = 226{,}000 \text{ in.-lb}$$

The difference between the flange-design bolt load and the total hydrostatic end force is obtained from Equation 11.75:

$$H_G = W - H = 160{,}500 - 130{,}500 = 30{,}000 \text{ lb}$$

The corresponding lever arm is computed from Equation 11.78:

$$h_G = \frac{C - G}{2} = \frac{35 - 32.75}{2} = 1.375 \text{ in.}$$

The moment from Equation 11.75 is

$$M_G = H_G \times h_G = 30{,}000 \times 1.375 = 41.200 \text{ in.-lb}$$

The difference between the total hydrostatic end force and the hydrostatic end force on the area inside of the flange is H_T from Equation 11.74,

$$H_T = H - H_D = 130{,}500 - 113{,}000 = 17{,}500 \text{ lb}$$

The corresponding lever arm given by Equation 11.79 is

$$h_T = \frac{h_D + h_G}{2} = \frac{2.0 + 1.375}{2} = 1.69 \text{ in.}$$

The moment is given by Equation 11.74 is

$$M_T = H_T \times h_T = 17{,}500 \times 1.69 = 29{,}600 \text{ in.-lb}$$

The summation of moment for the operating condition, M_0, is by Equation 11.76:

$$M_0 - M_D + M_G + M_T = 226{,}000 + 41{,}200 + 29{,}600 = 296{,}800 \text{ in.-lb}$$

Therefore, the operating moment is controlling, and

$$M_{max} = 296{,}800 \text{ in.-lb}$$

The flange thickness may now be computed from Equation 11.80:

$$t = \sqrt{(YM_{max})/(fB)}$$

$$K = \frac{A}{B} = \frac{36\frac{5}{8}}{30\frac{1}{4}} = 1.211$$

where K = the ratio of flange outside diameter
parameter Y = a function of K, obtained from Figure 11.32

$$Y = 10.25$$

Hence,

$$t = \sqrt{\frac{(10.25)\,(296{,}800)}{(15{,}000)\,(31)}} = 6.55 = 2.56 \text{ in.}$$

Thus, the designer should specify a $2\frac{3}{4}$ in. thick plate for the flange.

11.12 DESIGN PROCEDURE FOR INTEGRAL FLANGES

In accordance with the ASME code, integral hubbed flanges should have a hub taper equal to, or less than, 1:3. For a double taper, the hub slope adjacent to the weld should be equal to, or less than, 1:3 for a minimum length equal to 1.5 g_0 measured from the center of the weld. A single taper hub also may be used with a slope greater than 1:3 for a straight section, which has a minimum length of 1.5 g_0. The hub thickness, as a first trial, may be taken as equal to two times g up to 1.5 in., and 1.5 times g_0 for values exceeding 1.5 in. The flange loads, lever arms and bending moments can be evaluated, and constants, K, T, Z, Y and U determined. The value of g_1 may be determined from the following equation:

$$g_1 = 0.7\sqrt{M_0/fB} \tag{11.85}$$

where f = allowable stress (psi)

If g, determined from Equation 11.85, is less than g_0, then the tapered hub flange is not required, and g_1 (the thickness of the hub at the back of the flange) can be made equal to g_0 with h equal to the minimum required thickness for welding. If the computed g_1 is greater than g_0, a value for h may be determined from Figure 11.27 by assuming the factor f = 1.0. The stresses in the flange and hub for a selected flange thickness may be calculated with the use of g_1 and h values.

11.13 ADDITIONAL COMMENTS AND NOTES ON FLANGE DESIGN

The following factors should be considered when designing flanges:

1. All piping and vessel's nozzle flange templates must comply with the ASME code. However, in designing main joint flanges and heads there is no reason to follow the code in the flange templates. It is better to design the connection so that the distance between the gasket and bolt is at a minimum and, consequently, the bending moments, flange sizes and weight of the connection are held at a minimum.
2. The number of bolts must be a multiple of four.
3. In general it is desirable not to space the bolts on the principal axes.
4. The gasket should be entirely within the bolt circle.
5. The higher the internal pressure, the narrower the gasket should be. The more narrow the gasket, the less the bolt load for tightening the flange connection is required.
6. The flange bolts and studs must work only for tension. Nuts and bolt heads should not sit on fillets or sit crookedly on the stud or bolt. The nut must sit on the flange over its entire surface. In critical cases the flange surface should be faced. A less costly approach is to have the face under each nut reamed, but not deeper than 1-2 mm.

7. For flanges working under high temperatures or pressures, full-sized threaded bolts or bolts bored in the middle part to the inside diameter of thread are recommended in place of bolt studs. The increased elasticity of such studs decreases the danger of rupture as a result of stress concentrations.

8. The studs and flanges working under high temperatures must be fabricated from heat-resistant and creep-resistant steels.

9. The use of bolts and studs with diameters no less than 16 mm is recommended because thin bolts may be damaged easily during screwing up.

10. It is generally better to use fewer bolts of larger diameter since by increasing the bolt diameter the allowable stress is increased.

11. The spacing between bolts should not be less than 2.25 d to account for their tightening with a socket wrench, and no less than 3 d for a single-ended wrench. The normal spacing is 3.5–4 d. For tightness consideration the spacing should not exceed 5 d.

12. The outside diameter of a flange is taken equal to the bolt circle diameter plus 2 d bolts +5–10 mm.

13. The hub eases considerably under flange work. Hence, it is better to increase the thickness of a hub rather than the thickness of the flange.

14. The average hub thickness is taken from 0.5 to 0.75 of the flange thickness.

15. The hub slope should be from 1:2.5 to 1:4.

16. The fillet between the hub and the flange should not be too small for flanges fabricated from brittle materials. The same applies to flanges loaded with cyclic and reversed forces.

17. Flanges for vessels processing volatile, toxic and flammable substances must be designed for pressures above 16 atm. Flanges for atmospheric equipment must be designed for a pressure of no less than 2.5 atm.

NOMENCLATURE

A_B bolt area (in.^{2})
$B_s(max)$ maximum bolt spacing for tight joint (in.)
D_o,D_i outer and inside diameters of apparatus (m)
d bolt diameter (in.)
d_o,d_i outside, inside gasket diameters (in.)
E modulus of elasticity (N/m^2)
F area (m^2)
f stress correction factor
G mean gasket diameter (in.)
g_1 hub thickness (in.)
K coefficient in Equation 11.17
ℓ lever arm of moment (m)

m	gasket factor (see Table 11.7)
p	internal pressure (psi)
R	radius (m)
t	flange thickness (in.)
t_g	groove depth (mm)
W	bolt load (lb_f)
y	yield stress (psi)
ξ	relative elongation
$\tilde{\mu}$	Poisson' ratio
ν	angle (◦)
σ_{allow}	allowable bolt stress (psi or kg/cm^2)
ψ	parameter defined by Equation 11.31

REFERENCES

1. Brownell, L. E., and E. H. Young. *Process Equipment Design* (New York: John Wiley & Sons Inc., 1959).
2. Timoshenko, S. *Strength of Materials*, Part 2 (New York: Van Nostrand Reinhold Co., 1940).
3. Zahkarov, A. A. "Design of Flange Piping and Flanges in Cylindrical Vessels," *Macshgiz, M.* (1958).
4. Waters, E. O. W. Westrom, D. B. Roscheim, F. S. G. Williams and P. Westrom. "Formulas for Stresses in Bolted Flange Connections," *Trans. ASME* (April 1937).
5. Waters, E. O., and J. H. Taylor. "Strength of Pipe Flanges," *Trans. ASME* 49:531-542 (1927).
6. Waters, E. O., D. B. Roscheim, P. Weström and F. S. G. Williams. "Development of General Formulas for Bolted Flanges," Taylor Forge and Pipe Works Pub., Chicago (1949).

CHAPTER 12

THREADED CONNECTIONS

12.1 INTRODUCTION TO DESIGN PRINCIPLES

Threaded connections are commonly used in a variety of process equipment for joining smaller-sized, high-pressure pipes (diameters (d) typically less than 20 mm), pipes in service of moderate and low pressures and temperatures ($d \leqslant 32$ mm), and other preferentially metal parts (flanges, caps, nipples, etc.).

Such joints do not pose any special fire or explosion hazards, cause damage to valve seats, nor allow plugging of small orifices by permitting particles to enter into the piping system. They are not appreciably affected by changes in temperature. Major disadvantages are that they result in a weakening of the joint and increase the potential for corrosion.

Components consist of threaded male and female ends screwed together to provide satisfactory service without excessive cost. The proper manner in which the connection should be threaded is illustrated in Figure 12.1.

Normally, ¾ in./ft of thread length is recommended. The length of effective thread, F, can be determined from the following formula:

$$F = (0.80D + 6.8)\widetilde{P} \tag{12.1}$$

The unified thread form shown in Figure 12.2 is now the standard in the United States, Canada and Great Britain. This form is used for practically all threaded fasteners in the United States; therefore, no other thread types will be discussed.

Pitch diameters for the thread given in Figure 12.1 may be determined as follows:

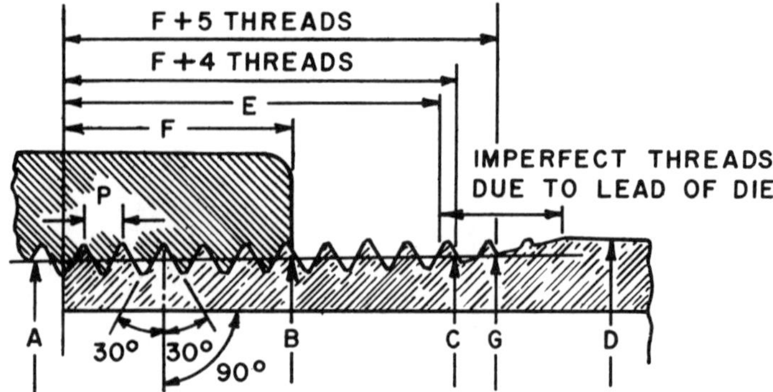

Figure 12.1. American standard taper pipe thread: A. pitch diameter at end of male thread; B. pitch diameter at end of female thread; C. maximum pitch diameter of straight male locknut thread; D. outside diameter of pipe; E. length of effective thread; F. normal engagement by hand between male and female thread; G. minimum pitch diameter of straight female locknut thread; P. pitch of thread.

$$A = D - (0.050D + 1.1)\widetilde{P} \tag{12.2}$$

$$B = A + 0.0625F \tag{12.3}$$

where A = pitch diameter of thread at end of connection
B = pitch diameter of thread at gauging notch
D = outside diameter of pipe
F = normal engagement by hand between external and internal threads
p = pitch of thread

The pitch of the thread is the distance the axis will advance in one revolution, that is, 1/n, where n is the number of threads per inch.

The taper of the thread is 1 in 16 as measured on the diameter, that is, there is a change in diameter. Threaded connections may be divided into three groups:

1. Pipe to pipe. This type of joint is required where it is necessary to install a section of pipe between pieces of equipment, valves or fittings.
2. Pipe to fittings. This connection is required to permit change in direction or size of piping, or to provide a means for branch connections to the system.
3. Pipe to equipment or valves. This type is required for proper operation of the system.

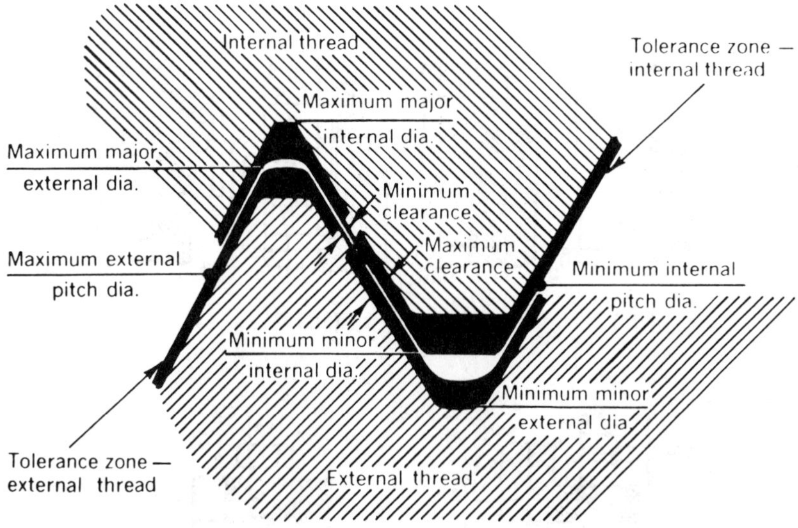

Figure 12.2. Unified thread form.

Pipe to pipe threaded connections are made with either couplings or screwed flange unions. Typical designs of detachable threaded connections used in manufacturing chemical equipment are illustrated in Figure 12.3.

Screwing parts I, II, IV and V (Figure 12.3) is accomplished with tensile strength indicating the axial load; designs III and VI are screwed without pulling. Devices should be provided for that will prevent rotation (for example, hexagonal, flattened spots for a wrench, etc.) in all connections where there is an axial load.

12.2 SCREWED FITTINGS

Screwed fittings include elbows, tees, crosses, 45° ells, Y-branches or laterals, return bends and reducers in various patterns and weights. Equipment and valves generally require joints similar to those used with fittings that are tapped for American Standard pipe threads. For normal requirements, couplings are threaded with right-hand threads for the full length and have plain ends. When the connection is made up, the ends of the pipe should not touch (Figure 12.4).

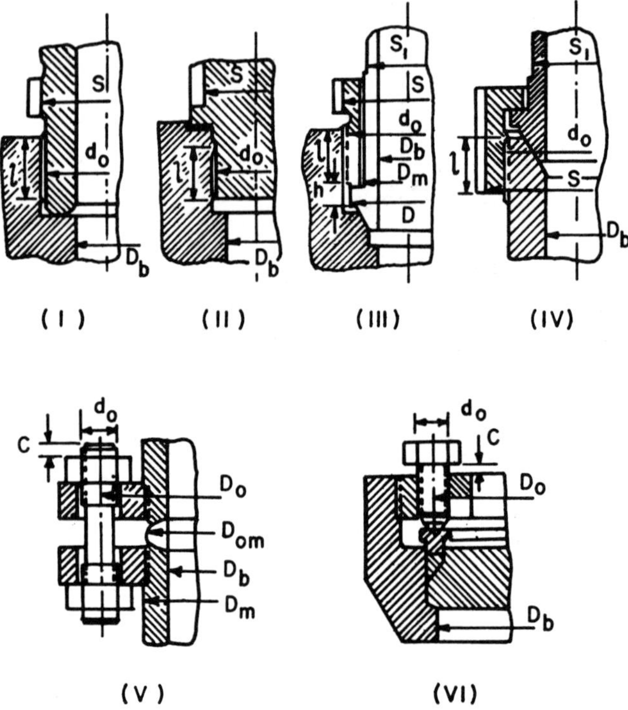

Figure 12.3. Main typical designs of threaded connections used for chemical equipment.

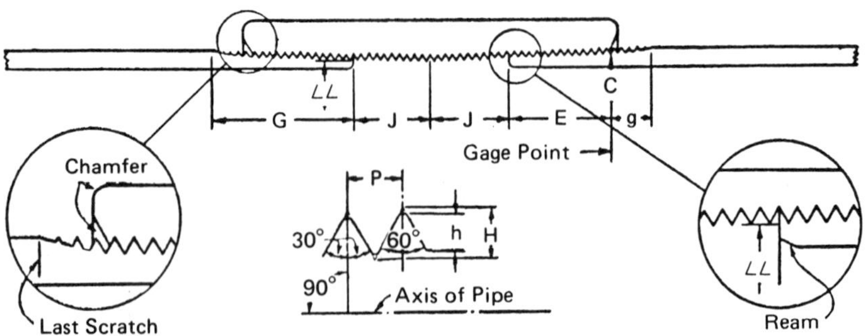

Figure 12.4. Thread specifications for coupling.

When it is necessary to install a piece of pipe close to the space between two pipe sections already in place, without using a union or flanges, one end of the filler piece must be threaded left-hand. The coupling is threaded with one end right-handed and the other with a left-handed thread. This makes it possible to screw the coupling on the ends of the two adjacent pieces at the same time, making the connection tight.

Couplings operating under high pressure for services in which shocks are frequent and for use on API line pipe, have the ends of the couplings recessed to cover the imperfect threads on the pipe. This is illustrated in Figure 12.5. While this design does not strengthen the joint, it does protect the thread against dirt and moisture and, therefore, reduces the danger of corrosion.

Unions consist of three pieces and, where used for connecting two adjacent pieces of pipe, have female ends threaded for American Standard pipe threads. The collar piece normally has a straight thread. The joint between two adjacent sections may be designed for use with a gasket or for metal-to-metal contact.

The application of unions to a typical layout is given in Figure 12.6.

Unions without gaskets may have beveled male and female adjoining surfaces with seats made of all iron, all steel, iron or steel to bronze, or bronze to bronze.

Unions with one male end and one female end provide a simple means of making a joint to equipment that must be disconnected frequently and is not equipped with flanged connections.

When selecting material for threaded parts, one should choose a higher hardness and strength for the male part. This can be achieved by choosing different materials or by corresponding heat treatment of the same material.

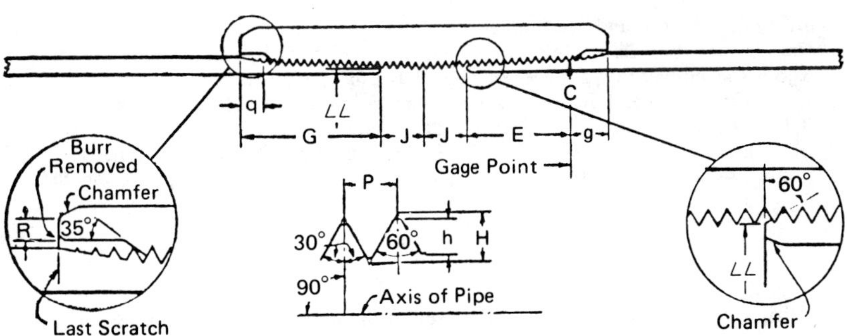

Figure 12.5. Couplings for high-pressure service.

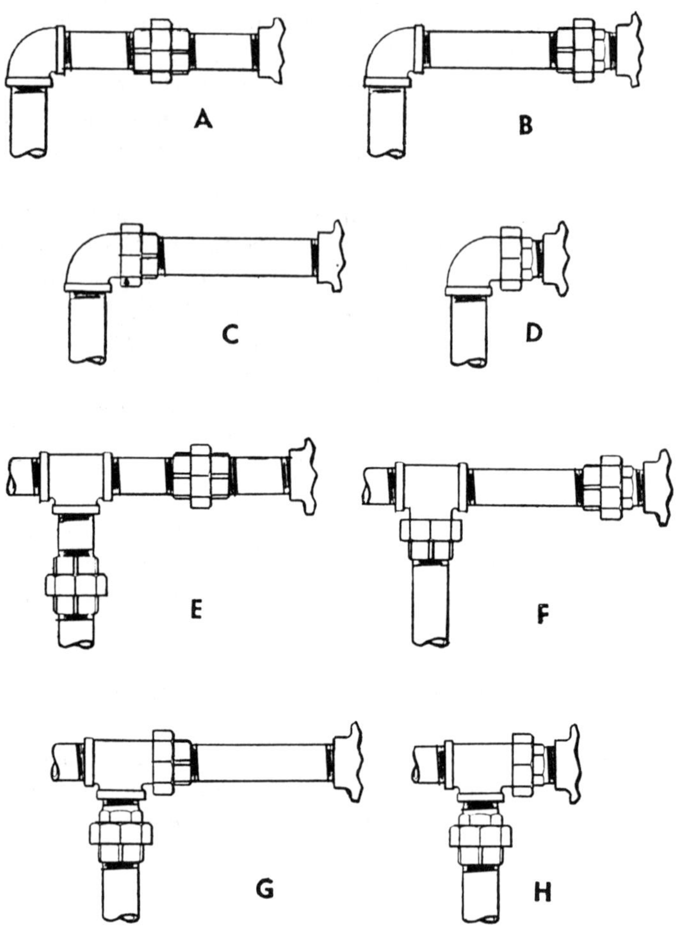

Figure 12.6. Elbow and tee connections with union joints: A. 1 piece of pipe, 2 fittings, 5 pipe threads, 6 joints assembled; B. 2 pieces of pipe, 2 fittings, 1 pipe thread, 5 joints assembled; C. 2 pieces of pipe, 1 fitting, 3 pipe threads, 4 joints assembled; D. 1 piece of pipe, 1 fitting, 1 pipe thread, 3 joints assembled; E. 5 pieces of pipe, 3 fittings, 8 pipe threads and 10 joints assembled; F. 3 pieces of pipe, 2 fittings, 4 pipe threads, 7 joints assembled; G. 3 pieces of pipe, 2 fittings, 4 pipe threads, 7 joints assembled; H. 2 pieces of pipe, 2 fittings, 2 pipe threads, 6 joints assembled.

Usually, the diameter of a thread is selected based on design considerations. Threads are selected from the standards with a definite pitch, t, and designed for the proper length of screwing.

Sources for stresses that may be imposed on male or female parts of threaded connections are the external load encountered in service, the load developed to make the connection tight, and the load involved in the tightening process. These loads produce the following stresses:

1. A bearing stress is caused by the male threads bearing on the threads of the female part of the connection.

2. A shear stress in the threads is produced, that is, a tendency to strip the threads. A full-depth thread has its minor diameter associated with the male, while the female part has a nearly equal diameter; the standard unified thread is based on a 75% thread depth. When tightened, a regular female part drilled out to only 50% of a full-depth thread will break the male part rather than strip the threads of the female part. A standard thread will develop the full tensile strength of the part when engaged at a distance of about one-half the diameter of the male part of the threaded connection. The thickness of the female part is almost equal to the diameter of the male part. Thus, the shear stress in the threads generally will not cause problems.

3. A shear stress in the body of male part is produced when it is used in shear. This becomes a simple problem of selecting the male part that has sufficient area in shear.

4. A tensile stress is produced in the male part of the threaded connection. If the male part is used in an application such that there is no load due to tightening and only a static applied load, the problem reduces to proper selection of a male part, with adequate area at the root diameter of the thread. If the male part is tightened to the extent that a tensile stress is developed that is almost equal to the yield strength, any load that induces a stress less than this will have no effect on the male part.

5. A torsion stress in the male part is produced due to tightening. As a male part is tightened, the tensile load that is induced forces the threads of the male and female parts together, thus increasing the friction. Consequently, to overcome this friction and to develop the tension load it is necessary to apply a torque, which will disappear when tightening is completed.

The required torque to produce a given tension load is

$$T = CDP \tag{12.4}$$

where T = torque
C = coefficient = 0.2
D = diameter of a male part
P = tension load in male part

12.3 DESIGN OF THREADED PARTS

The detailed design of threaded parts must be based on the requirements for bearing, bending and shearing stresses as follows:

The design length of screwing, ℓ', is determined on the basis of the bending strength for the male part:

$$\ell' \geqslant \frac{KP'(d_o - d_i)}{d_i t \sigma_{b,allow}}, \quad m \tag{12.5}$$

and for the female part:

$$\ell' \geqslant \frac{KP'(d_o - d_i)}{d_o t \sigma_{b,allow}}, \quad m \tag{12.6}$$

where P' = design load on the thread (N)
d_o = outside diameter (size) of thread (m)
d_i = inside diameter of thread (m)
K = coefficient: for metric thread, $K = 0.623$; for pipe cylindrical thread, $K = 0.685$

The design screwing length, ℓ', is determined from the condition of bearing strength for male and female parts of threaded connections as follows:

$$\ell' \geqslant \frac{1.27P't}{(d_o^2 - d_i^2)\sigma_{c,allow}} \tag{12.7}$$

The allowable bearing stress may be assumed. For connections in which there is no axial load during screwing (threaded flanges etc.), use the recommendations outlined in Table 12.1.

The value of allowable stress, depending on kind of stress, strength characteristics of construction material, design temperature of the part and service temperature, is determined from the following formulas:

$$\sigma_{allow} = \frac{\sigma_T^t}{n_T}\eta, \quad N/m^2 \tag{12.8}$$

$$\sigma_{allow} = \frac{\sigma_y^t}{n_y}\eta, \quad N/m^2 \tag{12.9}$$

$$\sigma_{allow} = \frac{\sigma_f^t}{n_f}, \quad N/m^2 \tag{12.10}$$

Table 12.1. Recommended Design Formulas for Threaded Connections[a]

Material	Temperature (°C)	Design formula				
		σ_{allow}	$\sigma_{c,allow}$	$\sigma_{b,allow}$	τ_{allow}	$\tau_{sh,allow}$
Carbon Steels	<420	12.8 and 12.9				
	⩾420	12.8, 12.9, 12.10 and 12.11				
Alloy Steels	<470	12.8 and 12.9				
	⩾470	12.8, 12.9, 12.10 and 12.11			$\approx 0.6\sigma_{allow}$	
Alloy Austenite Steels	<550	12.8 and 12.9				
	⩾550	12.8, 12.9, 12.10 and 12.11				
Nonferrous Metals and Their Alloys		12.8 and 12.9				
Cast Irons		12.8	12.12	12.13		
Quartz Glass		12.8	12.12	12.13	$\approx 0.8\sigma_{allow}$	
Polyvinyl Chloride Plastic		12.11	$1.2\sigma_{allow}$	$1.9\sigma_{allow}$	$1.8\sigma_{allow}$	$0.7\sigma_{allow}$

[a]The assumed value is taken as the smaller one from the above formulas.

$$\sigma_{allow} = \frac{\sigma_{cr}^{t}}{n_{cr}}, \quad N/m^2 \tag{12.11}$$

$$\sigma_{c,allow} = \frac{\sigma_{c}^{t}}{n_{T}}, \quad N/m^2 \tag{12.12}$$

$$\sigma_{b,allow} = \frac{\sigma_{b}^{t}}{n_{T}}, \quad N/m^2 \tag{12.13}$$

where n = correction coefficient, as given in Table 12.2

For connections in which axial loads arise during screwing (nipples, plugs, etc.), use the following formula:

$$\sigma_{c,allow} \leqslant 0.15\sigma_y, \quad N/m^2 \tag{12.14}$$

These equations assume σ_y of the material of the male part is no greater than that of the female part. In all cases, the design value is the greater value of ℓ', as computed from formulas 12.5, 12.6, 12.7, 12.15 and 12.16.

The screwing length, independent of design, should assume $\ell \geqslant t$.

Table 12.2. Values of Correction Coefficient

Operating Group	Operating Conditions	Operating Class of Apparatus I	II
1	Assemblies and parts heated with flame flue gases, open electric heaters	0.75	0.9
2	Assemblies and parts not heated, insulated from heating sources of Group I	0.9	1.0

Example 12.1

Determine the length of screwing for the following data: The shell and the flange are fabricated from the same material ($\sigma_{b.allow} = \sigma_{c.allow} = 233 \times 10^6$ N/m^2; $\tau_{sh} = 140 \times 10^6$ N/m^2); $d_o = 0.36$ m; $d_i = 0.3568$ m; t = 0.004; $P' = P_b = 351 \times 10^4$ N; h = 0.088 m.

Solution

The design screwing length is determined from the bending strength condition (Equation 12.5):

$$\ell' = \frac{KP'(d_o - d_i)}{d_i t \sigma_{b.allow}} = \frac{0.623 \times 0.351 \times 10^4 (0.36 - 0.3568)}{0.3568 \times 0.004 \times 233 \times 10^6}$$

$$= 0.021 \text{ m}$$

The design screwing length is determined from the bearing stress conditions (Equation 12.7):

$$\ell' = \frac{1.27P't}{(d_o^2 - d_i^2)\sigma_{c.allow}} = \frac{0.318 \times 351 \times 10^4}{0.623 \times 0.3568 \times 140 \times 10^6} = 0.0358 \text{ m}$$

The design screwing length is determined from shearing stress conditions (Equation 12.15):

$$\ell' = \frac{0.318P'}{Kd_i\tau_{sh}} = \frac{0.318 \times 351 \times 10^4}{0.623 \times 0.3568 \times 140 \times 10^6} = 0.0358 \text{ m}$$

The final design value is $\ell' = 38.2$ mm; we shall assume $\ell = 40$ mm (10 t).

Example 12.2

Determine the sizes of the main parts of the threaded connection (refer to Figure 12.2, type III) with obturation designed as in Example 10.7 for the following conditions: D = 0.028 m; the material of the nipple is steel ($\sigma_{b.allow}$ = 300 X 10^6 N/m^2); the material of the pressing screw is steel (σ_y = 500 X 106 N/m^2; $\sigma_{b.allow}$ = 250 X 106 N/m^2; τ_{sh} = 150 X 10^6 N/m^2); the material of the shell is stronger than that of the nipple and pressing screw. From Example 10.7 we have D_n = D_k = 0.025 m; p′ = 125 X 10^6 N/m^2; p_n' = 6800 N.

Solution

The design load from the internal pressure of fluid at k = 1 is determined from the following formula:

$$P_c' = 0.96\,KD_{np}^2 p' = 0.96 \times 1 \times 0.025^2 \times 125 \times 10^6 = 7.5 \times 10^4 \text{ N}$$

The design axial force acting on the connection is

$$P' = P_c' + P_n' = 7.5 \times 10^4 + 0.68 \times 10^4 = 8.18 \times 10^4 \text{ N}$$

The design outside diameter of the supporting surface of the pressing screw, D_2', is determined from the bearing strength condition from Equation 12.14:

$$\sigma_{sh,allow} = 0.15\,\sigma_y = 0.15 \times 500 \times 10^6 = 75 \times 10^6 \text{ N/m}^2$$

$$D_2' = \sqrt{\frac{P'}{0.8\,\sigma_{sh}} + D_n^2} = \sqrt{\frac{8.18 \times 10^4}{0.8 \times 75 \times 10^6} + 0.028^2} = 0.0462 \text{ m}$$

Assume D_2 = 46 mm. The design height, h′, of the nipple is determined from the bending strength condition:

$$h' = \sqrt{\frac{3P'(D_2 - D_1)}{\pi D_n\,\sigma_{b,allow}}} = \sqrt{\frac{3 \times 8.18 \times 10^4(0.046 - 0.028)}{\pi \times 0.028 \times 300 \times 10^6}} = 0.0129 \text{ m}$$

Assume h = 13 mm. The diameter of thread d_i = 48.1 mm. The design screwing length from Equation 12.7 (the bearing strength condition) is

$$\ell' = \frac{1.27P't}{(d_o^2 - d_i^2)\sigma_{sh,allow}} = \frac{1.27 \times 8.18 \times 10^4 \times 0.003}{(0.052^2 - 0.0481^2)75 \times 10^6}$$

$$= 0.01065 \text{ m}$$

The design screwing length is determined from the shearing strength condition as computed from Equation 12.15:

$$\ell' = \frac{0.318P'}{Kd_i\tau_{sh,allow}} = \frac{0.318 \times 8.18 \times 10^4}{0.623 \times 0.0481 \times 150 \times 10^6}$$

$$= 0.00577 \text{ m}$$

The final design value is $\ell' = 10.65$ mm. Assume $\ell = 8t = 8 \times 3 = 24$ mm.

NOMENCLATURE

A	pitch diameter of thread at end of connection (m)
B	pitch diameter of thread at gauging notch (m)
C	coefficient in Equation 12.4
D	outside diameter (m)
d_o,d_i	outside and inside diameters of thread (m)
K	thread coefficient (see Equations 12.5 and 12.6)
ℓ'	design length of screwing (m)
n	number of threads per inch
P	tension load (N)
p'	design load on thread (N)
$\tilde{p}$	pitch of thread
T	torque (N-m)
t	pitch
η	correction coefficient (see Table 12.2)
τ_{sh}	yield point (Pa)

REFERENCES

1. Parr, R. E. *Principles of Mechanical Design* (New York: McGraw-Hill Book Co., 1970).
2. Laugher, V. H., and A. D. Hargan. *Handbook of Fastening and Joining Metal Parts* (New York: McGraw-Hill Book Co., 1956).
3. York, J. E. *Methods of Joining Pipe* (New York: The Industrial Press, 1949).
4. *ASME Boiler and Pressure Vessels Code,* Section 8, Division 1 (New York: American Society of Mechanical Engineers, 1977).

CHAPTER 13

DESIGN OF COVERS

In contrast to heads that are nondetachable entities from shells, covers are designed as removable parts and, in a great number of cases, are locked hermetically to the shell of an apparatus, manhole, etc.

Covers provide convenience of assembly, permit loading and unloading of an apparatus during operation, and allow for repairs, inspections, etc. They may be located almost any place on an apparatus: from above, below or on its sides. The shape of covers may be round, rectangular or profile. Round covers are most common because of their ease in fabrication.

Examples of cover designs are given in Figure 13.1. According to the method of connection to equipment, covers are divided into three groups: flanged, bow and bayonet.

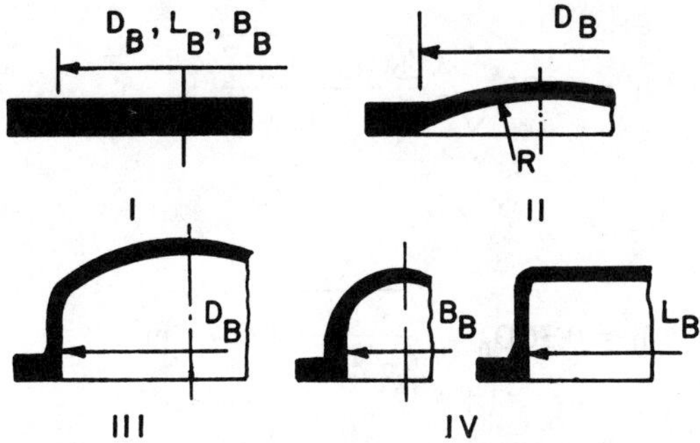

Figure 13.1. Examples of cover designs: I-flat cover (round, rectangular, profile); II-spherical (round); III-elliptical (round); IV-cylindrical (rectangular).

13.1 FLANGED COVERS

Flanged covers are widely used in process equipment. Connections of flanged covers are made by bolting, using a corresponding obturation.

13.1.1 Flat Flanged Covers

Flat round covers are commonly used; however, in some cases there is the need to apply rectangular and, less frequently, profile designs. In welded and cast apparatuses operating during filling operations, top covers of any shape are fabricated flat, independent of their size. However, large-sized covers should be reinforced with ribs. Lower and side round flat covers with $D \leqslant 600$ mm usually are applied at hydrostatic pressure, $p \leqslant 0.1$ MN/m^2. Round flat covers are recommended for apparatuses operating under pressure and vacuum under the following conditions:

- in welded apparatuses at $D < 400$ mm for $p <$ MN/m^2 and vacuum
- in forged apparatuses at $D \leqslant 800$ mm for $\geqslant 10$ MN/m^2

Rectangular covers are fabricated flat independent of pressure. These covers, when large and under elevated pressures, should be reinforced with ribs. Examples of designs of flanged flat covers are given in Figure 13.2.

13.1.2 Design of Round Flat Flanged Covers Operating Under Pressure

The design height of the cover at the sealing place, h′, is determined from the following formulas:

For types II–IV,

$$h' = 0.875\sqrt{\frac{P_b'(D_b - D_n)Z}{D_n \sigma_{b,allow} Z'}}, \quad m \tag{13.1}$$

For types V–VIII,

$$h' = 0.36 D_n \sqrt{\frac{P}{\sigma_{b,allow}}} + C, \quad m \tag{13.2}$$

The design height of the middle part of the cover, h′, is determined for all types from the following formula:

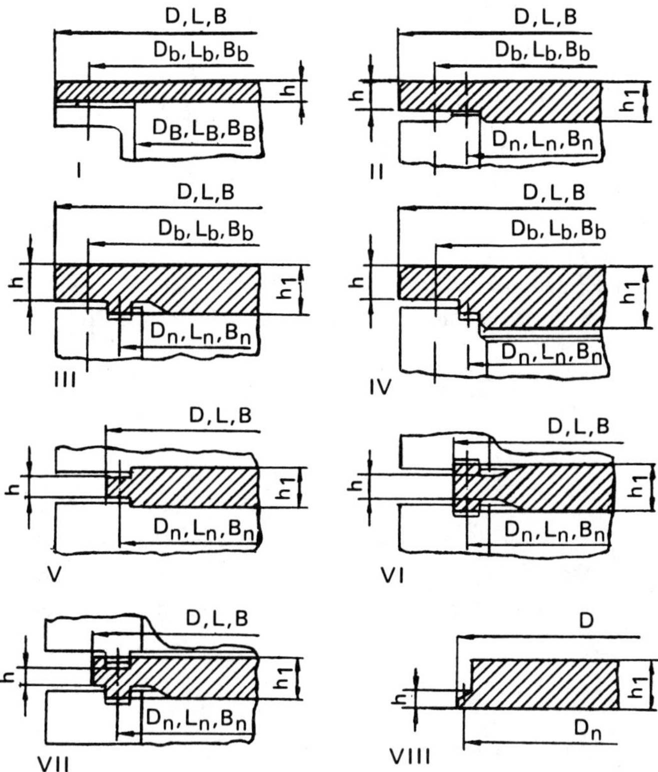

Figure 13.2. Examples of designs of flat flanged covers: I–IV-solid round and rectangular flanged covers; V–VII-round and rectangular covers inserted between the flanges; VIII-round cover inserted between flanges for high-pressure operation.

$$h_1' = D_n \sqrt{\frac{KP}{\sigma_{b,allow}\,\psi}} + C \ , \quad m \tag{13.3}$$

where P_b' = the design tensile force in bolts as determined from Equation 11.67

N, Z' and Z = the design and assumed number of bolts, respectively

K = a coefficient, depending on the design of the cover

ψ = a coefficient accounting for the weakening of the cover by openings

C = the corrosion allowance (m)

The value of coefficient K is assumed as follows:

For type I, K = 0.18; for types V–VIII, K = 0.3; for types II–IV,

$$K = 0.3 + \frac{0.7P_b'(D_b - D_n)}{P_c' D_n Z'} \tag{13.4}$$

where P_c' is defined in Equation 11.66.

The value of the weakening coefficient, ψ, is determined from the following:

$$\psi = \frac{D_n - \Sigma_d}{D_n} \tag{13.5}$$

where Σ_d is the sum of the opening diameters on the cover with D_n (for welded pipes the opening diameter is assumed to be the inside diameter).

The design outside diameter of the cover, D', is determined as follows:

For types I–IV as for round outside flanges, from the following formula:

$$D' = D_b + 2d_b + 0.005 \quad , \quad \text{m} \tag{13.6}$$

where d_b = diameter of bolt
D_b = diameter of bolt circle

For types V–VIII, according to the assumed obturation (see Chapter 10).

Example 13.1

Determine the basic sizes of a flat flanged cover (see Figure 13.2, type IV) using the following data: $D = 0.4$ m; $D_n = 0.423$ m; $D_b = 0.485$ m; $P_b' = 510 \times 10^4$ N; $P_c' = 424 \times 10^4$ N; $d_b = 0.03$ m; $p = 20 \times 10^6$ N/m^2; $Z = 20$; $Z' = 18.8$; the material of the cover is steel ($\sigma_{b.allow} = 250 \times 10^6$ N/m^2), C = 2 mm; the cover has no openings in its middle part.

Solution

The design outside diameter of the cover, from Equation 13.6 is

$$D' = D_b + 2d_b + 0.005 = 0.485 + 2 \times 0.03 + 0.005 = 0.55 \text{ m}$$

Assume D = 550 mm. The design height of the cover at the sealing place, from Equation 13.1, is

$$h' = 0.875 \sqrt{\frac{P_b(D_b - D_n)Z}{D_n \sigma_{b,allow} Z'}}$$

$$= 0.875 \sqrt{\frac{510 \times 10^4 (0.485 - 0.423) 20}{0.423 \times 250 \times 10^6 \times 18.8}} = 0.0493 \text{ m}$$

Assume h = 50 mm. The coefficient K in Equation 13.3 is determined for Equation 13.4:

$$K = 0.3 + \frac{0.7P'_b(D_b - D_n)Z}{P'_c D_n Z'} = 0.3 + \frac{0.7 \times 510 \times 10^4 (0.485 - 0.423) \times 20}{424 \times 10^4 \times 0.423 \times 18.8}$$

$$= 0.4315$$

The design height of the middle part of the cover, from Equation 13.3, is

$$h' = D_n \sqrt{\frac{KP}{\sigma_{b,allow} \psi}} + C = 0.423 \sqrt{\frac{0.4315 \times 20 \times 10^6}{250 \times 10^6 \times 1}} + 0.002$$

$$= 0.081 \text{ m}$$

Assume h_1 = 80 mm.

13.1.3 Design of Rectangular Flat Flanged Covers Operating Under Pressure (refer to Figure 13.2, types I–VII)

The design height of the cover, h′, for type I is determined as s′ from Equation 6.84. The design height of the cover at the sealing place, h′, for other types is determined from the following formulas:

For types II–IV,

$$h' = 1.1 \sqrt{\frac{P'_b(L_b + B_b - L_n - B_n)Z}{(L_n + B_n)\sigma_{b,allow} Z'}} \tag{13.7}$$

For types V–VII,

$$h' = \frac{K_2 Bp}{\tau_{sh,allow}} \qquad (13.8)$$

where K_2 is determined from Figure 13.3.

The design height of the middle part of the cover is

$$h' = B_n \sqrt{\frac{kp}{\sigma_{b,allow}\,\psi}} + C \qquad (13.9)$$

where k = coefficient, depending on the cover design and the ratio of its sides
ψ = coefficient of weakening by the holes

The values of coefficient k are assumed:

For types V–VII $k = K_1$ from Figure 13.3.
For types II–IV,

$$k = K_1 + \frac{0.7p_b'(B_b - B_n)Z}{p_c' B_n Z'} \qquad (13.10)$$

where P_b', P_c', Z and Z' are the same as in Equation 13.4.

The value of coefficient ψ is determined from the following:

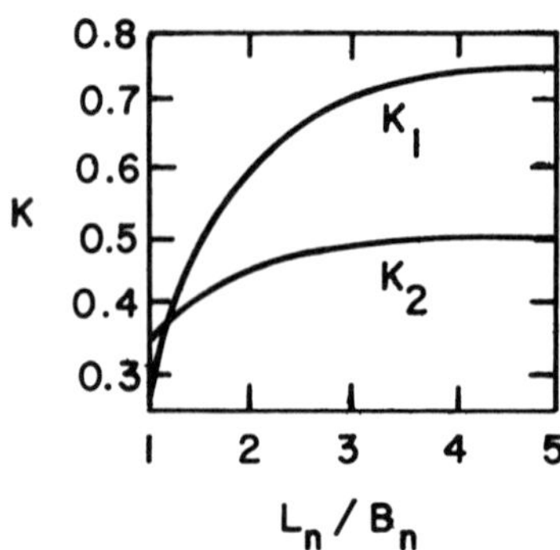

Figure 13.3. Values of coefficient k in Equations 13.8 and 13.10.

$$\psi = \frac{B_n - \Sigma d}{B_n} \tag{13.11}$$

where Σd = sum of diameters of holes on the length B_n, m.

The outside sizes of the cover L and B are determined as follows:

For types I–IV, from

$$L' = L_b + 2d_b + 0.005 \text{ m}$$

$$B' = B_b + 2d_b + 0.005 \text{ m}$$

For types V–VII, according to the assumed obturation (see Chapter 10).

Example 13.2

Determine the basic sizes of a flat rectangular flanged cover (refer to Figure 13.2, type II) using the following data: the material of the cover is steel ($\sigma_{b,allow}$ = 146 × 10^6 N/m^2); the cover has no openings (ψ = 1); C = 2 mm; p = 0.1 MN/m^2; P'_c = 8.12 × 10^4 N; Z = 38; Z′ = 21.8; L_n = 0.62 m; B_n = 0.42 m; L_b = 0.66 m; B_b = 0.46 m; L = 0.69 m; P'_b = 13 × 10^4 N.

Solution

The design height of the cover at the sealing place is determined from Equation 13.7:

$$h' = 1.1\sqrt{\frac{P'_b(L_b + B_b - L_n - B_n)Z}{(L_n + B_n)\sigma_{b,allow} Z'}}$$

$$= \sqrt{\frac{13 \times 10^4(0.66 + 0.46 - 0.62 - 0.42)38}{(0.62 + 0.42)\,146 \times 10^6 \times 21.8}} = 0.012 \text{ m}$$

Assume h = 12 mm. Coefficient K_1 from Figure 13.3 for L_n/B_n = 0.62/0.42 = 1.475; K_1 = 0.47. Coefficient k in Equation 13.9 is determined from Equation 13.10:

$$k = K_1 + \frac{0.7P'_b(B_b - B_n)Z}{P'_cB_nZ'} = 0.47 + \frac{0.7 \times 13 \times 10^4(0.46 - 0.42)38}{8.12 \times 10^4 \times 0.42 \times 21.8}$$

$$= 0.656$$

The design height of the middle part of the cover is obtained from Equation 13.9:

$$h' = B_n \sqrt{\frac{kp}{\sigma_{b,allow}\psi}} + C = 0.42 \sqrt{\frac{0.656 \times 0.1 \times 10^6}{146 \times 10^6 \times 1}} + 0.002$$

$$= 0.0109 \text{ m}$$

Assume from design considerations that $h_1 = 18$ mm.

13.1.4 Spherical Flanged Covers

Spherical flanged covers are only fabricated in round shapes and used as follows: (1) for welded apparatuses at $D \geqslant 400$ mm operating under pressure $p \leqslant 2.5$ MN/m^2 (in some cases up to 8 MN/m^2), vacuum and during filling operation (only lower and side covers at hydrostatic pressure higher than 0.1 MN/m^2); (2) for cast apparatuses, at any diameter independent of pressure; and (3) for forged apparatuses, at $D \geqslant 600$ mm operating under pressure $p \geqslant 10$ MN/m^2.

Examples of designs of flanged spherical covers are given in Figure 13.4.

Welded covers types I and II are comprised of spherical heads and flanges.

In the design of spherical flanged covers operating under pressure, the fol-

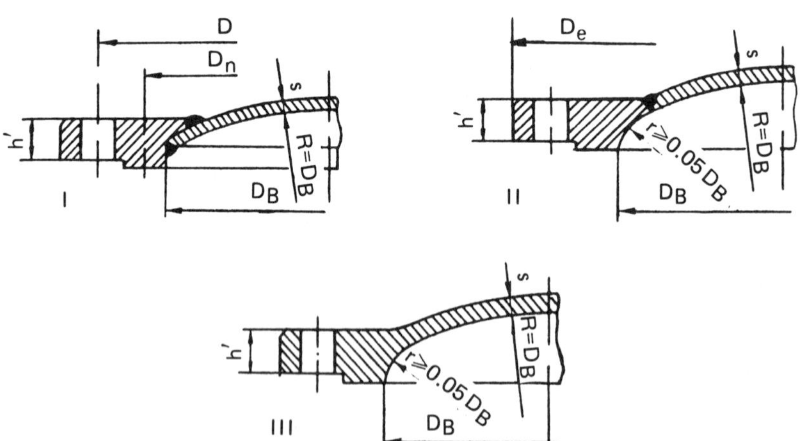

Figure 13.4. Examples of spherical flanged covers: I-welded at $p \leqslant 2.5$ MN/m^2; II-welded at $p > 2.5$ MN/m^2; III-forged and cast.

lowing should be determined: the design thickness of the spherical wall, S′, from Equation 6.20, and the design height of flat flange, h′, from

$$h' = 1.19 \sqrt{\frac{P_b'(D_b - D_n - d_b)(D_e + D)}{E^t ZZ' \eta \phi'_{allow}(D_e - D)}} \qquad (13.11B)$$

for plastic materials, where ϕ'_{allow} = allowable angle of flange distortion in circular direction: for gaskets similar to parotine, $\phi'_{allow} = 4 \times 10^{-4}$; for rubber gaskets, $\phi'_{allow} = 3 \times 10^{-3}$; for metal gaskets and without gaskets, $\phi'_{allow} = 2 \times 10^{-4}$.

For brittle materials use Equation 13.11C.

$$h' = \sqrt{\frac{P_b'[D_b - 0.5(D + D_n)]Z}{(D + D_n)\sigma_{b,allow} Z'}} \qquad (13.11C)$$

13.1.5 Elliptical Flanged Covers

The elliptical covers fabricated in round shape are used for welded apparatuses for $D \geqslant 400$ mm at $p > 1.6$–8 MN/m^2. The examples of welded elliptical flanged covers are given in Figure 13.5. The cover consists of an elliptical head and flange.

In the design of elliptical flanged covers operating under pressure (Figure 13.5), it is necessary to determine the design thickness of elliptical head, S′, from Equation 7.1, and the design height of the flange from Equations 13.11B or 13.11C, depending on the material and shape of the cover flange.

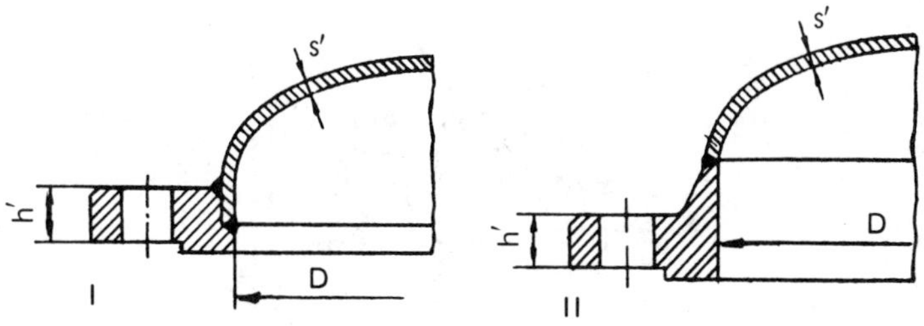

Figure 13.5. Elliptical welded flanged covers: I-welded steel covers at $p \leqslant 2.5$ MN/m^2; II-welded steel covers at $p > 2.5$ MN/m^2, welded of copper and aluminum at $p > 0.3$ MN/m^2.

13.1.6 Cylindrical Flanged Covers

The cylindrical covers fabricated in rectangular shape only are used mostly as collectors for apparatuses operating under pressure and vacuum and during filling.

Examples of flanged cylindrical rectangular covers are given in Figure 13.6. In design of flanged cylindrical covers operating under pressure (Figure 13.6), one should determine the following: the design thickness of the cylindrical cover walls, from Equation 6.20; the design thickness of flat walls, S_1, from Equation 6.84; and the design height of the flange, h', from

$$h' = 1.09 \sqrt{\frac{P_b' [L_b + B_b - 0.5(L + L_n + B + B_n)] Z}{(L + L_n + B + B_n) \sigma_{b,allow} Z'}}, \quad m \qquad (13.11D)$$

13.2 BOW COVERS

Bow covers are used mainly in cases in which it is necessary to gain quick and frequent access to the vessel's internals. Bow covers, as well as flanged covers, are fabricated in flat, spherical, elliptical and profile shapes. Their use

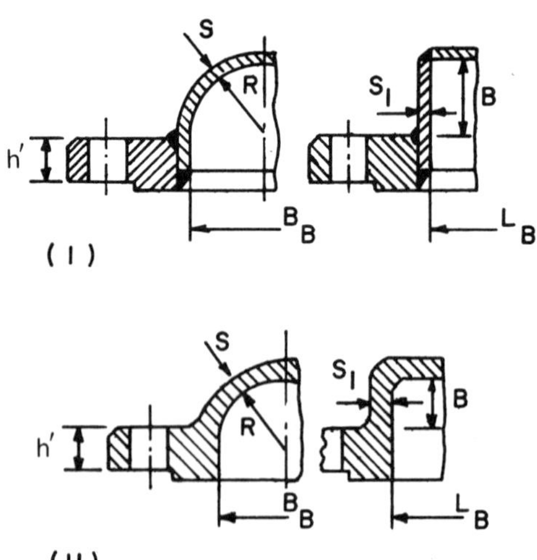

Figure 13.6. Examples of flanged cylindrical covers: I-welded; II-cast.

is limited to relatively low pressures and sizes. These covers are available as hinged, tilting and removable designs. An example of an oval manhole of carbon steel with internal removable bow cover with self-sealing is given in Figure 13.7.

The round covers usually have one bow, while rectangular and oval covers are provided with two bows, located in transverse direction.

The design load on a bow cover, P_b', is determined according to the data given in Chapter 11, depending on the size of a cover, fluid pressure, type of obturation and whether it is self-sealing. In the case of two bow covers, the distance between them is taken to be 0.5 L (where L is the greater size of the cover).

Flat bow covers are fabricated in two fashions: round and rectangular.

13.2.1 Design of Round Flat Bow Covers Operating Under Pressure (See Figures in Table 13.1)

The design height of the cover in the place of sealing, h′, is determined from the following:

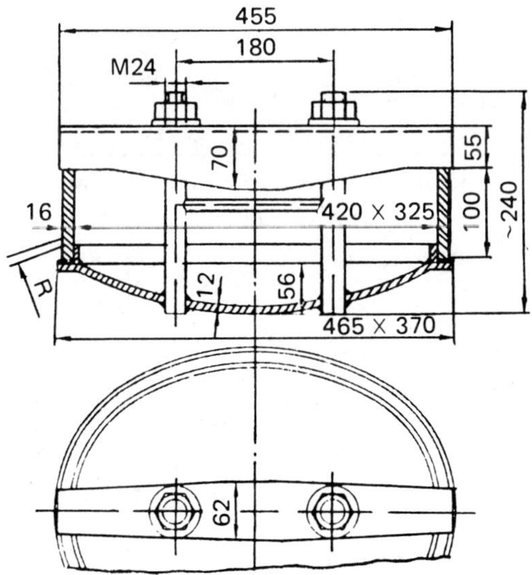

Figure 13.7. Design of oval manhole of carbon steel with internal removable bow cover, operating under pressure $p = 0.8\ MN/m^2$.

Table 13.1. Design Data for

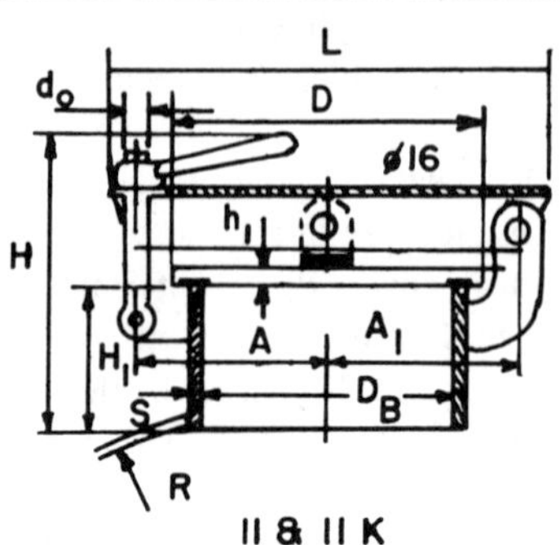

P_y, MN/m^2	D_B	D	H	H_1	A	A_1	L	ℓ	h_1
	mm								
Without Pressure	147 150	175	195	75	105	106	260	–	8
0.25	150	195	265	100	117.5	–	275	220	14
Without Pressure	250	280	225		156	160	365	–	10
0.25		300	350[c]	120	175	–	405	350	20
Without Pressure	400	430	290	140	236	238	530	–	12
0.25		460	455[d]		260	–	580	520	20

[a]Manholes of types II and III are fabricated for apparatuses of carbon steel.
[b]Manholes II K and III K are fabricated for apparatuses of high-alloy steel.
[c]340 for type III.
[d]440 for type III.

$$h' = 0.0366\sqrt{\frac{(2-\beta^2)P_b'}{\sigma_{b,allow}}} + C \qquad (13.12)$$

where P_b' = design load applied to the cover at its center and is equal to the corresponding design tensile force acting on the bolts (N)

$$\beta = \frac{\text{average diameter of sealing}}{\text{diameter of the circle where } P_b' \text{ acts}}$$

The design height of the middle part of the cover, h_1', is determined as

Round Flat Bow Covers[a,b]

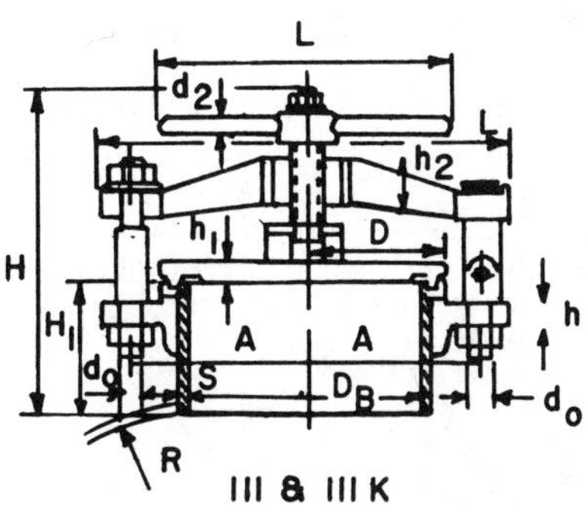

h	h_2	s	d_0	d_1	d_2	R	Type weight (kg)			
			mm				II	II K	III	III K
		6					–	6.74		
–	–		M16	–	–	⩾200			–	–
		4.5					6.3	–		
22	40	10		24 × 5	14		–	–	11.8	11.8
			M16			⩾250				
–	–	6		–	–		12.8	12.8	–	–
25	60	16	M21	32 × 6	16	⩾600	–	–	36.2	44.3
–	–	6	M20	–	–	⩾400	27.9	27.9	–	–
36	90	16	M30	44 × 8	20	⩾700	–	–	80.0	94.1

$$h_1' = 0.0505 \sqrt{\frac{(1.5 - 0.26\beta^2 - 1.95 \ln \beta) P_b'}{\sigma_{b,allow}}} + C \tag{13.13}$$

13.2.2 Design of Round Flat Bow Covers With Self-Sealing, Operating Under Pressure

The determination of design heights of such covers should be done as follows: h', from Equations 13.2 and 13.12; and h_1', from Equations 13.3 and 13.13. Larger values should be taken as the final design values.

13.2.3 Design of Rectangular Flat Bow Covers Operating Under Pressure

The design height of the middle part of the cover is determined from the following formula:

$$h_1' = 0.616 \sqrt{\frac{P_b'}{\sigma_{b,allow}}}\left[1 + k + \ell n \frac{(1+\tilde{\mu})B_n}{d}\right] + C \ , \quad m \qquad (13.14)$$

where $k = \dfrac{0.914}{1 + 1.6\beta^5} - 0.6$

$\beta = \dfrac{B_n}{L_n}$

d = diameter of the circle where P_b' acts

The design height of the middle part of the cover may be determined approximately from Equation 13.14, conditionally dividing the cover in two parts. The design should be done for the half similar to that of the cover with one bow.

In this case, the value P_b' in Equation 13.14 should be substituted by $0.5P_b'$ and, instead of B_n, the lesser from values 0.5 L_n or B_n should be chosen. The value β is assumed:

At

$$B_n \geqslant 0.5L_n \ , \quad \beta = \frac{L_n}{2B_n}$$

At

$$B_n < 0.5L_n \ , \quad \beta = \frac{2B_n}{L_n}$$

13.2.4 Spherical and Elliptical Bow Covers

Spherical bow covers in process apparatuses are used mostly in cast designs as bottoms and in welded equipment. The design wall thickness of such covers may be obtained from Equation 7.5, conditionally considering the load to be uniformly distributed over the surface of the spherical or elliptical wall ($P_e = (1.27/D^2)P_B'$,N/m^2).

In the center of the cover at the point of loading, the screw or the bow should be provided with a pad of thickness not less than 0.3 diameters of the cover. The pad should be rigidly connected with the wall of the cover.

13.2.4 Bow Device

A bow device intended for pressing down a flat rectangular cover (with two bows) is given in Figure 13.8.

The structural design of these parts is made using known formulas for the strength of materials and machine elements for basic load, p_b', acting in the middle of the cover.

Example 13.3

Determine the basic sizes of a flat rectangular cover and its bow device (Figure 13.8) for the following data: the material of the cover, brackets and clamp is steel ($\sigma_{b,allow} = \sigma_{c,allow} = 146 \times 10^6$ N/m^2); the bow, bolt and spindles are of steel ($\sigma_{T,allow} = \sigma_{b,allow} = 180 \times 10^6$ N/m^2); C = 2 mm; $\widetilde{\mu} = 0.3$; $P_b' = 13 \times 10^4$ N; L = 0.64 m; B = 0.44 m; $B_n = 0.42$.

Solution

We shall design bolt 2, Figure 13.8. The design cross-sectional area based on the internal diameter of the bolt thread is

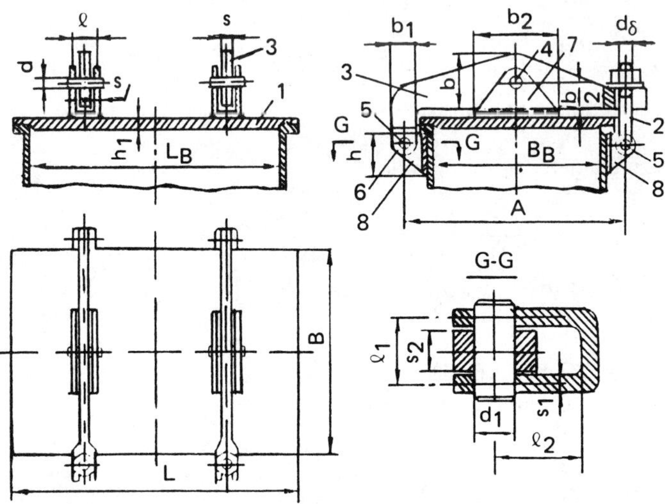

Figure 13.8. Flat rectangular cover with two bow devices: 1-cover; 2-bolt; 3-bow; 4-middle hinged spindle; 5-side spindles; 6-bow link; 7-clamp; 8-bracket.

$$F_b = \frac{P'_b}{4\sigma_{allow}} = \frac{13 \times 10^4}{4 \times 180 \times 10^6} = 181 \times 10^{-6} \text{ m}^2$$

Assume bolt diameter d = 20 mm (F_b = 220 × 10^{-6} m^2). For the bow 3: the bending moment at A = 0.48 m is

$$M = \frac{P'_b A}{2 \times 4} = \frac{13 \times 10^4 \times 0.48}{8} = 7800 \text{ N-m}$$

The design cross-sectional modulus for the middle part is

$$W' = \frac{M}{\sigma_{b,allow}} = \frac{7800}{180 \times 10^6} = 43.3 \times 10^{-6} \text{ m}^3$$

The design thickness at b/S = 5 is

$$S' = \sqrt[3]{\frac{6W'}{25}} = \sqrt[3]{\frac{6 \times 43.3 \times 10^{-6}}{25}} = 21.8 \times 10^{-3} \text{ m}$$

Assume S = 22 mm and b = 110 mm. The design of middle spindle 4: the bending moment at S_1 = 8 mm and ℓ = 36 mm:

$$M = \frac{P'_b(4\ell - S)}{2 \times 16} = \frac{13 \times 10^4(4 \times 0.036 - 0.022)}{2 \times 16} = 497 \text{ N-m}$$

The design diameter of the spindle is

$$d' = \sqrt[3]{\frac{M}{0.1\sigma_{b,allow}}} = \sqrt[3]{\frac{497}{0.1 \times 180 \times 10^6}} = 30.2 \times 10^3 \text{ m}$$

Assume d = 30 mm. Design of side spindle 5: the bending moment at s_1 = 8 mm; s_2 = 20 mm; ℓ_1 = 30 mm;

$$M = \frac{P'_b(4\ell_1 - S_2)}{4 \times 16} = \frac{13 \times 10^4(4 \times 0.03 - 0.02)}{4 \times 16} = 203 \text{ N-m}$$

The design diameter of the spindle is

$$d_1' = \sqrt[3]{\frac{M}{0.1\,\sigma_{b,allow}}} = \sqrt[3]{\frac{203}{0.1 \times 180 \times 10^6}} = 22.4 \times 10^3 \text{ m}$$

Assume $d_1 = 24$ mm. Design of bow link 6: the width of the link at $s_2 = 20$ mm and $d_1 = 24$ mm:

$$b_1' = \frac{P_b'}{4s_2\,\sigma_{allow}} + d_1 = \frac{13 \times 10^4}{4 \times 0.02 \times 180 \times 10^6} + 0.024$$

$$= 0.033 \text{ m}$$

Assume $b_1 = 36$ mm. Design of clamp 7: the bearing stress in the hole for a spindle is

$$\sigma = \frac{P_b'}{2d(2s_1)} = \frac{13 \times 10^4}{2 \times 0.034 \times 2 \times 0.008} = 135 \times 10^6 \text{ N/m}^2$$

Design of bracket 8 and the bolt: the bending moment at $\ell_2 = 26$ mm:

$$M = \frac{P_b'\ell_2}{4} = \frac{13 \times 10^4 \times 0.026}{4} = 845 \text{ N-m}$$

The design cross-sectional modulus at the manhole wall is

$$W' = \frac{M}{\sigma_{b,allow}} = \frac{845}{146 \times 10^6} = 5.8 \times 10^{-6} \text{ m}^3$$

The height of the bracket at $s_1 = 8$ mm is

$$h' = \sqrt{\frac{6W'}{2s_1}} = \sqrt{\frac{6 \times 5.8 \times 10^{-6}}{2 \times 0.008}} = 46.5 \times 10^{-3} \text{ m}$$

Assume h = 50 mm. The design of cover 1 is

$$\beta = \frac{L_n}{2B_n} = \frac{0.62}{2 \times 0.42} = 0.74$$

$$k = \frac{0.914}{1 + 1.6\beta^5} - 0.6 = \frac{0.914}{1 + 1.6(0.74)^5} - 0.6 = 0.074$$

The design height of the cover at $b_2 = 0.2$ m from Equation 13.13 is

$$h_1' = 0.616 \sqrt{\frac{P_b'}{2\sigma_{b,allow}} \left[1 + k + \ln \frac{(1 + \tilde{\mu})0.5 L_n}{0.5(b_2 + \ell_1 + S_1)}\right]} + C$$

$$= 0.616 \sqrt{\frac{13 \times 10^4}{2 \times 146 \times 10^6} \left[1 + 0.074 + \ln \frac{(1 + 0.3)0.5 \times 0.62}{0.5(0.2 + 0.036 + 0.008)}\right]} + 0.002$$

$$= 21.6 \times 10^{-3} \text{ m}$$

Assume $h_1 = 22.0$ mm.

13.3 BAYONET COVERS

Bayonet covers, as well as bow covers, are used in cases in which it is necessary for quick and frequent access to equipment internals. By contrast to bow covers, bayonet covers are fabricated only in round shapes and may be used for considerably higher pressures at inside diameters up to 3000 m.

Bayonet covers with special rubber packing operating under pressures $p \leqslant 1.6$ MN/m^2 or vacuum at normal and elevated temperatures of fluid are used in a variety of process equipment. Examples of bayonet connections for such covers are given in Figure 13.9. The design and size of rubber packing rings in these connections are given in Table 13.2. The obturation is achieved by pressing down the packing ring, where water, air or steam in the cavity is under pressures of 0.05–0.1 MN/m^2 higher than the operating pressure in the apparatus.

The relative rotation of parts with teeth around a common axis for locking (or unlocking) the bayonet connection of the cover is made by appropriate devices. This is illustrated in Figure 13.10.

The design of bayonet covers is similar to that of flanged covers. The elements of a bayonet connection are subjected to the load of the fluid pressure and are structurally designed by formulas for the strength of materials.

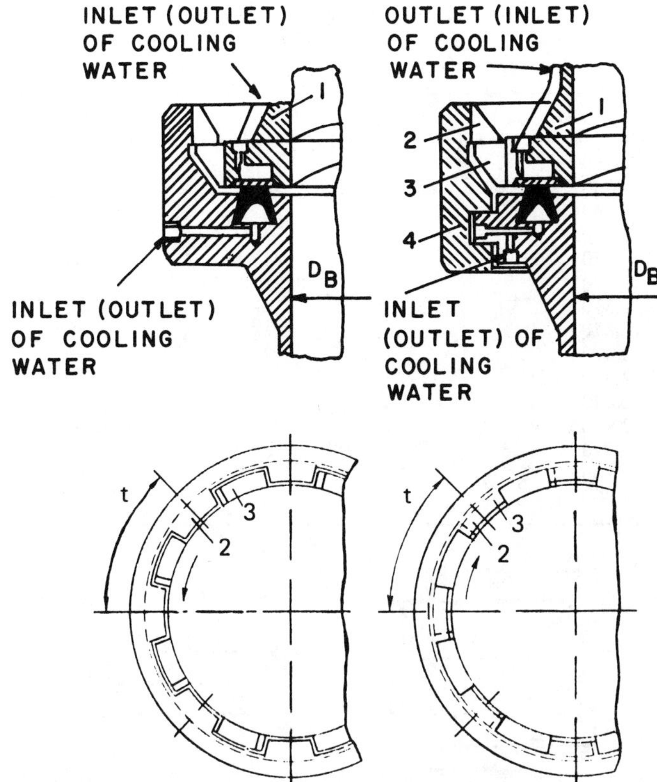

Figure 13.9. Examples of bayonet connections: I-with turning the cover; II-with turning of bayonet ring. 1-cover; 2-internal tooth; 3-external tooth; 4-bayonet ring (from two holes).

NOMENCLATURE

B_n	length (m)
C	corrosion allowance (m)
D	diameter (m)
D'	outer diameter of cover (m)
D_b	diameter of bolt circle (m)
d_b	bolt diameter (m)
h'	design height of sealing place (m)
K	coefficient in Equation 13.3

Table 13.2. Rubber Packing Rings for Bayonet Covers

D_a	D	H	B	h	b	R	P (bar)	Weight (kg)
		mm						
400	460						1,500	0.6
600	660	25	25	13	15	2	2,130	0.85
800	860						2,760	1.1
1,000	1,060						3,390	1.35
1,200	1,270						4,070	4.5
1,400	1,470						4,700	5.2
1,600	1,670	30	35	15	24	4	5,320	5.9
1,800	1,870						5,950	6.6
2,000	2,070						6,570	7.3
2,200	2,270						7,210	8.5
2,400	2,470						7,840	9.3
2,600	2,670						8,460	10.0
2,800	2,870	35	40	18	26	6	9.090	10.8
3,200	3,300						10,440	12.4
3,600	3,700						11,700	13.9

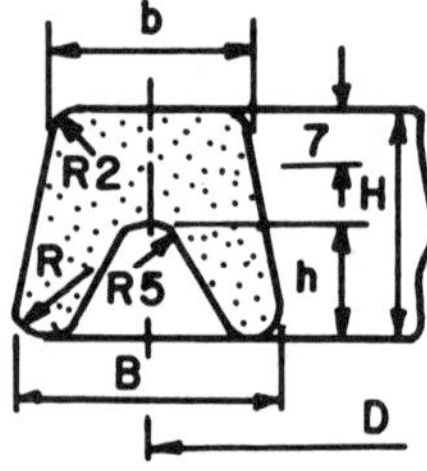

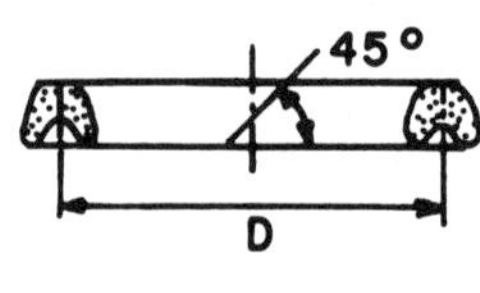

K_2	parameter in Figure 13.3
k	coefficient dependent on cover design and ratio of its sides (see Equation 13.9)
p	pressure (N/m^2)
P'_b	design tensile force on bolts (N)
P_c	critical pressure (N/m^2)
M	bending moment (N-m)
W'	cross-sectional modulus (m^3)
Z', Z	design or assumed number of bolts
β	ratio of average sealing to diameter of circle where P'_b acts
η	safety factor
μ	Poisson's ratio
$\sigma_{b,allow}$	allowable bolt stress (N/m^2)
$\tau_{sh,allow}$	yield point in shell (N-m)
ϕ'	allowable angle of flange distortion in circular direction (°)
ψ	coefficient accounting for weakening by openings

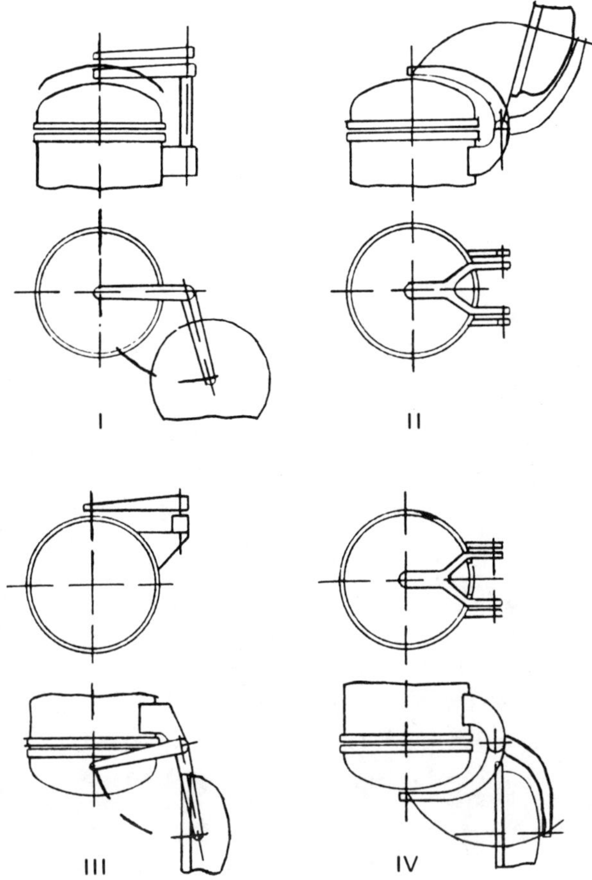

Figure 13.10. Schemes of devices for locking and unlocking bayonet covers: I, II-for top covers; III, IV-for side covers.

REFERENCES

1. Timoshenko, S. *Theory of Plates and Shells* (New York: McGraw-Hill Book Co., 1940).
2. *Boiler and Pressure Vessels Code,* Section 8, Division 1 (New York: American Society of Mechanical Engineers, 1977).
3. "Spravochnik mashinostroitelia," *Mashgiz, T3* (1955).
4. Kantorovich, Z. B. "Osovy rascheta khimicheskikh mashin i apparatov," *Mashgiz* (1960).

CHAPTER 14

DESIGN OF SUPPORTS

14.1 INTRODUCTION

Installation of process equipment on foundations or on special supporting structures is done with supports. Apparatuses with flat bottoms, operating during filling conditions, are installed directly onto the foundation. In all cases, the recommended average specific loads, q, on supporting surfaces are given in Table 14.1.

The support structures located between the vessel and the foundation may be divided into two groups: supports for vertical vessels and supports for horizontal vessels. Figure 14.1 gives typical designs of welded supports for vertical vessels.

Support types I–V are located underneath an apparatus, rigidly fixed with it and designated for a cylindrical apparatus. Support types VI–IX are located on the sides of an apparatus, and also rigidly fixed with it and designated for cylindrical and box-type apparatuses.

Support types I–III, VI and VII are unitized for supporting constructions, and support types IV, V, VIII and IX are separate lugs, the number of which should be no less than three for one apparatus. For small-sized apparatuses, in some cases it is possible to install only two lugs. In cast apparatuses the

Table 14.1. Average Values of Specific Loads on Supporting Surfaces

	q (MN/m^2)
Wooden Decking	≤2.0
Stonework	≤0.8
Concrete	≤2.0
Steel and Cast Iron Surface	≤100.0

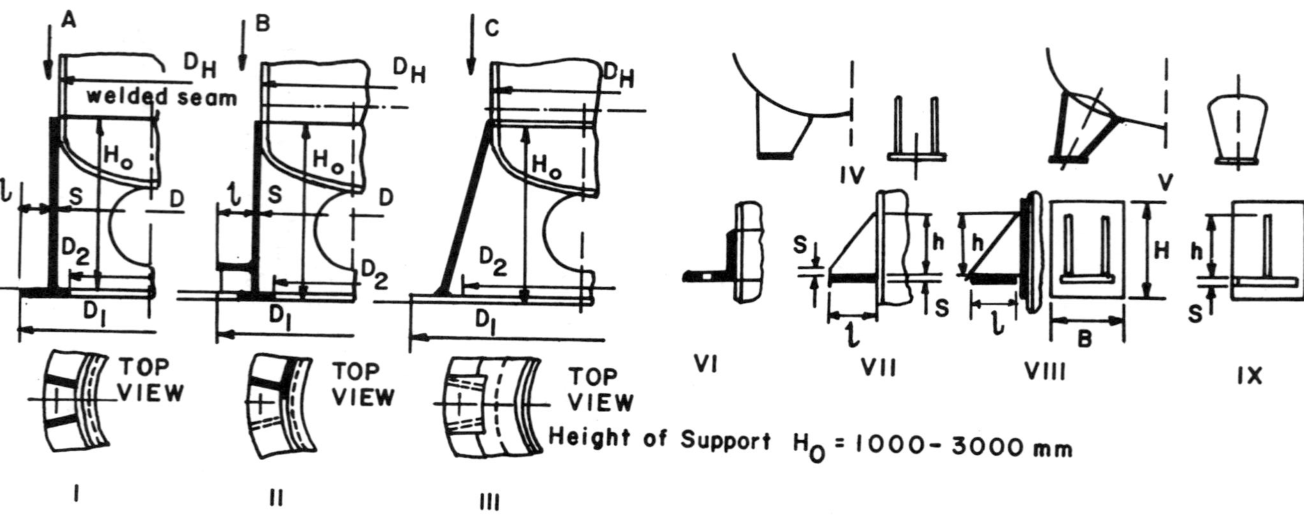

Figure 14.1. Examples of support designs for vertical apparatuses: I–III-for installation of apparatuses outside the building; IV–V-for installation of apparatuses on the floor; VI–IX-for suspended apparatuses.

supports are fabricated integral with a shell or the head. The support design in this case may be similar to types IV–IX.

The size and shape of supports depends basically on the amount and character of loads, material of the apparatus, its weight, as well as its location in space. If the vessel is to be installed on the floor inside a building, then support types IV or V are recommended. If the apparatus is to be suspended on the bearing structures between the intermediate floor, etc., then support types VI–IX are recommended. Skirt supports (types I–III) are used mainly for installation on foundations outside buildings, especially if the ratio of apparatus height to diameter is greater than 5.

If the apparatus is subjected to vibrations and dynamic forces such as machinery, then its supports are fabricated as a massive rigid frame. Its function is not only to transfer the weight of the apparatus on the bearing foundation surface, but with the use of anchor-bolts to involve all the foundation mass in absorbing dynamic forces and vibrations that take place during operation.

An alternative design may involve the use of lugs or brackets (Figure 14.1, types IV–IX) attached to the vessel and resting on columns or beams.

Many horizontal cylindrical apparatuses such as storage vessels, heat exchangers, etc., are commonly supported by saddle supports or cradles. The typical designs of welded supports of such apparatuses are shown in Figure 14.2.

Support types I and II are located underneath apparatuses and may be detachable (left-hand side) and welded to the apparatus (right-hand side). Support types III and IV are located on the sides of an apparatus and are welded to it. The number of support types I and II and doubled support types III and IV for an apparatus may be 2 or more, depending on the length of the apparatus.

Support types I and II (saddles) decrease the bending stresses in the wall of an apparatus as compared to the types III and IV—the lugs. In addition, they insulate the foundation from the apparatus. To decrease the bending stresses in the walls of the apparatus, the angle of the bearing arc of the shell on saddle should be no less than 120°.

Apparatuses used in hot service generally undergo elongation. If thermal elongation is not provided for, the walls may be subjected to thermal stresses:

$$\sigma = E\alpha(t_1 - t_2) \tag{14.1}$$

where E = modulus of elasticity
α = coefficient of thermal expansion
t_1 = temperature of cold apparatus walls
t_2 = temperature of walls at service conditions

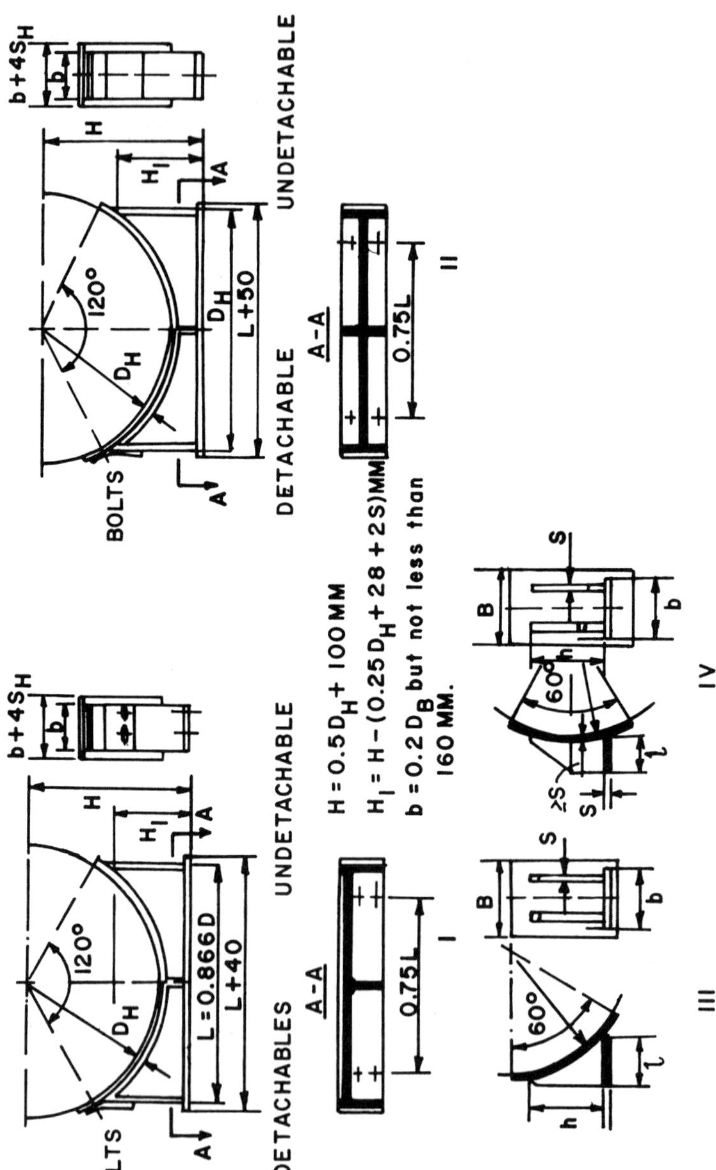

Figure 14.2. Examples of supports for horizontal cylindrical apparatuses: I, II-for installation on saddles; III, IV-for installation on lugs.

In application of detachable supports, one of the end supports of the apparatus is securely fixed while the others are provided with relative displacement of the apparatus along its axis due to thermal elongation. For this reason, oval holes are provided in the lugs on shells with which the apparatus is fastened with bolts to the support (Figure 14.2, element V). In these cases, all the supports are rigidly connected to the foundation. If there are inseparable supports, one of them is rigidly tightened to the foundation, whereas the others must have a free displacement relative to the foundation. This is achieved by inserting the support steel plates underneath, thus permitting the sliding of supports over them. To decrease friction, cylindrical rollers are inserted between the plate and the support (Figure 14.3); the number of rollers depends on the load accepted by the support.

14.2 DESIGN OF SUPPORTS FOR VERTICAL VESSELS

14.2.1 Skirt Supports

Tall vertical vessels usually are supported by skirts. Because a cylindrical skirt has an all-metal area located at a maximum distance (for a given diameter) from the neutral axis, the section modulus, W, is at a maximum and the induced stress for the metal involved is at a minimum. Thus, the cylindrical skirt is an economical design for the support of a tall vertical vessel.

The skirt may be welded directly to a bottom dished head, flush with the shell, or to the outside of the shell. If the skirt is welded flush with the shell,

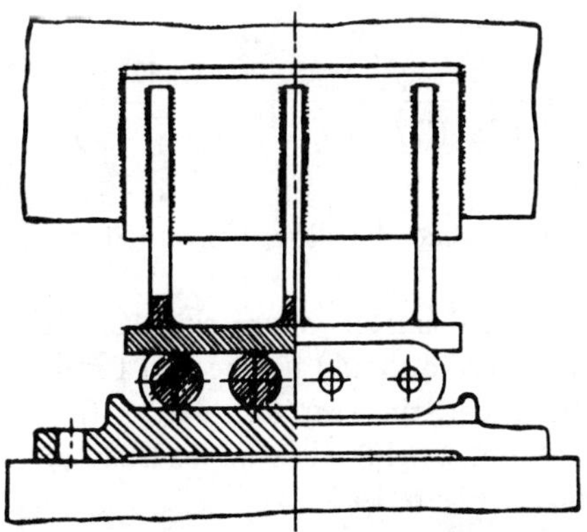

Figure 14.3. Saddle support with rollers.

the weight of the vessel in the absence of wind and seismic loads places the weld in compression. On the other hand, if the skirt is welded to the outside of the vessel, the weld seam is in shear.

The wall thickness of cylindrical skirt types I and II (Figure 14.1) and conical skirt type III (Figure 14.1) is selected from design considerations in the range S = 6 – 12 mm and then checked by the bending stress in the wall of the skirt as follows:

$$\sigma = \frac{GD + 4M_w}{\pi D_i^2 S} \leqslant \sigma_{b,allow}, \quad N/m^2 \tag{14.2}$$

where S = assumed wall thickness of the skirt (m)
G = the maximum load on the support from deadweight under conditions of operation and hydrostatic test
D_i = internal diameter of the support, which is (for support type III assume the lesser diameter) equal to the outside diameter of the shell, D (m)
M_w = wind-load moment (N-m)

In addition, the assumed wall thickness of the skirt should be checked for stability from formula 6.36.

The bottom of the skirt should be anchored securely to the concrete foundation by anchor bolts to prevent overturning from the bending moments induced by wind or seismic loads. The bending moment and the weight of the vertical vessel result in loading conditions on the concrete foundation, as shown in Figure 14.4. These induce a tensile load on the upwind anchor bolts and compressive load on the downwind anchor bolts.

The maximum compression stress on the surface of a bearing ring is

$$\sigma_{max} = \frac{G_{max}}{0.785(D_1^2 - D_2^2)} + \frac{M_w D_1}{0.1(D_1^4 - D_2^4)} \tag{14.3}$$

This stress should not exceed the allowable value for concrete, which is given in Table 14.1.

The outside and inside diameters of the skirt-bearing ring, D_1 and D_2, are taken from Figure 14.1 and should be checked for maximum bearing stress of the foundation under ring surface, σ_{max}, from formula 14.3. The diagrams of stresses from weight and wind loads, as well as a diagram of total stresses [1], are given in Figure 14.5.

If condition 14.3 is not fulfilled, the inside diameter, D_2, of the supporting ring should be decreased correspondingly. It is also recommended to add gusset plates inside the support.

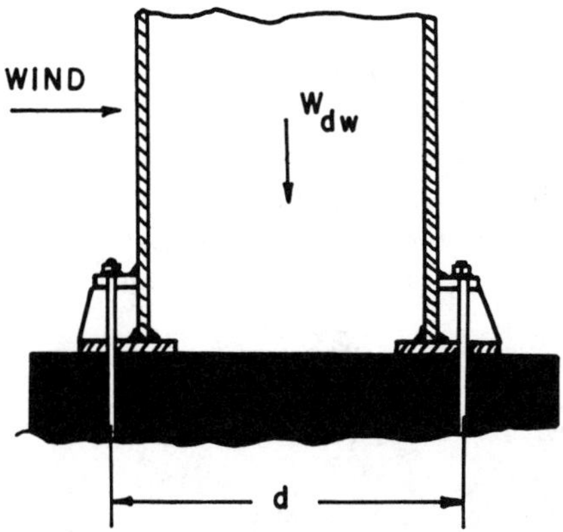

Figure 14.4. Sketch of loading on anchor bolts.

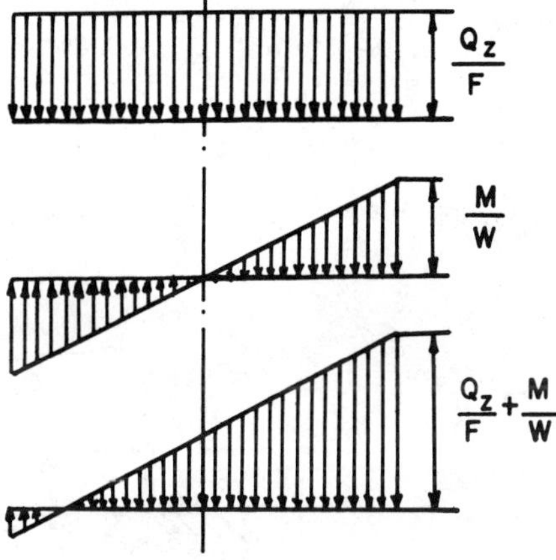

Figure 14.5. Stress diagrams from weight and wind loads.

The thickness of the skirt-bearing ring may be calculated from the following approximate procedure: The thickness of the bearing ring is determined by the compression load on the downwind side of the vertical vessel [2]. The minimum required width of the bearing ring was determined previously from Figure 14.1. The maximum compressive stress between the bearing ring and the concrete occurs at the outer periphery of the bearing ring. As shown in Figure 14.5, the compressive stress varies from a maximum to a lesser value at the junction of the skirt and the bearing ring. The value at the bolt circle may be used for simplicity in evaluating the required thickness of the bearing ring (Figure 14.6).

A bearing ring is assumed to be a uniformly loaded cantilever beam, with $\delta_{c(max.\ induced)}$ being the uniform load. The maximum bending moment for this beam takes place at the junction of the skirt and the bearing ring for a unit circumferential length (b = 1 cm), and is equal to

$$M_{max} = \sigma_c b \ell \left(\frac{\ell}{2}\right) = \frac{\sigma_c \ell^2}{2} \quad \text{(for b = 1)} \tag{14.4}$$

where ℓ = outer radius of bearing ring minus outer radius of skirt (from Figure 14.1)

The maximum stress in an elemental strip of unit width is

$$\sigma_{max} = \frac{M_{max}}{W} = \frac{6M_{max}}{bS_k^2} = \frac{3\sigma_c \ell^2}{S_k^2 \times 1} \tag{14.5}$$

where S_k = bearing-ring thickness

Letting $\sigma_{max} = \sigma_{c.allowable}$ and solving for S_k, we obtain

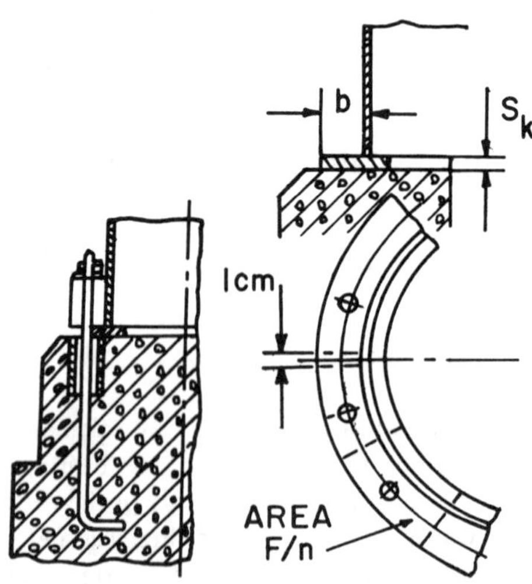

Figure 14.6. Sketch for calculating bearing ring thickness.

$$S_k = \sqrt{\frac{\sigma M_{max}}{\sigma_c}} = 1.73 \sqrt{\frac{\sigma_{max}}{\sigma_c}} \quad (14.6)$$

The thickness of the bearing ring, S_k, as calculated from Equation 14.6 is usually rounded off to the next larger standard thickness of the plate. The computed value of S_k must not be less than 12 mm.

If gussets are applied to stiffen the bearing ring, the loading conditions on the section of the ring between two gussets may be assumed to work similarly to that of a rectangular uniformly loaded plate ring, with two opposite edges simply supported by the gussets, the third edge joined to the shell and the fourth edge free [3].

Timoshenko [1] determined deflections and bending moments for bearing rings with gussets. Some of these values are given in Table 14.2.

Reviewing Table 14.2, the following conclusions become obvious: In the case in which $l/b = 0$ (no gussets or gusset spacing $b = \infty$), the bending moment reduces to Equation 14.4 and the thickness of the bearing ring is obtained by Equation 14.6. When l/b is equal to, or less than, $^3\!/_2$, the maximum bending moment occurs at the junction with the shell because of cantilever action. If l/b is greater than $^3\!/_2$, the maximum bending moment takes place at the middle of the free edge. The bearing ring thickness for rings with gussets is calculated from the bending moments:

$$F_{s(act)} = NF_B \geqslant F_{s(min)} \quad (14.7)$$

where N = number of bolts
$F_{s(act)}$ = actual area of bolting steel used (in.^{2})
$F_{s(min)}$ = minimum area of bolting steel (in.^{2})
F_B = root area of the bolt (in.^{2}) (Table 14.3)

Table 14.2. Maximum Bending Moments in a Bearing Plate with Gussets [1,2]

ℓ/b	$M_x\begin{pmatrix}x = b/2\\ y = \ell\end{pmatrix}$	$M_y\begin{pmatrix}x = b/2\\ y = 0\end{pmatrix}$
0	0	$-0.500\,\sigma_c\ell^2$
½	$0.0293\,\sigma_c b^2$	$-0.319\,\sigma_c\ell^2$
1	$0.0972\,\sigma_c b^2$	$-0.119\,\sigma_c\ell^2$
1½	$0.123\,\sigma_c b^2$	$-0.124\,\sigma_c\ell^2$
2	$0.131\,\sigma_c b^2$	$-0.125\,\sigma_c\ell^2$
4	$0.133\,\sigma_c b^2$	$-0.125\,\sigma_c\ell^2$
∞	$0.133\,\sigma_c b^2$	$-0.125\,\sigma_c\ell^2$

Table 14.3. Bolt Data [4]

Bolt Size (in.^{2})	Standard Thread		8-thread Series		Bolt Spacing[a]		Minimum Radial Distance, R (in.)	Edge Distance, ϵ (in.)	Nut Dimension (across flats) (in.)	Maximum Fillet Radius, r (in.)
	No. of Threads	Root Area	No. of Threads	Root Area	Minimum B_S (in.)	Preferred (in.)				
1/2	13	0.126	No. 8		1 1/4	3	13/16	5/8	7/8	1/4
5/8	11	0.202	thread		1 1/2	3	15/16	3/4	1 1/16	5/16
3/4	10	0.302	series		1 3/4	3	1 1/8	13/16	1 1/4	3/8
7/8	9	0.419	below 1″		2 1/16	3	1 1/4	15/16	1 7/16	3/8
1	8	0.551	8	0.551	2 1/4	3	1 3/8	1 1/16	1 5/8	7/16
1 1/8	7	0.693	8	0.728	2 1/2	3	1 1/2	1 1/8	1 13/16	7/16
1 1/4	7	0.890	8	0.929	2 13/16	3	1 3/4	1 1/4	2	9/16
1 1/2	.6	1.294	8	1.405	3 1/4		2	1 1/2	2 3/8	5/8
1 3/4	5	1.744	8	1.980	3 3/4		2 1/4	1 3/4	2 3/4	5/8
1 7/8	5	2.019	8	2.304	4		2 3/8	1 7/8	2 15/16	5/8
2	4 1/2	2.300	8	2.652	4 1/4		2 1/2	2	3 1/8	11/16
2 1/2	4	3.715	8	4.292	5 1/4		3 1/16	2 3/8	3 7/8	13/16
3	1	5.621	8	6.321	6 1/4		3 5/8	2 7/8	4 5/8	15/16

[a]B_S = center-to-center distance between bolts, inches.

For short vessels in which the calculated thickness of a bearing plate is ½ in. or less, a steel angle rolled to fit the outside of the skirt may be lap-welded, as shown in Figure 14.7.

If the design thickness of the bearing plate is ½–¾ in., a single-ring bearing may be applied, as shown in Figure 14.8. If the designed bearing-plate thickness is ¾ in. or greater for the design in Figure 14.8, a bolting "chair" can be applied, as shown in Figure 14.9. Table 14.4 presents a maximum number of chairs inserted in a skirt of a given diameter [2]. When we check the

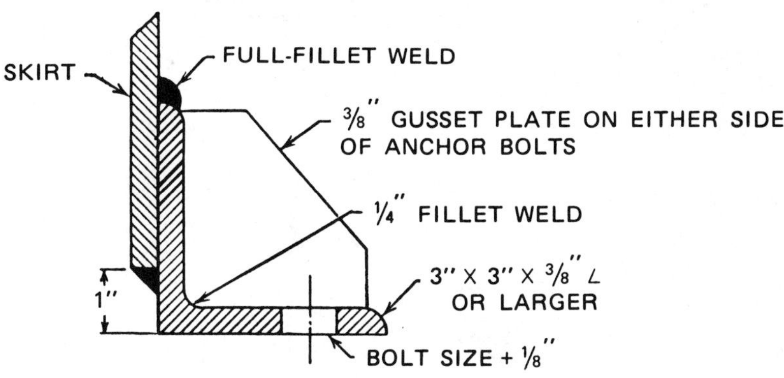

Figure 14.7. Rolled-angle bearing plate.

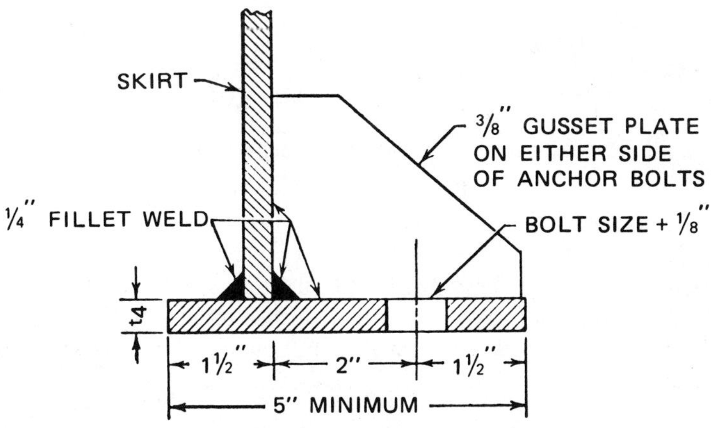

Figure 14.8. Single-ring beam plate with gussets.

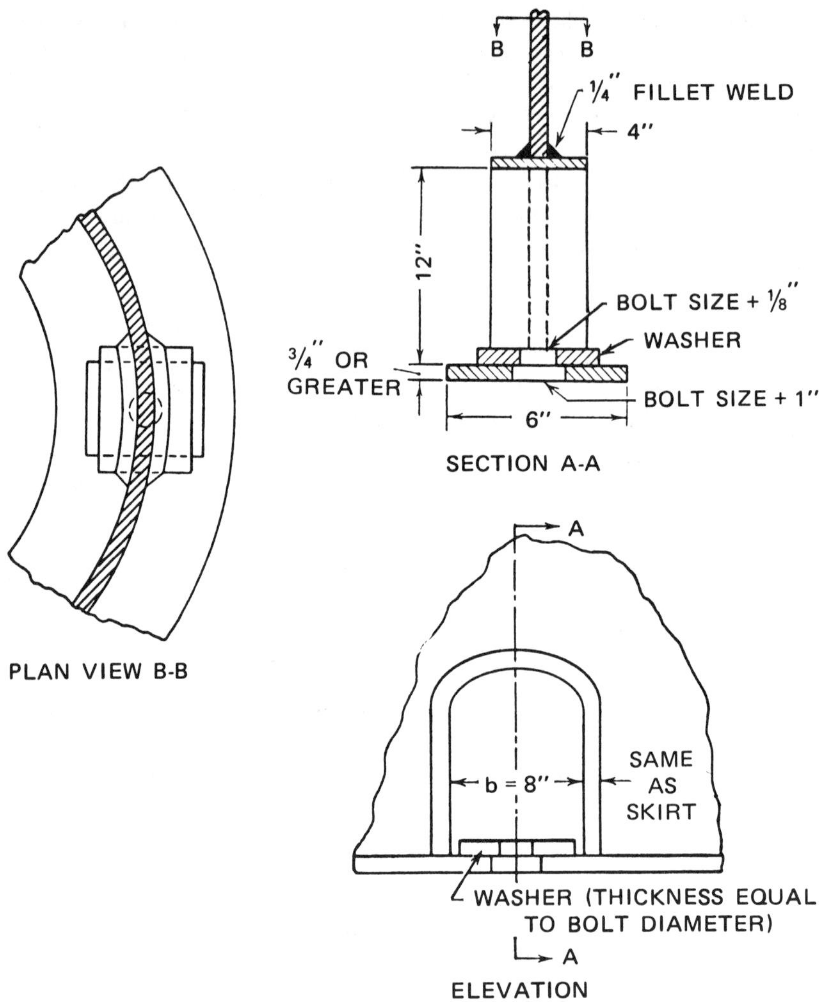

Figure 14.9. Centered anchor-bolt chair.

bearing-plate thickness for a centered chair, the plate inside the stiffeners may be assumed to work as a concentrated loaded beam with fixed ends.

The concentrated load, P, is equal to maximum bolting stress times the bolting area:

$$P = \sigma_s \times F_b \tag{14.8}$$

Table 14.4. Maximum Number of Centered Chairs in Various-Size Vessel Skirts [2]

Skirt Diameter (ft)	No. of Chairs
3	4
4	8
5	8
6	12
7	16
8	16
9	20
10	24

where σ_s = maximum induced stress, in bolts
F_b = root area of anchor bolts
P = maximum bolt load

The maximum bending moment in the bearing plate inside the chair takes place at the upwind dead center, located at or near the bolt where the cross-sectional area is minimum and the moment is

$$M_{max} = \frac{P \times b}{8} \tag{14.9}$$

where M_{max} = maximum bending moment
b = spacing inside chairs

Considering that the hole in the bearing plate reduces the effective beam width of the plate, the designed bearing-plate thickness inside the chair will be

$$S_{ch} = \sqrt{\frac{6M_{max}}{(S_w - d_h)\sigma_{allow}}} \tag{14.10}$$

where S_{ch} = bearing-plate thickness
S_w = bearing-plate width
d_h = bolt-hole diameter in bearing plate
σ_{allow} = allowable stress

The bending moment in the bearing plate between chairs will be controlling and can be calculated from Table 14.2. The thickness of the bearing plate outside the stiffeners may be evaluated from Equation 14.6.

If the design number exceeds values given in Table 14.4, external bolting chairs may be used (Figure 14.10). The proportions for the chair are given in detail b of Figure 14.10.

The maximum load on the compression ring at the top of an external chair takes place on the upwind side, where the reaction of the bolts produces a compression load. The compression ring is assumed to work as a rectangular plate bounded by the two gusset plates, the skirt and the outside of the plates. The bolt load is assumed to be distributed uniformly over a circular area equal to the bolt area [3]. The compression ring welded to skirt and gusset plates (Figure 14.11) provides additional rigidity, compensating the lack of support on the fourth side.

Timoshenko [3] hypothesized that the plate works as a sheet, freely supported on four sides with a concentrated load acting as a uniformly distributed load over a circular area of radius, ℓ. In reference to the "plan" view of Figure 14.10, with y in radial direction and x in the circumferential direction, the maximum bending moments, M_y and M_x, are determined by the following equations:

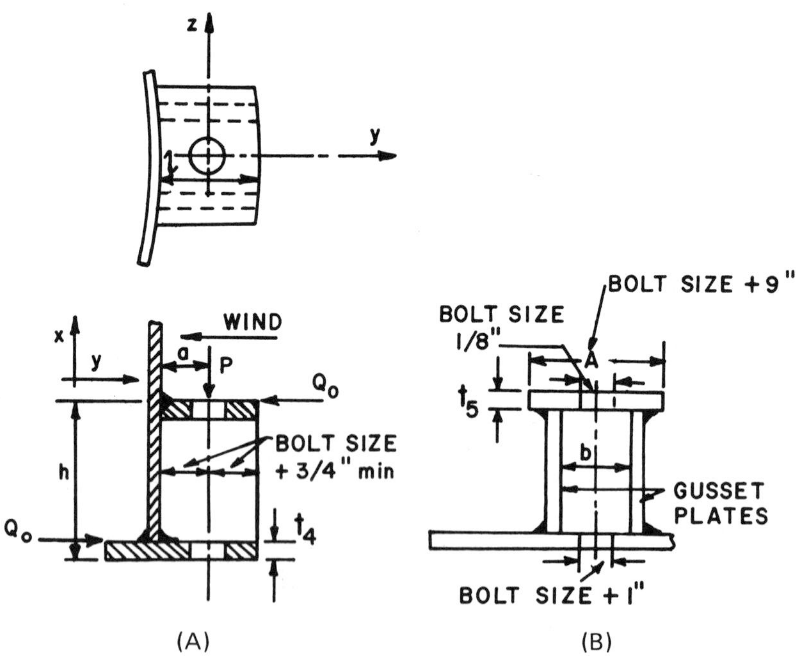

Figure 14.10. External bolting chair.

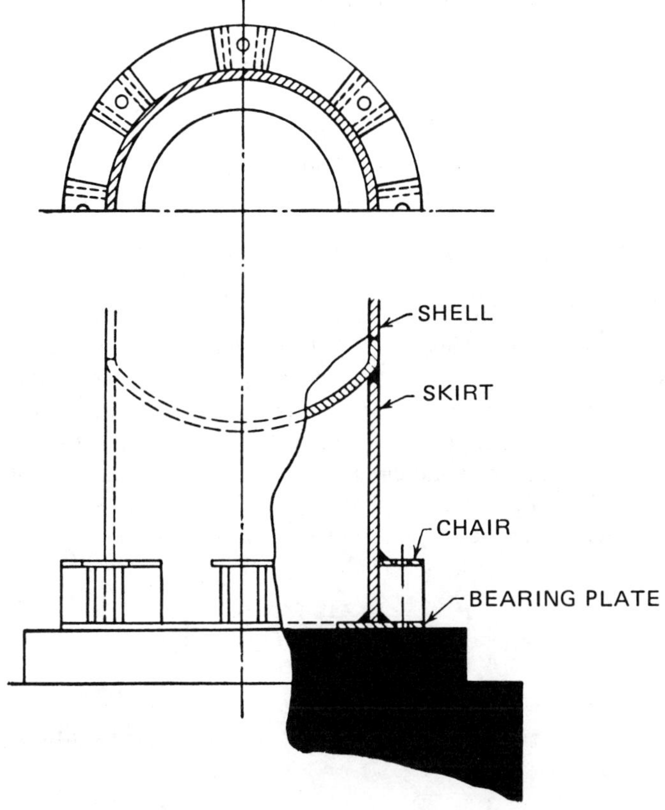

Figure 14.11. Vessel skirt with external bolting chairs.

$$M_y = \frac{P}{4\pi}\left[(1+\tilde{\mu})\ell n\left(\frac{2\ell \sin\left(\frac{\pi a}{\ell}\right)}{\pi e}\right)+1\right]-\left[\frac{\gamma_1 P}{4\pi}\right] \tag{14.11}$$

$$M_x = \frac{P}{4\pi}\left[(1+\tilde{\mu})\ell n\left(\frac{2\ell \sin\left(\frac{\pi a}{\ell}\right)}{\pi e}\right)+1\right]-\left[(1-\tilde{\mu}-\gamma_2)\frac{P}{4\pi}\right] \tag{14.12}$$

where M_y = maximum bending moment along radial axis
M_x = maximum bending moment along circumferential axis
P = maximum bolt load on upwind side

$\tilde{\mu}$ = Poisson's ratio (0.30 for steel)
a = radial distance from outside of skirt to bolt circle
ℓ = radial distance from outside the skirt to outer edge of compression plate
e = radius of action of concentrated load = one-half the distance across flats of bolting nut
γ_1, γ_2 = constants given in Table 14.5

In using Table 14.5 for b/ℓ values less than 1.0, invert b/ℓ and rotate the axes 90°.

If we compare Equations 14.11 and 14.12 for the case in which b = unity using Table 14.5, we obtain $M_x = M_y$. For all cases in which $b/\ell > 1$, $M_y > M_x$ and, consequently, M_y is controlling. After calculating the size of the bolt and selecting the width of the bearing plate, bolt-circle diameter and gusset spacing, the dimensions, a, b, e and 1 are fixed. Thus, the maximum bending moments in radial and circumferential directions may be calculated using formulas 14.11 and 4.12.

If a = $\ell/2$ and M_y is controlling, formula 14.11 reduces to

$$M_y = \frac{P}{4\pi}\left[(1+\tilde{\mu})\,\ell n\,\frac{2\ell}{\pi e} + (1-\gamma_1)\right] \tag{14.13}$$

The maximum stress in the compression ring as a strip of unit width is

$$\sigma_{max} = \frac{6M_y}{S_{cr}^2} \tag{14.14}$$

And if σ_x is assumed to be σ_{allow},

$$S_{cr} = \sqrt{\frac{6M_y}{\sigma_{allow}}} \tag{14.15}$$

A continuous ring usually is preferred in cases in which the spacing of external chairs becomes so small that the compression plates approach a continuous ring [2]. This is shown in Figure 14.12.

The maximum load on the continuous compression ring takes place on the upwind side of the vessel similar to the compression plate. This load induces a bending stress in the compression ring, and the vertical gussets transfer this stress to the bearing plate. It is assumed that each section of the ring between

Table 14.5. Constants for Moment Calculation in Compression Ring [3]

γ_1	γ_2	b/ℓ
0.565	0.135	1.0
0.211	0.085	1.4
0.125	0.057	1.6
0.073	0.037	1.8
0.042	0.023	2.0
0	0	∞

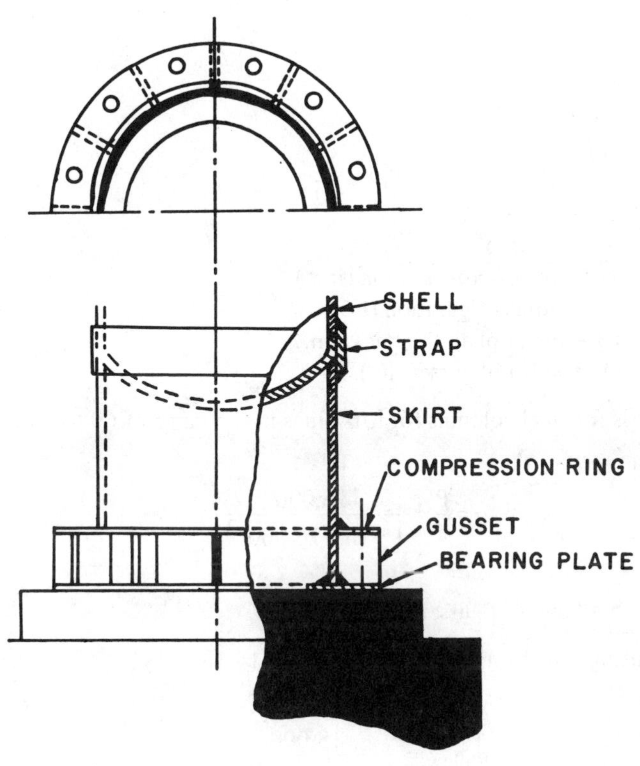

Figure 14.12. Skirt with continuous compression ring and strap.

the gussets works as a rectangular plate bounded by the two gusset plates—the shell and the outer ring. The bolt load is assumed to be a uniformly distributed load acting over the area of the bolt.

Consequently, in this case it is possible to use the approach for determining the thickness of compression plates, that is, to apply Equations 14.11, 14.12 or 14.13 and Table 14.5 and Equation 14.15.

In developing the equation for determining the thickness of gusset plates for compression rings, it is assumed that the gussets are evenly spaced alternately between bolts. Also, in this case the gusset plates may be considered a vertical column.

The moment of inertia of the gusset about the axis having the least radius of gyration [2] is

$$I = \frac{\ell S_2^3}{12} = ar^2 = \ell S_2 r^2 \tag{14.16}$$

or

$$r^2 = \frac{S_2^2}{12} \tag{14.17}$$

where a = area of cross section (in.^{2})
r = radius of gyration (in.)
S_2 = gusset-plate thickness (in.)
ℓ = width of gusset (in.)

The stress for steel columns for h/r values in the range of 60 to 200 is

$$\sigma = \frac{P}{a} = \frac{18{,}000}{1 + (h^2/18{,}000 r^2)} \tag{14.18}$$

where h = height of gusset (in.)

Substituting Equation 14.17 into Equation 14.18 gives

$$\sigma_{allow} = \frac{18{,}000}{1 + (h^2/1500\, S_2^2)} \tag{14.19}$$

The allowable stress, σ_{allow}, in Equation 14.19 must be

$$\sigma_{allow} = \frac{\text{Bolt Load}}{2\ell S_2} = \frac{P}{a} \tag{14.20}$$

Substituting Equation 14.20 into Equation 14.19, we obtain

$$\frac{\text{Bolt Load}}{2\ell S_2} = \frac{18{,}000}{1 + (h^2/1500 S_2)}$$

or

$$36{,}000\ell S_2^2 - (\text{Bolt Load})S_2 - \frac{h^2(\text{bolt load})}{1500} = 0 \qquad (14.21)$$

Solving Equation 14.21 for S_2 gives

$$S_2 = \text{Bolt Load}\left[\frac{1 + \sqrt{1 + (96\ell h^2/\text{bolt load})}}{72{,}000\ell}\right] \qquad (14.22)$$

The use of external bolting chairs or a compensation ring results in a loading condition that produces a reaction in the skirt. The design thickness of the skirt to resist this reaction is determined from the following formula [2]:

$$S_1' = 1.76\left(\frac{Pa}{mh\,\sigma_{allow}}\right)^{2/3} r^{1/3} \qquad (14.23)$$

where S_1' = skirt thickness required to resist reaction of external chairs or compression ring (in.)
r = radius of skirt (in.)
m = 2A (see Figure 14.10) or 2bs (two-bolt spacings)
P = maximum bolt load (lb)
a = radial distance from outside of skirt to bolt circle (in.)
h = gusset height (in.)

Vertical vessels that have to be installed outside the building should be checked for wind stability. The wind-load moment for vessels with height $H \leqslant 20$ is determined from the following formula [5]:

$$M_w = 0.5\, K_1 K_2 q_w H^2 D_0,\ \text{N-m} \qquad (14.24)$$

where H = height of the vessel (m)
D_0 = outside diameter of the vessel (m)
K_1 = aerodynamic coefficient: for cylindrical vessels, $K_1 = 0.7$; for box-type vessels, $K_1 = 1.4$
K_2 = dynamic coefficient (considered only for vessels with $H/D \geqslant 5$; in this case, $K_2 = 2$)
q_w = specific wind load; the recommended value is $10^3\ N/m^2$

The design of anchor-bolts for supports should be based on an empty apparatus, that is, the most unfavorable case for overturning. The minimum foundation-bearing stress under supporting surface σ_{min} is determined from the following formula:

$$\sigma_{min} = \frac{G_1}{0.785(D_1^2 - D_2^2)} - \frac{M_w D_1}{0.1(D_1^2 - D_2^4)} \tag{14.25}$$

where G_1 = deadweight of the empty vessel (N)

If σ_{min} is negative, then the apparatus is unstable and it becomes necessary to insert anchor-bolts. The total conditional design load on the anchor-bolts, P_b', assuming the case as a maximum load, is determined approximately from the following formula:

$$P_b' = 0.785(D_1^2 - D_2^2)\sigma_{min} \tag{14.26}$$

The design number of bolts, Z', is determined from the following formula:

$$Z' \geqslant \frac{1.2P_b'}{KF_b\sigma_y^t\eta} \tag{14.27}$$

where F_b = cross-sectional area of assumed bolt by the inside diameter of thread

K is a coefficient given in Table 14.6.

If σ_{min} is greater than zero, the apparatus is stable. In this case, it is necessary to determine the stability coefficient, K_{st}, from the following formula:

$$K_{st} = 0.42\frac{G_1 D}{M_w} \tag{14.28}$$

Table 14.6. Values of Coefficient K in Equation 14.27

	d_b (mm)							
	≤10	12	14	16	18	20	22	24
K	0.3	0.4	0.5	0.6	0.7	0.8	0.9	1.0

At $K_{st} \geqslant 1.5$ there is no need to insert anchor-bolts. At $K_{st} < 1.5$ the number of bolts is taken to be four or eight, depending on the diameter of the apparatus.

Example 14.1

From the following data, determine the main sizes of the support of type II (Figure 14.1) for a vertical cylindrical apparatus of carbon steel that is to be installed outside a building: the outside diameter of the apparatus D = 1.02 m; the apparatus height H = 12 m; the weight of the apparatus with liquid $G = 12 \times 10^4$ N; the weight of the empty apparatus $G_1 = 3 \times 10^4$ N; the height of the support $H_0 = 1$ m; the material of the support is steel ($\sigma_{b.allow} = 146 \times 10^6$ N/m^2; $\sigma_y = 240 \times 10^6$ N/m^2; $E = 205 \times 10^9$ N/m^2); the support rests on a concrete foundation.

Solution

The wind overturning moment at $K_1 = 0.7$; $K_2 = 2$ and $q_w = 10^3$ N/m^2 is determined from formula 14.24:

$$M_w = 0.5 \times K_1 \times K_2 \times q_w \times H^2 D_0 = 0.5 \times 0.7 \times 2 \times 10^3 \times 12^2 \times 1.02$$

$$= 10.3 \times 10^4 \text{ N M}$$

Assume a skirt wall thickness S = 8 mm, and diameter 1.02 m. The bending stress in the skirt wall is checked from formula 14.2:

$$\sigma = \frac{GD + 4M_w}{\pi D^2 S} = \frac{12 \times 10^4 \times 1.02 + 4 \times 10.3 \times 10^4}{\pi (1.02)^2 \times 0.008}$$

$$= 20.5 \times 10^6 \text{ N/m}^2 < \sigma_{b.allow} = 146 \times 10^6 \text{ N/m}^2$$

Checking the stability of the skirt,

For

$$\frac{D}{S} = \frac{1.02}{0.008} = 127.5, \text{ from Figure 6.12}$$

We have

$$K_1 = 1.6 \;;\quad K_2 = 6.2$$

The coefficient K is determined from formula 6.37:

$$K = \frac{2\,M_w}{0.25 \times G \times D + M_w} = \frac{2 \times 10.3 \times 10^4}{0.25 \times 12 \times 10^4 \times 1.02 + 10.3 \times 10^4}$$

$$= 1.54$$

The condition of stability is checked from formula 6.36:

$$\sqrt{\frac{1}{E}\left[K_1\left(1 - \frac{K}{2}\right) + 0.125\ Kk_2\right]\left(G + 4\,\frac{M_w}{D}\right)}$$

$$= \sqrt{\frac{1}{205 \times 10^9}\left[1.6\left(1 - \frac{1.54}{2}\right) + 0.125 \times 1.54 \times 6.2\right]\left(12 \times 10^4 + 4\,\frac{10.3 \times 10^4}{1.02}\right)}$$

$$= 0.002 \text{ m} < S = 0.008 \text{ m}$$

Hence, the condition of stability is fulfilled. From Figure 14.1 we determine

$$D_1 = D + 2s + 2\ell = 1.02 + 2 \times 0.008 + 2 \times 0.07 \approx 1.18 \text{ m}$$

The inside diameter of the bearing ring is

$$D_2 = D - 2 \times 0.03 = 1.02 - 2 \times 0.03 = 0.96 \text{ m}$$

The maximum bearing stress of the foundation under the supporting surface is determined from formula 14.3:

$$\sigma_{max} = \frac{G}{0.785(D_1^2 - D_2^2)} + \frac{M_w D_1}{0.1(D_1^4 - D_2^4)}$$

$$= \frac{12 \times 10^4}{0.785(1.18^2 - 0.96^2)} + \frac{10.3 \times 10^4 \times 1.18}{0.1(1.18^4 - 0.96^4)}$$

$$= 1.43 \times 10^6 \text{ N/m}^2 < q_{allow} = 2.2 \times 10^6 \text{ N/m}^2$$

The design thickness of the bearing ring, $\ell = 0.07$ m, is determined from formula 14.6:

$$S_k' = 1.73\ell \sqrt{\frac{\sigma_{max}}{\sigma_{b.allow}}} = 1.73 \times 0.07 \sqrt{\frac{1.43 \times 10^6}{146 \times 10^6}} = 0.012 \text{ m}$$

Assume $S_k = 12$ mm. The minimum foundation bearing stress under the support surface is determined from formula 14.25:

$$\sigma_{min} = \frac{G_1}{0.785(D_1^2 - D_2^2)} - \frac{M_w D_1}{0.1(D_1^4 - D_2^4)}$$

$$= \frac{3 \times 10^4}{0.785(1.18^2 - 0.96^2)} - \frac{10.3 \times 10^4 \times 1.18}{0.1(1.18^4 - 0.96^4)} = -1.04 \times 10^6 \text{ N/m}^2$$

that is, the apparatus is unstable and it is necessary to fix it with anchor bolts.

The total conditional design load on the anchor bolts is determined from formula 14.26:

$$P_b' = 0.785(D_1^2 - D_2^2)\sigma_{min} = 0.785(1.18^2 - 0.96^2)\,1.04 \times 10^6$$

$$= 38.4 \times 10^4 \text{ N/m}^2$$

The design number of anchor bolts is determined from formula 14.27 at $K = 1$; $F_b = 317 \times 10^{-6}$ m^2 ($d_b = 24$ mm); $\eta = 1$:

$$Z' = \frac{1.2 P_b'}{K F_b \sigma_y^t \eta} = \frac{1.2 \times 38.4 \times 10^4}{1 \times 317 \times 10^{-6} \times 240 \times 10^6 \times 4} = 6.05$$

Assume $Z = 8$.

Example 14.2 [2]

An external chair will be designed for a column 8 ft, 0 in. in diameter having 12 bolts 1½ in. in diameter with a calculated induced stress of 17,500 psi. The bolt circle diameter is 8 ft, 6 in., and the outside diameter of the bearing plate is 9 ft, 0 in. The gusset height is 12 in. Determine the thickness of the compression plate.

Solution

From Figure 14.10,

$$A = 9 \text{ in.} + (1\tfrac{1}{2}) = 10\tfrac{1}{2} \text{ in.}$$

$$b = 8 \text{ in.} + (1\tfrac{1}{2}) = 9\tfrac{1}{2} \text{ in.}$$

By Table 14.3, the root area of bolt is

$$A_b = 1.405 \text{ in.}^2$$

The bolt load by Equation 14.8 is

$$P = \sigma_s F_b = (17{,}500)\,(1.405) = 24{,}600 \text{ lb}$$

From Figure 14.10,

$$a = \frac{(8 \text{ ft. } 6 \text{ in.}) - (8 \text{ ft. } 0 \text{ in.})}{2} = 3 \text{ in.}$$

$$a' = \frac{(9 \text{ ft. } 0 \text{ in.}) - (8 \text{ ft. } 0 \text{ in.})}{2} = 6 \text{ in.}$$

From Table 14.3.

$$\ell = \frac{\text{nut dimension across flats}}{2} = \frac{2.375}{2} = 1.188 \text{ in.}$$

The compression-plate thickness is

$$\frac{b}{\ell} = \frac{9.5}{6} = 1.58$$

Interpolation from Table 14.5 gives

$$\gamma_1 = 0.134$$

Substituting into Equation 14.13 gives

$$M_y = \frac{P}{4\pi}\left[(1+\tilde{\mu})\,\ell n\,\frac{2\ell}{e\pi} + 1 - \gamma_1\right]$$

$$= \frac{24{,}600}{4\pi}\left[1.3\,\ell n\left(\frac{2 \times 6}{\pi \times 1.188}\right) + 1 - 0.134\right] = 4667 \text{ in.-lb}$$

Substituting into Equation 14.15 with $\sigma_{allow} = 17{,}500$ psi gives

$$t_s = \sqrt{\frac{6M_y}{\sigma_{allow}}} = \sqrt{\frac{6 \times 4667}{17{,}500}} = 1.265$$

Therefore, use 1.5-in. plate for the compression plate.

Example 14.3

A proposed fractionation column is 10 ft in diameter and 150 ft high and rests on a foundation of 3000 psi-strength concrete. The proposed bearing plate under the skirt extends 6½ in. out from the skirt with a bolt circle of 3¼ in. Outside the skirt, 24 bolts, 2½ in. in diameter, are to be used with a continuous compression ring. The gusset height is 12 in. Determine the required thickness of the skirt to resist the reaction of the bolting ring. The maximum induced stress in the bolts is 17,450 psi, and the maximum allowable stress in the skirt is 20,000 psi.

Solution (Figure 14.10)

$$a = 3\tfrac{1}{4} \text{ in.}$$

$$m = \frac{2\pi(126.5)}{24} = 33.2 \text{ in.}$$

$$\sigma_{allow} = 20{,}000 \text{ psi}$$

$$h = 12 \text{ in.}$$

$$P = 17{,}450 \times 3.72 \text{ in.}^2/\text{bolt} \; ; \quad r = 60 \text{ in.}$$

By Equation 14.23,

$$t = 1.76\left(\frac{Pa}{mh\,\sigma_{allow}}\right)^{2/3} r^{1/3}$$

$$= 1.76\left[\frac{17{,}450 \times 3.72 \times 3.25}{33.2 \times 12 \times 20{,}000}\right]^{2/3} (60)^{1/3}$$

$$= 0.607 \text{ in. or } 5/8 \text{ in.}$$

14.2.2 Lug Supports

Brackets, or lugs (Figure 14.1, types VI–IX), offer many advantages over other types of supports. They can absorb diametrical expansions, are easily attached to the apparatus and easily leveled or shimmed in the field.

Due to eccentricity of a lug support, compressive, tensile and shear stresses are induced in the wall of the apparatus. The tensile and compressive stresses are combined with pressure stresses circumferentially and longitudinally. The forces induced by these shear stresses are relatively small and can be neglected [6].

Lug supports are very favorable for thick-walled vessels because the thick wall has a considerable moment of inertia and, therefore, is capable of absorbing flexural stresses due to the eccentricity of the loads.

The eccentricity of loads applied to thin-walled vessels is not convenient unless the proper reinforcements are used and sufficiently tightened to the vessel (Figure 14.13).

If a vessel with lugs is located outside, the wind load as well as the dead-weight load should be taken into account in determining P. Because lug-

Figure 14.13. Insertion of reinforcing pad between the lug and thin-walled vessel.

supported vessels are usually smaller in height than their skirt-supported vessels, wind loads may be a minor consideration.

The highest compressive stresses in the supports occur on the leeward side when the vessel is full (because dead load and wind load are additive). The highest tensile stresses are induced on the windward side when the vessel is empty. The dead load in this case is subtracted from the wind load. As a result, stresses on the leeward side are the determining factors in the design of supports. The maximum total compression load is

$$P = \frac{4 P_w (H - L)}{n D_{bc}} + \frac{\Sigma W}{n} \tag{14.29}$$

where P_w = total wind load on exposed surfaces
H = height of vessel above foundation
D_{bc} = diameter of anchor-bolt circle
n = number of supports
ΣW = weight of empty vessel plus weight of liquid and other dead load

The lugs that are of essentially the same design as that shown in Figure 14.10 for external chairs are shown in Figure 14.14. These lugs use to their advantage the axial stiffness and strength of the cylindrical shell to absorb the bending stresses from the supports. The bottom and top plates have continuous welded seams because the maximum compressive and tensile stress takes place in these two plates, respectively.

These seams and the intermittent welds of the vertical gussets to the shell carry the vertical shear load. The load P (Figure 14.15) on the column has a lever arm "a," and this moment is resisted by the couple Q_0h (Figure 14.14) acting at the center lines of the top and bottom horizontal stiffening plates; the reaction is established in the shell, giving rise to axial and circumferential bending stresses, as in the case of bolting loads on external chairs [7].

The axial stress [2] is defined by

$$\sigma_{axial} = \frac{6M_0}{t^2} = \frac{-\beta^3 P a r^2}{(1 - \tilde{\mu}^2) m h} \tag{14.30}$$

where P = maximum total compression load (from Equation 14.29)
r = radius of gyration
$\tilde{\mu}$ = Poisson's ratio
m = effective arc width, taken as equal to 2A (Figure 14.14) with a and h determined from Figure 14.14:

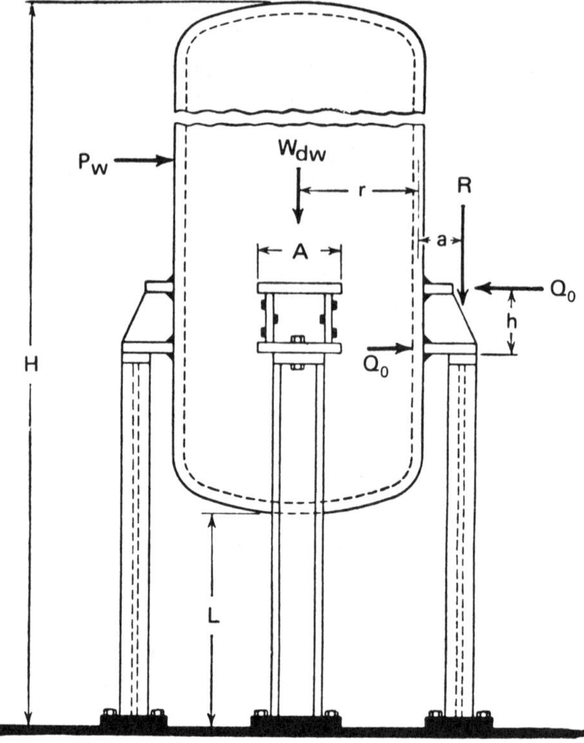

Figure 14.14. Sketch of vessel on four-lug supports with horizontal-plate stiffeners.

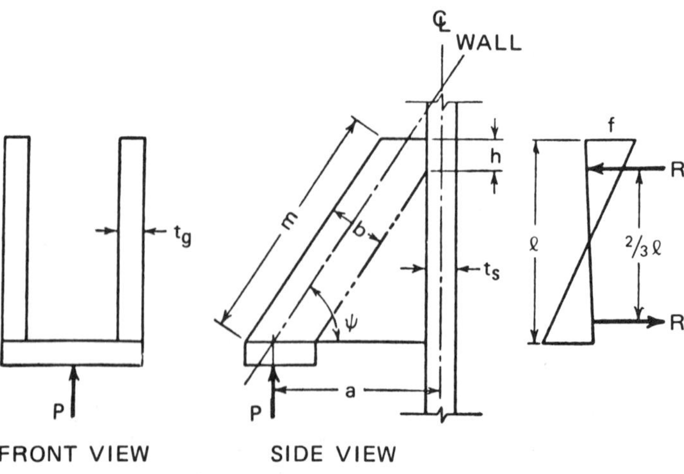

Figure 14.15. Sketch of lug in which only the vertical gussets are welded to the shell.

$$\beta = \sqrt[4]{\frac{3(1-\tilde{\mu}^2)}{r^2t^2}} \tag{14.31}$$

For lugs fabricated of two vertical gussets welded to the shell (without horizontal plates welded to shell, types VII–IX, Figure 14.1), the design thickness of a gusset S′ is determined from the following formula [5]:

$$S' = \frac{2.24G}{KZ\sigma_{b.allow}\ell} \tag{14.32}$$

where G = load on support (N)
K = coefficient depending on the ratio ℓ/s
ℓ = arm of the support (assumed from design consideration) (m)

An approximate value of 0.6 for coefficient K can be assumed for preliminary calculations. If the value of S′ obtained from formula 14.1 is greater than $\ell/13$, then the design value of S′ is final. Otherwise, the value of K should be decreased and S′ recalculated and the subsequent ℓ/s checked from Figure 14.16. The thickness of the supporting part should not be less than S. If the support is welded to the shell, the strength of the welded seams should correspond to the following condition:

$$G \leqslant 0.7 L_s \kappa \tau_{sh}, \quad N \tag{14.33}$$

where κ = fillet of a welded seam
L_s = total length of a welded seam

If, in the cylindrical vessel $(S - C)/D_b < 0.02$, the supporting lugs of types IV and V (Figure 14.1) should be reinforced by pads with the thickness S_p equal to that of the lug gusset.

Example 14.4

From the following data, determine the main sizes of a lug type VIII (Figure 14.1) for a vertical cylindrical vessel supported on four lugs: the load on one lug, $G = 10 \times 10^4$ N; the material of the lug is steel ($\sigma_{b.allow} = 146 \times 10^6$ N/m^2); the number of gussets in a lug Z = 2; lug extension $\ell - 0.25$ m, lugs are supported on a wooden pad ($q - 2 \times 10^6$ N/m^2); the shell thickness of the vessel $S_1 = 24$ mm.

Solution

Assume that the ratio of lug extension to the height of its gusset is $\ell/h = 0.5$. Then,

$$h = \frac{\ell}{0.5} = 0.5 \text{ m}$$

The design thickness of the lug gusset is determined from formula 14.32:

$$S' = \frac{2.24G}{KZ\sigma_{b,allow}\,\ell} = \frac{2.24 \times 10 \times 10^4}{0.6 \times 2 \times 146 \times 10^6 \times 0.25} = 0.0051 \text{ m}$$

The ratio $\ell/13 = 0.25/13 = 0.0192 > S' = 0.0051$; therefore, we decrease the value of K 0.25. At this value we read from Figure 14.16 $\ell/s = 22.5$. Recalculating s' we obtain

$$s' = 0.0051\,\frac{0.6}{0.25} = 0.0123 \text{ m} > \frac{0.25}{22.5} = 0.011 \text{ m}$$

Assume s = 14 mm. Assume the length of the bearing plate for the lug $\ell_1 = 0.23$ m; its thickness s = 14 mm. The design thickness of the bearing plate is

$$b' = \frac{G}{\ell_1 q} = \frac{10^5}{0.23 \times 2 \times 10^6} = 0.218 \text{ m}$$

Assume b = 0.22 m. From Table 14.7 at load $G = 8 \times 10^4$ N per lug and apparatus shell thickness $s_1 = 24$ mm, it can be seen that there is no need to insert a pad. However, considering the presence of a greater load $G = 10 \times 10^4$ N and taking into account the data of Table 14.7, we assume the pad with the following sizes:

$$s_H = 10 \text{ mm}, \quad H = 700 \text{ mm}, \quad B = 500 \text{ mm}$$

The gussets are welded to the pad all around with a fillet $\kappa = 8$ mm. The total length of the seam should be:

$$L_s = 4(h + s) = 4(0.5 + 0.014) = 2.056 \text{ m}$$

The strength of the welded seam at $\tau_{sh} = 80 \times 10^6$ N/m^2 is checked from formula 14.33:

$$G = 10 \times 10^4 \text{ N} < 0.7\, L_s \kappa \tau_{sh}$$
$$= 0.7 \times 2.056 \times 0.008 \times 80 \times 10^6 = 92 \times 10^4$$

Thus, the strength of the seam is provided.

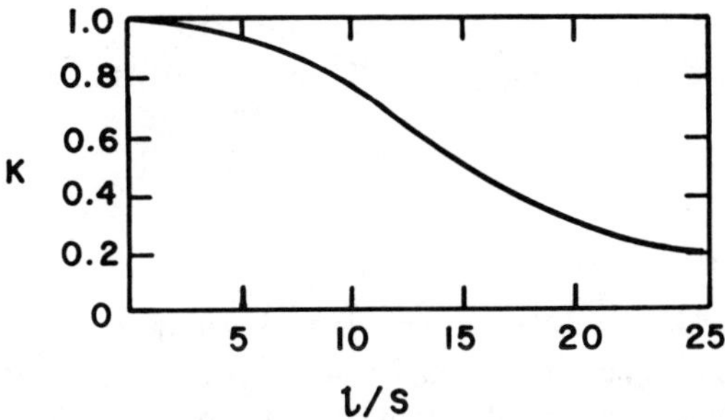

Figure 14.16. Diagram for determining coefficient K in Equation 14.32.

Table 14.7. Pad for Lugs (mm)

Allowable Load on Lug ($G \times 10^{-4}$ N)	Minimum Thickness of Wall Without Need of a Pad	Minimum Thickness of the Wall of an Apparatus	H	B	S_H
2.5	8	4	260	140	6
			460	320	8
4.0	12	8	365	200	6
6.0	20	5	740	550	10
		10	550	340	8
		14	460	230	6
8.0	24	10	750	530	10
		14	500	360	6
		20	490	250	

14.3 DESIGN OF SUPPORTS FOR HORIZONTAL VESSELS

In all cylindrical apparatuses resting on saddle supports horizontally, bending stresses arise in the wall of the shell resulting from the weight of an apparatus and its contents.

When resting on saddle supports such as shown in Figure 14.17, horizontal vessels behave as beams. In designing for this situation, it is necessary to determine the critical compression stress in the shell cross sections induced by the supports, which results in the loss of local stability (that is, formation of wrinkles and hollows in the compression zone).

It is also necessary to check the stability of the cylindrical shell under the action of concentrated loads of the supports. It is important, therefore, to determine the most favorable sizing between supports, rigidity of cross section of the vessel and its diameter. Determining and checking these parameters is important to the structural design of horizontal cylindrical apparatuses.

The determination of total stresses in cross sections of a shell may be done as follows:

$$\sigma_x = \sigma_1 + \sigma_b \tag{14.34}$$

σ_1 is determined only for vessels working under vacuum; in case of internal pressure, σ_x is assumed to be equal to σ_b, where σ_1 is the stress determined from the formulas given in Chapter 6 and σ_b is the bending stress for apparatuses behaving as beams. The stress σ_x must fulfill the following condition:

$$\sigma_x \leqslant \sigma_{allow} \tag{14.35}$$

where σ_{allow} = allowable tensile stress under the operating temperature

Also, the following criterion should be fulfilled:

$$\sigma_x \leqslant \frac{\sigma_{crit}}{\eta} = \sigma_{allow} \tag{14.36}$$

where σ_{crit} = critical compression stress when loss of stability is possible in the compression zone of the shell

η = safety factor

The critical stress of a bending vessel when crumpling is possible is determined from the following formula

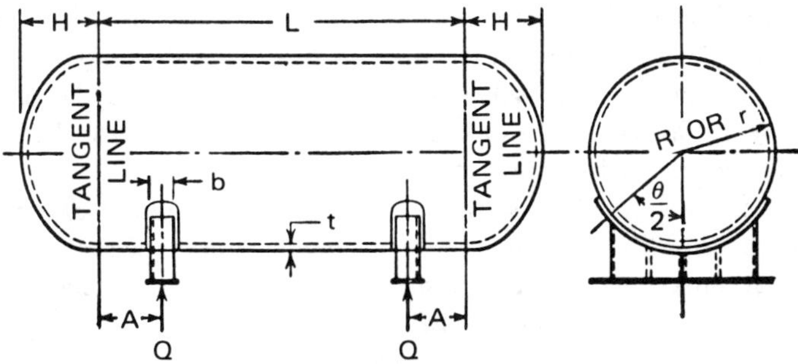

Figure 14.17. Sketch of a horizontal vessel with two saddle supports.

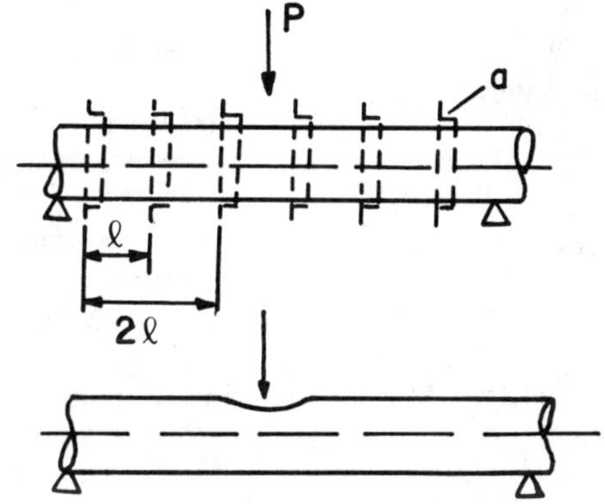

Figure 14.18. Formation of a hollow.

$$\sigma_{crit} = 1.37E\left(\frac{s'}{R}\right)^2 \tag{14.37}$$

where E = modulus of elasticity (dependent on temperature)
s′ = wall thickness, without corrosion allowance
R = outside diameter of the vessel.

The critical length ℓ_{crit} in Figure 14.18, for which a hollow section is formed during the loss of stability, is

$$\ell_{crit} = 1.009\pi R \tag{14.38}$$

In stability design of thin-walled large-diameter vessels, or in the case of saddles located away from the head (A/R > ½), the shell alone may not resist the load and the allowable stresses ($\sigma_{allow} = \sigma_{crit}/m$) may become very small. Ring stiffeners are then attached to the shell to alleviate the load on the shell. Figure 14.19 shows two recommended types of internal ring stiffeners, and Figure 14.20 shows the corresponding types of external ring stiffeners. Because the maximum stress in compression in the shell is reduced by internal operating or test pressures, the inside ring stiffener is more favorable from a strength standpoint.

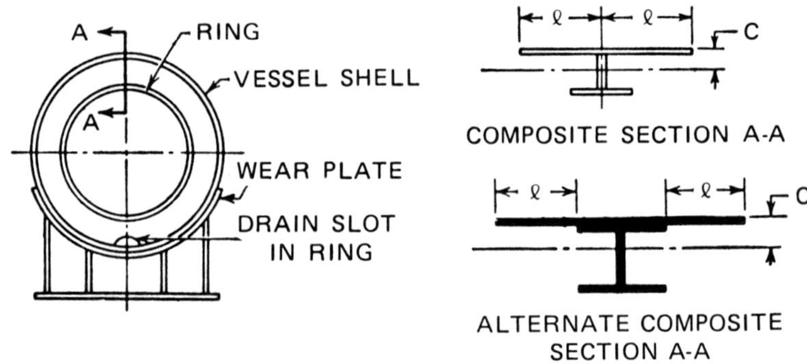

Figure 14.19. Examples of internal stiffening rings.

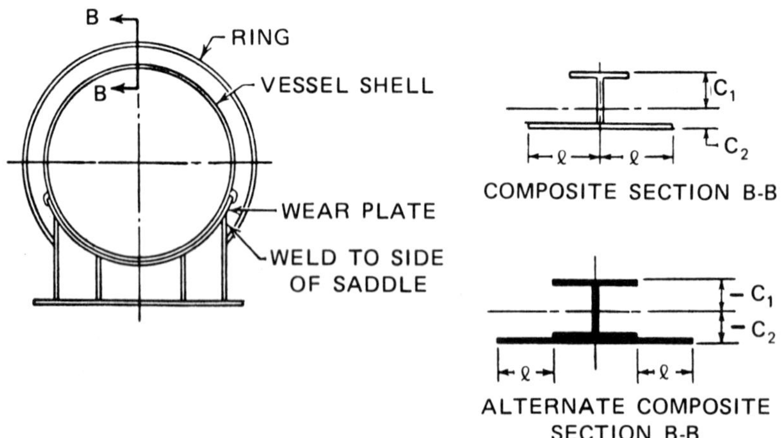

Figure 14.20. Examples of external stiffening rings.

An external ring stiffener is less desirable because the maximum stress may be either compression in the outer flange or tension in the vessel shell due to the total weight of an apparatus.

The distance between ring stiffeners is determined by calculation. For example, it is unreasonable to insert rings at the distance that is equal to, or exceeding, the critical stress (formula 14.35) because such ring spacing will not increase the stability of an apparatus, even if the rings have very high rigidity.

The design of ring stiffeners is reduced to the determination of the space between the rings and calculation of the rigidity of a ring stiffener.

The distance ℓ (Figure 14.18) should be less than $\ell_{crit} = 1.009\pi R$ and should be chosen in accordance with the following formula:

$$\sigma'_{crit} = 0.0605\,ES^2\left(\frac{9}{8}\frac{\ell^2}{R^4} + \frac{7}{6}\frac{\pi^4}{\ell^2}\right) \tag{14.39}$$

Here, σ'_{crit} is equal to critical stress when an apparatus reinforced with ring stiffeners can obtain hollows. The allowable stress, σ''_{allow}, in extreme fibers of an apparatus should be designed with a minimum safety factor of 2.5 for an apparatus without ring stiffeners:

$$\sigma''_{allow} \leqslant \frac{\sigma_{crit}}{2.5} \tag{14.40}$$

and for an apparatus reinforced with ring stiffeners,

$$\sigma''_{allow} \leqslant \frac{\sigma'_{crit}}{2.5} \tag{14.41}$$

As follows from Equation 14.39, a decrease in ℓ will increase the critical stress.

The cross section of a ring stiffener is determined from the following equation:

$$\sigma'_{crit} = 0.0605\,ES^2\left[4.5\frac{\ell^2}{R^4} + \frac{7}{6}\frac{\pi^4}{(2\ell)^2}\right] + 4.36\frac{\ell}{SR_r^3R}\,EI_2 \tag{14.42}$$

where R_r = radius of a ring stiffener through its gravity center (cm)
I_2 = moment of inertia of ring cross section (cm^4)

Consider the case in which a cylindrical apparatus is rested on two or three saddle supports such as those shown in Figure 14.21. One must take into account that in case more supports are used, the calculation of a bending moment will be similar to the use of the three-moment equation with a new number of supports as a continuous beam of constant rigidity and equal spans.

A cylindrical vessel with dished closures at the ends may be treated as an equivalent cylinder having a length equal to (L + ⁴⁄₃ H), where L is the distance between the tangent lines of the vessel and H is the depth of a dished closure (see Figure 14.17). This approximation assumes that the weight of the head and the fluid contained in the head are equal to two-thirds of the weight of a cylinder of length H and the fluid contained in it. This approximation is valid for hemispherical heads and elliptical dished heads and can be demonstrated by the use of the following formula for determination of the volume in cubic feet contained within the head, not including the straight-flange portion:

$$V = 0.000076 d_i^3 \tag{14.43}$$

where d_i = inside diameter of vessel (in.)
V = volume of elliptical dished head to straight flange (ft^3)

As an example, for an elliptical closure for a 100-in.-diameter vessel,

$$V = (0.000076)(100)^3 = 76 \text{ ft}^3$$

The depth of the dish, from Figure 7.1, is ID = 25 in. The volume of a cylinder 100 in. in diameter and 25 in. deep is 114 ft^3. The ratio of the volume of the head to the volume of the cylinder is ⁷⁶⁄₁₁₄, or ⅔.

For determination of the maximum bending moment, the weight of the fluid and the vessel may be considered to be a uniform load equal to the total weight divided by the equivalent length:

$$p = \frac{2Q}{L + \frac{4}{3} H} \tag{14.44}$$

where p = uniform load (kg/m)
L = length of vessel, tangent to tangent (m)
Q = total load per saddle (kg) = total weight divided by two
H = depth of head (m)

Figure 14.22 gives a scheme of the loaded conditions of the vessel with two and three saddle supports.

For a vessel with two saddle supports according to scheme (A), the reaction of supports is

$$P_A = P_B \ , \quad N \tag{14.45}$$

and the design bending moment is

$$M = 0.0215\ GL_{eq} \ , \quad N\text{-m} \tag{14.46}$$

For a vessel with three saddle supports according to scheme (B), reaction of supports is

$$P_A = P_D = 0.323\ G \ , \quad N \tag{14.47}$$

$$P_B = 0.354\ G \ , \quad N \tag{14.48}$$

The design bending moment for all three supports is

$$M = 0.0105\ GL_{eq} \ , \quad N\text{-m} \tag{14.49}$$

where G = total weight of the vessel and its content (N)
L_{eq} = equivalent length of the vessel (m)

The equivalent length of the vessel, if it has two equal heads, is

$$L_{eq} = L_c + 2\ L_h \tag{14.50}$$

where L_c = length of cylindrical shell (m)
L_h = length of head as an equivalent cylinder

The equivalent length of the head, L_h, in the vessel filled with fluid is determined from the following formula:

$$L_h = \frac{G_h + 10V_h\rho_f}{7.85[\rho(D_o^2 - D_i^2) + \rho_f D_i^2]} \ , \quad m \tag{14.51}$$

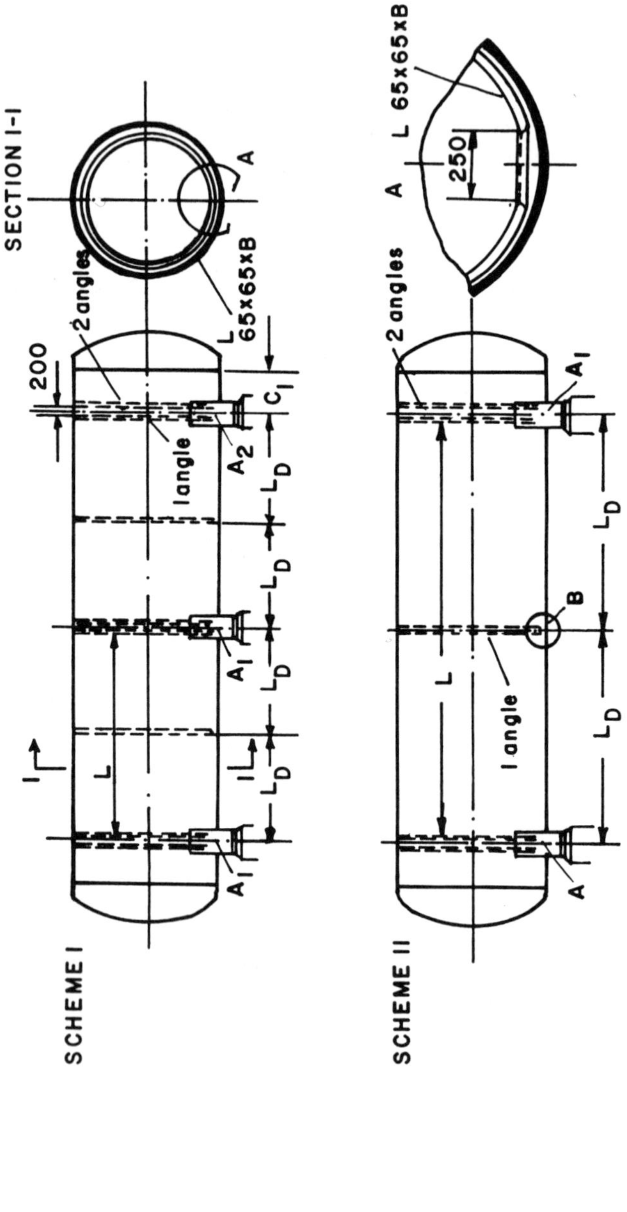

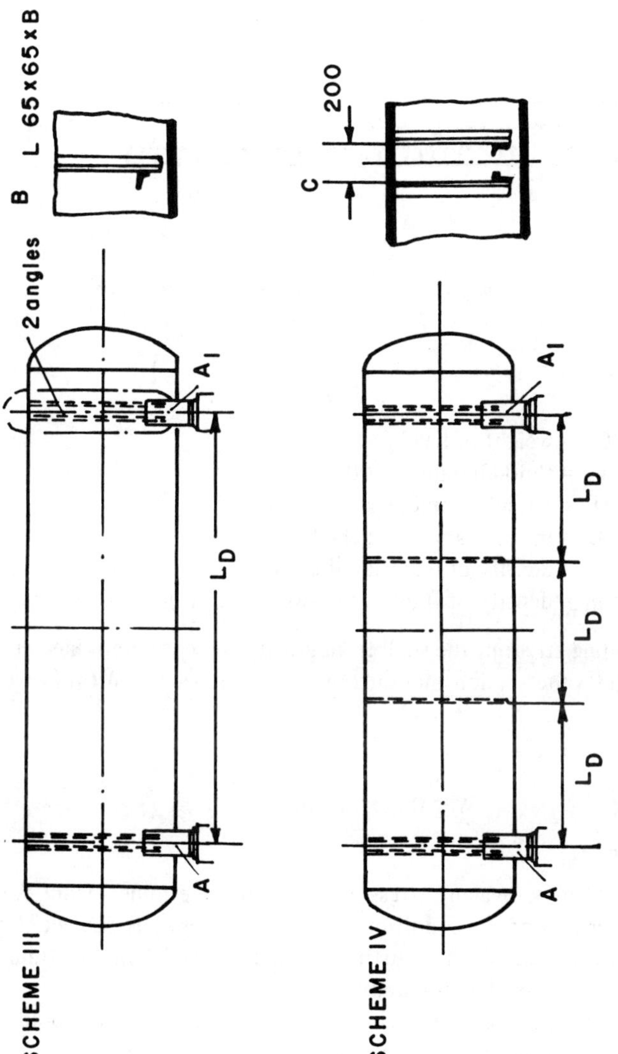

Figure 14.21. Sketches of horizontal vessels with two and three saddle supports.

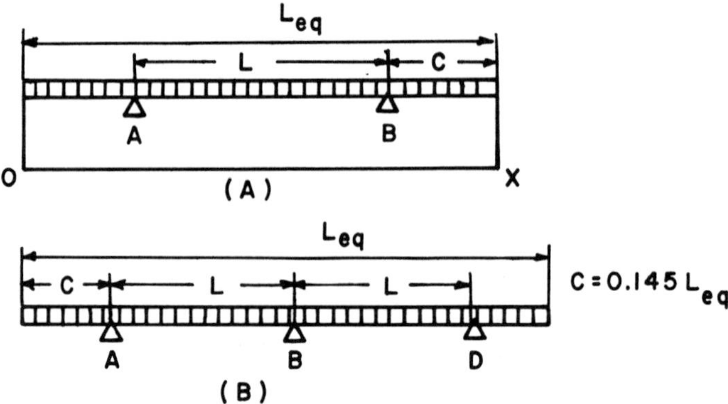

Figure 14.22. Loads and reactions on saddles.

where G_h = weight of head (N)
V_h = fluid contained in the head
D_o = outside diameter of shell
D_i = inside diameter of shell
ρ = density of material of shell (kg/m^3)
ρ_f = density of fluid in the vessel (kg/m^3)

The bending stress in the shell from gravity force is determined from formula 6.32, where the section modulus of shell cross section W (at $(s - c) < 0.1D_i$) is equal to

$$W \simeq 0.8D_i^2(s - c) \ , \quad m^3 \tag{14.52}$$

The value of equivalent stress in the shell determined from formula 6.30 should correspond to condition 6.34. If the vessel is subjected to external pressure, P_e, it should be substituted for P. In addition, the thickness of the shell should be checked for stability from formula 6.36.

The bending stress in the shell from support reaction should be checked from the following formula:

$$\sigma = \frac{0.02\ PD_0}{W'} \leqslant \sigma_{b,allow} \ ; \quad N/m^2 \tag{14.53}$$

where P = reaction of support (N)
W' = section of modulus of design cross section of wall element over the support about the x-axis passing through the center of gravity of the section parallel to the vessel axis (m)

The section modulus of the abovementioned section of wall element (Figure 14.23a) is determined from the following formula:

$$W' = \frac{[b + 8(s - c)]\,(s - c)^2}{6}, \quad m^3 \tag{14.54}$$

If condition 14.53 is not fulfilled, the shell under the support should be reinforced with the pad. The section modulus (consisting of a wall element and a pad) about the x-axis going through the gravity center parallel to vessel axis should be

$$W \geqslant \frac{0.02\, PD_0}{\sigma_{b,allow}}, \quad m^3 \tag{14.55}$$

It is recommended to assume the following pad thicknesses: At $4\sigma_{b,allow} \geqslant \sigma$ (from formula 14.53) $> \sigma_{b,allow}$, $S_H = S$; at σ (from formula 14.53) $> 4\,\sigma_{b,allow}$, $S_H > S$. S_H is determined by trial and error based on condition 14.55 being fulfilled.

The design section modulus of the combined section of a shell reinforced with a pad, W' (Figure 14.23b), is determined from the following formula:

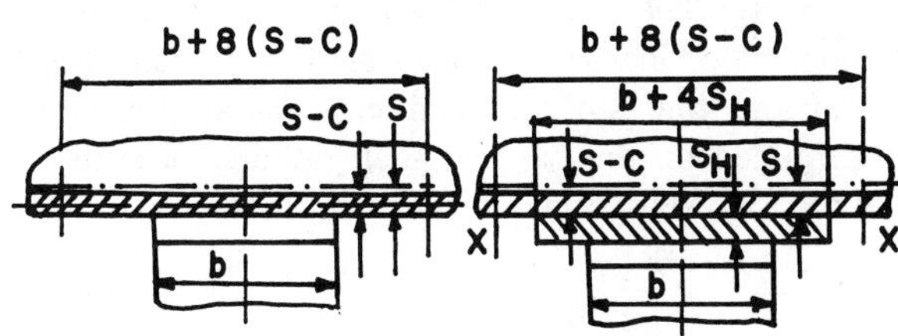

Figure 14.23. Design sections of wall of horizontal cylindrical vessel under the support in structural and stability design: (a) nonreinforced section of the wall; (b) reinforced with a pad section of the wall.

$$W' = \frac{J'_c + J'_H + F'_{sh}[S_H + 0.5(s - c) - y]^2 + F'_p(y - 0.5s)^2}{y} \qquad (14.56)$$

where F'_{sh} = design area of cross section of the shell (m^2)
F'_p = design area of cross section of wear plate (m^2)
J'_{sh} = moment of inertia of the area F_{sh} about the axis going through its center of gravity parallel to vessel's axis
J'_p = moment of inertia of the area F'_{p1} about the axis going through its center of gravity parallel to vessel's axis (m^4)
y = distance from lower surface of wear plate to the axis going through the gravity center of the area $F'_w + F_p$ parallel to the vessel's axis

The values in Equation 14.56 have the following values:

$$F'_{sh} = [b + 8(s - c)](s - c), \quad m^2 \qquad (14.57)$$

$$F'_p = (b + 4\, S_H)\, S_H, \quad m^2 \qquad (14.58)$$

$$J'_{sh} = \frac{F'_{sh}(s - c)^2}{12}, \quad m^4 \qquad (14.59)$$

$$J'_p = \frac{F'_p S_H^2}{12}, \quad m^4 \qquad (14.60)$$

$$y = \frac{F'_{sh}[S_H + 0.5(s - c)] + 0.5 F'_p S_H}{F'_{sh} + F'_p}, \quad m \qquad (14.61)$$

It is not recommended to use wear plates with $S_H > 1.6S$. If the wear plate does not conform with condition 14.55, then ring stiffeners should be provided near supports either from the outside or inside, depending on the design. The cross sections of ring stiffeners are illustrated in Figure 6.17.

The design moment of inertia of a cross section, J′, consisting of a cross-sectional area of the ring stiffener and cross-sectional area of the shell element, is determined on the basis of shell stability near the support (at $n_{stab} = 5$) from the following formula:

$$J' \geqslant \frac{0.208\, P D_0^2}{E^t \sin \alpha/2}, \quad m^4 \qquad (14.62)$$

where α = angle of contact (usually $\alpha = 120°$)

To determine the moment of inertia, J′, use Equation 14.62, where the design cross-sectional area of the shell element is calculated from formula 14.57. If lug types III and IV (Figure 14.2) are used, the ratio of their console to the gusset height should be assumed to be $\ell/h = 0.5$. The calculation of gusset thickness is made from formula 14.32.

Example 14.5

Check the strength and stability of a shell of a cylindrical vessel with elliptical heads and with three saddle supports (Figure 14.22B) from the following data: $D_o = 2.02$ m; $D_i = 2.0$ m; $s - c = 8$ mm; $L = 10.9$ m; $\psi = 0.95$; Head: $D_i = 2$ m; $h = 0.55$ m; $G_h = 0.36 \times 10^4$ N; $V_h = 1.17\ m^3$. The material of the shell and heads is steel ($E = 205 \times 10^9\ N/m^2$; $\sigma_y = 240 \times 10^6\ N/m^2$); $\sigma_{b,allow} = 146 \times 10^6\ N/m^2$; $\rho = 7.85 \times 10^3\ kg/m^3$. Fluid $p = 1 \times 10^6\ N/m^2$; $\rho = 1 \times 10^3\ kg/m^3$. Weight of the vessel filled with fluid $G = 35 \times 10^4$ N.

Solution

The design is made for the middle support having the greatest load from reaction of support, which is determined from formula 14.48:

$$P_B = 0.354G = 0.354 \times 35 \times 10^4 = 12.4 \times 10^4\ N$$

The equivalent length of head is determined from Equation 14.51:

$$L_h = \frac{G_h + 10 V_h \rho_f}{7.85[\rho(D_o^2 - D_i^2) + \rho_f D_i^2]}$$

$$= \frac{0.36 \times 10^4 + 10 \times 1.17 \times 10^3}{7.85[7.85 \times 10^3(2.02^2 - 2.0^2) + 10^3 \times 2^2} = 0.42\ m$$

The equivalent vessel's length is determined from Equation 14.50:

$$L_{eq} = L_c + 2\ L_h = 10.9 + 2 \times 0.42 = 11.74\ m$$

The distance between supports (see Figure 14.22) is

$$\ell = 0.5(L_{eq} - 2c) = 0.5(11.74 - 2 \times 0.145 \times 11.74) = 4.16\ m$$

The design bending moment from gravity force is determined from Equation 14.49:

$$M = 0.0105\,GL_{eq} = 0.0105 \times 35 \times 10^4 \times 11.74 = 4.32 \times 10^4 \text{ N-m}$$

The bending stress in the shell from gravity force is determined from Equations 14.52 and 6.32:

$$\sigma_1 = \frac{M}{W} = \frac{M}{0.8D_i^2(s - c)}$$

$$= \frac{4.32 \times 10^4}{0.8 \times 2^2 \times 0.008} = 1.69 \times 10^6 \text{ N/m}^2$$

Equivalent stress in the shell from additional loads at $\sigma = 0$ and $\tau = 0$ is determined from formula 6.30:

$$\sigma_{eq} = \sqrt{(\sigma + 0.8\sigma_1)^2 + 3\tau^2} = 0.8\sigma_1 = 0.8 \times 1.69 \times 10^6$$

$$= 1.35 \times 10^6 \text{ N/m}^2$$

Substituting values into condition 6.34:

$$\sqrt{1.2\sigma_{allow}^2 - \frac{[D_i + (s - c)]^2\,p^2}{4(s - c)^2\,\psi^2}}$$

we will obtain

$$\sqrt{1.2(146 \times 10^6)^2 - \frac{(2.0 + 0.008)^2(1 \times 10^6)^2}{4 \times 0.008^2 \times 0.95^2}}$$

$$= 90 \times 10^6 \text{ N/m}^2 > \sigma_{eq} = 1.36 \times 10^6 \text{ N/m}^2$$

that is, the shell is strong.

Checking the shell stability from formula 6.36, the coefficient K at axial load P = 0 from Equation 6.37 will be K = 2, and coefficient K_2 from Figure 6.12 at

$$\frac{D_i}{s - c} = \frac{2.0}{0.008} = 250 \ ; \quad K_2 = 6.7$$

Eliminating the values that are equal to zero in Equation 6.36, we obtain

$$\sqrt{\frac{0.125\, KK_2\, 4M}{E^t D_i}} = \sqrt{\frac{0.125 \times 2 \times 6.7 \times 4 \times 4.32 \times 10^4}{205 \times 10^9 \times 2.0}}$$

$$= 0.84 \times 10^{-3} \text{ m} < (s - c) = 8 \times 10^{-3} \text{ m}$$

that is, the shell is stable. We select the width of the support from Figure 14.2:

$$b = 0.2\, D_i = 0.2 \times 2.0 = 0.4 \text{ m}$$

The section modulus of the design section of the shell under the support from formula 14.54 is

$$W' = \frac{[b + 8(s - c)]\,(s - c)^2}{6} = \frac{[0.4 + 8 \times 0.008] \times 0.008^2}{6}$$

$$= 4.95 \times 10^{-6} \text{ m}^3$$

The bending stress in the shell from the reaction of support (from Equation 14.53) is

$$\sigma = \frac{0.02\, P_B D_0}{W'} = \frac{0.02 \times 12.4 \times 10^4 \times 2.02}{4.95 \times 10^{-6}}$$

$$= 1010 \times 10^6 \text{ N/m}^2 > \sigma_{b,allow} = 146 \times 10^6 \text{ N/m}^2$$

This is an indication that we need to insert the wear plate between the shell and the saddle support.

The required section modulus of reinforcement of the shell element under the support is determined from Equation 14.55:

$$W = \frac{0.02\, P_B D_0}{\sigma_{b,allow}} = \frac{0.02 \times 12.4 \times 10^4 \times 2.02}{146 \times 10^6} = 34.3 \times 10^{-6} \text{ m}^3$$

Because $4\sigma_{b,allow} = 4 \times 146 \times 10^6 = 584 \times 10^6\ N/m^2 < \sigma = 1010 \times 10^6\ N/m^2$, we assume the thickness of the wear plate is $s_H = 1.6 \times S = 1.6 \times 10 = 16$ mm. The design area of the shell cross section is calculated from Equation 14.57:

$$F'_{sh} = [b + 8(s - c)](s - c) = [0.4 + 8 \times 0.008] \times 0.008$$
$$= 37 \times 10^{-4}\ m^2$$

The design area of the wear plate cross section is calculated from Equation 14.58:

$$F'_p = (b + 4\,S_H)S_H = (0.4 + 4 \times 0.016) \times 0.016 = 74 \times 10^{-4}\ m^2$$

The moment of inertia of area F'_{sh} is calculated from Equation 14.59:

$$J'_{sh} = \frac{F'_{sh}(s - c)^2}{12} = \frac{37 \times 10^{-4} \times 0.008^2}{12} = 1.97 \times 10^{-8}\ m^4$$

The moment of inertia of the area F'_p is calculated from Equation 14.60:

$$J'_p = \frac{F'_p S_H^2}{12} = \frac{74 \times 10^{-4} \times 0.016^2}{12} = 15.8 \times 10^{-8}\ m^4$$

The distance from the lower surface of wear plate to the gravity center of the area $F'_{sh} + F'_p$ (from the formula 14.61) is

$$y = \frac{F'_{sh}[S_H + 0.5(s - c)] + 0.5\,F'_p S_H}{F'_{sh} + F'_p}$$

$$= \frac{37 \times 10^{-4}[0.016 + 0.5 \times 0.008] + 0.5 \times 74 \times 10^{-4} \times 0.016}{(37 + 74) \times 10^{-4}}$$

$$= 0.012\ m$$

The design section modulus of the shell section reinforced with a wear plate (from Equation 14.56) is

$$W' = \frac{J'_{sh} + J'_p + F'_{sh}[S_H + 0.5(s - c) - y]^2 + F'_p(y - 0.5\,S_H)^2}{y}$$

$$= 44.4 \times 10^{-6}\ m^3 > 34.3 \times 10^{-6}\ m^3$$

that is, the wear plate provides strength.

Example 14.6

Determine for Example 14.3 the sizes of the ring stiffener installed inside the shell near the supports instead of wear plates. The design of the ring stiffener should be made according to Figure 14.24. The shell contact angle with the saddle support is $\alpha = 120°$.

Solution

The design moment of inertia of combined cross section (from Equation 14.62) is

$$J' = \frac{0.208 P_B D_0^2}{E^t \sin\alpha/2} = \frac{0.208 \times 12.4 \times 10^4 \times 2.02^2}{205 \times 10^9 \sin 60°} = 59 \times 10^{-8}\ m^4$$

Assume angle 63 X 63 X 6.

$$J_2 = 2 \times 27.1 \times 10^{-8} = 54.2 \times 10^{-8}\ m^4$$

$$F_2 = 2 \times 7.28 \times 10^{-4} = 14.56 \times 10^{-4}\ m^2$$

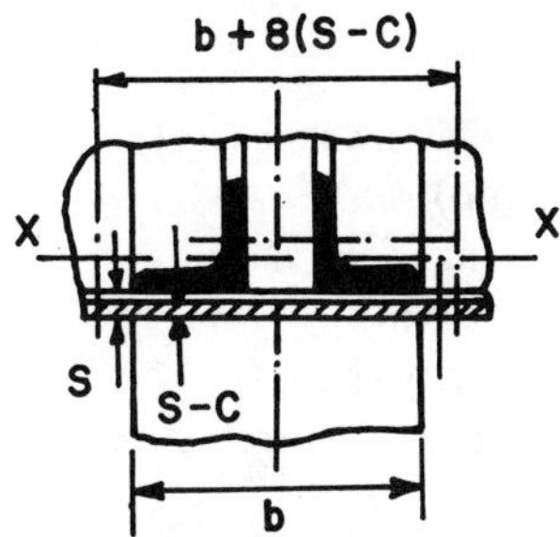

Figure 14.24. Sketch for Example 14.6.

From Example 14.3 we have $J_{sh} = 1.97 \times 10^{-8}$ m^4 and $F_{sh} = 37 \times 10^{-4}$ m^2. The distance y is obtained from the following relationship (see Figure 14.24):

$$F'_{sh} = [y - (s - c)\,0.5] = F_2(y_1 - y)$$

or

$$y = \frac{0.5 F'_{sh}(s - c) + F_2 y_1}{F'_{sh} + F_2}$$

$$y_1 = s + 17.8 = 10 + 17.8 = 28.8 \text{ mm} = 0.0278 \text{ m}$$

Hence,

$$y = \frac{0.5 \times 37 \times 10^{-4} \times 0.008 + 14.56 \times 10^{-4} \times 0.0278}{(37 + 14.56) \times 10^{-4}}$$

The moment of inertia of combined section is

$$J = J_{sh} + J_2 + F_{sh}[y - 0.5(s - c)]^2 + F_2(y_1 - y)^2$$

$$= 1.97 \times 10^{-8} + 54.2 \times 10^{-8} + 37 \times 10^{-4}[0.0107 - 0.5 \times 0.008]^2$$

$$+ 1456 \times 10^{-4}(0.0278 - 0.0107)^2 = 115.3 \times 10^{-8} \text{ m}^4 > J'$$

$$= 59.2 \times 10^{-8} \text{ m}^4$$

NOMENCLATURE

a	area (in^2.) or radial distance (in.)
b	gusset spacing (in.)
C	corrosion allowance (m)
d_h	bolt-hole diameter (m)
E	modulus of elasticity (N/m^2)
e	radius of action of concentrated load (m)
F	area (in^2.)
F_b	cross-sectional area of bolt (m^2)
G	maximum load on support (N)

H	vessel height (m)
h	height of gusset (in.)
I	moment of inertia (kg-m^2)
J	moment of inertia (m^4)
L	coefficient in Equation 14.27 (see Table 14.6)
K_1	aerodynamic coefficient (see Equation 14.24)
K_2	dynamic coefficient (see Equation 14.24)
K_{st}	stability coefficient
L	length
ℓ	difference between outer radius of bearing ring and outer radius of skirt (m)
$M_{x,y}$	bending moment (N-m/m)
M_w	wind-load moment (N-m)
P	bolt load (N)
Q	total load (kg)
q	specific load (N/m^2)
r	radius of gyration (in.)
r_1, r_2	constants in Equations 14.11 and 14.12 (see Table 14.5)
S	wall thickness (mm or m)
S_k	bearing ring thickness (mm)
t	temperature (°C)
V	volume (ft^3)
W	section modulus (m^3)
Z'	number of bolts
α	coefficient of thermal expansion, (m/m-°K) or angle of contact (°)
η	safety factor
κ	fillet of welded seam (in.)
ρ	density (kg/m^3)
σ	bending stress (N/m^2)
$\sigma_{b,allow}$	allowable bending stress (N/m^2)
σ_{max}	maximum compression stress (N/m^2)
σ_s	bolt stress (N/m^2)

REFERENCES

1. Timoshenko, S. *Strength of Materials,* Part II (New York: Van Nostrand Reinhold Co., 1940).
2. Brownell, L. E., and E. H. Young. *Process Equipment Design* (New York: John Wiley & Sons, Inc., 1959).
3. Timoshenko, S. *Theory of Plates and Shells* (New York: McGraw-Hill Book Co., 1940).

4. "Modern Flange Design," *Taylor Forge and Pipe Works Bull.* 502 (1950).
5. Domashnev, A. D. "Konstoruiovanie i Raschety Khimicheskikh apparatov," *Mashgiz* (1961).
6. "Penstock Analysis and Stiffener Design," Boulder Canyon Project, Final Reports, Part V, Technical Investigations Bulletin 5, U.S. Department of Interior, Bureau of Reclamation (1940).
7. Wolosewick, F. E. "Supports for Vertical Pressure Vessels, Part II," *Pet. Refiner* 30 (8): 101 (1951).
8. Zick, L. P. "Stresses in Large Horizontal Cylindrical Pressure Vessels on Two Saddle Supports," *Welding J. (N.Y.), Res. Suppl.* 30: 435-S (1951).

APPENDIX

TABLES OF RECOMMENDED SI UNITS AND CONVERSION FACTORS

The tables contained in this Appendix were reprinted from API Publication 2564–*Manual of Petroleum Measurement Standards*: Chapter 15, "Guidelines for the Use of the International System of Units (SI) in the Petroleum and Allied Industries," 2nd edition (December 1980), pp. 7-31, courtesy of the American Petroleum Institute, 2101 L Street, Northwest, Washington, DC.

NOTES ON USE OF TABLES

Metric units recommended for general use are given under the heading "API preferred metric unit." Other units that also may be needed are shown in the "other allowable" column. Preferred units do not preclude the use of other multiples or submultiples, as the choice of such unit-multiple is governed by the magnitude of the numerical value.

Notation used conforms to SI practice, i.e., groups of three digits to the left or right of the decimal marker are separated by spaces. No commas or other triad spacers are used. E-notation is used for convenience because it is a standard method of display in many calculators and because of the inability of computers to print out or transmit superscripts. An asterisk (*) denotes that all the succeeding digits would be zeroes. If a conversion factor ends in zero but does not have an asterisk, then any subsequent digits would not necessarily be zeroes.

For example, then

$$3.056\,0 \text{ E} + 00 = 3.056\,0 \times 10^{0} = 3.056\,0$$

$$3.056^{*}\text{E} - 01 = 3.056\,000 \times 10^{-1} = 0.305\,600\,0$$

$$9.290\,304 \text{ E} + 02 = 9.290\,304 \times 10^{2} = 929.030\,4$$

Table I. Nomenclature for Conversion Tables

Symbol	Name	Quantity	Definition: In Terms of Other Units	Definition: In Terms of Base Units	Type of Unit
A	Ampere	Electric current	–	–	Base
a	Annum (year)	Time	365 d	$3.153\,600 \times 10^7$s	Allowable
bar	Bar	Pressure	10^5 Pa	10^5 kg/(m·s^2)	Allowable
Bq	Becquerel	Activity (of a radionuclide)	–	1 s^{-1}	Derived
C	Coulomb	Quantity of electricity, electric charge	–	1 A·s	Derived
°C	Degree Celsius	Celsius temperature	–	–	Derived
cd	Candela	Luminous intensity	–	–	Base
cP	Centipoise	Dynamic viscosity	1 mPa·s	10^{-3} kg/(m·s)	Allowable
cSt	Centistokes	Kinematic viscosity	1 mm^2/s	10^{-6} m^2/s	Allowable
d	Day	Time	24 hr	8.640×10^4s	Allowable
F	Farad	Capacitance	1 C/V	1 s^4 A^2/(m^2·kg)	Derived
g	Gram	Mass	–	10^{-3} kg	Allowable (submultiple of base unit)
Gy	Gray	Absorbed dose, specific energy imparted, kerma, absorbed dose index	1 J/kg	1 m^2/S^2	Derived
h	Hour	Time	60 min	3.6×10^3 s	Allowable
H	Henry	Inductance	1 Wb/A	1 m^2·kg/(s^2·A^2)	Derived
ha	Hectare	Area	–	10^4 m^2	Allowable
Hz	Hertz	Frequency (of periodic phenomenon)	–	1 s^{-1}	Derived
J	Joule	Work, energy, quantity of heat	1 N·m	1 m^2·kg/s^2	Derived
K	Kelvin	Thermodynamic temperature	–	–	Base
kg	Kilogram	Mass	–	–	Base
kn	Knot	Velocity	1.852 km/h	5.144 444 m^2/s	Allowable
L	Liter	Volume	1 dm^3	10^{-3} m^3	Allowable

Table I, continued

Symbol	Name	Quantity	Definition: In Terms of Other Units	Definition: In Terms of Base Units	Type of Unit
lm	Lumen	Luminous flux	–	1 cd·sr	Derived
lx	Lux	Illumination	$1\ lm/m^2$	$1\ cd \cdot sr/m^2$	Derived
m	Meter	Length	–	–	Base
min	Minute	Time	–	60 s	Allowable
mol	Mole	Amount of substance	–	–	Base
N	Newton	Force	–	$1\ m \cdot kg/s^2$	Derived
naut. mi	Nautical mile	Distance	1.852 km	1.852×10^3 m	Allowable
Pa	Pascal	Pressure	$1\ N/m^2$	$1\ kg/(m \cdot s^2)$	Derived
r	Revolution	Angular displacement	360°	2π rad	Allowable
rad	Radian	Plane angle	–	–	Supplementary
S	Siemens	Conductance	1 A/V	$1\ s^3 \cdot A^2/(m^2 \cdot kg)$	Derived
s	Second	Time	–	–	Base
sr	Steradian	Solid angle	–	–	Supplementary
Sv	Sievert	Dose equivalent	1 J/kg	$1\ m^2/s^2$	Derived
t	Metric ton	Mass	1 Mg	10^3kg	Allowable
T	Tesla	Magnetic flux density	$1\ Wb/m^2$	$1\ kg/(s^2 \cdot A)$	Derived
V	Volt	Electric potential, potential difference, electromotive force	1 W/A	$1\ m^2 \cdot kg/(s^3 \cdot A)$	Derived
W	Watt	Power, radiant flux	1 J/s	$1\ m^2 \cdot kg/s^3$	Derived
Wb	Weber	Magnetic flux	1 V/s	$1\ m^2 \cdot kg/(s^2 \cdot A)$	Derived
Ω	Ohm	Electric resistance	1 V/A	$1\ m^2 \cdot kg/(s^3 \cdot A^2)$	Derived
°	Degree	Plane angle	–	$\pi/180$ rad	Allowable
′	Minute	Plane angle	(1/60)°	$2.908\ 882 \times 10^{-4}$ rad	Allowable
″	Second	Plane angle	(1/60)′	$4.848\ 137 \times 10^{-6}$ rad	Allowable

Table II. Recommended SI Units and Conversion Factors

Quantity	SI Unit	Customary Unit	Metric Unit: API Preferred	Metric Unit: Other Allowable	Conversion Factor (multiply quantity expressed in customary units by factor to get metric equivalent)	
Space, Time						
Length	m	naut. mi	km		1.852*	E+00
				naut. mi	1	
		mi	km		1.609 344*	E+00
		mi (U.S. statute)	km		1.609 347	E+00
		Chain	m		2.011 684	E+01
		Rod	m		5.029 210	E+00
		Fathom	m		1.828 804	E+00
		m	m		1	
		yard	m		9.144*	E–01
		ft	m		3.048*	E–01
		ft (U.S. survey)	m		3.048 006	E–01
		link	m		2.011 684	E–01
		in.	mm		2.54*	E+01
				cm	2.54*	E+00
		cm	mm		1.0*	E+01
				cm	1	
		mm	mm		1	
		mil	μm		2.54*	E+01
		micron (μ)	μm		1	

Table II, continued

Quantity	SI Unit	Customary Unit	Metric Unit: API Preferred	Metric Unit: Other Allowable	Conversion Factor (multiply quantity expressed in customary units by factor to get metric equivalent)
Space, Time (continued)					
Surface Texture	m	μin.	μm		2.54* E–02
		nm	μm		1.0* E–03
Length/Length	m/m	ft/mi	m/km		1.893 939 E–01
Length/Volume	m/m^3	ft/U.S. gal	m/m^3		8.051 964 E+01
		ft/ft^3	m/m^3		1.076 391 E+01
		ft/bbl	m/m^3		1.917 134 E+00
Length/Temperature	m/K	see Temperature, Pressure, Vacuum			
Area	m^2	mi^2	km^2		2.589 988 E+00
		mi^2 (U.S. statute)	km^2		2.589 998 E+00
		ha	m^2		1.0* E+04
		acre	ha		4.046 873 E–01
				m^2	4.046 873 E+03
		sq chain	m^2		4.046 873 E+02
		sq rod	m^2		2.529 295 E+01
Area		yd^2	m^2		8.361 274 E–01
		ft^2	m^2		9.290 304* E–02
		ft^2 (U.S. survey)	m^2		9.290 341 E–02

Quantity	SI unit	Customary unit	Metric unit	Other allowable unit	Conversion factor	Exponent
		in.^{2}	mm^2		6.451 6*	E+02
				cm^2	6.451 6*	E+00
		cm^2	mm^2		1.0*	E+02
				cm^2	1	
		mm^2	mm^2		1	
Area/Volume	m^2/m^3	ft^2/in.^{3}	m^2/cm^3		5.669 291	E–03
Area/Mass	m^2/kg	cm^2/g	m^2/kg		1.0*	E–01
Volume, Capacity	m^3	m^3	km^3		4.168 182	E+00
		acre·ft	m^3		1.233 489	E+03
				ha·m	1.233 489	E–01
		m^3	m^3		1	
		yd^3	m^3		7.645 549	E–01
		bbl (42 U.S. gal)	m^3		1.589 873	E–01
		ft^3	m^3		2.831 685	E–02
			dm^3	L	2.831 685	E+01
		Can. gal	m^3		4.546 09*	E–03
			dm^3	L	4.546 09*	E+00
		U.K. gal	m^3		4.546 092	E–03
			dm^3	L	4.546 092	E+00
		U.S. gal	m^3		3.785 412	E–03
			dm^3	L	3.785 412	E+00
		L	dm^3	L	1	
		U.K. qt	dm^3	L	1.136 523	E+00
		U.S. qt	dm^3	L	9.463 529	E–01
		U.K. pt	dm^3	L	5.682 615	E–01
		U.S. pt	dm^3	L	4.731 765	E–01

Table II, continued

Quantity	SI Unit	Customary Unit	Metric Unit: API Preferred	Metric Unit: Other Allowabie	Conversion Factor (multiply quantity expressed in customary units by factor to get metric equivalent)
Space, Time (continued)					
Volume, Capacity	m^3	U.K. fl oz	cm^3		2.841 308 E+01
		U.S. fl oz	cm^3		2.957 353 E+01
		in.^{3}	cm^3		1.638 706 E+01
		mL	cm^3		1
Volume/Length (linear displacement)	m^3/m	bbl/in.	m^3/m		6.259 342 E+00
		bbl/ft	m^3/m		5.216 119 E–01
		ft^3/ft	m^3/m		9.290 304* E–02
		U.S. gal/ft	dm^3/m	L/m	1.241 933 E+01
Volume/Mass	m^3/kg	see Density, Specific Volume, Concentration, Dosage			
Plane Angle	rad	rad	rad		1
		deg (°)	rad		1.745 329 E–02
				°	1
		min (′)	rad		2.908 882 E–04
				′	1
		sec (″)	rad		4.848 137 E–06
				″	1

Solid Angle	sr	sr	sr		1	
Time	s	million years (MY)	Ma		1	
		yr	a		1	
		wk	d		7.0*	E+00
		d	d		1	
		h	h		1	
				min	6.0*	E+01
		min	s		6.0*	E+01
				h	1.666 667	E–02
				min	1	
		s	s		1	
		millimicrosecond	ns		1	
Mass, Amount of Substance						
Mass	kg	U.K. ton (long ton)	Mg	t	1.016 047	E+00
		U.S. ton (short ton)	Mg	t	9.071 847	E–01
		U.K. cwt	kg		5.080 235	E+01
		U.S. cwt	kg		4.535 924	E+01
		kg	kg		1	
		lb	kg		4.535 924	E–01
		oz (troy)	g		3.110 348	E+01
		oz (avdp)	g		2.834 952	E+01
		g	g		1	
		grain	mg		6.479 891	E+01
		mg	mg		1	
		μg	μg		1	

Table II, continued

Quantity	SI Unit	Customary Unit	Metric Unit		Conversion Factor (multiply quantity expressed in customary units by factor to get metric equivalent)
			API Preferred	Other Allowable	
Mass, Amount of Substance (continued)					
Mass/Length	kg/m	see Mechanics			
Mass/Area	kg/m^2	see Mechanics			
Mass/Volume	kg/m^3	see Density, Specific Volume, Concentration, Dosage			
Mass/Mass	kg/kg	see Density, Specific Volume, Concentration, Dosage			
Amount of Substance	mol	ft^3 (60°F, 1 atm)	kmol		1.195 29 E–03
		ft^3 (60°F, 14.73 $lbf/in.^2$)	kmol		1.198 06 E–03
		m^3 (0°C, 1 atm)	kmol		4.461 53 E–02
		m^3 (15°C, 1 atm)	kmol		4.229 28 E–02
		m^3 (20°C, 1 atm)	kmol		4.157 15 E–02
		m^3 (25°C, 1 atm)	kmol		4.087 43 E–02
Heating Value, Entropy, Heat Capacity					
Heating Value (mass basis)	J/kg	Btu/lb	MJ/kg		2.326 000 E–03
			kJ/kg	J/g	2.326 000 E+00
				kW·h/kg	6.461 112 E–04
		cal/g	kJ/kg	J/g	4.184* E+00
		cal/lb	J/kg		9.224 141 E+00

Heating Value (mole basis)	J/mol	kcal/g mol	kJ/kmol		4.184*	E+03
		Btu/lb mol	MJ/kmol		2.326 000	E–03
			kJ/kmol		2.326 000	E+00
Heating Value (volume basis-solids and liquids)	J/m^3	therm/U.S. gal	MJ/m^3	kJ/dm^3	2.787 163	E+04
			kJ/m^3		2.787 163	E+07
				$kW{\cdot}h/dm^3$	7.742 119	E+00
		therm/U.K. gal	MJ/m^3	kJ/dm^3	2.320 798	E+04
			kJ/m^3		2.320 798	E+07
				$kW{\cdot}h/dm^3$	6.446 660	E+00
		therm/Can. gal	MJ/m^3	kJ/dm^3	2.320 799	E+04
			kJ/m^3		2.320 799	E+07
				$kW{\cdot}h/dm^3$	6.446 663	E+00
		Btu/U.S. gal	MJ/m^3	kJ/dm^3	2.787 163	E–01
			kJ/m^3		2.787 163	E+02
				$kW{\cdot}h/m^3$	7.742 119	E–02
		Btu/U.K. gal	MJ/m^3	kJ/dm^3	2.320 800	E–01
			kJ/m^3		2.320 800	E+02
				$kW{\cdot}h/m^3$	6.446 660	E–02
		Btu/Can. gal	MJ/m^3	kJ/dm^3	2.320 799	E–01
			kJ/m^3		2.320 799	E+02
				$kW{\cdot}h/m^3$	6.446 663	E–02
		Btu/ft^3	MJ/m^3	kJ/dm^3	3.725 895	E–02
			kJ/m^3		3.725 895	E+01
				$kW{\cdot}h/m^3$	1.034 971	E–02
		$kcal/m^3$	MJ/m^3	kJ/dm^3	4.184*	E–03
			kJ/m^3		4.184*	E+00
		cal/mL	MJ/m^3		4.184*	E+00
		ft·lbf/U.S. gal	kJ/m^3		3.581 692	E–01

Table II, continued

Quantity	SI Unit	Customary Unit	Metric Unit: API Preferred	Metric Unit: Other Allowable	Conversion Factor (multiply quantity expressed in customary units by factor to get metric equivalent)	
Heating Value, Entropy, Heat Capacity (continued)						
Heating Value (volume basis–gases)	J/m³	cal/mL	kJ/m³	J/dm³	4.184*	E+03
		kcal/m³	kJ/m³	J/dm³	4.184*	E+00
		Btu/ft³	kJ/m³	J/dm³	3.725 895	E+01
Specific Entropy	J/(kg·K)	Btu/(lb·°R)	kJ/(kg·K)	J/(g·K)	4.186 8*	E+00
		cal/(g·K)	kJ/(kg·K)	J/(g·K)	4.184*	E+00
		kcal/(kg·°C)	kJ/(kg·K)	J/(g·K)	4.184*	E+00
Specific Heat Capacity (mass basis)	J/(kg·K)	kW·h/(kg·°C)	kJ/(kg·K)	J/(g·°C)	3.6*	E+03
		Btu/(lb·°F)	kJ/(kg·K)	J/(g·°C)	4.186 8*	E+00
		kcal/(kg·°C)	kJ/(kg·K)	J/(g·°C)	4.184*	E+00
Molar Heat Capacity	J/(mol·K)	Btu/(lb mol·°F)	kJ/(kmol·K)	J/(g·°C)	4.186 8*	E+00
		cal/(g mol·°C)	kJ/(kmol·K)	J/(g·°C)	4.184*	E+00
Temperature, Pressure, Vacuum						
Temperature (absolute)	K	°R	K		5/9	
		K	K		1	
Temperature (traditional)	K	°F	°C		(°F − 32)/1.8	
		°C	°C		1	

Temperature (difference)	K	°F	K	°C	5/9
		°C	K	°C	1
Temperature/Length (geothermal gradient)	K/m	°F/100 ft	mK/m		1.822 689 E+01
Length/Temperature (geothermal step)	m/K	ft/°F	m/K		5.486 4* E–01
Pressure	Pa	atm (14.696 lbf/in.^{2} or 760 mm Hg at 0°C)	MPa		1.013 250* E–01
			kPa		1.013 250* E+02
				bar	1.013 250* E+00
		bar	MPa		1.0* E–01
			kPa		1.0* E+02
				bar	1
		at (kgf/cm^2) (technical atmosphere)	Mpa		9.806 650* E–02
			kPa		9.806 650* E+01
				bar	9.806 650* E–01
		lbf/in.^{2} (psi)	MPa		6.894 757 E–03
			kPa		6.894 757 E+00
				bar	6.894 757 E–02
		in Hg at 60°F	kPa		3.376 85 E+00
		in Hg at 32°F	kPa		3.386 38 E+00
		in H_2O at 39.2°F	kPa		2.490 82 E–01
		in H_2O at 60°F	kPa		2.488 4 E–01
		mm Hg at 0°C (torr)	kPa		1.333 22 E–01
		cm H_2O at 4°C	kPa		9.806 38 E–02
		lbf/ft^2(psf)	kPa		4.788 026 E–02
		μm Hg at 0°C	Pa		1.333 22 E–01
		μbar	Pa		1.0* E–01
		dyn/cm^2	Pa		1.0* E–01

Table II, continued

Quantity	SI Unit	Customary Unit	Metric Unit		Conversion Factor (multiply quantity expressed in customary units by factor to get metric equivalent)
			API Preferred	Other Allowable	
Temperature, Pressure, Vacuum (continued)					
Vacuum, Draft	Pa	in Hg at 60°F	kPa		3.376 85 E+00
		in H_2O at 39.2°F	kPa		2.490 82 E–01
		in H_2O at 60°F	kPa		2.488 4 E–01
		mm Hg at 0°C (torr)	kPa		1.333 22 E–01
		cm H_2O at 4°C	kPa		9.806 38 E–02
Liquid Head	m	ft	m		3.048* E–01
		in.	mm		2.54* E+01
Pressure Drop/Length	Pa/m	psi/ft	kPa/m		2.262 059 E+01
		psi/100 ft	kPa/m		2.262 059 E–01
		psi/mi	kPa/km		4.284 203 E+00
Density, Specific Volume, Concentration, Dosage					
Density (gases)	kg/m^3	lb/ft^3	kg/m^3		1.601 846 E+01
Density (liquids)	kg/m^3	lb/U.S. gal	kg/m^3		1.198 264 E+02
				kg/dm^3	1.198 264 E–01
		lb/U.K. gal	kg/m^3		9.977 633 E+01
				kg/dm^3	9.977 633 E–02
		lb/ft^3	kg/m^3		1.601 846 E+01
				kg/dm^3	1.601 846 E–02

		g/cm^3	kg/m^3		1.0*	E+03
				kg/dm^3	1	
		kg/L	kg/m^3		1.0*	E+03
		°API	kg/m^3		Use tables	
Density (solids)	kg/m^3	lb/ft^3	kg/m^3		1.601 846	E+01
				kg/dm^3	1.601 846	E–02
Specific Volume (gases)	m^3/kg	ft^3/lb	m^3/kg		6.242 796	E–02
				dm^3/kg	6.242 796	E+01
Specific Volume (liquids)	m^3/kg	ft^3/lb	m^3/kg		6.242 796	E–02
				dm^3/kg	6.242 796	E+01
		U.K. gal/lb	m^3/kg		1.022 242	E–02
				dm^3/kg	1.022 242	E+01
		U.S. gal/lb	m^3/kg		8.345 404	E–03
				dm^3/kg	8.345 404	E+00
Molar Volume	m^3/mol	L/g mol	$m^3/kmol$		1	
		ft^3/lb mol	$m^3/kmol$		6.242 796	E–02
Specific Volume (clay yield)	m^3/kg	bbl/U.S. ton	m^3/Mg	m^3/t	1.752 535	E–01
		bbl/U.K. ton	m^3/Mg	m^3/t	1.564 763	E–01
Yield (shale distillation)	m^3/kg	bbl/U.S. ton	dm^3/Mg	dm^3/t	1.752 535	E+02
		bbl/U.K. ton	dm^3/Mg	dm^3/t	1.564 763	E+02
		U.S. gal/U.S. ton	dm^3/Mg	dm^3/t	4.172 702	E+00
		U.S. gal/U.K. ton	dm^3/Mg	dm^3/t	3.725 627	E+00
Concentration (mass/mass)	kg/kg	wt %	kg/kg		1.0*	E–02
			g/kg		1.0*	E+01
		wt ppm	mg/kg		1	

Table II, continued

Quantity	SI Unit	Customary Unit	Metric Unit: API Preferred	Metric Unit: Other Allowable	Conversion Factor (multiply quantity expressed in customary units by factor to get metric equivalent)
Density, Specific Volume, Concentration, Dosage (continued)					
Concentration (mass/volume)	kg/m³	lb/bbl	kg/m³	g/dm³	2.853 010 E+00
		g/U.S. gal	kg/m³		2.641 720 E–01
		g/U.K. gal	kg/m³		2.199 692 E–01
		lb/1000 U.S. gal	g/m³	mg/dm³	1.198 264 E+02
		lb/1000 U.K. gal	g/m³	mg/dm³	9.977 633 E+01
		gr/U.S. gal	g/m³	mg/dm³	1.711 806 E+01
		lb/1000 bbl	g/m³	mg/dm³	2.853 010 E+00
		mg/U.S. gal	g/m³	mg/dm³	2.641 720 E–01
		gr/100 ft³	mg/m³		2.288 352 E+01
		gr/ft³	mg/m³		2.288 352 E+03
Concentration (volume/volume)	m³/m³	bbl/bbl	m³/m³		1
		ft³/ft³	m³/m³		1
		bbl/(acre·ft)	dm³/m³	L/m³	1.288 923 E–01
		U.K. gal/ft³	dm³/m³	L/m³	1.605 437 E+02
		U.S. gal/ft³	dm³/m³	L/m³	1.336 806 E+02
		mL/U.S. gal	dm³/m³	L/m³	2.641 720 E–01
		mL/U.K. gal	dm³/m³	L/m³	2.199 692 E–01
		Vol %	m³/m³		1.0* E–02
		Vol ppm	cm³/m³		1
			dm³/m³	L/m³	1.0* E–03

		U.K. gal/1000 bbl	cm^3/m^3		2.859 406	E+01
		U.S. gal/1000 bbl	cm^3/m^3		2.380 952	E+01
Concentration (mole/volume)	mol/m^3	lb mol/U.S. gal	$kmol/m^3$		1.198 264	E+02
		lb mol/U.K. gal	$kmol/m^3$		9.977 633	E+01
		lb mol/ft^3	$kmol/m^3$		1.601 846	E+01
		std ft^3(60°F, 1 atm)/bbl	$kmol/m^3$		7.518 18	E–03
Concentration (volume/mole)	m^3/mol	U.S. gal/1000 std ft^3 (60°F/60°F)	$dm^3/kmol$	L/kmol	3.166 93	E+00
		bbl/million std ft^3 (60°F/60°F)	$dm^3/kmol$	L/kmol	1.330 11	E–01

Facility Throughput, Capacity

Throughput (mass basis)	kg/s	million lb/yr	Mg/a	t/a	4.535 924	E+02
		U.K. ton/yr	Mg/a	t/a	1.016 047	E+00
		U.S. ton/yr	Mg/a	t/a	9.071 847	E–01
		U.K. ton/d	Mg/d	t/d	1.016 047	E+00
				t/h, Mg/h	4.233 529	E–02
		U.S. ton/d	Mg/d	t/d	9.071 847	E–01
				t/h, Mg/h	3.779 936	E–02
		U.K. ton/h	Mg/h	t/h	1.016 047	E+00
		U.S. ton/h	Mg/h	t/h	9.071 847	E–01
		lb/h	kg/h		4.535 924	E–01
Throughput (volume basis)	m^3/s	bbl/d	m^3/a		5.803 036	E+01
			m^3/d		1.589 873	E–01
			m^3/h		6.624 471	E–03
		ft^3/d	m^3/h		1.179 869	E–03
			m^3/d		2.831 685	E–02

Table II, continued

Quantity	SI Unit	Customary Unit	Metric Unit: API Preferred	Metric Unit: Other Allowable	Conversion Factor (multiply quantity expressed in customary units by factor to get metric equivalent)
Facility Throughput, Capacity (continued)					
		bbl/h	m^3/h		1.589 873 E–01
		ft^3/h	m^3/h		2.831 685 E–02
		U.K. gal/h	m^3/h		4.546 092 E–03
				L/h	4.546 092 E+00
		U.S. gal/h	m^3/h		3.785 412 E–03
				L/h	3.785 412 E+00
		U.K. gal/min	m^3/h		2.727 655 E–01
				L/min	4.546 092 E+00
		U.S. gal/min	m^3/h		2.271 247 E–01
				L/min	3.785 412 E+00
Throughput (mole basis)	mol/s	lb mol/h	kmol/h		4.535 924 E–01
				kmol/s	1.259 979 E–04
Pipeline Capacity	m^3/m	bbl/mi	m^3/km		9.879 013 E–02
Flowrate					
Flowrate (mass basis)	kg/s	U.K. ton/min	kg/s		1.693 412 E+01
		U.S. ton/min	kg/s		1.511 975 E+01
		U.K. ton/h	kg/s		2.822 353 E–01
		U.S. ton/h	kg/s		2.519 958 E–01

		million lb/d	kg/s		5.249 912	E+00
		U.K. ton/d	kg/s		1.175 980	E–02
		U.S. ton/d	kg/s		1.049 982	E–02
		million lb/yr	kg/s		1.438 332	E–02
		U.K. ton/yr	kg/s		3.221 864	E–05
		U.S. ton/yr	kg/s		2.876 664	E–05
		lb/s	kg/s		4.535 924	E–01
		lb/min	kg/s		7.559 873	E–03
		lb/h	kg/s		1.259 979	E–04
Flowrate (volume basis)	m^3/s	bbl/d	dm^3/s		1.840 131	E–03
		ft^3/d	dm^3/s		3.277 413	E–04
		bbl/h	dm^3/s		4.416 314	E–02
		ft^3/h	dm^3/s		7.865 791	E–03
		U.K. gal/h	dm^3/s		1.262 803	E–03
		U.S. gal/h	dm^3/s		1.051 503	E–03
		U.K. gal/min	dm^3/s		7.576 820	E–02
		U.S. gal/min	dm^3/s		6.309 020	E–02
		ft^3/min	dm^3/s		4.719 474	E–01
		ft^3/s	dm^3/s		2.831 685	E+01
Flowrate (mole basis)	mol/s	lb mol/s	kmol/s		4.535 924	E–01
		lb mol/h	kmol/s		1.259 979	E–04
		million scf/sd	kmol/s		1.383 449	E–02
Flowrate/Length (mass basis)	kg/(s·m)	lb/(s·ft)	kg/(s·m)		1.488 164	E+00
		lb/(h·ft)	kg/(s·m)		4.133 789	E–04
Flowrate/Length (volume basis)	m^2/s	U.K. gal/(min·ft)	m^2/s	$m^3/(s·m)$	2.485 833	E–04
		U.S. gal/(min·ft)	m^2/s	$m^3/(s·m)$	2.069 888	E–04

Table II, continued

Quantity	SI Unit	Customary Unit	Metric Unit		Conversion Factor (multiply quantity expressed in customary units by factor to get metric equivalent)
			API Preferred	Other Allowable	
Flowrate (continued)					
		U.K. gal/(h·in.)	m^2/s	$m^3/(s \cdot m)$	4.971 667 E–05
		U.S. gal/(h·in.)	m^2/s	$m^3/(s \cdot m)$	4.139 776 E–05
		U.K. gal/(h·ft)	m^2/s	$m^3/(s \cdot m)$	4.143 055 E–06
		U.S. gal/(h·ft)	m^2/s	$m^3/(s \cdot m)$	3.449 814 E–06
Flowrate/Area (mass basis)	$kg/(s \cdot m^2)$	$lb/(s \cdot ft^2)$	$kg/(s \cdot m^2)$		4.882 428 E+00
		$lb/(h \cdot ft^2)$	$kg/(s \cdot m^2)$		1.356 230 E–03
Flowrate/Area (volume basis)	m/s	$ft^3/(s \cdot ft^2)$	m/s	$m^3/(s \cdot m^2)$	3.048* E–01
		$ft^3/(min \cdot ft^2)$	m/s	$m^3/(s \cdot m^2)$	5.08* E–03
		U.K. gal/(h·in.^{2})	m/s	$m^3/(s \cdot m^2)$	1.957 349 E–03
		U.S. gal/(h·in.^{2})	m/s	$m^3/(s \cdot m^2)$	1.629 833 E–03
		U.K. gal/(min·ft^2)	m/s	$m^3/(s \cdot m^2)$	8.155 621 E–04
		U.S. gal/(min·ft^2)	m/s	$m^3/(s \cdot m^2)$	6.790 972 E–04
		U.K. gal/(h·ft^2)	m/s	$m^3/(s \cdot m^2)$	1.359 270 E–05
		U.S. gal/(h·ft^2)	m/s	$m^3/(s \cdot m^2)$	1.131 829 E–05
Flowrate/Pressure Drop (productivity index)	$m^3/(s \cdot Pa)$	bbl/(d·psi)	$m^3/(d \cdot kPa)$		2.305 916 E–02

Energy, Work, Quantity of Heat, Power						
Energy, Work, Quantity of Heat	J	quad	EJ		1.055 056	E+00
			TW·h		2.930 711	E+02
		therm	MJ		1.055 056	E+02
			kJ		1.055 056	E+05
				kW·h	2.930 711	E+01
		U.S. tonf·mi	MJ		1.431 744	E+01
		hp·h	MJ		2.684 520	E+00
			kJ		2.684 520	E+03
				kW·h	7.456 999	E–01
		ch·h or CV·h	MJ		2.647 796	E+00
			kJ		2.647 796	E+03
				kW·h	7.354 99	E–01
		kW·h	MJ		3.6*	E+00
			kJ		3.6*	E+03
		Chu	kJ		1.899 101	E+00
				kW·h	5.275 280	E–04
		Btu	kJ		1.055 056	E+00
				kW·h	2.930 711	E–04
		kcal	kJ		4.184*	E+00
		cal	kJ		4.184*	E–03
		ft·lbf	kJ		1.355 818	E–03
		J	kJ		1.0*	E–03
		lb·ft^2/s^2(ft·pdl)	kJ		4.214 011	E–05
		erg	J		1.0*	E–07

Table II, continued

Quantity	SI Unit	Customary Unit	Metric Unit: API Preferred	Metric Unit: Other Allowable	Conversion Factor (multiply quantity expressed in customary units by factor to get metric equivalent)
Energy, Work, Quantity of Heat Power (continued)					
Impact Energy	J	kgf·m	J		9.806 650* E+00
		ft·lbf	J		1.355 818 E+00
Work/Length	J/m	U.S. tonf·mi/ft	MJ/m		4.697 322 E+01
Surface Energy	J/m^2	erg/cm^2	mJ/m^2		1.0* E+00
Power	W	quad/yr	EJ/a		1.055 056 E+00
				GW	3.345 561 E+01
		erg/a	TW		3.170 979 E–27
			GW		3.170 979 E–24
		million Btu/h	MW		2.930 711 E–01
		ton of refrigeration	kW		3.516 853 E+00
		Btu/s	kW		1.055 056 E+00
		kW	kW		1
		hydraulic horse-power-hhp	kW		7.460 43 E–01
		hp (electric)	kW		7.46* E–01
		hp (550 ft·lbf/s)	kW		7.456 999 E–01
		ch or CV	kW		7.354 99 E–01
		Btu/min	kW		1.758 427 E–02
		ft·lbf/s	kW		1.355 818 E–03

		kcal/h	W		1.162 222 E+00
		Btu/h	W		2.930 711 E–01
		ft·lbf/min	W		2.259 697 E–02
Power/Area	W/m^3	Btu/(s·ft^2)	kW/m^2		1.135 653 E+01
		cal/(h·cm^2)	kW/m^2		1.162 222 E–02
		Btu/(h·ft^2)	kW/m^3		3.154 591 E–03
Heat Flow Unit-hfu (geothermics)		μcal/(s·cm^2)	mW/m^2		4.184* E+01
Heat Release Rate, Mixing Power	W/m^3	hp/ft^3	kW/m^3		2.633 414 E+01
		cal/(h·cm^3)	kW/m^3		1.162 222 E+00
		Btu/(s·ft^3)	kW/m^3		3.725 895 E+01
		Btu/(h·ft^3)	kW/m^3		1.034 971 E–02
Heat Generation Unit-hgu (radioactive rocks)		cal/(s·cm^3)	μW/m^3		4.184* E+12
Cooling Duty (machinery)	W/W	Btu/(bhp·h)	W/kW		3.930 148 E–01
Mass Fuel Consumption	kg/J	lb/(hp·h)	mg/J	kg/MJ	1.689 659 E–01
				kg/(kW·h)	6.082 774 E–01
Volume Fuel Consumption	m^3/J	m^3/(kW·h)	dm^3/MJ	mm^3/J	2.777 778 E+02
				dm^3/(kW·h)	1.0* E+03
		U.S. gal/(hp·h)	dm^3/MJ	mm^3/J	1.410 089 E+00
				dm^3/(kW·h)	5.076 321 E+00
		U.K. pt/(hp·h)	dm^3/MJ	mm^3/J	2.116 809 E–01
				dm^3/(kW·h)	7.620 512 E–01

Table II, continued

Quantity	SI Unit	Customary Unit	Metric Unit: API Preferred	Metric Unit: Other Allowable	Conversion Factor (multiply quantity expressed in customary units by factor to get metric equivalent)
Energy, Work, Quantity of Heat Power (continued)					
Fuel Consumption (automotive)	m^3/m	U.K. gal/mi	$dm^3/100$ km	L/100 km	2.824 811 E+02
		U.S. gal/mi	$dm^3/100$ km	L/100 km	2.352 146 E+02
		mi/U.S. gal	km/dm^3	km/L	4.251 437 E–01
		mi/U.K. gal	km/dm^3	km/L	3.540 060 E–01
Mechanics					
Velocity (linear), Speed	m/s	knot	km/h		1.852 E+00
				knot	1
		mi/h	km/h		1.609 344* E+00
		m/s	m/s		1
		ft/s	m/s		3.048* E–01
				cm/s	3.048* E+01
				m/ms	3.048* E–04
		ft/min	m/s		5.08* E–03
				cm/s	5.08* E–01
		ft/h	mm/s		8.466 667 E–02
				cm/s	8.466 667 E–03
		ft/d	mm/s		3.527 778 E–03
				m/d	3.048* E–01

Quantity	SI unit	Customary unit	Metric unit	Other unit	Factor	Exponent
		in./s	mm/s		2.54*	E+01
				cm/s	2.54*	E+00
		in./min	mm/s		4.233 333	E–01
				cm/s	4.233 333	E–02
Velocity (angular)	rad/s	r/min	rad/s		1.047 198	E–01
		r/min		r/min	1	
		r/s	rad/s		6.283 185	E+00
		r/s		r/s	1	
		deg/min	rad/s		2.908 882	E–04
		deg/s	rad/s		1.745 329	E–02
Reciprocal Velocity	μs/m	μs/ft	μs/m		3.280 840	E+00
Acceleration (linear)	m/s^2	ft/s^2	m/s^2		3.048*	E–01
		gal (cm/s^2)	m/s^2		1.0*	E–02
Acceleration (angular)	rad/s^2	rad/s^2	rad/s^2		1	
		rpm/s	rad/s^2		1.047 198	E–01
Corrosion Rate	mm/a	in./yr (ipy)	mm/a		2.54*	E+01
Momentum	kg·m/s	lb·ft/s	kg·m/s		1.382 550	E–01
Force	N	U.K. tonf	kN		9.964 016	E+00
		U.S. tonf	kN		8.896 443	E+00
		kgf (kp)	N		9.806 650*	E+00
		lbf	N		4.448 222	E+00
		N	N		1	
		pdl	mN		1.382 550	E+02
		dyn	mN		1.0*	E–02

Table II, continued

Quantity	SI Unit	Customary Unit	Metric Unit: API Preferred	Metric Unit: Other Allowable	Conversion Factor (multiply quantity expressed in customary units by factor to get metric equivalent)
Mechanics (continued)					
Bending Moment, Torque	N·m	U.S. tonf·m	kN·m		2.711 636 E+00
		kgf·m	N·m		9.806 650* E+00
		lbf·ft	N·m		1.355 818 E+00
		lbf·in.	N·m		1.129 848 E–01
		pdl·ft	N·m		4.214 011 E–02
Bending Movement, Length	N·m/m	lbf·ft/in.	N·m/m		5.337 866 E+01
		kgf·m/m	N·m/m		9.806 650* E+00
		lbf·in./in.	N·m/m		4.448 222 E+00
Moment of Inertia	kg·m^2	lb·ft^2	kg·m^2		4.214 011 E–02
		in.^{4}	cm^4		4.162 314 E+01
Stress	Pa	U.S. tonf/in.^{2}	MPa	N/mm^2	1.378 951 E+00
		kgf/mm^2	MPa	N/mm^2	9.806 650* E+00
		U.S. tonf/ft^2	MPa	N/mm^2	9.576 052 E–02
		lbf/in.^{2} (psi)	MPa	N/mm^2	6.894 757 E–03
		lbf/ft^2 (psf)	kPa		4.788 026 E–02
		dyn/cm^2	Pa		1.0* E–01
Yield Point, Gel Strength (drilling fluid)		lbf/100 ft^2	Pa		4.788 026 E+01

Mass/Length	kg/m	lb/ft	kg/m		1.488 164	E+00
Mass/Area Structural Loading, Bearing Capacity (mass basis)	kg/m^2	U.S. ton/ft^2	Mg/m^2	t/m^2	9.764 855	E+00
		lb/ft^2	kg/m^2		4.882 428	E+00
Modulus of Elasticity	Pa	$lbf/in.^2$ (psi)	MPa	N/mm^2	6.894 757	E–03
Section Modulus	m^3	$in.^3$	cm^3		1.638 706	E+01
Coefficient of Thermal Expansion	$m/(m \cdot K)$	in./(in.°F)	$mm/(mm \cdot K)$	$mm/(mm \cdot °C)$	5.555 556	E–01
Transport Properties						
Diffusivity	m^2/s	ft^2/s	mm^2/s		9.290 304*	E+04
		cm^2/s	mm^2/s		1.0*	E+02
		ft^2/h	mm^2/s		2.580 64*	E+01
Thermal Resistance	$K \cdot m^2/W$	$°C \cdot m^2 \cdot h/kcal$	$K \cdot m^2/kW$		8.604 208	E+02
		$°F \cdot ft^2 \cdot h/Btu$	$K \cdot m^2/kW$		1.761 102	E+02
Heat Flux	W/m^2	$Btu/(h \cdot ft^2)$	kW/m^2		3.154 591	E–03
Thermal Conductivity	$W/(m \cdot K)$	$cal/(s \cdot cm^2 \cdot °C/cm)$	$W/(m \cdot K)$	$W/(m^2 \cdot °C/m)$	4.184*	E+02
		$Btu/(h \cdot ft^2 \cdot °F/ft)$	$W/(m \cdot K)$	$W/(m^2 \cdot °C/m)$	1.730 735	E+00
		$kcal/(h \cdot m^2 \cdot °C/m)$	$W/(m \cdot K)$	$W/(m^2 \cdot °C/m)$	1.162 222	E+00
		$Btu/(h \cdot ft^2 \cdot °F/in.)$	$W/(m \cdot K)$	$W/(m^2 \cdot °C/m)$	1.422 279	E–01
		$cal/(h \cdot cm^2 \cdot °C/cm)$	$W/(m \cdot K)$	$W/(m^2 \cdot °C/m)$	1.162 222	E–01
Heat Transfer Coefficient	$W/(m^2 \cdot K)$	$cal/(s \cdot cm^2 \cdot °C)$	$kW/(m^2 \cdot K)$		4.184*	E+01
		$Btu/(s \cdot ft^2 \cdot °F)$	$kW/(m^2 \cdot K)$		2.044 175	E+01
		$cal/(h \cdot cm^2 \cdot °C)$	$kW/(m^2 \cdot K)$		1.162 222	E–02
		$Btu/(h \cdot ft^2 \cdot °F)$	$kW/(m^2 \cdot K)$		5.678 263	E–03
		$Btu/(h \cdot ft^2 \cdot °R)$	$kW/(m^2 \cdot K)$		5.678 263	E–03
		$kcal/(h \cdot m^2 \cdot °C)$	$kW/(m^2 \cdot K)$		1.162 222	E–03

Table II, continued

Quantity	SI Unit	Customary Unit	Metric Unit: API Preferred	Metric Unit: Other Allowable	Conversion Factor (multiply quantity expressed in customary units by factor to get metric equivalent)
Transport Properties (continued)					
Volumetric Heat Transfer Coefficient	$W/(m^3 \cdot K)$	$Btu/(s \cdot ft^3 \cdot °F)$	$kW/(m^3 \cdot K)$		6.706 611 E+01
		$Btu/(h \cdot ft^3 \cdot °F)$	$kW/(m^3 \cdot K)$		1.862 947 E–02
Surface Tension	N/m	dyn/cm	mN/m		1.0* E+00
Viscosity (dynamic)	Pa·s	$lbf \cdot s/in.^2$	mPa·s	cP	6.894 757 E+06
		$lbf \cdot s/ft^2$	mPa·s	cP	4.788 026 E+04
		$kgf \cdot s/m^2$	mPa·s	cP	9.806 650* E+03
		$dyn \cdot s/cm^2$	Pa·s		1.0* E–01
			mPa·s	cP	1.0* E+02
		P	Pa·s		1.0* E–01
			mPa·s	cP	1.0* E+02
		cP	mPa·s	cP	1.0* E+00
Viscosity (kinematic)	m^2/s	ft^2/s	mm^2/s	cSt	9.290 304* E+04
		$in.^2/s$	mm^2/s	cSt	6.451 6* E+02
		ft^2/h	mm^2/s	cSt	2.580 64* E+01
		m^2/h	mm^2/s	cSt	2.777 778 E+02
		cm^2/s	mm^2/s	cSt	1.0* E+02
		St	mm^2/s	cSt	1.0* E+02
		cSt	mm^2/s	cSt	1.0* E+00

Permeability	m^2	D	μm^2	D	1.0*	E+00
		mD	μm^2	D	1.0*	E–03
Electricity, Magnetism						
Admittance	S	S	S		1	
Capacitance	F	μF	μF		1	
Capacity, Storage Battery	C	A·h	kC		3.6*	E+00
Charge Density	C/m^3	C/mm^3	C/mm^3		1	
Conductance	S	S	S		1	
		℧	S		1	
Conductivity	S/m	S/m	S/m			
		m℧/m	mS/m		1	
Current Density	A/m^2	A/mm^2	A/mm^2		1	
		$A/in.^2$	A/mm^2		1.550 003	E–03
Displacement	C/m^2	C/cm^2	C/cm^2		1	
Electric Charge	C	C	C		1	
Electric Current	A	A	A		1	
Electric Dipole Moment	C·m	C·m	C·m		1	
Electric Field Strength	V/m	V/m	V/m		1	
Electric Flux	C	C	C		1	
Electric Polarization	C/m^2	C/cm^2	C/cm^2		1	

Table II, continued

Quantity	SI Unit	Customary Unit	Metric Unit: API Preferred	Metric Unit: Other Allowable	Conversion Factor (multiply quantity expressed in customary units by factor to get metric equivalent)
Electricity, Magnetism (continued)					
Electric Potential	V	V	V		1
		mV	mV		1
Electromagnetic Moment	$A \cdot m^2$	$A \cdot m^2$	$A \cdot m^2$		1
Electromotive Force	V	V	V		1
Flux of Displacement	C	C	C		1
Frequency	Hz	cycles/s	Hz		1
Impedance	Ω	Ω	Ω		1
Linear Current Density	A/m	A/mm	A/mm		1
Magnetic Dipole Moment	Wb·m	Wb·m	Wb·m		1
Magnetic Field Strength	A/m	A/mm	A/mm		1
		oersted	A/m		7.957 747 E+01
Magnetic Flux	Wb	mWb	mWb		1
		maxwell	nWb		1.0* E+01
Magnetic Flux Density	T	mT	mT		1
		gauss	T		1.0* E–04
		gamma	nT		1

Magnetic Induction	T	mT	mT	1
Magnetic Moment	$A \cdot m^2$	$A \cdot m^2$	$A \cdot m^2$	1
Magnetic Polarization	T	mT	mT	1
Magnetic Potential Difference	A	A	A	1
Magnetic Vector Potential	Wb/m	Wb/mm	Wb/mm	1
Magnetization	A/m	A/mm	A/mm	1
Modulus of Admittance	S	S	S	1
Modulus of Impedance	Ω	Ω	Ω	1
Mutual Inductance	H	H	H	1
Permeability	H/m	μH/m	μH/m	1
Permeance	H	H	H	1
Permittivity	F/m	μF/m	μF/m	1
Potential Difference	V	V	V	1
Quantity of Electricity	C	C	C	1
Reactance	Ω	Ω	Ω	1
Reluctance	H^{-1}	H^{-1}	H^{-1}	1
Resistance	Ω	Ω	Ω	1
Resistivity	$\Omega \cdot m$	$\Omega \cdot cm$ $\Omega \cdot m$	$\Omega \cdot cm$ $\Omega \cdot m$	1 1

Table II, continued

Quantity	SI Unit	Customary Unit	Metric Unit: API Preferred	Metric Unit: Other Allowable	Conversion Factor (multiply quantity expressed in customary units by factor to get metric equivalent)
Electricity, Magnetism (continued)					
Self Inductance	H	mH	mH		1
Surface Density of Charge	C/m^2	mC/m^2	mC/m^2		1
Susceptance	S	S	S		1
Volume Density of Charge	C/m^3	C/mm^3	C/mm^3		1
Acoustics, Light, Radiation					
Acoustical Energy	J	J	J		1
Acoustical Intensity	W/m^2	W/cm^2	W/m^2		1.0* E+04
Acoustical Power	W	W	W		1
Activity (of a radionuclide)	Bq	curie	Bq		3.7* E+10
Illuminance, Illumination	lx	footcandle (fc)	lx		1.076 391 E+01
		lm/ft^2	lx		1.076 391 E+01
Irradiance	W/m^2	W/m^2	W/m^2		1
Light Exposure	lx·s	fc·s	lx·s		1.076 391 E+01

Luminance	cd/m^2	cd/m^2	cd/m^2	1	
		footlambert (fL)	cd/m^2	3.426 259	E+00
Luminous Efficacy	lm/W	lm/W	lm/W	1	
Luminous Exitance	lm/m^2	lm/m^2	lm/m^2	1	
		lm/ft^2	lm/m^2	1.076 391	E+01
Luminous Flux	lm	lm	lm	1	
Luminous Intensity	cd	cd	cd	1	
Quantity of Light	lm·s	talbot	lm·s	1	
Radiance	W/(m^2·sr)	W/(m^2·sr)	W/(m^2·sr)	1	
Radiant Energy	J	J	J	1	
Radiant Flux	W	W	W	1	
Radiant Intensity	W/sr	W/sr	W/sr	1	
Radiant Power	W	W	W	1	
Sound Pressure	Pa	Pa	Pa	1	
Wavelength	m	Å	nm	1.0*	E–01

INDEX